2012 Proceedings of the European Solid-State Device Research Conference

(ESSDERC 2012)

Bordeaux, France
17 – 21 September 2012

IEEE Catalog Number: CFP12543-PRT
ISBN: 978-1-4673-1707-8

**Copyright © 2012 by the Institute of Electrical and Electronic Engineers, Inc
All Rights Reserved**

Copyright and Reprint Permissions: Abstracting is permitted with credit to the source. Libraries are permitted to photocopy beyond the limit of U.S. copyright law for private use of patrons those articles in this volume that carry a code at the bottom of the first page, provided the per-copy fee indicated in the code is paid through Copyright Clearance Center, 222 Rosewood Drive, Danvers, MA 01923.

For other copying, reprint or republication permission, write to IEEE Copyrights Manager, IEEE Service Center, 445 Hoes Lane, Piscataway, NJ 08854. All rights reserved.

***This publication is a representation of what appears in the IEEE Digital Libraries. Some format issues inherent in the e-media version may also appear in this print version.

IEEE Catalog Number: CFP12543-PRT
ISBN 13:   978-1-4673-1707-8
ISSN: 1930-8876

Additional Copies of This Publication Are Available From:

Curran Associates, Inc
57 Morehouse Lane
Red Hook, NY 12571 USA
Phone: (845) 758-0400
Fax: (845) 758-2633
E-mail: curran@proceedings.com
Web: www.proceedings.com

2012 Proceedings of the European Solid-State Device Research Conference (ESSDERC 2012)

Bordeaux, France
17-21 September 2012

IEEE Catalog Number: CFP12543-POD
ISBN: 978-1-46731-707-8

Table of Contents

A1L-A   PLENARY: Biopsychically Inspired Cognitive Control

Date: Tuesday, August 28, 2012
Time: 09:00 - 10:00
Room: Casino
Chair: Jerzy Sasiadek

Biopsychically Inspired Cognitive Control for Autonomous Agents Based on Motivated Learning ....N/A
J. Jim Zhu[2], Xudan Xu[1]
[1]Beihang University, China; [2]Ohio University, United States

---

A1L-A   JOINT PLENARY: Future IC Technology

Date: Tuesday, September 18, 2012
Time: 09:40 - 10:40
Room: Amphitheater A
Chair: Yann Deval; *IMS Lab*

Future Silicon Technology 1
Kinam Kim
Samsung Electronics Co., LTD., Korea, South

A3L-A   JOINT PLENARY: The Industrialization of the Silicon Photonics: Technology Road Map & Applications

---

Date: Tuesday, September 18, 2012
Time: 14:00 - 15:00
Room: Amphitheater A
Chair: Jean-Baptiste Begueret; *IMS Lab*

The Industrialization of the Silicon Photonics: Technology Road Map and Applications 7
Maurizio Zuffada
STMicroelectronics, Italy

B1L-A   JOINT PLENARY: Facing the Challenges of 450mm Manufacturing

---

Date: Wednesday, September 19, 2012
Time: 08:15 - 09:30
Room: Amphitheater A
Chair: Yann Deval; *IMS Lab*

Facing the Challenges of 450mm Manufacturing ....No Paper
Maria Marced
TSMC, United States

B4L-A   JOINT PLENARY: Solid-State and Bio Systems Interface

Date: Wednesday, September 19, 2012
Time: 14:00 - 15:00
Room: Amphitheater A
Chair: Thomas Skotnicki; *STMicroelectronics*

Solid-State and Biological Systems Interface 14
Nan Sun[3], Yong Liu[2], Ling Qin[1], Guangyu Xu[1], Donhee Ham[1]
[1]Harvard University, United States; [2]IBM T. J. Watson Research Center, United States; [3]University of Texas at Austin, United States

Table of Contents

C1L-A JOINT PLENARY: Advancing Very Large Scale High Performance Heterogeneous Integration

Date: Thursday, September 20, 2012
Time: 08:30 - 09:30
Room: Amphitheater A
Chair: Didier Belot; *STMicroelectronics*

Advancing High Performance Heterogeneous Integration Through Die Stacking 18
Liam Madden, Ephrem Wu, Namhoon Kim, Bahareh Banijamali, Khaldoon
Abugharbieh, Suresh Ramalingam, Xin Wu
Xilinx, Inc., United States

C4L-A JOINT PLENARY: Graphene for Microelectronics: Can it make a difference?

Date: Thursday, September 20, 2012
Time: 14:00 - 15:00
Room: Amphitheater A
Chair: Thomas Zimmer; *IMS Lab*

Graphene for Microelectronics: Can It Make a Difference? 25
Max C. Lemme
KTH Royal Institute of Technology, Sweden

A5L-A ESSDERC PLENARY: Nanostructure Devices for Logic & Memory and Beyond

Date: Tuesday, September 18, 2012
Time: 16:30 - 17:30
Room: Amphitheater A
Chair: Carlos Mazure; *Soitec*

Nanostructure Devices for Logic and Memory and Beyond 28
Sandip Tiwari
Cornell University, United States

B5L-A ESSDERC PLENARY: Modeling Circuits with Spins & Magnets for All-spin Logic

Date: Wednesday, September 19, 2012
Time: 15:10 - 16:10
Room: Amphitheater A
Chair: Sorin Cristoloveanu; *Grenoble INP*

Modeling Circuits with Spins and Magnets for All-Spin Logic 36
Behtash Behin-Aein[1], Angik Sarkar[2], Supriyo Datta[2]
[1]GlobalFoundries Inc., United States; [2]Purdue University, United States

C5L-A ESSDERC PLENARY: Organic Complementary Circuits

Date: Thursday, September 20, 2012
Time: 15:10 - 16:10
Room: Amphitheater A
Chair: Cor Claeys; *IMEC*

**Organic Complementary Circuits - Scaling Towards Low Voltage and
Submicron Channel Length** 41
Hagen Klauk
Max Plank Institute for Solid State Research, Germany

Table of Contents

B3L-A   JOINT ESSDERC/ESSCIRC Session on Compact Modeling

Date: Wednesday, September 19, 2012
Time: 11:00 - 12:20
Room: Amphitheater A
Chair(s):   Hervé Jaouen; *STMicroelectronics*
Christian Enz; *CSEM*

BSIM - Industry Standard Compact MOSFET Models 46
Yogesh Singh Chauhan, Sriramkumar Venugopalan, Mohammed Ahosan Ul Karim,
Sourabh Khandelwal, Navid Paydavosi, Pankaj Thakur, Ali Niknejad, Chenming Hu
University of California, Berkeley, United States

**Evaluation of the BSIM6 Compact MOSFET Model's Scalability in 40nm
CMOS Technology** 50
Maria-Anna Chalkiadaki[1], Anurag Mangla[1], Christian C. Enz[2], Yogesh Singh
Chauhan[3], Mohammed Ahosan Ul Karim[3], Sriramkumar Venugopalan[3], Ali
Niknejad[3], Chenming Hu[3]
*[1]École Polytechnique Fédérale de Lausanne, Switzerland; [2]École Polytechnique
Fédérale de Lausanne / Centre Suisse d'Electronique et de Microtechnique,
Switzerland; [3]University of California, Berkeley, United States*

4-Port Isolated MOS Modeling and Extraction for mmW Applications 54
Benjamin Dormieu[2], Patrick Scheer[2], Clement Charbuillet[2], Sebastien Jan[2], Francois
Danneville[1]
[1]IEMN, France; [2]STMicroelectronics, France

C3L-A   Joint ESSDERC/ESSCIRC Session on Variablity / Reliability

Date: Thursday, September 20, 2012
Time: 11:00 - 12:20
Room: Amphitheater A
Chairs: François Marc; *University Bordeaux 1*
Angel Rodriguez Vázquez; *IMSE-CNM*

**Variability Aware Cell Library Optimization for Reliable Sub-Threshold
Operation** 58
Tobias Gemmeke, Maryam Ashouei
Holst Center / imec, Netherlands

Advancements on Reliability-Aware Analog Circuit Design 62
Bertrand Ardouin[3], Jean-Yves Dupuy[1], Jean Godin[1], Virginie Nodjiadjim[1], Muriel
Riet[1], François Marc[2], Gilles Amadou Koné[2], Sudip Ghosh[2], Brice Grandchamp[2],
Cristell Maneux[2]
[1]III-V Lab, France; [2]Université de Bordeaux, France; [3]XMOD Technologies, France

A2L-E   High Mobility Devices

Date: Tuesday, September 18, 2012
Time: 11:00 - 12:20
Room: Room E2
Chairs: Emmanuel Dubois; *IEMN - UMR CNRS*
Nadine Collaert; *IMEC*

Design Challenges for Nano-Scale Devices 69
Marc Belleville, Olivier Thomas, Alexandre Valentian, Fabien Clermidy
CEA-LETI, France

Table of Contents

Study of Carrier Transport in Strained and Unstrained SOI Tri-Gate and Omega-Gate Si-Nanowire MOSFETs 73
Masahiro Koyama[2], Mikaël Cassé[1], Remi Coquand[3], Sylvain Barraud[4], Hiroshi Iwai[6], Gérard Ghibaudo[5], Gilles Reimbold[3]
[1]CEA LETI, France; [2]CEA-LETI, France; [3]CEA-LETI-MINATEC, France; [4]CEA-LETI-MINATEC and CEA-INAC, France; [5]IMEP-LAHC, MINATEC, INPG, France; [6]Tokyo Institute of Technology, Japan

Stability and Performance Optimization of InGaAs-OI and GeOI Hetero-Channel SRAM Cells 77
Vita Pi-Ho Hu, Ming-Long Fan, Pin Su, Ching-Te Chuang
National Chiao Tung University, Taiwan

A2L-F	**High-k Dielectrics and Applications**

Date: Tuesday, September 18, 2012
Time: 11:00 - 12:20
Room: Room F1
Chairs: Emmanuel Augendre;
Anton Bauer; *Fraunhofer*

Two-Step Annealing Effects on Ultrathin EOT Higher-K (K = 40) ALD-HfO2 Gate Stacks 81
Yukinori Morita, Shinji Migita, Wataru Mizubayashi, Meishoku Masahara, Hiroyuki Ota
Advanced Industrial Science and Technology, Japan

Thin Germanium Dioxide Film with a High Quality Interface Formed in a Direct Neutral Beam Oxidation Process 85
Akira Wada[1], Seiji Samukawa[1], Rui Zhang[2], Shinichi Takagi[2]
[1]Tohoku University, Japan; [2]University of Tokyo, Japan

(100)- and (110)-Oriented nMOSFETs with Highly Scaled EOT in La-Silicate/Si Interface for Multi-Gate Architecture 89
Takamasa Kawanago, Kuniyuki Kakushima, Parhat Ahmet, Yoshinori Kataoka, Akira Nishiyama, Nobuyuki Sugii, Kazuo Tsutsui, Kenji Natori, Takeo Hattori, Hiroshi Iwai
Tokyo Institute of Technology, Japan

CMOS Compatible ALD High-K Double Slot Grating Couplers for on-Chip Optical Interconnects 93
Maziar M. Naiini, Christoph Henkel, Gunnar B. Malm, Mikael Östling
KTH Royal Institute of Technology, Sweden

A2L-G	**Emerging Device Modeling**

Date: Tuesday, September 18, 2012
Time: 11:00 - 12:20
Room: Room F2
Chairs: Massimo Rudan; *Università di Bologna*
Cristell Maneux; *IMS Bordeaux*

Transport in Amorphous Materials with Applications to Phase-Change Memories 97
Carlo Jacoboni[1], Enrico Piccinini[2], Fabrizio Buscemi[2]
[1]Università degli Studi di Modena e Reggio Emilia, Italy; [2]Università di Bologna, Italy

Geometry Based Resistance Model for Phase Change Memory 101
K. C. Kwong, Philip K. T. Mok, Mansun Chan
Hong Kong Univeristy of Science and Technology, Hong Kong

Table of Contents

Drain-Conductance Optimization in Nanowire TFETs 105
Elena Gnani, Susanna Reggiani, Antonio Gnudi, Giorgio Baccarani
Università di Bologna, Italy

A4L-E	**Variability**

Date: Tuesday, September 18, 2012
Time: 15:10 - 16:10
Room: Room E2
Chairs: Hervé JAOUEN; *STMicroelectronics*
           Ray Hueting

Comprehensive Statistical Comparison of RTN and BTI in Deeply Scaled MOSFETs by Means of 3D 'Atomistic' Simulation 109
Salvatore Maria Amoroso, Louis Gerrer, Stanislav Markov, Fikru Adamu-Lema,
Asen Asenov
University of Glasgow, United Kingdom

Statistical Variability in 14-nm Node SOI FinFETs and its Impact on Corresponding 6T-SRAM Cell Design 113
Xingsheng Wang[2], Binjie Cheng[2], Andrew Brown[1], Campbell Millar[3], Asen Asenov[2]
[1]*Gold Standard Simulations Ltd., United Kingdom;* [2]*University of Glasgow, United Kingdom;* [3]*University of Glasgow, Gold Standard Simulations Ltd., United Kingdom*

Sensitivity-Based Investigation of Threshold Voltage Variability in 32-nm Flash Memory Cells 117
Valentina Bonfiglio, Giuseppe Iannaccone
Università degli Studi di Pisa, Italy

A4L-F	**Advanced FETs**

Date: Tuesday, September 18, 2012
Time: 15:10 - 16:10
Room: Room F1
Chairs: Giorgio Baccarani; *University of Bologna*
           Max Lemme; *KTH*

Scaling of Trigate Nanowire (NW) MOSFETs Down to 5 nm Width: 300 K Transition to Single Electron Transistor, Challenges and Opportunities 121
Veeresh Deshpande[2], Sylvain Barraud[2], Xavier Jehl[2], Romain Wacquez[2], Maud
Vinet[2], Remi Coquand[1], Benoit Roche[2], Benoit Voisin[2], François Triozon[1], Christian
Vizioz[2], Lucie Tosti[2], Bernard Previtali[2], Pierre Perreau[2], Thierry Poiroux[2], Marc
Sanquer[2], Olivier Faynot[1]
[1]*CEA-LETI-MINATEC, France;* [2]*CEA-LETI-MINATEC and CEA-INAC, France*

Active Strain Modulation in Field Effect Devices 125
Tom van Hemert, Raymond Hueting
Universiteit Twente, Netherlands

v

Table of Contents

A4L-G   Thin-Film Transistors

Date: Tuesday, September 18, 2012
Time: 15:10 - 16:10
Room: Room F2
Chairs: Peter Ashburn; *Southampton University*
       Ryoichi Ishihara; *TU Delft*

Static and Low Frequency Noise Characterization of Densely Packed CNT-TFTs 129
Min-Kyu Joo[1], Un Jeong Kim[4], Dae-Young Jeon[1], So Jeong Park[1], Mireille Mouis[1],
Gyu-Tae Kim[3], Gérard Ghibaudo[2]
*[1]IMEP-LAHC, France; [2]IMEP-LAHC, MINATEC, INPG, France; [3]Korea University,
Korea, South; [4]Samsung Advanced Institute of Technology, Korea, South*

Mechanically Flexible Double Gate a-IGZO TFTs 133
Niko Münzenrieder, Christoph Zysset, Thomas Kinkeldei, Luisa Petti, Giovanni A.
Salvatore, Gerhard Tröster
ETH Zurich, Switzerland

Top-Down Fabricated ZnO Nanowire Transistors for Application in Biosensors 137
Suhana Mohamed Sultan, Kai Sun, Maurits de Planque, Peter Ashburn, Harold
Chong
University of Southampton, United Kingdom

B2L-E   Novel Thin Film Integration

Date: Wednesday, September 19, 2012
Time: 09:40 - 10:40
Room: Room E2
Chairs: Jurriaan Schmitz; *University of Twente*
       Simon Deleonibus; *CEA*

Manufacturing Aspects of an Ultra-Thin Chip Technology 141
Evangelos Angelopoulos, Muhammad Al-Shahed, Wolfgang Appel, Stefan Endler,
Saleh Ferwana, Christine Harendt, Mahadi-Ul Hassan, Horst Rempp, Martin
Zimmermann, Joachim Burghartz
Insitute for Microelectronics Stuttgart - IMS CHIPS, Germany

**Epitaxial Growth of Large-Area p+n Diodes at 400 °C by Aluminum-Induced
Crystallization** 145
Agata Sakic, Lin Qi, Tom Scholtes, Johan van der Cingel, Lis Nanver
Technische Universiteit Delft, Netherlands

**Current-Voltage Characteristics of Vertical Diodes for Next Generation
Memories** 149
Hokyun An[1], Kongsoo Lee[1], Yoongoo Kang[1], Seonghoon Jeong[1], Wonseok Yoo[1],
Jaejong Han[1], Bonghyun Kim[1], Hanjin Lim[1], Seokwoo Nam[1], Gitae Jeong[1], Hokyu
Kang[1], Chilhee Chung[2], Byoungdeog Choi[3]
*[1]Samsung Electronics, Semiconductor R&D Center, Korea, South; [2]Semiconductor
R&D Center, Samsung Electronics Co., Korea, South; [3]Sungkyunkwan University,
Korea, South*

Table of Contents

B2L-F Tunneling Devices

Date: Wednesday, September 19, 2012
Time: 09:40 - 10:40
Room: Room F1
Chairs: Francois Andrieu; *CEA-Leti*
         Jean-Pierre Colinge; *Tyndall*

Si Tunneling Transistors with High on-Currents and Slopes of 50 mV/dec Using Segregation Doped NiSi2 Tunnel Junctions 153
Lars Knoll[1], Qing-Tai Zhao[1], Stefan Trellenkamp[1], Anna Schäfer[1], Konstantin Bourdelle[2], Siegfried Mantl[1]
[1]Forschungszentrum Jülich, Germany; [2]SOITEC, France

A Comparative Analysis of Tunneling FET Circuit Switching Characteristics and SRAM Stability and Performance 157
Yin-Nien Chen, Ming-Long Fan, Vita Pi-Ho Hu, Ming-Fu Tsai, Chia-Hao Pao, Pin Su, Ching-Te Chuang
National Chiao Tung University, Taiwan

Tunnel FET with Non-Uniform Gate Capacitance for Improved Device and Circuit Level Performance 161
Cem Alper, Luca De Michielis, Nilay Dagtekin, Livio Lattanzio, Adrian M. Ionescu
École Polytechnique Fédérale de Lausanne, Switzerland

B2L-G MEMS / OTFT

Date: Wednesday, September 19, 2012
Time: 09:40 - 10:40
Room: Room F2
Chairs: Piotr Grabiec; *Inst. of Electron Technology*
         Fiodor Sizov; *National Academy of Sciences of Ukraine*

From FinFET to Nanowire ISFET 165
Michal Zaborowski, Daniel Tomaszewski, Piotr Dumania, Piotr Grabiec
Institute of Electron Technology, Poland

Micro- and Nano-Link Ultra-Low Power Heaters for Sensors 169
Alfons Groenland, Elizaveta Vereshchagina, Alexey Kovalgin, Rob Wolters, Han Gardeniers, Jurriaan Schmitz
Universiteit Twente, Netherlands

High Performance Printed N and P-Type OTFTs for Complementary Circuits on Plastic Substrate 173
Stephanie Jacob[1], Mohammed Benwadih[1], Jacqueline Bablet[1], Isabelle Chartier[1], Romain Gwoziecki[1], Sahel Abdinia[3], Eugenio Cantatore[3], Lidia Maddiona[4], Francesca Tramontana[4], Giorgio Maiellaro[5], Luigi Mariucci[2], Giuseppe Palmisano[5], Romain Coppard[1]
[1]CEA-LITEN, France; [2]CNR-IMM, Italy; [3]Eindhoven University of Technology, Netherlands; [4]STMicroelectronics, Italy; [5]Università di Catania, Italy

Table of Contents

B3L-F High-frequency Transistors

Date: Wednesday, September 19, 2012
Time: 11:00 - 12:20
Room: Room F1
Chairs: Nathalie Malbert; *IMS - Bordeaus*
          Gilles Dambrine; *IEMN*

A Gate-Last In0.53Ga0.47As Channel FinFET with Molybdenum Source/Drain Contacts 177
Xingui Zhang, Hua Xin Guo, Xiao Gong, Yee-Chia Yeo
National University of Singapore, Singapore

Complementary RF-LDMOS Transistors Realized with Standard CMOS Implantations 181
Andreas Mai, Holger Rücker
IHP, Germany

TCAD Degradation Modeling for LDMOS Transistors 185
Susanna Reggiani[2], Gaetano Barone[2], Elena Gnani[2], Antonio Gnudi[2], Stefano Poli[1], Ming-Yeh Chuang[1], Weidong Tian[1], Rick Wise[1]
[1]Texas Instruments Incorporated, United States; [2]Università di Bologna, Italy

Pulsed I(V) - Pulsed RF Measurement System for Microwave Device Characterization with 80ns/45GHz 189
Mario Weiß, Sébastien Fregonese, Marco Santorelli, Amit Kumar Sahoo, Cristell Maneux, Thomas Zimmer
Université de Bordeaux, France

B3L-G DRAMs and SRAMs

Date: Wednesday, September 19, 2012
Time: 11:00 - 12:20
Room: Room F2
Chairs: Andreas Schenk; *ETH Zurich*
          Isodiana Crupi; *CNR - IMM Catania (Italy)*

Novel Deep Trench Buried-Body-Contact (DBBC) of 4F2 Cell for Sub 30nm DRAM Technology 193
Youngseung Cho[1], Yoosang Hwang[2], Huijung Kim[2], Eunok Lee[2], Soojin Hong[2], Hyunwoo Chung[2], Daeik Kim[2], Jiyoung Kim[2], Yongchul Oh[2], Hyeongsun Hong[2], Gyo-Young Jin[2], Chilhee Chung[2]
[1]Samsung Electronics Co., LTD., Korea, South; [2]Semiconductor R&D Center, Samsung Electronics Co., Korea, South

Z²-Fet Used as 1-Transistor High-Speed DRAM 197
Jing Wan[3], Cyrille Le Royer[2], Alex Zaslavsky[1], Sorin Cristoloveanu[3]
[1]Brown University, United States; [2]CEA-LETI-MINATEC, France; [3]IMEP-INPG/Minatec, France

A 5.61 pJ, 16 Kb 9T SRAM with Single-Ended Equalized Bitlines and Fast Local Write-Back for Cell Stability Improvement 201
Qi Li, Bo Wang, Tony Kim
Nanyang Technological University, Singapore

An Advanced Statistical Compact Model Strategy for SRAM Simulation at Reduced VDD 205
Plamen Asenov[2], Dave Reid[1], Scott Roy[2], Campbell Millar[3], Asen Asenov[2]
[1]Gold Standard Simulations Ltd., United Kingdom; [2]University of Glasgow, United Kingdom; [3]University of Glasgow, Gold Standard Simulations Ltd., United Kingdom

Table of Contents

B7L-E Mobility Characterization and Parameter Extraction in Advanced MOSFETs

Date: Wednesday, September 19, 2012
Time: 16:30 - 18:10
Room: Room E2
Chairs: Henryk Przewlocki; *Inst. of Electron Technology*
          Stefan Bengtsson; *Chalmers University of Technology*

Multibranch Mobility Characterization: Evidence of Carrier Mobility Enhancement by Back-Gate Biasing in FD-SOI MOSFET 209
Carlos Navarro[6], Noel Rodriguez[5], Luca Donetti[5], Akiko Ohata[4], Francisco Gamiz[5], François Andrieu[1], Olivier Faynot[1], Claire Fenouillet-Berangerand[2], Sorin Cristoloveanu[3]
[1]CEA-LETI-MINATEC, France; [2]CEA-LETI-MINATEC / STMicroelectronics, France; [3]IMEP-INPG/Minatec, France; [4]Osaka City University, Japan; [5]Universidad de Granada - CITIC, Spain; [6]Universidad de Granada - CITIC / IMEP-Minatec, Spain

The Role of the Temperature on the Scattering Mechanisms Limiting the Electron Mobility in Metal-Oxide-Semiconductor Field-Effect-Transistors Fabricated on (110) Silicon-Oriented Wafers 213
Philippe Gaubert, Akinobu Teramoto, Shigetoshi Sugawa, Tadahiro Ohmi
Tohoku University, Japan

New Parameter Extraction Method Based on Split C-V for FDSOI MOSFETs 217
Imed Ben Akkez[4], Antoine Cros[5], Claire Fenouillet-Beranger[1], Frederic Boeuf[5], Quentin Rafhay[2], Francis Balestra[2], Gérard Ghibaudo[3]
[1]CEA-LETI/STMicroelectronics (Crolles), France; [2]IMEP-LAHC, France; [3]IMEP-LAHC, MINATEC, INPG, France; [4]IMEP-LaHc/STMicroelectronics, France; [5]STMicroelectronics, France

Methodology for Extracting the Characteristic Capacitances of a Power MOSFET Transistor, Using Conventional on-Wafer Testing Techniques 221
Christoph Kerner, Ivan Ciofi, Thomas Chiarella, Stefaan Van Huylenbroeck
IMEC, Belgium

B7L-F Advanced Photodetectors

Date: Wednesday, September 19, 2012
Time: 16:30 - 18:10
Room: Room F1
Chairs: Denis Flandre; *UC Lovain*
          Lorenzo Faraone; *Univ. Western Australia*

A Gate Modulated Avalanche Bipolar Transistor in 130nm CMOS Technology 226
Robert Henderson, Eric A. G. Webster, Richard J. Walker
University of Edinburgh, United Kingdom

Low-Noise and Large-Area CMOS SPADs with Timing Response Free from Slow Tails 230
Danilo Bronzi[2], Federica Villa[2], Simone Bellisai[2], Bojan Markovic[2], Simone Tisa[2], Alberto Tosi[2], Franco Zappa[2], Sascha Weyers[1], Daniel Durini[1], Werner Brockherde[1], Uwe Paschen[1]
[1]Fraunhofer IMS, Germany; [2]Politecnico di Milano, Italy

Extreme Temperature 4H-SiC Metal-Semiconductor-Metal Ultraviolet Photodetectors 234
Wei-Cheng Lien[3], Albert P. Pisano[3], Dung-Sheng Tsai[1], Jr-Hau He[1], Debbie G. Senesky[2]
[1]National Taiwan University, Taiwan; [2]Stanford University, United States; [3]University of California, Berkeley, United States

Table of Contents

A Silicon Photomultiplier with >30% Detection Efficiency from 450-750nm and 11.6µm Pitch NMOS-Only Pixel with 21.6% Fill Factor in 130nm CMOS..........................238
Eric A. G. Webster[2], Richard J. Walker[2], Robert Henderson[2], Lindsay A. Grant[1]
[1]*STMicroelectronics, United Kingdom;* [2]*University of Edinburgh, United Kingdom*

B7L-G	**Analog/Low Power Devices**

Date: Wednesday, September 19, 2012
Time: 16:30 - 18:10
Room: Room F2
Chairs: Kazu Ishimaru; *Toshiba*
Thomas Ernst; *CEA-LETI*

Addressing Healthcare Challenges Using Semiconductor Technology Tools and Approaches... No paper
Peter Peumans
IMEC, Belgium

Low-Power DRAM-Compatible Replacement Gate High-k/Metal Gate Stacks 242
Romain Ritzenthaler[1], Tom Schram[1], Erik Bury[2], Jerome Mitard[1], Lars-Ake Ragnarsson[1], Guido Groeseneken[2], Naoto Horiguchi[1], Aaron Thean[1], Alessio Spessot[3], Christian Caillat[3], Vidya Srividya[3], Pierre Fazan[3]
[1]*IMEC, Belgium;* [2]*IMEC - Katholieke Universiteit Leuven, Belgium;* [3]*Micron Technology, Belgium*

On the UTBB SOI MOSFET Performance Improvement in Quasi-Double-Gate Regime...246
Valeriya Kilchytska[2], Denis Flandre[2], François Andrieu[1]
[1]*CEA-LETI-MINATEC, France;* [2]*Université catholique de Louvain, Belgium*

C2L-E	**Emerging Devices**

Date: Thursday, September 20, 2012
Time: 09:40 - 10:40
Room: Room E2
Chairs: Stephen Hall; *University of Liverpool*
Anthony O'Neill; *Newcastle University*

An Integration Approach for Graphene Double-Gate Transistors......................................250
Sam Vaziri[1], Anderson Smith[1], Christoph Henkel[1], Mikael Östling[1], Max C. Lemme[1], Grzegorz Lupina[2], Gunther Lippert[2], Jaroslaw Dabrowski[2], Wolfgang Mehr[2]
[1]*KTH Royal Institute of Technology, Sweden;* [2]*Leibniz-Institut für Innovative Mikroelektronik, IHP, Germany*

MTJ-Based Implication Logic Gates and Circuit Architecture for Large-Scale Spintronic Stateful Logic Systems..254
Hiwa Mahmoudi, Viktor Sverdlov, Siegfried Selberherr
Technische Universität Wien - IUE, Austria

Resistive Switching Memory Using Titanium-Oxide Nanoparticle Films 258
Emanuele Verrelli[2], Dimitris Tsoukalas[2], Pascal Normand[3], Nikos Boukos[3], Alistair H. Kean[1]
[1]*Mantis Deposition Ltd., United Kingdom;* [2]*National Technical University of Athens, Greece;* [3]*NCSR Demokritos, Greece*

Table of Contents

C2L-F   Characterization of Aging and Failure Mechanisms

Date: Thursday, September 20, 2012
Time: 09:40 - 10:40
Room: Room F1
Chairs: Montserrat Nafria; *UAB*
Joachim Burghartz; *Institut für Mikroelektronik Stuttgart*

An Array-Based Chip Lifetime Predictor Macro for Gate Dielectric Failures in Core and IO FETs..262
Pulkit Jain[2], John Keane[1], Chris Kim[2]
[1]Intel Corporation, United States; [2]University of Minnesota, United States

Unified Characterization of RTN and BTI for Circuit Performance and Variability Simulation 266
Nuria Ayala, Javier Martin-Martinez, Rosana Rodriguez, Montse Nafria, Xavier Aymerich
Universitat Autònoma de Barcelona, Spain

Kink Effect Characterization in AlGaN/GaN HEMTs by DC and Drain Current Transient Measurements..270
Laurent Brunel[2], Nathalie Malbert[2], Arnaud Curutchet[2], Nathalie Labat[2], Benoit Lambert[1]
[1]United Monolithic Semiconductors, France; [2]Université de Bordeaux, France

C3L-F   Resistive Memories

Date: Thursday, September 20, 2012
Time: 11:00 - 12:20
Room: Room F1
Chairs: Andrea Lacaita; *Politecnico di Milano*
Fernanda Irrera; *Università Roma La Sapienza*

Random Telegraph Signal Noise Properties of HfOx RRAM in High Resistive State..274
Francesco Maria Puglisi[2], Paolo Pavan[2], Andrea Padovani[2], Luca Larcher[2], Gennadi Bersuker[1]
[1]SEMATECH, United States; [2]Università degli Studi di Modena e Reggio Emilia, Italy

On the Impact of Ag Doping on Performance and Reliability of GeS2-Based Conductive Bridge Memories 278
Elisa Vianello[2], Carlo Cagli[2], Gabriel Molas[2], Emeline Souchier[2], Philippe Blaise[2], Catherine Carabasse[2], Guillaume Rodriguez[2], V. Jousseaume[2], Barbara De Salvo[3], Florian Longnos[1], Faiz Dahmani[1], Pascal Verrier[1], Damien Bretegnier[1], Jacques Liebault[1]
[1]Altis Semiconductor, France; [2]CEA-LETI, France; [3]CEA-LETI-MINATEC, France

Analysis of the Effect of Cell Parameters on the Maximum RRAM Array Size Considering Both Read and Write...282
Leqi Zhang[2], Stefan Cosemans[1], Dirk Wouters[2], Guido Groeseneken[2], Malgorzata Jurczak[1]
[1]IMEC, Belgium; [2]IMEC - Katholieke Universiteit Leuven, Belgium

Table of Contents

Carbon-Doped Ge2Sb2Te5 Phase-Change Memory Devices Featuring Reduced Reset Current and Power Consumption..................286
Quentin Hubert[1], Carine Jahan[1], Alain Toffoli[1], Gabriele Navarro[1], Sandhya Chandrashekar[1], Pierre Noé[1], Véronique Sousa[1], Luca Perniola[1], Jean-François Nodin[1], Alain Persico[1], Sylvain Maitrejean[1], Anne Roule[1], Ewen Henaff[1], Magali Tessaire[1], Paola Zuliani[3], Roberto Annunziata[3], Gilles Reimbold[1], Georges Pananakakis[2], Barbara De Salvo[1]
[1]*CEA-LETI-MINATEC, France;* [2]*IMEP - LAHC, France;* [3]*STMicroelectronics, Italy*

C3L-G   Quantum Transport

Date: Thursday, September 20, 2012
Time: 11:00 - 12:20
Room: Room F2
Chairs: Denis Rideau; *STMicroelectronics*
          David Esseni; *University of Udine*

Transport Properties of Strained Silicon Nanowires..................290
Yann Michel Niquet[3], Christophe Delerue[2], Viet Hung Nguyen[3], Christophe Krzeminski[2], François Triozon[1]
[1]*CEA-LETI-MINATEC, France;* [2]*IEMN, France;* [3]*SP2M, France*

Tin Nanowire Field Effect Transistor..................294
Lida Ansari[3], Giorgos Fagas[2], James C. Greer[1]
[1]*Tyndall National Institute, Ireland;* [2]*Tyndall National Institute, University College Cork, Ireland;* [3]*University College Cork, Tyndall National institute, Ireland*

Effects of Disorder on Transport Properties of Extremely Scaled Graphene Nanoribbons..................298
Mirko Poljak[1], Emil Song[1], Minsheng Wang[1], Tomislav Suligoj[2], Kang Wang[1]
[1]*University of California, Los Angeles - DRL, United States;* [2]*University of Zagreb - FER-ZEMRIS, Croatia*

C6L-E   GaN-based Power Switches

Date: Thursday, September 20, 2012
Time: 16:30 - 17:50
Room: Room E2
Chairs: Gaudenzio Meneghesso; *University of Padova*
          Tetsuya Suemitsu; *Tohoku University*

High Temperature Behaviour of GaN-on-Si High Power MISHEMT Devices..................302
Dirk Wellekens, Rafael Venegas, Xuanwu Kang, Mohammed Zahid, Tian-Li Wu, Denis Marcon, Puneet Srivastava, Marleen Van Hove, Stefaan Decoutere
IMEC, Belgium

High Voltage Low Ron In-situ SiN/Al0.35GaN0.65/GaNon-Si Power HEMTs Operation Up to 300 °C..................306
Abel Fontserè[1], Amador Pérez-Tomás[1], Phillipe Godignon[1], Jose Millán[1], John M. Parsey[2], Peter Moens[2]
[1]*IMB-CNM-CSIC, Spain;* [2]*On-Semiconductors, Belgium*

Critical Gate Module Process Enabling the Implementation of a 50A/600V AlGaN/GaN MOS-HEMT..................310
Sameh Khalil[2], Rongming Chu[2], Ray Li[2], Danny Wong[2], Scott Newell[2], Xu Chen[2], Mary Chen[2], Daniel Zehnder[2], Samuel Kim[2], Andrea Corrion[2], Brian Hughes[2], Karim Boutros[2], Chandra Namuduri[1]
[1]*General Motors, United States;* [2]*HRL Laboratories, LLC, United States*

Table of Contents

Scaling of InAlN/GaN Power Transistors...314
Daniel Piedra[2], Hyung-Seok Lee[2], Tomas Palacios[2], Xiang Gao[1], Shiping Guo[1]
[1]IQE RF LLC, United States; [2]Massachusetts Institute of Technology, United States

---

C6L-F   Semi-classical Transport

Date: Thursday, September 20, 2012
Time: 16:30 - 17:50
Room: Room F1
Chairs: Bernd Meinerzhagen; *TU Braunschweig*
          Tibor Grasser; *Vienna University of Technology*

**Deterministic Simulation of 3D and Quasi-2D Electron and Hole Systems in
SiGe Devices**...318
Christoph Jungemann[1], Anh-Tuan Pham[3], Sung-Min Hong[2], Bernd Meinerzhagen[4]
*[1]RWTH Aachen University, Germany; [2]Samsung, United States; [3]Synopsys Inc.,
United States; [4]Technische Universität Braunschweig, Germany*

**A Multi-Subband Monte Carlo Study of Electron Transport in Strained SiGe N-
Type FinFETs**...322
Daniel Lizzit[1], Pierpaolo Palestri[1], David Esseni[1], Francesco Conzatti[2], Luca Selmi[1]
[1]Università degli studi di Udine, Italy; [2]University of Udine, Italy

Electron Transport in Germanium Junctionless Nanowire Transistors.................326
Pedram Razavi, Giorgos Fagas, Isabelle Ferain, Ran Yu, Samaresh Das
Tyndall National Institute, University College Cork, Ireland

C6L-G   Low Frequency Noise in Next Generation FET Devices

Date: Thursday, September 20, 2012
Time: 16:30 - 17:50
Room: Room F2
Chairs: Paolo Pavan; *Universita degli Studi di Modena e Reggio Emilia*
          Gunnar Malm; *KTH Royal Institute of Technology*

**Low-Frequency Noise Assessment of the Transport Mechanisms in SiGe
Channel Bulk FinFETs**...330
Tommaso Romeo[1], Luigi Pantisano[1], Eddy Simoen[1], Raymond Krom[1], Mitsuhiro
Togo[1], Naoto Horiguchi[1], Jerome Mitard[1], Aaron Thean[1], Guido Groeseneken[2], Cor
Claeys[1], Felice Crupi[3]
*[1]IMEC, Belgium; [2]IMEC - Katholieke Universiteit Leuven, Belgium; [3]Università della
Calabria, Italy*

**Impact of Front-Back Gate Coupling on Low Frequency Noise in 28 nm FDSOI
MOSFETs**...334
Christoforos Theodorou[1], Eleftherios Ioannidis[2], Sebastien Haendler[4], Nicolas
Planes[4], Franck Arnaud[4], Jalal Jomaah[2], Charalabos Dimitriadis[1], Gérard Ghibaudo[3]
*[1]Aristotle University of Thessaloniki, Greece; [2]IMEP-LAHC, MINATEC, France;
[3]IMEP-LAHC, MINATEC, INPG, France; [4]STMicroelectronics, France*

**On the Correlation Between the Retention Time of FBRAM and the Low-
Frequency Noise of UTBOX SOI nMOSFETs**..338
Eddy Simoen[1], Marc Aoulaiche[1], Anabela Veloso[1], Gosja Jurczak[1], Cor Claeys[1],
Abraham Luque Rodríguez[3], Juan Antonio Jiménez Tejada[3], Luciano Mendes
Almeida[4], Maria Gloria C. Andrade[4], Christian Caillat[2], Pierre Fazan[2]
*[1]IMEC, Belgium; [2]Micron Technology, Belgium; [3]Universidad de Granada, Spain;
[4]Universidade de São Paulo, Brazil*

Effect of Substrate Bias on Frequency Dependence of MOSFET Noise Intensity.................342
Kenji Ohmori, Ranga Hettiarachchi, Keisaku Yamada
University of Tsukuba, Japan

Welcome to ESSDERC 2012

On behalf of the Organizing Committees of ESSDERC 2012, it is our pleasure to welcome you to the 42th European Solid-State Device research conference. ESSDERC 2012 runs in parallel to his sister conference ESSCIRC 2012, covering all aspects of modern solid-state systems, circuits and devices at a single event. The increasing level of integration for system-on-chip design made available by advances in silicon technology is stimulating more than ever before the need for deeper interaction among technologists, device experts and circuits and system designers. As a participant at ESSDERC and ESSCIRC, you will have the opportunity to learn of the latest advances in these fields, and to meet those who have dared, pioneered and succeeded.

The conferences are to be held at the Bordeaux Convention Center, ideally situated next to the ring road, close to the city center and just 10 minutes from Bordeaux-Merignac International Airport. Bordeaux's downtown is very close by and, from there visitors will find Bordeaux's major monuments and shops just a few short steps away.

This year, a total of 130 submissions originating from 28 countries were received for ESSDERC including 83 papers coming from Europe, 37 from Asia-Pacific and 10 from North-America. This is a proof of the truly international nature of ESSDERC. The Technical Program Committee with about 110 world-class experts from academia and industry selected 67 papers for oral presentations. Seven session invited papers have been brought in with especially two joint ESSDERC/ESSCIRC sessions: the first one is dedicated to compact modeling, the second is dedicated to reliability and variability. Twelve plenary presentations by outstanding guest speakers complete the program by focusing on highly relevant topics selected by the Technical Program Committees of both conferences. In addition to the conference programs, a pre-conference day with introductory tutorials and a post-conference day with workshops showcasing work currently being carried out by European research consortia will also be held.

We would like to thank the Steering Committee of ESSDERC/ESSCIRC for giving us the opportunity to organize this event.

The conference has been organized by members of the Institute IMS of Bordeaux, the University of Bordeaux 1, the CNRS and IPB (Institute Polytechnique de Bordeaux). We would like to thank the authorities of these institutions for their support and for allowing us to devote part of our time to the organization.

We have been extremely fortunate to have the help of an exceptional team of volunteers of the Organizing Committee and the Technical Program Committee, who have all worked very hard. We are hugely indebted to all these volunteers. Our warm thanks to all of them for their dedication, enthusiasm and professionalism.

Last but not least, we would like to express our greatest appreciation to all the authors who submitted papers to the conference and to all delegates, tutorial lecturers and plenary speakers who have travelled to Bordeaux to interact and share their thoughts during the conference.

Enjoy ESSDERC/ESSCIRC 2012 conference and your visit to Bordeaux. We hope to see you all back here more often.

Welcome, Bienvenue!

Yann Deval
General Chair – ESSDERC/ESSCIRC 2012

Thomas Zimmer, Thomas Skotnicki
TPC chairs – ESSDERC 2012

Committees

Organization Committee

Conference Chair	Yann Deval IMS Laboratory
ESSCIRC TPC Chairs	JB Begueret IMS Laboratory
	Didier Belot STMicroelectronics
ESSDERC TPC Chairs	Thomas Zimmer IMS Laboratory
	Thomas Skotnicki STMicroelectronics
Tutorials & Workshops Chairs	Thierry Taris IMS Laboratory
	Cristell Maneux IMS Laboratory
Industrial relationship Chairs	Doug Smith SMSC
	Patrice Gamand NXP
Publicity Chair	Domine Leenaerts NXP
Finance Chair	Magali De Matos IMS Laboratory
Publication Chairs	Eric Kerhervé IMS Laboratory
	Victor Dupuy IMS Laboratory
Local Arrangement Chairs	Olivier Mazouffre IMS Laboratory
	Nathalie Malbert IMS Laboratory
Secretariat	Christine Bogdan IMS Laboratory

Steering Committee

Sorin Cristoloveanu (Chair)	ENSERG-IMEP
Ralf Brederlow (Vice-Chair)	Texas Instruments
Cor Claeys (Executive secretary)	IMEC
Roberto Bez	Numonyx
Klaas Bult	Broadcom Corporation
Franz Dielacher	Infineon
Athanasios Dimoulas	NCSR Demokritos
Christian Enz	CSEM
Stephen Hall	Univ. of Liverpool
Kazunari Ishimaru	Toshiba
Carlos Mazure	SOITEC
Gaudenzio Meneghesso	Univ. of Padova
Doris Schmitt-Landsiedel	TU Munich
Thomas Skotnicki	STMicroelectronics
Hannu Tenhunen	KTH Stockholm
Roland Thewes	TU Berlin

ESSDERC Technical Program Committee

Ahopelto, Jouni	*VTT*	Faraone, Lorenzo	*U. Western Australia*
Andrieu, Francois	*CEA-Leti*	Fischetti, Massimo	*U. Massachussets*
Armstrong, Mervin	*QU Belfast*	Flandre, Denis	*UC Louvain*
Asenov, Asen	*U.Glasgow*	Franssila, Sami	*Aalto/VTT*
Ashburn, Peter	*Southampton*	Galy, Philippe	*ST Microelectronics Crolles*
Augendre, Emmanuel	*CEA-LETI*		
Baccarani, Giorgio	*U. Bologna*	Gamiz, Francisco	*U. Granada*
Baldi, Livio	*Numonyx*	Ghibaudo, Gerard	*IMEP*
Bauer, Anton	*Fraunhofer*	Godoy, Andres	*U. Granada*
Bawedin, Maryline	*University of Montpellier*	Grabiec, Piotr	*Institute of Electron Technology, Warsaw*
Beck, Romuald	*TU-Warsaw*	Grasser, Tibor	*TU-Wien*
Bengtsson, Stefan	*Chalmers*	Guiducci , Carlotta	*EPFL*
Bez, Roberto	*Numonyx*	Hall, Steve	*U. Liverpool*
Boeck, Josef	*Infineon*	Haspeslagh, Luc	*IMEC*
Burghartz, Joachim	*IMS*	Heinemann, Bernhard	*IHP*
Cimalla, Volker	*Fraunhofer, IAF; TU Imenau*	Hellström, Per-Erik	*KTH*
		Hijzen, Erwin	*NXP*
Colinge, Jean-Pierre	*Tyndall*	Huang, Ru	*Peking University*
Collaert, Nadine	*IMEC*	Hueting, Ray	*U. Twente*
Crupi, Isodiana	*Center MATIS CNR-IMM, Catania*	Iannaccone, Giuseppe	*U. Pisa*
		Ioannou, Dimitri	*George Mason University*
Dambrine, Gilles	*U. Lille*		
De Keersgieter, An	*IMEC*	Ionescu, Adrian	*EPFL*
De Meyer, Kristin	*IMEC*	Irrera, Fernanda	*U.Roma 'La Sapienza'*
De Salvo, Barbara	*CEA-LETI*		
De Wolf, Ingrid	*IMEC*	Ishihara, Ryoichi	*TU Delft*
Deleonibus, Simon	*CEA-LETI*	Ishimaru, Kazunari	*Toshiba*
Dubois, Emmanuel	*U. Lille*	Iwai, Hiroshi	*Tokyo Institute of Technology*
Ernst, Thomas	*CEA-LETI*		
Esseni, David	*U. Udine*	Jaouen, Herve	*STMicroelectronics*

JAUD, Marie Anne	*LETI*	Pozzovivo, Gianmauro	*Infineon*
Kreupl, Franz	*Qimonda*	Przewlocki, Henryk	*TU-Warsaw*
Krishnamohan, Tejas	*U.Stanford*	Rideau, Denis	*STMicroelectronics, Crolles*
Lacaita, Andrea	*Politecnico Milano*		
Lee, Sung-Young	*Samsung*	Rodriguez, Noel	*U. Granada*
Lemme, Max	*KTH*	Rudan, Massimo	*U. Bologna*
Liu, Ran	*Fudan*	Sangiorgi, Enrico	*U. Bologna*
Lombardo, Salvatore	*CNR*	Schenk, Andreas	*ETHZurich*
Lorenz, Jurgen	*Fraunhofer*	Schmitz, Jurriaan	*U. Twente*
Lukasiak, Lidia	*TU-Warsaw*	Selmi, Luca	*U. Udine*
Malbert, Nathalie	*IMS Lab,*	Shin, Changhwan	*University of Seoul*
,	*University Bordeaux*	Sizov, Fiodor	*Lashkaryov Institute of Semiconductor Physics, NAS of Ukraine*
Malm, Gunnar	*KTH*		
Maneux, Cristell	*IMS Lab, University Bordeaux*		
Masahara, Meishoku	*AIST*	Skotnicki, Thomas	*STMicroelectronics*
Mazure, Carlos	*SOITEC*	Soree , Bart	*U. Antwerp and IMEC*
Meinerzhagen, Bernd	*TU-Braunschweig*		
Meneghesso, Gaudenzio	*Universita' di Padova*	Steiner-Vanha, Ralph	*Sensirion*
		Suemitsu, Tetsuya	*Tohoku University*
Millan, Jose	*CNM Barcelona*	Sun, Yanning	*IBM*
Monfray, Stephane	*STMicroelectronics*	Tringe, Joseph	*Lawrence Livermore National Laboratory*
Moroz, Victor	*Synopsys*	,	
Nafria, Monserrat	*UA Barcelona*	Tsoukalas, Dimitris	*NTUA*
Nassiopoulou, Androula	*Demokritos*	Uchida, Ken	*Tokyo Institute Technol*
Nazarov, Alexei	*ISP-Kiev*		
Oldiges, Phil	*IBM*	Ueda, Tetsu	*Panasonic*
Olsson, Jörgen	*Uppsala University*	Wachutka, Gerhard	*TU-Munich*
O'Neill, Anthony	*U. Newcastle*	Woltjer, Reinout	*NXP*
Palacios, Tomas	*MIT*	Wu, DongPing	*Fudan*
Pandini, Davide	*STMicroelectronics*	Yoshimura, Katsunobu	*Aichi Science and Technology Foundation*
Paulasto-Kröckel, Mervi	*Aalto*		
Pavan, Paolo	*Università di Modena e Reggio Emilia*	Zimmer, Thomas	*IMS, U. Bordeaux 1*

Sponsors

Platinum

Silver

Regular

Future Silicon Technology

Dr. Kinam Kim
SAIT, Samsung Electronics Co.
Yongin-City, Gyunggi-Do, Rep. of Korea

Abstract—**Dimensional scaling will continue in Si CMOS technology which will extend to beyond 10nm. Key challenges for dimensional scaling and expansion of silicon-based technologies as well as research directions will be reviewed in traditional semiconductor applications such as DRAM, NAND Flash, logic as well as advanced devices including STT-MRAM, ReRAM and reconfigurable logic. Furthermore, other areas where Si technologies play import roles will be presented including power electronics, solid-state lighting as well as DNA sequencing and medical imaging.**

I. INTRODUCTION

The remarkable evolution of IT devices was enabled by the fast advancing silicon technology. The performances of devices such as CPU and memory improved tremendously over the past 20 years. The CPU performance improved 2400 times during this period, while DRAM and NAND Flash improved 1000 and 32000 times, respectively. And the mobile network has been getting faster by 840 times.

During the same period, the semiconductor market grew from sixty billion dollars in 1991 to over three hundred billion dollars in 2011 [1]. Over 67% of the current market has been driven by the advances of silicon scaling technologies as explained by Moore's Law. Silicon scaling technology, which is expected to reach sub-10nm, will continue to contribute to the growth of the semiconductor market in the future.

In addition to having an impact on the IT industry, silicon technology is bringing innovation to areas such as energy, health and medical applications: ultra fast DNA sequencing, extremely compact/efficient medical imaging devices, and energy efficient low cost lighting among others. These applications, fueled by Si technologies, could eventually take up a major portion of the semiconductor market in the next decade. This paper will attempt to present current work in emerging areas as well as conventional Si technologies..

II. MAINSTREAM SILICON TECHNOLOGY

A. Towards sub-20 nm DRAMs

DRAM density has been doubled every 18 months through scaling of the critical dimension, and it is now nearing the 20nm node as shown in Figure 1. Major challenges in further scaling, even down to sub-10nm, will require device innovations in terms of cell storage capacitors, cell array transistors (CAT) and patterning processes [2].

The cell capacitor structure has changed through generations in order to maintain the minimum cell capacitance needed to provide an adequate sensing signal and to meet retention time specifications [3]. In the current

DRAM generation, the capacitor is a unified simple cylinder-type structure of an extremely high aspect ratio (AR). For sub-10nm DRAMs, however, structural innovation may no longer be expected due to physical limitations in accommodating complex 3D structures.

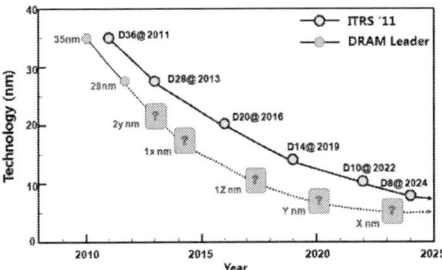

Figure 1. ITRS & DRAM Leader DRAM technology roadmap

At sub-10nm, the distance between electrodes becomes about 10nm, requiring the physical thickness of the storage electrodes and the dielectrics to be less than 5nm, This is very difficult to realize. But this limiting cell capacitor could be compensated by technological innovations in the CAT design. The CAT has to be designed for extremely low leakage current in order to obtain a high sensing signal.

The leakage current of the storage node is caused by the gate induced drain leakage, sub-threshold, as well as junction leakage of CAT. Data retention time, which is determined by the amount of stored charges and time dependent charge losses of the storage node, can be improved by suppressing it as shown in the evolution of CAT: PCAT (planar CAT) to RCAT (recessed CAT) to BCAT (buried CAT) to a new CAT like VCAT (vertical CAT) in the future [2, 4].

Furthermore, current sensing capability could be enhanced by reducing the bit line capacitance, which could be achieved by vertical scaling of the bit line electrodes with lower resistivity (ρ) metals and gap filling with lower-k dielectrics between the bit lines.

Nevertheless, patterning process technologies such as lithography and etching must also be developed further. Currently, the major concern is the productivity related to such technologies as EUV, double patterning lithography, and high aspect ratio deep capacitor hole etching. EUVL's laser source power and instability, in particular, are crucial concerns in continuing the scaling trend. The slow progress of EUVL technology will result in the need for expensive double or quadruple patterning technologies (DPT or QPT) for several critical layers for the 22nm node and beyond [3].

Technology push for smaller DRAM cells will continue while post DRAM devices and materials are being developed, such as a spin transfer torque (STT)-MRAM that uses the magnetic tunnel junction (MTJ) as the resistive storage element. As the STT-MRAM consists of one MTJ and one transistor, it can have similar densities with a higher operation speed and superior endurance as compared to its DRAM counterpart [5,6,7]. So far, spin torque switching speed of a few nanoseconds and endurance over 10^{15} cycles has been demonstrated. Scaling beyond 10 nm will be possible once the perpendicular magnetic anisotropy materials and vertical transistor stacking become possible [8,9,10]. Further materials cell property optimization as well as related process developments need to be addressed for STT-MRAM to be realized, i.e. tunneling magnetoresistance over 200%, thermal stability factor ($\Delta = K_u V / k_B T$) over 60 for 10 years data retention, and spin torque switching current density less than $1MA/cm^2$ must be simultaneously satisfied [10].

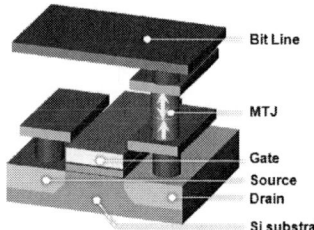

Figure 2. STT-MRAM structure where the magnetic tunnel junction (MTJ) shows two different resistance states depending on its magnetic configuration.

B. Beyond sub-20 nm NAND flash

The boom in mobile devices such as smart phones, tablets, etc, has increased the demand for NAND Flash memory. The conventional floating gate (FG) type NAND Flash cell technology is currently at the 20nm node [11]. Furthermore, sub-20nm cell using QPT has also been demonstrated [12] using the word line (WL) air-gap technology in order to overcome coupling interferences. However, as device technology enters the sub-20nm region, cell-to-cell coupling interferences, the numbers of stored electrons, WL to WL breakdown, endurance and data retention have become major issues[3, 13]. Although circuit technologies including parallel programming, shadow programming and extended ECC were successful in overcoming the cell-to-cell coupling down to the 20nm node, they would not be effective in overcoming issues for beyond the 20nm node planar NAND Flash because coupling interference ratio is inversely proportional to the design rule for 2D structures. Thus, as the design rule decreases, the coupling ratio will increase, reaching the allowed design limit, which is about 5 at around 20nm.

Unlike the 2D structured memories, the coupling interference ratio of the 3D VNAND remains far below the critical design limit even beyond the sub-10nm region, as shown in Figure 3. This is because the bit line coupling interference is almost eliminated due to a gate-all around structure. Additionally, the word line to word line coupling interference is reduced by the increased spacing between them. A wide programming window with a better erase speed

and endurance can be obtained by implementing a damascened metal gate and a TANOS (TiN-AlO-Nitride-Oxide-Si) cell, i.e. TCAT (TerabitCell Array Transistor).

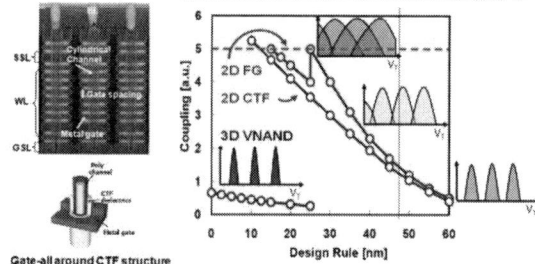

Figure 3. Gate-all around CTF structure and 3D NAND with gate-all around CTF structure. the coupling interference ratio of 3D VAND is much lower than the planar ones. The coupling interference ratio defines the charge build up which occurs while an adjacent cell is programming, it should not exceed the allowed design limit which is about 5.

Further research will be necessary in order to achieve the required read margin and write speeds, which could be negatively affected by a low mobility and high leakage of channel poly-Si. To reduce the leakage current, poly silicon has to be thin enough to be fully depleted, and its grain size has to be enlarged in order for its performance to be comparable to that of the single crystalline Si. Process technologies such as high aspect ratio channel-hole formation, bowing and leaning free stacks will also require further research.

Post-NAND devices are also being researched; one such being is the resistive RAM (ReRAM) for its structural simplicity. ReRAM is characterized by a resistive material acting as the memory element and a control element acting as a switch, resulting in a cross-point $4F^2$. This $4F^2$ ReRAM cell can be vertically stacked to obtain very high densities.

Although, there is not a clear solution for the control as well as the memory elements, there has been some progress in understanding the switching mechanism of the memory element [14]. Based on this understanding, suitable material has been used to design a ReRAM device, which showed that even under multi level cell operation, a sufficient sensing window and stable switching could be obtained with an excellent endurance, greater than 10^7 cycles. In addition, the on-off uniformity was better than 4.5 sigma(σ) with an estimated retention time of greater than 20 years.

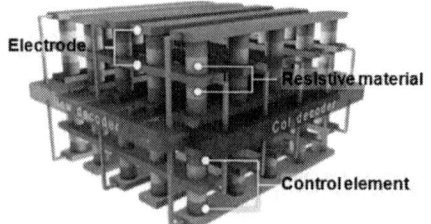

Figure 4. ReRAM with a cross-point structure with a resistive material acting as the memory element.

C. Logic technology

Up to the 20nm node, performance improvements and short channel effect (SCE) reduction could be obtained by scaling the gate dielectrics, since gate equivalent oxide thickness (EOT) is inversely proportional to the gate capacitance (~1/EOT). However, below 20nm, EOT will decrease sharply because in order to suppress SCE, channel doping concentration has to be increased rapidly for planar transistors. The EOT constraints can be relaxed if a multi-gate structure is used. In Figure 5, the orange curve shows better gate controllability allowing further scaling [15].

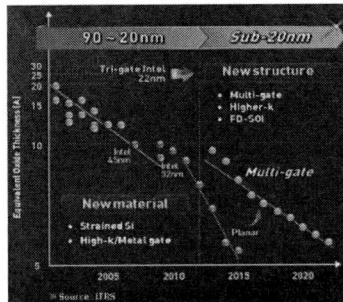

Figure 5. EOT scaling trends are shown. Due to the difficulty in controlling the SCE, a sharp decrease in EOT trend is inevitable for the coming nodes. However, historical trend can be reverted back in the case of multi-gate transistors [15]

Currently, the 3D transistor is moving from R&D to production. As of last year, Intel released the 22nm FINFET transistor that showed 26% faster operation at a lower voltage as compared to the 22nm planar transistor [16]. There still remain many challenges such as patterning, layout, design and control, etc.

Even though Si CMOS can go down to the sub 10nm regime, power will remain a critical issue. Alternative channel transistor materials are being considered in order to achieve high performance with lower power consumption. As an option for sub-10nm node, high mobility III-V channel transistor is being considered because they have less effective masses than Si, and thus, they can have higher on-currents at lower operating voltages. Recently, Intel announced the development of InGaAs multi gate quantum well FET [16]. And researchers at Purdue University also suggested a gate-all around structure using the same materials [17].

D. Advanced devices

Real-time Hybrid Reconfigurable Logic (RHRL)

Reconfigurable logic can provide significant flexibility in circuit design, a possibility of hardware revision or update as well as reduction of hardware development times, all of which are very attractive to many hardware and software engineers. However, device footprint, power consumption and slow operation speeds easily outweigh the above benefits, barely being noticeable in the logic market where the major players are ASIC or ASSP.

Recently, reconfigurable logic, including reconfigurable computing, is drawing a lot of attention as the demand for flexibility or multi-standard in various application fields is increasing. Reconfigurable logic can also relieve the software overhead in the processing units through a proper hardware organization [18]. Figure 6 is an example of a multi-functional and real-time-reconfigurable logic, where it is unnecessary to have bulky circuits such as look-up table and interconnect circuits, seen in conventional SRAM-based reconfigurable logic. External ROM or flash memory is also required in the conventional device in order to store the operation-code, which requires a large footprint and much power.

On the other hand, RHRL utilizes fast nonvolatile memories such as ReRAM and MRAM as logic setter that allows several advantages for a real-time reconfigurable logic. First, RHRL is compatible to CMOS processes; and more importantly, it changes functions in real-time. It can also perform much wider functions with the reduced area as compared to current ROM and SRAM based ones. Thus, RHRL will likely be useful for multi-standard applications, e.g. radio frequency communications in 3G, 4G, WiFi, etc., with a single chip.

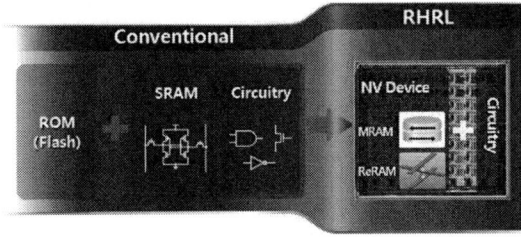

Figure 6. Real-time hybrid reconfigurable logic (RHRL) compared to the SRAM based reconfigurable logic

III. INNOVATIONS BASED ON SILICON TECHNOLOGIES

A. Silicon technology and optoelectronics

Optical interconnect

Si processing technology can integrate photonics and electronics on a Si die, enabling ultrahigh speed and low power consumption for data transmissions using optics [19]. Optical interconnect is an emerging technology for connectivity to overcome the speed bottleneck of electrical interconnects. Due to its high cost, use of optical connectivity has been limited to overseas communications and enterprise networks. However, with the decrease of manufacturing costs, optical interconnect in rack-to-rack and board-to-board levels can be observed, as shown in Figure 7 [20].

An optical interconnect system is comprised of Si optical transceivers for data conversion and waveguides or optical fibers for data transmission. Si photonics is a crucial platform to envisage optical transceivers that interconnect the optical and the electrical signals. Optical transceivers requiring the integration of silicon based optical and electrical building blocks can fully be supported with the CMOS processing

978-1-4673-1707-8/12 $31.00 © 2012 IEEE

technology. With cost effective Si photonics by 2015 [21], ultrahigh speed data communications in inter-chip and intra-chip levels may be implemented.

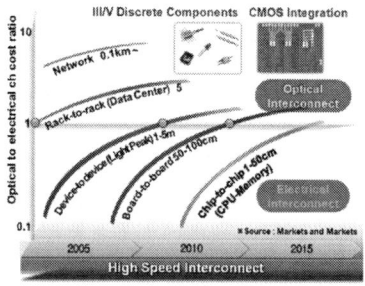

Figure 7. A trend of the high speed optical interconnects [21].

CMOS Image Sensor (CIS)

Recently, CIS and charge-coupled devices (CCD) have been the subject of high interest because of their diverse applications, especially in mobile phones, PC cameras, and digital still cameras (DSC). Even though CIS and CCD have different operating mechanism, they face similar problem due to their almost identical pixel structure. When the pixel size decreases, the amount of light reaching the photodiode also decreases resulting in the degeneration of the device performance. As a solution, wavelength selective organic photoconductive film and vertically stacked structure has been proposed [22, 23]. As reliability and performance of the color-selective organic photodiode are swiftly enhanced, it is possible that vertically stacked new image sensor can be realized in the near future [24].

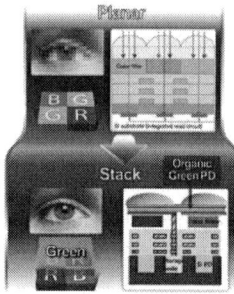

Figure 8. The comparison of planar CIS and stacked organic CIS

B. Silicon technology in the health industry

DNA sequencing

Fluorescence based detection on glass has been a well established method in DNA Sequencing. Recently an innovative sequencing device using Si- technology has been introduced [25].

Instrumental costs can be lowered by using a CMOS sensor and natural nucleotides, building blocks of a DNA molecule, instead of expensive laser scanning systems and fluorescent labeled nucleotides, respectively. After putting sample DNA into a well of the CMOS sensor, reagent solution with one of four nucleotides (adenine, guanine, cytosine and thymine) is introduced alternately into the well. Then DNA can be identified by detecting an electric signal, which is generated when protons, the by-product of DNA synthesis, are accumulated on the gate of CMOS sensor. Sequencing throughput can be increased by parallelizing the sequencing process, i.e. integrating large number of sensors on a chip. It has been announced that a person's entire DNA may be read in a day for $1,000 by the end of 2012 [26].

Medical Imaging – Ultrasound on Si

Currently, the transducer for ultrasonic probe is based on PZT (lead zirconate titanate) material. However, one must go through the meticulous processes of mechanically grinding it down to a designed thickness and dicing it to a specific pitch and number of elements, while maintaining the control wires attached to this transducer one-by-one. Therefore, the low yield and large deviation in performance between the elements are inevitable in the final product. The process is twice as difficult as the transducer is arranged in the form of two dimensional arrays.

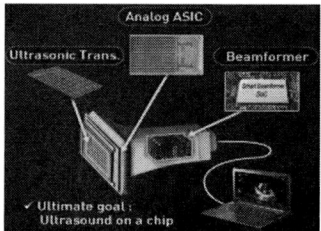

Figure 9. cMUT with control ASIC onto a single chip can be further integrated with a compact beam former into a probe.

On the other hand, an ultrasonic transducer based on the silicon technology, which is called capacitive micromachined ultrasonic transducer (cMUT), allows a higher performance with wider bandwidth. The true impact of cMUT-based probe design, however, lies in the possibility of integration with control ASIC. The integration with ASIC shortens the analog signal pass that increases the signal quality and enhances controllability. Flexibility of the aperture size and form factor can also be achieved as cMUT can be tiled.

X-ray detector on Si

The paradigm of x-ray detectors is being shifted by silicon technology. The film type x-ray detectors have been replaced with digital detectors, namely flat panel detectors (FPD), which are used in general radiography, radio/fluoroscopy and mammography systems. The signal can be detected indirectly, using scintillators that convert incident x-ray into visible light, which is detected by a p-i-n photodiode in a-Si TFT array. Or, it can be detected directly by using photoconductors that convert the incident x-ray into electron-hole pairs, yielding a superior image quality [27]. However, FPDs cannot count x-ray photons because of their charge integration detection

978-1-4673-1707-8/12 $31.00 © 2012 IEEE

method, which does not allow energy discrimination. As a consequence, multi-energy imaging is not possible.

X-ray energy detection is desirable because it enables the imaging of internal organs consisting of self tissues in the body. Detectors based on Si CMOS are capable of directly counting photons and discriminating their energy levels [28, 29].

Conventional photon counting detectors (PCD) can only take a full field x-ray image with multiple scanning, due to the areal limitation of PCD structures. If a large areal PCD is realized, it can create full field images without having to scan. With a large areal PCD, radiation exposure as well as patients' discomfort may be considerably reduced. Large areal x-ray PCDs on Si are expected to bring innovation to x-ray systems.

C. Silicon technology for energy technology

GaN LED on Si

Si technology is being introduced in the form of GaN on large scale Si wafers for solid-state lighting. GaN LED grown on a Si substrate instead of a sapphire substrate could accelerate the adoption of LED lighting due to its low cost. However, high quality GaN has been difficult to grow on Si. Recent advances show that GaN grown on a Si is comparable to the one grown on sapphire [30].

LED structures on top of a robust n-type GaN template on 8-inch diameter silicon substrates has been obtained, which has a low dislocation density and a 7 um-thick template without crack even at a sufficient Si doping condition. This high crystalline quality n-GaN has been obtained by optimizing the stress compensation and the dislocation reduction layers [31]. The measured dislocation density was about $2\sim3\times10^8/cm^2$ which is comparable to that to GaN grown on sapphire. Figure 10 shows a light emission image of 8-inch InGaN/GaN LED. The median forward voltage and the median output power at 350 mA (19.84 A/cm^2) of the vertical LED chips on Si fabricated from InGaN/GaN LED on 8-inch Si substrate were about 3.2 V and 484 mW, respectively, which was similar to that of a 4-inch LED with the same chip size [32].

Figure 10. A measured output power and operating current distribution of the vertical LED chips fabricated from InGaN/GaN LED on 8" Si wafer [32].

Power devices on Si

Today power electronic systems are very important for generating and converting energy in a sustainable way. Power semiconductor devices are one of the key components for energy-efficient power electronic systems, which require high voltage components with better performance in terms of static and dynamic loss reduction, current density, operating voltage, and operating temperatures. While Si power devices have been widely used for existing applications, the improvement of Si power device is hindered by the fundamental intrinsic materials properties. GaN-based high electron mobility transistors (HEMTs) have long been considered as remarkable candidates for high-power and high-frequency applications due to their high carrier mobility and high breakdown voltage characteristics [33]. Epitaxial growth of high quality GaN devices on large diameter Si substrates is one of the attractive approaches to reduce production costs, and thus GaN devices on Si can replace currently used Si-based power devices.

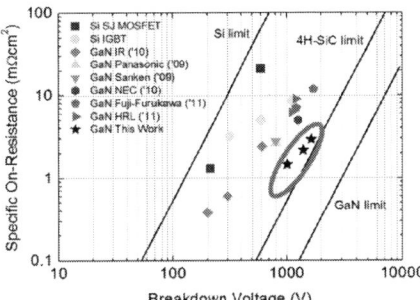

Figure 11. Specific on-resistance versus device breakdown voltage. The star in a red circle represents the most recent p-GaN HEMT devices results.

Normally-off HEMT devices with p-GaN/AlGaN/GaN heterostructure on Si substrate have been demonstrated. This device shows not only a high threshold voltage of 3 V, but also a low gate leakage current. The buffer and the device breakdown voltages exceed 1600 V, and specific on-state resistance is 2.9mΩ cm^2, which are very close to 4H-SiC limit as shown in Figure 11. The calculated figure of merit is 921 MV^2/Ωcm^2, which is the highest value reported for GaN E-mode devices [34].

Advances in GaN on Si based power devices, including hetero-epitaxial growth technology on silicon, will increase the demand for GaN devices in applications such as photovoltaics and electric vehicles.

IV. CONCLUSION

With the introduction of EUV Lithography in combination with multiple patterning for around 20nm, DRAM will be able to continue scaling down even to sub-10nm. And STT-MRAM is a strong post-DRAM candidate. 3D NAND flash will be introduced for sub-20nm node, with the possibility of being replaced by ReRAM in the future; but it will have to overcome process issues such as high aspect ratio deep hole etching. The main stream of 20nm logic will still be bulk-planar devices. However, due to the severe short channel effects, 3D multi-gate transistors using advanced materials, such as III-V channel materials, could be a promising option for future logic devices. Furthermore, RHRL, optical interconnect and advanced CIS, will enable high speed computing and smart IT devices.

978-1-4673-1707-8/12 $31.00 © 2012 IEEE

Silicon technology will offer low-cost/high performance solution to biomedical and energy applications. For instance, slow and expensive DNA sequencing can be done in one day for $1,000; bulky ultrasound probe can be made compact; x-ray imaging could detect internal organs; solid state lighting could be economical and renewable energy could become more efficient thanks to high performing silicon devices.

V. ACKNOWLEDGMENT

I thank Seong-Ho Cho, Chang-Jung Kim, Hojung Kim, Jun-Youn Kim, Kwang-Seok Kim, Kyu-sik Kim, Suhyeon Kim, Yougn-Bae Kim, Changho Lee, Y. Yvette Lee, Jae-Joon Oh, Seungjoo Park and Hyung-Jae Shin for their comments for this paper.

VI. REFERENCES

[1] iSupply market research, 2011

[2] K. Kim, U-In Chung, Y. W. Park, J.Y. Lee, J.H. Yeo and D.C. Kim, "Extending the DRAM and FLASH memory technologies to 10nm and beyond", Proceeding to the SPIE Advanced Lithography Conference, San Jose, Ca., February 2012.

[3] K. Kim, "From the future Si technology perspective: Challenges and opportunities", Technical Digest-International Electron Devices Meeting (IEDM), pp. 1–9, 2010.

[4] J. Lee, D. Ha, K. Kim, "Novel cell transistor using retracted Si3N4-liner STI for the improvement of data retention time in gigabit density DRAM and beyond ", IEEE Transactions on Electron Devices, vol. 48(6), pp. 1152–1158, June 2001.

[5] M. Durlam, Y. Chung, M. DeHerrera, B.N Engel, G. Grynkewich, B. Martino, B. Nguyen, J. Salter, P. Shah, and J.M. Slaughter, "MRAM Memory for Embedded and Stand Alone Systems", Proceedings 2007 IEEE International Conference on Integrated Circuit Design and Technology, ICICDT , art. no. 4299546 , pp. 75-78, 2007.

[6] K. Lee, S.H. Kang, "Development of embedded STT-MRAM for mobile System-on-Chips," IEEE Transactions on Magnetics, vol. 47, no. 1, pp. 131–135, 2011.

[7] S. C. Oh, J. H. Jeong , W. C. Lim, W. J. Kim, Y. H. Kim, H. J. Shin, et al., "On-axis scheme and novel MTJ structure for sub-30nm Gb density STT-MRAM", Technical Digest-International Electron Devices Meeting (IEDM), pp.12.6.1–12.6.4, 2010.

[8] H. Yoda, T. Kishi, T. Nagase, M. Yoshikawa, K. Nishiyama, E. Kitagawa, T. Daibou, et al., "High efficient spin transfer torque writing on perpendicular magnetic tunnel junctions for high density MRAMs," Current Applied Physics, vol. 10, e87–89, 2010.

[9] J. J. Nowak, R. P. Robertazzi, J. Z. Sun, G. Hu, David W. Abraham, et al., "Demonstration of Ultralow Bit Error Rates for Spin-Torque Magnetic Random-Access Memory With Perpendicular Magnetic Anisotropy", IEEE Magnetics Letters, vol.2 , pp.3000204, 2011.

[10] T. Kawahara, K. Ito, R. Takemura, H. Ohno, "Spin-transfer torque RAM technology: Review and prospect," Microelectronics Reliability, vol. 52(4), pp. 613–627, April 2012.

[11] K. Prall, K. Parat, "25nm 64Gb MLC NAND technology and scaling challenges", Technical Digest-International Electron Devices Meeting (IEDM), pp. 5.2.1–5.2.4, December 2010.

[12] J. Hwang, J. Seo, Y. Lee, S. Park, J. Leem, J. Kim, et al., "A middle-1X nm NAND flash memory cell (M1X-NAND) with highly manufacturable integration technologies,", Technical Digest-International Electron Devices Meeting (IEDM), p. 9.1.1–9.1.4, December 2011.

[13] K. Kim, "Future memory technology: challenges and opportunities," Proceedings of technical program(VLSI-TSA), pp. 5-9, April 2008.

[14] M.J. Lee, C.B. Lee, D.S. Lee, S.R. Lee, M. Chang, J.H. Hur, et al., "A fast, high-endurance and scalable non-volatile memory device made from asymmetric Ta$_2$O$_5$−x/TaO$_2$−x bilayer structures", Nature Materials, vol. 10, pp. 625–630, 2011.

[15] B.H. Lee, et al., "Gate stack technology for nanoscale devices", Materials Today, 9, pp. 32, 2006.

[16] M. Radosavljevic, G. Dewey, D. Basu, J. Boardman, B. Chu-Kung, J.M. Fastenau et al., "Electrostatics improvement in 3-D tri-gate over ultra-thin body planar InGaAs quantum well field effect transistors with high-K gate dielectric and scaled gate-to-drain/gate-to-source separation," Technical Digest-International Electron Devices Meeting (IEDM), pp.33.1.1–33.1.4, December 2011.

[17] J.J. Gu, Y.Q. Liu, Y.Q. Wu, R. Colby, R.G. Gordon, P.D. Ye, "First experimental demonstration of gate-all-around III–V MOSFETs by top-down approach," Technical Digest-International Electron Devices Meeting (IEDM), pp.33.2.1–33.2.4, December 2011.

[18] Gartner, November 2011.

[19] D. Miller, "Device Requirements for Optical Interconnects to Silicon Chips," Proceedings of the IEEE, vol. 97, no. 7, pp. 1166-1185, 2009.

[20] Optical to electrical channel cost ratio, Markets and markets 2009.

[21] M. Haurylau, G. Chen, H. Chen, J. Zhang, N. A. Nelson, D. H. Albonesi, et al., "On-Chip Optical Interconnect Roadmap: Challenges and Critical Directions," IEEE Journal of Selected Topics in Quantum Electronics, vol. 12(6), pp. 1699-1705, November 2006.

[22] S. Aihara, Y. Hirano, T. Tajima, K. Tanioka, N. Abe, N. Saito, et al., "Wavelength selectivities of organic photoconductive films: Dye-doped polysilanes and zinc phthalocyanine/tris-8-hydroxyquinoline aluminum double layer," Applied Physics Letters , vol. 82(4), pp. 511-513, 2003.

[23] S. Takada, M. Ihama, M. Inuiya, "CMOS image sensor with organic photoconductive layer having narrow absorption band and proposal of stack type solid-state image sensors", Proc, SPIE, vol. 60680A, 2006.

[24] Y. Higashi, K. S. Kim, H. G. Jeon, M. Ichikawa, "Enhancing spectral contrast in organic red-light photodetectors based on a light-absorbing and exciton-blocking layered system," Journal of Applied Physics, vol. 108(3), pp.034502-034502-5, August 2010.

[25] J. M. Rothberg, W. Hinz, T.M. Rearick, J. Schultz, W. Mileski, M. Davey, et al., "An integrated semiconductor device enabling non-optical genome sequencing", Nature, vol. 475, pp. 348–352, July 2011.

[26] http://www.lifetechnologies.com/

[27] S. I. Kim, S. W. Kim, J. C. Park, Y. Kim, S. W. Han, H. K. Kim, et al., "Highly sensitive and reliable X-ray detector with HgI2 photoconductor and oxide drive TFT," IEEE International Electron Devices Meeting (IEDM), pp. 14.2.1-14.2.4, December 2011.

[28] M. Danielsson, "Photon counting is an intuitive way to detect x-rays, which by nature are digital and have a color spectrum," Diagnostic Imaging Europe, Vol. 25, No. 7, November 2009.

[29] T. G. Schmidt, "Optimal "image-based" weighting for energy-resolved CT," Medical Physics, vol. 36(7), pp. 3018, June 2009.

[30] J. -Y. Kim, Y. Tak, J. W. Lee, H. –G. Hong, S. Chae, H. Choi, et al., "Highly efficient InGaN/GaN blue LED grown on Si (111) substrate", Invited, CLEO in Baltimore, USA, May 2011.

[31] Y. Tak, J. W. Lee, J. -Y. Kim, H.-G. Hong, S. Chae, B. Min, et al., "Metal-organic vapor phase epitaxy of high quality GaN layers with a crack-free thickness exceeding 10um on Si (111) substrate", in proc. IWN 2010, Florida, A3.6, 2010.

[32] J. -Y. Kim, Y. Tak, J. Kim, H. –G. Hong, S. Chae, J. W. Lee, et al., "Highly efficient InGaN/GaN blue LED on 8-inch Si (111) substrate", Invited, SPIE photonic west, 8262, 8262-47, USA, 2012.

[33] M. Yanagihara, Y. Uemoto, T. Ueda, T. Tanaka, and D. Ueda, "Recent advances in GaN transistors for future emerging applications", Phys. Status Solid A 206, 1221, 2009.

[34] I. Hwang, H. Choi, J. Lee, H. Choi, J. Kim, J. Ha, et al., "1.6KV, 2.9 mΩ cm^2 Normally-off p-GaN HEMT Device", International Symposium on Power Semiconductor Devices and ICs, 2012.

The industrialization of the Silicon Photonics: technology road map and applications

Maurizio Zuffada

STMicroelectronics

Milan, Italy

maurizio.zuffada@st.com

Abstract— The R&D maturity level reached by the Silicon Photonics technology envisions a clear road map fitting the needs and the challenges of the future ICT (Information Communication Technology) systems and services. Four key applications will drive the evolution: intensive computing, broadband communication, mass storage and consumer multimedia. The Silicon Photonics technology, ported in the 300 mm silicon wafer fabs, together with the most advanced package techniques, will offer a low cost manufacturing infrastructure for 3D heterogeneous system integration. Starting from a summary of the peculiarities of this technology, the unprecedented level of miniaturization and power consumption reduction, this paper will address several disruptive applications: starting from active optical cables, going to optical modules, to inter-chips communications, up to intra-chip applications, it will be shown how the Silicon Photonics technology will progressively evolve from 40 Gbps to 3200 Gbps rate, from 20 pJ/bit to 1 pJ/bit dissipation, from 4 \$/Gbps to 0.04 \$/Gbps cost and from 250 mm^3/Gbps to 0.05 mm^3/Gbps volume. Furthermore, a picture on how industry is preparing for the massive adoption of Silicon Photonics, by exploiting the existing large scale semiconductor manufacturing environment, will be provided.

I. INTRODUCTION

Data communications through metallic interconnections (cables, back-planes, boards and Integrated Circuits) show relevant attenuation and electrical crosstalk even at few GHz of bandwidth and relatively short distances [1]. Intensive computing applications in data centers are being more and more limited [2]. In particular, power consumed for I/O data transfer and for system cooling are increasing at too fast pace. While metallic substrates supporting electrical data transfer are unable to offer a reliable long term solution, photonics, up to date used for long distances, is a good candidate also for applications entailed at short distances, by taking advantage of the huge bandwidth and low energy consumption. The push toward ever increasing data rates coupled with low power consumption suggests integration of optics together with electronics on silicon giving rise to Silicon Photonics, the technology discussed in this paper. The chart, shown in fig.1, synthesizes the foreseen electrical limits versus data communication distances. Traditional hybrid photonics, based

on the 2D/3D assembly of III-V photonic devices with the traditional silicon electronics, may address problems of electrical interconnects; nevertheless it will not fill the gap for photonics electronics integration. The main limit of the hybrid solution comes from the miniaturization required by the long term intra-chip communication. The emerging Silicon Photonics technology promises a low cost solution by allowing close proximity integration of photonics with electronics.

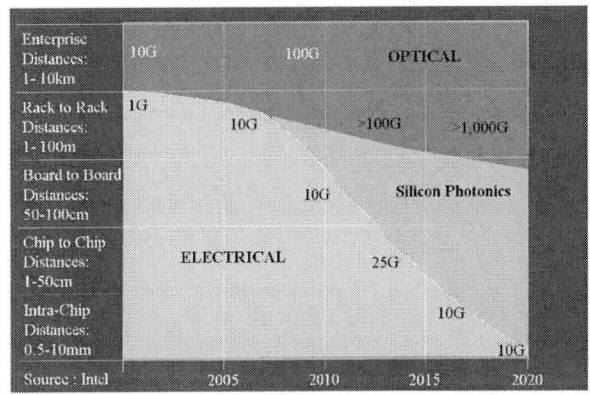

Figure 1 – Electrical Data Rate limits for different data communication distances

This article is organized in four Sections. **Section II** will describe an electro optical transceiver based on the Silicon Photonics technology and its characteristics. The four key figures of merit, namely: bandwidth, energy, size and cost are introduced through an example. **Section III** describes the necessary breakthroughs for the evolution of Silicon Photonics in order to address application specifications. Main technological challenges to be addressed and solved are also discussed. **Section IV** introduces a possible Road Map vision of the Silicon Photonics Technology and its applications. Finally, **Section V** provides a summary and draws the main conclusions.

978-1-4673-1707-8/12 \$31.00 © 2012 IEEE

II. SILICON PHOTONICS ELECTRO-OPTICAL TRANSCEIVER

The Silicon Photonics CMOS process starts from silicon-on-insulator (SOI) wafer. The buried oxide and the epitaxial layer thicknesses are chosen and optimized according to the range of the selected operating wavelengths. SOI is required for the confinement of the light in the optical waveguides. First process steps are instrumental to realize the photonic structures inside the silicon epitaxial layer. There are six basic photonic structures necessary to build up an optical transceiver: two types of Grating Couplers (GC), the single mode waveguides, the beam splitter, the modulator and the Photo-Detector (PD). The GC structures allow the coupling of the light, either from the laser source or the I/O single mode fiber arrays with the silicon waveguides.

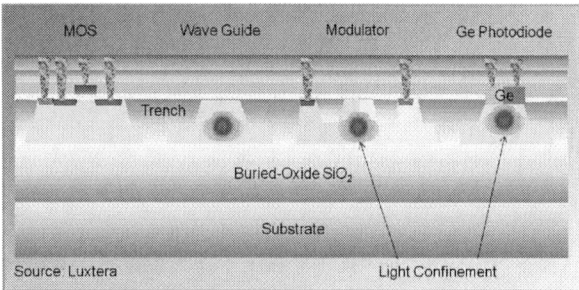

Figure 2 – CMOS Silicon Photonics process cross-section

As shown in fig.2, the light is confined and guided into the silicon strip layer, thanks to the difference of the two refractive index of the silicon epi-core (n Si $\cong$ 3.5) with respect to the cladding silicon oxide (n SiO$_2$ $\cong$ 1.5). Selective P-type and N-type implantations form the P-N and the P-I-N junctions as well as the diffused resistors, essential to implement optical phase modulators. After the definition of the CMOS transistors, a pure germanium layer is selectively grown over small areas of the waveguides. These small Ge islands enable the monolithic integration of high-performance photodetectors within the silicon waveguides. The input/output light is coupled to the Silicon Photonics die through the Grating Couplers [3] which are passive optical I/O structures, shown in fig.3. The GC's are designed to operate with tilted incident light angle; this improves the efficiency of the coupling. The GC's are diffractive (periodic) structures etched in the silicon material. Vertical optical interfaces allow very low loss optical coupling from/to the surface of the chip. Low loss is enabled by the special design of the structure so that the light coupled to the diffractive structure is "mode matched" with a single mode optical fiber. Vertical optical interfaces have significant advantages being fully compatible with the CMOS process and the single mode fibers assembly. Another big advantage offered by GC's is in the testing of the optical circuits at wafer level with high speed optical wafer probing. Because of the strong dependence from the signal polarization state, Single Polarization Grating Couplers are used to couple the light from the (polarized) light source into the die and the light from the die into the single mode fiber. Instead, the Polarization Splitting Grating Couplers are used to couple the light from standard single mode fiber into the silicon wave guides. These couplers are more complex and consist of a tapered bi-dimensional diffractive structure. The Single Polarization Grating Couplers are used in transmission direction when a laser source, which has a defined polarization state, is directly coupled to a waveguide (without the transit of the light through an optical fiber which normally rotates the polarization randomly).

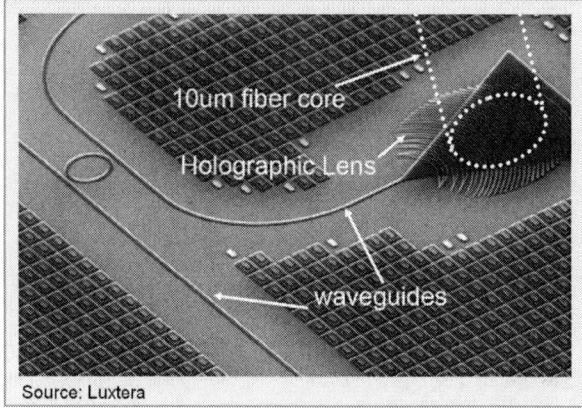

Source: Luxtera

Figure 3 –SPGC - Single polarization grating coupler

The Polarization Splitting Grating Couplers are used in the receiver direction because the polarization state of the received light signal is not defined. Propagating mode in silicon waveguides is characterized by the transverse electric (TE) field. The typical WG losses are below 0.5 dB/cm. The beam splitters are usually cascaded to distribute optical power over many waveguides. The commonly used electro-optical modulator is based on the Mach-Zehnder interferometer, whose structure is shown schematically in fig. 4. The narrowband input field E_{in} enters in the MZ through a single-mode waveguide, then it is split into two equal parts by a directional coupler. The phase velocity of the light in the silicon waveguides is equal to c/n_g, where n_g is the effective group refractive index that depends on temperature and dopants concentration in the silicon core. By controlling the reverse biasing of the depletion region in the P-N junctions of the two arms, the c/n_g phase velocity of each arm can be varied and, over the MZ length, the optical phase shifted. After two symmetrical paths, the CW beams are recombined using the same directional coupler structure to produce the output field E_{out}. The phase of the balanced 50/50 CW light in the two arms is controlled so to achieve a relative phase shift either equal to 0 or π. When the two arms are combined together, the light interferes constructively or destructively and hence the light is OOK (On Off Keing) modulated at the exiting waveguides. Usually only one output is used, the other output has the inverted optical data. In order to set the operating point of the MZI modulator, a precise static offset control of the phase difference between the two arms is necessary. This is possible by means of a low frequency feedback loop. The phase mismatch compensation can be realized with a relatively short time constant, in the order of ms, therefore the phase modulators can be based either on the carrier plasma dispersion effect (PIN type) or on the thermo-optical effect. In the carrier plasma dispersion effect the variation of the waveguide refractive index is caused by carrier injection while in the thermo-optical effect the index variation is caused by local heating.

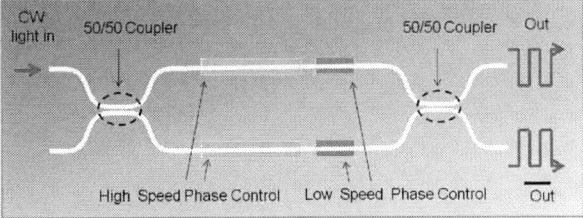

Figure 4 – Mach-Zehnder Interferometer

The Ge photodetectors are P-N diodes, biased at a reverse voltage of about -1 ÷ -2 V. These Ge diodes are very small micro-strips (see fig.5), about 1μm wide and 10 μm long, with a thickness of about 300-500 nm, built on top of the silicon waveguide. Thanks to the higher refractive index of the germanium, the light, while traveling along the z axis, is totally absorbed by the Germanium strip. The absorbed photons generate hole-electron pairs inside the depletion p-n region, which produce the photo current. Typical values of Ge photodiode responsivity, for wavelengths from 1300 nm to 1550 nm, range from 0.7 A/W to 1 A/W. Being these devices very small, their parasitic capacitance is below 10 fF, ideal to achieve a large BW at the receiver side. The typical BW (-3dB) of available Ge photodiodes are in excess of 30 GHz [4]. The electro-optical transceiver, shown in fig.6, is the fundamental sub-system through which the electrical data are up-down converted into optical signals. In this section we will limit our analysis to an electro-optical transceiver that is exchanging the data with an IC placed at a distance of few centimeters through serial ports at a given Data Rate. On the TX side, a continuous wave light beam, provided by an external laser source at a given wavelength, is coupled to a Silicon Photonics waveguide, through a GC.

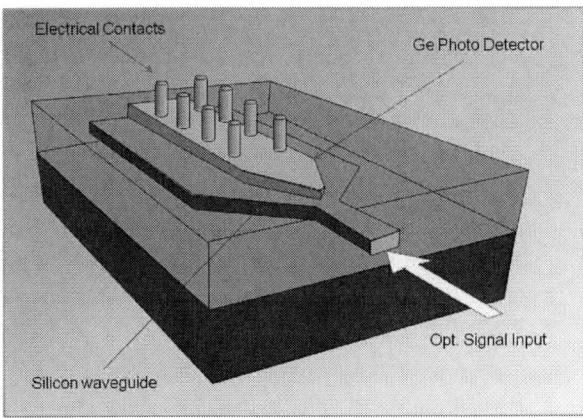

Figure 5 – PIN wave guide Ge Photo Detector

Then, the beam is split by a factor of 2 or 4 depending on the number of links/fibers the optical module has to be compatible to. After the splitters the light beam enters in the high speed MZ modulator. The serial data, received by the input buffer, are equalized and resynchronized (if necessary). The NRZ

digital inputs modulate the light beam phase delay in the two arms of the MZ. As a consequence, the output of the MZ shows an amplitude modulated beam with two level of light power, P1 and P0. The ratio P1/P0 represents the Extinction Ratio, while the difference (P1-P0) represents the OMA (Optical Modulation Amplitude). Today CW lasers electrical power consumption, operating at wavelength between 1310nm and 1550nm, sustaining an optical output power of 13dBm, in nominal conditions, shows a compliance voltage of about 1.4 V with a drive current of about 80 mA, hence the electronics power spent for the driver is close to 150 mW. The receiver part of the electro optical transceiver, shown in fig.6, includes an external SM fiber which is coupled to a PSGC (Polarization Splitting Grating Coupler). The PSGC couples the input modulated beam light to the silicon waveguides and to the Ge PD. The PSGC ensures that no penalty will be paid for the optical signal energy carried on the two polarization modes.

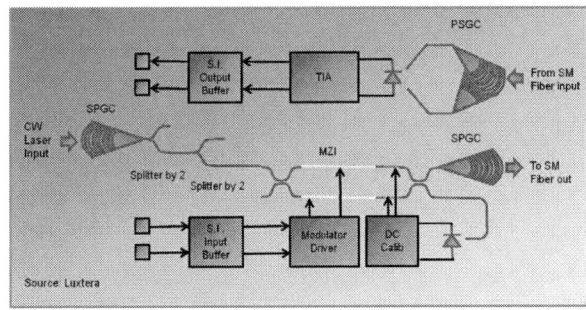

Figure 6 – Silicon Photonics Electro Optical transceiver functional block diagram

The electrical current signal, produced by the Ge PD, will be equal to the OMA input power multiplied by the PD responsivity. For a given BER (Bit Error Rate), the OMA sensitivity at the receiver is proportional to the total rms noise at the input of the TIA, and inversely proportional to photodetector responsivity. The OMA sensitivity reached with the Silicon Photonics at 10Gbps has been shown to be better than -22dBm with a BER< 10^{-12}. These results have been obtained with a TIA-LA implemented in CMOS 130 nm with power consumption below 25 mW and a size of 0.9 mm^2 [5]. A monolithic Silicon Photonics electro-optical transceiver (4 x 10Gbps) has been commercialized, since 2010, by Luxtera within Active Optical Cable applications. The Silicon Photonics transceiver has a size below 40 mm^2 and is assembled in a QSFP housing with a total power consumption of 780 mW. The QSFP housing contains the Silicon Photonics IC (see fig.7), the laser source and the fiber input/output array plus the voltage regulators and few discrete electronic components. This electro-optical transceiver represents the reference starting point for the next Section II, setting the following performances: a DR of 10 Gbps/link, an energy per bit below 20 pJ/bit, a form factor of 40 Gbps/10,000 mm^3 and a reference cost in the order of 4 $/Gbps

978-1-4673-1707-8/12 $31.00 © 2012 IEEE

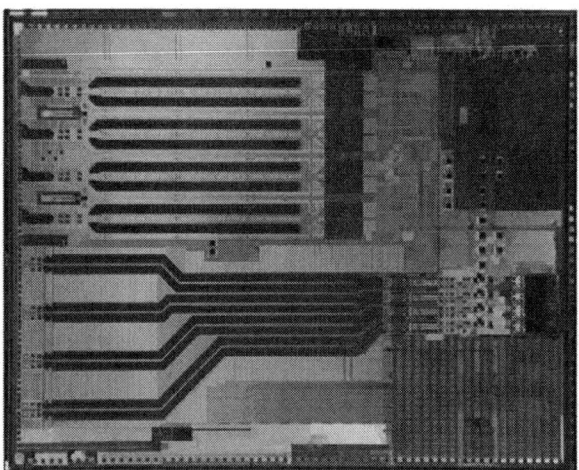

Source : Luxtera

Figure 7 – Luxtera's Silicon Photonics Transceiver

III. SILICON PHOTONICS BREAKTHROUGHS

The Silicon Photonics technology, analyzed in Section I, has been developed on 200 mm wafers with a lithography and metrology process node of 130 nm. By porting such a technology into 300 mm wafers with lithography and metrology process node of 65 nm, an increase of the data rate per link of a 3x factor can be expected. This improvement is due to CMOS electronics which allows to scale down the parasitics and to increase the f_T of the transistors by 3x (f_T=180GHz). The 300 mm, 65 nm process environment will also improve the matching of the photonic devices as well as the losses due to the light's scattering. Matching and reduced losses will also imply less required laser output power. The die size will not significantly scale down because the chip is dominated by the photonics device occupancy and their constraints, however with some optimization, the total transceiver die size could be squeezed to about 35 mm^2 according to the Pareto block diagram shown in fig.8. The 65 nm CMOS transistors will also allow some improvement in the electronic MZI driver's efficiency. Based on these assumptions we could reasonably predict that a monolithic Silicon Photonics electro-optical transceiver, similar to the reference described in section I, ported in a 300mm wafer CMOS 65 nm SOI process, will achieve a total aggregate BW approaching 4 x 30G, with an energy per bit below 12 pJ/bit (compatible with the QSFP size and power dissipation capability). This technology will be matured for massive production in three years time frame offering a transceiver cost below 1.33 $/Gbps. These characteristics will allow to increase the Silicon Photonics market share of the active optical cables and enter in new application fields such as the optical back-planes for board to board communications. Meanwhile some high end inter-chip interconnections based on Silicon Photonics electro optical interoperability (Silicon Photonics interposer) could also occur.

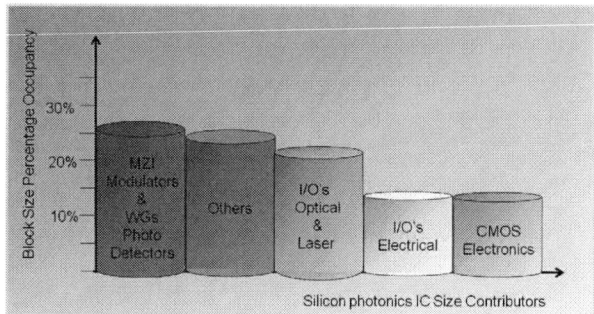

Figure 8 – 2015 Pareto of the Transceiver size

The 12 pJ/bit are distributed according to the Pareto diagram shown in fig.9. In order to make significant advances in the third generation of the Silicon Photonics technology, it will be necessary to introduce a new class of modulators (based on the micro ring resonators) and the Wavelength Division Multiplexing (WDM). The micro ring modulators have already been demonstrated and well characterized [6]. Micro rings are optical structures [7], 200 times smaller in size with respect to the today MZI. These devices are suitable to realize DWDM (Dense Wavelength Division Multiplexing) filters mux's and demux's (multiplexer, demultiplexer) as well as optical switches (see fig.10 and fig.11).

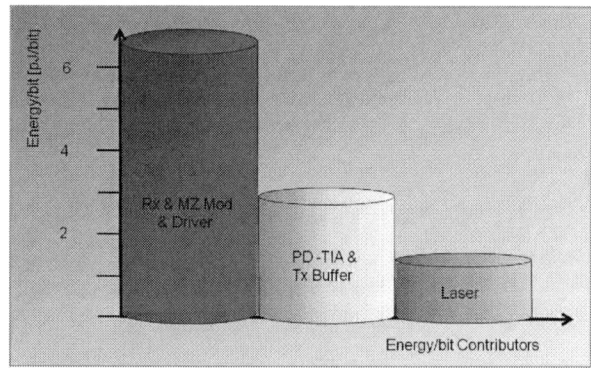

Figure 9 – Silicon Photonics 2015: Pareto of the energy/bit

The main challenge of these devices is in the thermal stabilization, because their optical parameters, like the resonant wavelength, are very sensitive to the temperature variations. Their maximum driving voltage swing (1V) is compatible with modern CMOS technology and the ER is in excess of 4 (6dB). By using a CMOS 28 nm electronics, and assuming an f_T=300GHz, a modulation speed of 50 Gbps/link per wavelength could be sustainable. This new class of modulators will dramatically reduce the modulation energy per bit because both the capacitance and the voltage will scale by 24x and 2.5x respectively when compared to the MZI structure. Thus, at the same DR the energy/bit will decrease 150 times.The modulator size will become negligible (0.002

978-1-4673-1707-8/12 $31.00 © 2012 IEEE

mm²/micro-ring). We can imagine to use 4 lasers each with a different wavelength spaced by a Δλ (a coarse spacing could mitigates the thermal constraints) among each other. These improvements allow to increase the transceiver aggregate bandwidth to 800 Gbps on 4 links in an estimated transceiver size of about 52 mm².

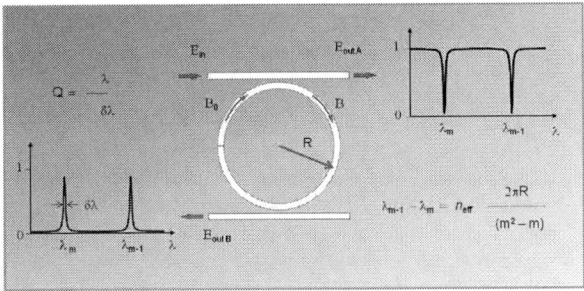

Figure 10 – Ring resonator schematic and frequency response

This third generation of the Silicon Photonics technology could be mature for massive production in 2018. For this generation, by using 4 lasers, we can estimate a total energy per bit from 2 to 2.5 pJ/bit, still being compatible with the same module form factor and power constraints. Based on these assumptions, the third generation could achieve a cost scaling factor of 6.66x, by allowing 0.32 $/Gbps. By changing the fiber in/out assembly from vertical to horizontal, we could also expect the module thickness to scale down by a factor 2. Thus an opto transceiver module of 800 Gbps could fit in about 5000 mm³. This last size compression could be achieved in two ways: either by modifying the end finishing of the fiber block, realizing a 45 ° mirror that couples the SM fiber light to the GC (this will also eliminate the fibers curvature issue), or by using the low-loss inverted taper coupler for silicon-on-insulator ridge waveguides [8], as represented in fig. 12. So, in 2018, we could imagine the Silicon Photonics pervading the on board inter-chip applications as well as gain market share in the small form factor 2D-3D opto-asic applications. When this will happen, a further growth in volumes will occur being fully compliant with 300 mm wafers fabs. The third generation of the Silicon Photonics technology will get a Pareto diagram (energy/bit) close to the one shown in fig. 13. It is important to notice that the Silicon Photonics transceiver size will be now dominated by the optical I/Os, the lasers micro-packages and the electrical I/Os. The fourth generation of the Silicon Photonics technology will find its applications in the intra–chip communications [6]. Thus, for this generation, the Silicon Photonics will be used to realize an optical communication layer capable to satisfy the required bandwidth for intra-chip communications with almost zero latency. The integration of the multi-wavelength lasers sources must be introduced in this generation by solving all the issues related to the realization of low cost, highly integrated external optical DWDM optical sources. The electrical I/O's will consist of very dense vias connecting the CMOS electronics to the photonics devices. The communications, at distances inside 1mm radius, at clock frequency between 3-5GHz, could remain based on traditional metals, while the wide bandwidth interconnections among clusters of processors and memories will be based on a Silicon Photonics layer.

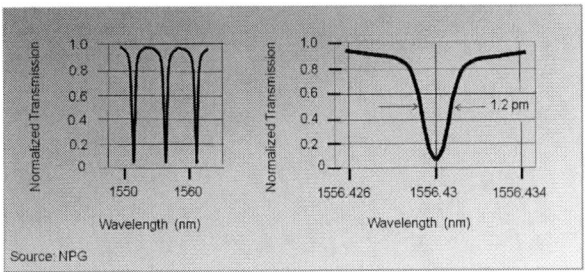

Figure 11 – Transmission spectrum of a 40-μm-radius micro ring resonator and zoomed-in spectrum of an optical mode with a 1.2 pm line width

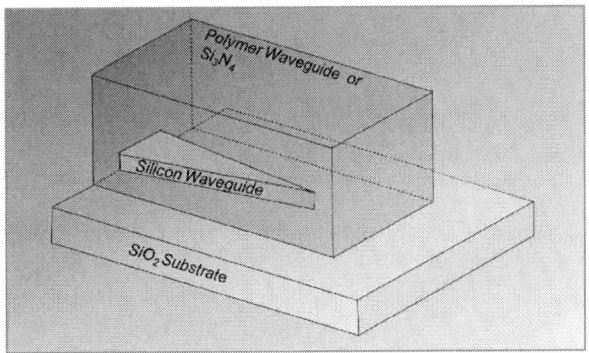

Figure 12 – Schematic drawing of the nano-taper coupler

That layer could be back end post processed after the wafer to wafer molecular bonding [9]. Let's imagine to interconnect 16 clusters with a clock of 4 GHz and I/O data of 64 bits. For each cluster we need an aggregate bidirectional bandwidth in excess of 256 Gbps. The numbers of nodes will be 120, to stay in a sustainable transceiver power budget of 20W we need to achieve an energy/bit below 700 fJ/bit. Advanced R&D Labs [10] have already shown hybrid micro-ring lasers with electrical power consumption 3x times lower than present DFB lasers. The electrical and optical characteristics of these micro-lasers are shown in fig. 14. This may be the way to integrate the lasers and simultaneously scale down the laser energy/bit below a value of 250 fJ/bit. The intra-chip communications will spare both the GC and SPGC losses by allowing to share more links/laser and to achieve better SNR. That will relax the photodiode and TIA noise constraints. Thus a reduction of the power consumption of 3x may be expected from less requirements of TIA and LA (Limiter Amplifier), while the output buffer will simply disappear. This fourth generation of Silicon Photonics could be envisaged in massive production by 2021. The total volume, included the 3D package constraints, has been estimated to be below 150 mm³. At this time, the Silicon Photonics could finally enter in the more traditional high volume silicon VLSI domain.

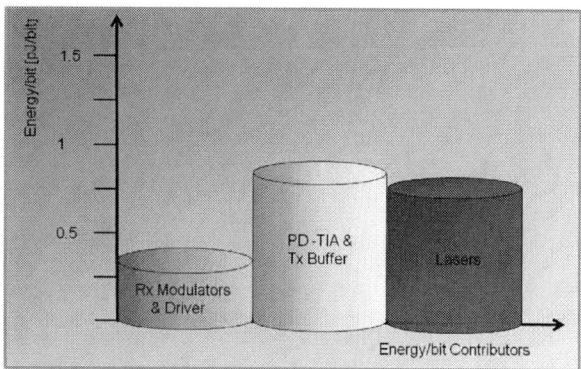

Figure 13 – Energy/bit Pareto diagrams of the Silicon Photonics Third Generation

The single Silicon Photonics transceiver, based on 16 μ-ring lasers, each with a power of 50 mW, could achieve 800 Gbps/WG (3200 Gbps on 4 + 4 wave guides) by fitting a total estimated size of 12.5 mm².

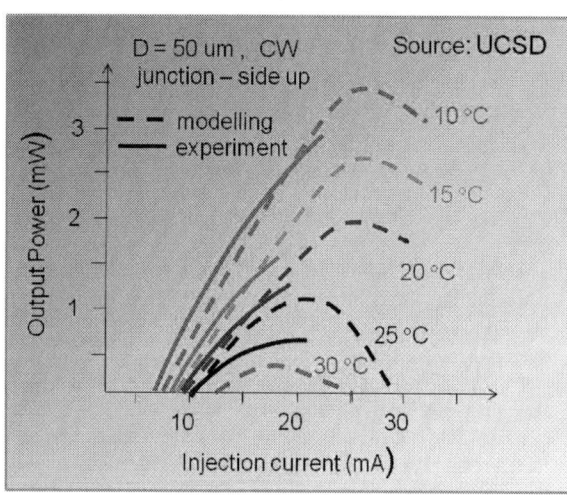

Figure 14 – UCSD Hybrid μ-ring laser

The cost/Gbps of an equivalent transceiver, after molecular bonding and post processing, could be estimated in 0.04 $/Gbps. The main challenges will be in the thermal IC control that will need a temperature stabilized inside T_{ref} +/- 10 °K, and the wafer to wafer alignment and planarization. On the temperature issue, advanced R&D labs are already working on the fluidic cooling [11] and SoD (Silicon on Diamond) platform [12]. Other issues to be addressed are related to EO VLSI packages. As a conclusion of this section, we can summarize that by 2021 a fourth generation of the Silicon

Photonics technology could be ready for the massive production by allowing an energy/bit of 600-800 fJ/bit, a total aggregate bandwidth of 3200 Gbps with and a cost below 0.04 $/Gbps in a total size of about 150 mm³.

IV. SILICON PHOTONICS ROADMAP

Based on the assessment of the today Silicon Photonics technology, described in section I, and the new class of photonics devices already demonstrated in the R&D labs, we could envision a Silicon Photonics technology road map as synthesized in fig. 15. The reference generation is based on the assessment of the today technology built on a 200mm SOI wafers with CMOS 130nm. This technology is already in production by meeting its main application in the active optical cables. The reference key characteristics are summarized in a total aggregate bandwidth of 40 Gbps on 4 links, with energy per bit below 20 pJ/bit. The Silicon Photonics chip is integrated in a QSFP module form factor with a total cost close to 4 $/Gbps. Based on these characteristics and by porting this technology on 300mm SOI wafers, with CMOS 65nm , we can expect to improve the four key figures of merit by a factor 3x. Thus, we can increase the aggregate bandwidth and improve the optical losses. We expect the second generation to be mature for industrial application in 2015. The driving applications of this generation can be envisaged in HPC and modern data centers where the intensive computing and huge routing are strongly requiring wide bandwidth at low power and low latency. The volumes are expected to grow with the consequent reduction of the overall cost. The cost is scaled down by the aggregate growth and by the energy reduction that is indirectly impacting the cost of maintenance and cooling. The technology breakthrough offered by the micro ring modulators will dramatically scale down both the modulator size and the modulators energy. The micro ring resonators will also be very efficient for the introduction of the coarse wavelength division multiplexing. We expect this generation to be mature for massive production in 2018. By thinking to this generation, associated with an improved electronics (CMOS 28 nm), we could foresee an aggregate of 800 Gbps with energy per bit below 3pJ/bit. Thanks to these breakthroughs the third generation could dramatically decrease both the cost and the form factor size. Among the features offered by the Silicon Photonics the low latency will also play a key role by allowing that generation to offer big advantages in the chip to chip applications. Here we can think to a Silicon Photonics interposer capable to scale down the multi layers electrical board size of almost 5 times. If this will happen we could also expect a consistent volume growth of the Silicon Photonics market (10MU/Y, 20kWs/Y). Based on the recent R&D demos, we could forecast that another key breakthrough will allow to integrate the multi-wavelength micro-ring-lasers by allowing the usage of the dense wavelength division multiplexing. Two other 3D technologies, as the wafer to wafer molecular bonding and the fluidic cooling or black diamond, could be the enablers for the Silicon Photonics fourth generation. This generation, that we estimate ready for massive production in 2021, could enable the intra-chip interconnections with the characteristics shown in the table of fig. 15. When this will happen, the traditional semiconductors silicon volumes will finally occur and the intra-chip communications, today based on copper, will be achieved by the multi-λ integrated nano-photonics.

Year	2012	2015	2018	2021
Application	Active Cables	Active Cables Opto Modules	Opto Modules 2D/3D Opto Asics	Opto Asics Opto Chip
Technology Breakthrough	Reference	CMOS 65 nm Electronics	CMOS 28nm μ-Ring-Resonators WDM(4 λ)	CMOS 15nm μ-Lasers DWDM (16 λ) 3D PK & Cool
Data Rate per link	10 Gbps	30 Gbps	200 Gbps	800 Gbps
Aggregate Bandwidth	40 Gbps	120 Gbps	800 Gbps	3200 Gbps
Parallel Link (4x) — Estimated Energy/bit	< 20 pJ/bit	< 12 pJ/bit	< 3 pJ/bit	< 1 pJ/bit
Parallel Link (4x) — Estimated Size/module	10.000 mm³	10.000 mm³	5,000 mm³	150 mm³
Parallel Link (4x) — Estimated cost/module	4 $/Gbps	1.33 $/Gbps	0.32 $/Gbps	0.04 $/Gbps

Figure 15 – Silicon photonics road map

V. CONCLUSIONS

This paper, starting from an analysis of a mature Silicon Photonics technology and the most advanced R&D demonstrators and by introducing a new class of micro and nano photonics devices, provides a vision of a possible road map, up to 2021. Though this road map is based on assumptions and challenges that must be further verified and validated, the Silicon Photonics technology is showing features capable to provide an answer to the increasing demand of short distance, wide bandwidth, low power and low cost wired communications. Volumes are expected to increase while the distances are becoming shorter from tens of meters to millimeters. More the distances will scale and more the Silicon Photonics technology will become efficient and competitive versus copper and hybrid photonics. Finally we have shown how the technology road map is also consistent with the required miniaturization for the efficient communications at intra-chip level.

VI. AKNOWLEDGMENTS

This paper has benefited from the cooperation and technical discussion with the members of my team: G. Chiaretti, A. Fincato, A. Pallotta and E. Temporiti, as well as many people inside and outside STMicroelectronics. A particular thank to P.De Dobbelaere [Luxtera] and S.Rotolo for proof-reading this article and providing comments to improve the final version.

References

[1] R.Beausoleil: "A nanophotonics interconnect for High-Performance many-core computation" IEEE LEOS NEWSLETTER June 2008.

[2] G.Astfalk : "Why optical communications and why now?" Journal of Applied Physics Vol.95, issue 4, pag.933-940, June 2009 & October 23, 2011.

[3] A.Mekis et al. : "Grating-Coupler-Enabled CMOS Photonics" published on IEEE Journal of Selected Topics in Quantum Electronics Vol. 17,NO.3, May/June 2011

[4] T.Yin & Co: "31GHz Ge n-i-p waveguide photodetectors on Silicon on Insulator substrate" Optics Express Vol.15, No.21 17 Oct. 2007.

[5] D.Kucharsky & Co: " 10Gb/s 15 mW optical receiver with integrated Germanium Photodetector and Hybrid Inductor Peakingin 0.13um SOI CMOS Technology" ISSCC 2010.

[6] R. Beausoleil : " Large-Scale Integrated Photonics for high performance interconnects " ACM Vol.7, No.2, Article 6, published : May 2011.

[7] Nature Photonics Volume: 5, Pages: 770–776 2011.

[8] [O.C.Vol. 283, Issue 19, 1 October 2010, Pages [3678–3682].

[9] L.Fulbert & J.M. Fedeli: "Photonics-Electronics Integration on CMOS", ESSCIRC 2011.

[10] Di Liang: "Electrically-pumped compact hybrid silicon microring lasers for optical interconnects", Optics Express Vol.17, No.22 26 Oct. 2009.

[11] IBM tutorial, OFC 2012.

[12] Di Liang: " Silicon On Diamond Platform" IEEE Photon. Techn. Lett. 23, 657(2011)

978-1-4673-1707-8/12 $31.00 © 2012 IEEE

Solid-State and Biological Systems Interface

Nan Sun[¶,*], Yong Liu[#], Ling Qin[§], Guangyu Xu[§], and Donhee Ham[§,*]

[¶]Department of Electrical and Computer Engineering, University of Texas at Austin, Austin, TX, USA
[#]IBM T. J. Watson Research Center, Yorktown Heights, NY, USA
[§]School of Engineering and Applied Sciences, Harvard University, Cambridge, MA, USA
[*]Emails: nansun@mail.utexas.edu, donhee@seas.harvard.edu

Abstract—Solid-state electronic devices can be engineered to detect and manipulate biological molecules and cells by using electric or magnetic interactions. The integrated circuits, which can contain a large number of such devices, may then potentially be developed into low-cost chip-scale platforms to perform bioanalytical tasks in a multiplexed manner for applications in biology, biotechnology, and personalized medicine. This paper reviews some recent developments in this solid-state electronic and biological systems interface.

I. INTRODUCTION

It is well known that field-effect transistors can be arranged to operate as biomolecular sensors [1-6]. Imagine an integrated circuit, where a top metal connected to a transistor's gate is surface treated with a thin dielectric layer and then with single-stranded DNA molecules of a known sequence, which one may call probe DNA (Fig. 1). The thin dielectric layer is exposed to an electrolyte containing single-stranded DNA molecules of varying sequences. If these include DNA molecules whose sequence is complementary to the probe DNA sequence, binding will occur at the dielectric-electrolyte interface. As DNA molecules are negatively charged in typical buffer condition, the binding event will alter the transistor's conductance, offering the basis for electrical DNA detection. This ion-sensitive field effect transistor configuration [1-4] can be also used to detect other types of biological molecules such as proteins, by surface treating the device with the antibodies that can specifically bind to target proteins. With bottom-up nanoscale transistors, femto-molar sensitivity and single virus particle detection have been demonstrated [5,6].

As illustrated with the ion-sensitive field effect transistor example, solid-state electronic devices can be engineered to execute biomolecular sensing. Not only the charge-based electric sensing, but also spin-based, magnetic sensing is possible [7,8]. Furthermore, the electronic devices can be used to manipulate the motion of individual biological cells using electric or magnetic fields [9,10]. As a large number of electronic devices can be integrated inexpensively, integrated circuits have been envisioned as low-cost chip-scale multiplexed bioanalytical platforms for applications in biology, biotechnology, and personalized medicine. For example, instead of the optical DNA microarray requiring a bulky and expensive optical scanner [11], one can imagine developing a low-cost all-electronic DNA microarray that can be used like a memory stick with a personal computer. While optics-based bio-analytical systems will continue to be used broadly into the foreseeable future and although electronic bioanalytical chips still require much improvement in fidelity and yield, the latter devices have attractive features such as low-cost chip-scale operation and easy accessibility, and developing them represents a growing branch of engineering. The recent exciting development of the CMOS integrated circuit for DNA sequencing [12] (not to be confused with DNA array) shines positive light on this direction; this work employs a large array of ion-sensitive field effect transistors, with each transistor designed to detect the local pH change that arises from the nucleotide incorporation.

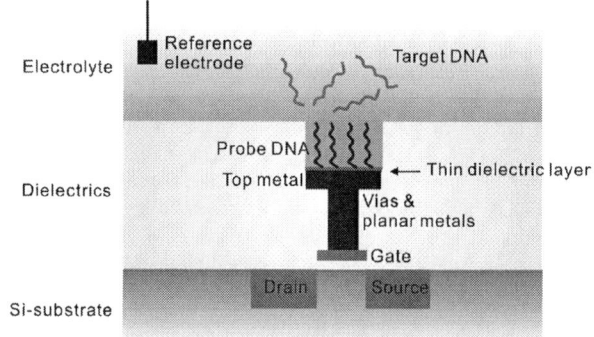

Fig. 1: Ion-sensitive field-effect transistor.

This invited paper is *not* a new technical contribution, but *reviews* some recent developments in this interdisciplinary field of solid-state electronic and biological systems interface, in the hope for providing a perspective on the on-going scene in this field. This review will include some of our own works that have been published elsewhere. Images that appear from here on are adopted from these prior publications of ours.

II. BIOLOGICAL CELL MANIPULATION

The ability to control the motion of individual biological cells can be useful in single cell interrogation, cell mechanics

978-1-4673-1707-8/12 $31.00 © 2012 IEEE

study, cell sorting, and designer tissue assembly. Biological cells can be manipulated on the integrated circuit, by using the electric or magnetic fields it generates.

Non-uniform electric fields exert a force on electric dipoles, thus, charge-neutral dielectric objects like biological cells, and move them towards where the field is stronger (if the background electrolyte's dielectric constant is larger than that of the objects, the movement will be towards where the field is weaker). This dielectrophoresis can be performed on top of an integrated circuit containing an array of micro-electrodes made out of top metals; by controlling the voltage of each electrode independently, the array can produce a variable electric field pattern, with which one can manipulate cells suspended above the chip in a microfluidic environment [10]. In order to prevent charged particles such as ions from accumulating to electrodes and screening their effect, AC voltages are applied to the electrodes. With the resulting time-varying electric fields, the charged particles will oscillate back and forth, not accumulating onto electrodes; by contrast, despite the field's periodic polarity change, charge-neutral biological cells will move in one local direction determined by the field magnitude gradient, as opposed to oscillating back and forth. The integrated circuit executing dielectrophoresis may be thought of as electric tweezers.

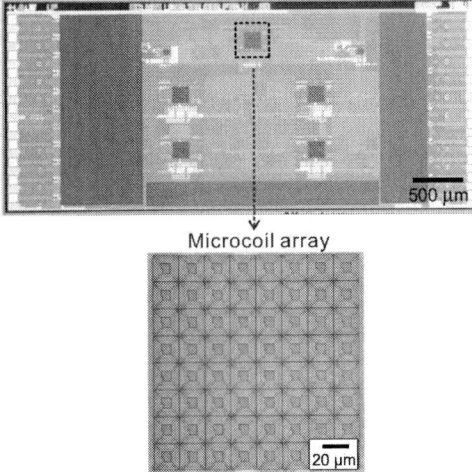

Fig. 2: CMOS magnetic manipulator. Modified reprint from our paper, *IEEE J. Solid-State Circ.*, **41**, 1471 (2006) [9]. © IEEE.

Integrated circuits can be used as magnetic tweezers, too. Non-uniform magnetic fields exert force on magnetic dipoles, thus can move cells attached to magnetic particles (most cells are non-magnetic; more accurately, they are not strongly magnetic enough to receive appreciable magnetic force). An array of planar microcoils made from top metals in an integrated circuit produces a variable magnetic field pattern, by controlling each coil's current individually. This variable field pattern can actuate magnetic-particle-bound cells.

Fig. 2 is a microcoil array CMOS chip we developed a number of years ago [9]. Each coil produces a magnetic field of about 30 G at the coil center on the chip surface with a *dc* current of 20 mA, and can pull a magnetic particle or a magnetic-particle-bound cell into its center. Fig. 3 shows a

manipulation of an 8.5-μm-diameter magnetic particle, from our prior work [9]; as the magnetic field peak position is moved by successively turning on and off the microcoils along a particular path, the particle is transported along that path on the chip surface, suspended inside a microfluidic chamber fabricated on top of the CMOS chip. Fig. 4, again from [9], shows an example manipulation of two bovine capillary endothelial cells, which have engulfed antibody-treated 250-nm diameter magnetic nanoparticles; these two cells are joined together by appropriately operating the microcoil array in the CMOS chip.

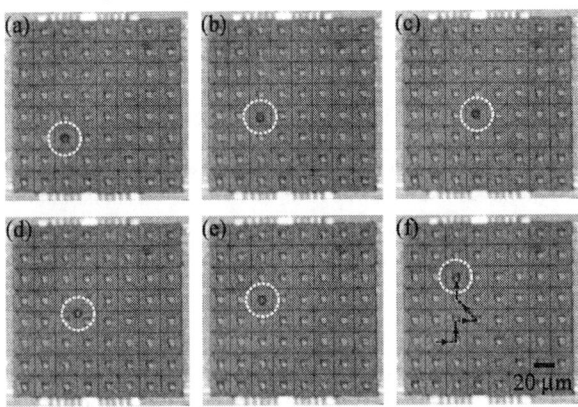

Fig. 3: Magnetic particle manipulation. Reprint from our earlier paper, *IEEE J. Solid-State Circ.*, **41**, 1471 (2006) [9]. © IEEE.

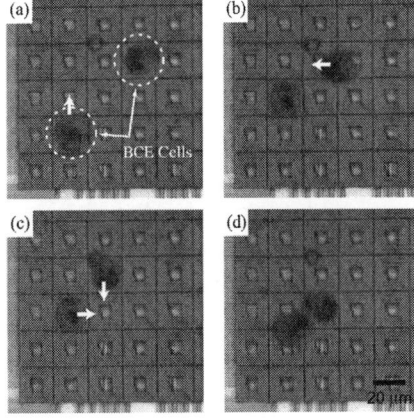

Fig. 4: Magnetic manipulation of bovine capillary endothelial cells. Reprint from our earlier publication, *IEEE J. Solid-State Circ.*, **41**, 1471 (2006) [9]. © IEEE.

The electric and magnetic manipulation of biological cells on the CMOS chips is limited to two dimensions, lacking vertical control, due to the planar nature of integrated circuits. Despite this limitation, the individual cell manipulation with integrated circuits may still offer opportunities in facilitating cytometry, cell mechanics study, high-precision cell sorting, and layer-by-layer designer tissue assembly.

We now turn to biomolecular sensing, where the planar nature of integrated circuits imposes less limitations. In fact, biomolecular sensing with integrated circuits may have far broader applications and impacts.

III. BIOMOLECULAR DETECTION

The ion-sensitive field effect transistor discussed at the start of this review detects biomolecules by way of charge sensing. Biomolecular detection with integrated circuits can be also done by resorting to spin-based magnetic sensing; e.g., we recently developed CMOS biomolecular sensors using nuclear magnetic resonance (NMR) [7,8]. The charge-based electric sensing and spin-based magnetic sensing contrast each other in an interesting way; the former has a larger signal, but also has a larger noise due to irrelevant ions in the electrolyte; the latter has a smaller signal, but also has a smaller noise, for the ions in the electrolyte have no first-order effect on magnetic interactions.

We illustrate how spin-based biomolecular sensing works, using our CMOS NMR system as an example. While rigorous description of NMR can entail quantum mechanics, classical picture suffices here. The nuclei of hydrogen atoms, which are single protons, are tiny magnets due to their spin, and interact with magnetic fields. Apply a static magnetic field B_0 produced from a permanent magnet or an electromagnet to a sample of water, which contains a large number of hydrogen atoms. After a while, the hydrogen proton magnetic moments line up preferentially along the static magnetic field B_0 to minimize the overall potential energy; not all proton magnetic moments line up along the static magnetic B_0 due to thermal agitation, but their vector sum per unit volume (*i.e.*, magnetization) does. Now we add an RF magnetic field $B_1(t)$ perpendicularly to the static magnetic field B_0 by wrapping a coil around the water sample and injecting an RF current into the coil. If the RF field's frequency is tuned into a value given by $f_0 = B_0 \times \gamma/(2\pi)$ [γ: proton gyromagnetic ratio; its numerical value is such that B_0 of 1 T gives f_0 of 42.6 MHz], the magnetization resonantly absorbs energy from the RF field, which is NMR. The absorbed energy increases the overall potential energy of protons; equivalently, the magnetization originally aligned with the static magnetic field B_0 is then rotated away from B_0, enlarging the angle θ it makes with B_0.

If the RF field is turned off when θ becomes, say, 90°, the magnetization will stay in the θ=90° plane orthogonal to B_0 for a while, before it eventually sheds its potential energy and lines back up with B_0. With the non-zero θ, the magnetization undergoes a precession motion about B_0, for B_0 exerts a torque on it and alters its angular momentum associated with the proton spin. The precession frequency is also f_0. This precession will produce a periodically varying magnetic flux across the aforementioned coil, inducing a voltage across the coil. This reception-phase voltage may be referred to as NMR signal, although NMR actually occurred during the excitation phase.

The NMR signal of the pure water after the 90° excitation is an exponentially damped sinusoid, and the characteristic relaxation time is called T_2 (Fig. 5, top). This damping occurs, as each individual proton precession is randomly disturbed by nearby protons (spin-spin interaction), developing what is like the oscillator phase noise. Hence, precessions of a large number of protons grow increasingly phase incoherent with one another, leading to the exponential decay of the overall magnetization. The damping also occurs due to the gradual

potential energy loss of the magnetization, which reduces θ. Since the phase decoherence due to spin-spin interaction occurs faster than the energy relaxation, T_2 is largely governed by the former effect.

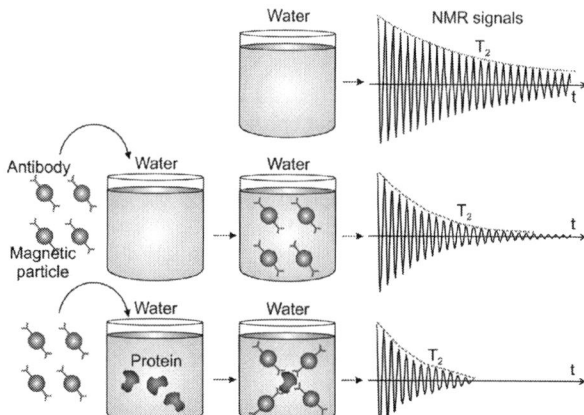

Fig. 5: NMR-based biomolecular sensing [13]. The illustration is a modified reprint from our earlier publication, *IEEE J. Solid-State Circ.*, **44**, 1629 (2009) [7]. © IEEE.

Consider detecting a certain type of proteins in a biological sample, which contains a large number of water molecules. Nanoscale magnetic particles coated with antibodies that can specifically bind to the target proteins are introduced into the sample. If target proteins are absent (Fig. 5, middle), the magnetic particles stay mono-dispersed. These magnetic particles constantly move around, creating fluctuating magnetic fields. These perturb precessions of individual protons, increasing the phase noise beyond that due to the basic spin-spin interactions. Consequently, the phase coherence is lost more rapidly, reducing T_2. In the presence of the target proteins (Fig. 5, bottom), the magnetic particles self-assemble into larger local clusters, which are even more effective in accelerating phase decoherence [13], leading to an even smaller T_2. In sum, by monitoring the NMR relaxation time, target proteins can be detected [13].

Conventional NMR relaxometry systems are heavy, bulky, and expensive. E.g., a state-of-the-art commercial benchtop NMR relaxometer weighs over 100 kg. This is because they use large-sized magnets to produce a highly homogenous static magnetic field. Even for the same static field strength, a more uniform field yields a larger NMR signal, hence the use of the large-sized magnets. Through a series of developments, we miniaturized NMR relaxometry systems significantly, e.g., Figs. 6, 7 [7,8]. The key to this miniaturization was opting for much smaller magnets (the magnet in Fig. 6 is comparable to a hamburger in size; the magnet in Fig. 7 is as small as a ping-pong ball), and then designing highly sensitive CMOS RF transceivers to deal with the NMR signal substantially weakened by the small-sized magnets. The 0.1-kg NMR system of Fig. 7 is three orders of magnitude lighter and smaller than the commercial benchtop system, yet two orders of magnitude more spin-mass sensitive. Since the ping-pong ball size magnet severely degrades the NMR signal, this system of Fig. 7 uses an off-chip high-quality solenoidal coil (hidden inside the magnet in Fig. 7), not to further degrade the signal. The NMR system of Fig. 6 integrates the coil in the CMOS chip together with the RF transceiver; here, biological samples are placed directly on the coil portion of the CMOS chip for on-chip detection of target biomolecules. As the on-chip coil degrades the NMR signal, not to further degrade it, this system of Fig. 6 uses the larger, hamburger sized magnet. Yet, this system still weighs only 1.5 kg, and is 60 times more spin mass-sensitive than the commercial system.

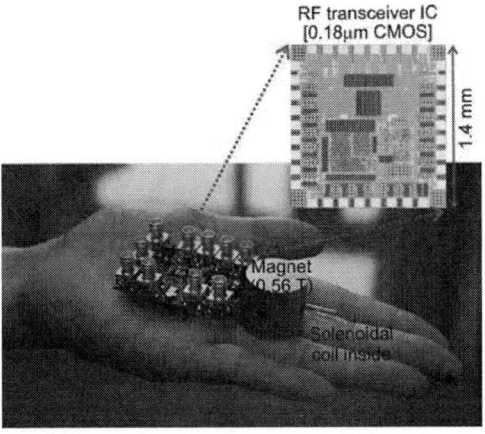

Fig. 6: A CMOS NMR biomolecular sensor. Reprint from our earlier paper, *IEEE J. Solid-State Circ.*, **46**, 342 (2011) [8]. © IEEE.

Fig. 7: Another CMOS NMR biosensor. Reprint from our prior paper, *IEEE J. Solid-State Circ.*, **46**, 342 (2011) [8]. © IEEE.

With these CMOS NMR systems, we demonstrated the detection of avidin, hCG, and human bladder cancer cell [7,8] (multiplexed measurements were demonstrated in [14]). These CMOS NMR systems, implemented in the low-cost, hand-held platform, are examples of the solid-state and bio systems interface using magnetic interactions. We find them also interesting in that they showcase how CMOS RF integrated circuits can be used for biomolecular sensing, beyond their traditional wireless applications.

IV. ANOTHER AVENUE: CMOS-NEURON INTERFACE

Another interesting avenue to explore is the CMOS-neuron interface [15-17]. Understanding how the neuronal network in the brain codes information is one of the most celebrated problems in all of science. The challenge is colossal. To help make a step forward in tackling this problem, a variety of

approaches have been proposed. One approach is to interface a CMOS integrated circuit with interconnected neurons to interrogate them simultaneously [15-17] *in vivo, ex vivo,* or *in vitro.* In principle, the density of CMOS circuits promises sufficient spatio-temporal resolution in massively parallel recording at least with two-dimensional neuronal networks. Nonetheless, massively parallel recording of a large number of neurons with single-cell resolution has yet to be achieved, let alone its biologically relevant interpretation in connection with functions and behaviors. Developing the CMOS-neuron interface offers exciting opportunities at the intersection of devices, circuits, electrochemistry, and neurobiology.

REFERENCES

[1] P. Bergveld, "The future of biosensors," Sens. Actuators A, vol. 56, pp. 65-73, 1996.

[2] J. Bausells, J. Carrabina, A. Errachid, and A. Merlos, "Ion-sensitive field-effect transistors fabricated in a commercial CMOS technology," Sens. Actuators B Chem. vol. 57, pp. 56–62, 1999.

[3] M. Milgrew, P. Hammond, and D. Cumming, "The development of scalable sensor arrays using standard CMOS technology," Sens. Actuators B Chem. vol. 103, pp.37-42, 2004.

[4] M. Milgrew, and D. Cumming, "Matching the transconductance characteristics of CMOS ISFET arrays by removing trapped charge," IEEE Trans. Electron Devices, vol.55, pp.1074–1079, 2008.

[5] F. Patolksy, B. Timko, G. Zheng, and C. M. Lieber, "Nanowire-based nano-electronic devices in the life sciences," MRS Bulletin, vol. 32, pp. 142-149, 2007.

[6] F. Patolsky, G. Zheng, O. Hayden, M. Lakadamyali, X. Zhuang, and C. M. Lieber, "Electrical detection of single viruses," Proc. Nat. Acad. Sci. vol. 101, pp. 14017-14022, 2004.

[7] N. Sun, Y. Liu, H. Lee, R. Weissleder, and D. Ham, "CMOS RF biosensor utilizing nuclear magnetic resonance," IEEE J. Solid-State Circuits, vol. 44, pp. 1629-1643, 2009.

[8] N. Sun, T. J. Yoon, H. Lee, W. Andress, R. Weissleder, and D. Ham, "Palm NMR and 1-Chip NMR," IEEE J. Solid-State Circuits, vol. 46, pp. 342-352, 2011.

[9] H. Lee, Y. Liu, R. M. Westervelt, and D. Ham, "IC/microfluidic hybrid system for magnetic manipulation of biological cells," IEEE J. Solid-State Circuits, vol. 41, pp. 1471-1479, 2006.

[10] T. P. Hunt, D. Issadore, and R. M. Westervelt, "Integrated circuit / microfluidic chip to programmably trap and move cells and droplets with dielectrophoresis," Lab on a Chip, vol. 8, pp. 81-87, 2007.

[11] M. Schena et al, "Quantitative monitoring of gene expression patterns with a complementary DNA microarray," Science, vol. 270, pp. 467-470, 1995.

[12] J. M. Rothberg et al, "An integrated semiconductor device enabling non-optical genome sequencing," Nature, vol. 475, pp. 348-352, 2011.

[13] J. M. Perez, L. Josephson, T. O'Loughlin, D. Hoegeman, and R. Weissleder, "Magnetic relaxation switches capable of sensing molecular interactions," Nature Biotechnology, vol. 20, pp. 816-820, 2002.

[14] H. Lee, E. Sun, D. Ham, and R. Weissleder, "Chip-NMR biosensor for detection and molecular analysis of cells," Nature Medicine, vol. 14, pp. 869-874, 2008.

[15] B. Eversmann et al, "A 128 times; 128 CMOS biosensor array for extracellular recording of neural activity," IEEE J. Solid-State Circuits, vol. 38, pp. 2306-2317, 2003.

[16] F. Heer et al, "CMOS microelectrode array for bidirectional interaction with neuronal networks," IEEE J. Solid-State Circuits, vol. 41, pp. 1620-1629, 2006.

[17] U. Frey et al, "Switch-matrix-based high-density microelectrode array in CMOS technology," IEEE J. Solid-State Circuits, vol. 45, pp. 467-482, 2010.

Advancing High Performance Heterogeneous Integration Through Die Stacking

Liam Madden, Ephrem Wu, Namhoon Kim, Bahareh Banijamali, Khaldoon Abugharbieh, Suresh Ramalingam and Xin Wu

Xilinx, Inc.

2100 Logic Drive, San Jose CA 95124

bahare.banijamali@xilinx.com

Abstract

This paper describes the industry's first heterogeneous Stacked Silicon Interconnect (SSI) FPGA family (3D integration). Each device is housed in a low-temperature co-fired ceramic (LTCC) package for optimal signal integrity. Inside the package, a heterogeneous IC stack delivers up to 2.78Tb/s transceiver bandwidth. The resulting bandwidth is approximately three times that achievable in a monolithic solution. Mounted on a passive silicon interposer with through-silicon vias (TSVs), the heterogeneous IC stack comprises FPGA ICs with 13.1-Gb/s transceivers and dedicated analog ICs with 28-Gb/s transceivers. Optimization took place concurrently on multiple facets of the design which were necessary to successfully implement the 3-D integration. In particular, this paper outlines the choices that were made in terms of package substrate material and interposer resistivity in order to optimize 28Gb/s system channel characteristics. These choices were validated through extensive electrical simulation and test chip correlation. In addition, this paper describes the design and timing verification of inter-die interconnects, an area that the electronic design automation industry had not yet fully addressed. This paper further describes 3D thermal-mechanical modeling and analysis for package reliability. The modeling was performed to address package coplanarity issues and stresses imposed by the interposer on the active dice, the low-k dielectric material, the micro-bumps and the C4 attach. The results indicate heterogeneous stacked-silicon (3D) integration is a reliable method to build very high-bandwidth multi-chip devices that exceed current monolithic capabilities.

Introduction

Since the advent of integrated circuit technology in 1958, the industry has focused primarily on monolithic integration. Unfortunately, due to physical and economic issues, the vast majority of high performance analog chips, high density memory chips, and high performance digital chips are each build on separate technologies. Therefore, in order to deliver optimum system performance, power and cost, it is desirable to integrate multiple different die, each using its own optimized technology, in a single package. In this implementation the system optimization was achieved through integrating digital FPGA die and separate high speed analog SerDes die on a silicon interposer. The underlying interposer technology allows up to 10,000 inter-chip connections with signaling power levels approaching intra-die connections. This paper describes the underlying technology and challenges encountered in implementing a heterogeneous FPGA and SerDes which delivers a previously unattainable off chip bandwidth of 2.78Tb/s.

3D Integration and its Scalability

The XC7VH580T [1] is the first commercial FPGA built with heterogeneous SSI, and is the first in a family of three FPGAs with 28Gb/s transceivers. The device consists of a passive silicon interposer and three active die: an 8 x 28Gb/s transceiver IC (GTZ-IC) and two FPGA ICs known as Super Logic Regions (SLR), see Figure 1. Two additional products are derived from these building blocks: The XC7VH290T with one fewer SLR and the XC7VH870T with one additional GTZ-IC and one additional SLR, see Figure 2. The XC7VH870T consists of 16 28Gb/s transceivers and 72 13.1Gb/s transceivers, which allows the delivery of the previously unattainable bandwidth of 2.78Tb/s.

The enabling technologies which deliver this scalability include a silicon interposer with four high-density (~1um pitch) interconnects, through silicon vias (TSVs), and fine-pitch micro-bumps as shown in Figure 3.

These technologies make it possible to manufacture FPGAs that offer bandwidth and capacity exceeding that of the largest possible monolithic FPGA die with the manufacturing and time-to-market advantages of smaller die. The integrated dice appear to the designer as a single, ultra-high-capacity, monolithic FPGA.

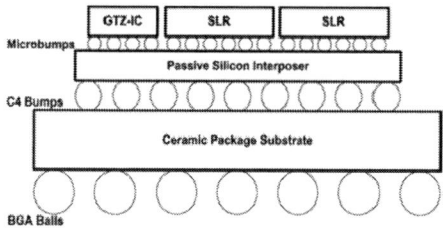

Fig. 1 *Conceptual cross-sectional view of XC7VH580T*

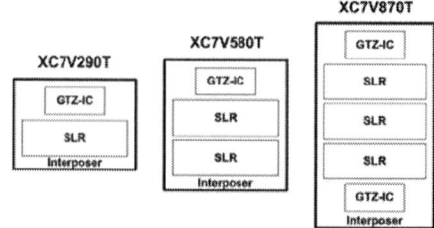

Fig. 2 *Bird's eye view of heterogeneous FPGAs (not to scale)*

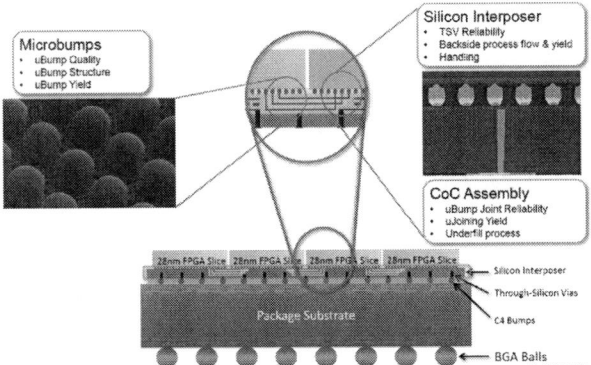

Fig. 3 *Stacked Silicon Interconnect cross section*

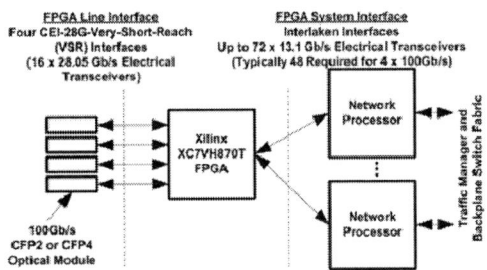

Fig. 5 *4 x 100Gb/s line card*

Applications

The Xilinx XC7VH580T FPGA targets high-density dual 100Gb/s networking applictions with CFP2 or CFP4 optical modules [2] on the front plate of a networking card. The FPGA consists of two main datapath interfaces: the line interface and the system interface. The former connects the FPGA with optical modules while the latter connects the FPGA with network processors, Figure 4. The FPGA line interface consists of eight CEI-28G-Very-Short-Reach (VSR)[1] transceivers, each operating at up to 28.05Gb/s. Thus, the XC7VH580T supports up to 224.4 Gb/s of full-duplex line-side traffic at the physical layer. The system interface typically operates under the Interlaken protocol [3]. At the physical layer, the XC7VH580T supports up to 48 13.1Gb/s transceivers for the system interface, or 628.8Gb/s of full-duplex transceiver bandwidth. The aggregate full-duplex transceiver bandwidth of this FPGA is thus 853.2Gb/s. A second version of the FPGA, the XC7VH870T, supports 4 x 100Gb/s applications, Figure 5.

Interposer Signal Integrity

Although the silicon interposer is only a hundred microns thick, failure to consider high frequency effects of the TSVs will degrade the rise/fall time, increase crosstalk, increase noise injection, and cause significant performance degradation of the signal transmitted through the high speed channel. In addition, die to die signals which pass through the microbumps and laterally through the fine pitch metallization also need careful consideration.

This section summarizes signal integrity studies for two types of off-die signals in 3D designs, Type I and Type II signals. A Type I signal connects one or more active ICs and a package pin. A Type II signal supports inter-die communication only and is not connected to any package pins. For clarity, Type I signals are electrically connected to TSVs in the silicon interposer whereas Type II signals are not.

Signal "A" in Figure 6 is a Type I signal. Typically, multiple micro-bumps are connected to one C4 bump to ensure adequate electro-migration margin. Signal "B" is a Type II signal from IC 2 to IC 1, whereas signal "C" is a Type II signal with a fan-out of two from IC 1 to both IC 2 and IC 3.

The subsections below discuss signal integrity related to the 2x100Gb/s datapath in the GTZ-IC.

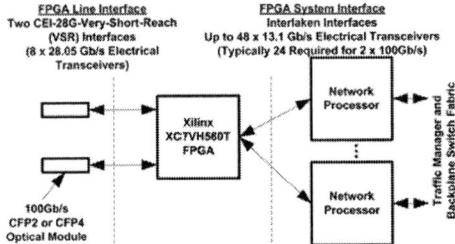

Fig. 4 *2 x 100Gb/s line card*

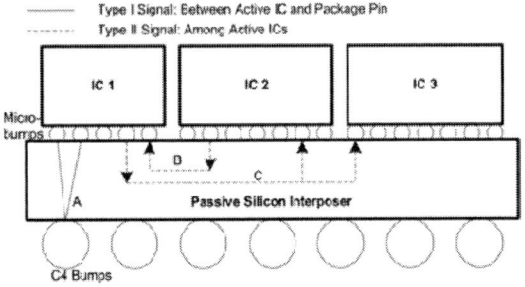

Fig. 6 *Type I and type II signals*

a) Type I Signals

Figure 7 illustrates the 2x100Gb/s datapath in a GTZ-IC. In the ingress direction, it receives eight differential serial streams at up to 28.05Gb/s each, de-serializes them with serial-in-parallel-out (SIPO) blocks, and sends the parallel data and necessary clocks to the SLR. Likewise in the egress

---

[1] The OIF CEI-28G-VSR is a standard electrical interface under development. It is for interoperability between CFP2 or CFP4 optical modules and electronic components. Unlike CFP optical modules, which divide the nominal 100Gb/s full-duplex bandwidth at the electrical interface among ten 11.2Gb/s transceivers, the electrical interface of a CFP2 and a CFP4 consists of four transceivers, each operating up to 28.05Gb/s.

978-1-4673-1707-8/12 $31.00 © 2012 IEEE

direction, the GTZ-IC receives parallel data from the SLR, serializes the traffic with parallel-in-serial-out (PISO) blocks, and outputs the resulting differential serial streams.

The GTZ Type I signals are in the analog supply domain, which has its own power and ground metal planes. The parallel Type II signals are in the digital supply domain, sharing the digital and ground planes with the SLR. Analog power and ground decoupling capacitors are distributed on the die, on the interposer, and on the ceramic package substrate to ensure optimal signal integrity at 28Gb/s.

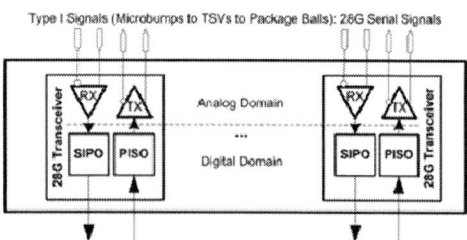

Fig. 7 *Datapath and power domains of GTZ-IC*

1) Silicon Interposer TSV Modeling and Optimization
A full 3D EM field solver is used to model the silicon TSV interposer accurately over a wide frequency range. A broadband S-parameter model is generated with an upper frequency limit of 50GHz. The interposer consists of TSVs, four metal layers for die-to-die connections, micro-bumps for die to interposer connections and C4 bumps for interposer to package connections. All components are incorporated into the model to predict actual interposer performance in the system. A silicon interposer test vehicle is fabricated for measurement and verification. Various TSV test structures are characterized across a wide frequency range. The TSV is less than 100um tall and is quite difficult to measure. De-embedding microprobe effects is critical for accurate frequency domain vector network analyzer measurements. As a first step, a microprobe calibration is performed to move the reference plane to the end of microprobe. Depending on the metallization of the probe tip and the surface of the pad, a residual, uncompensated resistive impedance term on the order of 10~100 mOhm can still be present. The flexing or over-travel of the microprobe can also result in an uncompensated, residual inductance of as much as 100pH. To overcome these parasitics a 2-port VNA technique is used. [4]

The 2-port VNA technique eliminates the residual contact resistance and inductance. However, mutual magnetic coupling between probes still exists if probes are located face to face in close proximity. To improve high frequency measurement accuracy, 90 degree orthogonal probing is performed to cancel the magnetic field coupling from the two probes as shown in Figure 8. This results in improved measurement especially for higher frequencies (>5GHz).

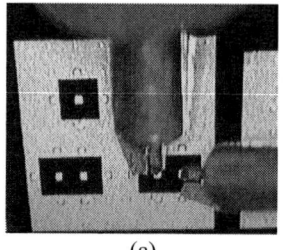 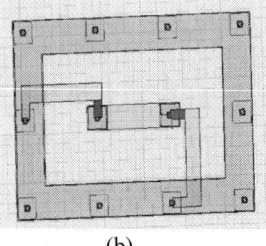

(a)   (b)

Fig. 8 *90 degree 2-port VNA orthogonal probing method. (a) measurement (b) simulation setup*

Since the focus is on multi-gigabit operation within the interposer, the model must support and have a good agreement for frequencies up to tens of gigahertz. The measured data is compared with simulation results. Figure 9 shows the comparative results of effective capacitances which were extracted from the test structures. Two different silicon substrates are fabricated to compare the substrate resistivity effects. One test interposer sample has a 10 ohm-cm silicon resistivity representative of a typical commercial process and the other test interposer sample has a 20 ohm-cm silicon resistivity.

In Figure 9, the dotted traces represent the measured data and the solid traces represent the simulated data. Green and pink traces represent effective capacitance and conductance values over frequency from the 10 ohm-cm silicon resistivity substrate. Blue and red traces represent effective capacitance and conductance over frequency from the 20 ohm-cm silicon resistivity substrate. We observe good agreement between the measured and simulated results. The effective capacitance decreases as a function of frequency (slow wave model to quasi-TEM mode). This effect can be modeled as resistance and capacitance in the silicon. The measured effective capacitance from the 10 ohm-cm silicon resistivity substrate was about 205.2fF at low frequency and 19.4fF at high frequency. The measured effective capacitance from the 20 ohm-cm silicon resistivity substrate was about 115.8fF at low frequency and 12.4fF at high frequency. Both low and high frequency capacitance are important electrical parameters because they directly affect data eye opening, power consumption and delay.

A second structure is fabricated to evaluate insertion and return loss for serially connected TSV's. S21 insertion loss and S11 return loss are measured and compared with simulation as shown in Figure 10. The measured insertion loss at the Nyquist frequency (14GHz) is -0.414dB from the 20 ohm-cm silicon substrate and -0.822dB from 10 ohm-cm silicon substrate, respectively. The high resistivity silicon substrate provides lower loss over the entire frequency range as shown in this figure.

Based on these measurements and simulation results, we conclude that a 20 ohm-cm high resistivity silicon substrate is the best choice for very high speed signaling applications such as the 28Gb/s Serdes.

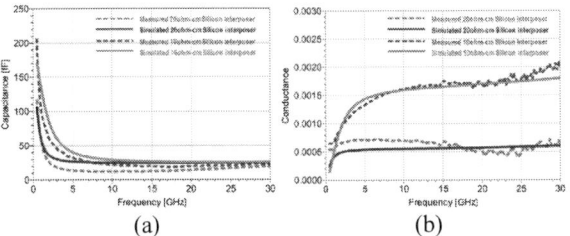

(a)   (b)

Fig. 9 *(a) Effective capacitance correlation[fF] (b) Effective conductance correlation;*

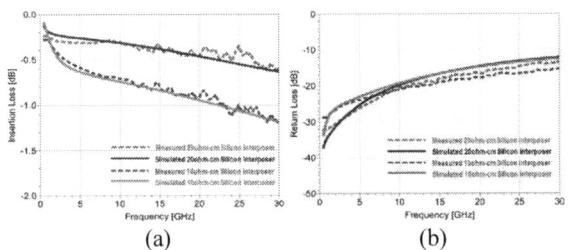

(a)   (b)

Fig. 10 *(a) Insertion Loss correlation [dB] (b) Return Loss correlation [dB]*

2) Package Design for Ultra Low Loss

After the silicon interposer is optimized, the package is designed to minimize channel loss. The package substrate can be a major source of losses due to discontinuities, dielectric loss, skin effect and surface roughness loss due to narrower trace widths. A Low Temperature Co-Fired Ceramic (LTCC) package substrate is chosen to minimize loss and discontinuities. The LTCC package has a lower loss tangent dielectric material when compared with an organic substrate. In addition, relatively wide traces can be used to control impedance. The ceramic package also has the advantage of less copper loss including skin effect loss and surface roughness loss.

3) Embedded Capacitors Design in the Silicon Interposer to Minimize SSO Noise

Once signal insertion/reflection loss and crosstalk are optimized, the power plane coupling noise including simultaneous switching noise (SSN) has to be considered. The ceramic package provides excellent signal integrity in the signal channel but at the same time has a disadvantage in power integrity due to the long vertical power connections. Higher inductance in the package coupled with the on-die capacitance creates higher plane impedance (and therefore noise) at the resonant frequency. To minimize the power plane noise due to higher plane impedance, a finger capacitor constructed from metal layers in the silicon interposer is used. Approximately $0.375nF/mm^2$ can be achieved using the finger capacitors. Since the metal finger capacitor location is much closer to the die, it is much more effective at decoupling than package capacitors. Time domain SPICE simulations are performed to see the power plane noise effect. Figure 11 shows one example of worst case simultaneous switching noise simulated at the resonant frequency. The coupling noise amplitude waveform at the package ball side with the interposer finger capacitance is compared to the noise amplitude waveform without the interposer finger

capacitance. Power noise is significantly reduced by the finger capacitor in the silicon interposer as shown in Figure 11.

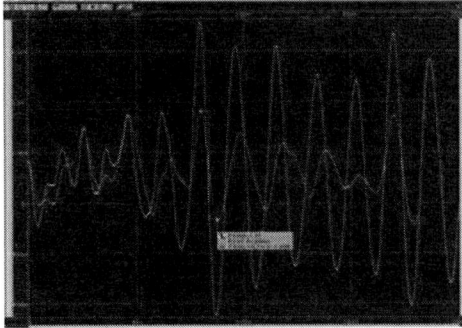

Fig. 11 *Simultaneous switching noise simulation in system without interposer capacitor (in yellow) and with interposer capacitor (in blue)*

4) Channel Simulation

Full channel analysis is performed including an optimized silicon interposer model on top of a low loss package substrate and a PCB model. The simulation data is then compared to the measured 28Gb/s eye diagram and shows very good correlation as shown in Figure 12. The measured eye amplitude is 923mVp-p and the measured total jitter is 6.25ps at BER 10^{-12}. The measured Random jitter is 230fs which is 0.175UI of total jitter. The proposed optimized channel system enables high performance signaling suitable for 28Gb/s Serdes FPGA products.

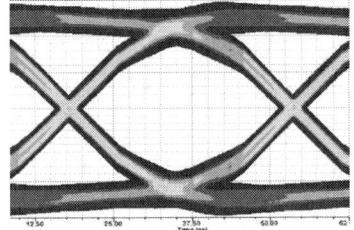

(a)   28Gb/s Simulated Eye Diagram

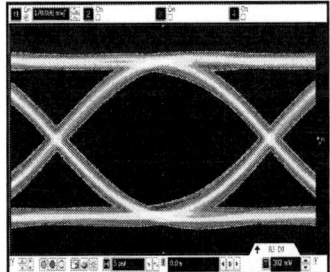

(b)   28Gb/s Measured Eye Diagram

Fig. 12 *System Level Eye Diagram Comparison*

b) Type II Signals

1) Silicon Interposer Signal Routing

The silicon interposer lateral routing consists of four layers of metal: one layer for redistribution, one for ground reference, and two for signal routing. The ground reference layer separates the two signal-routing layers. In each signal-routing layer, every signal wire runs parallel to a grounded side shield.

Figure 13 illustrates the effect of these grounded side shields on a victim wire. Eleven interposer wires of 4.5mm in a single metal layer on top of a ground reference layer are simulated. The victim wire is at the center of the eleven-wire bus. The interposer wires are modeled with 3D RLC elements. The simulation consists of ten aggressors transitioning with a range of delay values relative to the victim. Figure 13(a) and Figure 13(b) illustrate the victim waveform without side shields under odd-mode and even-mode coupling, respectively. Figure 13(c) and Figure 13(d) show the same simulation with side shields inserted into the eleven-wire bus to eliminate overshoots and undershoots.

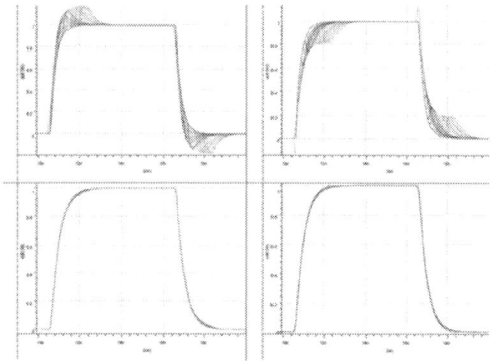

Fig. 13. *(a) Odd-Mode Coupling without Side Shields. (b) Even-Mode Coupling without Side Shields. (c) Odd-Mode Coupling with Side Shields. (d) Even-Mode Coupling with Side Shields.*

2) Inter-Die Timing Verification

Inter-die routing requires rigorous timing analysis to verify adequate set-up and hold time.

There are over 4000 Type II signals between the GTZ-IC and its neighboring SLR. Figure 14 illustrates the wire length distribution of these Type II signals. All but five signals are routed on traces no more than 6mm in length, while 85% of the traces are less than 3.75mm in length. Clocking between the GTZ-IC and its neighboring SLR is system-synchronous. Careful balancing of clock networks on these two pieces of silicon across process, voltage, and temperature is necessary. Static timing analysis can verify inter-die timing efficiently. However, during XC7VH580T development, static timing analysis tools did not accept RLC interconnect models. As a result, inter-die propagation delay and transition time values were calibrated between SPICE simulations with 3D RLC interconnect models and static timing analysis with RC interconnect models. The layout ensures that Type II signaling in the XC7VH580T is not in transmission-line mode and static timing analysis with RC-based interconnects suffices.

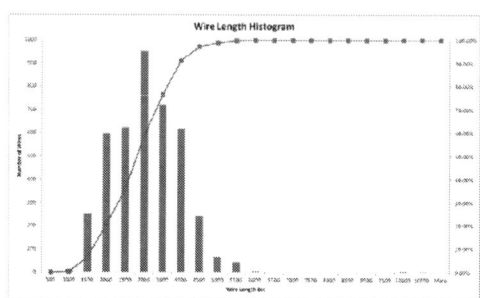

Fig. 14 *Wire Length Histogram for Type II Signals between GTZ-IC and Neighboring SLR*

3) Interposer and Active Device Assembly

Figure 15(a) shows the XC7VH580T interposer in a ceramic package substrate. Figure 15(b) illustrates two SLRs and one GTZ-IC mounted on an interposer face-down. Both the interposer and the package contain additional decoupling capacitors. In particular, the interposer provides additional decoupling for the GTZ analog power domain.

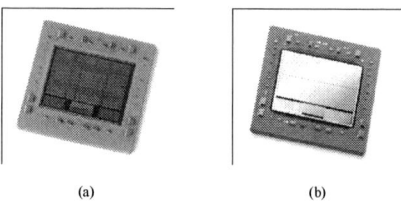

(a) (b)

Fig. 15 *XC7VH580T in Package Substrate. (a) Interposer Only. (b) GTZ-IC and Two SLRs Mounted Face-Down on Interposer.*

3D Thermo-Mechanical Simulation and Analysis

This work relies heavily on thermo-mechanical modeling to verify long-term package/interposer reliability. 3D models are constructed using the commercial software ABAQUS to perform package thermo-mechanical simulation and analysis.

Owing to the symmetry of the package, only one half is modeled to increase computational efficiency. The package is assumed to be stress-free at an underfill cure temperature of 150C. Furthermore, sub-modeling techniques are used. A detailed model for the ubumps, TSV and Redistribution Layers (RDL), C4 solder, low-k dielectric material, substrate solder mask and metallization greatly affects the accuracy and rate of convergence of the solution. Therefore, detailed local models are solved at selected locations on the global model.

The material properties of package compounds, Young's modulus and CTE, are temperature dependent. Table 1 shows these properties at room temperature.

Table 1. *Material properties of package compounds at room temperature*

Materials	Young's modulus (GPa)	CTE (ppm)
LTCC ceramic	75	12.3
Si	150	3
ubump UF	6	38
C4 UF	8.5	32
Copper	130	17
Sn Ag solder	51	22.4
TIM1	5.8e-3	-
Eut solder	31.5	25.3

Table 2. *Warpage simulation results at room temperature*

Max warpage	Max die stress
Package: 55um	Logic slices: 102MPa
Interposer: 52um	SerDes die: 86Mpa

a) Package Warpage

In this section, 3D non-linear thermo-mechanical global models are built. No printed circuit board is considered here and the focus is on warpage of the package under thermal stress. In the warpage model, the package is stress free at an underfill cure temperature of 150C, and out of plane displacement is obtained after cooling to room temperature.

Different core materials (organic low-CTE vs. ceramic), substrate sizes and heat spreader designs (forged-lid, stamped-lid, copper and AlSiC lid materials) are investigated in order to optimize warpage and low-k stress. Simulations show that after chip-on-chip (COC) attach and subsequent ubump underfill, the interposer warps when returned to room temperature. In addition, after the COC is packaged it continues to warp during thermal stressing. However, introducing an AlSiC heat spreader decreases the warpage significantly by balancing the Coefficient of Thermal Expansion (CTE) mismatch within the package. The CTE of AlSiC is close to that of the LTCC ceramic substrate. Table 2 and Figure16 show the warpage and die stress simulation results.

b) ubump and C4 Solder Reliability

In order to study lead free ubumps and eutectic C4 solders in detail and simulate strains, inelastic energy and fatigue to a high level of accuracy, detailed local models were built for a Thermal Cycling (TC) test. The ubump and C4 local models include one ubump-pitch or C4-pitch portion of the package at the corner where inelastic energy is higher under TC.

In sub-modeling, at each time point, displacements calculated from the global model are applied to the corresponding points at the boundaries of the local model. The local model is then solved for the applied boundary conditions and loads. In the global model, the package is assumed to be stress-free at an underfill cure temperature of 150C and stresses, inelastic strains and energy are obtained after cooling the package from the underfill cure temperature to room temperature and through temperature cycles of -55C/+125C.

The simulation is run for 3 temperature cycles. The results shown in Figure 17 and Table 3 indicate that for the given package and TSV interposer attributes and metallization the ubumps and C4 bumps undergo acceptable reliable solder inelastic strain and fatigue.

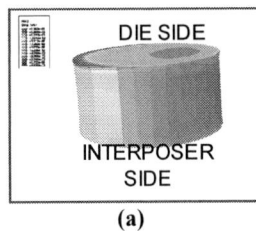 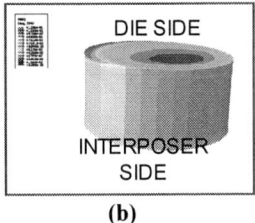

Fig. 17 *ubump inelastic strains after three TC cycles at 125C (a) Logic slices (b) SerDes die*

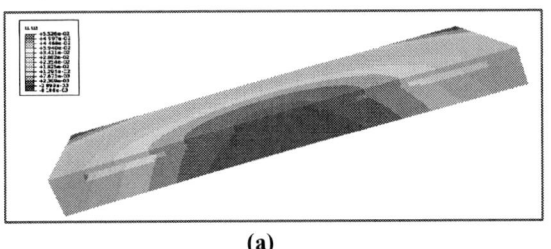

(a)

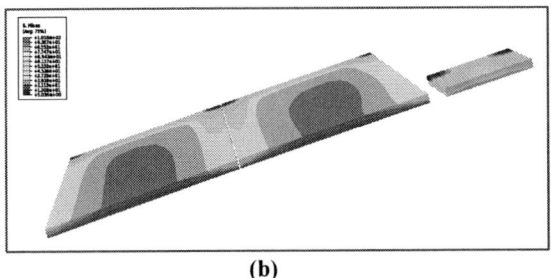

(b)

Fig. 16 *Simulation results at 25C (a) warpage (b) die stress*

Table 3. *Summary of solder fatigue study*

	After 3 cycles	
	@ 125C	@ -55C
ubump solder inelastic strains	0.13	0.39
ubump solder inelastic energy	3.4	13.3
C4 solder inelastic strains	0.09	0.2

85C and 100C when every logic chip is performing at the same power. Simulation results confirm that a heatsink is required to cool down the package and for the selected passive heatsink the FPGA package is thermally reliable and meets the thermal specs.

Advanced Thermal Study

3D non-linear models are constructed using the commercial software Flotherm to perform the package thermal simulation and analysis.

An accurate measurement or estimation of power is required to drive the simulation. The Xilinx Power Estimator (XPE) tool is used to estimate the FPGA power [5]. The tool is typically used in the pre-design and pre-implementation phases of a project. XPE assists with architecture evaluation and device selection and helps in selecting the appropriate power supply and thermal management components which may be required for an application. XPE considers the design's resource usage, toggle rates, I/O loading, and many other factors which it combines with the device models to calculate the estimated power distribution. The device models are extracted from measurements, simulation, and/or extrapolation.

The package is assumed to be initially at ambient temperatures of 55C, 85C and 100C. A JEDEC 2s2p board is considered here.

The material properties of package compounds, conductivity and specific heat, are temperature dependent. At room temperature, conductivity is shown in Table 4.

Table 4. *Material properties of package compounds at room temperature*

Materials	Conductivity W/mK)
LTCC ceramic	2.0
Prepeg	0.35
Si	117.5
Copper	389
Solder mask	0.24
TIM1	0.92

In order to study heat transfer to a high level of accuracy and without the need to model every ubump, C4 bump and BGA ball, a representative conductivity and specific heat of ubumps/ubump underfill, C4 bumps/C4 underfill and BGA level are calculated using composite material property equations. In addition, considering TSV pitch, attributes and material properties, equivalent TSV silicon interposer conductivity and specific heat are estimated.

The 3D non-linear model is solved for applied boundary conditions and corresponding power. The package is assumed to be initially at ambient temperature, and temperatures, temperature gradients and thermal resistances are obtained after the system reaches a steady-state condition.

Table 5 demonstrates thermal simulation results for the optimized design for 3 different ambient conditions of 55C,

Table 5. *Thermal Simulation Results*

Virtex7 DEVICE	Power (Watt)	Ta (C)	JB (C /W)	JC (C /W)	Eff. JA (C /W)	Max Junction temp. (C)
No HS	36	55	4.2	0.07	7.6	297.2
No HS	37	85			7.4	338.4
No HS	44	100			6.9	355.2
w *HS @ 250 LFM	36	55			0.6	81.2
w *HS @ 250 LFM	37	85			0.6	107.1
w *HS @ 250 LFM	44	100			0.6	130.3
*HS = Xilinx reference heatsink						

Conclusions

A heterogeneous FPGA and transceiver solution with up to 2.78Tb/s bandwidth is detailed. The design methodologies to minimize signal insertion loss, reflection loss, crosstalk and power coupling noise in the system are presented. Full channel analysis is performed including an optimized silicon interposer model on top of a low loss package substrate and a PCB model. The simulation data when compared to the measured 28Gb/s eye diagram shows excellent correlation. The proposed optimized channel system enables high performance signaling at 28Gb/s. Furthermore, 3D thermal and mechanical simulations are run to study the thermal performance and mechanical reliability of the system. The optimized design delivers a Stacked Silicon Interconnect implementation with acceptable thermal performance, warpage/coplanarity and highly reliable micro bumps and C4 bumps.

Acknowledgments

Thanks to Sanjiv Stokes, Suzanne Yiu, Oliver Huang, Daniel Wu, Jack Carrel, Chris Wyland and Hoa Do.

References

[1] Xilinx, Inc., "7 Series FPGAs Overview," http://www.xilinx.com/support/documentation/data_sheets/ds180_7Series_Overview.pdf, May 5, 2012.

[2] The CFP MSA, "CFP MSA 100G Roadmap and Applications," http://www.cfp-msa.org/Documents/CFP_MSA_roadmap.pdf, November 2010.

[3] The Interlaken Alliance, "Interlaken Protocol Specification, v1.2," http://www.interlakenalliance.com/Interlaken_Protocol_Definition_v1.2.pdf, October 2008.

[4] Agilent Application Note, "Ultra-low Impedance Measurements Using 2-port Measurement"

[5] http://www.xilinx.com/power

978-1-4673-1707-8/12 $31.00 © 2012 IEEE

Graphene for Microelectronics:
Can it make a difference?

Max C. Lemme
KTH Royal Institute of Technology
School of Information and Communication Technology
Electrum 229, 16440 Kista, Sweden
lemme@kth.se

INTRODUCTION

Benchmarking figures for graphene show remarkable properties like ballistic conductance over several hundred nanometers or charge carrier mobilities of several 100.000 cm^2/Vs [1, 2]. When graphene is integrated and processed, however, defects in the graphene and its dielectric environment dominate device performance [3, 4]. Furthermore, the lack of a band gap limits the applicability of graphene field effect transistors (GFETs) for logic applications. Yet, there are many options for graphene to make a difference in the future of microelectronics, many of which can be attributed to the "More than Moore" domain defined in the ITRS. These will be discussed in this talk.

MANUFACTURING TECHNOLOGY

The future industrial manufacturability of graphene devices, independent of their specific application, will depend on the availability of a scalable and technologically compatible graphene fabrication method. The state-of-the-art in graphene growth will therefore be discussed with respect to silicon technology. While chemical vapor deposition and epitaxy from silicon carbide both promise scalability, they are not (yet) fully compatible with silicon technology. Direct growth of graphene on insulating substrates would be a major step. Even though first results are available [5, 6], this technology is still at a very early stage. This has implications on potential entry points of graphene as an add-on to mainstream silicon technology, which will be discussed in the talk.

RF APPLICATIONS

On the device level, Radio frequency analog graphene FETs will be compared with current silicon CMOS technology, paying particular attention to requirements concerning carrier mobility and contact resistance. We find that graphene FETs slightly lag behind CMOS in terms of speed despite of their higher mobility. In addition, GFETs achieve their best performance only for narrow ranges of V_{DS} and I_{DS}, which must be carefully considered for the design of biasing circuitry [7].

As an alternative, a novel graphene-based hot electron transistor, the Graphene Base Transistor (GBT) will be introduced [8]. The GBT is based on a vertical arrangement of emitter, base and collector contacts (Fig. 1).

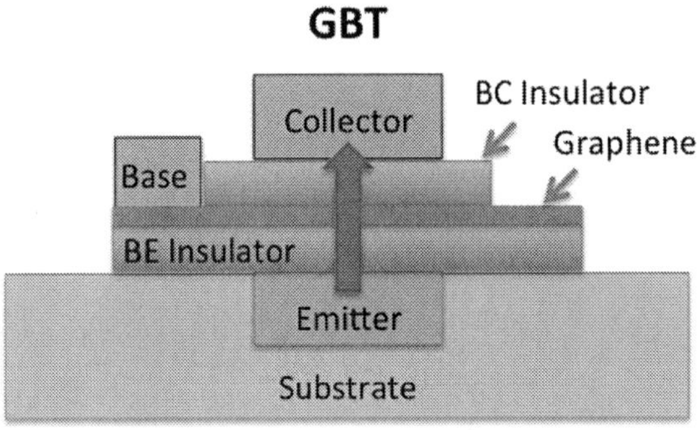

Fig. 1: Schematic of a Graphene Base Transistor. In the on-state, carriers tunnel through the emitter-base-insulator across the graphene base into the conduction band of the base-collector-insulator.

In the off-state, charge carriers face a dielectric barrier. In the on-state, the emitter-base diode injects hot electrons across the emitter-base-insulator (EBI) through Fowler-Nordheim tunneling. The carriers than cross the ultra-thin graphene base and enter the conduction band of the collector-base-insulator (BCI), as illustrated in Fig. 2. The GBT is predicted to achieve large I_{on}/I_{off} ratios if the BCI is design appropriately. It is also predicted to exhibit current saturation when the output voltage exceeds the value necessary to remove the tunneling barrier of the BCI. As the transit time depends mainly on the base thickness,

978-1-4673-1707-8/12 $31.00 © 2012 IEEE

monoatomic graphene films enable high switching speeds, potentially in the THz range. The large I_{on}/I_{off} ratio may also enable logic integration. A scalable, CMOS compatible fabrication scheme (Fig. 3) will be illustrated [9].

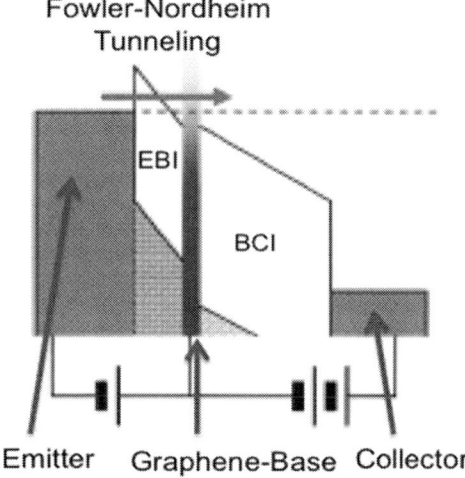

Fig. 2: Schematic band diagram for a graphene Base hot electron transistor in the on-state.

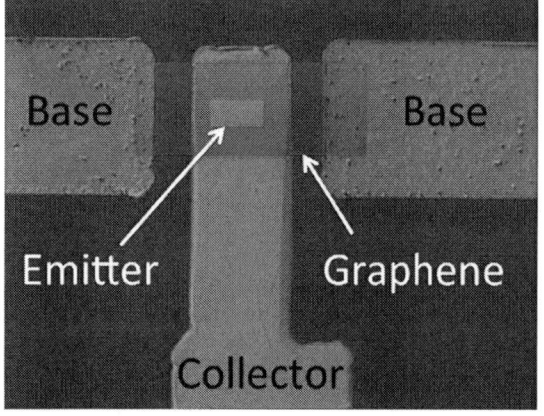

Fig. 3: Optical micrograph of a GBT. Emitter and graphene are enhanced for clarity.

OPTOELECTRONICS

The unique band structure of graphene with a linear dispersion relation from 0 up to approximately ±1eV makes it an interesting optoelectronic material. Photons can in principle generate electron-hole pairs nearly independent of their energy, i.e. from UV to THz frequencies. The talk will include results from double-gated graphene devices (Fig. 4) that allow tuning the graphene channel from bipolar to unipolar. This is explained with a model of the photothermal effect. As a result, the photoresponse becomes gate-activated in the visible light spectrum, suggesting applications such as broadband photodetectors and pixelized cameras in the visible and IR range as well as bolometers [10]. The high mobility in graphene could enable applications such as high speed photodetectors [11]. Finally, plasmonic nanostructures can enhance the light-matter interaction and hence the quantum efficiency of graphene-based optoelectronic devices [12, 13].

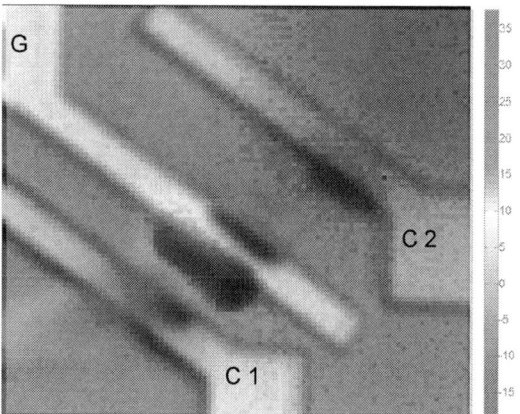

Fig. 4: Scanning electron microscope image of a gated graphene photodetector. The overlay shows photocurrent maps of positive (red) and negative currents (blue). Strong signals are observed at pn junctions near contacts (C1, C2) and gate (G) electrode.

NEMS

Its low mass and high mechanical stability suggest graphene for nanoelectromechanical systems (NEMS). Resonance frequencies of approximately 400GHz have been deduced from STM measurements of naturally occurring graphene nanomembranes [14]. In this context, a graphene-based pressure sensor will be discussed that utilizes a graphene membrane as an electromechanical transducer [15]. An SEM image of such a sensor is shown in Fig. 5 The sensitivity of graphene pressure sensors will be compare to state of the art silicon and carbon nanotube devices [16].

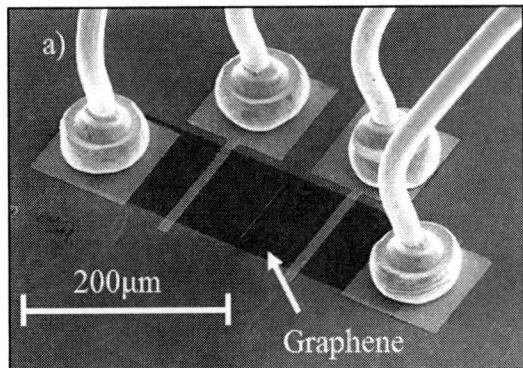

Fig. 5: Scanning electron microscope image of a graphene membrane based pressure sensor.

SUMMARY

Graphene has a large number of exceptional materials properties. Many reported benchmarking figures, however, are based on idealized measurement conditions and may not stand in realistic devices and applications. In addition, the lack of a " band gap results in extremely poor switching of graphene field effect transistors. Thus, the first wave of enthusiasm for post silicon "graphene electronics" has abated somewhat. Nevertheless, there is great promise in graphene, particularly for applications in the More than Moore domain. Some like RF analog devices, optoelectronics and NEMS are summarized in this contribution, but others include non-conventional switches like the Bilayer PseudoSpin Field-Effect Transistor (BiSFET) [17] or the Barristor [18].

REFERENCES

[1] S. V. Morozov, K. S. Novoselov, M. I. Katsnelson, F. Schedin, D. C. Elias, J. A. Jaszczak, and A. K. Geim, "Giant Intrinsic Carrier Mobilities in Graphene and Its Bilayer," *Physical Review Letters,* vol. 100, pp. 016602-4, 2008.

[2] K. I. Bolotin, K. J. Sikes, Z. Jiang, M. Klima, G. Fudenberg, J. Hone, P. Kim, and H. L. Stormer, "Ultrahigh electron mobility in suspended graphene," *Solid State Communications,* vol. 146, pp. 351-355, 2008.

[3] M. C. Lemme, T. J. Echtermeyer, M. Baus, and H. Kurz, "A Graphene Field-Effect Device," *IEEE Electron Device Letters,* vol. 28, pp. 282-284, 2007.

[4] S. Adam, E. H. Hwang, V. M. Galitski, and S. Das Sarma, "A self-consistent theory for graphene transport," *Proceedings of the National Academy of Sciences,* vol. 104, pp. 18392-18397, November 20, 2007 2007.

[5] G. Lippert, J. Dabrowski, M. C. Lemme, O. Seifarth, and G. Lupina, "Direct Graphene Growth on Insulator," *Physica Status Solidi B* vol. 248, pp. 2619–2622, 2011.

[6] U. Wurstbauer, T. Schiros, C. Jaye, A. S. Plaut, R. He, A. Rigosi, C. Gutiérrez, D. Fischer, L. N. Pfeiffer, A. N. Pasupathy, A. Pinczuk, and J. M. García, "Molecular beam growth of graphene nanocrystals on dielectric substrates," *Carbon,* 2012.

[7] S. V. S. Rodriguez, M. Ostling, A. Rusu, E. Alarcon, M.C. Lemme, "RF Performance Projections of Graphene FETs vs. Silicon MOSFETs," *ECS Solid State Letters,* 2012.

[8] W. Mehr, J. C. Scheytt, J. Dabrowski, G. Lippert, Y.-H. Xie, M. C. Lemme, M. Ostling, and G. Lupina, "Vertical Transistor with a Graphene Base," *IEEE Electron Device Letters,* vol. 33, pp. 691-693, 2012.

[9] S. Vaziri, A. D. Smith, C. Henkel, M. Östling, M. C. Lemme, G. Lupina, G. Lippert, J. Dabrowski, and W. Mehr, "An Integration Approach for Graphene Double-Gate Transistors," *ESSDERC,* 2012.

[10] J. Yan, M.-H. Kim, J. A. Elle, A. B. Sushkov, G. S. Jenkins, H. M. Milchberg, M. S. Fuhrer, and H. D. Drew, "Dual-gated bilayer graphene hot electron bolometer," *arxiv:1111.1202v1,* 2011.

[11] T. Mueller, F. Xia, and P. Avouris, "Graphene photodetectors for high-speed optical communications," *Nat Photon,* vol. 4, pp. 297-301, 2010.

[12] T. J. Echtermeyer, L. Britnell, P. K. Jasnos, A. Lombardo, R. V. Gorbachev, A. N. Grigorenko, A. K. Geim, A. C. Ferrari, and K. S. Novoselov, "Strong plasmonic enhancement of photovoltage in graphene," *Nat Commun,* vol. 2, p. 458, 2011.

[13] F. H. L. Koppens, D. E. Chang, and F. J. García de Abajo, "Graphene Plasmonics: A Platform for Strong Light–Matter Interactions," *Nano Letters,* vol. 11, pp. 3370-3377, 2011/08/10 2011.

[14] T. Mashoff, M. Pratzer, V. Geringer, T. J. Echtermeyer, M. C. Lemme, M. Liebmann, and M. Morgenstern, "Bistability and Oscillatory Motion of Natural Nanomembranes Appearing within Monolayer Graphene on Silicon Dioxide," *Nano Letters,* vol. 10, pp. 461-465, 2010.

[15] A. D. Smith, S. Vaziri, M. Östling, and M. C. Lemme, "Strain Engineering in Suspended Graphene Devices for Pressure Sensor Applications," in *13th International Conference on Ultimate Integration on Silicon (ULIS)* Grenoble, France, 2012.

[16] A. D. Smith, F. Niklaus, S. Vaziri, A. C. Fischer, M. Sterner, A. Delin, M. Östling, and M. C. Lemme, "Graphene Membrane-based Nanoelectromechanical Pressure Sensors," *submitted,* 2012.

[17] S. K. Banerjee, L. F. Register, E. Tutuc, D. Reddy, and A. H. MacDonald, "Bilayer PseudoSpin Field-Effect Transistor (BiSFET): A Proposed New Logic Device," *Electron Device Letters, IEEE,* vol. 30, pp. 158-160, 2009.

[18] H. Yang, J. Heo, S. Park, H. J. Song, D. H. Seo, K.-E. Byun, P. Kim, I. Yoo, H.-J. Chung, and K. Kim, "Graphene Barristor, a Triode Device with a Gate-Controlled Schottky Barrier," *Science,* vol. 336, pp. 1140-1143, June 1, 2012 2012.

978-1-4673-1707-8/12 $31.00 © 2012 IEEE

Nanostructure Devices for Logic and Memory and Beyond

Invited Paper

Sandip Tiwari
Cornell University
Ithaca, NY 14853, USA
st222@cornell.edu

Abstract—**After six decades of device size reduction and its efficient use through hierarchical design, the semiconductor area encounters two major conflicting currents: (a) quantum, stochastic (atomic and signal/noise) and other probabilistic effects with size reduction at the bottom and (b) thermodynamic consequences in the inefficiencies of the information engine as a large numbers of devices are assembled together hierarchically at the top. This is the central intellectual challenge when discussing the future of nanostructure devices and their use in information machines. We discuss the conceptual fundamentals of the small and the large that ties this scale change that exists in time, size, energy, and other dimensions of the machinery. From this, we derive ideas for devices, robustness, information efficiency, and performance under practical constraints so that the next six decades are just as fruitful and useful for the society.**

I. INTRODUCTION

As devices have shrunk, their variability and propensity to statically and dynamically fail continues to grow. Quantum mechanics' consequences also appear in multitude of forms, the most deleterious examples of which are the increase in static currents and threshold voltage for inversion in a transistor. In the deterministic digital style that we practice in electronics, i.e., always being correct, the cost in energy consumption per element has stopped decreasing because of the plethora of probabilistic variations in important properties and parasitics that increase and thus counter the density increase from shrinking of dimensions and of switching energy contribution from decreased potential.

Each device is relatively more inefficient, even if smaller, due to these conflicting considerations.

We employ each these minute components to build the chip as a more sophisticated information machine. Hierarchical design, i.e., creation of *de facto* interfaces between devices, circuits, their functions, and the system architecture and software that separates the concerns of these domains by abstracting the most critical attributes, also leads to large increases in inefficiencies because of the nonlinearities inherent in building interfaces. This tremendously successful approach is thus encountering enormous barriers.

This problem is a manifestation of the inherent complexity of large scale changes, in size, in time, in energy, and other dimensions: nanometer devices in a centimeter or larger system, picoecond devices in a millisecond human environment, femto-to-nanojoule components in milliwatt to megawatt machines. At nanoscale, a terascale of devices are potentially connectible in an integrated circuit. Scale changes of decades of decades abound. At such large numbers and such large scale changes, respect to thermodynamics/statistical mechanics is a necessity, not just in power and heat design, but also in finding operational principles that make high efficiency of the information engine feasible.

These observed inefficiencies are also a manifestation of two physical styles at work representing the two different ways that we understand and gainfully employ physical phenomena: "reductionism" as in scaling to pack higher and higher density, and "constructionism" [1] as in collecting to affect order; in this case via hierarchy. Constructionism's manifestation in physical systems is emergence – appearance of new properties, an order from disorder, such as mobility or bandgap or ferroelectricity that arises in a crystal. Emergence of simple physical laws is a property of complex systems that is manifest all around us in the physical world. The intensive properties of gaseous phases such as pressure, the melting and boiling temperatures of liquids, are reproducible even if the molecular blocks have probabilistically variant properties. Thermodynamics and statistical mechanic help us understand this and define the efficient paths, as Nicolas Léonard Sadi Carnot did in laying the foundation of second law of thermodynamics and in elucidating the reversible connection between useful work and heat in reversibility of a mechanical engine.

In the information engine, in discussing nanostructures and their use in electronics, it is necessary to include laws of large numbers together with the fundamentals of the smallest scale to draw meaningful conclusions – in particular in how could one achieve extremely low energy computation of decisions in the midst of uncertainty – a task that is naturally probabilistic, and how does it map best to the complex systems that we now

---

Support by IBM Research, DARPA, NSF, TeraNano, AFOSR and Mellowes endowment over the years has been critical to the thinking articulated here.

978-1-4673-1707-8/12 $31.00 © 2012 IEEE

create through nanoscale devices so that they become highly efficient engines.

The following is the author's attempt at connecting this scope.

II. PROBABILITIES, ERRORS & CONSEQUENCES FOR ENERGY

Logic and memory operate in a noisy environment with noisy signals. The noise can be thermal and also arises from a multitude of other statistical fluctuations. In this sense thermal noise and threshold voltage variations are probabilistic effects arising from different phenomena but resulting in similar energetic consequences in a logic switch or in a memory where the physical and the informational world come together. We look at these generic logic switches and memories in the following way to extract some general principles.

A. Switches and their Ensembles:

Take an ideal switch turning on or off a connection to a load, say another switch, represented as a simple capacitance C. The energy per operation is $U_i = P_i \cdot t_i$ relating power and switching time. For thermal noise arising from fluctuations, the mean thermal energy generated in the circuit is $\frac{1}{2}C\bar{v}^2 = \frac{1}{2}k_BT$. The "0" and "1" states, with these random fluctuations, is

$$P_0 = \frac{1}{\sqrt{2\pi}}\exp\left[-\frac{v^2C}{2k_BT}\right] \;\; and \;\; P_1 = \frac{1}{\sqrt{2\pi}}\exp\left[-\frac{(v-V)^2C}{2k_BT}\right] \quad (1)$$

which allows us to determine the error rates [2] (see Fig. 1) for a normalized switching threshold of $S = \frac{V}{2\sqrt{k_BT/C}}$ as

$$\epsilon = \frac{\epsilon_{01}+\epsilon_{10}}{2} = \frac{2}{\sqrt{2\pi}}\int_S^\infty \exp\left[-\frac{v^2C}{2k_BT}\right] d\left(\frac{V}{2\sqrt{k_BT/C}}\right). \quad (2)$$

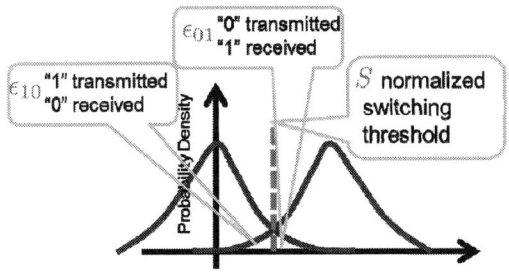

Figure 1. Errors in switched states due probabilistic distribution arising from noise and statistical fluctuations.

This relationship can be inverted to relate error rates in this ideal noisy system as seen in Fig. 2.

For this ideal switch, 1 error in 10 years (10^{-20}) corresponds to an energy consumption of $200 k_BT$. This relationship is general since the principle of $\frac{1}{2}k_BT$ of energy per degree of freedom is a general conclusion of equipartition of energy and the virial of Claussius - valid from quantum to macroscopic world for systems whose stability arises from harmonic behavior. This consequence holds true for magnetic

systems including spin systems, electrical systems, mechanical systems and their combinations.

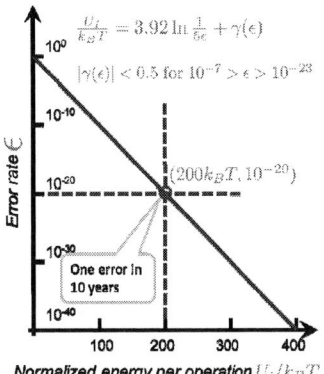

Figure 2. Error rate relationship to energy per operation in ideal capacitive switch in the presence of thermal noise.

One can introduce non-idealities to this and other probabilistic effects to make this calculation more realistic. The consequence of additional probabilistic effects is to increase the energy necessary to assure correctness. The general feature of these error-energy relationships is that the error probability has a relationship of $\epsilon \propto \exp\left(-\frac{\Delta U}{k_BT}\right)$. The Boltzmann factor represents the distribution probability of the ensemble of physical objects or of random noise which leads to the error. Error reduction requires larger and larger energy consumption for all the errors arising from random effects – thermal or other statistical sources and whose contributions are from independent sources. For example, in a CMOS inverter, this relationship is $\epsilon \propto \exp\left[-\frac{2(V_{DD}-V_T)}{\sigma(V_T)}\right]$ [3]. The statistical variations in threshold voltage, a random variation effect, causes errors in a similar form as thermal noise, and in the smallest of geometries, such as in sRAMs, this effect is larger than that due to thermal noise.

These random effects are independent and add, and one can see their consequence in circuit design. Consider designing a 100 million gates circuit. If one wants 90% yield for the ensemble, $0.9 = (1 - \epsilon)^{10^8}$, or $(V_{DD} - V_T) = 10.3 \cdot \sigma(V_T)$. In sRAMs, variances of ~40 mV are currently the norm, so an ~0.5 V excess voltage above the threshold is a necessity to achieve a 90% yield in a 100 million gate system. While defect correction can be applied in ordered systems, the penalty of correction in less-ordered system is tremendous [4] with the consequences of defects in interconnect lines most consequential. Defect correction at fine granularity extracts a very high system price in power, area and timing, and is really not a practical option.

B. Memory:

Memories are switches with stable states, so the previous discussion still holds. If it is a bistable system, the two energy

978-1-4673-1707-8/12 $31.00 © 2012 IEEE

minima are the desired states of the system around which fluctuations occur (see top part of Fig. 3). For example, magnetic domain switching, with magnetic/inductive energy as the degree of freedom, undergoes a flip in presence of thermal noise, whose error rate of 1 in 10 years requires that the minimum stored energy by of the order of $200k_BT$ if only one bit is used. If more are employed, much more energy is needed. A 10^{10} (10 Gbits) ensemble based on these arguments would need $\sim 300k_BT$ energy per bit for assuring 1 error in 10 years in the ensemble. Use of error correction techniques in disk drives allows one to achieve characteristics near this number at higher densities. Note that the viewpoint that time stability $T_s \approx \tau_s \, exp\left(-\frac{U}{k_BT}\right)$ where τ_s is an attempt time must also include the probabilistic distributions of attempt times in order to find the error rates in flipping erroneously.

Memories that we employ fall in two general classes (see Fig. 3):

(a) *Bistable Memories* described well by the description above and

(b) *Random Walk Memories* where the particle performs a very slow random walk in time and space.

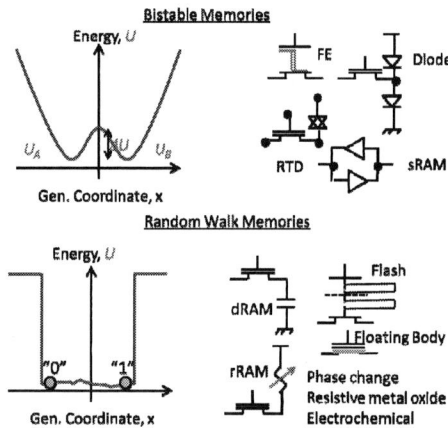

Figure 3. Memory classification based on operating principle of bistability or random walk (very very slow walk in transport of carriers and their annihilation).

The bistable memories have two (or more) low energy (electrical, magnetic, mechanical) states separated by a potential barrier whose minimum correspond to the two peaks of the distribution of Fig. 1. The random walk memories are quasi-stable memories depending as they do on the slowness of a degradation process (electron leakage from a capacitor in dRAM, or of Flash storage, or movement of ions in metal oxide electrochemical memories). As size reduces, the statistical effects in these random walk structures should not be truly random and subject to anomalies. A good example of this is the variable retention time of dRAMs[5] and its random telegraph behavior which may exhibit multiple time

constant phenomena depending on the different states that the system collapses to.

C. Example of Emergence in Logic and Memory:

What happens when one puts a large number of such structures together? There are many emergent properties but one that is extremely important and one that one works with in any design is the confluence of power, heat and speed. These are all tied to each other. If α is the activity factor, U the energy per operation, A the cross-section area, and Q the density of heat removal, equilibrium implies that the time constant of the system, $\tau = \frac{\alpha U}{QA}$. In one-dimensional heat spreading, so a planar array, the heat extraction capacity (Q) in silicon of 100 W/cm^2 implies a time constant of 5 ns. This is what sRAMs should be. If a device is isolated, rf or clock driver, and heat spreading is three-dimensional ($Q = 10^5$ W/cm^2), the time constant is about 5 ps which is what electronic devices are close to being capable of.

III. INFORMATION IN DIGITAL MACHINES

The energy input to the information machine (dU) translates to useful work (dW) and production of heat (dQ), i.e., $dU = dW + dQ$. Computers only process information. They do not create it. The amount of information content in the machine provided through the design of the physical hardware and the algorithms implemented sets the upper bound of the information hard- and soft-wired into the machine. One can estimate this. For n circuit elements, m terminals, and q terminals for input/output, grid and power, the total capacity is $N = nm + q$, and the total possible arrangements N^N. The information content of this collection is $I = N \log_2 N$.

For $n = 250 \cdot 10^9$, a nano-compatible number of elements with average FI+FO of $m = 4$, and $q = 1000$, $N \approx 10^{12} \approx 2^{39.9}$ with $I \approx 4 \cdot 10^{13}$ bits. In entropy units, this is $S \approx 4 \cdot 10^{-3}$, a very small number. Only a very small fraction of all the states that are possible in random arrangements are accessible to the deterministic phase state space of the deterministic information machine. Szilard had an apt metaphor: *a computer is a translator, not a writer*. And one can see the inefficiencies when comparing analog based elements, e.g., digital-signal processors or analog state machines, to purely deterministic digital machines, a consequence of the limitations posed by this thermodynamic consideration of limiting the phase space accessible.

IV. IMPLICATIONS

This brief physical and thermodynamic view points out that digital information machines are quite inefficient thermodynamically. The hierarchical organization through the interfaces that have been so powerful in the success of this approach, also increase the inefficiencies. Movement of data also has its inefficiencies: transmission is an added burden between units that one designs to be the efficient within the deterministic architecture, and transmission is in a thermodynamic view an inefficient boundary/interface. The gains with lower energy in transmission such as through low voltage differential signaling, while important because of their large role in high performance computing, has only a fractional impact because of the inefficiencies that abound.

A. Implications of Nanoscale and Probabilistic Variations

Given the fundamentals of energy-error relationships, both logic and memory structures in a deterministic style of computing have limitations in energy that naturally arise from the statistical relationships. Independent of the form the device takes, these energy limitations are lower bounds as are emergent properties in time scales as a consequence for the deterministic approach. The form the nanoscale device takes may determine how the energy is incorporated, but equi-partition of energy of fluctuations still arises and applies. One may get closer to the limits, e.g. error correction techniques or sensitive measurement techniques allow one to get closer to these limits, but in turn this will extract a price in the time. Magnetic storage media is a good example; the price in energy and time is paid in the read head and integration times to recover signal. In our current efforts, device ideas and technology are all directed to achieving the smallest dimensions and unique characteristics (e.g., non-volatility versus volatility or suppression of leakage currents with less degradation in sub-threshold swing), but it misses the bigger picture of information and best ways to achieve efficiencies in it. An information system is more than the physical hardware, it includes the signals, the software, the algorithms, and buried in this is the content – the information that one wants to work with, the "useful work, dW".

B. Implications of Nature of Signal

Data can be encoded in various forms (charge, spin, polarization) in the particles (electrons, photons, plasmons, polaritons, etc.) and many of these forms are being investigated. In each of these forms, however, for deterministic approaches the above minimum energies still apply. Spin loses coherence through the statistical interactions (via interactions with other spins in surroundings, with field fluctuations, etc.) and in this, many of the proposals for spin-based logic, suffer from the same effects albeit over a shorter length scale than electrical signal loss in interconnects. This signal recovery is critical, and signal recovery requires transduction that consumes energy. Photons or photon interaction with mobile charge (plasmon) or with ions (polariton), like spin, provide unique effects that are of special purpose large use (e.g., storage from spin, enhanced non-linearities useful in antennas from plasmons, or terahertz radiation from polaritons), but photon interaction lengths are long, and for the coupled modes, statistical effects still apply.

V. COMPLEXITY AND INFORMATION

This discussion of lower bounds in deterministic approaches, and inefficiency in information processing, serves as a useful contrast to evolution's information processing which handles sensation, perception, motor control, memory, emotion, planning and decision-making. Human power consumption is about 125 W of which about 20 W is consumption by brain. Evolution's approach is a machinery that (a) has heterogeneity and specialization at each spatial scale, (b) employs number coding and computation together with error correction and signal-to-noise ratio redundancy, (c) provides trade-offs at level of neurons and networks, and (d) learns and adapts at each temporal scale, e.g., gene adapts to the mean while the synapse to the variance. In particular, in the more understood perception system, the observation is that the interactions are adapted to the statistical structure of the inputs and to the structure of downstream computations. For example, the retinal ganglion cell network has a response distribution with correlation that is a pairwise excitation and inhibition.

A. InfoMax Principle

Linsker [5] argues that a governing principle for efficient information utilization is maximum information preservation in the network that is formed. What this implies is that network connections of information processing in perceptual systems develop in such a way as to maximize the amount of information that is preserved when signals are transformed at each processing stage subject to certain constraints. This principle, InfoMax, by preserving information in the flow from input to output, maximizes the content and minimizes the spurious. The principle naturally leads to redundancy when the noise is high. When the noise is low, it naturally leads to variance maximization between the input and the output. By maximizing information preservation, it minimizes energy in a computational task.

B. Network of Information Flow

The flow of information, so that energy and time can be optimized for a statistical distribution of information content, origin and destination, is also a statistical mechanics problem of percolation. In an information machine environment, it is a problem with both short range and long range percolation. Probabilistic studies have shown that optimization points to a mixture of long range and short range connections [7] whose details are related to the heterogeneity of the problem.

C. Heterogeneity

Heterogeneity allows one to design at the smaller function scale efficient engines specific for the task – multipliers, adders, or others, one could imagine specific matrix multiplication engines that implement BLAS efficiently with a three-dimensional memory structure where nearest neighbors have the strongest statistical probability and hence shortest time to access. Heterogeneity also allows specific tuning of the resources to the task at hand.

D. Learning: Adaptation and Evolution

Efficient engines adapt to tasks, bicycles and cars change gear ratios, drilling machines speeds, cooking stoves heat flow. Similar tuning is needed in a more sophisticated way to improve energy efficiencies so that the high integration scales can be utilized effectively. Evolution, such as through dynamic configuring when one determines that tasks can be conducted better in hardware or software, would also provide more powerful capabilities beyond the static approach of today.

E. Accepting Inexactness

Nature and devices are non-deterministic. Working with a level of precision appropriate to the necessity is central to efficiencies. Much of complex computation, working with sensor data, modeling of geological, economic, social or other data, is with incomplete data using incomplete models. Using deterministic approaches to such non-deterministic problems is inherently brittle. Using probabilistic methods that accept the variability of devices and of the models and incorporate the principles of A through E of this section would, while being a significant departure from the past sixty years of practice, bring more robust and powerful decision and judgment tools for the next sixty years – from healthcare, robotics, and others that emphasize small form factor to high performance computation that emphasize high performance.

VI. DISCUSSION; AND BEYOND

While a larger number of possibilities exist in choices of devices, they are ultimately constrained in the present deterministic approaches by the energy required to reduce error, the energy required to transmit data around for manipulation, and by separations built in the hierarchy. These approaches have been enormously successful and would continue to as evolutionary progress in technological integration successfully tackles diverse tasks that information processing can serve – sensor integration, health diagnostics, and others. However, it is an increasingly inefficient approach both for complex problems that naturally rely on judgment calls and probabilistic and incomplete information and model. The present path is also subject to increasing static and dynamic yield issues. It is also an inefficient method for mobile applications where one wishes to stress ultra-low power. Additionally, a machine performing only a limited set of tasks is an efficient use of a large collection of resources that the information machine environment provides. At nanometers scale, terascale is within reach. The machine should be more powerful, even learn to improve at tasks without human intervention. These are all directions that deviate substantially from the von Neumann path followed to this point. The discussion of the previous section argues that several new principles can be incorporated to improve efficiencies, to work with information and knowledge, rather than just perform linear algebra on data.

The following illustrate a few examples that introduce technology themes that support the Section V principles and point new directions. It is the nature of research that these directions all have related issues, practical questions that may make the direction less compelling, but thinking and pursuing these subjects is important to getting beyond the current model that is entirely predicated on shrinking of dimensions, a path that is near the end.

A. Adaptation

Adaptation is not too complex to introduce into devices. For example, a transistor in its multi-gate form, finfet or back-gated double gate device, naturally lead themselves to a rapid threshold change if the second gate is allowed to be independent whose potential can be manipulated via feedback without a penalty in basic device speed. Functions implemented with such device structures will allow energy

and speed tuning to the desired objective. Fig. 4 gives an example of a multiplier critical path and the operational characteristics that could be obtained in different regimes should adaptation be a design principle employed. These examples are for a 40 nm gate length [8] technology.

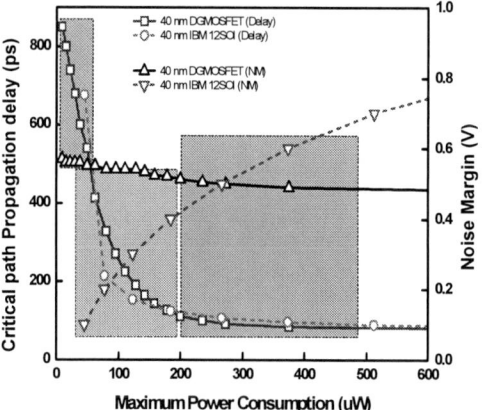

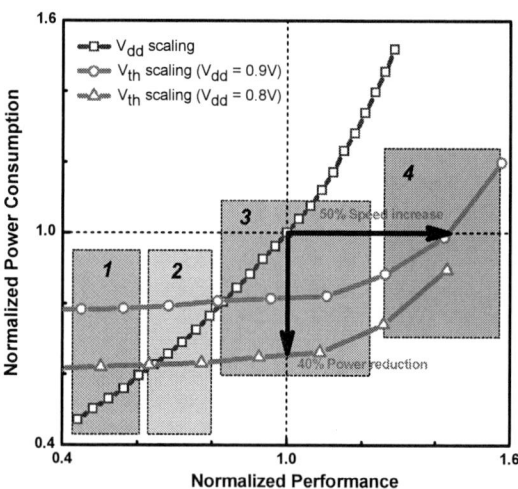

Figure 4. Critical path delay in a 32 bit multiplier and characteristics of an 8 bit Radix-4 multiplier block. Both simulated charactristics showing the utility of adaptation are at 40 nm technology node.

B. Inexact Computation

Does the noisy world require computation to be always correct? No. Numerous real world applications, certainly those utilizing signal processing or much of real world analytics can do well with reduction in energy so long as one can specify the exactness required. This in turn allows us to manipulate the energy in most significant to least significant bits of a computation to deliver the necessary preciseness. Fig. 5 shows the energy-speed trade-off in presence of variances, for a conditional carry select a conditional sum select (CCS-CSS) adder's 8 bit block [9]. Note the exponential relationship between error and energy that we had earlier argued for is reflected in the power dependence of this figure. This inexact approach can be expanded to nearly any functional block that is employed in current architectures.

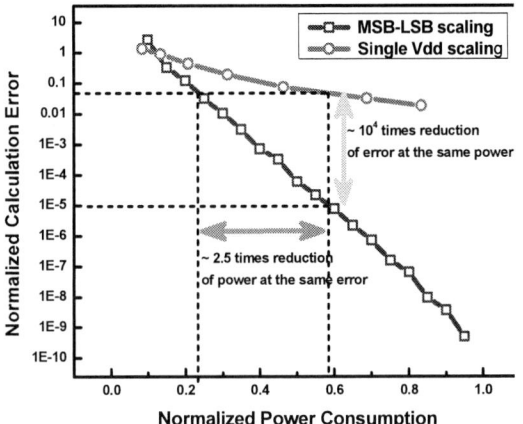

Figure 5. Error-power consumption relationship when the energy in put to most significant bits to least significant bit is adapted to achieve a speficied level of correctness in a CCS-CSS adder's 8 bit block.

C. Dynamic Configurability

Dynamic configurability, one where the chip itself adapts to tasks is a subject that has been discussed at length in the past several years and examples now exist which show its efficacy. Fig. 6 shows work from Vahid et al. [10] where dynamic profiling is employed to map tasks that are found to be repeated to hardware (in this case an FPGA island on the chip). Such approaches lead to reduction both in energy consumption and delays.

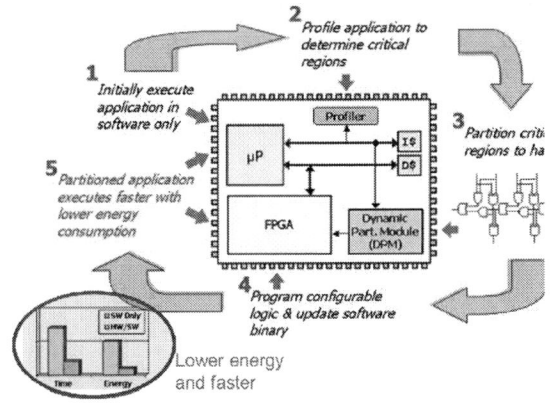

Figure 6. Error-power consumption relationship when the energy in put to most significant bits to least significant bit is adapted to achieve a speficied level of correctness.

Such dynamic configurability is also of necessity in nanometers environment because of the dynamic degradation

as well as failures to be expected, and the stochastic variability.

Dynamic configurability points to an urgent need of the chip environment. We do not have a compact method for dynamic configuring. FPGAs are both area and power consuming and are volatile because of the use of sRAMs. sRAMs have the significant attribute of voltage level compatibility with configuring as the storage and the sense function are both merged within the six transistor cell. Other memories that may compete in speed – dynamic RAM, magnetic, spin-torque, etc. - all require a sensing function which is counter to a direct manipulation of the switch with a voltage output from the memory element. The semiconductor environment, for all the newer approaches that are argued for in this article, require a memory element that is dense, is compatible with silicon electronics, and has a signal level compatibility that is direct and without sensing mediation. If it is non-volatile and fast, with low power consumption, it would have an enormous impact in the coming decades.

I point to three directions where such an element may arise from, each guided by the voltage correspondence argument.

D. Novel Effects: Phase Transitions

Phase transitions, such as of ferroelectricity, ferromagnetism, superconductivity are employed gainfully in semiconductor technology for a variety of specialized needs.

Phase transitions have this unique property that in a collective ensemble an emergent property appears that is quite reproducible, even largely invariant of dimensions, unless and until other energetic phenomena intervene. Below a critical temperature, polarizations align in a ferroelectric crystal, or the Cooper pair becomes the particle carrying current without energy loss below the threshold for superconductivity in low temperature superconductors. Such a property is highly desirable.

For ferroelectric materials, one can describe the free energy as

$$F(P,T) = \int \left[F(P=0) + \frac{1}{2} g_2 P^2 + \frac{1}{4} g_4 P^4 + \frac{1}{6} g_6 P^6 - \frac{1}{2} P \cdot E \right] d^3 r \quad (3)$$

If the 4th power coefficient dominates, and the coefficient is positive, this is a 1st order transition and if it is less than 0, it is a 2nd order transition (see Fig. 7). In the metal-insulator phase transition, small bandgap semiconductors undergo phase transition to metallic phases because of the electron-crystal interactions. In, e.g., the presence of lattice distortion, the valence electron goes from a localized (around the ion) to delocalized (bound only to the crystal) state, i.e., the material undergoes metal-to-insulator transition. If such an effect could be utilized, one would obtain a memory that is non-volatile and fast. Some examples of materials where memory effects have been observed are VO_2 and $SmNiO_3$, in each of which, however, polarization induced effects also exist. Fig. 7 shows characteristics of the phase transition in VO_2 due to electron injection, and memory in a VO_2 floating gate structure [11], where no charge injection in the floating gate occurs, just a phase transition dominated by the ferroelectric effect.

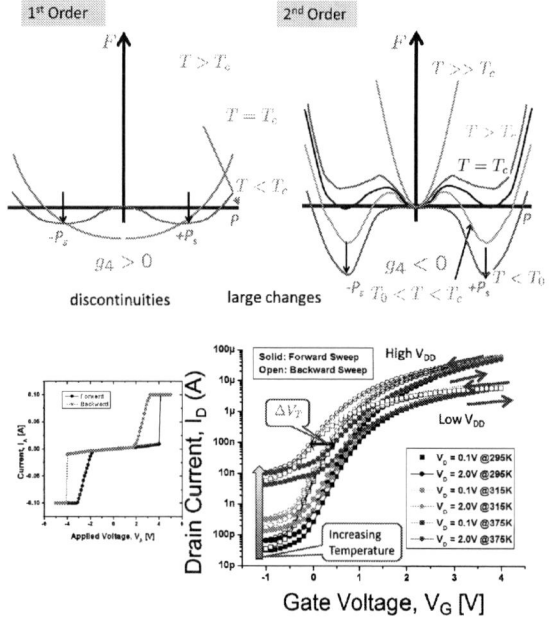

Figure 7. Order in phase transitions. This is an example for ferroelectric effect. In the bottom is shown electron injection induced metal-insulator phase transition in VO_2. On the right is shown a ferroelectric dominated phase-transition memory utilizing VO_2 as a floating gate.

If a fast non-volatile phase transition memory can be realized, it will be low energy, and a configuring switch could be realizable using just two such memory elements replacing a six transistor configuring switch dynamic cell. Such a structure would be fast and useful in implementing numerous of the principles articulated in Section V and VI.

E. Novel Effects: Shape Memories and Mechanical Elements

The utilization of mechanical elements in mainstream technologies is relatively minimal. It is a form in which storage of energy can be quite substantial. It is therefore of possible utility in nanostructures, again with low energy and non-volatile properties potentially at nanoscale. We point to two examples here. Fig. 8 shows stress-strain characteristics of a one way effect shape memory alloy. NiTi is one example of such a shape memory alloy that undergoes a pseudoplastic deformation, with nearly 10% strain and therefore enormous storage of energy, but then it can also recover. Starting at origin in the martensitic phase when large stress is applied, deformation occurs and the structure is in a high strain in plastic deformation, Raising of temperature moves the structure to austenitic phase and relieves the strain. At this point dropping the temperature completes the loop. The figures shows how in a device form, one might deform to set an "on" state, and then by passing current that heats, return this back to "off" state, i.e., achieving the function of a

compact memory. This figure also shows a torsion switch [12] that is a fast nanoseconds switch because torsion modes are higher in frequency than bending modes. Such a switch when combined with a transistor and utilizing stiction, provides another way of making a fast configuring element.

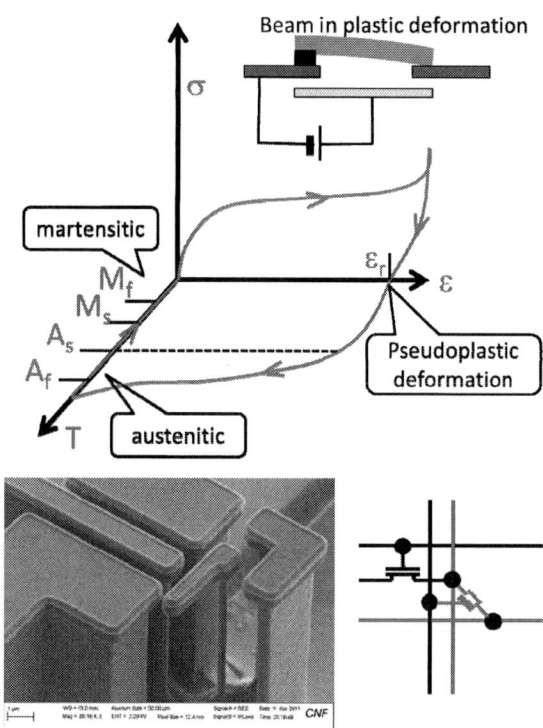

Figure 8. Shape memory alloys and their use in a relay memory. The bottom figure shows an electromechanical torsion switch employable as a configuring element using transistor based configuration actuation. This arrangement can be potentially non-volatile through the use of stiction.

VII. SUMMARY

This paper argues the following: (a) the lowest energies in deterministic, always correct, switches and memories is of the order of few hundred $k_B T$ whose magnitude is a function of the scale of integration and the principles employed, and (b) the error-energy relationships and the scale of integration with hierarchy are non-linearly connected and lead to vast energy inefficiencies. As integration levels increase, this disconnection inherent in the vast scale changes of the information machine make the current evolutionary path of integration and scaling increasingly challenging. However, there are numerous principles that connect the probabilistic world that we inhabit with the large numbers of elements that we employ for information processing that are enormously beneficial and that we do not currently leverage. The next sixty years, like the past sixty years, would be just as exciting and as useful for the society if we pursue the principles of (a) heterogeneity with efficient specialized functions/engines, (b) adaptation, (c) scales suitable to the efficiency of the function, (d) learning incorporated within the chip environment, (e) evolution, (f) maximization of information preservation, (g) information flow networks that incorporate the approach of (f) and utilize the statistical nature of the information within the data. There are numerous technology directions – in materials, devices, and their assembly, including 3D integration which was not discussed, that would open new territories for the information machines that we build. Information is physical, and the physical world is an information-centric world. Nanoscale science, engineering and technology can benefit the information machines tremendously, but the larger steps in progress will come from constructionism rather than reductionism.

ACKNOWLEDGMENT

The thoughts of the author have evolved over the decades building on insights of numerous former colleagues from IBM, friends across disciplines, and from the technical community at large. Views of R. Landauer, C. Bennett, P. Solomon, H. Davidson, R. Linsker, long discussions with research students and their own work, some of them referenced here, and received wisdom from physicists, computer scientists, applied mathematicians, neuroscientists, and electrical engineers are reflected in the thinking. To all, a sincere thank you.

REFERENCES

[1] View reductionism as the approach to finding the laws that govern to a finer and finer depth, and constructionism as the emergence of certainty and order from collection's self or mediated organization.

[2] K. U. Stein, IEEE J. SSC, SC12 (1977).

[3] See C. Mead and L. Conway, "Introduction to VLSI Systems," Addision-Wesley (1990).

[4] This is the corrollary of Maxwell's Demon. For a discussion of this penalty, see A. Kumar and S. Tiwari, "Testing and Defect Tolerance: A Rent's Rule Based Analysis and Implications on Nanoelectronics," Proc. of IEEE International Symposium on Defect and Fault Tolerance in VLSI Systems, Oct., 280-288(2004)

[5] P. J. Restle, J. W. Park and B. F. Lloyd, "DRAM Variable Retention Time," IEEE Tech. Dig. of IEDM, 32.2.1 (1992)

[6] R. Linsker, "Local Synaptic Learning Rules Suffice to Maximize Mutual Information in a Linear Network," Neural Computation, **4**, 691(1992)

[7] M. Aizenman, H. Kesten and C. M. Newman, "Uniqueness of the Infinite Cluster and Continuity of Coonectivity Functions for Short- and Long-Range Percolation," Comm. Math. Phys., 111 505-532(1987)

[8] J. Y. Kim, P. Solomon and S. Tiwari, "Adaptive Circuit Design Using Independently Biased Back-Gated Double-Gate MOSFETS," IEEE Circuits and Systems I, 59 (4), Apr.., 806-819(2012)

[9] J. Y. Kim and S. Tiwari, "Inexact Computing using Probabilistic Circuits: Ultra Low-Power Digital Processing", submitted to ACM Journal of Emerging Technologies. Also see J. Y. Kim, S. Tiwari, "Inexact Computing for Ultra Low-Power Nanometer Digital Circuit Design," Tech. Digest of IEEE/ACM Symp. on Nanoarch, 24-31 (2011)

[10] F. Vahid, G. Stitt and R. Lysecky, "Warp Processing: Dynamic Translation of Binaries to FPGA Circuits," IEEE Computer, 41, 40-46(2008)

[11] S. H. Lee, M. Kim, J-W Lee, Z. Yang, S. Ramanathan, S. Tiwari, "VO_2 is also a Ferroelectric: Properties from Memory Structures," Tech. Dig. of IEEE NANO, WeP1T2.1 (2011)

[12] J. Rubin, R. Sundararaman, M. Kim and S. Tiwari, "A Low-Voltage Torsion Nanorelay," IEEE Electron Device Letters, V32, No.3, 414(2011)

Modeling circuits with spins and magnets for all-spin logic

Behtash Behin-Aein
Global Foundries
Sunnyvale, USA
Behtash.Behin-Aein@globalfoundries.com

Angik Sarkar and Supriyo Datta
School of ECE, Purdue University
West Lafayette, USA
datta@purdue.edu

Abstract—**This talk will summarize our work in the last three years exploring the possibility of spin-magnet circuits that utilize two key recent advances, namely (1) spin valve (SV) devices demonstrating spin injection into metals and semiconductors from magnetic contacts and (2) spin transfer torque (STT) devices demonstrating the switching of magnets by the injected spins. Utilizing an experimentally benchmarked model describing these phenomena we have shown the possibility of Boolean all-spin logic (ASL) circuits with intrinsic directionality or with directionality enforced through STT based Bennett clocking. We also discuss probabilistic ASL circuits and summarize the modeling framework that we have developed for the analysis of generic spin-magnet circuits.**

I. INTRODUCTION

It is generally recognized that there are major roadblocks to continued miniaturization of transistors, the most important one being power dissipation. Experts have urged the need to "reinvent the transistor" to reduce its operating voltage [1]. In our recent work we have proposed the possibility of using a network of spins and magnets (Fig.1) to implement all-spin logic (ASL) devices for information processing [2] that could operate at voltages V of the order of tens of millivolts [3].

This approach is based on two key recent advances namely (1) spin valve (SV) devices demonstrating spin injection into metals and semiconductors from magnetic contacts and (2) spin transfer torque (STT) devices demonstrating the switching of a second magnet by the injected spins. Magnets inject spin and spins turn magnets forming a closed "ecosystem" suggesting an all-spin approach to information processing without converting to charge at every stage.

Practical SV and STT devices (Fig.2) are typically magnetic tunnel junctions (MTJ's) having a 'vertical' structure with the two magnets separated by a spacer layer short enough for electrons to tunnel through. ASL devices would typically

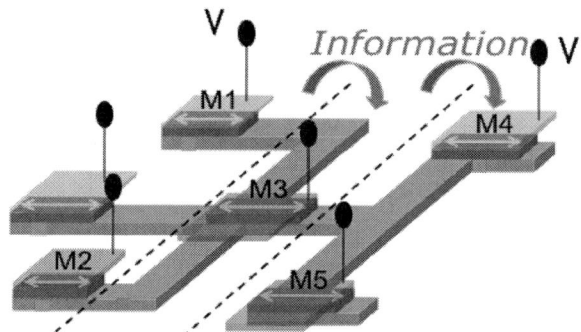

Figure 1 – The basic ASL device stores information in nanomagnets which communicate using spin currents just as a standard CMOS switch stores information as charge on capacitors which communicate through charge currents. The energy in either case comes from the power supply V. With ASL devices, V could be as low as tens of millivolts.

require a 'lateral structure' with magnets separated by relatively longer distances. Lateral SV devices have been studied for over twenty years (see for example [4]), but the signals are relatively small compared to MTJ's, while lateral STT devices are relatively rare (see for example [5,6]) reflecting the difficulty of generating spin currents large enough to switch magnets.

We have shown that both the spin-valve characteristics (resistance versus voltage) as well as the spin-torque characteristics (in-plane and out-of-plane torque versus voltage) of an MTJ can be described well using a model of the type shown in Fig.3 with spin transport described by a Non-equilibrium Green's Function (NEGF)-based model using a semi-empirical effective mass Hamiltonian (See for example, Datta *et.al.* [7]).

Supported by the Nanoelectronics Research Initiative

978-1-4673-1707-8/12 $31.00 © 2012 IEEE

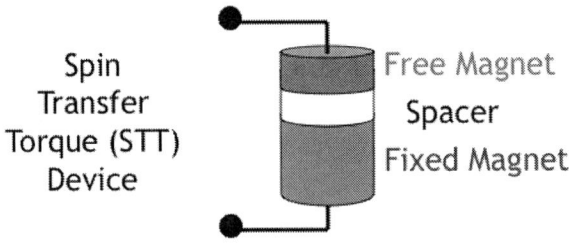

Figure 2 – Basic STT device with a fixed magnet separated from a free magnet by a short spacer layer.

Lateral devices too can be modeled using the same general scheme except that the structures are much longer and a direct solution of the NEGF equations is inconvenient. Instead we have developed a 4-component spin circuit model building on the work of Brataas, Bauer and co-workers [8] as described in [3], [9] and showed that it describes the experimental results in Ref.[5] quite well.

An all-spin logic device (ASLD) involves multiple free layers requiring multiple LLG's to be coupled to the spin transport model (Fig.4). This is in principle no different from STT devices with just one free layer having multiple domains each of which is like a separate "magnet". In ASL devices, however, the individual magnets are well separated physically making their direct dipolar and exchange interactions relatively small compared to the spin current mediated interaction, introducing a rich new set of issues and related physics.

II. ALL SPIN LOGIC CIRCUITS

An important aspect of this interesting physics was discussed at length in [9] in three magnet systems. It was predicted that three nanomagnets cascaded in a chain would behave as a chain of inverters (Fig 5) and if arranged in a ring should show spontaneous oscillations in the magnetization, if the supply voltage is positive, but not if it is negative. To see this, we first note that when the voltage is positive, each magnet tries to make its neighbors *anti-parallel* to itself, but if the voltage is negative, neighboring magnets want to be *parallel*. This is not obvious but a simple argument was presented in [9] which is repeated here for convenience.

The full LLG equation describing the dynamics of magnet 'i' is discussed later in the section on modeling. It includes external magnetic fields, anisotropy fields, thermal noise fields as well as spin currents generated by other magnets. For the moment let us consider just these spin currents. Assuming small damping factors α, we can write for magnets 1 and 2,

$$\frac{d\hat{m}_1}{dt} \sim -\hat{m}_1 \times \hat{m}_1 \times \vec{I}_{s,1}$$

$$\frac{d\hat{m}_2}{dt} \sim -\hat{m}_2 \times \hat{m}_2 \times \vec{I}_{s,2}$$

(1)

and note that the spin currents can often be approximated as [8]

$$\vec{I}_{s,1} = B_{11}\hat{m}_1 + B_{12}\hat{m}_2 + C_1 \hat{m}_1 \times \hat{m}_2$$

$$\vec{I}_{s,2} = B_{21}\hat{m}_1 + B_{22}\hat{m}_2 + C_2 \hat{m}_2 \times \hat{m}_1$$

(2)

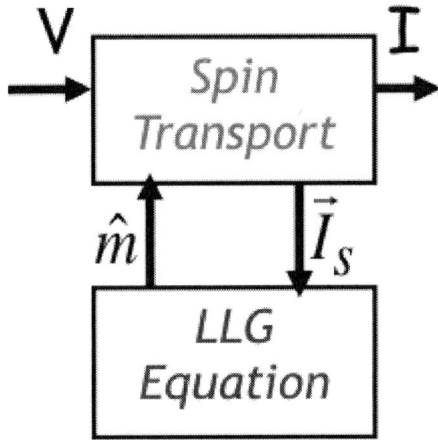

Figure 3 – Spin transfer torque (STT) devices typically require (a) a model for spin transport that calculates the charge and spin currents I, I_s for a given voltage V and magnetization m, and (b) a model for the magnetization dynamics (LLG equation) that calculates the magnetization for a given spin current.

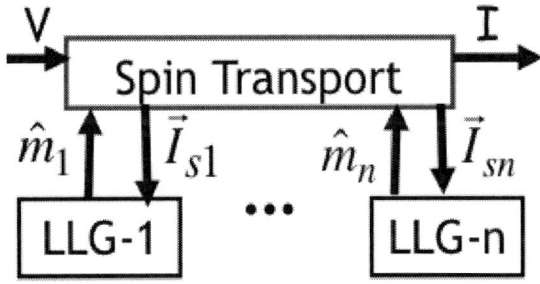

Figure 4 – In general, STT as well as ASL devices require multiple LLG equations, one for each free layer, to be coupled to the spin transport model.

Using (1) and (2) we can write

$$\frac{d}{dt}c \sim (B_{12} + B_{21})(1 - c^2)$$

(3)

where $c \equiv \hat{m}_1.\hat{m}_2$. Clearly from (3), c= ±1 represents steady states with dc/dt =0; c=+1 represents parallel magnets and c=-1 represents anti-parallel magnets. However, both steady states are not stable, only the one with a negative Jacobian corresponding to $(B_{12} + B_{21})c > 0$.

A negative V_{ss} makes a magnet inject electrons spin-polarized in the direction of the magnet making B_{12} and B_{21} positive so that c = +1, representing parallel magnets, the stable state. A positive V_{ss} makes B_{12} and B_{21} negative, so that c = -1 representing anti-parallel magnets becomes the stable state.

If we accept that neighboring magnets want to be parallel (anti-parallel) for negative (positive) voltages, then it is easy to see that with negative voltages all three magnets become parallel. With positive voltages, in a ring structure, there is no way to make all neighboring magnets anti-parallel and it leads to oscillations.

However, in a chain of magnets, the adjoining magnets want to orient anti-parallel to each other. For example, in Fig 5, magnet 1 was at m_z=-1, so magnet 2 switched from m_z=-1 to +1 to be antiparallel to magnet 1. This leads to magnet 3 switching to m_z=-1 so that the magnets are anti-parallel to each other.

The operation we just described is similar to the well-known chain of inverters composed of a chain of complementary

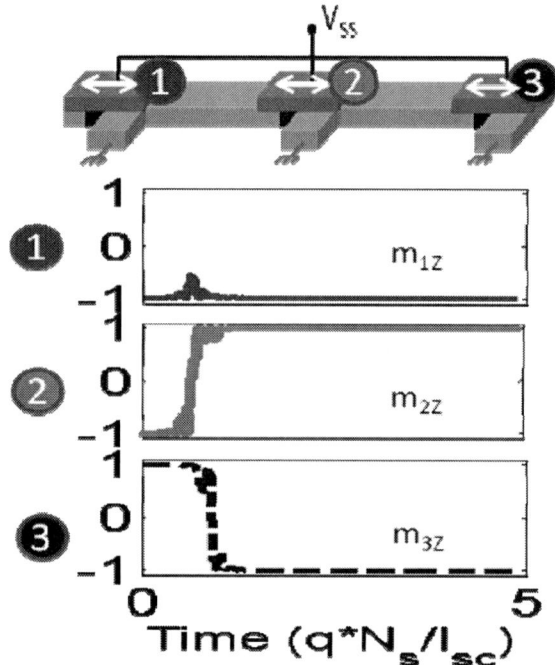

Fig 5- ASL Chain of inverters consisting of three magnets connected to a positive supply voltage. Information propagates from magnet 1→2→3.

metal oxide semiconductor (CMOS) inverters. Nevertheless, it is surprising that a chain of magnets would show this

behavior since it requires *directivity* whereby information flows preferentially from magnet 1 to 2 and onto 3 for nominally identical magnets. This directivity comes naturally with CMOS inverters whose input determines the output while the output has negligible influence on the input. But it is not common with magnets. The usual equilibrium interactions between magnets, whether dipolar or exchange, are reciprocal in nature: If magnet 1 affects magnet 2, then magnet 2 affects magnet 1 equally. It is the non-equilibrium nature of the spin-transfer torque interaction that allows it to be non-reciprocal in a properly designed structure: In Fig.5 m_1 affects m_2 more than m_2 affects m_1 because the ground terminal is located near m_1 as explained in [9]. The resulting non-reciprocity enables chain of inverters functionality and can lead to cascaded universal logic gates like the NAND gate [10].

III. PROBABILISTIC ALL SPIN LOGIC?

Information processing in multi-magnet circuits like the ASL chain of inverters is an interesting prediction but has not been demonstrated experimentally and seems difficult to implement at this time, since lateral STT devices are relatively rare. Instead a different class of devices that we could call probabilistic ASL, seems more promising [11].

The basic idea is based on the concept of Bennett clocking (see [2] and references therein). Consider the same chain of inverters structure as in Fig 5, but with a voltage V_{ss} that is below the critical value V_c needed to switch the magnet. Ordinarily nothing should happen, the magnets should continue in whatever state they happen to be in.

But suppose, we first use a large initial magnetic field to place all three magnets along their hard axis which is perpendicular to the easy axis that they normally like to align with. Then, the external bias is turned on and the external field is removed. In the absence of the field, the system of magnets relaxes from the neutral hard axis to some final state which we register.

A typical result [11] is shown in Fig.6, where we have used "u" for up to denote m_z = +1, "d" for down to denote m_z = -1. Here, we consider the same chain of inverters as in Fig 5; but with sub-threshold voltages characterized by 'r', the ratio between the applied bias and the threshold voltage. Evidently, when r is close to 1, only the anti-parallel states, *udu* and *dud* appear with high probability, similar to the results in Fig 5. It is quite surprising that the chain of inverters operation can be observed even below threshold using the probabilistic approach.

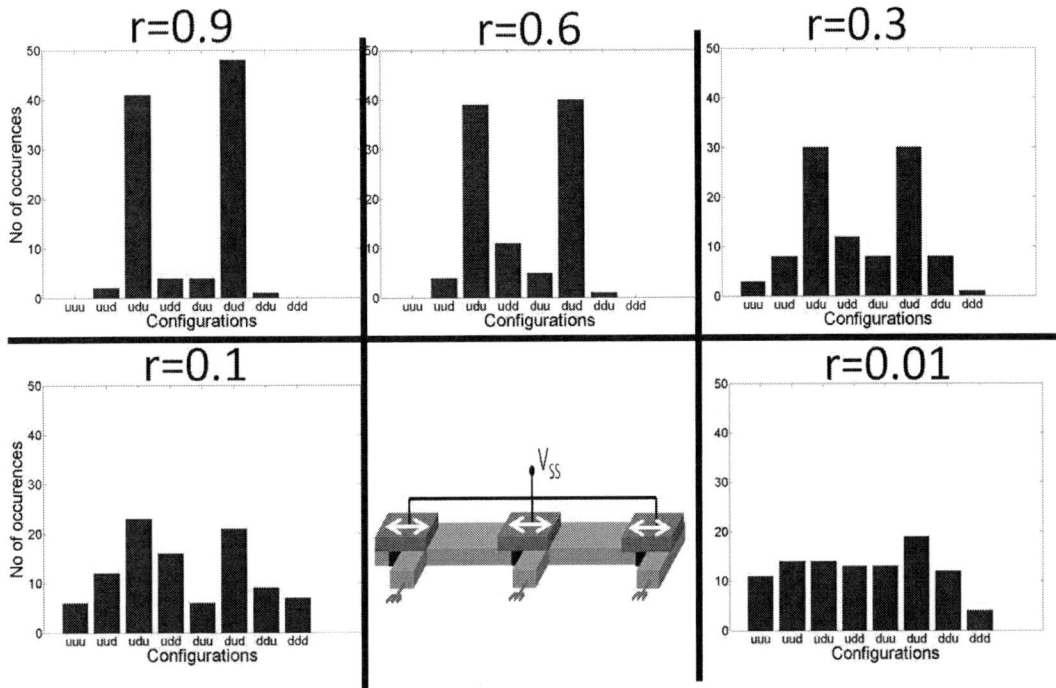

Fig 6. Probabilistic all spin logic? Cascadability of ASL can be demonstrated even with a bias below threshold. Consider the chain of inverters as in FIg 6, biased Below threshold (r the ratio of the applied bias to threshold). The anti-parallel states udu and dud appear with higher probability than other configurations. However, the probability goes with as r.

When r is scaled down from 1, the probability of occurrence of *udu* and *dud* states reduce and other states start appearing primarily due to thermal noise (Fig 6). However, it is interesting to see that *udu* and *dud* occur with significant probability even at r=0.1. This makes experimental demonstration of the effect easier, since this implies that it is possible to experimentally verify our predictions on multi-magnet all spin operations even with spin signals an order of magnitude lower than the threshold.

It is also important to note that the probability of occurrence of the most probable states (*udu* and *dud* in this case) decreases monotonically as r is decreased. It appears that the probability of occurrence of these states is a function of r, at a given temperature and magnetic field. In conservative systems, such a function would simply be the Hamiltonian of the system. However, the phenomenon of spin torque is non-conservative and an 'energy' function does not exist.

Nevertheless, from Eqn 3, it seems that the function $\sum_{i,j} (B_{12} + B_{ji})(\hat{m}_i \bullet \hat{m}_j)$ is being minimized. For example,

in Fig 6, the function is minimum when all $(\hat{m}_i \bullet \hat{m}_j)$ products are negative or the magnetizations are anti-parallel for adjacent magnets, as is observed in Fig 6. Is this function minimized in all the probabilistic experiments? If so, these probabilistic experiments could be used to solve hard optimization problems. We leave investigation of the scope of the function for future work.

IV. Modeling Probabilistic Spin-Magnet Circuits

The numerical results we presented in this paper are based on the coupled model of spin transport and magnetization dynamics (Fig.4) used in [9]. Each nanomagnet i is assumed to be monodomain and treated as a macrospin $\hat{m}_i$ whose dynamics is described by an LLG equation in presence of thermal noise

$$(1+\alpha_i^2)\frac{d\hat{m}_i}{dt} = -\hat{m}_i \times \left(\gamma(\vec{H}_i + \vec{h}_{fl}) + \hat{m}_i \times \vec{I}_{s,i} \right)$$
$$- \alpha_i \hat{m}_i \times \hat{m}_i \times \gamma(\vec{H}_i + \vec{h}_{fl}) + \alpha_i \hat{m}_i \times \vec{I}_{s,i} \quad (4)$$

γ being the magnitude of the gyromagnetic ratio, α_i, the Gilbert damping parameter, $\vec{I}_{s,i}$ is the spin current incident at magnet i, calculated using the full 4-component spin circuit model [9]. $\vec{H}_i$ represents the total magnetic field including external fields and internal anisotropy fields. We model the effect of thermal fluctuations by adding the fluctuating field $\vec{h}_{sl}$ with zero mean [2]:

$$\left\langle h_{fl,i}^{x_\alpha}(t)\right\rangle = 0 \qquad (5)$$

where the superscript x_α refers to the field's vector components.

The field is assumed to be Gaussian distributed with a standard deviation σ given by the fluctuation dissipation theorem:

$$\left\langle h_{fl,i}^{x_\alpha}(t) h_{fl,i}^{x_\beta}(t')\right\rangle = \delta_{ij}\delta(t-t')\sigma^2 \qquad (6)$$

$$\sigma^2 = \frac{\alpha}{(1+\alpha^2)}\frac{2k_B T}{\gamma M_s V} \qquad (7)$$

where T is the temperature that is set to 300K in our simulations for room temperature operations.

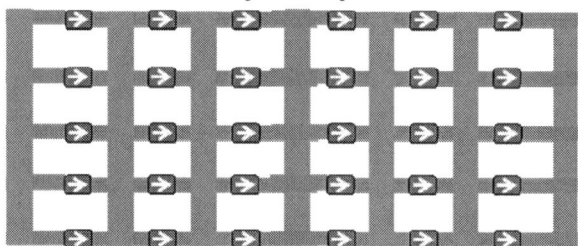

Fig 7. Vision for all-spin multi-magnet networks: an array of nanomagnets interacting to nearest neighbors via spin currents driven by applied bias on the magnets.

V. SUMMARY AND CONCLUSIONS

All-spin logic (ASL) represents a new approach to information processing where the roles of charges and capacitors in CMOS are played by spins and magnets. In this talk, we have presented multi-magnet simulations to show the possibility of constructing logic 'circuits' demonstrating all-spin behavior in networks of nanomagnets interacting via spin currents (Fig 7). However, spin-torque based nanomagnet switching in lateral structures is a challenging task, primarily because of the large spin signal required to switch a magnet. Hence, we have presented a different class of experiments that involve probabilistic switching of magnets at spin signals way below the threshold.

This class of probabilistic experiments, described in section III, seems closer to practical implementation since it can be carried out even when the voltage V_{ss} is significantly below the critical value needed for switching a magnet. But can it be useful? Further work is needed to establish specific applications, but one can envision nanomagnetic arrays (Fig 7) with applied voltages that can be used to control the effective interaction B_{ij} between magnets i and j which determine the final state distribution. The similarity to the new class of quantum computers aimed at solving hard optimization problems [see for example, 12, 13] seems worth investigating, but that is a different story for the future.

ACKNOWLEDGEMENTS

It is a pleasure to acknowledge the contribution of Srikant Srinivasan to sections I and II of this paper summarizing the results from Refs 3 and 9 which were coauthored by him. This work was supported by the Nanoelectronics Research Initiative (NRI) and the Institute for Nanoelectronics Discovery and Exploration (INDEX) Center and NSF Center for Science of Information (CSoI).

REFERENCES

[1] T. N. Theis and P. M. Solomon, "It's time to reinvent the transistor!," Science, vol. 327, no. 5973, pp. 1600-1601, Mar. 2010.

[2] B. Behinaein, D. Datta, S. Salahuddin and S.Datta," Proposal for an all-spin logic device with built-in memory", Nature Nanotechnology, **5**, 266 (2010).

[3] B. Behinaein, A. Sarkar, S.Srinivasan and S.Datta, "Switching energy and delay of spin logic devices," Appl. Phys. Lett. **98**, 123510 (2011).

[4] M. Johnson, "Optimized device characteristics of lateral spin valves", IEEE Trans. on Electron Devices, 54, 1024 (2007).

[5] T. Yang, K. Kimura and Y. Otani, "Giant spin accumulation and pure spin-current-induced reversible magnetization switching," *Nature Phys.*, vol. 4, pp.851- 854, Oct. 2008.

[6] J. Z. Sun, M. C. Gaidis, E. J. O'Sullivan, E. A. Joseph, G. Hu, D. W. Abraham, J. J. Nowak, P. L. Trouilloud, Yu Lu, S. L. Brown, D. C. Worledge, and W. J. Gallagher. "A three-terminal spin-torque-driven magnetic switch," *App. Phys. Lett.*, vol.95, pp. 083506.1-083506.3, Aug. 2009.

[7] D. Datta, B. Behinaein, S. Salahuddin and S.Datta, "Voltage Asymmetry of Spin-Transfer Torques," to appear in IEEE Trans. On Nanotechnology.

[8] A. Brataas and G. Bauer, "Non-collinear magnetoelectronics," *Phys. Rep.*, **427**, 157 (2006)

[9] S.Srinivasan, A. Sarkar, B. Behinaein and S.Datta, "All-spin logic device with inbuilt non-reciprocity," IEEE Trans. on Magnetics, **47**, 4026 (2011). See also book chapter "Modeling multi-magnet circuits" to appear in Handbook of Spintronics, eds. D.Awschalom, J.Nitta and Y.Xu.

[10] A.Sarkar, S. Srinivasan, B. Behin-Aein, S. Datta, "Modeling all spin logic: Multi-magnet networks interacting via spin currents, Proc IEEE IEDM, 2011

[11] B. Behin-Aein, A. Sarkar and S. Datta, unpublished.

[12] Johnson M.W. et al. Quantum annealing with manufactured spins, Nature **473**, 194 (2011).

[13] Britton J.W. et al. Engineered two-dimensional Ising interactions in a trapped-ion quantum simulator with hundreds of spins, Nature, Advance Online Publication 25 April 2012.

Organic Complementary Circuits - Scaling Towards Low Voltage and Submicron Channel Length

Hagen Klauk

Max Planck Institute for Solid State Research
Stuttgart, Germany
H.Klauk@fkf.mpg.de

Abstract—Organic thin-film transistors (TFTs) can usually be fabricated at temperatures of about 100 °C or less and thus on a variety of unconventional substrates, such as flexible plastics and paper. This makes organic TFTs potentially useful for the integration with organic light-emitting diodes (OLEDs) into bendable, rollable or foldable emissive active-matrix displays for next-generation mobile devices, provided the organic TFTs meet the stringent static and dynamic performance requirements of such displays.

I. INTRODUCTION

Organic thin-film transistors (TFTs) are of interest for a variety of flexible, large-area electronics applications in which individual switching devices or simple integrated circuits are distributed over large areas on mechanically flexible substrates, and in which the use of single-crystalline silicon transistors and circuits is technically or economically not feasible. Examples include bendable, rollable or foldable information displays [1], conformable sensor arrays [2], and plastic circuits [3,4].

In some of the more advanced applications anticipated for organic TFTs, such as the integrated row and column drivers of active-matrix organic light-emitting diode (AMOLED) displays [5], the TFTs will be required to control electrical signals of about 2.5 to 3 V (which is the supply voltage range of high-efficiency OLEDs [6]) at frequencies of up to about 10 MHz. Since the field-effect mobilities of organic TFTs are usually well below 10 cm²/Vs, these advanced applications demand aggressively scaled organic TFTs with lateral dimensions (channel length and parasitic gate-to-contact overlap) of about 1 μm or less.

II. HIGH-CAPACITANCE GATE DIELECTRICS

The first requirement for the realization of low-voltage flexible organic TFTs is a high-capacitance gate dielectric that can be produced at temperatures compatible with plastic substrates, i.e., at temperatures of about 100 °C or less. Assuming a maximum gate-source voltage of 3 V and a minimum charge-carrier density in the channel of 10^{13} cm^{-2},

the gate dielectric must have a capacitance per unit area of at least 600 nF/cm². Further assuming a permittivity of 5, this implies a thickness of less than 6 nm for the gate dielectric. Besides allowing the TFTs to operate with low voltages, a small gate-dielectric thickness provides the additional benefit of greatly suppressing a number of undesired short-channel effects (such as drain-induced barrier lowering (DIBL) and threshold-voltage roll-off) when the channel length is aggressively reduced with the goal of extending the cutoff frequency of the devices into the megahertz regime.

Over the past decade, several promising approaches for the realization of high-capacitance gate dielectrics at process temperatures not exceeding 100 °C for flexible, low-voltage organic TFTs have been developed. The most prominent examples are high-permittivity metal oxides (which are either grown by electrochemistry or deposited by physical vapor deposition or atomic layer deposition [7]), ultrathin insulating polymers (usually deposited by spin-coating from solution [8]), and organic self-assembled multilayers [9]. The most aggressive approach in terms of dielectric capacitance is the use of a solid-state electrolyte, such as a polyelectrolyte [10] or an ion gel [11], with capacitances exceeding 10 μF/cm². However, electrolytes are often limited by the slow response to transients which makes them less useful at high frequencies.

An alternative approach that is more suitable for high frequencies is a hybrid gate insulator consisting of an ultrathin aluminum oxide (AlO$_x$) layer (obtained by briefly exposing the surface of the aluminum gate electrodes to an oxygen plasma) and an alkylphosphonic acid self-assembled monolayer (SAM), obtained by exposing the aluminum oxide surface to a solution of the phosphonic acid molecules [12]. Depending on the plasma conditions and the chain length of the alkylphosphonic acid molecules, this hybrid AlO$_x$/SAM gate dielectric has a total thickness of about 4 to 6 nm and a capacitance of 700 to 900 nF/cm² [13]. Despite its small thickness and despite the fact that it is obtained at a temperature of less than 100 °C, the leakage current density through this hybrid dielectric is less than 10^{-5} A/cm² at an electric field of 5 MV/cm.

978-1-4673-1707-8/12 $31.00 © 2012 IEEE

Fig. 1 shows the cross-section, a transmission electron microscopy (TEM) image [14], photographs, and the electrical characteristics of a low-voltage TFT with an AlO_x/SAM gate dielectric fabricated by U. Zschieschang (MPI for Solid State Research). For this p-channel TFT, the small-molecule semiconductor 2,9-didecyldinaphtho[2,3-b:2',3'-f]thieno[3,2-b]thiophene (C_{10}-DNTT) that was recently developed by M. J. Kang and K. Takimiya at Hiroshima University [15] was employed. C_{10}-DNTT TFTs with a large channel length (L = 30 μm) fabricated on flexible polyethylene naphthalate (PEN) substrates exhibit field-effect mobilities of 4.3 cm^2/Vs, on/off ratios of 10^8, subthreshold swings of 68 mV/decade, and a signal propagation delay (measured using a 5-stage ring oscillator) of 25 μs per stage at a supply voltage of 3 V [16].

III. MOLECULAR CONTACT DOPING

In order to extend the frequency performance of organic TFTs into the megahertz regime, the channel length and the parasitic gate-to-contact overlap must be reduced to about 1 μm. This creates a new challenge: When the channel length of a field-effect transistor is reduced, the channel resistance decreases proportionally, which means that the relative influence of the contact resistance on the total device resistance (channel resistance + contact resistance) increases.

This increase of the relative influence of the contact resistance on the total resistance of the transistor causes the well-documented suppression and the distinct non-linearity of the drain current measured at small drain-source voltages, which can be observed in essentially all previous reports on organic TFTs with small channel lengths (i.e., L < 2 μm). To alleviate this undesired effect, area-selective contact doping using a strong molecular dopant with the aim of reducing the contact resistance has recently been introduced into organic TFTs [17-19].

To illustrate the beneficial effect of contact doping, Fig. 2 shows a comparison of the measured output characteristics of two organic p-channel TFTs, one without doping and one with a thin layer of the molecular dopant NDP-9 (developed by Novaled AG, Germany) inserted at the interface between the organic semiconductor (dinaphtho[2,3-b:2',3'-f]thieno[3,2-b]thiophene, DNTT) and the gold source and drain contacts [20]. Both TFTs, which were fabricated by F. Ante (MPI for Solid State Research) have a channel length of 150 nm. For these TFTs, the submicron transistor patterns were defined by electron-beam lithography, and for simplicity these devices were fabricated on silicon substrates, rather than on flexible plastic substrates. As can be seen in Fig. 2, the area-selective contact doping indeed greatly alleviates the contact resistance problem and helps to recover the desired linear current-voltage behavior at small drain-source voltages in organic TFTs with submicron channel length [20].

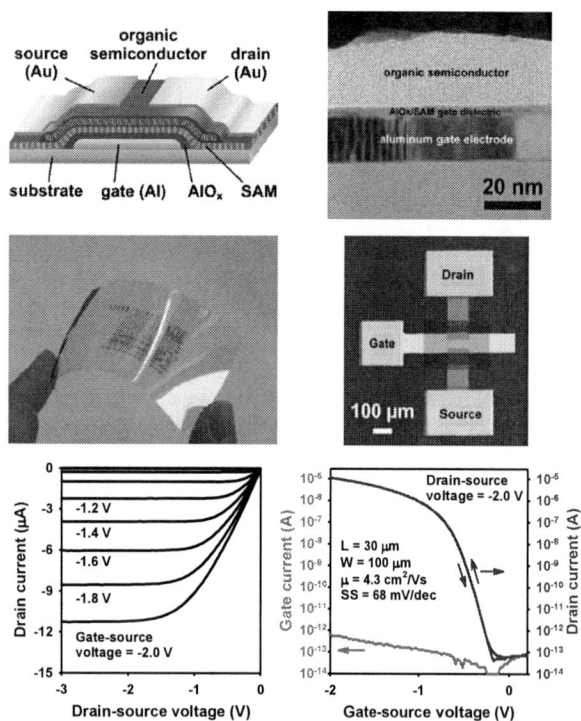

Figure 1. Schematic cross-section, cross-sectional transmission electron microscopy (TEM) image [14], photographs, and current-voltage characteristics of a long-channel C_{10}-DNTT TFT with a hybrid AlO_x/SAM gate dielectric fabricated on a flexible polyethylene naphthalate (PEN) substrate. The TFT has a channel length of 30 μm, a field-effect mobility of 4.3 cm^2/Vs, an on/off ratio of 10^8, a subthreshold swing of 68 mV/decade, and a signal propagation delay (measured using 5-stage ring oscillators) of 25 μs per stage at a supply voltage of 3 V. (See also reference [16].)

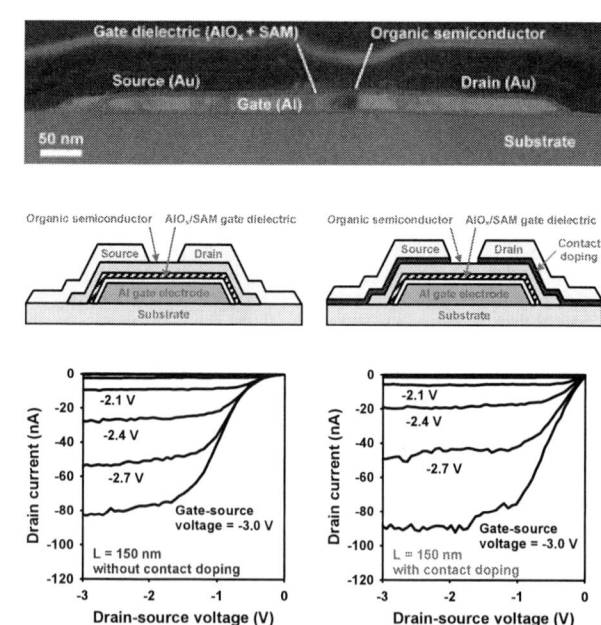

Figure 2. Output characteristics of two organic TFTs, one without contact doping (left) and one with a thin layer of the molecular dopant NDP-9 inserted at the interface between the organic semiconductor layer and the gold source and drain contacts (right). Both TFTs have a channel length of 150 nm and were fabricated on silicon substrates with the help of direct-write electron-beam lithography. A cross-sectional TEM image of the TFTs is also shown. (See also reference [20].)

978-1-4673-1707-8/12 $31.00 © 2012 IEEE

IV. HIGH-RESOLUTION SILICON STENCIL MASKS

Direct-write electron-beam lithography is obviously not suitable for large-area manufacturing. A far more promising approach for the routine fabrication of organic TFTs with lateral dimensions near (and possibly below) 1 μm on flexible, large-area substrates is the use of high-resolution silicon stencil masks [21-23].

Fig. 3 shows a scanning electron microscopy (SEM) image of a part of a silicon stencil mask, as well as a photograph, an SEM image and the measured output characteristics of a C_{10}-DNTT TFT with a channel length of 1 μm and a channel width of 5 μm fabricated on a flexible PEN substrate using a set of silicon stencil masks developed at IMS Chips (Stuttgart, Germany). This TFT has an effective field-effect mobility of 1.2 cm²/Vs and a width-normalized transconductance of 1.3 S/m. Also shown in Fig. 3 are the transfer characteristics of 16 DNTT TFTs, all with a channel length of 1 μm and a channel width of 10 μm, fabricated on a glass substrate, demonstrating the excellent parameter uniformity attainable with the stencil-mask process.

The main advantage of the reduced lateral dimensions that are achieved with the high-resolution stencil masks is the greatly improved dynamic performance of the TFTs; flexible C_{10}-DNTT p-channel TFTs with a channel length of 1 μm have a signal propagation delay of 420 ns per stage at a supply voltage of 3 V (see Fig. 4).

Another advantage of the reduced TFT dimensions is the smaller footprint. Fig. 5 shows a photograph and the measured transfer curve (after calibration) of a 6-bit current-steering digital-to-analog converter designed and characterized by T. Zaki, H. Richter and J. N. Burghartz at IMS Chips and fabricated on glass using DNTT TFTs with a channel length of 4 μm. The circuit occupies an area of just 2.6 × 4.6 mm² and operates with sampling rates of up to 100 kS/s [24,25].

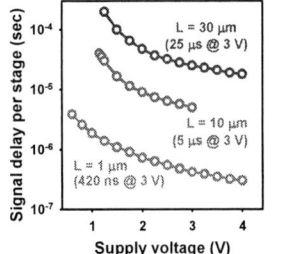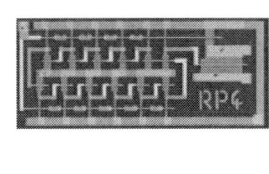

Figure 4. Signal propagation delay as a function of supply voltage of organic TFTs with various channel lengths fabricated on flexible PEN substrates and measured using ring oscillators. The photograph shows an 11-stage ring oscillator patterned using stencil masks.

V. ORGANIC COMPLEMENTARY CIRCUITS

Integrated digital or mixed-signal circuits, such as sensor and display drivers, identification tags and microprocessors, can in principle all be implemented using a unipolar circuit topology, i.e., with the exclusive use of organic p-channel TFTs [1,3,4,11,24-26]. However, in terms of noise margin and power consumption (and thus especially in view of mobile applications), complementary circuits provide substantial advantages over unipolar circuits. The development of organic n-channel TFTs with adequate performance and stability is therefore of significant interest.

The first requirement for the realization of an air-stable organic n-channel field-effect transistor is a semiconductor with a large electron affinity. A successful approach to organic semiconductors with large electron affinities is the substitution of hydrogen atoms with strongly electron-withdrawing moieties, such as fluorine, chlorine or cyano groups. Examples include hexadecafluorocopperperphthalocyanine (F_{16}CuPc [27]), heptafluorobutyl-dicyano-perylene tetracarboxylic diimide (PTCDI-$(CN)_2$-$CH_2C_3F_7$ [28]), heptafluorobutyl-dichloro-naphthalene tetracarboxylic diimide (NTCDI-Cl_2-$CH_2C_3F_7$ [29]), and poly{[N,N'-bis(2-octyldodecyl)-naphthalene-1,4,5,8-bis(dicarboximide)-2,6-diyl]-alt-5,5'-(2,2'-bithiophene)} (P(NDI2OD-T2) [30]), all of which have shown promising n-channel TFT characteristics.

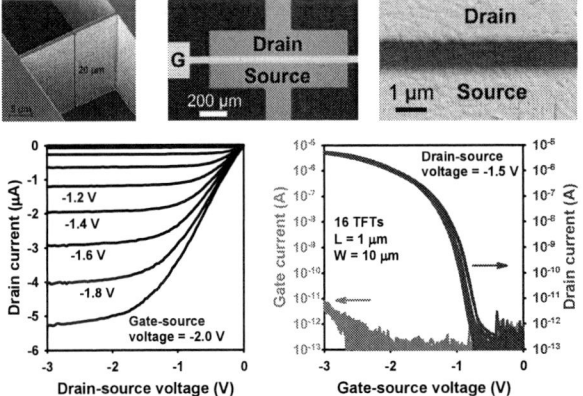

Figure 3. Top left: Scanning electron microscopy (SEM) image of a part of a silicon stencil mask. Top center, right: Photograph and SEM image of a TFT with a channel length of 1 μm fabricated with Si stencil masks. Bottom left: Output characteristics of a C_{10}-DNTT TFT with a channel length of 1 μm, a channel width of 5 μm, a mobility of 1.2 cm²/Vs, and a transconductance of 1.3 S/m fabricated on a flexible PEN substrate. Bottom right: Transfer characteristics of 16 DNTT TFTs with a channel length of 1 μm and a channel width of 10 μm fabricated on a glass substrate. (See also reference [22].)

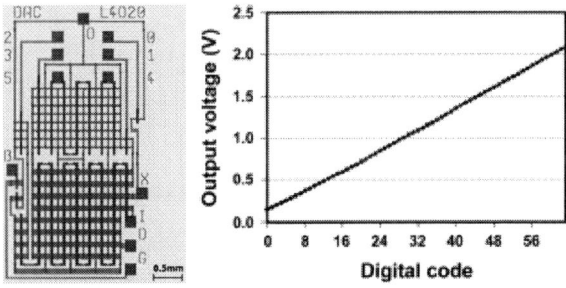

Figure 5. Photograph and measured transfer curve (after calibration) of a 6-bit current-steering digital-to-analog converter (DAC) fabricated on a glass substrate using DNTT TFTs with a channel length of 4 μm. The circuit occupies an area of 2.6 × 4.6 mm² and operates with sampling rates of up to 100 kS/s.
(See also references [24] and [25].)

978-1-4673-1707-8/12 $31.00 © 2012 IEEE

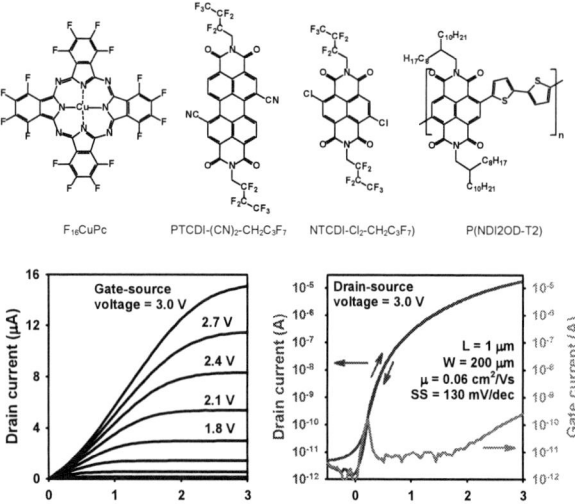

Figure 6. Top: Organic semiconductors that have shown promise for the realization of air-stable organic n-channel TFTs.
(See also references [27], [28], [29] and [30].)
Bottom: Measured output and transfer characteristics of an air-stable organic n-channel TFT based on heptafluorobutyl-dichloro-naphthalene tetracarboxylic diimide (NTCDI-Cl$_2$-CH$_2$C$_3$F$_7$). The TFT has a channel length of 1 µm, a field-effect mobility of 0.06 cm^2/Vs, an on/off ratio of 10^6, and a subthreshold swing of 130 mV/decade.

Fig. 6 shows the current-voltage characteristics of a TFT based on NTCDI-Cl$_2$-CH$_2$C$_3$F$_7$, recently developed by S.-L. Suraru and F. Würthner at the University of Würzburg, with a channel length of 1 µm. This TFT has an electron mobility of 0.06 cm^2/Vs, an on/off ratio of 10^6, a subthreshold swing of 130 mV/decade, and a transconductance of 0.09 S/m.

Fig. 7 shows a photograph and the measured transfer curve (after calibration) of a 6-bit C-2C analog-to-digital converter designed and characterized by W. Xiong and B. Murmann at Stanford University and fabricated with DNTT p-channel and F$_{16}$CuPc n-channel TFTs on glass. Due to the complementary design, the power consumption at a sampling rate of 100 S/s is just 3.6 µW (of which the comparator consumes 2.9 µW) [31].

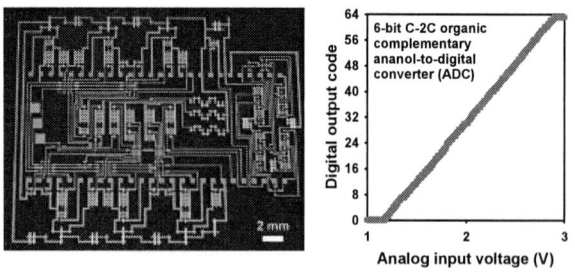

Figure 7. Photograph and measured transfer curve (after calibration) of a 6-bit C-2C analog-to-digital converter (ADC) fabricated in a complementary design with DNTT p-channel and F$_{16}$CuPc n-channel TFTs on a glass substrate. At a sampling rate of 100 S/s, the ADC has a power consumption of 3.6 µW.
(See also reference [31].)

ACKNOWLEDGMENTS

Ute Zschieschang (Max Planck Institute for Solid State Research), Frederik Ante (Max Planck Institute for Solid State Research), Reinhold Rödel (Max Planck Institute for Solid State Research), Ulrike Kraft (Max Planck Institute for Solid State Research), Mirsada Sejfić (Max Planck Institute for Solid State Research), Robert Hofmockel (Max Planck Institute for Solid State Research), Daniel Kälblein (Max Planck Institute for Solid State Research), Hyeyeon Ryu (Max Planck Institute for Solid State Research), Marko Burghard (Max Planck Institute for Solid State Research), Konstantin Amsharov (Max Planck Institute for Solid State Research), Claudia Kamella (Max Planck Institute for Solid State Research), Ulrike Waizmann (Max Planck Institute for Solid State Research), Marion Hagel (Max Planck Institute for Solid State Research), Bernhard Fenk (Max Planck Institute for Solid State Research), Wolfgang Winter (Max Planck Institute for Solid State Research), Klaus Kern (Max Planck Institute for Solid State Research), Martin Jansen (Max Planck Institute for Solid State Research), Kersten Hahn (Max Planck Institute for Metals Research), Peter van Aken (Max Planck Institute for Metals Research), Edwin Weber (TU Bergakademie Freiberg), Tarek Zaki (IMS Chips), Florian Letzkus (IMS Chips), Jörg Butschke (IMS Chips), Harald Richter (IMS Chips), Joachim N. Burghartz (IMS Chips), Alberto Salleo (Stanford University), Wei Xiong (Stanford University), Boris Murmann (Stanford University), Marcus Halik (University of Erlangen), Tatsuya Yamamoto (University of Hiroshima), Myeong J. Kang (University of Hiroshima), Kazuo Takimiya (University of Hiroshima), Masaaki Ikeda (Nippon Kayaku), Hirokazu Kuwabara (Nippon Kayaku), Tobias W. Canzler (Novaled), Ansgar Werner (Novaled), Jan Blochwitz-Nimoth (Novaled), Sabin-Lucian Suraru (University of Würzburg), Matthias Stolte (University of Würzburg), Frank Würthner (University of Würzburg), Susanne C. Martens (University of Heidelberg), Sonja Geib (University of Heidelberg), Lutz H. Gade (University of Heidelberg), Thomas Gessner (BASF), Jochen Brill (BASF), R. Thomas Weitz (BASF), Nis Hauke Hansen (University of Würzburg), Jens Pflaum (University of Würzburg), Tsuyoshi Sekitani (University of Tokyo), John E. Anthony (University of Kentucky), Takao Someya (University of Tokyo), Richard Rook (CADiLAC Laser), and William MacDonald (DuPont Teijin Films) are acknowledged for their countless and invaluable contributions.

REFERENCES

[1] M. Noda, N. Kobayashi, M. Katsuhara, A. Yumoto, S. Ushikura, R. Yasuda, N. Hirai, G. Yukawa, I. Yagi, K. Nomoto, and T. Urabe, "An OTFT-driven rollable OLED display," J. Soc. Inf. Display, vol. 19, p. 316, 2011.

[2] T. Someya, Y. Kato, T. Sekitani, S. Iba, Y. Noguchi, Y. Murase, H. Kawaguchi, and T. Sakurai, "Conformable, flexible, large-area networks of pressure and thermal sensors with organic transistor active matrixes," Proceedings of the National Academy of Sciences, vol. 102, p. 12321, 2005.

[3] K. Myny, S. Steudel, S. Smout, P. Vicca, F. Furthner, B. van der Putten, A.K. Tripathi, G. H. Gelinck, J. Genoe, W. Dehaene, and P. Heremans, "Organic RFID transponder chip with data rate compatible with electronic product coding," Org. Electronics, vol. 11, p. 1176, 2010.

978-1-4673-1707-8/12 $31.00 © 2012 IEEE

[4] K. Myny, E. van Veenendaal, G. H. Gelinck, J. Genoe, W. Dehaene, and P. Heremans, "An 8-bit, 40-instructions-per-second organic microprocessor on plastic foil," IEEE J. Solid-State Circ., vol. 47, p. 284, 2012.

[5] P. Schalberger, M. Herrmann, S. Hoehla, and N. Fruehauf, "A fully integrated 1-in. AMOLED display using current feedback based on a five-mask LTPS CMOS process," J. Soc. Inf. Display , vol. 19, p. 496, 2011.

[6] J. Birnstock, T. Canzler, M. Hofmann, A. Lux, S. Murano, P. Wellmann, and A. Werner, "PIN OLEDs - Improved structures and materials to enhance device lifetime," J. Soc. Inf. Display, vol. 16, p. 221, 2008.

[7] W. Y. Lin, R. Müller, K. Myny, S. Steudel, J. Genoe, and P. Heremans, "Room-temperature solution-processed high-k gate dielectrics for large area electronic applications," Org. Electronics, vol. 12, p. 955, 2011.

[8] M. E. Roberts, S. C. B. Mannsfeld, N. Queralto, C. Reese, J. Locklin, W. Knoll, and Z. Bao, "Water-stable organic transistors and their application in chemical and biological sensors," Proceedings of the National Academy of Sciences, vol. 105, p. 12134, 2008.

[9] S. A. DiBenedetto, D. Frattarelli, A. Facchetti, M. A. Ratner, and T. J. Marks, "Structure-performance correlations in vapor phase deposited self-assembled nanodielectrics for organic field-effect transistors," J. Am. Chem. Soc., vol. 131, p. 11080, 2009.

[10] L. Herlogsson, X. Crispin, S. Tiemey, and M. Berggren, "Polyelectrolyte-gated organic complementary circuits operating at low power and voltage," Adv. Mater., vol. 23, p. 4684, 2011.

[11] Y. Xia, W. Zhang, M. Ha, J. H. Cho, M. J. Renn, C. H. Kim, and C. D. Frisbie, "Printed sub-2 V gel-electrolyte-gated polymer transistors and circuits," Adv. Funct. Mater., vol. 20, p. 587, 2010.

[12] U. Zschieschang, M. Halik, and H. Klauk, "Microcontact-printed self-assembled monolayers as ultrathin gate dielectrics in organic thin-film transistors and complementary circuits," Langmuir, vol. 24, p. 1665, 2008.

[13] A. Jedaa, M. Burkhardt, U. Zschieschang, H. Klauk, D. Habich, G. Schmid, and M. Halik, "The impact of self-assembled monolayer thickness in hybrid gate dielectrics for organic thin-film transistors," Org. Electronics, vol. 10, p. 1442, 2009.

[14] T. Sekitani, U. Zschieschang, H. Klauk, and T. Someya, "Flexible organic transistors and circuits with extreme bending stability," Nature Mater., vol. 9, p. 1015, 2010.

[15] M. J. Kang, I. Doi , H. Mori , E. Miyazaki , K. Takimiya, M. Ikeda, and H. Kuwabara, "Alkylated dinaphtho[2,3-b:2′,3′-f]thieno[3,2-b]thiophenes (C_n-DNTTs): Organic semiconductors for high-performance thin-film transistors," Adv. Mater., vol. 23, p. 1222, 2011.

[16] U. Zschieschang, M. J. Kang, K. Takimiya, T. Sekitani, T. Someya, T. W. Canzler, A. Werner, J. Blochwitz-Nimoth, and H. Klauk, "Flexible low-voltage organic thin-film transistors and circuits based on C_{10}-DNTT," J. Mater. Chem., vol. 22, p. 4273, 2012.

[17] K. Tsukagoshi, K. Shigeto, I. Yagi, and Y. Aoyagi, "Interface modification of a pentacene field-effect transistor with a submicron channel," Appl. Phys. Lett., vol. 89, p. 113507, 2006.

[18] J. Li, X. W. Zhang, L. Zhang, K. U. Haq, X. Y. Jiang, W. Q. Zhu, and Z. L. Zhang, "Improving organic transistor performance through contact-area-limited doping," Solid-State Commun., vol. 149, p. 1826, 2009.

[19] S. P. Tiwari, W. J. Potscavage Jr., T. Sajoto, S. Barlow, S R. Marder, and B. Kippelen, "Pentacene organic field-effect transistors with doped electrode-semiconductor contacts," Org. Electronics, vol. 11, p. 860, 2010.

[20] F. Ante, D. Kälblein, U. Zschieschang, T. W. Canzler, A. Werner, K. Takimiya, M. Ikeda, T. Sekitani, T. Someya, and H. Klauk, "Contact doping and ultrathin gate dielectrics for nanoscale organic thin-film transistors," Small, vol. 7, p. 1186, 2011.

[21] F. Letzkus, J. Butschke, B. Höfflinger, M. Irmscher, C. Reuter, R. Springer, A. Ehrmann, and J. Mathuni, "Dry Etch Improvements in the SOI Wafer Flow Process for IPL Stencil Mask Fabrication," Microelectronic Eng., vol. 53, p. 609, 2000.

[22] F. Ante, D. Kälblein, T. Zaki, U. Zschieschang, K. Takimiya, M. Ikeda, T. Sekitani, T. Someya, J. N. Burghartz, K. Kern, and H. Klauk, "Contact resistance and megahertz operation of aggressively scaled organic transistors," Small, vol. 8, p. 73, 2012.

[23] K. Sidler, N. V. Cvetkovic, V. Savu, D. Tsamados, A. M. Ionescu, and J. Brugger, "Organic thin film transistors on flexible polyimide substrates fabricated by full-wafer stencil lithography," Sensors and Actuators A, vol. 162, p. 155, 2010.

[24] T. Zaki, F. Ante, U. Zschieschang, J. Butschke, F. Letzkus, H. Richter, H. Klauk, and J. N. Burghartz, "A 3.3 V 6b 100kS/s current-steering D/A converter using organic thin-film transistors on glass," 2011 IEEE International Solid-State Circuits Conference Technical Digest, p. 324, 2011.

[25] T. Zaki, F. Ante, U. Zschieschang, J. Butschke, F. Letzkus, H. Richter, H. Klauk, and J. N. Burghartz, "A 3.3 V 6-bit 100 kS/s current-steering digital-to-analog converter using organic p-type thin-film transistors on glass," IEEE J. Solid-State Circuits, vol. 47, p. 292, 2012.

[26] G. H. Gelinck et al., "Flexible active-matrix displays and shift registers based on solution processed organic transistors," Nature Mater., vol. 3, p. 106, 2004.

[27] Z. Bao, A. Lovinger, and J. Brown, "New air-stable n-channel organic thin film transistors," J. Am. Chem. Soc., vol. 120, p. 207, 1998.

[28] B. A. Jones, M. J. Ahrens, M. H. Yoon, A. Facchetti, T. J. Marks, and M. R. Wasielewski, "High-mobility air-stable n type semiconductors with processing versatility: Dicyanoperylene 3,4:9,10-bis(dicarboximides)," Angew. Chem. Int. Ed., vol. 43, p. 6363, 2004.

[29] J. H. Oh, S. L. Suraru, W. Y. Lee, M. Könemann, H. W. Höffken, C. Röger, R. Schmidt, Y. Chung, W. C. Chen, F. Würthner, and Z. Bao, "High-performance air-stable n-type organic transistors based on core-chlorinated naphthalene tetracarboxylic diimides," Adv. Funct. Mater., vol. 20, p. 2148, 2010.

[30] H. Yan, Z. Chen, Y. Zheng, C. Newman, J. R. Quinn, F. Dötz, M. Kastler, and A. Facchetti, "A high-mobility electron-transporting polymer for printed transistors," Nature, vol. 457, p. 679, 2009.

[31] W. Xiong, U. Zschieschang, H. Klauk, and B. Murmann, "A 3V 6b successive-approximation ADC using complementary organic thin-film transistors on glass," 2010 IEEE International Solid-State Circuits Conference Technical Digest, p. 134, 2010.

978-1-4673-1707-8/12 $31.00 © 2012 IEEE

BSIM – Industry Standard Compact MOSFET Models

Yogesh Singh Chauhan[*,†], Sriram Venugopalan[†], Mohammed A. Karim[†], Sourabh Khandelwal[†],
Navid Paydavosi[†], Pankaj Thakur[†], Ali M. Niknejad[†] and Chenming C. Hu[†]
[*]Department of Electrical Engineering, Indian Institute of Technology (IIT) Kanpur, India
[†]Department of Electrical Engineering and Computer Science, University of California Berkeley, CA-94720,
Email: yogesh@eecs.berkeley.edu, hu@eecs.berkeley.edu

Abstract—BSIM compact models have served industry for more than a decade starting with BSIM3 and later BSIM4 and BSIMSOI. Here we will briefly discuss the ongoing work on current and future device models in BSIM group. BSIM6 is the next generation bulk RF MOSFET Model which uses charge based core with physical models adapted from BSIM4. Model fulfills all symmetry tests and shows correct slopes for harmonics. The BSIM-CMG and BSIM-IMG are the surface potential based models for multi-gate MOSFETs. The BSIM-CMG model has been developed to model common symmetric double, triple, quadruple and surround gate MOSFET. The BSIM-IMG model has been developed to model independent double-gate MOSFET capturing threshold voltage variation with back gate bias. Models include all read device effects like SCE, DIBL, mobility degradation, poly depletion, QME etc.

I. INTRODUCTION

BSIM group started device modeling in 80's and since then has been working to advance the modeling for circuit designers. First two models BSIM1 and BSIM2 were the beginning of MOSFET modeling in the group. BSIM3 developed in early 90's was the full fledged model, which was selected as the industry standard model is 1996 by Compact Model Council [1]. BSIM4 was selected as industry standard compact MOSFET model in 2000 [1] and since then, its been used by major semiconductor companies and design houses. The beauty of BSIM4 lies in the flexibility to fit data from different technologies starting from 0.35um to 28nm in production today [2]. Around same time BSIM group also worked on developing compact model for SOI MOSFET covering both partially and fully-depleted SOI, which was standardized in 2003. The compact model consists of two main components - core model and real device models. The core is the ideal long channel model, which is threshold voltage based in case of BSIM4 [3], [4]. The real device models are the models used to capture the effects in real devices e.g. short channel effect, velocity saturation effect, quantum mechanical effect etc. The real device models of BSIM4 are physically derived expressions for different effects and have excellently captured the silicon data and provided accuracy in parameter extraction. Although BSIM4 is being used for all types of designs, Analog and RF designers have complained on symmetry issue in the model. To address this issue, BSIM group started BSIM6 development in late 2010. BSIM6 uses charge based core with all physical models adapted from BSIM4 and guarantees

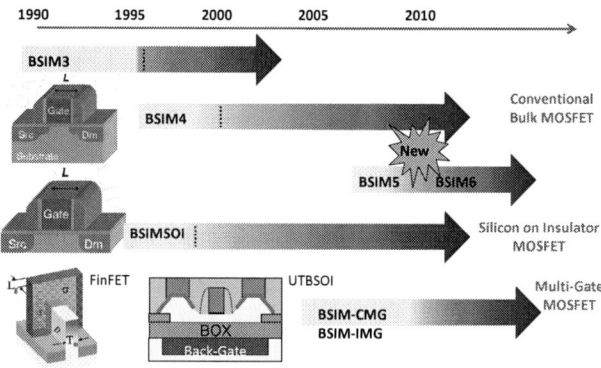

Fig. 1. Evolution of BSIM family of Compact MOSFET Models

symmetry around $V_{DS}=0$, thus providing correct slopes for harmonic balance simulation [5].

Although bulk and SOI technologies have successfully scaled till 28nm, off-state leakage has forced the industry to look for alternatives. FinFET and UTBSOI FET are the two multi-gate FET architecture already in production at sub-32nm nodes. Multiple gates surrounding the channel provide significantly better electrostatic control compared to conventional planar MOSFET [6], [7], [8] . BSIM group started its efforts on multi-gate modeling in early 2000. The BSIM-CMG and BSIM-IMG are the surface potential based physical compact models for multi-gate MOSFETs. The BSIM-CMG model has been developed to model common symmetric double, triple, quadruple and surround gate MOSFET. The Compact Model Council has selected BSIM-CMG as industry standard model for FinFET in March 2012. The BSIM-IMG model has been developed to model independent double-gate MOSFET capturing threshold voltage variation with back gate bias. BSIM-IMG is under consideration by CMC as a standard for Ultra-Thin-Body SOI technology, which is listed in the International Technology Roadmap for Semiconductors (ITRS) together with FinFET as the two successors to the planar CMOS. Here we will briefly report on BSIM models especially on BSIM6, BSIM-CMG and BSIM-IMG.

II. MULTI-GATE MOSFET

There are different flavors of multi-gate MOSFETs [6], [9]. The best known example is FinFET [6] which consists of a

978-1-4673-1707-8/12 $31.00 © 2012 IEEE

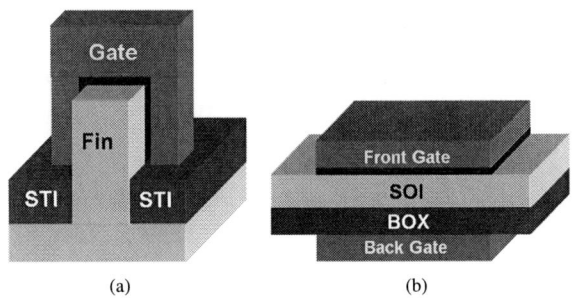

Fig. 2. Multi-Gate MOSFETs: (a) Triple-gate FinFET on Bulk Si (b) Planar Double-gate UTBSOI FET

thin silicon body (fin) and a gate wrapping around its top and two sides as shown in fig. 2(a). FinFETs can be made on either bulk or SOI substrates, creating the bulk FinFET or the SOI FinFETs. Double-gate MOSFETs may also be made as a planar device [7] such as planar SOI MOSFET with a thin buried oxide (BOX) as shown in fig. 2(b). A heavily-doped region in silicon under the BOX acts as the back-gate. Unlike front-gate, the back-gate is primarily used for tuning the device's threshold voltage (V_{th}). The back-gate can be used to dynamically raise or lower V_{th} circuit by circuit within a chip in response to the need for less leakage or more speed. Recently, there has been tremendous progress in improving UTBSOI performance by using very thin silicon body and thin BOX [8].

III. MULTI-GATE MOSFET MODELS

It is likely that more than one flavor of multi-gate MOSFETs will be used in production. Therefore the compact model should ideally cover as many of these flavors as possible. We have classified multi-gate MOSFETs into two main categories: independent multi-gate (IMG) and common multi-gate (CMG) MOSFETs. IMG refers to independent double-gate MOSFET with two separate gates. The front- and back-gate stacks are allowed to have different gate workfunctions, biases, dielectric thicknesses and materials e.g. UTBSOI. CMG refers to a special case where all the gates have identical workfunction, bias and dielectric thickness and material. Regular FinFETs and gate all-around MOSFETs fall into to this category. Two separate compact models BSIM-IMG and BSIM-CMG have been developed in a single framework for IMG and CMG devices, respectively.

A. BSIM-CMG MODEL

The BSIM-CMG is a surface potential-based model for a multi-gate MOSFET with undoped or doped body. The Poisson's equation with inversion carrier is perturbed by the body doping and a modification to the surface potential is derived. The analytical surface potential agrees well with TCAD double-gate device simulation for different doping concentration of the fin without any fitting parameter [10]. The core I-V model in BSIM-CMG is based on the drift-diffusion formulation without using the charge-sheet approximation. BSIM-CMG has been verified with measurements of cylindrical-gate

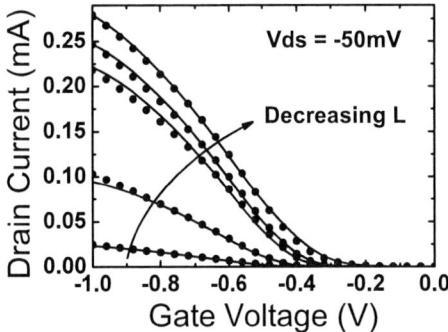

(a) $I_{ds} - V_{gs}$ for p-type FinFET ($V_{ds} = -50$mV)

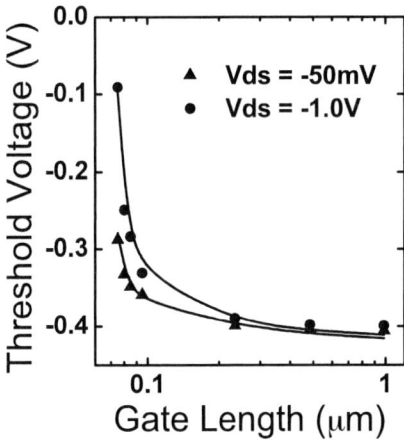

(b) $V_{th} - L$ for p-type FinFET

Fig. 3. Global extraction result of p-type SOI FinFETs. ($H_{fin} = 60$nm, $T_{fin} = 22$nm, $N_{fin} = 20$, $EOT = 2$nm, $L = 75$nm, 85nm, 90nm, 235nm and 1μm). Symbols: Measurement and lines: model.

FET and also on both SOI and bulk FinFET technologies [11], [12]. Fig. 3 (a) and (b) show $I_d - V_{gs}$ and threshold voltage roll-off for p-type FinFETs for different channel lengths showing excellent agreement between the model and data. A global extraction methodology has developed to fit devices for entire gate length array using single set of parameters.

B. BSIM-IMG MODEL

BSIM-IMG is also a surface-potential based model, i.e. the I-V and C-V are expressed in terms of electric potentials at the silicon/oxide interfaces. The front and back sides are asymmetric such as in UTBSOI transistor. We assume that the silicon body is lightly-doped as independent double-gate MOSFETs will likely have a lightly-doped body to minimize random dopant fluctuation. To get an analytical closed form solution of surface potential, we assumed that the inversion carrier density at the back surface is much smaller than that at the front surface. Thus, BSIM-IMG model is valid when the back-side surface is in depletion or weak inversion which

978-1-4673-1707-8/12 $31.00 © 2012 IEEE

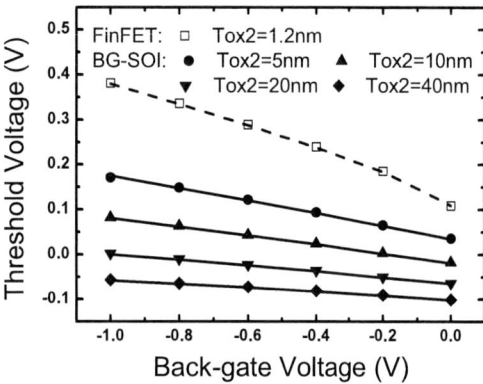

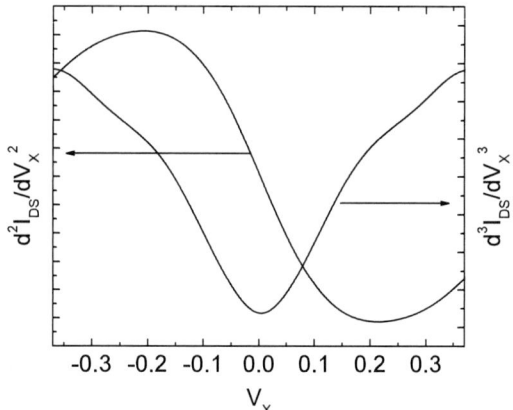

Fig. 4. Threshold voltage versus front-gate bias with varying back-oxide thickness. Solid lines and closed symbols: asymmetric structure ($T_{si} = 15$nm, $T_{ox1} = 1.2$nm, $V_{ch} = 0$, $\Phi_{g1} = 4.17$V, $\Phi_{g2} = 5.29$V); dashed lines and open symbols: symmetric structure ($T_{ox1} = T_{ox2} = 1.2$nm, $T_{si} = 15$nm, $\Phi_{g1} = 4.4$V, $\Phi_{g2} = 4.4$V; Symbols: TCAD; Lines: Model).

Fig. 6. Gummel Symmetry Test: $\frac{d^2 I_{DS}}{dV_X^2}$ and $\frac{d^3 I_{DS}}{dV_X^3}$ vs. $V_X = V_D$, where drain and source biases are swept in opposite direction.

IV. BSIM6 MODEL

To address symmetry issues in BSIM4, the BSIM group started BSIM6 development in 2010. The goal is to introduce a new core model to solve the symmetry issue while maintaining BSIM4's accuracy, speed, and above all, excellent user experience. BSIM6 has charge based core, which is derived from Poisson's solution for long channel MOSFET. The reason for choosing charge based core is due to its physical nature as well as accuracy along with computational efficiency. Several major improvements were done to ensure symmetry such as modified V_{ds} to V_{dsat} transition function, new velocity saturation model, updated channel length modulation, gate current model etc. As a result the model provides accurate results for modeling harmonic distortion [5]. As a successor of the BSIM4 model, it inherits a wide array of real-device effects, capturing all of the physical phenomena that appear in the state-of-the art bulk CMOS technologies. BSIM4 users can transition easily to BSIM6, whose parameters names are exactly the same as BSIM4 so that there is no learning required to make the transition. The BSIM6 is under evaluation at CMC to be the next bulk MOSFET model standard.

Fig. 6 shows the famous Gummel symmetry test results for BSIM6 model. The drain current and all its derivatives are continuous up to any order depending on the value of DELTA parameter, whose maximum value has been fixed to 0.5 to ensure that third derivative is always continuous. The charges associated with drain and source nodes are obtained using Ward-Dutton charge partitioning scheme. Fig. 7 shows the normalized capacitance plots, where it is shown that model behaves physically across bias range. It is evident that transcapacitances C_{gs} and C_{gd} overlap each other for $V_{ds} = 0$, which is an important condition for good compact model. Fig. 8 show the capacitance symmetry plots for BSIM6 model, which clearly demonstrate that model satisfies AC symmetry test [15]. It is must for a compact model to qualify this test for correct harmonics behavior. In fact, we had to

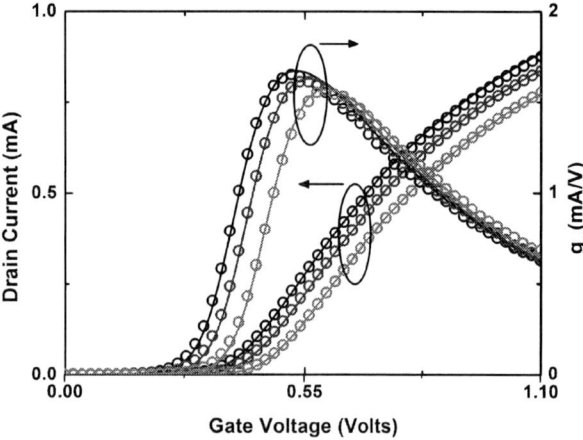

Fig. 5. $I_D - V_{fg}$ and $g_m - V_{fg}$ for different values of $V_{bg} (= 0, -0.2, -0.5V)$ at $V_{ds} = 50$mV for a short-channel device. ($W = 10\mu m$, $L = 30$nm, $T_{ox1} = 1.2$nm, $T_{Si} = 8$nm, and $T_{ox2} = 10$nm.)

is also the case with industrial UTBSOI devices [8]. The model has been tested with TCAD simulations for scalability across T_{si}, $Tox1$ and BOX thickness [10]. In Fig. 4, V_{th} is plotted versus back-gate bias (V_{bg}) for independent double-gate devices with both symmetric and asymmetric structures. Larger slope for thin back-oxide devices is due to the stronger coupling from the back side. Fig. 5 shows model validation on measured I_{ds} and g_m versus V_{gs} for different back-gate biases [13].

Self heating effect has been included in both BSIM-CMG and BSIM-IMG models and novel thermal network extraction methodology has been proposed recently [14].

978-1-4673-1707-8/12 $31.00 © 2012 IEEE

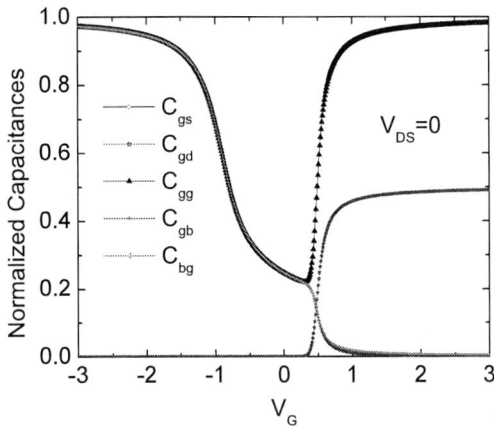

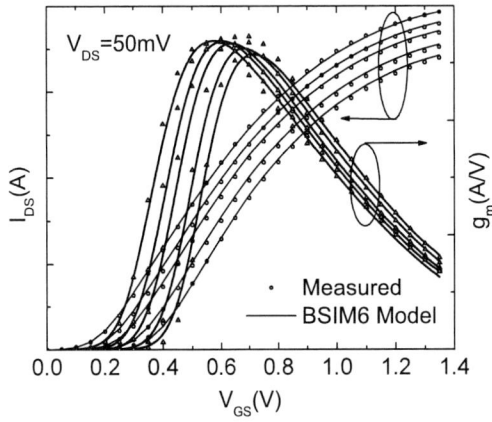

Fig. 7. Normalized capacitances from BSIM6 model. Quantum mechanical and poly depletion effects have been added in the model (not shown here).

Fig. 9. I_{ds} vs. V_{gs} for multiple body biases for medium channel device. Symbols: Measured Data, Lines: BSIM6 Model

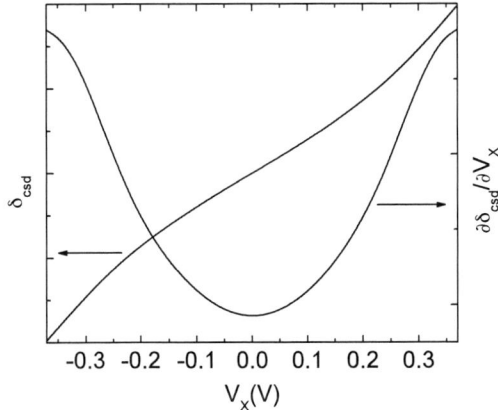

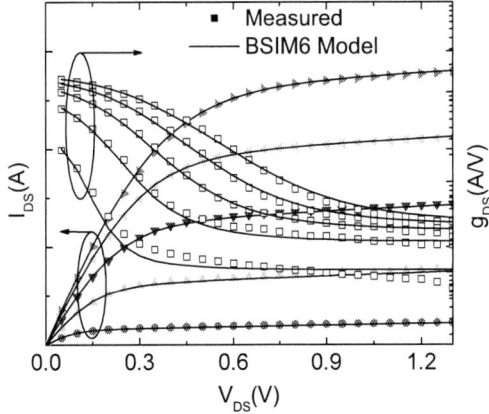

Fig. 8. AC Symmetry Test: $\delta_{csd} = \frac{(i_{s-}+i_{d-})+(i_{s+}-i_{d+})}{(i_{s-}-i_{d-})+(i_{s+}+i_{d+})} = \frac{C_{ss}-C_{dd}}{C_{ss}+C_{dd}}$ and its derivative vs. $V_X = V_D$, where drain and source biases are swept in opposite direction.

Fig. 10. I_{ds} vs. V_{gs} for multiple body biases for medium channel device. Symbols: Measured Data, Lines: BSIM6 Model

update the junction capacitance model taken from BSIM4 to satisfy this test. Fig. 9 and Fig. 10 show the model validation on measured characteristics from medium technology node. A detailed extraction for different lengths and widths using BSIM6 has been demonstrated in [16].

V. CONCLUSION

BSIM group has continuously delivered industry standard compact models for circuit designers. BSIM-CMG and BSIM-IMG models are production ready surface potential based models for multi-gate MOSFETs. BSIM6 Model is the next generation Bulk RF MOSFET Model which uses charge based core with all physical models adapted from BSIM4 model. BSIM6 is under standardization at Compact Model Council. All models fulfill quality tests e.g. Gummel symmetry and AC symmetry test. Models have been rigorously tested in DC,

small signal, transient and RF simulation and shows excellent convergence in circuit simulation.

REFERENCES

[1] Compact Model Council - http://www.geia.org/CMC-Introduction
[2] Chenming Hu and Christian Enz, GSA forum Article, March 2012.
[3] BSIM Models: http://www-device.eecs.berkeley.edu/bsim/
[4] Wiedong Liu and Chenming Hu, "BSIM4 and MOSFET Modeling for IC Simulation", World Scientific, 2011.
[5] Y. S. Chauhan et al., Workshop on Compact Modeling, 2012.
[6] X. Huang et al., IEEE IEDM, 67-70, 1999.
[7] I. Y. Yang et al., IEEE Trans. on Electron Devices, 1997.
[8] O. Faynot et al., IEEE IEDM, 2010.
[9] H.-S. Kim et al., Symposium on VLSI Technology, 1995.
[10] Darsen D. Lu, Ph.D. thesis, University of California Berkeley.
[11] M. V. Dunga et al., Symposium on VLSI Technology, 60-61, 2007
[12] S. Venugopalan et al., VLSI-TSA, 2011.
[13] S. Khandelwal et al., IEEE Trans. Electron Devices, 2012.
[14] M. A. Karim et. al., IEEE Electron Device Letters, 2012 (in press).
[15] C. C. McAndrew, IEEE Trans. Electron Devices, 2006.
[16] M.-A. Chalkiadakim et al., IEEE ESSDERC, 2012.

Evaluation of the BSIM6 Compact MOSFET Model's Scalability in 40nm CMOS Technology

M.-A. Chalkiadaki[*], A. Mangla[*], C.C. Enz [*][†], Y.S. Chauhan[‡],
M. A. Karim[‡], S. Venugopalan[‡], A. Niknejad[‡], C. Hu[‡]
[*] École Polytechnique Fédérale de Lausanne (EPFL), Switzerland
Email: maria-anna.chalkiadaki@epfl.ch
[†] Swiss Center for Electronics and Microtechnology (CSEM), Neuchtel, Switzerland
[‡] University of California, Berkeley, CA-94720

Abstract—**The aggressive downscaling of advanced bulk CMOS technologies demands MOSFET models that are able to describe accurately the behavior of devices accounting for all the physical phenomena. A reliable model should have the ability to handle all the different operating regions of the MOS transistor in the whole geometry range of one technology. Targeting to meet the aforementioned needs, the new charge-based compact model BSIM6 has been developed. In this article, as a first benchmarking of BSIM6, the model is evaluated for its scaling capabilities when a single set of parameters is used. The model is compared against a state-of-the-art 40nm CMOS technology. The results attest the model's scalability under all bias conditions, proving its reliability for nowadays complex IC designs.**

I. INTRODUCTION

Advanced Systems-on-Chip (SOCs) consist of millions of devices with the MOS transistor being the key element among them. The widely recognized advantages of bulk CMOS have resulted in it becoming the leading technology for microelectronics industry, and according to the latest 2011 ITRS roadmap [1] it will remain in the foreground for many more years to come. Due to the increasing complexity of modern integrated circuits (ICs), designers rely heavily on device models that would serve as an accurate interface between circuit design and fabrication.

Expecting to meet the demands posed by the rapid downscaling of CMOS technology and all the emerging physical phenomena, the BSIM family has introduced its latest member, BSIM6 as the new generation compact bulk MOSFET model. The limitations of the threshold-based approach adopted by its predecessors, have led to the transition to the physical charge-based analysis for the core of the BSIM6 model [2], [3]. This approach ensures the continuity of the model through all different levels of inversion which results from the continuous current-charge and charge-voltage relations. As a result, the model is able to capture the behavior of the MOS transistor even in the transition from weak to strong inversion; a known bottleneck for most of the traditional compact models. This property is valuable for advanced designs, where the need of low voltage supply leads MOSFETs to operate in moderate or even weak inversion.

This work was partly funded by the Swiss National Science Foundation Grand No. 200021-127241 and ENIAC project MIRANDELA.

Despite its multiple advantages, the downscaling of devices results also in increased short and narrow channel effects, which become dominant and must be accurately modeled. Consequently, one of the most important challenges for compact MOSFET models is their scaling property, namely their ability to accurately represent the behavior of the real devices across the whole range of W and L and for all bias conditions, using a single model parameter set without any binning.

State-of-the-art technologies need more complex models that subsequently demand more time to be developed. Taking also into account both the advancements of technology and the time required for a model to be adopted by the design community, the actual lifetime of a model is reduced significantly. Under the described conditions, it is crucial that the model is thoroughly benchmarked from its first steps to speed-up its development and adoption. This paper demonstrates the scaling properies of the new BSIM6 compact model in all regions of operation.

II. THE BSIM6 MODEL

The BSIM group started the development of BSIM6 in late 2010 expecting it to become the next standard model for bulk CMOS devices. In the core of BSIM6, a charge-based approach has been adopted. The main advantages of this approach are its physical nature, accuracy and computational efficiency. Unlike its predecessors that were source-referenced, BSIM6 is a bulk-referenced model and thus consistent with the existing symmetry between source and drain of real devices.

The core charge density equation of BSIM6 [4], is given by

$$2q_i + ln(q_i) + ln\left(\frac{4n}{\gamma}\left(\frac{n}{\gamma}q_i + \sqrt{\psi_p - 2q_i}\right)\right) = \psi_p - 2\phi_f - v_{ch} \quad (1)$$

where $q_i = -Q_i/2nC_{ox}U_T$ is the normalized inversion charge density – with Q_i being the inversion charge density, n the slope factor, C_{ox} the oxide capacitance per unit area, and $U_T = kT/q$ the thermodynamic voltage – and $\psi_p = \Psi_p/U_T$ is the normalized pinch-off surface potential. In earlier approaches [4], [5] the second log term of Eq. 1 was neglected in the evaluation of charge density at source and drain. In BSIM6, though, Eq. 1 has been solved analytically with respect to q_i, without using any approximations, ensuring better accuracy of the model for the entire bias range.

The normalized drain to source current, obtained using the well-known drift-diffusion model, is given by [6],

$$i_{\mathrm{ds}} = \frac{I_{\mathrm{DS}}}{I_{\mathrm{spec}}} = \frac{\left(q_{\mathrm{s}}^2 + q_{\mathrm{s}}\right) - \left(q_{\mathrm{d}}^2 + q_{\mathrm{d}}\right)}{\frac{1}{2}\left(1 + \sqrt{1 + \left(\lambda_{\mathrm{c}}\left(q_{\mathrm{s}} - q_{\mathrm{d}}\right)\right)^2}\right)} \quad (2)$$

where q_{s} and q_{d} are the normalized charge densities at source and drain end respectively. I_{spec} is the specific current defined as $I_{\mathrm{spec}} = 2n\beta U_{\mathrm{T}}^2$, where $\beta = \mu C_{\mathrm{ox}}\frac{W}{L}$ with μ being the mobility of the carriers. The denominator term in the above equation accounts for velocity saturation for short channel transistors; where $\lambda_c = \frac{2\mu_{\mathrm{eff}} U_{\mathrm{T}}}{v_{\mathrm{sat}} L_{\mathrm{eff}}}$ [6]. Note, that the drain charge density q_{d} is the effective charge density at drain, which is obtained using effective drain voltage accounting for V_{D} to V_{Dsat} transition [7].

BSIM6 inherits most of the expressions used to capture the higher order effects appearing in devices e.g. short channel effect, channel length modulation, velocity saturation etc. from the BSIM4 model. Although some parts of the model used to describe these effects are not electrically equivalent to BSIM4, the parameter names are kept identical for easier extraction based on the BSIM4 experience. During the BSIM6 development, extra effort was made in preserving model symmetry

around $V_{\mathrm{DS}} = 0$. The real device effect models were updated to ensure symmetry during DC and AC analyses. BSIM6 has been tested and satisfies all quality tests for compact MOSFET models [8]. Currently the model is under standardization at the CMC – Compact Model Council [9].

III. EVALUATION OF THE BSIM6 MODEL

For the evaluation of the scalability of BSIM6, along with its ability to represent accurately the behavior of MOS transistors in all regions of operation, the model was compared against measurements of an advanced 40 nm bulk CMOS technology from ST-Microelectronics. All the results presented in this paper are in form of normalized quantities. The drain current, as already discussed, is normalized with respect to I_{spec} while the gate transconductance G_{m} and output conductance G_{ds} are normalized with respect to the specific conductance $G_{\mathrm{spec}} = I_{\mathrm{spec}}/U_{\mathrm{T}}$. So, the normalized quantities of G_{m} and G_{ds} are then defined as $g_{\mathrm{m}} = G_{\mathrm{m}}/G_{\mathrm{spec}}$ and $g_{\mathrm{ds}} = G_{\mathrm{ds}}/G_{\mathrm{spec}}$ [6].

A. Parameter extraction procedure

Since the model maintains most of the parameter names of the BSIM4 model, almost the same step-by-step parameter extraction procedure as the one used in BSIM4 [10], with some minor adjustments for this improved model, was followed. First, geometry independent parameters were extracted using

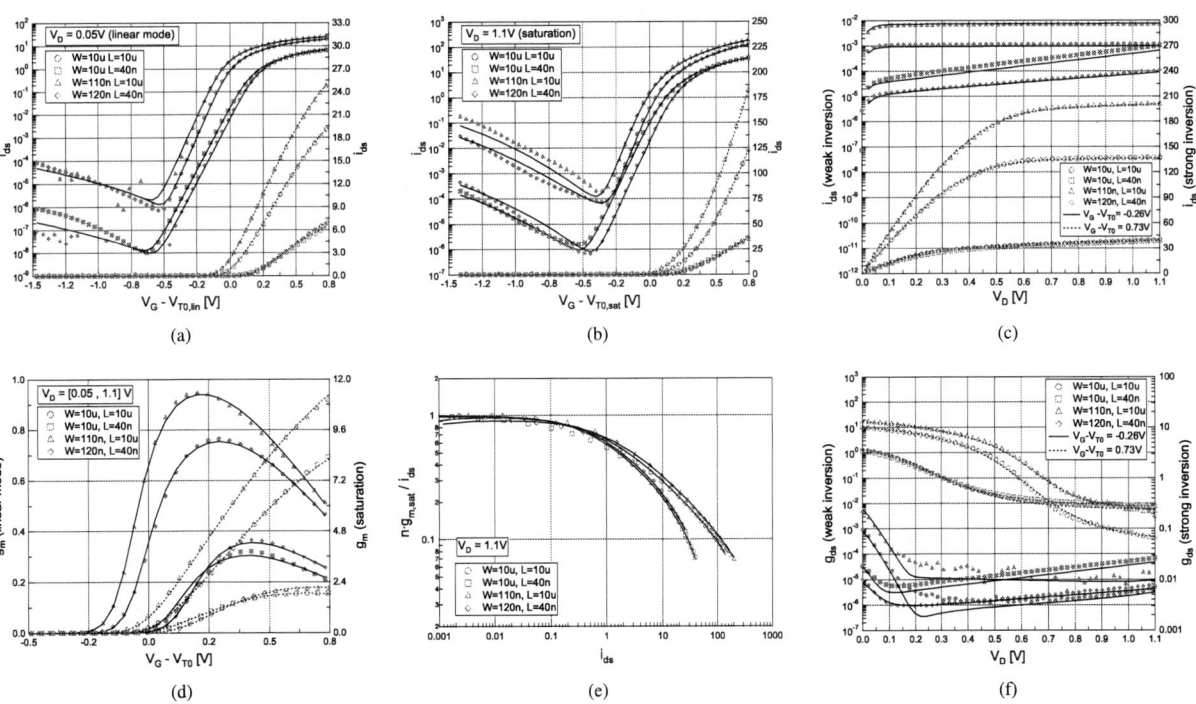

Fig. 1. DC analyses for the corner DUTs. (a) i_{ds} vs. $V_{\mathrm{G}} - V_{\mathrm{T0,lin}}$, $V_{\mathrm{D}} = 0.05V$ (linear mode), in linear and logarithmic scale, (b) i_{ds} vs. $V_{\mathrm{G}} - V_{\mathrm{T0,sat}}$, $V_{\mathrm{D}} = 1.1V$ (saturation), in linear and logarithmic scale, (c) i_{ds} vs. V_{D} for $V_{\mathrm{G}} - V_{\mathrm{T0}} = -0.26V$ (weak inversion) and $V_{\mathrm{G}} - V_{\mathrm{T0}} = 0.73V$ (strong inversion), (d) g_{m} vs. $V_{\mathrm{G}} - V_{\mathrm{T0}}$ for $V_{\mathrm{D}} = 0.05V$ (linear mode) and $V_{\mathrm{D}} = 1.1V$ (saturation), (e) $ng_{\mathrm{m,sat}}/i_{\mathrm{ds}}$ vs. i_{ds} for $V_{\mathrm{D}} = 1.1V$ (saturation), (f) g_{ds} vs. V_{D} for $V_{\mathrm{G}} - V_{\mathrm{T0}} = -0.26V$ (weak inversion) and $V_{\mathrm{G}} - V_{\mathrm{T0}} = 0.73V$ (strong inversion). Symbols stand for measurements and lines for BSIM6 model. Solid lines and symbols correspond to the left y-axis while the dashed lines and empty symbols correspond to the right y-axis.

978-1-4673-1707-8/12 $31.00 © 2012 IEEE

a large device. Then two sets of smaller-size devices, one set with constant wide channel but different channel lengths and one set with constant long channel but different channel widths, were used to extract parameters related to short and narrow channel effects respectively. To ensure accuracy of the extracted global model, the measurements of a wide range of device geometries were used.

B. Evaluation of the model for the corner DUTs

The evaluation of the scaling features of a model is meaningful only after the model's ability to describe accurately the overall behavior of at least one device, with all the involved physical phenomena, is shown. For consistency, the results of the model simulations for the fundamental DC analyses compared to measurements concerning the four corner DUTs, are presented. Fig. 1(a), (b) and (d) present i_{ds} and g_m vs. $V_G - V_{T0}$, in both linear and saturation modes, where V_{T0} is the threshold voltage of the specified region. Fig. 1(c) and (f), show the i_{ds} and g_{ds} vs. V_D for weak and strong inversion and last Fig. 1(e) shows the normalized transconductance efficiency $nG_m U_T / I_{DS}$ vs. I_{DS} in saturation, which is calculated after

$$\frac{nG_m U_T}{I_D} = \frac{\frac{nG_{m,sat}}{I_{spec}} U_T}{\frac{I_{DS}}{I_{spec}}} = \frac{ng_{m,sat}}{i_{ds}} \quad (3)$$

The results verify the correct behavior of the model across all regions of operation for the extreme geometries of the studied technology.

C. Evaluation of the model's scalability

A reliable model should be able to predict the drain current for all combinations of V_G and V_D, across W and L. In order to study this property of the model, the normalized currents in linear operation and in saturation for different levels of inversion were used in Fig. 2.

In addition, in order to highlight the behavior in weak inversion, the model was evaluated for its ability to predict the subthreshold slope (SS), given by $SS = \partial V_{GS}/\partial log(I_D)$ (Fig. 3). It can be observed that the model demonstrates a very good scalability across the entire width/length plane.

D. Study of the model's symmetry

Source and drain symmetry is a fundamental feature of an ideal MOSFET model. To check if this longitudinal symmetry is preserved by BSIM6 when the same parameter set extracted for this process is used, the Gummel symmetry test (GST) in weak and strong inversion was carried out [11]. The device that was selected is a short channel device where the symmetry of the model is more difficult to be preserved due to the prevailing short channel effects. The model was tested for the current $I_X = I_D - I_S$ vs. $V_X = V_D - V_S$, when $V_D = -V_S$, and its derivatives up to the 5th degree (Fig. 4). The quantities are normalized to their maximum values as $i_X = I_X/max(I_X)$ and $\frac{\partial^n i_X}{\partial V_X^n} = \frac{(\partial^n I_X)/\partial V_X^n}{max(\partial^n I_X/\partial V_X^n)}$. The model exhibits a smooth behavior around $V_{DS} = 0$.

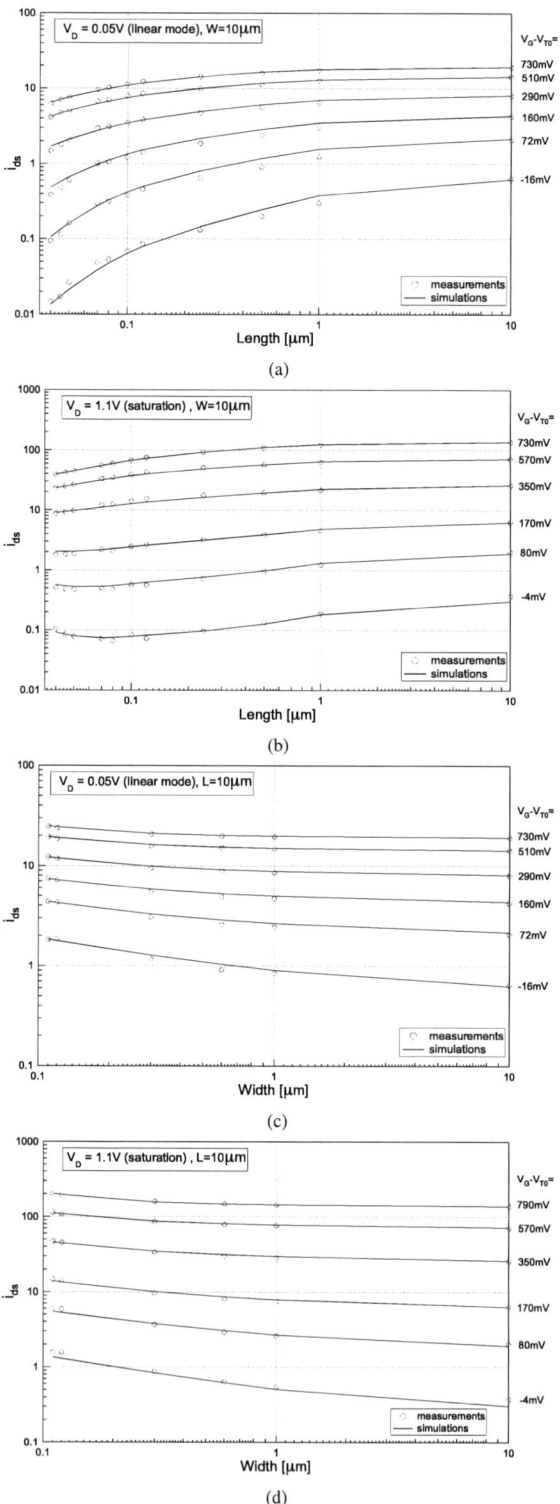

Fig. 2. Length and Width Scaling of inversion coefficient i_{ds} for various V_G. (a) i_{ds} vs L, $V_D = 0.05V$ (linear mode), (b) i_{ds} vs L, $V_D = 1.1V$ (saturation) (c) i_{ds} vs W, $V_D = 0.05V$ (linear mode), (d) i_{ds} vs W, $V_D = 1.1V$ (saturation).

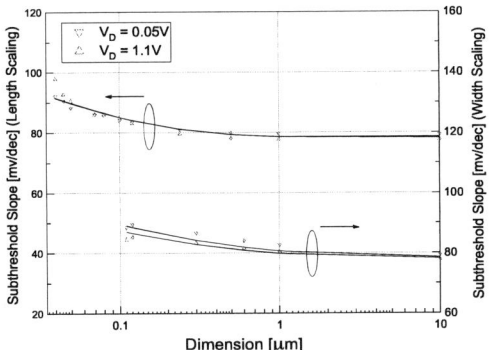

Fig. 3. Length and Width Scaling of Subthreshold Slope (SS) for $V_D = 0.05V$ (linear mode) and $V_D = 1.1V$ (saturation). In Length Scaling $W = 10\mu m$ while in Width Scaling $L = 10\mu m$. Symbols stand for measurements and lines for BSIM6 model.

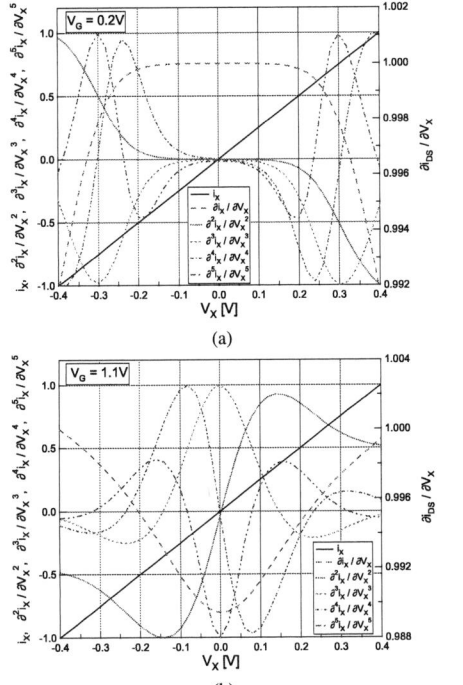

(a)

(b)

Fig. 4. Gummel symmetry test (GST) of BSIM6 model in (a) $V_G = 0.2V$ (weak inversion) and (b) $V_G = 1.1V$ (strong inversion). i_X vs. V_X and its partial derivatives up to 5th degree are shown.

E. Results

The results, when comparing model simulations and measurements, validate the excellent behavior of the BSIM6 model across W and L. The model has the ability to capture all the short and narrow channel effects that appear in state-of-the-art CMOS technologies. This ability is more obvious in Fig. 2(b) where the model can predict DIBL effect and velocity saturation in low and high levels of inversion respectively. It must also be highlighted that a single set of parameters is used for all the simulations, thus making binning unnecessary.

In addition, the model behaves accurately for all regions of operation, not excluding weak and moderate inversion. Furthermore, despite the complexity of the model, in order to incorporate all physical phenomena of modern processes, it still preserves its symmetry with respect to source-drain interchange even for higher order derivatives of the current.

IV. CONCLUSION

The increased requirements of IC designers for models that can precisely describe the behavior of modern MOSFET devices, pose a great challenge for model developers. The physical phenomena that come along with the downscaling of technologies must be accurately modeled. The new BSIM6 charge-based compact model shows excellent behavior in all regimes of operation and especially in the transition between subthreshold region and strong inversion as well as between linear operation and saturation. Without binning it is able to model the behavior of MOS devices, for a wide range of geometries. Furthermore, despite the complexity of the higher order effects that are predicted by the model, BSIM6 preserves its source-drain symmetry. This paper presented a first benchmark of BSIM6 against ST 40 nm CMOS technology measurements, in the early stages of the model's life cycle, which is a critical aspect, given the rapid progress of technology. The results are promising and show the potential of the model to become the next industrial standard MOSFET model.

ACKNOWLEDGMENT

The authors would like to thank ST-Microelectronics for providing the measurements and especially Didier Belot, Patrick Scheer and André Juge for all their kind support.

REFERENCES

[1] International Technology Roadmap for Semiconductors - ITRS, *Executive Summary*, http://www.itrs.net/Links/2011ITRS/2011Chapters/2011ExecSum.pdf, 2011.
[2] M. A. Maher, C. A. Mead, *A physical Charge-Controlled Model for MOS Transistors*, Advanced Research in VLSI, P. Losleben (ed.), MIT Press, Cambridge, MA, 1987.
[3] A. I. A. Cunha, M. C. Schneider, C. Galup-Montoro, *An MOS Transistor Model for Analog Circuit Design*, IEEE J. Solid-State Circuits, vol. 33, no. 10, pp. 1510-1519, 1998 .
[4] J.-M. Sallese, M. Bucher, F. Krummenacher, P. Fazan, *Inversion charge linearization in MOSFET modeling and rigorous derivation of the EKV compact model*, Solid-State Electronics, pp. 677-683, 2003.
[5] J. He, X. Xi, H. Wan, M. Dunga, M. Chan, A. M. Niknejad, *BSIM5: An advanced charge-based MOSFET model for nanoscale VLSI circuit simulation*, Solid-State Electronics, pp. 433-444, 2007.
[6] C. C. Enz, E. A. Vittoz, *Charge-Based MOS Transistor Modeling - The EKV Model for Low Power and RF IC Design*, 1st ed. John Wiley, 2006.
[7] K. Joardar, K. K. Gullapalli, C. C. McAndrew, M. E. Burnham and A. Wild, *An Improved MOSFET Model for Circuit Simulation*, IEEE Trans. Electron Devices, vol. 45, no. 1, pp. 134-148, 1998.
[8] C. C. McAndrew, *Validation of MOSFET model Source - Drain Symmetry*, IEEE Trans. Electron Devices, vol. 53, no. 9, pp. 2202-2206, 2006.
[9] Compact Model Council - http://www.geia.org/CMCIntroduction.
[10] T.-H. Morsed et al. *BSIM4v4.7 MOSFET Model User's Manual*, www-device.eecs.berkeley.edu/~bsim/Files/BSIM4/BSIM470/BSIM470_Manual.pdf, 2011.
[11] C. C. McAndrew, H. Gummel, K. Singhal, *Benchmarks for compact MOSFET models*, in Proc. SEMATECH Compact Models Workshop, Mar. 1995.

978-1-4673-1707-8/12 $31.00 © 2012 IEEE

4-Port Isolated MOS Modeling and Extraction for mmW Applications

B. Dormieu, P. Scheer, C. Charbuillet, S. Jan
STMicroelectronics
Crolles, France

F. Danneville
IEMN
Villeneuve d'Ascq, France

Abstract—**This paper reports on the extraction of the small-signal equivalent circuit of 28nm isolated RF MOS transistors using on-wafer 4-port S-parameter measurements up to 50GHz. It shows that modeling accuracy of RF MOS is significantly enhanced via a 4-resistance cross-type substrate network plus an isolation sub-network. In addition, the impact of substrate network on Mason gain is presented. Finally, the whole methodology is shown to be very promising to extract and model RF MOS in sub-threshold region for low power/high frequency applications.**

I. INTRODUCTION

Recent MOS transistors exhibit high f_t/f_{max} performances, making these devices good candidates for RF and mmW applications [1]. Besides, using MOS transistor for low-power applications becomes popular[2][3][4]. However, substrate coupling in this application domain may be not negligible anymore, especially impacting Mason gain, and in turn f_{max} [5]. Extracting components values for a given substrate network topology appears to be challenging. Methods based on classic common source configuration of the transistor [6] often need important approximations and are frequency limited. Another possible solution is to directly observe the substrate by at least one of the two ports [7][8][9], but information on f_t and f_{max} are then lost. According to the authors, 4-port measurements [10] are a very efficient way to extract substrate components, because they give a straightforward access to all electric paths between terminals. They also gain in interest with the increase of their frequency range [11], reaching now a domain where substrate coupling impacts device performances. Considering that MOS transistors are intrinsically 4-port devices, this paper aims to present an extraction methodology of a cross-type substrate network for 28nm High-k/Metal Gate (HKMG) transistors, with a buried deep-nwell layer. Indeed, the modeling of isolated MOSFETs is often neglected, although those devices are widely used in industry. For generality purpose, substrate model and extraction methodology presented in this paper are applied along a small-signal equivalent circuit. Nevertheless, any intrinsic model, such as compact models, could be used. This paper is organized as follows: first, 4-port structures and measurements setup are briefly presented, then substrate model and extraction method are detailed, finally the substrate model is evaluated in a practical case

through a comparison of measured and simulated Mason gain of a transistor in common source configuration.

II. MEASUREMENTS AND DE-EMBEDDING

A. Measurements

Test structures are 28nm HKMG nMOS transistors [12] with buried deep n-well (substrate isolation) layer. Three different geometries and topologies are investigated (gate length/gate width): M1:30nm/120μm, M2:90nm/120μm, M3:30nm/300μm. Measurements equipment is a PNA-X N5247A 4-port from Agilent with Cascade Infinity 67GHz GSGSG differential probes. S-parameters measurements are performed from 100MHz up to 50GHz and biasing voltages are consistent with standard transistor common source configuration: VG from 0 to 1V, VD = 0, 0.3V and 1V, VS=0V, VB=0V. Fig. 1 presents a 4-port device with access lines and ports configuration. To define the 4x4 S-parameters matrix, device terminals are numbered as follows: Gate 1, drain 2, source 3 and bulk 4.

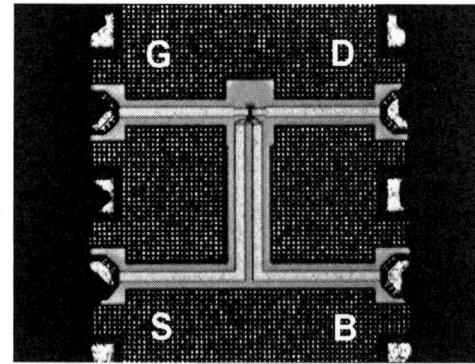

Fig. 1. Top photographic view of MOS device in 4-port characterization configuration.

B. De-embedding

Usual OPEN-SHORT technique is employed for the de-embedding procedure. Indeed, equation (1), well-known in the case of 2-port measurements, is still valid here.

$$Y_{DUT}=((Y_{MES}-Y_{OPEN})^{-1}-(Y_{SHORT}-Y_{OPEN})^{-1})^{-1} \quad (1)$$

$Y_{DUT}, Y_{MES}, Y_{SHORT}, Y_{OPEN}$ are the Y-parameters matrix of the de-embedded device, the measured device,

978-1-4673-1707-8/12 $31.00 © 2012 IEEE

the short and open structure, respectively. Comparison with 2-port devices with same topologies have been made and validated the method. In the following, S-parameters refer to the de-embedded S-parameters.

III. MODEL AND PARAMETERS EXTRACTION

A. Model description

A 4-terminal model of isolated 28nm HKMG MOS devices suitable for RF applications is shown on Fig. 2. Its equivalent schematic conventionally includes both intrinsic and extrinsic components. Intrinsic part is expected to describe the transistor effect, especially the channel behavior and is thus strongly bias dependent. Extrinsic part mainly contains parasitic elements which have not been removed by de-embedding procedures, such as gate resistance Rg and fringing capacitances, denoted here by C_{gsext} and C_{gdext}.

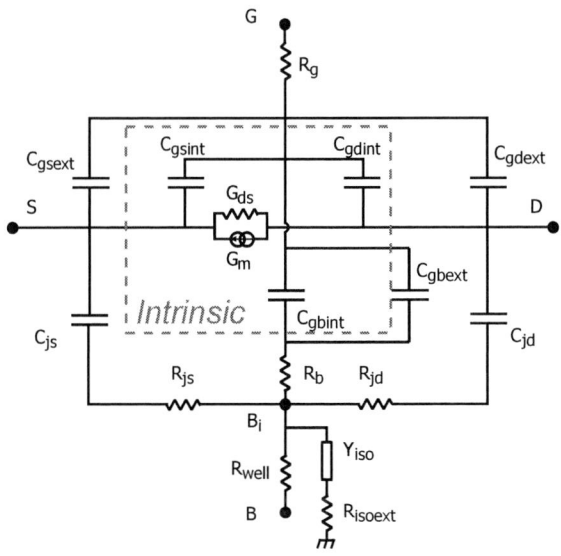

Fig. 2. Complete MOS equivalent schematic model.

Various substrate network topologies exist in literature. Considering the isolated MOS cross-section of Fig. 3 the four resistances cross-type substrate model, as found in PSP compact model [13] and shown on Fig. 2, is best suited. Still, to account for the isolation layer contribution, a sub-network needs to be added. According to the Fig. 3, following lumped components may be identified: an access resistance R_{isoext}, in series with a deep-nwell resistance, R_{iso}, and a pwell/deep nwell junction capacitance, C_{iso}.

Since the deep nwell layer is spreading under the pwell, R_{iso} and C_{iso} are actually distributed in nature. In RF and mmW domains, this must be considered to properly model the isolation layer impedance, Y_{iso}. Therefore, a simplified but still accurate version of model from [9] is chosen to describe the isolation layer.

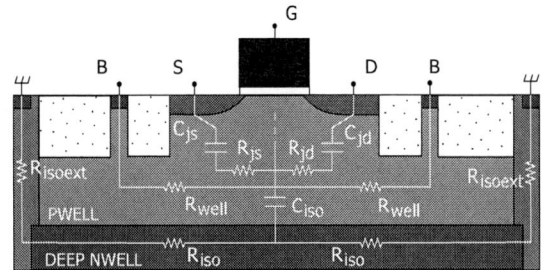

Fig. 3. Simplified MOS substrate cross section with lumped substrate equivalent network.

As a result, the distributed admittance Y_{iso} related to the isolation sub-network writes:

$$Y_{iso} = j\omega C_{iso} \frac{\tanh\left(\gamma_{iso}/2\right)}{\gamma_{iso}/2} \tag{2}$$

$$\text{with } \gamma_{iso} = \sqrt{j\omega C_{iso} R_{iso}} \tag{3}$$

B. Extraction

The extraction procedure begins with bias-independent extrinsic part. Capacitances C_{gsext} and C_{gdext} are directly extracted when the transistor is off (VG=VD=VS=VB=0V) using the branch admittances Y_{12} and Y_{13}. R_{well} is given by $Re(Z_{44} - Z_{14})$. In strong inversion regime (VG=1V and other ports DC biased at 0V), R_g is obtained from $R_{gg} = Re(1/Y_{11})$, neglecting channel contribution for devices with small gate length [14]. Simultaneously, C_{gbext} is found using the branch admittance Y_{14}, since C_{gbint} is null in that regime of inversion.

Other substrate parameters are extracted for null DC bias at all terminals (VG=VD=VS=VB=0). In depletion regime, the 4-terminal MOS model of Fig. 2 may be reduced to the schematic shown in Fig. 4. Thus, after removing access resistances R_g and R_{well}, all branches between ports may be conveniently extracted using the 4-port measurements. Therefore, an intrinsic admittance matrix $\mathbf{Y}'$ is defined by $\mathbf{Y}' = \mathbf{Z}'^{-1}$ where $\mathbf{Z}'$ writes:

$$\mathbf{Z}' = \begin{bmatrix} Z_{11} - R_g & Z_{12} & Z_{13} & Z_{14} \\ Z_{21} & Z_{22} & Z_{23} & Z_{24} \\ Z_{31} & Z_{32} & Z_{33} & Z_{34} \\ Z_{41} & Z_{42} & Z_{43} & Z_{44} - R_{well} \end{bmatrix} \tag{4}$$

Junction capacitances and resistances are extracted according to:

$$C_{jd/s} = \frac{1}{-Im(\frac{1}{Y_{jd/s}})\omega} \quad R_{jd/s} = Re(\frac{1}{Y_{jd/s}}) \tag{5}$$

$$\text{with } Y_{jd/s} = -Y'_{42/3} \tag{6}$$

978-1-4673-1707-8/12 $31.00 © 2012 IEEE

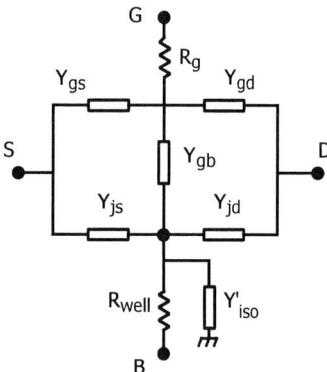

Fig. 4. Simplified MOS model in depletion regime

The benefit of removing R_g and R_{well} contributions to obtain accurate results is illustrated on Fig. 5. Indeed, junction capacitances and resistances are frequency independent when extracted using $\mathbf{Y}'$. Whereas it is not the case for results based on the original $\mathbf{Y}$ matrix.

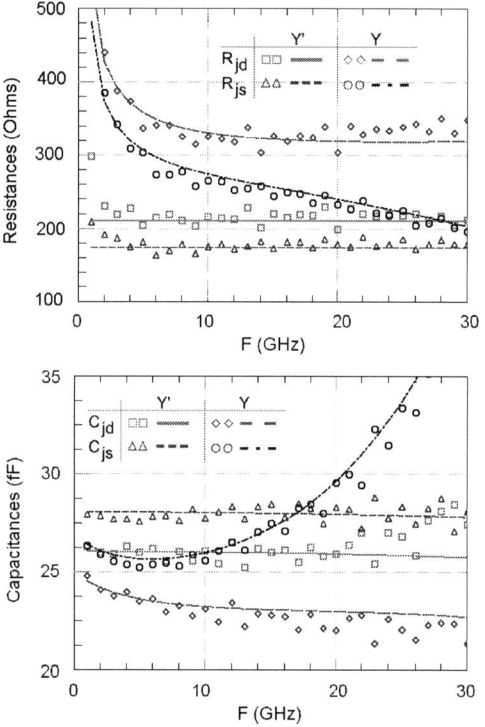

Fig. 5. Extracted and simulated Cjd, Cjs, Rjd, Rjs for the transistor M1, using the $\mathbf{Y}'$ matrix and the original $\mathbf{Y}'$ one, which contains access resistances R_g and R_{well}.

Isolation network is defined by the three parameters C_{iso}, R_{iso} and R_{isoext}. Since expression (2) involving C_{iso} and R_{iso} is complex, the latter can not be directly extracted. Therefore, the two following figures of merit are defined:

$$C'_{iso} = \frac{1}{-Im(\frac{1}{Y'_{iso}})\omega} \quad R'_{iso} = Re(\frac{1}{Y'_{iso}}) \quad (7)$$

where $Y'_{iso} = Y'_{44} + Y'_{41} + Y'_{42} + Y'_{43}$, according to Fig. 4.

Parameter C_{iso} is given by the value of C'_{iso} at low frequency, below 1 GHz in our case. This is due to the size of the deep n-well layer area, which leads to an large associated time constant $R_{iso} C_{iso}$, about 20 ns. Then, R_{iso} and R_{isoext} are extracted using an optimization procedure on the figure of merit R'_{iso}. A good starting value of R_{isoext} is given by the value at high frequency of R'_{iso}. Results of isolation parameters extraction are presented in Fig. 6.

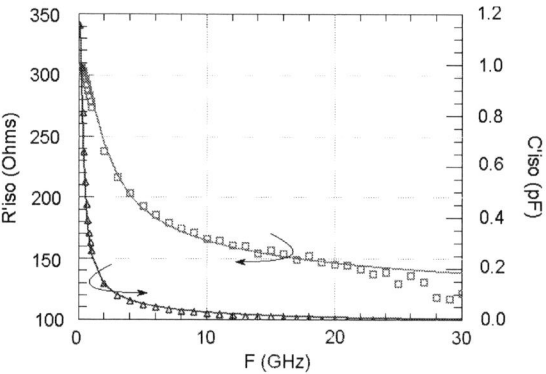

Fig. 6. Isolation network extraction: measured and simulated R'_{iso} and C'_{iso} for the transistor M1 versus the frequency.

Finally, the value of R_b is extracted from $Re(\frac{-1}{Y'_{14}})$ which gives the resistive path from the gate to the bulk. A summary of extracted substrate parameters for all transistors is given in TABLE I.

TABLE I
SUBSTRATE MODEL PARAMETERS FOR HIGH-K 28NM
BULK DEVICES

	R_{well}	R_{js}	R_{jd}	R_b	R_{iso}	R_{isoext}	C_{jd}	C_{js}	C_{iso}
	Ω						fF		
M1	25	174	211	1500	2504	98	26	28	1207
M2	27	160	210	290	1850	91	24	26	1207
M3	25	78	104	800	2324	55	70	74	1207

Once extrinsic elements are known, the intrinsic network may be extracted at each considered bias. Compared to standard 2-port configuration, actual values of parameters, especially capacitances C_{gs}, C_{gb} and channel conductance G_{ds} may be directly extracted without any hypothesis of symmetry or approximations (Fig. 7).

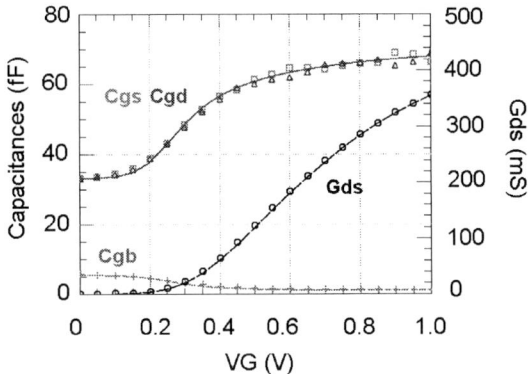

Fig. 7. Measured (symbols) and simulated (line) capacitances Cgs, Cgd, Cgb and channel conductance Gds of transistor M1 versus gate bias voltage VG, for VD=0V, f=1GHz.

IV. APPLICATION

As a direct application of the substrate parameters extraction, a MOS transistor with same geometry as M1 is measured in a conventional 2-port (common source) configuration and simulated with model of Fig. 2. Intrinsic and extrinsic parameters are extracted following traditional methodology. In addition, substrate model is determined by the previous 4-port study. Results on Mason gain (U) for several gate voltages are presented on Fig. 8.

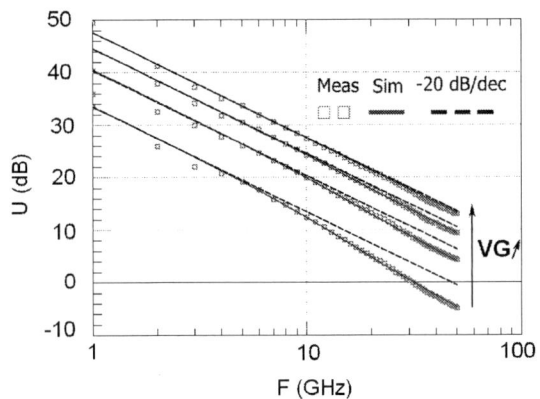

Fig. 8. Mason Gain (U) of transistor Lg=30nm, W=120μm, for VG from 0.1V to 0.7V, VD=1V.

At high VG, U presents a -20dB/decade slope and f_{max} may be thus extrapolated at any frequency. However, when gate bias decreases, substrate influence is not negligible anymore and U presents a higher and non-constant slope. Therefore, for weak and moderate inversion regimes, accurate substrate determination and modeling become crucial to correctly predict - and thus not overestimate - the device RF performances.

V. CONCLUSION

4-port measurements were shown to enable a simple and robust methodology to extract an equivalent schematic circuit for MOS device. As an illustration, the extraction of a cross-type substrate sub-network including distributed isolation was detailed. The method may be applied to extract any substrate network, especially in built-in compact model ones. Indeed, 4-port measurements allow a direct extraction of all model parameters with a single structure, at the expense of larger structure size. The effect of substrate coupling on f_{max} was also demonstrated.

REFERENCES

[1] (2011). ITRS, [Online]. Available: www.itrs.net.

[2] E. Vittoz, "Weak inversion for ultra low-power and very low-voltage circuits", *IEEE Asian Solid-State Circuit Conference*, pp. 127–132, Nov. 2009.

[3] A. Shameli, "Ultra-low power RFIC design using moderately inverted MOSFETs: an analytical/experimental study", *IEEE RFIC Symp.*, Jun. 2006.

[4] Y.-Z. Xiong, K. Kai, L Nan, and F. Lin, "Rf noise of 65-nm MOSFETs in the weak-to-moderate-inversion region", *IEEE EDL*, pp. 185–188, Feb. 2009.

[5] E. Bouhana, P. Scheer, S. Boret, D. Gloria, G. Dambrine, M. Minondo, and H. Jaouen, "Analyse and modeling of substrate impedance network in RF CMOS", *IEEE ICMTS*, pp. 65–70, 2006.

[6] I.-M. Kang and H. Shin, "Extraction method for cross-type substrate resistances of RF MOSFETs based on PSP model", *IEEE Electronics Letters*, vol. 46, May 2010.

[7] U. Mahalingam, S. Rustagi, and G. Samudra, "Direct extraction of substrate network parameters for RF MOSFET modeling using a simple test structure", *IEEE EDL*, vol. 27, pp. 130–132, 2006.

[8] J. Liu, L. Sun, Z. Yu, and M. Condon, "A new substrate model and parameter extraction method for DNW RF MOSFETs", *IEEE Int. Symp on Cicruits and Systems*, pp. 2478–2481, 2010.

[9] B. Dormieu, P. Scheer, C. Charbuillet, N. Kauffmann, and F. Danneville, "Mmw modeling of isolated mos substrate network through gate-bulk measurements", *IEEE RFIC Symp*, pp. 1–4, Jun. 2011.

[10] S.-D. Wu, G.-W. Hwang, and K.-H. Liao, "Modeling the substrate effect of rf mosfet's based on four-port measurement", *ARFTG Conference*, pp. 186–189, 2006.

[11] *2-Port and 4-Port PNA-X Network Analyzer N5247A Data Sheet and Technical Specifications*, Agilent Technologies, Feb. 2012.

[12] F. Arnaud *et al.*, "Competitive and cost effective high-k based 28nm cmos technology for low power applications", *IEEE Electronics Device Meeting*, pp. 1–4, Dec. 2009.

[13] *PSP 103.1 manual*, NXP Semiconductors, Apr. 2009.

[14] X. Jin, J.-J. Ou, Chen, C-H., W. Liu, M.-J. Deen, P.-R. Gray, and C. Hu, "An effective gate resistance model for cmos rf and noise modeling", *IEDM Tech. Digest*, pp. 961–964, 1998.

Variability Aware Cell Library Optimization for Reliable Sub-Threshold Operation

Tobias Gemmeke, Maryam Ashouei
Holst Center / imec
Eindhoven, The Netherlands
tobias.gemmeke@imec-nl.nl

Abstract—**Standard cell libraries are designed focusing on the best performance-area trade-off for a technology at nominal supply. Scaling supply voltages emphasizes the effects of systematic or random variation. We revisit existing approaches and present two new design points in standard CMOS that target variability hardened standard cells integrated into the digital design flow. They are optimized for dynamic and stand-by power, respectively. A speed-up from 1.4 to more than 3x is achieved on cell level. These gains are preserved in the example of an FIR filter while even improving in energy efficiency. The analysis and design has been performed in a low-power 40nm CMOS technology.**

Standard-cell Design; Nanometer Technologie; Near-threshold design; Sub-threshold Logic; Ultra-Low Power; Variability.

I. INTRODUCTION

Sub-threshold operation is an attractive approach for reducing the energy consumption in energy-constraint applications with limited performance requirements. The large improvement in energy efficiency comes at the cost of severe speed degradation and increased susceptibility to process variations. Today, most SoC designs in the sub-threshold regime rely on standard cell libraries that are optimized for performance-area trade-offs in the super-threshold region, which neglects energy savings potential a low voltage.

Although there is a limit on the minimum V_{DD} for which the circuit is still functional ($V_{DD,min}$), the voltage for the minimum energy per operation [1] often exceeds this limit as shown in Figure 1. The optimum supply voltage $V_{DD,opt}$ is reached when the energy savings due to voltage scaling are equal to the increase in leakage energy, which is governed by the two opposing trends: an exponential increase in cycle time and a super-linear decrease in leakage power. This point is dependent on multiple parameters including the activity factor, the logic depth, and the amount of memory in the design. With proper device sizing, $V_{DD,opt}$ can be lowered together with the minimum energy per operation.

Recent research has revisited the cell design optimizing for sub-threshold operation while addressing the performance, the variability, or both. In [2] and [3] a new cell library is proposed suitable for sub-threshold operation considering logical effort and reverse short channel effect, respectively. Standard cell design exploiting the inverse narrow width effect is proposed in [4]. Pass-gate-based cell design is presented in [5]. Schmitt-trigger logic is proposed in [6]. Most of these consider older processes featuring less process variability.

The design choices are severely impacted and sometimes even reversed when process variability is considered. For example, [7] showed that minimum sized devices are theoretically optimal for the sub-threshold region. The theory is invalidated taking device variability into account [8]. [9] presents a sizing methodology that considers threshold voltage variation in the threshold regime.

In this work, we propose specific design points for standard cells targeting two different use case scenarios. The design points combine the inverse narrow width effect and reverse short channel effect. The first scenario is a small always-on domain with very low activity, where the standby leakage power is dominant. The second scenario focuses on energy optimum point for wireless sensor nodes. It should be noted that the proposed design approach is targeting the threshold region and will not be optimum for the voltage ranges well above the threshold voltage. The two design points are compared with a commercial standard cell library in a low-power 40nm CMOS technology.

II. CELL OPTIMIZATION

A. Optimization Methodology

At supplies around the threshold voltage random variability plays a dominant role in circuit speed which can be captured with Monte-Carlo (MC) analysis. Here, 1400 points are used to quantify the effect of random variation around the typical process corner using foundry provided statistical models (unless otherwise noted: V_{DD}=0.3V, T=25C). Other process corners can be pushed to the typical corner by applying adaptive body biasing (ABB) [10].

Random variation of threshold voltages follows a Gaussian distribution. With the exponential dependency of the transistor current on the threshold voltage, this leads to a log-normal distribution of the on-current I_{on}. The relationship between I_{on} and circuit speed (in terms of propagation delay t_p) is shown in figure 2 (left) for the example of an inverter. Its reciprocal is following a linear trend as shown in figure 2 (right). Only for very fast t_p the effect of the input slope matters resulting in the drop off visible in the lower left corner. In the case of minimal

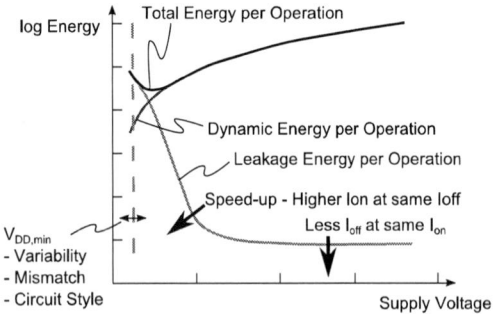

Figure 1. Improving on the energy optimal point.

978-1-4673-1707-8/12 $31.00 © 2012 IEEE

cycle time only slow samples matter, which can be modeled using a shifted hyperbola with fitting constants $t_{p,0}$ and m:

$$t_p = t_{p,0} + m \cdot 1/I_{on}. \quad (1)$$

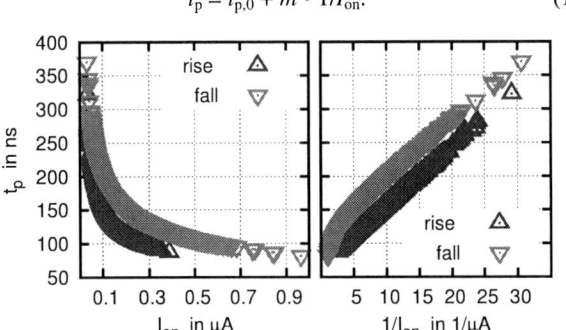

Figure 2. Mismatch distribution of propagation delay t_p vs. on-curents I_{on}.

Simulation results of various devices and input transition times confirm the model. Therefore, the following analysis considers I_{on}, only, as opposed to t_p. This reduces simulation time significantly as I_{on} is determined using basic DC analysis.

B. Gate Capacitance

Transistor sizing changes gate capacitance and therefore directly impacts dynamic energy. Figure 3 shows the ratio c of gate capacitance C_{gate} per gate area A_{gate} for NMOS and PMOS, and various device geometries. This ratio is normalized in the figure to the value at L_{min} and W_{min}. Both constants are varied: $L=[L_{min}=40\text{nm}, \; 2xL_{min}]$ and $W=[W_{min}=120\text{nm}, \; W_{min}+200\text{nm}]$.

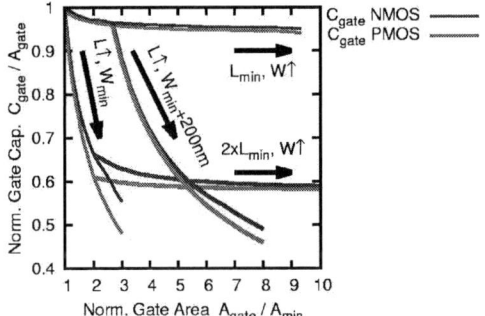

Figure 3. Normalized ratio of gate capacitance to gate area vs. various channel widths and lengths ($A_{gate}=L\cdot W$, $A_{min}=L_{min}\cdot W_{min}$).

The ratio c remains almost constant with scaled W. Contrarily, increases in L reduce the ratio due to dominant fringe capacitances [11]. The following optimization considers gate capacitance and device area. Hence, I_{on} is analyzed along with the ratio of I_{on} to C_{gate}.

C. Dependency of I_{on} on Parmeters L, W, and Finger Count

The impact of various nanometer effects on I_{on} is quantified using dedicated analysis for random dopant fluctuation, inverse narrow-width effect (INWE), and short-channel effect (SCE).

1) Random Dopant Fluctuation

According to [12], the amount of variation of the threshold voltage is inversely proportional to the area of the device:

$$\sigma_{th} \propto 1/\sqrt{WL} . \quad (2)$$

In the case of a Gaussian distribution, the term '$\mu \pm x \cdot \sigma$' is commonly used to refer to an x-sigma confidence interval with mean μ and standard deviation $x \cdot \sigma$. It contains a specific percentage of samples, e.g. $\mu \pm 3\sigma$ covers 99.7%. At supplies around the threshold voltage, the on-current is log-normally distribution as $I_{on} \propto e^{VDD-Vth}$. For the variation of such distribution the above relationship does not hold. Hence, the parameters μ and σ of a Gaussian distribution are fitted to logarithm of the on-currents, $\log I_{on}$. They define the boundaries of the confidence interval of I_{on}: $I_{\mu \pm 3\sigma} = 10^{\mu \pm 3\sigma}$ and the relative spread of currents: $I_{\mu \pm 3\sigma}/I_{\mu}$.

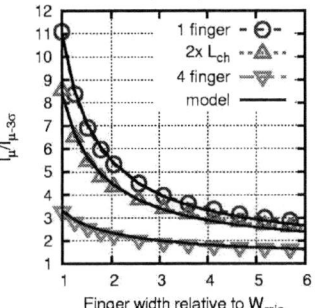

Figure 4. Inverse of the relative current spread vs. scaled gate area.

Figure 4 shows the inverse of this spread vs. the relative finger width, while scaling W, L, and the number of fingers. The lower current bound at μ-3σ corresponds to the slow side of the delay distribution (cf. eq. (1), fig. 2). MC Simulation results (symbols) are overlaid with the trend (model) based on scaled σ according to (2). Equivalent reductions in the spread are achieved by increasing the finger width, the finger count, or the channel length L. Please note that a change in variability of the threshold voltage leads to an above linear improvement in I_{on} due to its log-normal dependency on the threshold voltage. For example, by going from a 1-finger to 4 of equal finger size, the V_{th} variability improves by sqrt(4)=2-times, while the current spread improves by more than 3.7-times.

2) Inverse Narrow-Width Effect (INWE)

The typical corner on-current I_{TT} per unit width is shown for a PMOS in figure 5 as triangles. For narrow channels the current per unit width is significantly higher. This suggests that many narrow channels are more efficient than a single wide device. Small fingers feature high variability, which can equivalently be compensated either by up-sizing or changing the finger count as long as the effective area is constant.

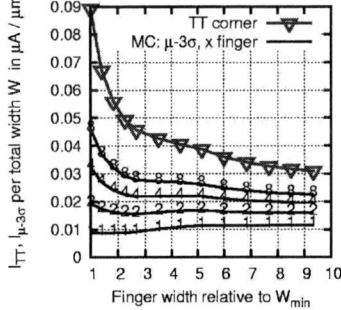

Figure 5. Current of PMOS devices at typical process corner with and without mismatch as fct. of device width and finger count.

The combined effect is indicated in figure 5 as black lines, which are based on the same MC simulation as figure 4. Here, the current is not normalized to its average. As expected, the current per total width is improved due the reduction of variability with increasing finger count.

3) Reverse Short Channel Effect (RSCE)

In contrast to super-threshold design, increases in channel length are considered not only for leakage reduction but also to boost performance at scaled supply voltages. Figure 6 (left) shows the current $I_{\mu-3\sigma}$ vs. sizing of the channel length for two different finger widths (W_{min}, W_{min}+200nm). For small increases in channel length the current is significantly reduced (despite the reduction in variability). Interestingly, the current recovers by further stretching the channel length. In small finger count configurations, $I_{\mu-3\sigma}$ even exceeds the level at L_{min}. As before, multi fingered devices are better. For example, the blue line of the 4 fingered device always remains above the 2 fingered device of increased channel width, although the latter has a larger total gate area.

The maximal current is important when driving high loads. Especially if the gate drives small capacitances like local interconnects, the current per gate capacitance becomes relevant as a higher ratio results in better performance at the same energy level. This ratio is shown on the right-hand-side of figure 6. Starting at L_{min} it features an even stronger drop-off. However, around 2.5xL_{min} the slope reaches a side maximum or becomes shallow before falling off again. In the following, two design points are selected:

1. Minimum channel length devices L_{min} that exploit the short channel effect, and

2. Long channel devices L_{long}=100nm that reduce leakage current at the cost of increased dynamic energy conversion.

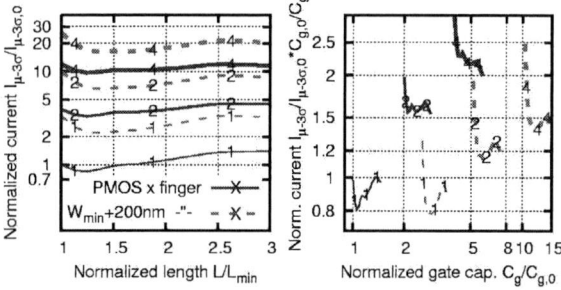

Figure 6. Normalized on current $I_{\mu\pm3\sigma}$ (normalized to $I_{\mu\pm3\sigma,0}$ featuring L_{min} and 1finger) vs. normalized channel length L (left), and normalized on current $I_{\mu\pm3\sigma}$ (normalized to gate capacitance) vs. increase in gate capacitance (right).

Finally, a corresponding analysis with respect to the channel width is performed. The results are shown in figure 7. Channel width is scaled for the 2 previously defined design points. For small number of fingers and small width, the output current of the long channel devices exceeds the L_{min} values. However, the output current per gate capacitance is never improved. The benefit of sizing widths appears to be very limited if compared to the increases in drive strength achieved by adding a finger.

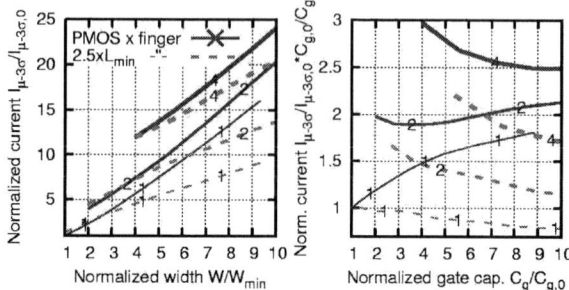

Figure 7. Increase in on-current vs. total channel width W (left) and gate capacitance C_{gate} (right).

The previously defined design points are therefore combined with minimum width devices, only. Drive strength and P-to-N ratio are adapted by the number of fingers used.

III. CELL DESIGN

Design for lowest energy at a given voltage relies on small device geometries. The drive strength used for local interconnects and logic is therefore selected to be similar to the smallest drive strength of the standard cell library. A reasonable P-to-N ratio is achieved using 2 NMOS and 2 PMOS devices in case of the L_{min} design point. For the L_{long} design, 3 PMOS fingers are combined with 2 NMOS fingers. On one side, the L_{min} design is targeted for designs of high switching activity when increases in gate capacitance have a considerable impact on energy conversion. On the other side, the L_{long} cells are suited for always-on domains of low switching activity when leakage power is dominating.

A single finger device is not used, as the impact of the predicted variability is considered excessive for minimal channel width. To further reduce variability, a regular design style is adopted, i.e. poly fingers are all of the same size and placed at identical pitch. Also, there is no metal 1 routing of the transistor gates to limit its impact on threshold voltage shift [13]. Additionally, diffusion sizing and spacing is kept as uniform as possible. Figure 8 shows inverters as examples for the two design points (b) L_{min} and (c) L_{long}. For comparison, it also features an inverter of a standard cell (a).

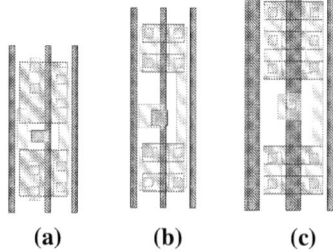

(a)　　　(b)　　　(c)

Figure 8. Standard cell inverter design: (a) super-threshold, (b) low-voltage, active energy oriented, and (c) low-voltage, static energy oriented.

Due to the regularity requirements and the use of fingers, the footprint of the cells increases. Based on a sample library the cell footprint overhead can be considerable, especially in the case of the L_{long} design point which shows a 2x increase (cf. table 1). Hence, this dedicated design point has to be considered carefully to prevent significant impact on the total chip area.

Table 1. Scaling of footprint as function of the design point.

	Reference	L_{min}	L_{long}
Area Ratio	1x	1.3x	2x

Finally, sequential elements are designed with equal drive in the feed-forward and -back paths to avoid issues at low voltages.

IV. RESULTS

As first step, the 2 design points are compared to a standard cell library using an inverter chain with different wire loads and extracted netlists. To highlight the benefits of the fingered approach, sub-types were created, that have the area between the fingers filled with diffusion (cf. figure 9, lower right) creating a much wider device of identical footprint.

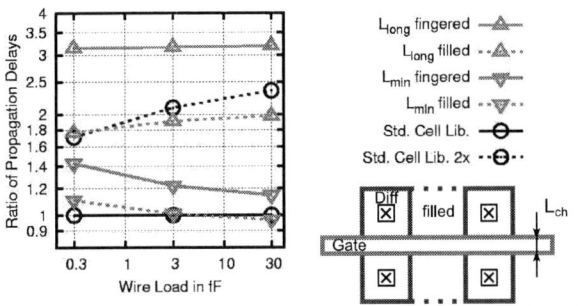

Figure 9. Propagation delay ratio $t_{p,std.\ cell}/t_{p,other}$ (0.4V, TT corner + mismatch: $\mu+3\sigma$, 25C) for various wire loads.

In figure 9 the ratio of propagation delays of the standard cell design to the alternative design is shown for different wire loads. The new design points achieve significant improvements in propagation delays even at a moderately scaled voltage of 0.4V (V_{DD} nominal: 1.1V) with a 1.3x for L_{min} and more than 3x for L_{min}. The filled variants show worse performance highlighting the benefit of fingered devices. Also shown is the performance of a standard cell inverter of identical footprint as the L_{long} cell. However, it does not reach the same performance level and exhibits much higher leakage due to the shorter channel length.

Finally, an FIR filter is used with the different cell designs. As the focus is not on FIR optimization, the already highly optimized, energy efficient architecture of [14] is used. The timing critical paths in the core logic are analyzed using MC simulations (1400 points, TT corner + mismatch, 0C). The leakage power is computed at 80C and fast process corner. The results are plotted in figure 10.

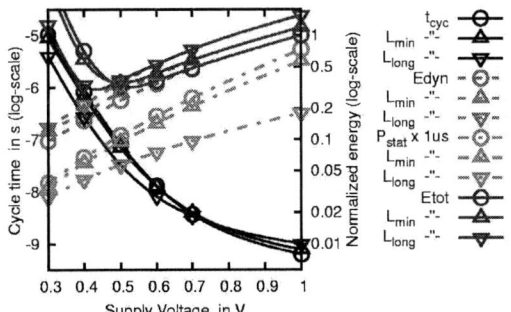

Figure 10. FIR performance using new design points. Energy is normalized to E_{tot} of standard cell design.

Despite the reduction in gate capacitance in the case of the L_{min} design, the dynamic power is increased. This is partially due to the larger cell footprint pushing the wire capacitances. On the other side leakage power is reduced. Apparently, the standby savings are far more pronounced for the L_{long} design point. The energy minimal point is a function of the switching activity. In this example it lays between 0.4 and 0.5V with a 15% switching activity. Overall, the L_{long} design point is the most appealing, especially for always on domains. Although, its minimal energy is only 5% below the standard cell library, it features a 3x faster cycle time enabling a significant leap in performance for ultra low power bio-medical applications.

V. CONCLUSION

New design points for standard cells operating around the energy optimal point have been developed based on a detailed analysis of variability, inverse narrow-width, and reverse short channel effects. This was combined with layout techniques to reduce systematic variability in nanometer technologies. The resulting cell library outperforms a standard library by more than 3x. The benefit of the cells has been confirmed in a FIR design combining higher energy efficiency with the 3x speed-up. The analysis and design was performed in a low-power 40nm CMOS technology using Monte-Carlo simulation together with extracted netlists.

REFERENCES

[1] S. Hanson, et al., "Ultra-Low Voltage Minimum-Energy CMOS," IBM Journal of Research and Development, Volume 50 Issue 4/5, July 2006.

[2] J. Keane, et al., "Subthreshold Logical Effort: A Systematic Framework for Optimal Subthreshold Device Sizing," In Proeedings of 43rd ACM/IEEE Design Automation Conference, pp. 425-428, 2006.

[3] T. H. Kim, et al., "Utilizing Reverse Short-Channel Effect for Optimal Subthreshold Circuit Design," IEEE Trans. VLSI Syst. 15(7), pp. 821-829, 2007.

[4] J. Zhou, et al., "A 40 nm inverse-narrow-width-effect-aware sub-threshold standard cell library," DAC, pp. 441-446. 2011.

[5] N. Reynders and W. Dehaene, "A 190mV supply, 10MHz, 90nm CMOS, Pipelined Sub-Threshold Adder using Variation-Resilient Circuit Techniques," ASSCC, pp. 113-116, 2011.

[6] N. Lotze, et al. ,"A 62mV 0.13μm CMOS standard-cell-based design technique using schmitt-trigger logic," ISSCC, pp. 340-342, 2011.

[7] B. Calhoun and A. Chandrakasan, "Characterizing and Modeling Minimum Energy Operation for Subthreshold Circuits," ISLPED 2004.

[8] J. Kwong and A. Chandrakasan, "Variation-Driven Device Sizing for Minimum Energy Sub-Threshold Circuit," ISLPED 2006: 8-13.

[9] Bo Liu, et al., "Process Variation Reduction for CMOS Logic Operating at Sub-threshold Supply Voltage", DSD 2011: 135-139.

[10] J.W. Tschanz, et al. , "Adaptive body bias for reducing impacts of die-to-die and within-die parameter variations on microprocessor frequency and leakage," IEEE Journal of Solid-State Circuits, vol.37, no.11, pp. 1396- 1402, Nov 2002.

[11] H. B. Bakoglu, "Circuits, Interconnections, and Packaging for VLSI," Reading, Addison-Wesley, 1990.

[12] M. Pelgrom, et al. "Matching properties of MOS transistors," European Solid-State Circuits Conference (ESSCIRC), pp.327-330, Sept. 1988.

[13] H. Tuinhout et al., "Effects of Metal Coverage on MOSFET Matching," IEDM, pp. 735-8, 1996.

[14] T. Gemmeke, M. Gansen, H.J. Stockmanns, T.G. Noll, "Design optimization of low-power high-performance DSP building blocks," IEEE Journal of Solid-State Circuits, vol.39, no.7, pp. 1131-1139, 2004.

Advancements on Reliability-Aware Analog Circuit Design

B. Ardouin
XMOD Technologies
Bordeaux, France
ardouin@xmodtech.com

J.-Y. Dupuy , J. Godin, V. Nodjiadjim, M. Riet
III-V Lab, joint lab between Bell Labs, Thales and CEA/Leti,
Marcoussis, France

F. Marc, G. A. Koné, S. Ghosh, B. Grandchamp, C. Maneux
Laboratoire IMS, Université de Bordeaux
France

Abstract—**This paper presents a new physics-based method for reliability prediction and modeling of Integrated Circuits (ICs). By implementing transistor degradation mechanisms via differential equations in the transistor compact model, the aging of the circuit can be simulated over (accelerated) time under real conditions. Actually, each transistor in the circuit integrates the voltage, current and temperature stress it suffers which results in (slowly) varying model parameters over time. Due to its straightforward implementation in commercial Computer Aided Design (CAD) flows, this method allows designers creating reliability-aware circuit architectures at an early stage of the design procedure, well before real circuits are actually fabricated. Application examples and results are presented for an InP/InGaAs DHBT process, but the universality of the method makes it suitable also for silicon based technologies such as CMOS and (SiGe) BiCMOS.**

I. INTRODUCTION

Today, technology decisions for IC fabrication are driven by both performance and reliability criterion. An accurate example could be find in the field of high speed optical communications systems working above 100 Gbit/s, especially when talking about submarine cable systems, it is easily understood that reliability is of utmost importance, due to maintenance difficulties. In the work towards improved robustness, engineers apply intensive (post IC fabrication) qualification procedures in order to predict and possibly guarantee operational reliability. Nevertheless, fabricated ICs failing qualification need design re-spin involving additional costs and delays; circuit reliability is improved from one generation to the other only at the cost of time-consuming feedback loops between reliability and design teams. Therefore, beside process reliability improvement done by technologists, it is highly desirable that designers create reliability aware circuit architectures such as gain or offset compensation loops providing transistor degradation compensation over circuit lifetime. In order to do so, modeling and simulation of transistors degradations is mandatory since designers can not only rely on their experience and intuition.

A few approaches have been proposed for circuit reliability simulation. Most of them mimics the first reliability simulation flows [1] where an electric simulator use standard transistors models to simulate the circuit, then the "aging manager" computes the evolutions of transistors parameters due to degradation and start another simulation with aged model transistors. In this flow, the aging manager is a separate software pre-processing and post-processing the electric simulation. It computes a cumulated stress as a function of voltages and currents in the circuit then extrapolates this stress to the final age and computes the corresponding aged model transistors parameters using a table obtained by aging experiment. Some later improvements of this methodology have been proposed including (i) multi-step aging simulation to reduce extrapolation errors, (ii) use of formulas instead of table for aged parameters computation, (iii) modification of electric model for degradation specific conduction phenomenon [2] or (iv) introduction of the aging manager in the simulation tool using specific commands. Moreover, circuit reliability simulation seems only have been performed on MOS silicon technologies.

In this paper, we present a physics based reliability simulation methodology. This methodology can be applied without a specific tool for aging and easily integrated in any design flow. It takes into account reliability by computing degradation laws directly in compact models used in all Berkeley-SPICE derivative simulators and includes a "virtual degradation acceleration mechanism" to simulate years of aging and picoseconds of electric behavior [3]. This methodology has been developed and applied to an InP/InGaAs DHBT process of III-V lab featuring an f_T and f_{MAX} of approximately 300 GHz. The principle can be summarized as follows: First the transistors are submitted to accelerated aging and degradation mechanisms and are

978-1-4673-1707-8/12 $31.00 © 2012 IEEE

analyzed using TCAD (Technology Computer Aided Design) simulator. Next, degradation laws are derived and implemented in the HiCuM [4] compact model. Nominal and aging model parameters are then extracted based on electrical measurements of the transistors before and after accelerated aging. Results are finally verified against post-aging measurements of a high-speed trans-impedance amplifier reported in [5].

It has to be noted that the methodology principles presented in this paper are not specific to III-V technologies nor to HBT transistors and can be applied to silicon-based technologies such as CMOS and (SiGe) BiCMOS provided the failure mechanisms to be modeled are governed by progressive degradation laws.

Section II of this paper presents the complete procedure which has been applied to derive the aging model. Part II-A presents a summary of the technology, part II-B presents the bias stress experiments, part II-C details the derivation of the aging laws, part II-D presents their verilogA implementation in HiCuM and part II-E shows the results of the model parameter extraction (before aging). Section III presents the validation of the aging model at circuit level for the case study of a transimpedance amplifier. Finally, Section IV gives an example of reliability aware circuit optimization. Conclusion and perspectives are drawn in section V.

II. TRANSISTOR AGING MODEL

A. Summary of III-V Lab InP DHBT technology

The Technology under investigation in this paper is a process under development as part of the ROBUST project [6], in which three different generations of InP/InGaAs DHBT technology have been realized (G0, G1 and G2). All results presented in this paper are related to generation 1 (G1).

HBT heterostructures are grown by Gas Source MBE on a 3-inch semi-insulated substrate. The structure includes a 130 nm thick composite collector with an InGaAs spacer, a 20 nm highly doped InP region and a low-doped InP layer (2.5×10^{16} cm^{-3}). A thin sub-collector InGaAs layer is also used for low thermal resistance. The base (~ 30nm) is compositionally graded and highly doped (C-doping $\sim 8 \times 19$ cm^{-3}) to minimize the base sheet resistance.

Submicron devices are fabricated using an all wet-etched self-aligned triple mesa technology. The effective width of the emitter contact is 0.5 μm and the emitter length varies from 5 to 10 μm. The base contact extends 0.3 μm on each side of the emitter and includes a plug for the connection. Both contacts are defined by electron-beam lithography because of the high alignment accuracy required for this technology. The other steps rely on optical lithography. Figure 1 shows a 0.5x5μm² effective emitter size DHBT microphotograph. After planarization with polyimide, the collector contacts are accessed through via holes while emitter and base contacts are opened by etching to interconnect the devices. The process also includes film resistors and MIM capacitors, as well as three interconnection metal levels for the circuits. Fabricated HBTs demonstrate a DC current gain above 25 and a common emitter breakdown voltage of 5 V. Maximum frequency performances f_T and f_{MAX} are around 300 GHz.

Figure 1. 0.5X5 μm² effective emitter size DHBT

B. Accelerated Aging Tests

Accelerated Aging tests (AAT) have been performed on three HBT geometries with effective emitter dimensions of 0.5x5 μm², 0.5x7 μm² and 0.5x10μm² (respectively labeled T5, T7 and T10). A total of five stress conditions (named as P1, P2, P3, P4 and P2') have been used. The collector current density was kept fixed at 400 kA/cm² for the first four stress conditions. The collector-emitter voltage (V_{CE}) was 1.5, 2, 2.5 and 2.7 V for P1, P2, P3 and P4 respectively. The corresponding junction temperatures T_j are 80, 92, 106 and 112°C where the ambient temperature was maintained at 30 °C. In order to separately observe the junction temperature and the current density impact on the HBT degradation mechanisms, we chose J_C=610 kA/cm² and V_{CE}=1.3 V for P2' condition so that P2' has the same junction temperature as P2. All these stress conditions are chosen such that all the bias points are in the Safe Operating Area (SOA) of the transistor. These current densities and emitter-collector voltages are chosen before the onset of avalanche. During aging tests, DC characterization has been performed on HBTs of the three geometries at 1, 2, 4, 8, 16, 24, 72, 144, 250, 500, 750, 1000, 1250, 1500, 2000 and 3000 hours. As it can be observed in Figure 2, after 2500 h hours of temperature accelerated test, device degradations are visible. At medium current densities (i.e. before the onset of high current effects), a moderate increase of the collector current and an important increase of the base current (resulting in a global current gain degradation) are observed. Moreover at high current densities, a moderate collector current decrease is observed, which has been attributed at first order to an increase of the emitter contact resistance. Although, an aging model has been derived also for the emitter resistance, it has been de-activated in this study for the sake of simplicity.

C. Derivation of Aging Laws

In order to get a deep insight into the physical failure mechanisms involved in the HBT aging process and to validate the proposed methodology within the frame of the ROBUST research project, 2D simulations have been performed using TCAD Sdevice tools. Nevertheless, from a practical point of view, the experimental results (i.e. the electrical characteristics) resulting of the applied bias stress are the only information required to formulate the degradation laws of the compact model parameters. This means that the used methodology can be applied in a fairly straightforward manner in an industrial environment.

978-1-4673-1707-8/12 $31.00 © 2012 IEEE

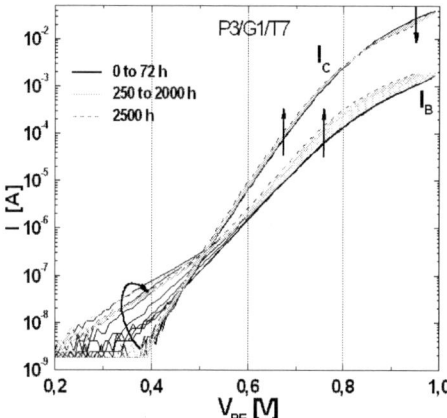

Figure 2. IC and IB degradation due to electrical and thermal stress for G1 T7 HBT under P3 bias condition

The TCAD simulations work realized within this project is briefly described below. Figure 3 presents the simulated 2D InP/InGaAs DHBT half structure. To achieve the required accuracy, TCAD simulations are performed using the hydrodynamic model equations using the Stratton formulation [7][8]. Before aging, the current gain is mainly limited by peripheral recombination simulated by the presence of traps on the emitter sidewall and on the extrinsic base surface. The trap locations are shown in Figure 3. These traps are defined using the same methodologies as [8][9][10][11] and have been calibrated on measurements before aging. Their origins are probably related to unpassivated level of missing atoms at the surface of semiconductors.

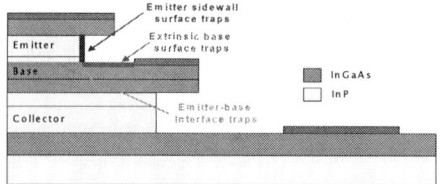

Figure 3 Simulated 2D structure with surface traps' locations.

The observed increase of both I_C and I_B (0.6V≤V_{BE}≤1V) due to the bias aging tests can be explained through the introduction of donor traps on the entire surface of the emitter-base junction (Figure 3). This new trap level is located at $E_T-E_V=0.83eV$ above the valence band energy. This level acts as recombination center in the Space Charge Region (SCR) of the E-B junction and therefore increases the base recombination current. But, it also introduces positive charges, changing the pinning of the conduction band at the E-B junction modifying the electron injection across the junction. The physical origin of this level can be due to the appearance of exotic impurity in emitter-base SCR. For example, very close donor level has been observed in InP originated from gold impurity [12]. Figure 4 represents the evolution of the trap density (determined from TCAD simulations [13]) as a linear function of stress time associated

with the considered trap level for all tested T7 devices sorted by their bias aging conditions. These TCAD simulation results correlated with the electrical measurements provide some guidance in the choice of the parameters for the modeling of aged devices.

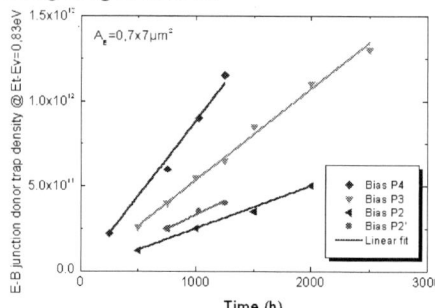

Figure 4. Evolution of trap density with time for various aging conditions (from TCAD simulations)

All modern bipolar transistor models describe the (intrinsic) base current i_{jBEi} and the transfer (collector) current i_{Tf} with equations of the form:

$$i_{jBEi} = I_{BEiS}\left[\exp\left(\frac{v_{B'E'}}{m_{BEi}V_T}\right)-1\right] + I_{REiS}\left[\exp\left(\frac{v_{B'E'}}{m_{REi}V_T}\right)-1\right] \quad (1)$$

$$i_{Tf} = \frac{I_S}{Q_{p,T}/Q_{p0}}\left[\exp\left(\frac{v_{B'E'}}{V_T}\right)-\exp\left(\frac{v_{B'C'}}{V_T}\right)\right] \quad (2)$$

In equation (1), I_{BEis} and I_{REis} are respectively the ideal and non-ideal intrinsic base saturation current parameters, m_{BEi} and m_{REi} are the ideality factors of the ideal and non-ideal part respectively, and V_T is the thermal voltage. In equation (2), I_S is the collector saturation current, $Q_{p,T}$ and Q_{p0} are respectively the weighted and zero bias hole charge in the transistor. $v_{B'E'}$ and $v_{B'C'}$ are the intrinsic B-E and B-C voltages. Here the HiCuM l2v2.30 [4] formalism has been used for consistency with the compact model we have selected in this work. Equations (1) and (2) have been used to model the transistor characteristics before and after aging and the I_{BEis} and I_S parameters have been extracted from all characteristics measured during the aging process. Their linear evolution with stress time (not shown here) is found to be very similar to that of the trap density determined from TCAD simulations, suggesting that (i) I_{BEis} and I_S follow the same law than the trap density and (ii) I_{BEis} and I_S can be advantageously taken as variables of the aging model for the implementation into a compact model. Actually, the trap density can be modeled using the physical description of a generation mechanism as described in equation (3):

$$\frac{dN_{trap}(t)}{dt} = G(T, J_C) \quad (3)$$

Where t is the time, T is the absolute temperature and J_C is the collector current density. According to the experimental results we can consider that I_{BEis} and I_S are linear functions of $N_{trap}(t)$. Therefore, we can define the constant increasing rates A_{IBEIS} and A_{IS} for I_{BEis} and I_S respectively. This gives the degradations laws for I_{BEis} and I_S shown in equation (4 which can be easily implemented in a compact model:

$$\frac{dI_{BEiS}(t)}{dt} = A_{IBEIS}(T, J_C) \quad , \qquad \frac{dI_S(t)}{dt} = A_{IS}(T, J_C) \qquad (4)$$

Figure 5 presents the extracted values of AIS and AIBEIS (normalized) as a function of 1/T for T7 and T10 at different stress conditions (P2, P3, P4). It has to be noted that the data suffer some scattering, which comes from experimental conditions of the aging test (due to the stress test bench intermediate characterization capabilities). Nevertheless, from a linear regression performed on the decimal logarithm of the data of Figure 5, it can be determined that both A_{IS} and A_{IBEIS} follow an Arrhenius law.

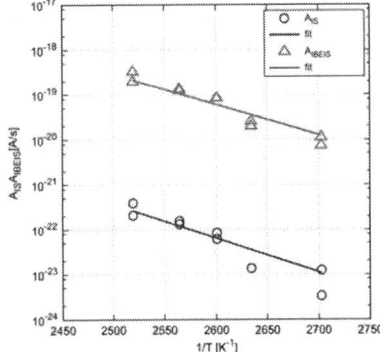

Figure 5. Normalized extracted values of A_{IS} and A_{IBEIS} as a function of 1000/T for T7 and T10 at different stress conditions (P2, P3, P4) at same J_C

Taking into account the dependence on the collector current density J_C (observed for the stress condition P'2, one can express A_{IS} and A_{IBEIS} by:

$$A_{IBEIS}(T, J_C) = B_{IBEIS}\left[\exp\left(\frac{-E_{IBEIS}}{k_B T}\right)\right] J_c^{\alpha_1} \qquad (5)$$

$$A_{IS}(T, J_C) = B_{IS}\left[\exp\left(\frac{-E_{IS}}{k_B T}\right)\right] J_c^{\alpha_2} \qquad (6)$$

The extracted values for E_{IBEIS} and E_{IS} are 1.34 eV and 1.5 eV respectively (note these values are related to the trap generation mechanism and do not represent the trap energy level).

D. VerilogA Implemetation of Aging laws in HiCUM L2v2.30

The HiCuM L2v2.30 model has been selected due to its ability to provide an accurate description of the InP/InGaAs DHBTs as a function of bias, frequency and temperature. The self-heating effect modeling is also crucial due to the dependence of the constant increasing rate parameters on junction temperature. Moreover, the version 2.30 of the HiCuM model introduces bias-dependent hole charge

weighting factors which are extremely important to accurately model medium current densities transconductance degradation observed in HBTs with aggressive bandgap engineering, such as those developed under the ROBUST project (although this formulation was initially introduced for SiGeC HBTs). It has also the advantage of being an internationally standardized model via the Compact Modeling Council (CMC) label, and is therefore available in all commercial simulators. A major advantage is also availability of the VerilogA code of the model.

From a practical point of view, implementing the differential forms of the aging laws in the VerilogA code of HiCuM (equation (4)) is fairly straightforward and consists in the introduction of fictive transistor nodes (in a way similar to the implementation of self-heating). Unfortunately, such implementation would result in unacceptable simulations times, since one would need to simulate several thousand hours to observe circuit degradations resulting in years of CPU effective time. The solution consists in introducing an accelerating time scale factor (ATSF) in the equations. One needs to pay attention though to keep the degradation laws time constants large enough with respect to the circuit's time constants. Similarly, it is also beneficial to scale in a similar manner the thermal (self-heating) time constant. A good rule of thumb is to keep the thermal time constant 100 times lower than aging laws time constants. Since degradation laws of the compact model are only activated for transient simulations, we have also implemented an automatic procedure (based on a Perl script) to collect the aging parameters of all transistors after an aging simulation and to re-introduce them in the circuit netlist for subsequent analysis of the aged circuit. With this procedure, one can run quasi-static analysis (such as AC, S-Parameters or harmonic balance simulations) of the circuit at any time of its life-cycle.

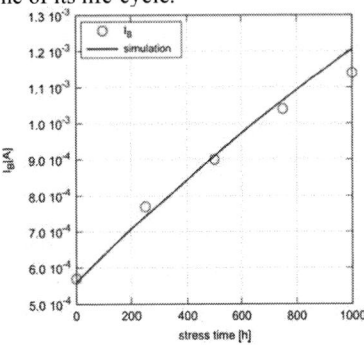

Figure 6. Comparison between measurements and simulations of the base current monitoring during T7 bias stress

During the applied bias stress, the transistor base current and base-emitter voltage have been monitored. The bias stress conditions can be reproduced in simulation in order to verify the degradations at the transistor level. For this, the parameter ATSF has been set to 36 10^{12} in order to simulate 1000 hours of stress in 100 ns of transient simulation time. Additionally, a special circuit is used in simulation to reproduce the forced collector current and forced collector-emitter voltage conditions applied by the test bench. The obtained results are shown in Figure 6 for the transistor T7 under bias condition P4. Good agreement is obtained between the measured and simulated base current, thus validating the aging model at the transistor level.

E. Parameter Extraction Before Aging

The HiCuM L2v2.30 compact model parameters have been extracted for the three transistor geometries (T5, T7 and T10) from on-wafer measurements performed before aging. DC and S-parameters measurements (up to 50 GHz) have been realized at 6 different temperatures and de-embedded from the pads parasitics. Details on the applied parameter extraction procedures can be found in [14] and [15].As it can be seen from Figure 7 the extracted model provides a very accurate description of the measured data, which highlights the ability of the HiCuM model to describe advanced III-V DHBT technology. It is worth noting that the self-heating effect is well modeled, which is very important since the transistor junction temperature (T_j) is of utmost importance for our aging model. Also, a good description of the S parameters of f_{MAX}, of the Mason's gain and of the unilateral gain can only be obtained when the base resistance value and junction capacitances partitioning across the base resistance is correctly determined. Additionally, simple equations have been introduced in the model cards (using the simulator macro language) in order to reproduce the process variations observed from one wafer to another. The first order parameters can be skewed to reproduce the transistor measurements of each produced wafer. This feature is useful when validating the model on circuit measurements coming from different wafers than the one used to build the model. Actually, for a technology being at the initial phase of its development (like the G1 generation investigated here), process variations can be significant enough to induce erroneous conclusions.

III. CASE STUDY WITH A TRANSIMPEDANCE AMPLIFIER

The circuit investigated to validate the methodology at the circuit level is a high-speed trans-impedance amplifier typically used in photo-receivers for optical communication systems. As reported in [5], it is composed of 24 transistors and consumes around 0.63 W. Its role is to amplify and convert a single-ended input photocurrent from a photodiode into a differential output voltage with 100-Ω differential impedance.

The circuit complete schematic is shown in Figure 8. The first stage of the circuit consists in a single-ended common-emitter amplifier with shunt-shunt resistive feedback through level shifters composed of diodes and an emitter follower. It is followed by a differential amplifier, the complementary input of which is provided by a reference voltage, generated by an internal automatic offset compensation loop. Within the functional range of the loop, all DC variations of the input signal or from inside the circuit should be cancelled.

Two versions of the circuit have been investigated which are referenced as TLIA and TLIA HF. The first one is a high gain version and the latter is lower gain version having a higher cutoff frequency. The S parameters of the two circuits have been measured up to 65 GHz before and after applying 1008 hours of stress at 70°C. These measurements are compared to the reliability-aware models (with model parameters re-centered to transistor measurements coming from the circuit wafers). First a transient simulation is performed (stop time of 100 ns) with an accelerated time scale factor of 36 10^{12} corresponding to 1000 hours. Note that the external capacitance value used in the offset compensation loop has to be adjusted in order to obtain a time constant compatible to a 100 ns simulation. Some examples of resulting transient output voltages are shown in Figure 10. Next the aging parameters are collected at the new model output nodes for each transistor instance and re-inserted for subsequent simulations. Quasi-static simulations (such as AC, S-parameters or harmonic balance) can now be run with model parameters reflecting aged transistors behavior. Figure 9 presents measurement and simulation comparisons for the TLIA and TLIA HF before and after 1008 h of bias stress at 70°C. As it can be seen the aged model is slightly pessimistic but provides fairly accurate predictions of the circuit degradations, given experimental uncertainty and statistical variations of a process under development.

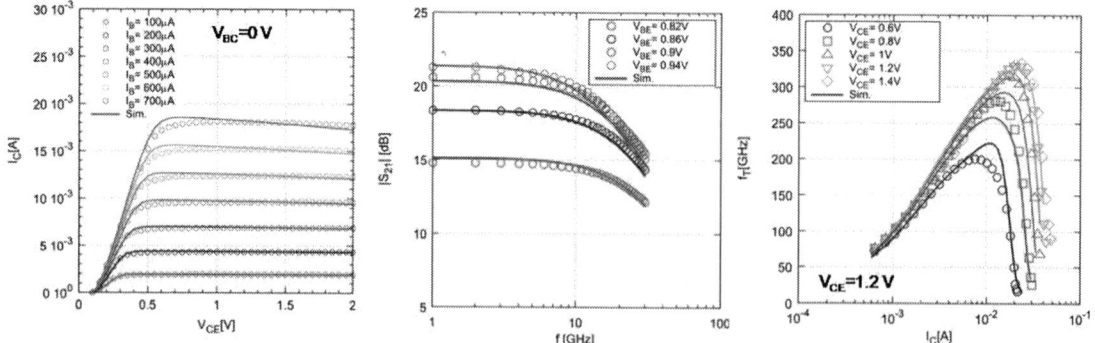

Figure 7. Comparison between measurements and simulations with the HiCuM model for T7. From left to right: output chracteristic, magnitude of the S_{21} parameter and transist frequency

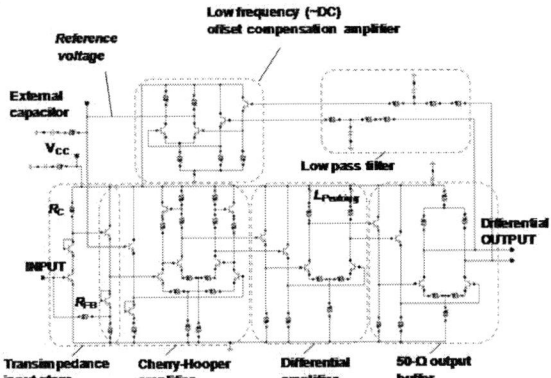

Figure 8. Transimpedance amplifier (TIA) schematic

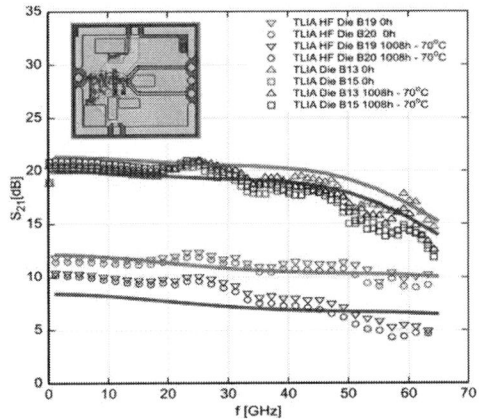

Figure 9. Comparison of measurements and simulations of $|S_{21}|$ before and after 1008 hours of stress at 70°C for the TLIA and TLIA HF.
Lines : corresponding simulation results.
Inset : TLIA top view

IV. RELIABILITY-AWARE CIRCUIT OPTIMIZATION

Based on the aging model presented in previous sections, a study has been carried out on the trans-impedance amplifier presented in section III.

It is well known in all differential-pair-based circuits that any DC asymmetry, often called "DC offset", causes signal distortion. A common way to compensate for this effect is to implement a so-called "DC offset compensation loop" [5]. As depicted in Figure 9, it can be as simple as a low-pass filtering of complementary outputs, feeding a DC amplifier generating a reference voltage driving the unused input of a single-ended-to-differential conversion part in the circuit. The loop is designed to ensure that the reference voltage (V_{REF} in Figure 9) is properly adjusted to cancel the output DC offset (DC difference between complementary outputs).

To illustrate how the presented novel aging model enables reliability-aware design optimization in that respect, we hereafter present 3 simulation cases, where a 3-GHz sinusoidal signal is applied to the circuit, the ambient temperature is 100°C and the transient response is simulated over 100 ns (equivalent to 1000 h here with ad-hoc accelerating factor).

A. Case 1: TIA without DC offset compensation loop

To emulate not using the DC offset compensation loop, the reference voltage V_{REF} is set to a fixed value. As shown in Figure 10-a, while at 0 ns, V_{REF} is correctly set so that the output offset equals zero, after a few ns V_{TIA} starts to drift, due to the aging of the transistors (especially the input transistor). As V_{REF} is fixed, V_{OUT} and V_{OUTb} start to drift in different directions causing DC offset. A loop is therefore needed to dynamically adjust V_{REF} according to V_{TIA}'s variations.

B. Case 2 : TIA with DC offset compensation loop; loop gain ≈ main amplifier gain

V_{REF} is now generated by the DC offset compensation loop, where the gain of the loop is set only considering typical voltage variations at the beginning of the circuit's lifetime. As shown in the simulated transient response (Figure 10-b), V_{REF} follows properly V_{TIA} in the very beginning of the simulation time, while it shrinks after a few tens of ns, causing again output DC offset. This due to the fact that the loop gain is not large enough compared to the main amplifier gain, so that the loop's efficiency depends on both. When main amplifier's gain decreases (observed on the decrease of V_{OUT} and V_{OUTb}'s amplitudes), due to the aging of transistors, the loop doesn't manage to make V_{REF} properly follow V_{TIA} any longer. Although this effect can be easily predictable qualitatively, only an accurate simulation with the aging model allows evaluating quantitatively its extent and deducing that the loop gain is not large enough.

C. Case 3 : TIA with DC offset compensation loop; loop gain >> main amplifier gain

The loop gain is now set to a large value, so that the loop's efficiency is insensitive to the main amplifier's gain. As shown in Figure 10-c, V_{REF} now properly follows V_{TIA}, resulting is an output DC offset close to zero over the whole simulation time. One can notice some oscillations in V_{REF}, resulting in oscillations of V_{OUT} and V_{OUTb}'s mean value. This simulation artifact is due to the reduced stability margin caused by the increased loop gain. While in reality this is easily compensated for by increasing the loop's time constant (using off-chip high-value capacitors), we had to artificially decrease it for the simulation to allow observing V_{REF} variations.

D. Discussion

We have illustrated in this section how the novel aging model allows improving the reliability of the presented TIA with quantitative assessment of the performances variations over time. In particular we have shown how it allows a better optimization of the DC offset compensation loop design.

Based on this new design methodology, architecture-level optimization can also be contemplated in the future. For instance, from the simulations shown in Figure 10, one can draw that a compensation for main amplifier's gain reduction is also needed. Adding such a "gain compensation loop" would relax the requirement on the gain of the DC offset compensation loop (helping improve the stability of the loop), since its insensitivity to main amplifier's gain would no longer be mandatory.

978-1-4673-1707-8/12 $31.00 © 2012 IEEE

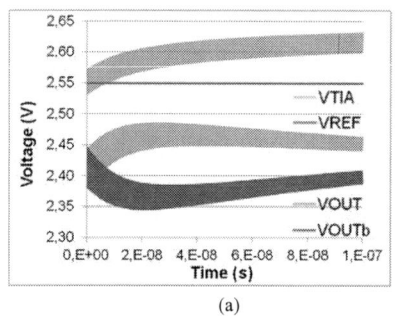 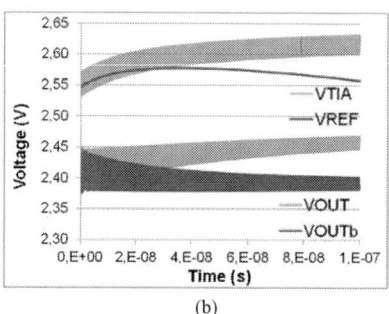 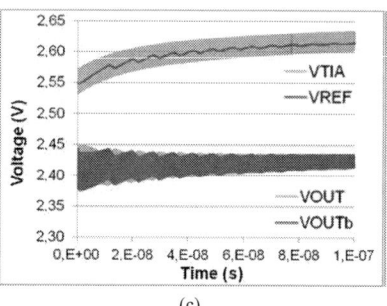

(a)   (b)   (c)

Figure 10. Aging transient simulation results
(a): TIA without DC offset compensation; fixed reference voltage
(b): TIA with DC offset compensation loop; loop gain ≈ main amplifier gain
(c): TIA with DC offset compensation loop; loop gain >> main amplifier gain

V. CONCLUSION AND PERSPECTIVES

We have presented a new methodology for the simulation of degradation mechanisms in IC technology, which consists in the implementation of degradation laws in standard compact models. The feasibility is demonstrated for a InP/InGaAs DHBT process, but the method can be applied also to silicon based technologies. TCAD simulations are used to understand the degradation mechanisms observed on the measurements of transistors submitted to accelerated bias aging stress. Degradation laws are then derived and implemented in the standard HiCuM model. Model parameters have been extracted for this process in order to simulate two variants of a transimpedance amplifier also submitted to aging process. Comparisons to post stress measurements show good agreement, thus validating the proposed approach at circuit level. Finally, simulations with the new model are used to experiment circuit optimization, demonstrating the powerful usage which can be made of the new model. The proposed procedure seems very promising, therefore extensive validation is now required on various stable technologies. Most difficulties of the approach have been understood therefore allowing straightforward investigations. Also the new models need to be intensively used by designers in order to find innovative utilization of this new reliability aware design capability.

ACKNOWLEDGMENT

The work presented in this paper is the result of a highly collaborative work, realized within the ROBUST research project supported by the ANR involving III-V lab, IMS, XMOD, OMMIC and IEMN. This work has been funded by the French National Agency for Research (ANR) via the ROBUST project.

REFERENCES

[1] C.Hu,"IC reliability simulation", IEEE J.Solid State Circuits, vol.27, March 1992, pp241-246

[2] M.M. Lunenborg et al. "PRESS – A reliability circuit simulator with built-in hot carrier degradation model" Conf.proc. ESREF, pp. 157-161, october 1993.

[3] F. Marc, B. Mongellaz, C. Bestory, H. Levi, Y. Danto. "Improvement of aging simulation of electronic circuits using behavioral modeling".

IEEE Transactions on Device and Materials Reliability, 6 :228–234, 2006

[4] M. Schröter, "High-frequency circuit design oriented compact bipolar transistor modeling with HICUM", IEICE Transactions on Electronics, Special Issue on Analog Circuit and Device Technologies, Vol. E88-C, No. 6, pp. 1098-1113, 2005

[5] J.-Y. Dupuy, F. Jorge, M. Riet, A. Konczykowska, J. Godin, "InP DHBT Transimpedance Amplifiers With Automatic Offset Compensation for 100 Gbit/s Optical Communications," in Proc. of the Microwave Integrated Circuits Conference (EuMIC), 2010 European , pp.341-344, 27-28 Sept. 2010.

[6] J. Godin, V. Nodjiadjim, M. Riet, P. Berdaguer, O. Drisse, E. Derouin, A. Konczykowska, J. Moulu, J.-Y. Dupuy, F. Jorge, J.-L. Gentner, A. Scavennec, T.K. Johansen, V. Krozer, "Submicron InP DHBT Technology for High-Speed High-Swing Mixed-Signal ICs," Compound Semiconductor Integrated Circuits Symposium, 2008. CSIC '08. IEEE, pp.1-4, 12-15 Oct. 2008.

[7] "Sentaurus device release. TCAD Sentaurus version X-2005, Synopsys; 2005."

[8] C. Maneux, N. Labat, N. Malbert, A. Touboul, Y. Danto, J-M. Dumas, J. L. Benchimol, M. Riet,, ''Experimental procedure for the evaluation of GaAs-based HBT's reliability'', Microelectronics Journal, Vol. 32, N°4, pp 357-371, 2001

[9] Tao NG et al. "Impact of surface state modelling on the characteristics of InP/GaAsSb/InP DHBTs", Solid State Electron, 2007;1(4):185–9.

[10] C. Maneux, M. Belhaj, B. Grandchamp, N. Labat, A. Touboul, ''Two dimensional DC simulation methodology for InP/GaAs0.51Sb0.49/InP heterojunction bipolar transistor'', Solid State Electronics, Vol. 49, N°6, pp. 956-964, 2005.

[11] Ruiz-Palmero JM et al. "Impact of surface traps on downscaled InP/InGaAs DHBTs", In: 7th conference on solid-state circuit and integrated-circuit technology, Beijing, China; 2004.

[12] V. Parguel et al, "Gold diffusion in InP" J. Appl. Phys. 62, p-824 (1987).

[13] G. A. Koné et al., "Reliability on submicron InP/InGaAs DHBT on accelerated aging tests under thermal and electrical stresses", Microelectronics Reliability 51 (2011) 1730-1735

[14] S. Fregonese, S. Lehmann, T. Zimmer, M. Schröter, D. Celi, B. Ardouin, H. Beckrich, P. Brenner, W. Kraus, "A Computationally Efficient Physics-Based Compact Bipolar Transistor Model for Circuit Design—Part II: Parameter Extraction and Experimental Results", IEEE Trans. Electron Dev., Vol. 53, pp. 287- 295, 2006.

[15] B. Ardouin, C. Raya, M. Schroter, A. Pawlak, D. Celi, F. Pourchon, K. Aufinger, T.F. Meister, T. Zimmer, T., "Modeling and parameter extraction of SiGe: C HBT's with HICUM for the emerging terahertz era", Proceeding of the Microwave Integrated Circuits Conference (EuMIC), Sept. 2010, pp. 25-28

Design Challenges for Nano-Scale Devices

Marc Belleville, Olivier Thomas, Alexandre Valentian, Fabien Clermidy
CEA, LETI, Minatec Campus
Grenoble, France
{Firstname.Lastname}@cea.fr

Abstract— **This paper presents an overview of the design challenges and solutions under development for Nano-scale technologies. Major applications requirements and nano-technologies design limitations are introduced. Adaptive techniques aiming to cope with variations and to track an optimal energy operating point are presented.**

I. INTRODUCTION

In the past, digital integrated circuits have benefitted a lot from the CMOS technology scaling, following Moore's law. Indeed, for an unchanged function, a new technology node used to bring performance improvements, as well as area and power savings. However, CMOS devices getting close to physical and technological limits, this progress is not as obvious and only materials, devices and IC design innovations allow keeping the pace.

This paper will describe, in the perspective of relevant applications, the main challenges Integrated Circuits designers have to face when using nano-scale technologies and the new techniques they need to use. It is organized as follows: in Section II, three main application drivers will be described with their consequences on circuits; Section III will briefly present the major limitations of nano-scale devices, in the circuit design perspectives; design solutions for Logic circuits will be described in Section IV, while Section V will focus on SRAM memories.

II. APPLICATIONS LANDSCAPE

A. Main Application drivers for nanoscale technologies

The wireless communication field has experienced an amazing growth and has been, in addition to the PC market, a strong driver of the semiconductor products. This trend will not decline in the future, as only such high volume markets will be able to access nano-scale technologies. A common way to illustrate and segment the communication market of the future is based on three tiers. The central one consists of a large number of huge data and computing centers. Connected to it, the second tier is made of billions of mobile devices. Finally, it is expected that, in the future, trillions of smart devices like sensors, or actuators, potentially interconnected via the Internet of Things (IoT), will create the third tier [1]. We will discuss in the following paragraphs the key requirements of each of those three fields.

B. Data and Computing Centers

Over time, computers have shown a constant exponential growth in performance; an improvement of 10x is observed every 3.6 years. Simple extrapolation of High Performance Computing trend suggests that Exascale performance (exaFLOPS) should be reached in 2018 with power requirement of up to 500 MW. Power consumption is a driving and limiting factor in supercomputing, and in order to be economically sustainable, a politico-economic threshold of 20 MW has been suggested as a boundary for next decade supercomputer [2]. Consequently, a substantial increase in energy efficiency is required by minimum a factor of 60. Thus data and computing centers are considering using mobile technologies, because of their performance/power advantages. In such systems, the key circuit issues are the power/energy efficiency, and the thermal constraints, at high performance.

C. Mobile Devices

Mobile communications set specific challenges to integrated electronics. Firstly, the algorithmic complexity of the digital communications is sharply increasing. It comes from the necessity to enhance the spectral efficiency of such systems. Indeed, as the usable spectral resources are very limited, it is mandatory to optimize its usage in order to limit the infrastructure cost while improving the data transmission bandwidth.

In addition to communication, mobile devices require more and more computing power to accommodate new Smartphone applications like HD Video [3]. However, as it is portable devices and as the battery technologies are only slowly progressing, this huge increase in computing power has to be achieved at a quasi constant power budget.

Technology scaling does not provide enough power savings, and has to come with complex circuit design techniques. A specificity of mobile applications is that the required computing power greatly varies over time (e.g. from standby mode to HD Video). Thus, the circuit has to be efficient in standby, as well as operating power, and design techniques like Power Gating (PG) and Adaptive Voltage and Frequency Scaling (AVFS) are required. In the next generation, it is expected that this will be exacerbated, leading to Ultra Wide Voltage Range (UWVR) (e.g. 1.2V to 0.4V)

978-1-4673-1707-8/12 $31.00 © 2012 IEEE

circuit operation. This drastically changes the design targets, which used to be VDD (Power supply) +/- 10%.

D. Sensor Nodes for the Internet of Things

The IoT aims to create smart environments and self-aware things for climate, food, energy, mobility, digital society and health applications [4]. Energy issues, from harvesting to conservation and usage, are central to the development of the IoT. As the total expected budget of such nodes is often well below 1mW, there is a need for new solutions having as an objective ultra low power devices, current devices seeming inadequate considering the processing power needed.

As such devices will operate frequently at a very low duty cycle, minimizing drastically the standby power is a primary issue. Concerning the operating power, in addition to dedicated architectures, computing efficiently at a very low supply voltage is the most promising path considering early demonstrations of near-threshold and sub-threshold processors [5][6].

E. Main Conclusions for Integrated Circuits

Whatever the application, permanently seeking an optimized energy efficiency is a major target. This has one main consequence on the circuit power supply voltage that can locally and dynamically be tuned between 0V and its maximum value, with several intermediate points, to: (1) dynamically switch off any unused sub-circuit, while keeping the sensitive data; (2) dynamically track the optimal point according to the on-going application requirements.

III. CHALLENGES LINKED TO NANO-SCALE TECHNOLOGIES

A. Power and Leakage

Power supply voltage reduction, which is required to cope with the devices aging and is very efficient to reduce the power consumption, has induced a lowering of the devices threshold voltages to keep high dynamic performances. This, combined with a difficult channel electrostatic control, leads to a large increase in leakage currents. This static power consumption, dramatic for circuits in sleep mode, is also now a significant portion of the total operating power. Although new technology options like High K/Metal gate and thin film devices like FDSOI or Finfet are helping a lot, specific design techniques are required in addition to meet the application targets.

B. Process variations and aging

Simultaneously, due to the nano-scale dimensions, physical and technological variations are increasing strongly. Random dopant fluctuations, edges roughness as well as metal gate granularity are examples of parameters that affect the devices electrical behavior. Random and local variations cannot be neglected anymore, compared to global ones [7].

A first consequence is that the minimum operating voltage of nano-scale circuits based on logic gates, SRAMs and DRAMs increases as devices are scaled down due to the larger variations of the MOSFETs electrical characteristics [8].

In addition, aging due to phenomena like NBTI/PBTI also affects this behavior. Finally, environmental parameters like the temperature and the power supply voltage variations also strongly impact the performances.

To cope with the uncertainty, designers used to consider worst case corners (process, temperature, voltage). But, in nano-scale technologies, simply increasing the design margin, to take into account the larger variations, is not effective anymore, as it hides the benefits of scaling. In addition, the number of different corners to consider, guaranteeing circuit functionality, moved from less than ten to sometimes close to a hundred (in particular for UWVR design). Statistical design approaches that are more accurate are currently considered [9].

Part of the variations comes from neighboring effects (e.g. stress induced) and lithography. To minimize this impact, regular layout rules (discretized width and spacings, layer uni-directionality, avoidance of jogs and notches) have been introduced [10], but it has not been achieved without an area penalty.

The following two sections will describe some design techniques that have been introduced to cope with those issues.

IV. LOGIC DESIGN IN NANO-SCALE TECHNOLOGIES

A. Circuits for Leakage Power

The most common technique used in advanced technologies to cut-off the leakage power of standard-cell-based logic blocs in non-active mode, relies on the introduction of High-VT power switches between the power supply (or the ground) and the logic (Figure 1. left) [11]. Those devices have to meet two conflicting requirements: a low RON to minimize the voltage drop in operation, and thus the impact on the performances; and a high ROFF to reduce the leakage current. Several design techniques can be used to maximize the ROFF/RON ratio: positive or negative Gate-to-Source overdrive; adaptive body biasing, applying a Body-to-Source forward bias in ON mode, and a backward bias in OFF mode. This latter can be advantageously extended to Backplane polarization in Ultra-Thin Body and Box (UTBB) FDSOI technology, exploiting its wide VT tunability.

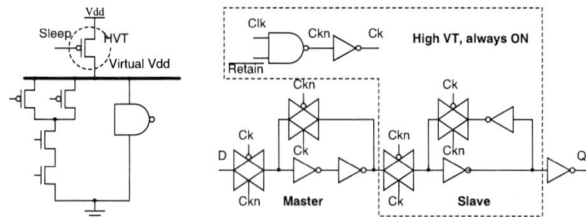

Figure 1. Standard power gating scheme (left) and retention flip-flop (right)

Power gating has to be associated with retention flip-flops to store internal states during Power-off (Figure 1. Right), and start-up circuits to control the dI/dt and the induced IR drop.

B. Low variability and Low voltage circuits

Optimizing standard cells library to cope with variability can be done either by selecting the less sensitive schematic

[12], and/or by proper sizing of the transistors. A second step of optimization can be done at the design synthesis step.

However, when designs are moving from the super-threshold to the near/sub-threshold regime, the effective/idle current ratio decreases drastically, degrading the logic cells noise margin and potentially leading to logic failures. Global and local process variations will further worsen this issue. Thus specific design techniques have to be used to guarantee circuit functionality over a wide voltage range [13][14].

A first approach is to remove, from a regular standard cell library, the cells that exhibit poor operating characteristics at low voltage. For instance, cells with a large number of parallel or stacked transistors are prohibited, as well as cells having internally transient driving conflicts between transistors, like in ratioed logic or bus keepers.

Transistors resizing is also used to balance the cells in low voltage and/or reduce the random variability impact. To cope with a global unbalance between NMOS and PMOS, which can change with the power supply voltage, tuning the threshold voltages by applying body biasing in bulk, or backplane biasing in FDSOI will further improve the minimum operating voltage and the performances.

C. Adaptive Circuits

Dynamic energy consumption largely comes from the charging and discharging of internal capacitances and can be reduced quadratically by lowering the supply voltage (VDD). Thus, providing only the minimum power supply voltage to meet the system targets is an efficient way to save energy. Considering that the system targets often vary overtime, and that the process and environmental variations strongly affect the performances, adaptive circuits are required to track optimal energy points (Figure 2.).

In a synchronous circuit, the clock frequency and the power supply voltage are the two main knobs. An efficient control of the threshold voltages of the transistors (e.g. via backplane polarization in FDSOI) will further extend the optimization capabilities (Figure 2.).

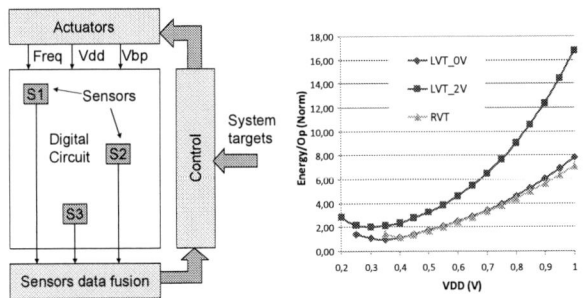

Figure 2. Adaptive circuit (left) and Optimum energy points in FDSOI -2 VT flavors, RVT & LVT, and 2 backplane biases for LVT (right)

Sensors are required to feed the servo-control, ranging from PVT (Process, Voltage, and Temperature) sensors, to timing fault [15](or warning [16]) and voltage droop sensors. A first step is to extract relevant information from the sensors, which can be simple ring oscillators easy to integrate in a digital circuit [17]. A second step is to define the control algorithms, which can range from a simple local feedback to high-level tasks remapping or rescheduling. In addition to energy optimization, a typical concern is to avoid, in heavy computing systems, thermal hotspots or thermal runaway (i.e. a positive feedback between leakage power and temperature that leads to an uncontrollable increase in temperature).

D. Regular circuits

Layer uni-directionality, fixed width and spacing are already widely used for the poly-gate lines of digital standard cells. Metal1 is expected to be concerned soon. Although this impacts the cell area, traditional transistor width sizing design methods are still usable with advanced 2D technologies like FDSOI. Moving to 3D technologies like Finfet worsens the sizing and layout issue, as all the devices must have in addition the same fixed width (the fin height).

Looking towards nanowire technologies, extremely regular logic structures, like PLAs (Programmable Logic Array) or Croosbars, have been revisited and proposed (e.g. Figure 3.).

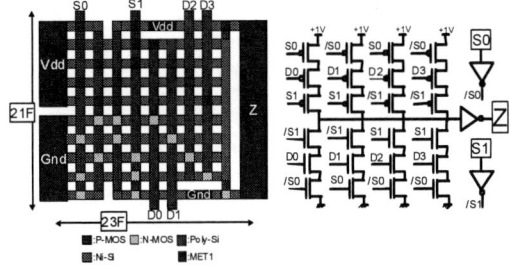

Figure 3. 4:1 MUX crossbar implementation and equivalent schematic [18]

V. SRAM DESIGN IN NANO-SCALE TECHNOLOGIES

Embedded memories are critical in CMOS ICs. Various memory alternatives (DRAM, Flip-Flops) exist and can be differentiated by their access time, energy efficiency, density and process integration. Among the different alternatives, SRAM is the most common embedded-memory option. Moderate speed and energy can be achieved for an adequate density and full CMOS integration compatibility.

As energy efficient IC design requires wide voltage range, SRAM must remain compatible with operating conditions. In the mean time, increasingly parallel architectures ask for more on chip memory to effectively share information across parallel processing units, governing ICs yield.

Designing ultra wide voltage and high capacity SRAM faces lot of challenges, originating from process electrical parameters to CAD methodologies.

Process variations in nano-scale devices affect all SRAM functionality parameters. They result in yield degradation mainly dominated by soft failures. Soft failures can be classified into three categories: failures in maintaining data stability (read-stability), failures in the time it takes the bitcells to generate a target bitline differential voltage (read-ability) and failures in the time to flip the bitcell (write-ability). Low operation voltage exacerbates soft failures, making SRAM bitcells unreadable or unwritable, while sense amplifiers offset increases and timing-control signals deviate. The limit on the minimum operating supply voltage (Vmin) is critical. Therefore, circuit designers must address the need of low voltage SRAM design through novel circuit design techniques in the periphery and the bitcell[19].

6 transistors (6T) SRAM bitcell has been the workhorse for embedded memory applications for decades. Alternative solutions abandon 6T bitcell to bypass the write-ability read-stability contention or read fails caused by a lack of bitcell read current. However, most of the 6T alternatives lead to an area overhead, require single-ended sensing instead of differential sensing and cannot interleave bitcell columns. Various circuit design solutions tried to solve the problems. Assist techniques, such as wordline boost, negative bitlines, VDD droop and GND boost are employed to expand the operating margins. The effectiveness of these techniques is related to the area and energy overhead.

Resilient low voltage memory design must consider bit density to limit application cost increase, which might reduce the range of applicability because of economic factors. Therefore a better understanding of failure cases is becoming a crucial issue. SRAM arrays can fail if one out of millions of bitcells fail. Crude Monte Carlo simulations can be used to measure the effect of variability. However, finding a rare failure event requires an unfeasible number of simulations. To cope with simulation time, most measurements of effectiveness use static metrics, assume that failure distribution is Gaussian and extrapolate how many σ (for example six σ) from the mean can be used to estimate failure rate. Those approaches come up with two main problems. First, static metrics do not accurately track read and write margins caused by non-Gaussian distribution. Second, these metrics considered infinite access time and do not account for transient effects. To break down the simulation barrier, importance sampling is used in [20] to determine the SRAM bit error rate (BER), based on the estimation of the most probable failure point (MPFP). This highly efficient and general methodology for a quick evaluation of SRAM design has been employed in [21]. The methodology not only estimates failure probability, but also tracks bitcell sensitivity to transistor strength through evaluation of the MPFP. The BER estimation reveals that Vmin is limited by write-ability and read-ability, but not read-stability.

In opposition to static write margin, write-ability is mainly limited by the high logic level node completion. Therefore, write-ability and read-ability can be improved by reinforcing the drivability of the pMOS and nMOS transistors, respectively.

In UTBB FDSOI technology, the V_T of the transistor can be adjusted by both the backplane doping type and polarization. In [21], a low voltage single p-well 6T bitcell architecture trading off soft failures is proposed.

VI. CONCLUSION

Circuit dynamic adaptability and Ultra Wide Voltage Operating range are two new design issues designers will have to address in nano-scale technologies, in addition to already established static leakage power and layout regularity. Proposed solutions will have to prove to be cost effective, considering area, robustness and test.

REFERENCES

[1] J. Rabaey. "Swarm Vision". Talk or presentation, BWRC Winter 2012 Retreat, 8, January, 2012, available at http://bwrc.eecs.berkeley.edu/php/pubs/pubs.php/1893.html

[2] P. M. Kogge et al, "ExaScale Computing Study: Technology Challenges in Achieving Exascale Systems," DARPA Information Processing Techniques Office, Washington, DC, pp. 278, September 28, 2008

[3] C.H. van Berkel, "Multi-Core for Mobile Phones" DATE, April 2009, Nice, France

[4] O. Vermesan et al., "Internet of Things Strategic Research Roadmap" available at www.internet-of-things-research.eu, 2011.

[5] R. Dreslinski et al., ''Near-threshold Computing: ReclaimingMoore's Law through Energy Efficient Integrated Circuits,'' Proc. IEEE, vol. 98, no. 2, Feb. 2010, pp. 253-256.

[6] V. Chandra, "Dependable Design in Nanoscale CMOS Technologies: Challenges and Solutions", WDSN'09, June 2009, Lisbon, Portugal

[7] K.A. Bowman, X. Tang, J.C. Eble, J.D.Meindl, "Impact of extrinsic and intrinsic parameter fluctuations on CMOS circuit performance", 2000, IEEE Journal of Solid-State Circuits 35 (8) , pp. 1186-1193

[8] K Itoh, M.Horiguchi, "Low-voltage scaling limitations for nano-scale CMOS LSIs", 2009, Solid-State Electronics 53 (4) , pp. 402-410

[9] D.Blaauw, K. Chopra, A. Srivastava, L.Scheffer, "Statistical Timing Analysis: From Basic Principles to State of the Art", IEEE Transactions on Computer-Aided Design of Integrated Circuits and Systems, April 2008, Vol. 27 , Issue: 4, pp. 589-607

[10] S. R. Nassif and K. J. Nowka, "Physical design challenges beyond the 22nm node", ISPD'10, March 2010, pp. 13-14.

[11] S. Mutoh et al., "1-V power supply high-speed digital circuit technology with multithreshold-voltage CMOS", IEEE Journal of Solid-State Circuits, Vol. 30 , Issue 8, 1995 , pp. 847-854

[12] B. Rebaud et al., "A comparative study of variability impact on static flip-flop timing characteristics", in Proc. IEEE ICICDT 2008, pp. 167-170.

[13] A. Wang, B.H. Calhoun and A.P. Chandrakazan, "Sub-threshold Design for Ultra Low-Power Systems", Springer, 2006

[14] S; Hsu et al., "A 280mV-to-1.1V 256b reconfigurable SIMD vector permutation engine with 2-dimensional shuffle in 22nm CMOS", ISSCC 2012, pp. 178-179

[15] D Blaauw et al., "Razor II: In Situ Error Detection and Correction for PVT and SER Tolerance", ISSCC 2008, pp. 400-401

[16] B. Rebaud, et al., "Timing slack monitoring under process and environmental variations: Application to a DSP performance optimization", 2011, Microelectronics Journal 42 (5) , pp. 718-732

[17] Vincent L., Maurine P., Lesecq S., Beigne E., "Embedding Statistical Tests for On-Chip Dynamic Voltage and Temperature Monitoring", DAC 2012, pp. 994-999

[18] P.-E. Gaillardon, M.H. Ben-Jamaa, F. Clermidy, I.O'Connor, "Evaluation of a crossbar multiplexer in a lithography-based nanowire technology", ISCAS 2011, pp. 2930-2933

[19] M. Qazi, M.-E. Sinangil, A-P. Chandrrakasan, "Challenges and Directions for Low-Voltages SRAM", Design & Test of Computers, Jan-Feb 2011, pp 32-43.

[20] L. Dolecek, M.Qasi,D. Shaha, A-P. Chandrakasan, "Breaking the simulation barrier: SRAM evaluation Through Norm Minimization", ICCAD, 2008, pp. 323-329.

[21] O. Thomas, et al., "6T SRAM Design for Wide Voltage Range in 28nm FDSOI", SOI Conference, 2012.

Study of Carrier Transport in Strained and Unstrained SOI Tri-gate and Omega-gate Si-Nanowire MOSFETs

M. Koyama[1,4], M. Cassé[1], R. Coquand[1,2,3], S. Barraud[1], H. Iwai[4], G. Ghibaudo[2], and G. Reimbold[1]

[1] CEA LETI, MINATEC Campus, 17 rue des Martyrs 38054 Grenoble Cedex 9, France,
[2] IMEP-LAHC, INPG-MINATEC, 3 Parvis Louis Neel, 38016 Grenoble Cedex 1, France,
[3] STMicroelectronics, 850 rue J. Monnet, 38926 Crolles, France,
[4] Frontier Research Center, Tokyo Institute of Technology, 4259, Nagatsuta, Midori-ku, Yokohama, 226-8502, Japan

Abstract— We report an experimental study of the carrier transport in long channel tri-gate (TG) and omega-gate (ΩG) Si nanowire (NW) transistors with cross-section width down to 10 nm. Electron and hole mobility have been measured down to 20 K. We discuss the influence of channel shape, channel width and strain on carrier mobility. In particular we have shown that transport properties are mainly driven by the relative contribution of the different inversion surfaces, without noticeable differences between TG and ΩGNWs. We have also demonstrated the effectiveness of an additional uniaxial strain in NWs down to 10nm width.

I. INTRODUCTION

Silicon nanowire (Si NW) MOSFETs are strongly attractive for future CMOS technology nodes, due to their good immunity to short channel effect, drain induced barrier lowering, and better electrostatic control (lower off-current and steeper subthreshold slope) [1]. These innovative devices thus offer advantages in downscaling and power consumption.

The reduced dimensions of the cross-section of nanowire below 20 nm can lead to the confinement of carriers and change the transport properties of NW devices [2]. Moreover the side-walls, with different crystallographic orientation, may also contribute to the total carrier mobility and modify the drain current [3]. Stress engineering can be used to further improve the device performance [2,4]. The combination of stress and NW architecture appears as a promising solution to scale further CMOS technology. The understanding of carrier transport in both NMOS and PMOS Si NWs, with additional uniaxial strain along the channel direction, is non trivial and important to investigate.

In this work, we present an experimental study of the carrier transport in rectangular (tri-gate) and Omega-shaped (Ω-gate) Si NW MOSFETs. In particular, we will focus on the influence the cross-section shape, width dimensions, and the effect of a uniaxial tensile strain. Carrier mobility has been measured on NMOS and PMOS NW transistors down to low temperature and its behavior is discussed in details.

II. DEVICES DESCRIPTION

Si NWs were fabricated using the top-down approach by optical lithography followed by a resist trimming process [5, 6]. [110]-oriented tri-gate (TG) and omega-gate (ΩG) NWs were fabricated on (100) unstrained- or strained-SOI (SOI and sSOI) wafers, with ~10 nm thickness. ΩG NW was formed with additional H_2 annealing (Fig.1). Strained NWs are obtained from sSOI substrate corresponding to ~1.4 GPa biaxial tensile stress. After etching, the biaxial strain reduces to a uniaxial tensile strain along the [110] channel direction, due to lateral relaxation [5].

The high-k/metal gate stack consists in 2 nm CVD HfSiON, 5 nm ALD TiN covered with 50 nm poly-Si. The total EOT varies from 12.5 Å for wide and TGNWs to 15 Å for ΩGNWs, as a probable consequence of the H_2 anneal.

N-type and P-type FET structures with varied widths were fabricated, from 10 μm (wide FET) down to 10 nm (NW FET). Measurements were done on 50 NWs in parallel in order to extract carrier mobility using the conventional split C-V technique. We focused our study to long channel transistors with channel length L=10 μm. In the following the nanowire width and height are defined as top-view width W and H (which is equal to the silicon film thickness t_{si}), as illustrated in Fig.1.

Structure of NW	Tri-gate	Omega-gate
	H W	H W
	10nm	10nm
SOI	H = 11 nm W = 10, 30 nm	H = 10 nm W = 23 nm
sSOI (σ = 1.4GPa)	H = 10.5 nm W = 16, 36 nm	H = 8.4 nm W = 33 nm
L = 10 μm, array of 50 channels		

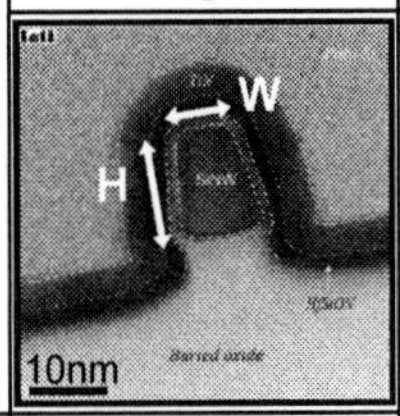

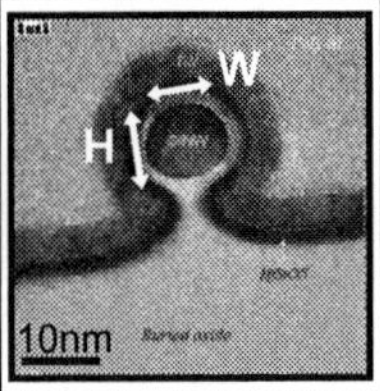

Fig. 1: Channel dimensions (height H and top-view width W) of SiNW MOSFETs and cross-sectional TEM images of the channel of TG (left) and ΩG (right) SiNW MOSFETs.

III. RESULTS AND DISCUSSION

Carrier mobility μ_{eff} has been measured as a function of temperature, from room temperature down to 20 K. Figure 2 shows the electron and hole mobility extracted as a function of inversion carrier density (N_{inv}) for SOI TGNW and wide FETs. For NWs, μ_{eff} was obtained through C_{gc} measurement which takes into account inversion carrier density in the whole structure (incl. side-walls and top surface). In NMOS case, the effective mobility of TGNW is degraded compared to wide FET, while the hole mobility is improved in high N_{inv} region in PMOS. In particular, the degradation in NMOS is larger at low temperature and high N_{inv} (above 0.4×10^{13} cm^{-2}). Moreover, the peak mobility μ_{max} appears at lower N_{inv} for NMOS TGNW ($\mu_{max} \sim 650$ cm^2/Vs at $N_{inv} \sim 0.2 \times 10^{13}$ cm^{-2} and T=100 K) compared to wide FETs ($\mu_{max} \sim 800$ cm^2/Vs at $N_{inv} \sim 0.6 \times 10^{13}$ cm^{-2} and T=100 K). On the other hand, there is no shift of the maximum mobility in PMOS case.

978-1-4673-1707-8/12 $31.00 © 2012 IEEE

The mobility improvement for PMOS and mobility degradation for NMOS is in good agreement with the increasing contribution of the (110)-oriented sidewalls as the transistor width is reduced [6]. Indeed the hole mobility in (110)/[110] channel is higher than the one in (100)/[110] channel [7]. In contrast the electron mobility is degraded in (110)/[110] channel.

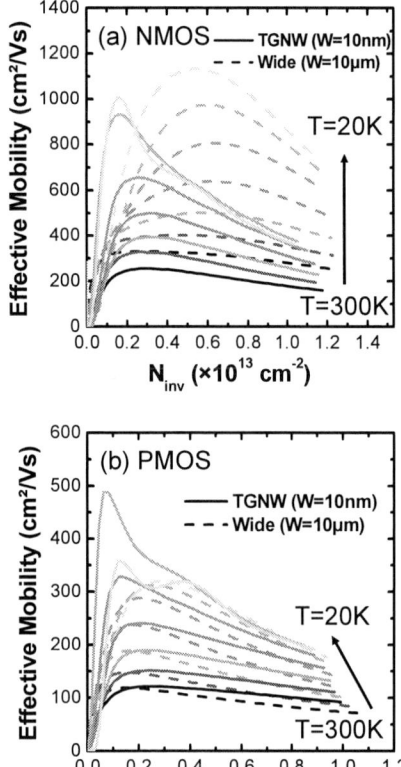

Fig. 2: Effective mobility extracted as a function of inversion carrier density (N_{inv}) on SOI TGNW and wide channel at different temperatures for (a) NMOS and (b) PMOS SOI TGNW-FETs.and wide MOSFETs. The channel width is 10 nm and 10 μm for TGNW and wide transistors respectively.

In Fig.3 we compare the electron and hole mobility as a function of N_{inv} for SOI TG and ΩG NW FETs with similar width. For PMOS the mobility of ΩGNW and TGNW are very similar in the whole range of N_{inv} and temperature, with a slightly higher mobility at low N_{inv} for ΩGNW. On the other hand, slight differences appear for NMOS at low temperature: the mobility is higher for ΩGNW at low N_{inv}, while it is deteriorated at high N_{inv} region.

These results show that the shape of the NWs (rectangular vs. semi-circular) has only a little influence on carrier transport for dimension as small as 10nm×10nm. The (110)-oriented surface can still be distinguished even in the semi-circular geometry of ΩGNW.

The difference at low N_{inv} could indicate a lower Coulomb scattering contribution (CS) in the case of ΩGNWs, due to the H_2 anneal process, in contrast with previous work [8].

In the following we discuss the contribution of top and side-wall surfaces on TGNW mobility, using the total mobility expressed as [3,6]:

$$\mu_{TG} = \mu_{top}^{(100)}\left(W/(2H+W)\right) + \mu_{side-wall}^{(110)}\left(2H/(2H+W)\right)$$

where μ_{top} and $\mu_{side\text{-}wall}$ are the mobility corresponding to each surface orientation ((100) for top surface and (110) for side-walls). Using this equation and making the reliable assumption that the (100)-oriented top mobility μ_{top} is given by the 10μm-wide FETs, we can de-correlate the mobility contributions of the top and the sidewalls for both N- and P-MOS TGNWs. The extracted contributions as a function of N_{inv} are shown in Fig.4. Side-wall mobility of both NMOS and PMOS is in good agreement with the referential mobility of Si(110) wide FET measured at 300 K, showing that the transport properties of TGNWs are governed by the independent inversion surfaces.

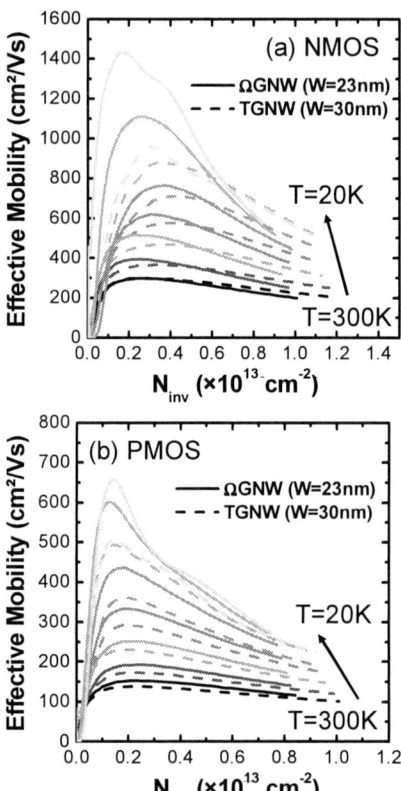

Fig. 3: Effective mobility as a function of inversion carrier density (N_{inv}) extracted at different temperatures for (a) NMOS and (b) PMOS SOI TG and ΩG SiNW-FETs. The channel width is 30 nm and 23 nm for TG and ΩG respectively.

At low temperature, phonon scattering disappears, and below 100 K, the mobility is only limited by CS at low N_{inv} and surface roughness (SR) scattering at higher N_{inv}. Figure 4 shows that for NMOS, the SR contribution in NW is drastically reduced compared to the wide (100) reference, in good agreement with data reported for (110)/[110] electrons in literature [7,9].

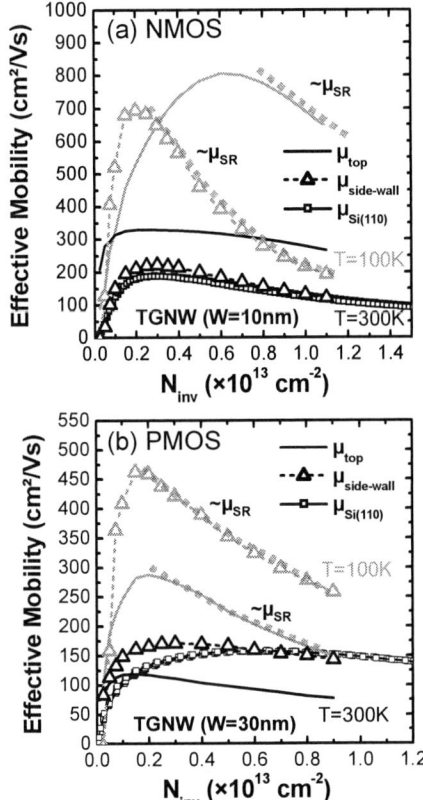

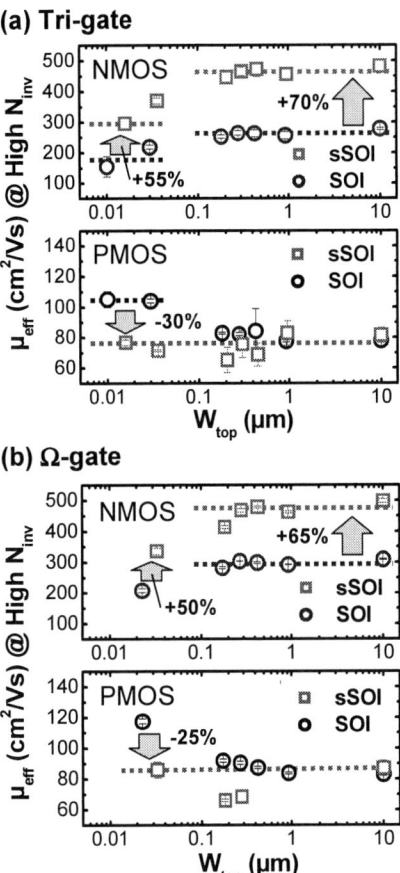

Fig. 4: Extraction of the top and side-wall mobility contributions in SOI TGNW FETs (μ_{top} and $\mu_{side-wall}$, see text for details) as a function of inversion carrier density (N_{inv}) on (a) NMOS with width of 10 nm and (b) PMOS with width of 30 nm. The mobility of Si(110) wide MOSFET ($\mu_{Si(110)}$) measured at 300 K is also shown as a reference.

Figure 5 shows the effective mobility extracted a high inversion carrier density as a function of channel width for all SOI and sSOI transistors.

For SOI transistors, the behavior is well explained by the contribution of top and side-wall mobility and the surface orientation dependence of the mobility (see table in Fig.6). The Si(110) side-walls are beneficial to hole transport parallel to [110], while Si(100) top surface is advantageous to electron transport. Therefore, for NMOS the mobility is degraded as top surface area decreases, *i.e.* as the width W decreases. On the other hand, the mobility is increased in PMOS as side-wall contribution increases. The hole mobility enhancement and the electron mobility degradation as W decreases is similar for TG and ΩG NWs as already noticed.

For sSOI devices the effect of strain for different surface orientations and stress conditions (biaxial vs. uniaxial) has to be taken into account to understand our results. Moreover as the width decreases, the tensile biaxial stress in wide transistor (100) plane changes to a uniaxial stress along the [110] channel direction (Fig.6b). Thus, a uniaxial tensile strain along [110] direction is expected to enhance electron mobility in both (100) (top surface) and (110) (side-wall) planes. The enhancement mainly results from the strain induced repopulation of Si conduction valleys, and the reduction of intervalley phonon scattering [9,10].

Fig. 5: Effective mobility as a function of channel width (W_{top}) for SOI and sSOI (a) TGNW and (b) ΩGNW MOSFETs at 300 K. The TGNW mobility was extracted at $N_{inv}=10^{13}$ cm^{-2}. In ΩGNW, the mobility was extracted at $N_{inv}=0.8\times10^{13}$ cm^{-2} for NMOS and at 0.7×10^{13} cm^{-2} for PMOS.

The mobility is clearly degraded as the width is narrowed in both SOI and sSOI NMOS. However it is worth noting that the electron mobility of strained NMOS devices is still enhanced for a given width, showing the efficiency of a uniaxial strain in NW, with up to +55% gain in μ_{eff} for W=10 nm. We noticed that both TG and ΩGNW exhibit roughly the same mobility improvement for the same width.

(a)	Strain	NMOS		PMOS	
	(transport //[110])	**(100)**	**(110)**	**(100)**	**(110)**
	no strain	0	-	0	+
	biaxial tensile	+		- / =	
	uniaxial tensile // [110]	++	++	--	-

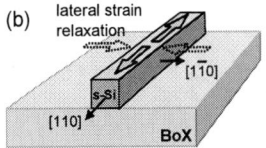

Fig. 6: (a) Table summarizing the effect of the stress on carrier transport along [110], for various stress configurations of interest here (mainly established from piezoresistive coefficients and results given in Refs.[9-16]). (b) Schematics of the strain relaxation occurring in NWs.

This shows that the strain relaxation is the same in both geometries, and that the piezoresistive properties are identical in TG and ΩGNWs despite the more complex inversion surface orientations of ΩGNW.

On the contrary, hole mobility is degraded by a tensile uniaxial stress along the channel especially for (100) surface [11]. In (110) surface the better mobility in unstrained case is counterbalanced by the uniaxial tensile strain. As a result, as the width is decreased the total hole mobility in sSOI NWs is no more improved as compared to wide transistors, and remain roughly constant with W variation. Again TG and ΩGNW exhibit roughly the same mobility for a given width, in agreement with NMOS results.

Finally, the effective mobility extracted at a constant high N_{inv} (resp. 0.7 and 0.8×10^{13} cm^{-2} for PMOS and NMOS) has been plotted as a function of temperature, for SOI and sSOI FETs (Fig.7). NMOS mobility improvement and PMOS mobility degradation in sSOI FETs are observed in the whole temperature range. Below 100 K, the phonon contribution becomes negligible, and the mobility at high N_{inv} saturates, as a consequence of the dominant SR limited contribution. For both type of carriers, we observe a similar SR term for NWs compared to wide FETs. Electron mobility is degraded on narrowest width in both SOI and sSOI FETs without correlation to channel shape, in agreement with a dominant (110) surface orientation for NWs. In contrast, hole mobility is degraded on strained NWs revealing a drastic change of SR contribution for tensely strained (110)/[110].

Above 100 K, the slope of the temperature dependent mobility is driven by phonon scattering, especially at moderate N_{inv} (typically around μ_{max}).

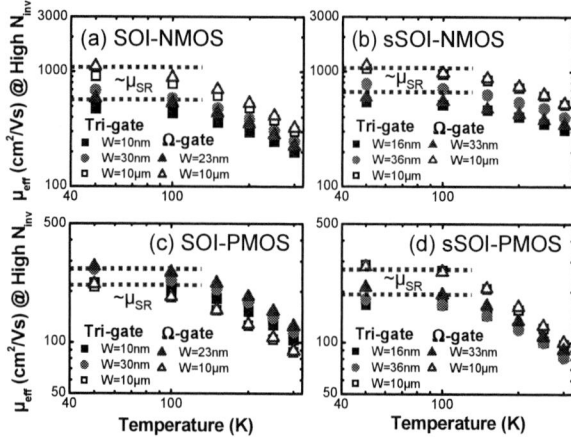

Fig. 7: Effective mobility extracted at high N_{inv} as a function of temperature in all MOSFETs: (a) SOI NMOS, (b) sSOI NMOS, (c) SOI PMOS, and (d) sSOI PMOS. The mobility was extracted at $N_{inv}=0.8 \times 10^{13}$ cm^{-2} for NMOS (a) and (b), and at $N_{inv}=0.7 \times 10^{13}$ cm^{-2} for PMOS (c) and (d).

Figure 8 shows the values of the power law exponent, γ, of the temperature dependent maximum mobility ($\mu_{max} \sim T^{-\gamma}$) extracted for all FETs studied. For SOI or sSOI, the values do not differ significantly for each structure: wide, TG or ΩGNW FETs. On the other hand, the temperature dependence changes significantly when comparing unstrained (SOI) and strained

(sSOI) devices, especially for NMOS. These results indicate that the temperature dependence of phonon-limited electron and hole mobility is mainly governed by the strain down to NW width of 10 nm.

Structure		NMOS		PMOS	
		SOI	**sSOI**	SOI	**sSOI**
Tri-gate w/o H$_2$ anneal	Wide (10µm)	0.95	**0.64**	1.00	**1.18**
	NW	0.94	**0.63**	0.98	**1.25**
Ω-gate with H$_2$ anneal	Wide (10µm)	1.05	**0.69**	1.03	**1.07**
	NW	1.05	**0.41**	1.12	**1.13**

Fig. 8: Values of the power law exponent γ of the temperature dependence of μ_{eff} extracted at peak mobility ($\mu_{max} \sim T^{-\gamma}$) for all devices studied.

Moreover no significant difference in T-dependence of μ_{eff} can be observed between TG and ΩG NWs, revealing again no significant influence of cross-sectional shape of NW down to 10 nm width.

IV. CONCLUSION

We have carefully studied the carrier transport in strained and unstrained NWs with rectangular and rounded cross-section shape. We found that: (i) the transport properties in TGNWs are well described by the separate contribution of inversion surfaces, for rectangular sections as small as 10nm×10nm. (ii) ΩGNWs exhibit the same mobility behavior, despite a more complex geometry with multiple surface orientations, for top view width down to 23 nm. (iii) Lower Coulomb scattering is observed in ΩGNWs, as possible consequence of the H$_2$ anneal process. (iv) Additional uniaxial strain obtained from sSOI starting substrate is effective in NWs, independently of the NW geometry, and can thus be exploited to enhance NMOS performances in NW devices.

ACKNOWLEDGMENT

This work has been carried out in the frame of the ST/IBM/LETI joint program.

REFERENCES

[1] I. Ferain, C. Colinge, and J.-P. Colinge, Nature 479, p.310 (2011)
[2] M. Baykan et al., J Appl. Phys. 108, 093716 (2010)
[3] J. Chen et al., Symp. VLSI Tech., p.32 (2008)
[4] C.W. Liu et al., IEEE Circ. and Dev. Mag. 21, p.21 (2005)
[5] R. Coquand et al., Symp. VLSI Tech., to be published (2012)
[6] R. Coquand et al., Proc. ULIS conf. (2012)
[7] S. Takagi et al., IEEE TED, p.2363 (1994)
[8] K. Tachi et al., IEDM Tech. Dig., 34.4.1 (2010)
[9] M. V. Fischetti et al., J. Appl. Phys. 94, p.1079 (2003)
[10] Y. Sun et al., Strain Effect in Semiconductors, Springer&Verlag (2010)
[11] G. Sun et al., J.Appl. Phys. 102, 084501 (2007)
[12] K. Uchida et al., IEDM Tech.Dig. (2005) and IEDM Tech. Dig. (2006)
[13] H. Irie et al., IEDM Tech. Dig., p.225 (2004)
[14] T. Mizuno et al., IEEE TED 52, p.367 (2005)
[15] A. Teramoto et al., IEEE TED 54, p.1438 (2007)
[16] B. Mereu et al., J. Appl. Phys. 100, 014504 (2006)

Stability and Performance Optimization of InGaAs-OI and GeOI Hetero-channel SRAM Cells

Vita Pi-Ho Hu, Ming-Long Fan, Pin Su and Ching-Te Chuang

Department of Electronics Engineering & Institute of Electronics, National Chiao Tung University, Hsinchu, Taiwan
Email: vitabee.ee93g@nctu.edu.tw

Abstract—InGaAs-OI and GeOI SRAM cells using optimized threshold voltage (Vt) design to enhance the intrinsic variation immunity of high-performance (super-threshold) and low-voltage (near-/sub-threshold) 6T SRAM cells are presented. For low-voltage SRAMs operating at low Vdd, low-Vt design shows smaller variability while the design trade-off between performance and leakage should be considered. For high-performance SRAMs operating at high Vdd, high-Vt design shows smaller variability. Moreover, compared with the SOI SRAMs with high-Vt design, InGaAs-OI/GeOI hetero-channel SRAM cells with high-Vt design exhibit improved Read/Write stability and performance, and maintain comparable RSNM variations for the high-performance SRAM applications.

I. INTRODUCTION

III-V semiconductors are promising candidates as channel materials for future CMOS technology due to their high mobility. Lower immunity to Short-Channel Effects (SCE) [1, 2] could be overcome by using Ultra-Thin-Body (UTB) device architecture [3, 4, 5]. SRAM cells with III-V MOSFETs have been studied for improving stability and performance [6, 7]. However, the higher leakage and variability may limit the use of III-V MOSFETs in SRAM applications.

In this paper, the stability, variability, performance and cell leakage of III-V-OI and GeOI SRAM cells are analyzed under two scenarios (InGaAs-OI(NFET)/GeOI(PFET) and GeOI NFET/PFET) and compared with the SOI SRAM cells. For low-voltage (near-/sub-threshold) SRAM cells, [8] shows that the stability and variability of SRAM cells can be improved by using lower threshold voltage (Vt) device. However, the impact of Vt design on the variability of high-performance (super-threshold) SRAM cells remains to be examined. In this paper, Vt design and optimization are used to improve the variability of high-performance and low-voltage III-V-OI and GeOI SRAM cells. We show that III-V-OI and GeOI SRAM cells with optimized Vt design exhibit improved stability and performance, and comparable variability compared with the SOI SRAM cells.

The UTB MOSFETs used in this study has 25 nm gate length (L_g) and raised source/drain structure with thin Buried Oxide (BOX). (Fig. 1(a)). The device parameters are listed

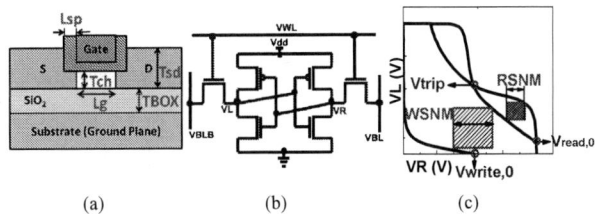

Fig. 1. (a) The schematic of a UTB MOSFET with thin BOX and raised source/drain structure. (b) The schematic of a 6T SRAM cell. (c) The definitions of RSNM and WSNM.

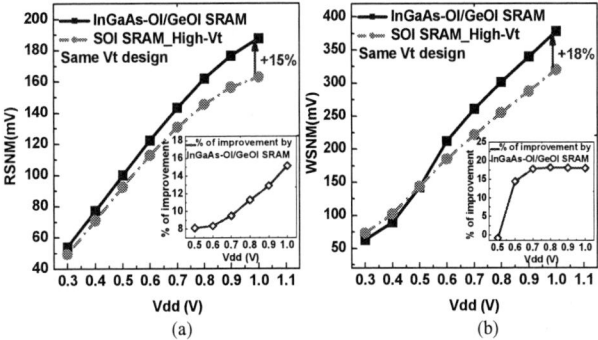

Fig. 2. InGaAs-OI/GeOI SRAM cell shows (a) larger RSNM and (b) larger WSNM than the (same-) high-Vt SOI SRAM cells at high Vdd.

below: channel thickness (T_{ch}) = 5 nm, EOT = 0.7 nm with high-k gate dielectric (HfO$_2$, permittivity = 22), buried oxide thickness (T_{BOX}) = 10 nm, spacer length (L_{sp}) = 10 nm, raised source/drain thickness (T_{sd}) = 22.5 nm, channel doping concentration (N_{ch}) = 1E16 cm^{-3}, and source/drain doping concentration (N_{sd}) = 5.5E19 cm^{-3} for GeOI and 1E20 cm^{-3} for InGaAs-OI and SOI MOSFETs respectively. The band-to-band tunneling and mobility model are calibrated with experimental data [5, 9, 10, 11] to accurately assess the power-performance of UTB GeOI and InGaAs-OI devices using atomistic TCAD mixed-mode simulations [12]. UTB MOSFETs are designed with two Vt values (low-Vt = 0.2V and high-Vt = 0.4V) to investigate the variability of SRAM cells and the impact of III-V-OI and GeOI devices on the stability, variability, and performance of SRAM cells.

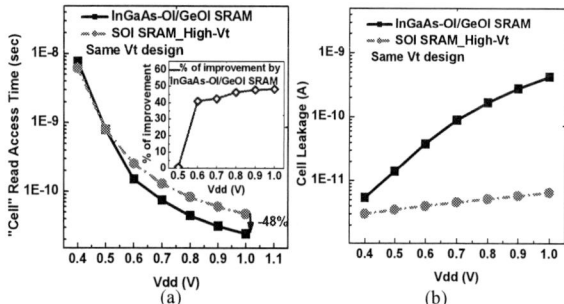

Fig. 3 (a) InGaAs-OI/GeOI SRAM cell shows 48% improvement in Read time than the (same-) high-Vt SOI SRAMs at high Vdd. (b) InGaAs-OI/GeOI SRAM cell shows larger cell leakage than the (same-) high-Vt SOI SRAMs due to band-to-band tunneling.

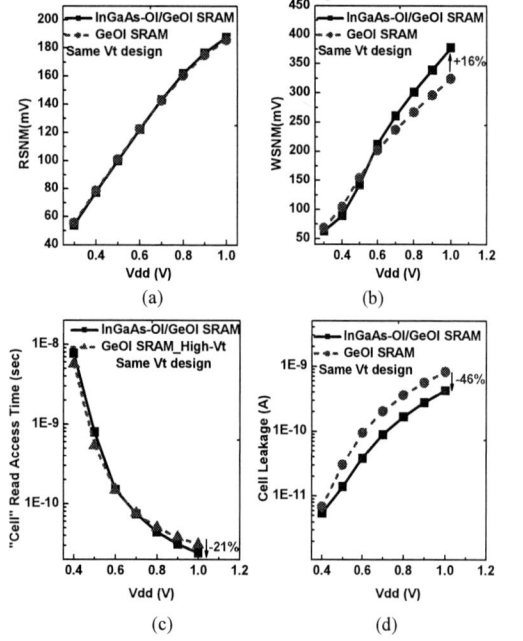

Fig. 4. InGaAs-OI/GeOI SRAM cell shows larger WSNM, comparable RSNM, smaller Read time and cell leakage at high Vdd compared with the GeOI SRAMs.

II. INGAAS-OI(NFET)/GEOI(PFET) SRAMS AND GEOI SRAMS

Fig. 1(c) shows the definitions of Read/Write Static Noise Margin (RSNM/WSNM). Vread,0 is determined by the pass-gate and pull-down NFETs, and Vtrip is determined by the pull-up PFET and pull-down NFET. Fig. 2(a) and 2(b) show that the RSNM and WSNM comparisons between InGaAs-OI/GeOI SRAMs and SOI SRAM cells with (same-) high-Vt design. As can be seen, the InGaAs-OI/GeOI SRAMs show 15% and 18% improvements in RSNM and WSNM at Vdd = 1V respectively compared with the high-Vt SOI SRAMs. The insets of Fig. 2(a) and 2(b) show the improvement in RSNM and WSNM over SOI SRAM cell by using InGaAs-OI/GeOI SRAM cells. The InGaAs-OI/GeOI SRAMs with stronger pull-down NFET device show smaller Read disturb voltage (Vread,0) and larger RSNM at Vdd = 1V.

Fig. 3(a) shows the "Cell" Read access time comparisons between InGaAs-OI/GeOI and SOI SRAMs. The "Cell" Read access time is analyzed for 64 cells per bit-line. A capacitive load is added onto each bit-line to account for the capacitance of wires and the connected devices. The "Cell" Read access time is defined as the time required for developing 0.1Vdd bit-line differential voltage after the word-line is activated during a Read operation. Read access time depends on the Read current through pass-gate and pull-down transistors. The InGaAs-OI/GeOI SRAMs with stronger pass-gate and pull-down transistors exhibit 48% improvement in "Cell" Read access time compared with the high-Vt SOI SRAMs at Vdd = 1V. Due to smaller bandgap, the cell leakage of InGaAs-OI/GeOI SRAMs dominated by band-to-band tunneling is larger than that of SOI SRAMs as shown in Fig. 3(b). As Vdd scales from 1V to 0.4V, the cell leakage of InGaAs-OI/GeOI SRAM cells shows larger reduction than that of SOI SRAM cells since the band-to-band tunneling leakage is very sensitive to Vdd [13]. The band-to-band tunneling leakage may be reduced by using underlap device design [14, 15].

Fig. 4 shows the RSNM, WSNM, "Cell" Read access time and cell leakage comparisons between InGaAs-OI/GeOI SRAMs and GeOI SRAMs with (same-) high-Vt design. (The GeOI PFET is identical for these two scenarios.) Compared with GeOI NFET, InGaAs-OI NFET shows larger drive current due to its higher mobility. As can be seen, the InGaAs-OI/GeOI SRAMs with stronger NFET show comparable RSNM, larger WSNM and smaller "Cell" Read access time. In other words, InGaAs-OI/GeOI SRAM cells with stronger NFET shows larger improvement in WSNM than in RSNM. InGaAs-OI NFET also shows smaller band-to-band tunneling leakage due to its larger band-gap compared with the GeOI NFET. Therefore, the InGaAs-OI/GeOI SRAMs show smaller cell leakage than the GeOI SRAMs.

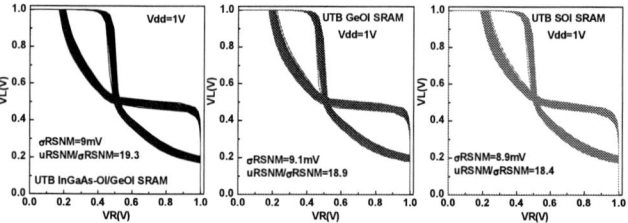

Fig. 5 InGaAs-OI/GeOI SRAM cell and GeOI SRAM cell with high-Vt design show comparable σRSNM compared with the (same-) high-Vt SOI SRAM. For LER: correlation length=20nm, rms amplitude=1.5nm.

Fig. 5 shows that the σRSNM of InGaAs-OI/GeOI, GeOI and SOI SRAMs with (same-) high-Vt design considering gate Line-Edge Roughness (LER) at Vdd = 1V. For lightly-doped UTB MOSFETs, LER has been known as the dominant intrinsic variation source [16, 17]. To assess the LER, the line edge patterns were derived using the Fourier synthesis approach [17] with correlation length = 20nm, rms amplitude = 1.5nm [18]. μRSNM is defined as the mean of RSNM, and σRSNM is defined as the standard deviation of RSNM. It can be seen that InGaAs-OI/GeOI and GeOI SRAM cells with high-Vt design show comparable σRSNM compared with the (same-) high-Vt SOI SRAM cell at Vdd = 1V.

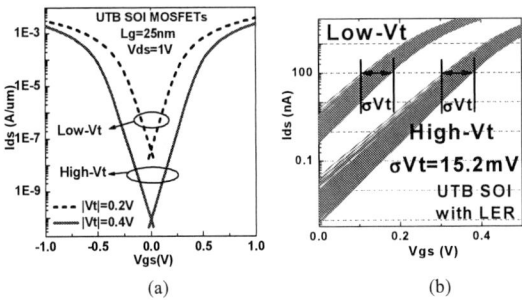

(a)   (b)

Fig. 6. (a) UTB SOI MOSFETs with high- and low-Vt design. (b) The σVt's due to LER are comparable (15.2mV) for the high- and low-Vt device

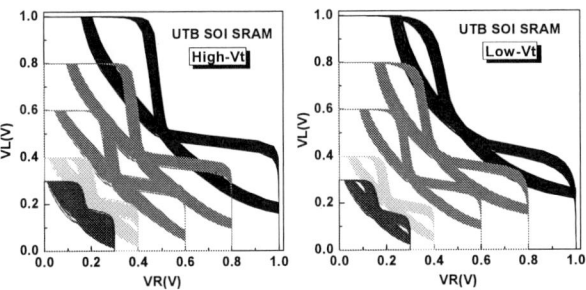

Fig. 7 High-Vt UTB SOI SRAMs show larger RSNM variation at low Vdd and larger RSNM at high Vdd than the low-Vt UTB SOI SRAMs.

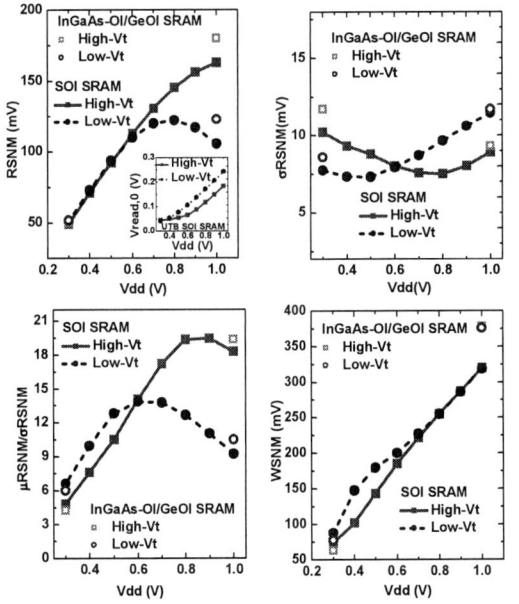

Fig. 8. High-Vt UTB SOI SRAMs show larger RSNM and smaller σRSNM at high Vdd than the low-Vt one. Low-Vt UTB SOI SRAMs show lower σRSNM and larger WSNM at low Vdd.

III. IMPACT OF VT DESIGN ON LOW-VOLTAGE (NEAR-/SUB-THRESHOLD) AND HIGH-PERFORMANCE (SUPER-THRESHOLD) SRAM CELLS

Device variability caused by continued technology scaling makes diminished Static Noise Margin (SNM) a serious

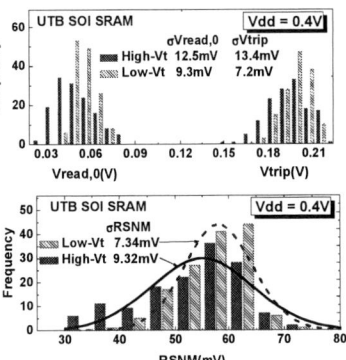

Fig. 9. At Vdd = 0.4V, high-Vt UTB SOI SRAM cell shows larger σVread,0, σVtrip and σRSNM than the low-Vt one.

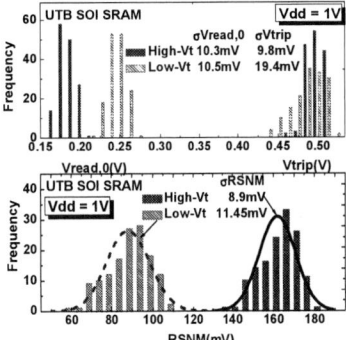

Fig. 10. At Vdd = 1V, low-Vt UTB SOI SRAM cell shows larger σVtrip and larger Vread,0, therefore, larger σRSNM and smaller RSNM.

problem for SRAM design. Here, we show that the Vt design can be used to improve the variability and stability of high-performance and low-voltage SRAMs. Fig. 6(a) shows I_{ds}-V_{gs} characteristics for UTB SOI MOSFETs with high- and low-Vt design. The Vt is adjusted by varying the work function [19]. Notice that the σVt's due to LER are comparable (15.2mV) for the high- and low-Vt device shown in Fig. 6(b).

Fig. 7 shows the RSNM considering LER for high-Vt and low-Vt UTB SOI SRAMs. High-Vt and low-Vt SOI MOSFETs with comparable Vt variation (Fig. 6(b)) show different impacts on RSNM variation at high Vdd and low Vdd. Fig. 8 shows that for low-voltage SRAMs operating at low Vdd, UTB SOI and InGaAs-OI/GeOI SRAM cells with low-Vt design show smaller σRSNM, larger μRSNM/σRSNM and larger WSNM than those with high-Vt design. However, for high-performance SRAMs operating at high Vdd, UTB SOI and InGaAs-OI/GeOI SRAM cells with high-Vt design show larger RSNM, smaller σRSNM and larger μRSNM/σRSNM than those with low-Vt design.

Fig. 9 illustrates why low-voltage SRAMs (Vdd = 0.4V) show smaller variability with low-Vt design. For low-voltage SRAMs operating at low Vdd, low-Vt devices show smaller drain current variation as its operation region moves slightly into the super-threshold region. Therefore, low-voltage SRAMs with low-Vt design show smaller σVread,0, σVtrip, and σRSNM than that with high-Vt design. In Fig. 10, for high-performance SRAMs operating at high Vdd (Vdd = 1V),

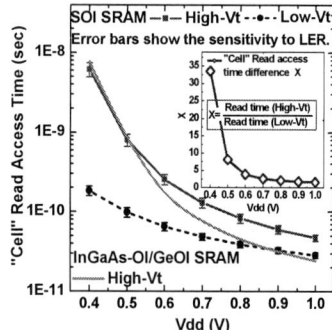

Fig. 11. The "Cell" Read access time degradation of high-Vt UTB SOI SRAM cell becomes smaller as Vdd increases.

Table 1. Design suggestion for low-voltage and high-performance SRAMs.
(↑ : Improve, ↓ : Degrade, − : Comparable)

SRAM		Devices Used	RSNM	σRSNM	μ/σ RSNM	WSNM	"Cell" Read Access Time	Cell Leakage
Low voltage (Near-/Sub-Vt)	Compared with high-Vt	Low Vt	−	↑	↑	↑	↑	↓
	Design Suggestion	• Use low Vt and well-controlled SCE devices to improve variability • Trade-off between performance and leakage						
High Performance (Super-Vt)	Compared with low-Vt	High Vt	↑	↑	↑	−	↓	↑
	Design Suggestion	• Use high Vt devices to improve stability/variability • III-V-OI devices to improve performance						

SRAMs with high-Vt design show smaller Vread,0 and larger RSNM than that with low-Vt design because the pass-gate transistor is very sensitive to Vt. At Vdd = 1V, SRAMs with low-Vt design show larger σVtrip and σRSNM. As can be seen in Fig. 10 (top figure), the Vtrip of low-Vt SRAMs ranges from 0.44 to 0.51V, while the Vtrip of high-Vt SRAMs ranges from 0.48-0.51V. With balanced N/PFET(same |Vt| for N/PFET), NFET is stronger than PFET in the super-threshold region due to more significant contribution of mobility to drain current, while their strength would be comparable in the subthreshold region. Therefore, under the bias condition for determining Vtrip, low-Vt pull-up and pull-down devices would operate more/deeper into super-threshold region, resulting in smaller Vtrip tail and larger σVtrip.

Fig. 11 shows that the high-Vt SOI SRAMs show larger "Cell" Read access time than the low-Vt SOI SRAMs, and the "Cell" Read access time difference between high-Vt and low-Vt SRAMs becomes smaller as Vdd increases as shown in the inset. Compared with low-Vt design, SOI SRAMs with high-Vt design show improvement in variability while sacrificing the "Cell" Read access time. However, the InGaAs-OI/GeOI SRAMs with high-Vt design show much smaller "Cell" Read access time than the UTB SOI SRAM cells with (same-) high-Vt design at Vdd = 1V. In other words, for high-performance SRAMs at Vdd=1V, InGaAs-OI/GeOI SRAMs with high-Vt design simultaneously offer smaller variability and better performance over the (same-) high-Vt UTB SOI SRAM. Table 1 summarizes the design suggestions for low-voltage and high-performance SRAMs.

IV. CONCLUSION

The stability, variability, performance and cell leakage of InGaAs-OI/GeOI and GeOI SRAM cells are analyzed and compared with the SOI SRAM cells. For low-voltage SRAMs, low-Vt design shows smaller variability, larger cell leakage and smaller "Cell" Read access time, and the design trade-off between performance and leakage should be considered. For high-performance SRAMs, SOI SRAMs with high-Vt design show smaller variability and degraded performance. Using InGaAs-OI(NFET)/GeOI(PFET) SRAMs with high-Vt design improves the stability and performance while maintaining comparable RSNM variations for high-performance SRAM applications.

ACKNOWLEDGMENT

This work was supported in part by the National Science Council of Taiwan under Contracts NSC 100-2628-E-009-024-MY2, in part by the Ministry of Economic Affairs in Taiwan under Contract 100-EC-17-A-01-S1-124, and in part by the Ministry of Education in Taiwan under ATU Program.

REFERENCES

[1] T. Skotnicki et al., "How can high mobility channel materials boost or degrade performance in advanced CMOS," *Symp. VLSI Tech.*, pp. 153-154, 2010.

[2] Y.-S. Wu et al., "A Closed-Form Quantum Dark Space Model for Predicting the Electrostatic Integrity of Germanium MOSFETs with High-k Gate Dielectric," *IEEE TED*, vol. 59, no. 3, pp. 530-535, 2012.

[3] V. P.-H. Hu et al., "Investigation of Electrostatic Integrity for Ultra-Thin-Body Germanium-On-Nothing (GeON) MOSFET," *IEEE TNano*, vol. 10, no. 2, pp. 325-330, 2011.

[4] S. H. Kim et al., "High performance Extremely-thin Body III-V-On-Insulator MOSFETs on a Si substrate with Ni-InGaAs metal S/D and MOS Interface Buffer Engineering," *Symp. VLSI Tech.*, pp. 58-59, 2011.

[5] L. Hutin et al., "GeOI pMOSFETs Scaled Down to 30-nm Gate Length With Record Off-State Current," *IEEE EDL*, vol. 31, no. 3, pp. 234-236, 2010.

[6] S. Oh et al., "Vaibility Study of All-III-V SRAM for Beyond-22-nm Logic Circuits," *IEEE EDL*, vol. 32, no. 7, pp. 877-879, 2011.

[7] J.-P. Kulkarni et al., "Technology Circuit Co-Design for Ultra Fast InSb Quantum Well Transistors," *IEEE TED*, vol. 55, no. 10, pp. 2537-2545, 2008.

[8] V.P.-H. Hu et al., "Impact of Work Function Design on the Stability and Performance of Ultra-Thin-Body SOI Subthreshold SRAM, " *ESSDERC*, pp. 145-148, 2009.

[9] E. Batail et al., "Localized ultra-thin GeOI: An innovative approach to germanium channel MOSFETs on bulk Si substrates", *IEDM Tech. Dig.*, pp. 397-400, 2008.

[10] M. Poljak et al., "Features of Electron Mobility in Ultrathin-Body InGaAs-On-Insulator MOSFETs down to Body Thickness of 2 nm," *IEEE SOI conf.*, 2011.

[11] V.P.-H. Hu et al., "Band-to-Band-Tunneling Leakage Suppression for Ultra-Thin-Body GeOI MOSFETs Using Transistor Stacking," *IEEE EDL*, vol. 33, no. 2, pp. 197-199, 2012.

[12] *Sentaurus TCAD, C2009-06 Manual.*

[13] V. P.-H. Hu et al., "Comprehensive Analysis of UTB GeOI Logic Circuits and 6T SRAM Cells considering Variability and Temperature Sensitivity," *IEDM Tech. Dig.*, pp. 753-756, 2011.

[14] S.K. Gupta et al., "Exploration of device-circuit interactions in FinFET-based memories for sub-15nm technologies using a mixed mode quantum simulation framework: Atoms to systems," *IEDM Tech. Dig.*, pp. 757-760, 2011

[15] V. P.-H. Hu et al., "Evaluation of Static Noise Margin and Performance of 6T FinFET SRAM Cells with Asymmetric Gate to Source/Drain Underlap Devices," *IEEE SOI Conference*, pp. 74-75, 2010.

[16] T. Ohtou et al., "Impact of Parameter Variations and Random Dopant Fluations on Short-Channel Fully Depleted SOI MOSFETs with Extremely Thin BOX," *IEEE EDL*, vol. 28, no. 8, pp. 740-742, 2007.

[17] A. Asenov et al., "Intrinsic parameter fluctuations in decananometer MOSFETs introduced by gate line edge roughness," *IEEE TED*, vol. 50, no. 5, pp. 1254-1260, 2003.

[18] A. Dixit et al., "Impact of Stochastic Mismatch on Measured SRAM Performance of FinFETs with Resis/Spacer-Defined Fins: Role of Line-Edge-Roughness," IEDM Tech. Dig., 27.8, 2006.

[19] J. Pery et al., "A VFB tunable Single Metal Single Dielectric approach using As I/I into TiN/HfO2 for 32nm node and beyond," *VLSI-TSA*, pp. 57-58, 2009.

978-1-4673-1707-8/12 $31.00 © 2012 IEEE

Two-step annealing effects on ultrathin EOT higher-k (k = 40) ALD-HfO$_2$ gate stacks

Yukinori Morita, Shinji Migita, Wataru Mizubayashi, Meishoku Masahara, and Hiroyuki Ota

Green Nanoelectronics Center (GNC), Nanoelectronics Research Institute,
National Institute of Advanced Industrial Science and Technology (AIST)
Tsukuba West SCR, 16-1 Onogawa, Tsukuba, Ibaraki 305-8569, JAPAN
e-mail address: y.morita@aist.go.jp

Abstract—We fabricated ultrathin HfO$_2$ gate stacks of very high permittivity by atomic layer deposition (ALD) and novel two-step post-deposition annealing (PDA) technique. First, no-cap PDA degasses residual contaminations in ALD layer, and second, Ti-cap PDA enhances permittivity of HfO$_2$ by generating cubic crystal phase without SiO$_2$ interfacial layer growth. Using these techniques, the dielectric constant of the ALD-HfO$_2$ can be enhanced to ~40, and a 0.3 nm equivalent oxide thickness is obtained.

I. INTRODUCTION

For equivalent oxide thickness (EOT) scaling in the latest device technology, the increase in the dielectric constant (k) of a gate insulator material has been a promising solution ever since "high-k" dielectrics were adopted in a gate stack structure. In the sub-0.5-nm EOT regime, however, the physical thickness of a high-k HfO$_2$ (k ~ 13–20) gate insulator needs to be as small as ~2 nm with a direct-contact architecture with no interfacial layer (IL). [1-4] This physical thickness is close to the direct-tunneling limit of the gate leakage current. This implies that for further EOT scaling, further enhancement of the permittivity of high-k dielectrics (higher-k) is needed. [5-10]

In our previous study, we developed an oxygen-controlled cap post-deposition annealing (cap PDA) technique, which realized a k>40 HfO$_2$ gate stack without a SiO$_2$ IL. [10,11] A high-permittivity "cubic" phase HfO$_2$ was generated by abrupt and short-time annealing with a Ti-cap layer. The Ti-cap layer stabilizes the metastable cubic phase and absorbs excess oxygen in the HfO$_2$ layer during annealing, suppressing SiO$_2$ IL growth.

In this study, we fabricate ultrathin higher-k gate stacks by a novel two-step PDA technique consisting of no-cap and Ti-cap annealing for HfO$_2$ dielectric layer prepared by atomic layer deposition (ALD). As a result, the dielectric constant of ALD-HfO$_2$ increases to k ~ 40 over a wide range of HfO$_2$ physical thickness, and HfO$_2$ gate stacks of extremely thin EOT (~0.3 nm) are realized.

II. EXPERIMENTAL SETUP

Fig.1 shows the concept of permittivity enhancement by the cap PDA on TiN/HfO$_2$/Si gate stacks. [8,10,11] The HfO$_2$ layers were deposited by HfO$_2$ target sputtering. By abrupt and short-time annealing with a TiN-cap layer, amorphous HfO$_2$ crystallizes into a high-permittivity cubic phase. X-ray diffraction (XRD) spectra obtained before and after TiN-cap PDA clearly show that the amorphous phase transforms into the cubic phase. (**Figs.1(c) & (d)**). However, the results of cross-sectional transmission electron microscopy (X-TEM) show that a thick SiO$_2$ IL grows simultaneously. (**Figs.1(a) & (b)**) This is caused by oxygen released from HfO$_2$ during crystallization.

To suppress IL growth, we chose a Ti-cap layer instead of the TiN in this study. Ti can absorb up to 30 at.% oxygen without forming Ti oxide. By controlling Ti and HfO$_2$ thicknesses and PDA temperature, SiO$_2$ IL growth can be

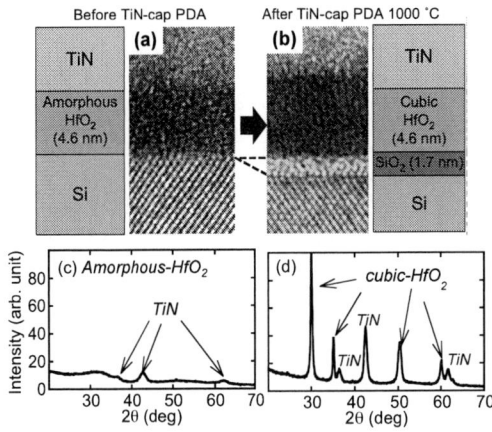

Fig. 1 Concept of k-enhancement by the cap PDA method. (a) and (b) show the X-TEM images of HfO$_2$ gate stacks obtained before and after TiN-cap PDA, respectively. The gate stack structures are also shown. (c) and (d) show XRD spectra for TiN-cap HfO$_2$ before and after cap-PDA, respectively. By abrupt and short-time annealing with a cap layer, amorphous HfO$_2$ crystallizes into the cubic phase. The dielectric constant of cubic HfO$_2$ is increased to ~50. However, a SiO$_2$ IL grew simultaneously because of the oxygen released during crystallization.

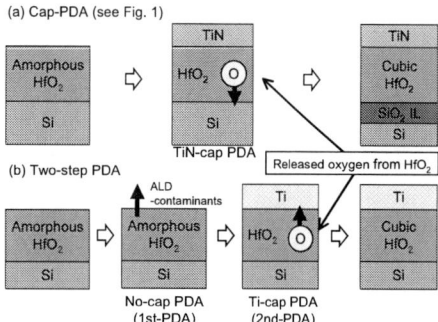

Fig.2 Concept of IL suppression for cubic phase generation in the present study. In (a), TiN barely absorbs oxygen, and then, oxygen released from HfO₂ oxidizes the Si substrate. In (b), no-cap PDA removes contaminants in ALD-HfO₂ layer, and Ti-cap PDA crystallizes HfO₂ into cubic phase without IL formation.

suppressed. In addition, another issue relating to PDA for ALD-HfO₂ layer exists: The high-k films deposited by ALD contain considerable amount of fragments from ALD precursors. These ALD-related contaminants severely degrade dielectric performance, and thus, an appropriate annealing process for degassing is needed. We used a combination of PDAs consisting of first no-cap PDA and second Ti-cap PDA. **Fig.2** shows a comparison between (a) cap-PDA and (b) two-step oxygen-controlled PDA in the present experiment. The Ti-cap PDA crystallizes HfO₂ into cubic phase, and suppresses SiO₂ IL growth reducing oxygen during crystallization.

Fig.3 shows the process flow of a higher-k HfO₂ gate stacks in this study. The wafers were treated with diluted HF to remove the SiO₂ sacrificial oxide. ALD-HfO₂ was directly deposited on the OH-terminated Si surface to form a Si–O–Hf direct contact interface [3] and then short-time PDA was performed in vacuum condition. After the deposition of Ti and 5 nm TiN gate electrodes, the second short-time PDA was performed with Ti cap. For the higher-k gate stack formation, the short-time annealing is the key to generate a metastable cubic HfO₂. [8,12] **Fig.3(b)** shows the time dependence of the specimen temperature for thermal treatments at 910 °C. The maximum and average ramp rates were 86 and 57 °C/s, respectively. These values are sufficiently large for generating the cubic HfO₂ phase. After

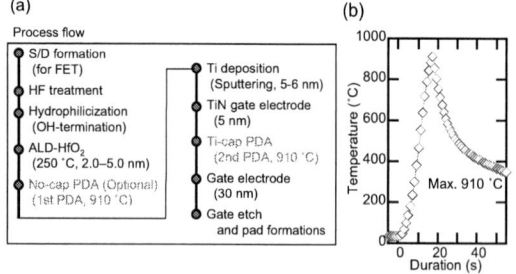

Fig.3 (a) Sample preparation procedure for this experiment. The HfO₂ layer is directly deposited on OH-terminated Si without a SiO₂ IL. (b) Time dependence of specimen temperature for 910 °C thermal treatments.

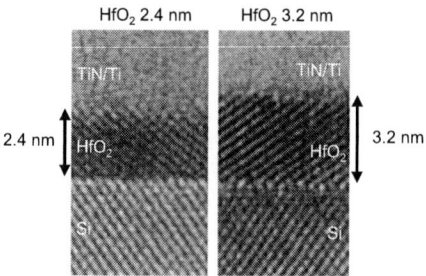

Fig.4 X-TEM images of the TiN/Ti/ALD-HfO₂/Si gate stacks after Ti-cap PDA at 910 °C. No SiO₂ IL is observed, and the HfO₂ layers crystallize in both the 2.4 and 3.2 nm cases.

the device fabrication, the electrical properties of the gate stacks were measured.

We note another effect of such a cap material with large oxygen absorbability, which is "IL scavenging." [1,2,10] This effect decomposes the SiO₂ IL and effectively reduces the EOT in the high-k gate stacks. In our previous study, it was difficult to control SiO₂ decomposition, and Si-Hf silicidation by excess SiO₂ decomposition was unavoidable. Therefore, in this experiment, we chose thermal treatment conditions and Ti thickness carefully to avoid the IL scavenging effect.

III. RESULTS

First, we confirmed the physical effect of Ti-cap for the ALD-HfO₂ gate stacks. (**Fig.4**) By the Ti-cap PDA for TiN/Ti/HfO₂/Si stacks, no SiO₂ IL was formed in both (a) and (b) specimens, showing the effective suppression of SiO₂ IL growth. **Fig.5** shows (a) front-side and (b) back-side secondary ion mass spectroscopy (SIMS) analysis results for the 2.4 nm HfO₂ gate stack, confirming the depth profiles of (a) O, Si, and (b) Ti. After Ti-cap PDA at 910 °C, the accumulation of O atoms in the TiN/Ti layer was confirmed. These O atoms were considered the residuals released from HfO₂, and the Ti-cap layer effectively absorbed O atoms. On the other hand, no Si accumulation in the TiN/Ti layer was observed. In our previous study, the Si atoms from the SiO₂ IL decomposed by the IL scavenging effect migrated into the TiN/Ti layer. The absence of Si accumulation in the present

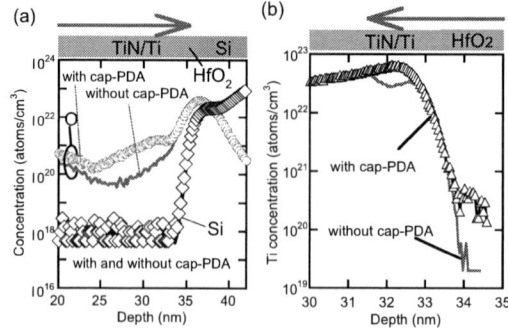

Fig.5 (a) Front-side and (b) back-side SIMS analysis results for the TiN/Ti/ALD-HfO₂/Si stacks with Ti-cap PDA, confirming the distribution of O, Si, and Ti atoms. The PDA temperature is 910 °C. The measurement directions (arrows) and stack structures are superimposed on the figures.

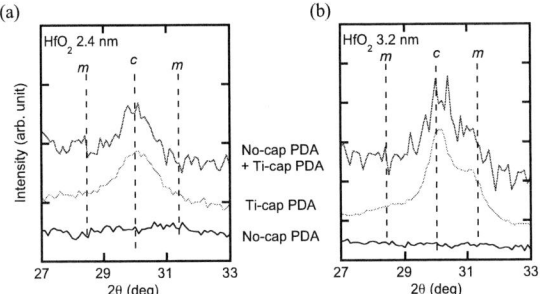

Fig.6 In-plane XRD analysis results for 2.4 and 3.2 nm ALD-HfO$_2$ after three types of PDA combinations at 910 °C. The indexed c and m peaks indicate the cubic and monoclinic phases of HfO$_2$, respectively.

result indicates that SiO$_2$ IL decomposition was suppressed. In (b), ~1% Ti migration into HfO$_2$ layer was observed.

Fig.6 shows in-plane X-ray diffraction (XRD) spectra for the (a) 2.4 and (b) 3.2 nm HfO$_2$ stacks. The HfO$_2$ layers treated only by no-cap PDA were amorphous. As for the Ti-cap PDA case, the cubic phase was generated in 2.4 nm HfO$_2$, and the cubic and monoclinic phases coexisted in 3.2 nm HfO$_2$. After two-step PDA, the shape of the spectra appeared to be similar to those of Ti-cap PDA instead of an insertion of no-cap PDA before Ti-cap PDA.

Fig.7 shows the C–V characteristics of TiN/Ti/HfO$_2$/Si gate stacks. The data were frequency-corrected with 500 kHz and 1 MHz. In (a) (Ti-cap PDA), the 2.4-nm-thick HfO$_2$ gate stack showed a very large capacitance value. The ideal C–V curve for 0.25 nm EOT is also plotted in the figure, which is simulated by considering the quantization capacitance at the high-k/Si interface. [13] The ideal C–V curve shows good agreement with that for the 2.4-nm-thick HfO$_2$ gate stack. This implies that the very large accumulation capacitance corresponds to the very small EOT (0.25 nm). The impedance–frequency (Z–f) measurements of the 0.25-nm-EOT TiN/Ti/HfO$_2$/Si gate stack result also indicates the validity of the 0.25 nm EOT in the present study. (**Fig.8**) [11] In (b), the capacitance for 3.2 nm HfO$_2$ gate stack only annealed by the Ti-cap PDA drastically increased with two-step PDA.

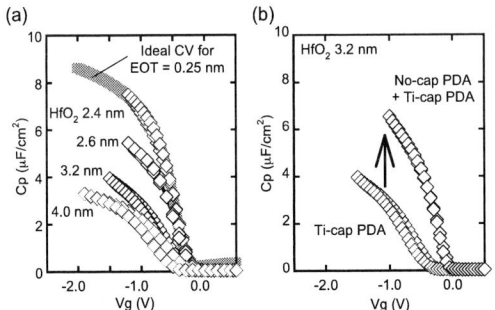

Fig. 7 C–V curves of TiN/Ti/ALD-HfO$_2$/Si nMOS gate capacitors. The data are frequency-corrected with 500 kHz and 1 MHz. The Ideal C–V curve is simulated by MIRAI-ACCEPT. (a) Variation of capacitance curve for HfO$_2$ thickness changes with the Ti-cap PDA. (b) Effect of insertion of no-cap PDA before Ti-cap PDA for C–V. The HfO$_2$ thickness is 3.2 nm.

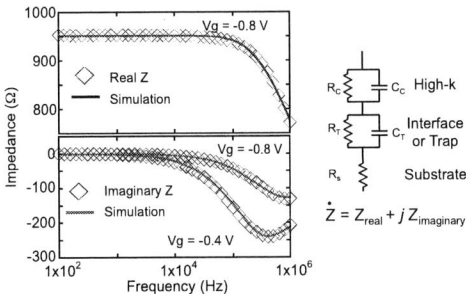

Fig. 8 Z-f measurements of TiN/Ti/ALD-HfO$_2$/Si gate stacks of 0.25 nm EOT. Simulated curves are also plotted on the basis of the three-component device model shown on the right-hand side. The real and imaginary data agree with the simulated curves, indicating validity of 0.25 nm EOT in present experiment.

Fig.9 shows the relationship between HfO$_2$ physical thickness (T$_{Phys}$) and EOT for TiN/Ti/HfO$_2$/Si gate stacks. In the case of the Ti-cap PDA (1st PDA), the very small EOT abruptly increased at around T$_{Phys}$ = 2.6 nm. The dielectric constant (k) was estimated to be 40 below T$_{Phys}$ = 2.6 nm and 18 above it, from the slope of fitting lines. The very high dielectric constant of around 40 possibly originated from the cubic phase. On the other hand, permittivity degraded at larger T$_{phys}$. In case of the standard long-time PDA (700 °C, 20 s) process, the EOT values increased and a k value of ~13 was estimated. This was caused by generation of low permittivity monoclinic phase, although Ti-cap PDA was added after standard PDA. The monoclinic phase is obviously stable and cannot be transformed into cubic phase even if the Ti-cap PDA is added. The best performance of the EOT in this measurement was obtained for two-step PDA. From the slope of the fitting line, a k value of ~40 was estimated for 2.4, 3.2, and 4.0 nm of HfO$_2$ layers.

IV. DISCUSSION

As shown in **Fig.6**, HfO$_2$ of T$_{Phys}$ = 2.4 nm crystallized into the cubic phase with Ti-cap PDA at 910 °C. The high permittivity cubic phase in dielectric layer may be the origin of very small EOT. In the case of 3.2 nm HfO$_2$, the monoclinic phase simultaneously grew with the cubic phase. The dielectric constant degraded at T$_{Phys}$ over 2.6 nm HfO$_2$. (**Fig.9**) Does the cogeneration of the low-permittivity

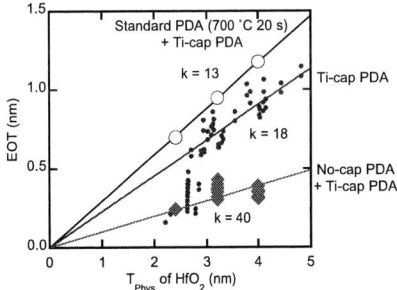

Fig. 9 EOT-T$_{Phys}$ plot for the TiN/Ti/ALD-HfO$_2$/Si gate stacks prepared by three types of PDA combinations.

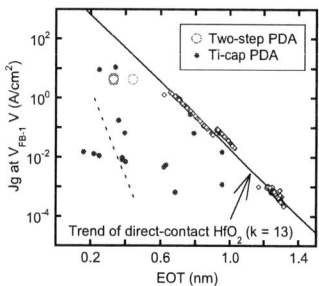

Fig. 10 Jg–EOT$_s$ plot for the TiN/Ti/ALD-HfO$_2$/Si gate stacks prepared by Ti-cap (closed dots) and two-step (open circles) PDAs. The trend of direct-contact HfO$_2$ gate stacks (k = 13) is also plotted for comparison. [4]

monoclinic phase result in the drop of the dielectric constant with Ti-cap PDA?

Insertion of the no-cap PDA before Ti-cap PDA suppressed degradation of permittivity, as shown in **Fig.9**. This implies that no-cap PDA is a key process for permittivity enhancement. However, the no-cap PDA itself did not generate the cubic phase. The HfO$_2$ layers annealed only by the no-cap PDA were amorphous.

The XRD spectra after two-step PDA show no considerable difference from those of Ti-cap PDA. Of course the generation of the cubic phase would be one main reason for permittivity enhancement in the two-step PDA. One possible origin of permittivity degradation on Ti-cap PDA is attributed to the ALD-related impurities in HfO$_2$ layers. The Ti-cap layer blocks desorption of the ALD-related contaminants. These fragments could prevent the condensation of ALD films and the growth of the cubic phase crystal domain at the initial stage of Ti-cap PDA. A small T$_{Phys}$ (2.4 nm) with less impurities would not suppress film condensation.

The insertion of no-cap PDA improves the permittivity of the HfO$_2$ layer. However, the insulating performance, at the same time, is affected by no-cap PDA. (**Fig.10**) The Jg value appears to be degraded after two-step PDA compared with that after the Ti-cap PDA. Defects relating to the excess oxygen desorption are generated during no-cap PDA process at 910 °C, which may be the origin of a larger Jg. To take advantage of the higher-k HfO$_2$, each PDA parameter must be optimized.

V. SUMMARY

We demonstrated extremely scaled higher-k ALD-HfO$_2$ gate stacks. The enhancement of permittivity was conducted by combination of no-cap PDA, which removes ALD-contaminants from films, and Ti-cap PDA for crystallization of ALD-HfO$_2$ into cubic crystallographic phase. The SiO$_2$ IL was also effectively suppressed by oxygen absorption effect of Ti capping layer. The dielectric constant of HfO$_2$ was effectively increased to ~40 for wider range of HfO$_2$ physical thickness, and thinnest EOT of ~0.3 nm was achieved. The ultrathin gate stack technology for sub-0.5 nm EOT regime will be furthermore important for not only the ultrascaled

MOSFET application but also novel devices such as a tunnel-FET.

ACKNOWLEDGMENT

Technical support was provided by the ICAN, AIST. The work is supported by a grant by JSPS through the First Program, "Development of Core Technologies for Green Nanoelectronics" initiated by CSTP. This work is also based on prior research in the MIRAI Project.

REFERENCES

[1] C. Choi, C.Y. Kang, S.J. Rhee, M.S. Abkar, S.A. Krishna, M.H. Zhang, H. Kim, T. Lee, F. Zhu, I. Ok, S. Koveshnikov, and J.C. Lee, "Fabrication of TaN-gated ultra-thin MOSFETs (EOT < 1.0nm) with HfO$_2$ using a novel oxygen scavenging process for sub 65nm application," Symp. VLSI Tech. Dig., pp. 226-227, 2005.

[2] T. Ando, M.M. Frank, K. Choi, C. Choi, J. Bruley, M. Hopstaken, M. Copel, E. Cartier, A. Kerber, A. Callegari, D. Lacey, S. Brown, Q. Yang, and V. Narayanan, "Understanding Mobility Mechanisms in Extremely Scaled HfO$_2$ (EOT = 0.42 nm) Using Remote Interfacial Layer Scavenging Technique and V-t-tuning Dipoles with Gate-First Process," IEDM Tech. Dig., pp. 394-397, 2009.

[3] Y. Morita, A. Hirano, S. Migita, H. Ota, T. Nabatame, and A. Toriumi, "Impact of Surface Hydrophilization prior to Atomic Layer Deposition for HfO$_2$/Si Direct-Contact Gate Stacks," Appl. Phys. Express, vol. 2, p. 011201, 2009.

[4] Y. Morita, S. Migita, N. Taoka, W. Misubayashi, and H. Ota, "Oxygen-Terminated Si Surface for Atomic Layer Deposition and its Impact on Interfacial Electrical Quality of sub-nm-EOT high-k Gate Stacks," Ext. Abstr. SSDM, pp. 52-53, 2009.

[5] K. Kita, K. Kyuno, and A. Toriumi, "Permittivity increase of yttrium-doped HfO$_2$ through structural phase transformation," Appl. Phys. Lett., vol. 86, p. 102906, 2005.

[6] M. Suzuki, M. Tomita, T. Yamaguchi, and N. Fukushima, "Ultra-thin (EOT = 3 angstrom) and low leakage dielectrics of La-alminate directly on Si substrate fabricated by high temperature deposition," IEDM Tech. Dig., pp. 445-448, 2005.

[7] H. Arimura, N. Kitano, Y. Naitou, Y. Oku, T. Minami, M. Kosuda, T. Hosoi, T. Shimura, and H. Watanabe, "Excellent electrical properties of TiO$_2$/HfSiO/SiO$_2$ layered higher-k gate dielectrics with sub-1?nm equivalent oxide thickness," Appl. Phys. Lett., vol. 92, p. 212902, 2008.

[8] S. Migita, Y. Watanabe, H. Ota, H. Ito, Y. Kamimuta, T. Nabatame, and A. Toriumi, "Design and demonstration of very high-k (k > 50) HfO$_2$ for ultra-scaled si CMOS," Symp. VLSI Tech. Dig., pp. 152-153, 2008.

[9] M.M. Frank, S. Kim, S.L. Brown, J. Bruley, M. Copel, M. Hopstaken, M. Chudzik, and V. Narayanan, "Scaling the MOSFET gate dielectric: From high-k to higher-k? (Invited Paper)," Microelectron. Eng., vol. 86, pp. 1603-1608, 2009.

[10] Y. Morita, S. Migita, W. Mizubayashi, and H. Ota, "Direct-contact higher-k HfO$_2$ gate stacks by oxygen-controlled cap-post deposition annealing," Jpn. J. Appl. Phys., vol. 50, p. 10PG01, 2011.

[11] Y. Morita, S. Migita, W. Mizubayashi, and H. Ota, "Extremely scaled (~0.2 nm) equivalent oxide thickness of higher-k (k = 40) HfO$_2$ gate stacks prepared by atomic layer deposition and oxygen-controlled cap post-deposition annealing," Jpn. J. Appl. Phys., vol. 51, p. 02BA04, 2012.

[12] Y. Nakajima, K. Kita, T. Nishimura, K. Nagashio, and A. Toriumi, "Experimental Demonstration of Higher-k Phase HfO$_2$ Through Non-Equilibrium Thermal Treatment," ECS Trans., pp. 203-212, 2010.

[13] N. Yasuda, H. Ota, T. Horikawa, T. Nabatame, H. Satake, A. Toriumi, Y. Tamura, T. Sasaki, and F. Ootsuka, "Reliable Extractions of EOT and V$_{fb}$ in Poly-Si Gate High-k MISFETs through Advanced Modeling of Gate and Substrate Capacitances," Ext. Abstr. SSDM, pp. 250-251, 2005.

Thin Germanium Dioxide Film with a High Quality Interface Formed in a Direct Neutral Beam Oxidation Process

Akira Wada and Seiji Samukawa

Institute of Fluid Science, Tohoku University
Micro System Integration Center, Tohoku University
2-1-1 Katahira, Aoba-ku, Sendai, Miyagi 980-8577, Japan
Samukawa@ifs.tohoku.ac.jp

Rui Zhang and Shinichi Takagi

School of Engineering
The University of Tokyo
2-11-16 Yayoi, Bunkyo-ku, Tokyo 113–8656, Japan

Abstract—Germanium dioxide (GeO_2) thin film with a high-quality interface was directly formed by using a damage-free and low-temperature neutral beam oxidation (NBO) process. GeO_2 film with little suboxide could be formed even at a low substrate temperature of 300°C because of the extremely low activation energy (E_a) oxidation resulting from bombardment with energetic oxygen neutral-beams of 5 eV. A high-quality GeO_2/Ge interface with an low interface state density (D_{it}) of less than 1×10^{11} $cm^{-2}eV^{-1}$ was created by combining the NBO process with a hydrogen (H) radical native oxide removal treatment.

I. Introduction

Germanium (Ge) complementary metal oxide semiconductor (CMOS) devices have been widely investigated for their potential as next-generation size-scaled silicon (Si)-CMOS devices. However, Ge-FET mobility is easily degraded by the instability of the dielectric/Ge interface the instability of the dielectric/Ge interface caused by the poor thermal stability of Ge and germanium dioxide (GeO_2) films. Ge monoxide (GeO) desorption occurs at the interface during high temperature processes exceeding 400°C [1,2]. The result is in an increase in surface roughness and interfacial state density (D_{it}). Recently, several Ge pMOSFETs with a GeO_2/Ge interface formed by thermal oxidation were reported to have low D_{it} [3,4]. However, these GeO_2 films had very thick (more than 20 nm) equivalent oxide thicknesses (EOTs). It is difficult to scale down the EOT and maintain a low D_{it} at the same time. Furthermore, thin GeO_2 is so weak that it easily degrades the GeO_2/Ge interface [5].

To suppress the degradation of dielectric/Ge interfaces that occurs with thin GeO_2 films, we developed a low-

temperature and damage-free NBO process. In this study, we used this process to form thin GeO_2 films and investigated the resulting film's quality by analyzing its composition. We were able to create a high-quality GeO_2/Ge interface with an extremely low D_{it} of less than 1×10^{11} $cm^{-2}eV^{-1}$ for Au/Al_2O_3/GeO_2/Ge MOS capacitors.

II. Experimental Method

Neutral beam (NB) is an advanced technique to achieve damage-free processing [6,7]. A typical NB apparatus consists of plasma and process chambers that are separated by a silicon aperture (Fig. 1). An H radical treatment can also be performed in an L/L chamber. The Si aperture can effectively neutralize charged particles and eliminate UV photons when the plasma passes through it. Therefore, the surface oxidation can proceed with neutral beam oxidation (NBO) without any damage from charged particles or high-energy photons in the plasma. The details of the NBO system are described elsewhere [7,8].

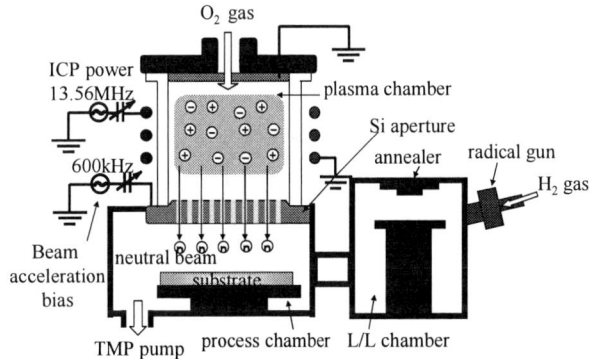

Figure 1. Neutral beam apparatus

This works was supported in part by the Creation of Innovation Centers for Advanced Interdisciplinary Research Areas Program, Japan.

A p-type Ge (100) substrate was used for surface oxidation. Thin GeO$_2$ films were formed by using the damage-free, direct NBO process at a substrate temperature of 300°C. The plasma was generated by inductively coupled plasma (ICP; 13.56 MHz), and the power was 500 W. The energy of the oxygen NB (which ranged from 1 to 10 eV) was precisely controlled by changing the beam acceleration bias voltage and the pressure in the process chamber (Fig. 1).

We compared the oxide removal performance of the H radical treatment with that of a wet treatment [diluted hydrofluoric acid (DHF-2%)]. For the H radical treatment, 40 sccm of hydrogen gas was introduced into the radical generator. A 190-W microwave (2.45 GHz) was applied to the generator to generate the H radicals, which then irradiated the sample Ge substrate at 280°C for 30 minutes. The DHF treatment was performed for 2 minutes at room temperature, followed by a rinse in deionized water.

We fabricated MOS capacitors to evaluate the D$_{it}$ at the GeO$_2$/Ge interface; the process flow of the fabrication is shown in Fig. 2. After direct formation of extremely thin GeO$_2$ film using the NBO process, Al$_2$O$_3$ film was deposited by using atomic layer deposition (ALD) at 300°C [5]. After high-k film deposition, post-deposition annealing (PDA) was performed at 400°C in N$_2$ ambient for 30 min. An Au electrode and an Al back contact were then deposited by thermal evaporation.

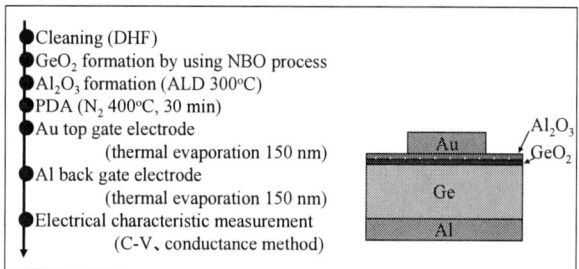

Figure 2. Process flow for fabricating Ge MOS capacitors

III. RESULTS AND DISCUSSION

First, the native oxide removal process was investigated. Because native oxide is easily formed in air for Ge, a dry process using H radicals was chosen. Since the chamber for the H radical treatment is connected to the NBO apparatus and the sample can be transported into the NBO chamber without breaking the vacuum (Fig. 1), undesired growth of native oxide does not occur. Figure 3 shows the XPS spectra of the Ge substrate before and after the H radical or DHF (2%) treatment. The native oxide peak (about 33 eV) was almost completely removed after the H radical treatment, whereas it still remained after the DHF treatment. Prior to performing the NBO process, we believed that the native oxide might completely disappear because the sample was transported into the NBO chamber after the H radical treatment without breaking the vacuum. Under this condition, we measured the

surface roughness of the Ge substrate after native oxide removal by using atomic force microscopy (AFM) (Fig. 4). The surface remained atomically flat even after the H radical treatment. These results indicate that H radical treatment is an effective process for removing native oxide.

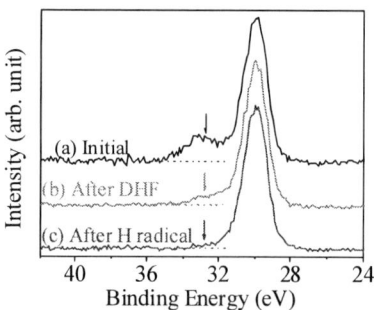

Figure 3. Ge 3d region of XPS spectra before and after native oxide removal

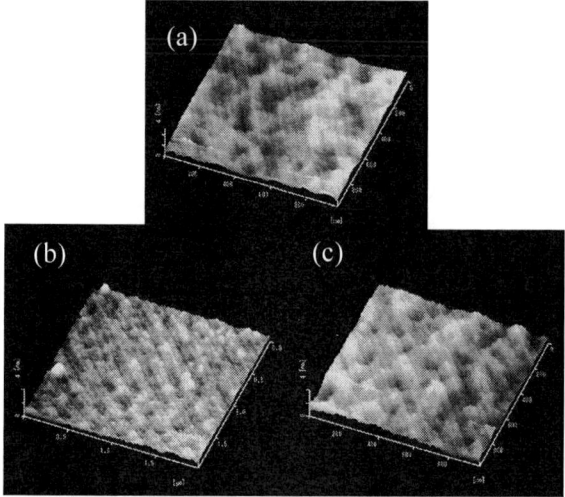

Figure 4. AFM images of Ge substrate: (a) initial condition, (b) after H radical treatment, (c) after DHF treatment. RMS values were respectively 0.29, 0.19, and 0.31.

Figure 5 shows the oxidation rate as a function of the energy of oxygen NB. From Fig. 5(a), we can see that the oxidation rate at the beam energy of 10 eV was almost independent of substrate temperature when the film thickness was less than about 2.5 nm. The oxidation rate in this region is linear with respect to the neutral-beam irradiation time, which limits the surface reaction [9]. Conversely, at the beam energy of 1 eV, the oxidation rate was slightly dependent on temperature when the film thickness was less than about 2.5 nm, as shown in Fig. 5(b). To investigate what effect the beam energy had on the oxidation reaction, we calculated activation energy (E$_a$) for the oxidation reaction (Fig. 6). The figure also shows the activation energy for radical oxidation, which is reported to be about 0.03 eV [10]. It should be noted that E$_a$ of the oxidation reaction in the NBO process is much lower than the values reported for the conventional thermal (1.73 eV) [11] and radical oxidation processes. Specifically, we achieved a very low E_a (below 0.005 eV) when oxygen NBs with energies of 5 and 10 eV were used. These results indicate that oxide film growth with NBO does not depend on the substrate temperature, and therefore, it is possible to achieve low-temperature oxidation.

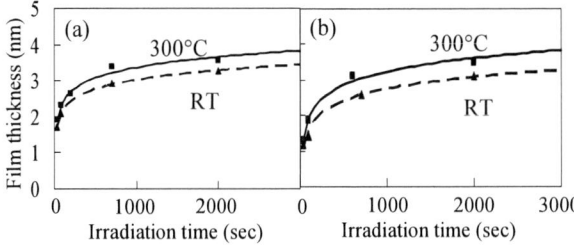

Figure 5. Oxidation rate at beam energy of (a) 10 eV and (b) 1 eV.

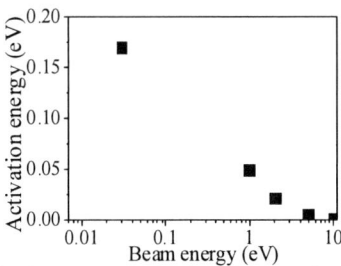

Figure 6. Activation energy as a function of energy of oxygen atoms. The activation energy for radical oxidation (0.03 eV) is also shown.

Figure 7 shows a cross-sectional transmission electron microscopy (TEM) image of a GeO_2 thin film formed by NBO. The interface between the GeO_2 and Ge substrate was very flat, without any roughness or lattice defects. Figure 8 shows the XPS spectra of GeO_2 thin films (3-nm thick) formed using NBO at 1, 5, and 10 eV. The spectra were fitted to peaks of Ge^0(elemental Ge $3d_{5/2}$, 29.3 eV), Ge^{1+} (Ge_2O $3d_{5/2}$, 30.1 eV), Ge^{2+} (GeO $3d_{5/2}$, 31.1 eV), Ge^{3+} (Ge_2O_3 $3d_{5/2}$, 31.9 eV), and Ge^{4+} (GeO_2 $3d_{5/2}$, 32.7 eV). For the GeO_2 thin films formed with NBO, the area of GeO_x (suboxide) was 15.3% at 1 eV, 7.7% at 5 eV, and 8.8% at 10 eV. Figure 9 shows the calculated compositions of the suboxide-peak and dioxide-peak areas. The suboxide compositions formed by NBO at a beam energy of 5 eV were much smaller than those previously reported for

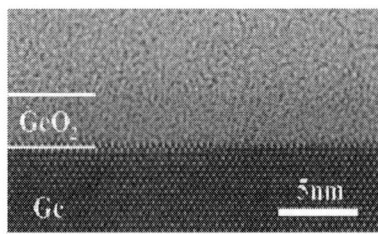

Figure 7. TEM image of a 3-nm-thick GeO_2 film formed by NBO.

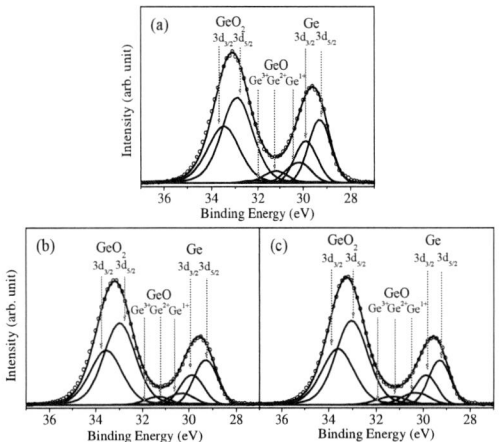

Figure 8. XPS spectra of GeO2 thin films (3 nm thick) formed using NBO at beam energies of (a) 1, (b) 5, and (c) 10 eV.

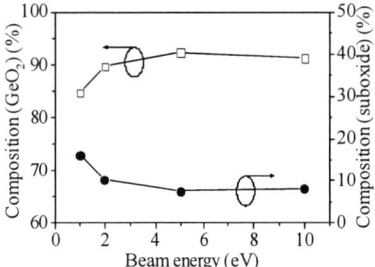

Figure 9. Dependence of germanium oxide composition (%) on beam energy, as derived from fitting analysis of XPS spectra.

thermally oxidized GeO_2 films [12]. This indicates that thanks to the low activation energy, our NBO process created high-quality GeO_2 thin films with little suboxide even at low temperature.

On the basis of these results, we fabricated Ge MOS capacitors with thin GeO_2 films using NBO at an energy of 5 eV and a temperature of 300°C. The electrical characteristics were evaluated in terms of the capacitance-voltage (C-V) of the Ge MOS capacitors, as shown in Fig. 10. The native oxide removal treatments yielded almost the same curves. The EOT of $Al_2O_3/GeO_2/Ge$ gate stacks with /without H radical native oxide treatment (i.e., only DHF treatment without H radical treatment) were 5.5/ 6.7 nm. The difference in the EOTs was caused by native oxide remaining in the stacks formed without the H radical treatment. The D_{it} values were calculated by using a low-temperature conductance method. Figure 11 shows the D_{it} of $Au/Al_2O_3/GeO_2/Ge$ MOS capacitors. D_{it} without the H radical treatment was higher than with the H radical treatment. In the case of $Au/Al_2O_3/GeO_2/Ge$ MOS capacitors without the H radical treatment, because the native oxide contained a large amount of sub-oxide, the GeO_2/Ge interface had a high defect density. D_{it} was less than 1×10^{11} cm^{-2} near the mid-gap in all of the GeO_2 films made with the direct NBO process and H

978-1-4673-1707-8/12 $31.00 © 2012 IEEE   87

radical native oxide removal treatment. This result suggested that a high-quality GeO_2/Ge interface requires the native oxide to be completely removed before NBO. Figure 12 shows the calculated D_{it} versus the EOT thickness. The Ge oxidation process yielded higher quality compared with the thermal oxidation process [3,4] and direct plasma oxidation process [5,13] even for a thin EOT.

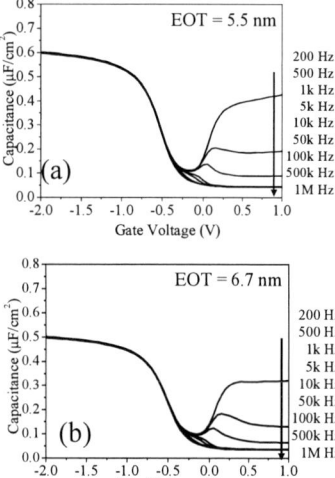

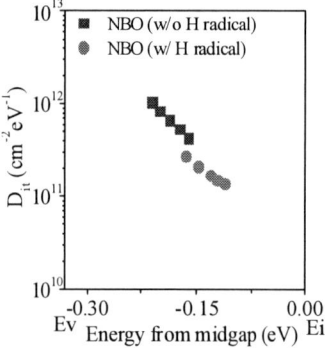

Figure 10. C-V characteristics of Au/Al_2O_3/GeO_2/Ge MOS capacitors fabricated using NBO at 5 eV beam energy (a) with H radical treatment and (b) without H radical treatment. The EOT of Al_2O_3/GeO_2/Ge gate stacks are (a) 5.5 and (b) 6.7 nm, respectively.

Figure 11. Calculated D_{it} of Au/Al_2O_3/GeO_2/Ge MOS capacitors. The 3-nm-thick GeO_2 films were formed using the NBO process with and without H radical treatment.

IV. CONCLUSION

We investigated GeO_2 thin films formed at low temperature (300°C) with a combination of an H radical treatment and our newly developed NBO process. The H

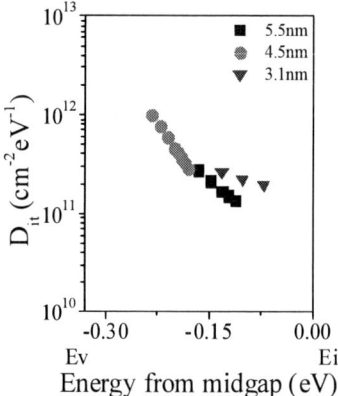

Figure 12. Calculated D_{it} of NBO-fabricated Au/Al_2O_3/GeO_2/Ge MOS capacitor structure versus EOT.

radical treatment enabled removal of the native oxide without causing any roughness. The NBO process's extremely low activation energy (<0.005 eV) enabled high-quality GeO_2 thin films with little suboxide to be formed even at low temperature. We used the process to form high-quality GeO_2 film with a low interfacial state density (D_{it}) at the GeO_2/Ge interface for Al_2O_3/GeO_2/Ge gate stacks having thin EOTs. The results of our experimental fabrication demonstrate the outstanding potential of NB technology for forming high-quality GeO_2 thin films.

REFERENCES

[1] K. Prabhakaran, F. Maeda, Y. Watanabe, and T. Ogino, Appl. Phys. Lett., **76** (2000) 2244.

[2] K. Kita, S. Suzuki, H. Nomura, T. Takahashi, T. Nishimura, and A. Toriumi, Jpn. J. Appl. Phys., **47** (2008) 2349.

[3] Y. Nakakita, R. Nakane, T. Sasada, H. Matsubara, M. Takenaka, and S. Takagi, Tech. Dig. IEDM (2008) p. 877.

[4] T. Nishimura, C. H. Lee, T. Tabata, S. K. Wang, K. Nagashio, K. Kita, and A. Toriumi, Appl. Phys. Express **4**, 064201 (2011).

[5] R. Zhang, T. Iwasaki, N. Taoka, M. Takenaka, and S. Takagi, J. Electrochem. Soc. **158**, G178 (2011).

[6] S. Sakamoto, K. Ichiki, and S. Samukawa, J. Appl. Phys., **40** (2001) 779.

[7] M. Yonemoto, T. Ikoma, T. Kano, K. Endo, T. Matsukawa, M. Masahara, and S. Samukawa, Jpn. J. Appl. Phys., **48** (2009) 04C007.

[8] T. Ikoma, C. Taguchi, S. Fukuda, K. Endo, H. Watanabe, and S. Samukawa, Ext. Abstr. Solid State Devices and Materials, (2006) 448.

[9] B. E. Deal and A. S. Grove, J. Appl. Phys., **36**, (1965) 3770.

[10] T. Nakano, N. Sadeghi, and R. A. Gottscho, Appl. Phys. Lett., **58** (1991) 458.

[11] M. Kobayashi, G. Thareja, M. Ishibashi, Y. Sun, P. Griffin, J. McVittie, P. Pianetta, K. Saraswat, and Y. Nishi, J. Appl. Phys., **106**, (2009) 104117.

[12] A. Molle, M. Bhuiyan, G. Tallarida, and M. Fanciulli, Appl. Phys. Lett., **89** (2006) 083504.

[13] R. Zhang, T. Iwasaki, N. Taoka, M. Takenaka, and S. Takagi, VLSI, pp. 56, 2011.

(100)- and (110)-oriented nMOSFETs with highly Scaled EOT in La-silicate/Si Interface for Multi-gate Architecture

T. Kawanago[1], K. Kakushima[2], P. Ahmet[1], Y. Kataoka[2], A. Nishiyama[2], N. Sugii[2], K. Tsutsui[2], K. Natori[1], T. Hattori[1] and H. Iwai[1]

[1] Frontier Research Center, Tokyo Institute of Technology, 4259, Nagatsuta, Midori-ku, Yokohama 226-8502, Japan,
[2] Interdisciplinary Graduate School of Science and Engineering, Tokyo Institute of Technology,
Tel.: +81-45-924-5847, E-mail: kawanago.t.ab@m.titech.ac.jp

Abstract— **This paper reports on detailed comparison between (100)- and (110)-oriented nMOSFETs with direct contact of La-silicate/Si interface structure for expansion to multi-gate architecture including FinFETs, trigate FETs, and nanowire FETs. Scaled EOT of 0.73 nm for (110)-oriented nMOSFETs has been achieved as well as (100)-oriented nMOSFETs. Although the large interface state density originating from (110) orientation was observed, fairly nice interfacial property was obtained from (110)-oriented nMOSFETs at scaled EOT region. Moreover, larger interface state density in (110) orientation did not affect on V_{th} instability. It was found that V_{th} shift of nMOSFETs is mainly caused by bulk trapping of electron in La-silicate as well as Hf-based oxides.**

I. INTRODUCTION

The scaling in an equivalent oxide thickness (EOT) with high-k/metal gate stacks is continuously important to suppress not only the gate leakage current but also the short-channel effect and threshold voltage variability in advanced MOSFETs [1, 2]. Although a SiO_2 interfacial layer (IL) is typically formed to recover from the interfacial property or reduced effective mobility [3], the removal of SiO_2-IL is essential for reduction of the EOT in high-k/metal gate stacks. Several groups have been reported that an EOT of 0.5 nm can be achieved with Hf-based oxides in directly contact with Si by sophisticated methods to inhibit or scavenge SiO_x-based interfacial layer growth [4]. Moreover, an EOT of 0.62 nm accompanied by superior interfacial property has been reported with direct contact La-silicate/Si interface structure and effective electron mobility of 155 cm^2/Vsec has been demonstrated [5]. Since the conventional planar FETs is significantly challenging from a viewpoint of gate controllability, multi-gate architectures such as FinFETs, trigate FETs or nanowire FETs should be implemented because of its better short channel immunity and reduced random dopant fluctuation [6]. Therefore, direct contact high-k/Si structure is thought to be combined with multi-gate architectures. In this context, (110) plane is usually utilized as the side-surface in FinFETs or nanowire FETs [7]. Since reactively-formed La-silicate dielectrics can easily provide the direct contact high-k/Si interface, it is thus of great interest to investigate the electrical characteristics of MOSFETs fabricated on (110) orientation with direct contact of high-k/Si

structure. The object in this study is extrapolation for multi-gate architectures with direct contact high-k/Si interface based on the experimental findings.

II. DEVICE FABRICATION

La_2O_3 was deposited on HF-last Si (100)- and (110)-oriented substrate by e-beam evaporation in an ultra-high vacuum chamber, followed by *in-situ* W (tungsten) metal deposition by RF sputtering. The channel direction is parallel to <110>. nMOSFETs were fabricated by gate last process using source and drain pre-formed p-Si substrates with a substrate doping concentration of 3×10^{16} cm^{-3} [5]. TiN and Si were deposited on W metal by RF sputtering for scaled EOT. The details had been reported elsewhere [5]. The thickness of W, TiN and Si were 5 nm, 10 nm and 100 nm, respectively. Post-metallization annealing was performed at 800 °C for 30 min in forming gas ambient ($H_2:N_2$ = 3%:97%) with various gate structures to form La-silicate at La_2O_3/Si interface [5]. Al was deposited on the source/drain region and back side of the substrate as a contact. Finally, recovery annealing was performed at 420 °C for 30 min in forming gas ambient. Device fabrication process is summarized in **Figure 1**. Thermally-grown SiO_2 nMOSFETs were also prepared as references.

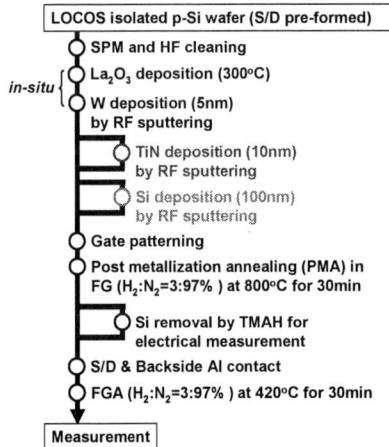

Figure 1. Fabrication process for nMOSFETs.

This study was supported by New Energy and Industrial Technology Development Organization (NEDO).

III. RESULTS AND DISCUSSION

First, silicate reaction at La_2O_3/Si interface was investigated by evaluating the EOT. Previous study has revealed that the oxidation rate of Si substrate is dependent on the surface orientation [8]. **Figure 2** shows the dependence of EOT on (100) and (110) orientation with various gate structures. The oxygen partial pressure can be controlled by Si deposition because Si layer can prevent the oxygen diffusion from atmosphere, enabling the scaled EOT [5]. The EOT is decreased by lowering the oxygen partial pressure. Moreover, no significant difference between (100) and (110) orientation can be confirmed. This result suggests that the scaled EOT can be obtained by controlling oxygen partial pressure regardless of Si substrate orientation. Moreover, the formation rate of La-silicate is found to be almost identical irrespective of substrate orientation unlike with the oxidation of Si substrate. One of the origins for same EOT is that the silicate reaction at interface is attributed to oxygen radical [9]. Deposited Si layer can provide EOT of 0.75 nm and 0.73 nm for (100) and (110) orientation, respectively.

Next, the interface state density was evaluated by charge pumping method [10]. It is well known that available bonds in (110) orientation is higher than that in (100) orientation, leading to larger interface state density in (110) orientation [8]. **Figure 3** shows the charge pumping current as a function of frequency. The results of nMOSFETs with SiO_2 are also shown in **Figure 3**. nMOSFETs fabricated on (110) orientation shows the higher interface state density irrespective of gate dielectrics. These results reflect the difference of surface orientation. The interface state density of 9.5×10^{10} cm^{-2}eV^{-1} can be obtained on (100) orientation with La-silicate gate dielectrics, while larger interface state density on (110) orientation is clearly confirmed. **Figure 4** shows the I_d–V_g characteristics of nMOSFETs by adjusting the threshold voltage (V_{th}) to compare the sub-threshold slope (SS). SS of (100)-oriented MOSFET is 65 mv/dec, indicating the superior interfacial property. On the other hand, 78 mV/dec of SS is observed from (110)-oriented MOSFET. The degradation of SS is likely caused by the large interface state density as previously shown in **Figure 3**. From **Figure 3** and **Figure 4**, reduced interface state density on (110) orientation is one of the challenging issues.

Figure 5 shows the gate-channel capacitances (C_{gc}–V) of nMOSFETs with La-silicate gate dielectrics on (100) and (110) orientation at frequency of 1MHz measured by split C–V method [11]. The C_{gc}–V characteristic on (110) orientation is shifted by 100 mV to negative direction compared to that on (100) orientation, while no hysteresis in C_{gc}–V characteristics are clearly observed for both (100)- and (110)-oriented nMOSFETs. In the case of SiO_2/Si interface, the fixed charges are typically assumed to be located at the SiO_2/Si interface. Moreover, the dependence of the density of fixed charges is the same as interface state density, namely the density of fixed charges on (110) orientation is larger than that on (100) orientation [8]. The existence of fixed charges at interface affects on the threshold voltage of MOSFETs. Based on the study of SiO_2/Si interface, it is speculated that shift in C_{gc}–V characteristics on (110) orientation may be attributed to the presence of fixed charges at interface [7].

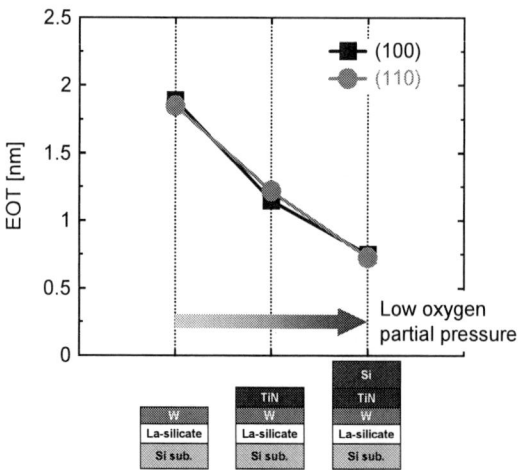

Figure 2.   Dependence of EOT on (100) and (110) orientation.

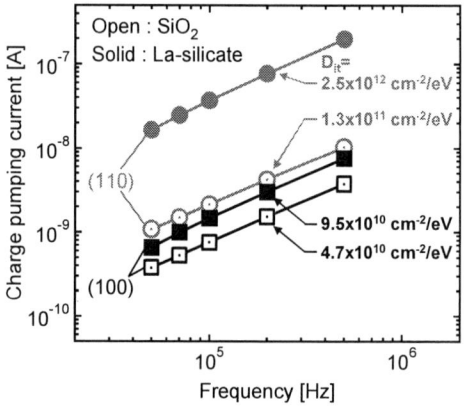

Figure 3.   Charge pumping current for (100)-and (110)-oriented nMOSFETs with La-silicate. The results of SiO_2 nMOSFETs are also shown.

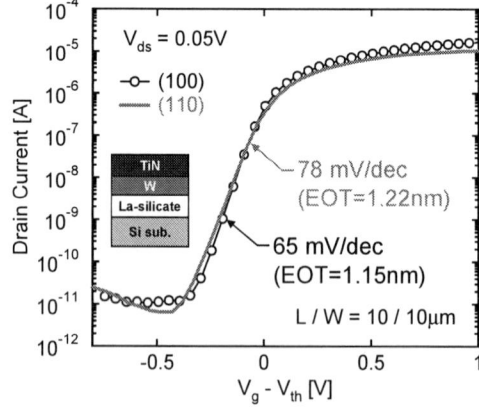

Figure 4. I_d–V_g characteristics of nMOSFETs by adjusting the threshold voltage (V_{th}).

Figure 6 shows the comparison of frequency dispersion in C_{gc}–V characteristics between (100) and (110) orientation. Frequencies used to measure C_{gc}–V characteristics were 100 kHz, 500 kHz, and 1 MHz, respectively. In C_{gc}–V characteristics for (100) orientation, no frequency dispersion can be clearly observed as shown in **Figure 6 (a)**, indicating the superior interfacial property at EOT of 0.75 nm. On the other hand, C_{gc}–V characteristics for (110)-oriented nMOSFET shows the frequency dispersion, suggesting the larger amount of interface state density compared with that for (100)-oriented nMOSFETs in **Figure 6 (b)**. However, the frequency dispersion in **Figure 6 (b)** is not so severe expected from the results of charge pumping measurements as shown in **Figure 3**. Although the frequency dispersion should be suppressed, fairly nice C_{gc}–V characteristics of (110)-oriented nMOSFET was obtained at EOT of 0.73 nm.

Subsequently, the potential barrier of gate dielectrics was evaluated by measuring the gate leakage current with carrier separation method [12]. The electrons injected form inversion layer are measured by the source/drain terminals of the nMOSFETs while the current sensed at substrate terminal correspond to the current flow of the holes. Since the physical thickness of La-silicate is less than 3 nm, the direct tunneling current is thought to be mainly dominating. Therefore, these current flows are dependent on the band offset at dielectrics and substrate interface. **Figure 7** shows the comparison of current component for (100)- and (110)-oriented nMOSFETs. No significant difference in current component between (100)- and (110)-oriented nMOSFETs is clearly observed. In addition, breakdown voltages for (100)- and (110)-oriented nMOSFETs were 2.85 V and 2.9 V, respectively. These results mean that the bulk property of reactively-formed La-silicate gate dielectrics and potential barrier against the Si substrate are almost identical regardless of substrate orientation. Besides, the gate leakage current is four orders of magnitude lower than that of ITRS roadmap.

Figure 8 shows the effective electron mobility for both (100)-and (110)-oriented nMOSFETs with La-silicate. The effective electron mobility of nMOSFETs with SiO_2 is also shown. The channel direction is parallel to <110>. It has been reported that the electron mobility of (110)-oriented nMOSFETs is lower than that of (100)-oriented nMOSFETs [13]. This lower mobility is mainly associated with the smaller energy split between 4-fold valleys and 2-fold valleys as well as the larger DOS of lower-energy valleys in (110) surface orientation [13]. As shown in **Figure 8**, the electron mobility of (110)-oriented nMOSFETs is reduced regardless of gate dielectrics. This result is consistent with previous study originating from fundamental properties in (110) surface orientation [13]. On the other hand, electron mobility with La-silicate is degraded compared to nMOSFETs with SiO_2. This is attributed to the inherent mechanisms for high-k gate dielectrics, such as remote Coulomb scattering (RCS) due to fixed charges in high-k, remote phonon scattering (RPS) caused by low-energy optical phonon modes in high-k gate dielectrics and remote surface roughness scattering (SRS) [6]. One of the interesting features regarding La-silicate dielectrics is that the electron mobility is close to the electron mobility of nMOSFETs with SiO_2 at high N_{inv} region.

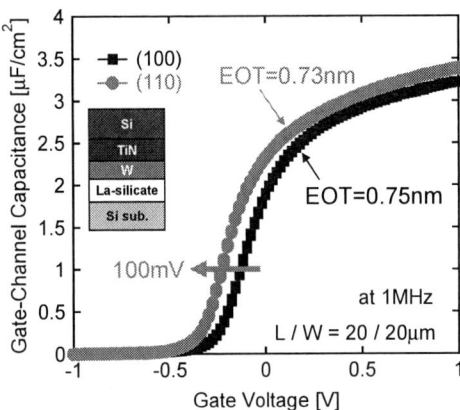

Figure 5. Gate-channel capacitances (C_{gc}–V) of nMOSFETs with La-silicate gate dielectrics on (100) and (110) orientation.

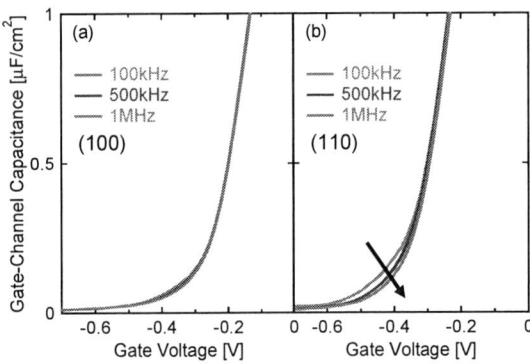

Figure 6. Comparison of frequency dispersion in C_{gc}–V characteristics for (a) (100) and (b) (110) orientation, respectively.

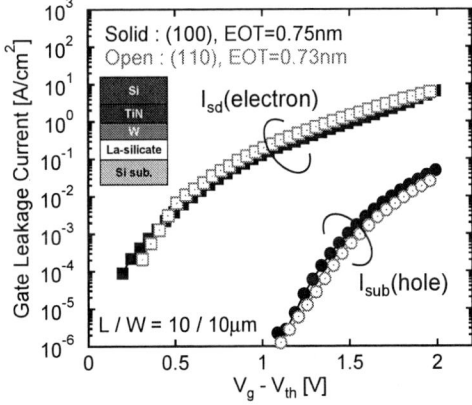

Figure 7. Comparison of current component for (100)- and (110)-oriented nMOSFETs by carrier separation method.

978-1-4673-1707-8/12 $31.00 © 2012 IEEE

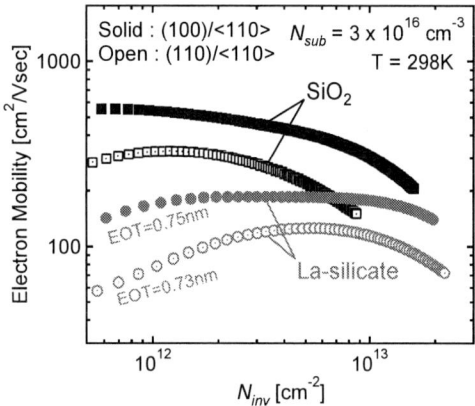

Figure 8. Effective electron mobility for (100)-and (110)-oriented nMOSFETs with La-silicate. The effective electron mobility of SiO$_2$ nMOSFETs are also shown.

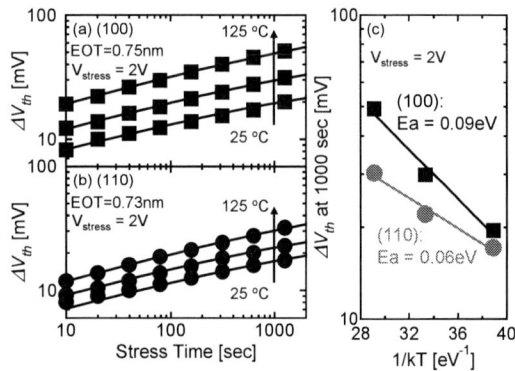

Figure 9. Time evolution of the V$_{th}$ shift for (a) (100) and (b) (110) orientation, respectively. (c) Arrhenius plot of the V$_{th}$ shift at 1000 sec.

Finally, V$_{th}$ instability of nMOSFETs was investigated under positive gate bias stress. **Figure 9 (a)** and **(b)** show the time evolution of the V$_{th}$ shift (ΔV_{th}). The gate bias stress was set to be 2 V. The V$_{th}$ shift is increased with increasing stress time and measured temperature. A model with a stretched exponential equation was utilized for fitting V$_{th}$ shift versus stress time [14]. Although the activation energy (Ea) of V$_{th}$ shift for (100) orientation is slightly larger than that for (110) orientation as shown in **Figure 9 (c)**, these low Ea are thought to be direct tunneling of electron from substrate into the La-silicate, causing the V$_{th}$ shift due to electron trapping similar to previous study with Hf-based oxides (Ea=0.08eV) [15].

IV. CONCLUSIONS

The electrical characteristics of (100)- and (110)-oriented nMOSFETs has been investigated with direct contact La-silicate/Si interface structure. The scaled EOT of (110)-oriented nMOSFETs can be attained by controlling oxygen partial pressure as well as (100)-oriented nMOSFETs. Small interface state density on (110) orientation is one of the challenging issues, while fairly nice C$_{gc}$–V characteristic for (110)-oriented nMOSFET has been demonstrated at EOT of 0.73 nm. Potential barrier at La-silicate and the Si substrate interface are almost identical irrespective of surface orientation. Bulk trapping in La-silicate is found to be main origin responsible for the V$_{th}$ shift even with larger interface state density on (110) orientation. These experimental findings are useful information for fabrication and designing high-performance and/or low-power multi-gate architectures including FinFETs, trigate FETs, and nanowire MOSFETs because this study successfully demonstrated the capability of direct contact La-silicate/Si interface structure for not only (100) but also (110) orientation.

REFERENCES

[1] K. Henson et al., "Gate Length Scaling and High Drive Currents Enabled for High Performance SOI Technology using High-κ/Metal Gate," in IEDM Tech. Dig., pp.645, (2008).

[2] K. J. Kuhn, "Reducing Variation in Advanced Logic Technologies: Approaches to Process and Design for Manufacturability of Nanoscale CMOS," in IEDM Tech. Dig., pp.471, (2007).

[3] S. Saito et al., "Unified Mobility Model for high-k Gate Stacks," in IEDM Tech. Dig., pp.797, (2003).

[4] T. Ando et al., "Understanding Mobility Mechanisms in Extremely Scaled HfO$_2$ (EOT 0.42 nm) Using Remote Interfacial Layer Scavenging Technique and Vt-tuning Dipoles with Gate-First Process," in IEDM Tech. Dig., pp.423, (2009).

[5] T. Kawanago et al., "EOT of 0.62nm and High Electron Mobility in La-silicate/Si Structure based nMOSFETs Achieved by Utilizing Metal Inserted Poly-Si Stacks and Annealing at High Temperature," IEEE Trans. Electron Devices, vol. 59, no. 2, pp. 269, Feb. (2012).

[6] M. M. Frank, "High-k/metal gate innovations enabling continued CMOS scaling," in Proc. 41th European Solid-State Device Research Conference, pp. 25, (2011).

[7] M. Saitoh et al., "Understanding of Short-Channel Mobility in Tri-Gate Nanowire MOSFETs and Enhanced Stress Memorization Technique for Performance Improvement," in IEDM Tech. Dig., pp.780, (2010).

[8] S. M. Sze: Physics of Semiconductor Devices (Wiley, New York, 1981) 2nd ed.,

[9] K. Kakushima et al., "Characterization of flatband voltage roll-off and roll-up behavior in La$_2$O$_3$/silicatre gate dielectric," Solid-State Electron. 54, pp.720, (2010).

[10] G. Groeseneken et al., "A Reliable Approach to Charge-Pumping Measurements in MOS Transistors," IEEE Trans. Electron Devices, vol. 31, no. 1, pp. 42, January (1984).

[11] J. R. Hauser, "Extraction of Experimental Mobility Data for MOS Devices," IEEE Trans. Electron Devices, Vol. 43, no. 11, pp. 1981, (1996).

[12] R. Iijima et al., "Intrinsic Effects of the Crystal Orientation Difference between (100) and (110) Silicon Substrates on Characteristics of High-k/Metal Gate Metal–Oxide–Semiconductor Field-Effect Transistors," Jpn. J. Appl. Phys. vol. 50, 061503 (2011).

[13] K. Uchida et al., "Carrier Transport and Stress Engineering in Advanced Nanoscale Transistors From (100) and (110) Transistors To Carbon Nanotube FETs and Beyond," in IEDM Tech. Dig., pp.569, (2008).

[14] S. Zafar et al., "A Comparative Study of NBTI and PBTI (Charge Trapping) in SiO$_2$/HfO$_2$ Stacks with FUSI, TiN, Re Gates," in VLSI Symp. Tech. Dig., pp. 23, (2006).

[15] S. Pae et al., "BTI Reliability of 45 nm High-k + Metal-Gate Process Technology," Proc. IRPS. pp. 352, (2008).

CMOS Compatible ALD High-k Double Slot Grating Couplers for On-Chip Optical Interconnects

Maziar M. Naiini, Christoph Henkel, Gunnar B. Malm, Mikael Östling

Integrated Devices and Circuits, School of Information and Communication Technology
KTH Royal Institute of Technology, P.O. Box 229, SE-16440 Kista, Sweden
[mamn |chenkel | gunta | ostling]@kth.se

Abstract— Silicon-on-insulator(SOI) novel on-chip grating couplers for double slot high-k waveguides are experimentally demonstrated. The devices were fabricated with standard CMOS process technology. The grating couplers were designed for the best performance at the C-band communication range. Two thin layers of aluminum oxide formed the slot region of the waveguide. The high-k layers were deposited using the atomic layer deposition (ALD) method. A reliable process was realized by etching the structures to the buried oxide. Effect of the top oxide cladding layer on the efficiency was studied. The grating couplers had a measured efficiency of 22% at 1.55μm wavelength. This efficiency is competitive to other results reported by other groups.

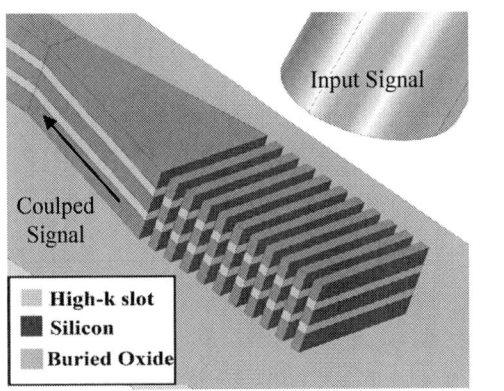

Figure 1. Schematic view of the device. Yellow regions represent the ALD high-k slot layers. Blue regions represent the amorphous-Si and the grey layer the buried silicon oxide layer.

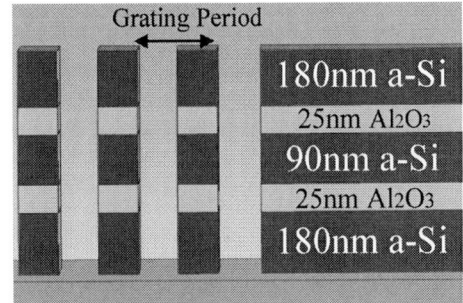

Figure 2. Cross-section of the grating structure and the thin film specifications of the waveguide stack.

I. INTRODUCTION

The ever increasing demand of high speed data transfer rate interconnects has made the need of optical solutions evident. This technology push will not only affect the servers but also the personal computers. This technology is built by a variety of components including the light source, low loss waveguides, low consumption light modulators and efficient photo detectors. All-silicon active optical devices have not yet shown a best- in-class performance because of the limitations of silicon, and some of the components are required to be manufactured with alternative materials integrated directly with the CMOS platform[1]. Horizontal slot technology is a low cost alternative that can take advantage of the alternative materials while being compatible with the SOI technology. Light modulators and detectors using slot waveguides have been introduced and demonstrated with non-linear filling polymers [2,3]. An electrically pumped light generator with a ring resonator cavity has been introduced [4]. Yet coupling to the waveguides remains a major issue to sub-wavelength size waveguides [5].

Previously, highly reproducible grating couplers for single and double slot waveguides were designed [6,7]. Experimental fabrication and characterization of the single slot silicon dioxide slot waveguides were also demonstrated [8]. In this work double slot high-k structures are practically studied. Increasing the number of slots can increase the total fraction

of the optical power in the slot regions (higher modal confinement) [9]. Aluminum oxide (Al_2O_3) is chosen as the high-k material. The Al_2O_3 films are deposited using the atomic layer deposition (ALD) method with a high refractive index and thickness uniformity. The quality of these films is suitable for photonics applications [10]. Fig.1 shows a schematic view of the grating coupler and the slot waveguide

978-1-4673-1707-8/12 $31.00 © 2012 IEEE

structure. The material stack is prepared by multiple deposition steps. Amorphous silicon layers are deposited with LPCVD at 480°C with the disilane gas. The detailed specification of the stack layering and the thicknesses is depicted in Fig. 2. The fabricated devices are optically characterized and the results are post-processed for the measurement of the coupling efficiency. Other grating coupler designs for the horizontal slot waveguides have been studied [11,12], which do not benefit from a etch stop layer. In these devices the efficiency will vary by the variations in the etch depths. The achieved practical coupling efficiency is competitive to the mentioned designs. Nanotapers have also been studied by other groups [13,14]. These structures have high coupling efficiency but the photonic chips should be diced and the facade of the waveguides needs a polishing treatment to lower the scattering.

This work is organized in three sections: In section (II) the fabrication flow is explained with details. In section (III) the performance and characteristics of the measured devices is presented and finally in (IV) the achievements are summarized.

II. FABRICATION FLOW

A. Material Stack Depostion

As illustrated by Fig. 2 the waveguide structure was implemented by five deposition steps. The structure was layered upon a thermally grown silicon dioxide layer with the thickness of 1.2μm. Three layers of amorphous silicon were deposited using the LPCVD method at a low temperature of 480°C. This can only be established if the disilane gas is used as the source. The silane gas has a higher molecule cracking temperature of 560°C.

High-k dielectric layers were deposited with an 8-inch ALD BENEQ TFS200. Al_2O_3 layers were deposited at 200°C with a trimethylaluminum (TMA) precursor. After the

deposition of each layer the thickness was measured by means of spectral ellipsometry. Measurements were done with a UVSEL ER HORIBA. Both the thickness and the refractive index of the layers were inspected. The final material stack is shown in Fig. 3. Each layer has the same color code as in Fig1 and Fig2. SEM measurements show the successful deposition of two slot layers with similar thicknesses. It is also evident that the layers are smooth and thus suitable for photonic applications. The completed waveguide stack was covered by silicon dioxide hardmask that was deposited by the PECVD method at 300°C. The thickness of the hardmask was about 400nm to protect the structures while etching. The remaining of the hardmask layer after etching serves as the waveguide top cladding layer.

B. Patterining and Etching

The wafers were coated by an SPR 700 resist with the thickness of 1.2μm. The structures were patterned by an i-line reticle stepper. The designed mask consisted of grating periods ranging from 970nm to 1060nm. The optimum exposure rate was experimentally found by means of a dose-to-clear matrix on a wafer with the same material stack.

The patterned features were dry etched using an Applied Materials P5000 cluster tool. The etching was done in six stages. Firstly, the hardmask was etched and the photoresist was striped to avoid the fencing while the silicon etching. The amorphous silicon layers were etched with an HBr/Cl$_2$/He/O$_2$ chemistry and the Al_2O_3 layers were etched with a BCl$_3$/Cl$_2$/N$_2$ chemistry. Each amorphous silicon etching process is stopped by the endpoint detection method.

C. Device Final Cladding

The grating couplers can be covered with a thick SiO$_2$ layer as a protective layer and an index matching material. At the grating sites this layer will fill the etched teeth. In this work non-cladded devices were also prepared and characterized. Afterwards a 500nm SiO$_2$ was deposited by PECVD at 300°C and the devices were characterized. The impact of this process on the optical characteristics of the grating couplers was further investigated.

III. RESULTS AND DISCUSSION

A. Characerization Setup Configuration

An in house optical measurement setup was developed in which the angle of incidence is tunable. The measurements were conducted with 17° tilted fibers. The transmission in the waveguides with the lengths of 600μm, 1000μm, and 1400μm were measured and the attenuation per unit length was calculated. The total attenuation was excluded from the original transmission spectrums to characterize the efficiency of the grating couplers. In order to reduce the reflections at the fiber ends, utilization of index matching gels is reported [15]. In the measurements this method was not used because of the issues that can be caused by the gel while electrical probing in case of active devices.

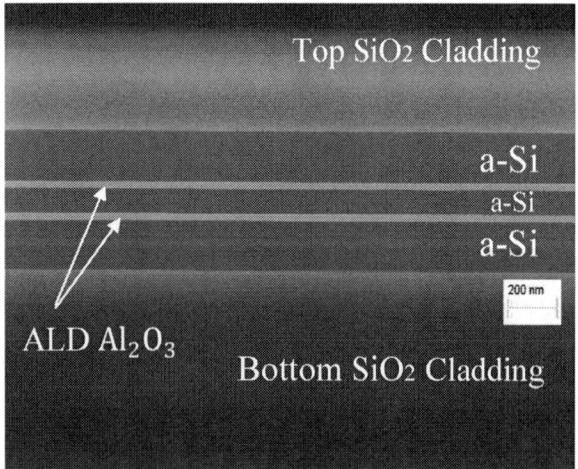

Figure 3. SEM image of the material stack. The layers are colorized as in Fig. 1 and Fig. 2.

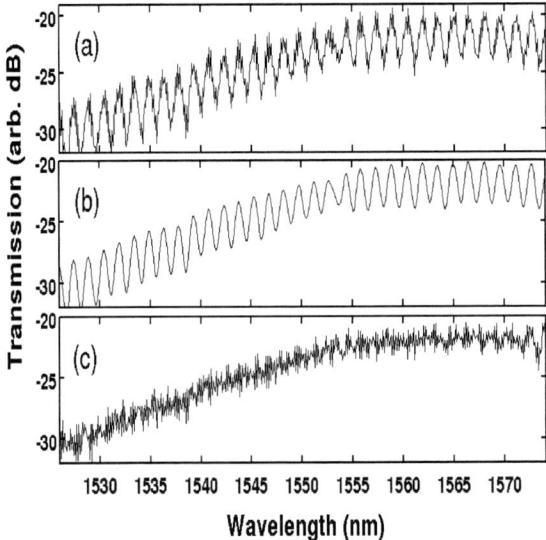

Figure 4. Transmission spectrum excluding the waveguide propagation loss (a) the primitive signal (b) filtered signal with a low pass filter (c) a band reject filtered signal .

B. Coupling Efficiencies

A measured transmission spectrum is demonstrated in Fig. 4(a). The primitive spectrum was analyzed by means of Fourier transforms and filters.

With this analysis the corresponding resonance cavity was calculated. Filters were applied to characterize the transmission signal. Fig. 4(b) is the result of a low pass filter and the frequency of the fluctuation is 0.64(1/nm). The resonance cavity size of this fluctuation is smaller than the waveguide length. This resonance is caused by the reflections at the fiber tips and not by the internal reflections. Further, a band reject filter was applied to the signal to filter the mentioned fluctuation. The resulting spectrum is shown in Fig. 4(c). Specifications of the filter are shown in Fig. 5. The

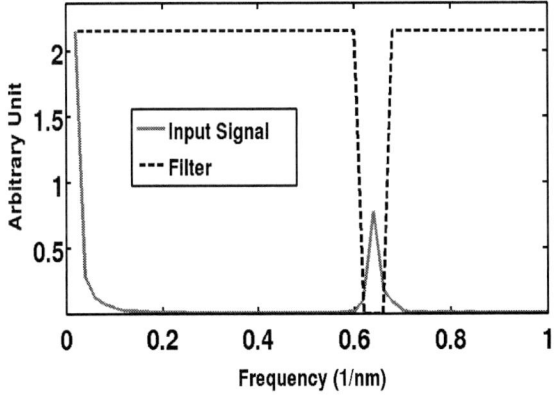

Figure 5. Fourier transform of the primitive spectrum and the applied band reject filter. The filter has a center frequency of 0.64 (1/nm) that corresponds to the fluctuation with a period of 1.6nm in the spectrum. The bandwidth of the filter is 0.06 (1/nm).

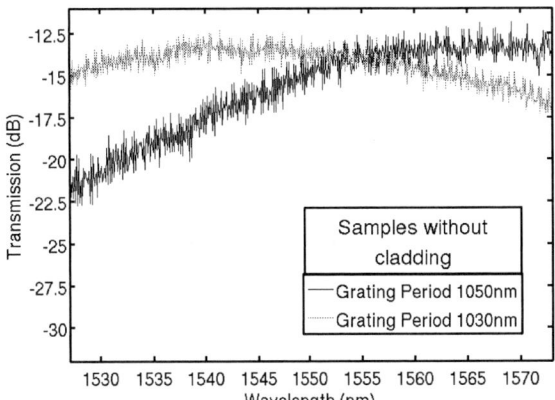

Figure 6. The transmission spectrum excluding the loss in the waveguides for the grating couplers without the final cladding (filling).

grating couplers were further characterized using this filter. Fig. 6 shows the transmission for gratings without the cladding layer. The gratings with a larger grating period had a larger peak wavelength. The peak transmission includes the insertion loss of two grating couplers. The insertion loss per grating coupler was calculated assuming that the devices are reciprocal. The insertion loss of the non-cladded grating couplers were measured 6.6dB. The measured couplers had a 2.5dB bandwidth of 60nm. The transmission spectrum of the cladded samples are shown in Fig. 7. The insertion loss measured for the grating couplers with the final cladding layer was 6.5dB and the 2.5dB bandwidth of the couplers was 48nm. Addition of the cladding layer has a major effect on the peak wavelength of the spectrum. Grating couplers of available sizes were characterized. The results are summarized in the Fig. 8. Every point represents the suitable grating period for the peak coupling wavelength. The coupling efficiency depends on the top oxide thickness. The dependency trend was studied with simulations and the results are summarized in

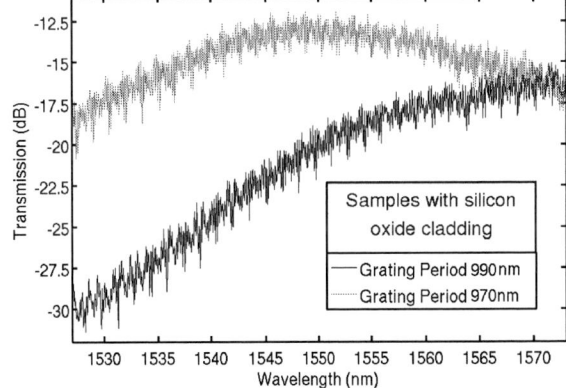

Figure 7. The transmission spectrums for grating couplers with the final oxide cladding layer excluding the total attenuation in the waveguide.

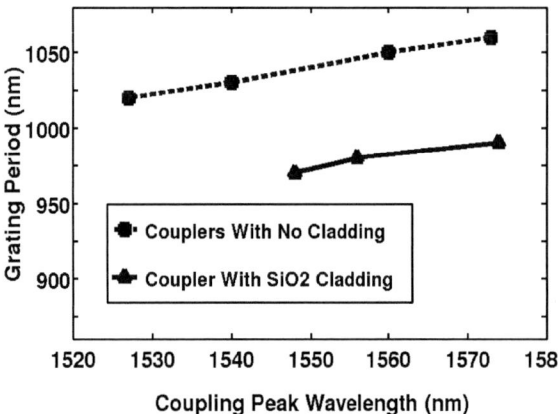

Figure 8. Suitable grating periods vs. coupling peak wavelength. Both sample

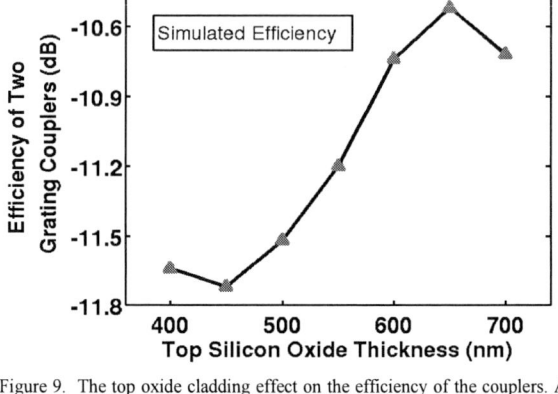

Figure 9. The top oxide cladding effect on the efficiency of the couplers. A 1.2dB modulation was calculated.

Fig. 9. The efficiency modulation for a pair of grating couplers is 1.2dB. The efficiency of the characterized couplers was 0.8dB lower than the simulated devices.

IV. CONCLUSIONS

Grating couplers for double slot high-k waveguides are successfully fabricated and characterized. The fabrication process is highly reproducible and reliable. The devices are manufactured in a wafer scale process with the standard CMOS aspects. The fabricated devices are suitable solutions for the SOI on-chip optical integrated circuits. The high-k deposition temperature is non destructive to the underlying layers. The etching depth of the gratings was well controlled by the material contrast resulting in homogenous devices. The measured signal spectrums were analyzed with Fourier transforms and filters. The couplers have a maximum coupling efficiency of 22%. Devices with and without a final cladding layer were characterized. The efficiency of the couplers depends on the thickness of the top layer as well as the bottom cladding layer.

V. AKNOWLEDGEMENT

This financial support by the ERC advanced grant OSIRIS is greatly acknowledged.

References

[1] T. Baehr-Jones, T. Pinguet, P. Lo Guo-Qiang, S. Danziger, D. Prather, and M. Hochberg, "Myths and rumours of silicon photonics," Nature Photonics, vol. 6, no. 4, pp. 206-208, Mar. 2012.

[2] T. Baehr-Jones et al., "Optical modulation and detection in slotted Silicon waveguides," Optics Express, vol. 13, no. 14, p. 5216, Jul.2005.

[3] M. Gould et al., "Silicon-polymer hybrid slot waveguide ring-resonator modulator," Optics Express, vol. 19, no. 5, p. 3952, Feb.2011.

[4] C. A. Barrios and M. Lipson, "Electrically driven silicon resonant

light emitting device based on slot-waveguide," Optics Express, vol.13, no. 25, p. 10092, Dec. 2005.

[5] D. Feng, B. J. Luff, and M. Asghari, "Recent advances in manufactured silicon photonics," in Proceedings of SPIE, 2012, vol. 8265, no. 1, pp. 826507-826507-9.

[6] M. M. Naiini, G. Malm, and M. Ostling, "Fully etched grating couplers for atomic layer deposited horizontal slot waveguides," in Ulis 2011 Ultimate Integration on Silicon, 2011, pp. 1-4.

[7] M. M. Naiini, C. Henkel, G. B. Malm, and M. Ostling, "ALD high-k layer grating couplers for single and double slot on-chip SOI photonics," in 2011 Proceedings of the European Solid-State Device Research Conference (ESSDERC), 2011, pp. 191-194.

[8] M. M. Naiini, C. Henkel, G. B. Malm, and M. Ostling, "ALD high-k layer grating couplers for single and double slot on-chip SOI photonics," accepted for publication in the Solid Sate Electronics Journal.

[9] R. Sun et al., "Horizontal single and multiple slot waveguides: optical transmission at λ = 1550 nm," Optics Express, vol. 15, no. 26, p.17967, 2007.

[10] K. Solehmainen, M. Kapulainen, P. Heimala, and K. Polamo,"Erbium-Doped Waveguides Fabricated With Atomic Layer Deposition Method," IEEE Photonics Technology Letters, vol. 16, no. 1, pp. 194-196, Jan. 2004.

[11] A. M. P. Fievre, T. Liu, and R. R. Panepucci, "Nanotaper coupler for the horizontal slot-waveguide," in Proceedings of SPIE, 2007, vol. 6645, no. 1, p. 66450E-66450E-8.

[12] D. Vermeulen et al., "High-efficiency fiber-to-chip grating couplers realized using an advanced CMOS-compatible Silicon-On-Insulator platform," Optics Express, vol. 18, no. 17, p. 18278, Aug. 2010.

[13] J. V. Galan, P. Sanchis, J. Blasco, and J. Marti, "Study of High Efficiency Grating Couplers for Silicon-Based Horizontal Slot Waveguides," IEEE Photonics Technology Letters, vol. 20, no. 12, pp.985-987, Jun. 2008.

[14] J. V. Galan et al., Vertical grating couplers for silicon sandwiched slot waveguides. IEEE, 2008, pp. 105-107.

[15] V. R. Almeida, R. R. Panepucci, and M. Lipson, "Nanotaper for compact mode conversion," Optics Letters, vol. 28, no. 15, p. 1302, Aug. 2003.

Transport in Amorphous Materials with Applications to Phase-Change Memories

Carlo Jacoboni
Dipartimento di Scienze Fisiche, Informatiche e Matematiche
Università di Modena e Reggio Emilia
Via Campi 213/A, I-41125 Modena, Italy
Istituto Nanoscienze CNR-S3
Email: carlo.jacoboni@unimore.it

Enrico Piccinini and Fabrizio Buscemi
ARCES Research Center
Università di Bologna
Via Toffano 2/2, I-40125 Bologna, Italy
Email: enrico.piccinini@unimore.it
fabrizio.buscemi@unimore.it

Abstract—Trap-limited conduction is here analyzed with special attention to chalcogenide materials used for phase-change memories. After a short review of existing theories, new developments are presented, based on very general microscopic assumptions. Electric field, carrier concentration, and electron temperature variable along the device, as well as diffusion and Poisson self-consistency are considered. The results account for and interpret all main experimental findings in PCM cells.

I. INTRODUCTION

A general theory of transport in amorphous materials was originally developed by Mott considering hopping events (see, for example, [1], [2]). This theory represents a milestone in the analysis of amorphous materials and has widely been applied since the 1960s. Nevertheless, classes of amorphous materials, such as chalcogenide glasses, feature an electrical switching behavior, i.e., a sudden change in resistivity by several orders of magnitude when a threshold field is reached, that cannot be explained by the original Mott theory. Such behavior is of great interest for the technology of phase-change memories (PCM) [3], so that in the last decade the study of electron transport in amorphous materials raised more interest.

Thus, a number of alternative models have been developed to explain the S-shaped Negative Differential Resistivity (SNDR) in such materials. After some brief considerations on trap-limited conduction in Sect. II, existing theories are shortly reviewed in Sect. III. In Sect. IV some recent developments resulting from a collaboration between the Universities of Bologna and of Modena and Reggio Emilia will be presented.

II. REDUCED MOBILITY AND HOPPING CONDUCTION

Let us consider, first, electrons travelling in a semiconductor conduction band with an applied field F and a mobility $\mu_\circ$, in absence of traps. On the other hand, if traps are present, which capture electrons with an average trapping time τ_b and release them with an average detrapping time τ_t (see Fig. 1), it is immediate to see that the electron mobility is reduced to

$$\mu = \mu_\circ \frac{\tau_b}{\tau_b + \tau_t} \ . \tag{1}$$

If $\tau_t \gg \tau_b$, electrons spend most of their time in the traps, and the above reduces to $\mu_\circ \tau_b / \tau_t$. In this last case the drift

velocity becomes

$$v_d = \mu F = \mu_\circ F \frac{\tau_b}{\tau_t} = \frac{l}{\tau_t} \ , \tag{2}$$

where l is the average distance traveled between consecutive detrapping and trapping events. Thus, only the average distance traveled by the electrons in each transition and the waiting time between successive flights are relevant, as in hopping trasport.

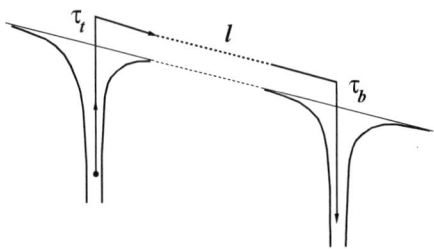

Fig. 1.   When the time τ_t spent by the carriers inside the traps is much longer than the time τ_b spent to move from one trap to another, the trap-limited mobility is characterized by two parameters: the mean flight length l and the "detrapping" time τ_t, as for hopping conduction.

Strictly speaking, we refer to hopping conduction when electrons jump from one site to another without entering the conduction band. Such transitions may occur if the two traps are close enough and the energy barrier between them is not too high. They may consist in a direct tunneling, or may be thermally assisted or thermally activated, as shown in Fig. 2.

In the chalcogenide materials considered here, the temperature and field dependences of the conductance show that transport occurs mainly via thermally-activated trap-limited mobility, with electrons moving from one site to another in the band, or, at least, above the band mobility edge [4], [5]. An ideal PCM cell is sketched in Fig. 3. It may occur that two traps are close enough to allow for direct tunneling, but a Monte Carlo simulation [6] has shown that in such a case carriers jump with high frequency back and forth between the two centers which become equivalent to a single trap. These motions contribute to the high-frequency noise, but not to conduction.

978-1-4673-1707-8/12 $31.00 © 2012 IEEE

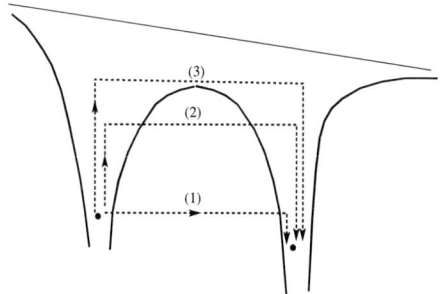

Fig. 2. Hopping may occur through a direct tunneling (1), thermally-assisted tunneling (2) or thermal activation (3). The thin line on top indicates the band mobility edge.

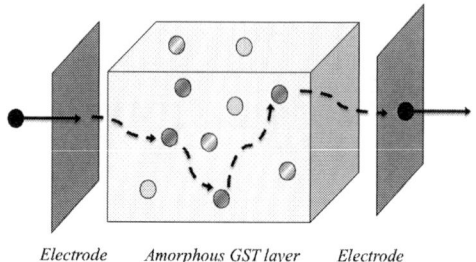

Electrode   Amorphous GST layer   Electrode

Fig. 3. Schematics of an ideal PCM cell made of an amorphous chalcogenide layer sandwiched between two planar metallic electrodes. Carriers from the injecting contact (left) move in the band by means of successive detrapping and trapping events until they reach the collecting contact (right). The different colors of the dots refer to traps of different kind and charge state.

III. AVAILABLE THEORIES

The first model proposed to explain the SNDR behavior in chalcogenides was developed by Adler and coworkers in the early 1980s [7]. According to this model, conduction is sustained by electrons and holes, and both have their respective traps. Some defects can act as traps for electrons or holes according to their charge state. When the field reaches a threshold value, following a complex reaction kinetics, which includes also generation of pairs by impact ionization, a large number of carriers is able to neutralize a large part of such centers; electrons can now move more freely, and a low resistivity state is set up through the generation of highly conductive filaments.

The ideas of Adler have been questioned in the last ten years. The main skepticism was about the size of the filament in the high-conductivity state, whose radius is expected in the micrometer range. More recent experimental investigations have shown that the switching behavior is present also in devices with smaller cross-sections. These arguments gave origin to alternative interpretations.

The model proposed by Lacaita and coworkers [8] still relies on traps and impact ionization, but transport is described within the more common framework of semiconductor bands. Impact ionization is the main ingredient of the model, where trap states act like pillars of a bridge between the valence and the conduction bands. At low currents, impact ionization and recombination via trap centers balance each other. As the current increases, the generation rate increases and traps are more and more filled. The recombination process, instead, is weakly dependent on the bias, so that a critical point is reached where recombination cannot balance generation any longer. The only way to re-establish a steady balance is obtained through a reduction of the bias, thus of the generation rate along the device. Since the currents are high, this condition must be accompanied by an increased concentration of carriers with high mobility in the conduction band.

Two arguments can be opposed to this model. The first deals with the generation mechanism: though favored by trap states within the band gap, carriers must still acquire a large energy before ionizing, which would made be possible by a mean free-path longer than that expected in PCM. Next,

impact ionization does not account for the activation energy resulting from experimental data. The latter consideration on the activation energy opens the way to two other models, that are currently still debated.

The model proposed by Karpov and coworkers, [9], [10] stems from the system free-energy balance and brings back the idea of crystalline filaments surrounded by an amorphous matrix proposed by Adler and Ovshinsky [7]. Crystalline nuclei can form inside the chalcogenide amorphous matrix due to local energy dissipation, and a resulting local stronger field favors the growth of these initial nuclei. As the switching point is approached, a crystalline filament spans across the whole device shunting the electrodes. The presence of highly-conductive crystalline filaments surrounded by an amorphous matrix makes a large increase in the current possible, without a significant effect on (or a small reduction of) the potential drop.

The last model to be mentioned in this short review, due to Ielmini and coworkers [4], [5], is the one most acknowledged today. Contrary to the Karpov model, where the energy is used to activate the nucleation process, in the Ielmini model the activation energy refers directly to the transport mechanism, being the energy required by a carrier to reach the band edge. According to this theory, transport can be described by means of a trap-limited conduction scheme, as illustrated in Fig. 1. Given a pair of traps, since an applied field bends the band edge, the forward and the backward particle fluxes are influenced by the field with opposite exponential dependences. If the current, thus the field, is low, the conventional Ohmic behavior is found. For higher fields, the exponential dependence of the energy barrier on the field makes one flux much larger than the other one, and the exponential regime of the $I(V)$ characteristic is attained. The energy gain from the field is counterbalanced by a relaxation process, but, eventually, a heated carrier distribution is found sufficiently far from the injecting contact. When the heating is large enough, carriers move with higher mobility. This fact separates the device into two regions: the short "off" region near the injecting contact, where the local field is high and carriers heat up, and the "on" region where hot carriers move with higher mobility in presence of a lower local field. Above the threshold point, an increase in the current makes the carriers in the on region hotter, and their mobility increases giving origin to a reduction of the potential drop.

IV. RECENT DEVELOPMENTS

A. Advances in hot-carrier trap-limited conduction

The model proposed by the Bologna-Modena group is an evolution of the work of Ielmini and coworkers. Starting from Ielmini's idea of hot carriers, the authors have derived an enhanced model for trap-limited conduction, reworking the analytical elaboration from scratch. Some of the original simplifying assumptions have been eliminated, and diffusion and electrostatic self-consistency have been introduced. The model has been derived for simplicity in the one-dimensional case, z being the coordinate parallel to the current, but can be easily extended to higher dimensions.

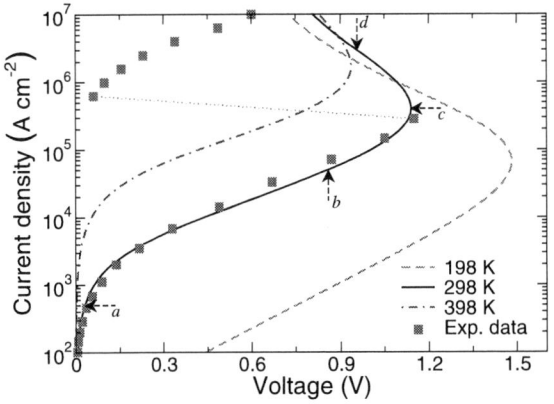

Fig. 4. $J(V)$ characteristics at different temperatures and comparison with experimental results [5] for a 40 nm-long device with a 1000 nm^2 cross-sectional area. The arrows show the position on the $J(V)$ curve of four significant points that are analyzed in Fig. 5. The upper part of the experimental characteristic is not described by the model because the material crystallized due to Joule-heating. The parameters used in the numerical calculations are $\Delta z = 7$ nm, $\tau_0 = 1.2 \cdot 10^{-14}$ s, $\tau_R = 7.8 \cdot 10^{-14}$ s, $n_T = 6.8 \cdot 10^{19}$ cm^{-3}, $\Delta E_G = 0.68$ eV.

Three physical quantities, namely the electric field $F(z)$, the concentration of the carriers $n(z)$ and the associated mean energy $\langle E(z) \rangle$ are investigated. The two latter quantities are conveniently described as moments of the distibution function $f_F(E, z)$, where the equilibrium Fermi level E_F and the lattice temperature T_L have been replaced by the quasi-Fermi level $E_n(z)$ and the electron temperature $T_e(z)$, respectively. The introduction of a z-dependent carrier concentration also enables carrier diffusion, which has properly been accounted for. Following Eq. (2) and the reasoning in [4], one can define the velocity of a carrier with energy E as:

$$v_{\rightleftharpoons}(E, z) = \frac{\Delta z}{\tau_0} \exp\left[-\frac{E_C - E \pm qF(z)\Delta z/2}{k_B T_L}\right] \quad (3)$$

where $\Delta z, E_C, \tau_0$ and k_B are the average traveling distance in a detrapping-trapping event, the band edge, a characteristic

time constant, and the Boltzmann constant, respectively. The arrows/signs indicate the two opposite directions of motion.

The method requires the simultaneous solution of three differential equations that are obtained by proper integrations over the distribution function $f_F(E, z)$ in the energy region filled with traps. For simplicity, a uniform density of trap states in the band gap is assumed.

1) *Carrier concentration and Poisson equation*

$$\frac{dF(z)}{dz} = -\frac{q}{\varepsilon}[n(z) - n_{eq}] \quad (4)$$

where $n(z) = \int \gamma f_F(E, z)\, dE$ and n_{eq} is the equilibrium carrier concentration, with γ the density of states.

2) *Carrier flux equation*

$$\frac{J}{q} = -n(z)\left[\langle v_{\rightarrow}(z)\rangle - \langle v_{\leftarrow}(z)\rangle\right] + \\ + \frac{\Delta z}{2}\frac{d}{dz}\left\{n(z)\left[\langle v_{\rightarrow}(z)\rangle + \langle v_{\leftarrow}(z)\rangle\right]\right\} \quad (5)$$

where the average velocities $\langle v_{\rightleftharpoons}(z)\rangle$ have been obtained as $\langle v_{\rightleftharpoons}(z)\rangle = \frac{1}{n(z)}\int \gamma f_F(E, z)v_{\rightleftharpoons}(E, z)\, dE$.

3) *Energy balance equation*

$$\frac{d}{dz}\left\{n(z)\left[\langle P_{\rightarrow}(z)\rangle - \langle P_{\leftarrow}(z)\rangle\right]\right\} + \\ - \frac{\Delta z}{2}\frac{d^2}{dz^2}\left\{n(z)\left[\langle P_{\rightarrow}(z)\rangle + \langle P_{\leftarrow}(z)\rangle\right]\right\} = \\ = JF(z) - \frac{n(z)\left[\langle E(z)\rangle - \langle E_\circ(z)\rangle\right]}{\tau_R} \quad (6)$$

where the average energy of the carrier distribution $\langle E(z)\rangle$ and the energy fluxes $\langle P_{\rightleftharpoons}(z)\rangle$ have been defined as $\langle E(z)\rangle = \frac{1}{n(z)}\int \gamma f_F(E, z)(E - E_V)\, dE$ and $\langle P_{\rightleftharpoons}(z)\rangle = \frac{1}{n(z)}\int \gamma f_F(E, z)(E - E_V)v_{\rightleftharpoons}(E, z)\, dE$. In eq. (6), τ_R is the energy relaxation time, and $\langle E_\circ(z)\rangle$ represents the "target energy" for the relaxation process, namely the energy that the population $n(z)$ would have if at thermal equilibrium with the lattice.

Results. The $J(V)$ characteristics obtained with the numerical solution of the three equations above are shown in Fig. 4 for different temperatures, and compared to the experimental results reported in [5]. The parameters used in the calculations are listed in the caption of the figure.

Fig. 5 shows the profiles of the quantities of interest along the device. These quantities may help to understand the shape of the $J(V)$ characteristics shown in Fig. 4. It is important to recall that the experimental characteristics show SNDR regions, so that the model must be current-driven. At low currents the field is essentially uniform, and the carrier temperature is close to the lattice temperature. As a consequence, the carrier concentration is constant. This description also holds for most of the subthreshold region, where the most important effect is simply a spatially-uniform growth of the electric field. However, as the current increases and approaches the switching point, the profiles show different values in the regions close to the injecting contact (off region) and to the collecting contact (on region). This is because

978-1-4673-1707-8/12 $31.00 © 2012 IEEE

in the off region the power provided by the electric field is partially used to heat up the carriers, since the energy-relaxation process is not effective enough when carriers have the lattice temperature. As $T_e(z)$ increases, electron move faster and, in order to maintain the prescribed current, $n(z)$ decreases. When $T_e(z)$ is sufficiently high, the dissipation restores a spatially-homogeneous situation where $n(z)$ recovers the charge-neutrality value and the field $F(z)$ is constant. During this process, the electron temperature undergoes an overshoot, particularly evident above threshold. Above threshold, higher currents induce higher electron temperatures and therefore lower fields in the on region, as in the Ielmini model.

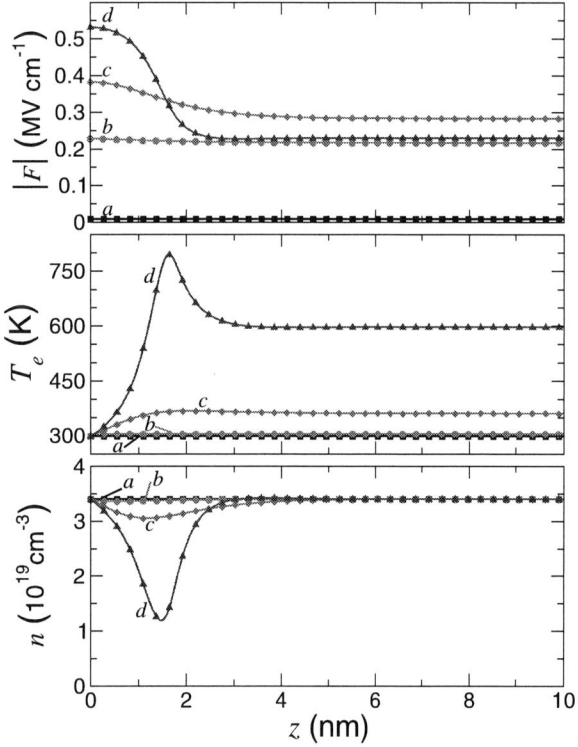

Fig. 5. Electric field (top), carrier temperature (middle), and carrier concentratration (bottom) for the four current densities indicated in Fig. 4: *(a)* $5 \cdot 10^2$ A/cm², *(b)* $5 \cdot 10^4$ A/cm², *(c)* $4 \cdot 10^5$ A/cm² and *(d)* $3 \cdot 10^6$ A/cm². The z axis has been truncated at 10 nm to highlight the off region.

B. Cooperative ionization

Recently, Rudan *et al.* [11] suggested that electron-electron interaction can play an important role in determining the snap-back behavior in the $I(V)$ characteristic of amorphous chalcogenide materials. According to this model, electrons in low-energy states move by hopping between traps while the ones in higher-energy extended states move above the band edge. The former carriers have a smaller mobility than the latter. Trap-to-band transitions are induced mainly by the cooperative effect of Coulomb interactions of a number of band electrons with a trap electron. Quantum-mechanics investigations showed that such a cooperative effect depends

strongly on the current density [12]. This dependence and the difference in mobility between the trap and band electrons are critical for the occurrence of the switching phenomenon in amorphous chalcogenides. In the snap-back region, the voltage increase across the material due to an increasing current is indeed overcome by a competitive effect, that is, the decrease in the same voltage induced by the sudden increase in the material conductivity.

V. Conclusion

Several theories have been proposed to explain the conductivity behavior of chalcogenide glasses used in PCM device technology. Recent advances reduce the number of *ad hoc* hypotheses and approximations within the theoretical approach and provide sets of equations manageable to study a variety of experimental conditions.

Measurements on nanostructures reveal, however, that the properties of individual devices may be influenced by the specific distribution of the existing defects. Thus, their performances may show a significative variance (see, e.g., [13]). For such a reason a Monte Carlo simulation program is being developed by the Bologna-Modena group. A previous work on these lines [14] is being modified in order to implement the features of the theory presented above.

Due to the technological relevance of phase-change materials we expect that more experimental results will contribute to enlight definitively the microscopic bases of transport in amorphous materials.

Acknowledgment

The authors would like to thank Rossella Brunetti, Andrea Cappelli, Fabio Giovanardi, Daniele Ielmini, Massimo Rudan, and Gianpaolo Spadini for many useful discussions.

We gratefully acknowledge the Intel Corporation support under the contract 3477131/2011.

References

[1] N. Mott and E. Davis, *Electronic Processes in Noncrystalline Materials*, Oxford: Clarendon Press, 1961.
[2] B. Shklovskii and A.L. Efros, *Electronic properties of Doped Semiconductors*, Berlin: Springer, 1984.
[3] D.Kau *et al.*, *IEDM Tech. Digest*, p. 617 2009.
[4] D. Ielmini and Y. Zhang, J. Appl. Phys., vol. 102, p. 054517, 2007.
[5] D. Ielmini, Phys. Rev. B, vol. 78, p. 035308, 2008.
[6] F. Buscemi, E. Piccinini, M. Rudan, R. Brunetti, and C. Jacoboni, Fluctuation and Noise Letters, in press.
[7] D. Adler, M.S. Shur, M. Silver, and S.R. Ovshinsky, J. Appl. Phys., vol. 51, p. 3289, 1980.
[8] A. Pirovano, A.L. Lacaita, A. Benvenuti, F. Pellizzer, and R. Bez, IEEE Trans. Electron Devices, vol. 51,p. 452, 2004.
[9] V.G. Karpov, Y.A. Kryukov, S.D. Savransky, and I.V. Karpov, Appl. Phys. Lett., vol. 90, p. 123504, 2007.
[10] M. Simon, N. Nardone, V.G. Karpov, and I.V. Karpov, J. Appl. Phys., vol. 108 ,p. 064514, 2010.
[11] M. Rudan, F. Giovanardi, E. Piccinini, F. Buscemi, R. Brunetti, and C. Jacoboni, IEEE Trans. Electron Devices, vol. 58, p. 4361, 2011.
[12] F. Buscemi, E. Piccinini, F. Giovanardi, M. Rudan, R. Brunetti, and C. Jacoboni, *Proc. of SISPAD 2011*, p. 67, 2011.
[13] S. Roy and A. Asenov, Science, vol. 309, p. 388, 2005.
[14] F. Buscemi, E. Piccinini, R. Brunetti, M. Rudan, and C. Jacoboni, J. Appl. Phys., vol. 106, p. 103706, 2009.

Geometry Based Resistance Model for Phase Change Memory

K. C. Kwong[*], Philip K. T. Mok and Mansun Chan

Department of ECE, Hong Kong University of Science and Technology,
Clear Water Bay, Kowloon, Hong Kong
*Email: eehk@ust.hk

Abstract—A phase change memory resistance model accounting for the geometry of SET and RESET state is developed. The resistance of the memory cell with different dimensions and boundary conditions is solved using conformal mapping including the current crowding effect. When combining with a proper thermal heating model, the read resistance at different degree of crystallization can be predicted, which is important for multi-bit storage simulation. The model has been verified by numerical simulation with different cell geometry and programming current magnitude. The model calculation result is verified by the numerical simulation.

I. INTRODUCTION

Phase change memory (PCM), as a candidate for next generation non-volatile memory (NVM) [1], in addition to storing "0" and "1", more recent studies also reveal the possibility to use PCM to store multi-bit data in one cell [2]. The data storage of PCM is achieved by changing the cell resistance by thermal heating and the ON/OFF resistance ratio of the high resistance state (RESET) and low resistance state (SET) is a core parameter of PCM to determine the error margin. The introduction of multi-bit storage requires intermediate states between complete SET and RESET states [3] and makes the accurate determination of the intermediate state resistances important [4]. However, there is very little study on the resistance of PCM cell as a function of geometry and programming conditions. Some resistance models have been proposed [5], but the complete geometry and the current crowding effects have not been studied.

In this work, a resistance model for PCM cell is developed which includes the current crowding effect due to the cell geometry using conformal mapping. From the geometrical configuration of the amorphous region of the PCM cell resulting from thermal heating induced with different input current pulse, the intermediate resistance between SET and RESET can also be determined. Extensive numerical simulation has been used to validate the model.

II. SET STATE RESISTANCE MODEL

Fig. 1 shows the generic structure of the PCM cell. In the SET state, the PC material in crystalline state is placed between the top and bottom electrode with different widths. Current flows from the top electrode to the bottom electrode with a non-uniform current path resulting in current crowding. To calculate the equivalent resistance with the complex boundary condition, conformal mapping is used to transform the boundary conditions to a simpler form in a different co-ordinate system. Due to the symmetry of the PCM structure, conformal mapping is applied to only half of the cell as shown in Fig. 2. The corresponding new dimension can be expressed as:

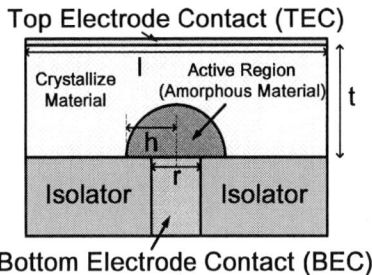

Figure 1. Generic PCM structure labelled with dimension

SET State
(a) z=x+iy (b) w=u+iv

(a1) (a2) (a5) (a3) (a4)
(b1) (b2) (b4) (b3)

Mapping Function:
$$sn(w, L_{SET}') = sn[(\frac{2z}{l}) \cdot K_z, k_z] \cdot \frac{k_z}{L_{SET}}, \quad \& \quad \frac{K_z}{K_z'} = \frac{l}{2t}$$

(c)

Side	Dimension	B.C.	Side	Dimension	B.C.
a1	l	$V=V_0$	b1	l_{SET}'	$V=V_0$
a2	t	$\partial V/\partial x=0$	b2	t_{SET}'	$\partial V/\partial u=0$
a3	r	$V=0$	b3	l_{SET}'	$V=0$
a4	l-r	$\partial V/\partial y=0$	b4	t_{SET}'	$\partial V/\partial u=0$
a5	t	$\partial V/\partial x=0$			

Figure 2. (a) SET state PCM before conformal mapping (b) SET state PCM after conformal mapping (c) A list of dimension and boundary condition of different side in Fig. 2(a) and Fig. 2(b)

$$l_{SET}' = \int_0^{\frac{\pi}{2}} \frac{d\theta}{\sqrt{1-\{k_z \cdot sn[(\frac{r}{l})K_z, k_z]\}^2 \cdot \sin^2\theta}} \quad (1)$$

$$t_{SET}' = \int_0^{\frac{\pi}{2}} \frac{d\theta}{\sqrt{k_z^2 \cdot sn^2[(\frac{r}{l})K_z, k_z] \cdot \sin^2\theta}} \quad (2)$$

where l_{SET}' and t_{SET}' are the transformed half width and thickness of the cell respectively, K_z, K_z' are the complete elliptical integral of the first kind for modulus k_z and k_z'. After performing the

978-1-4673-1707-8/12 $31.00 © 2012 IEEE

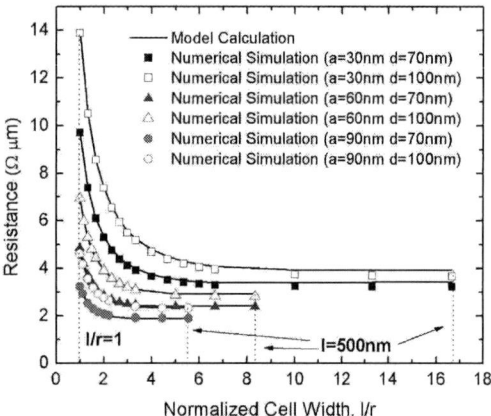

Figure 3. Comparison between model calculation and numerical simulation of SET state PCM resistance with different normalized cell width

conformal transformation, a simple resistance equation can be applied calculated according to Fig. 2:

$$R_{SET} = \frac{1}{2} \cdot \rho_{cry} \cdot \frac{t_{SET}'}{l_{SET}'} \qquad (3)$$

where ρ_{cry} is the resistivity of the crystalline material [6]. The calculated resistance in Fig. 2 is multiplied by half in equation (3) to include both left and right half of the structure.

Fig. 3 shows the comparison between model calculation and numerical simulation with different cell thicknesses and electrode widths. As the width of one electrode increases, the resistance decreases as in normal resistor until a threshold is reached. Further increase in the width of the longer electrode cannot reduce the resistance due to current crowding in the shorter electrode, which agrees with the physics.

III. RESET STATE RESISTANCE MODEL

PCM in RESET state consists of a portion of PC material in amorphous state defined as active region. The resistivity of amorphous PC material depends on the crystal fraction of the material determined by the temperature history of the cell. Its value is usually several orders of magnitude higher than the PC material in crystalline state. The shape of the active region depends on the quenching time of the cooling process and in general has a form of hemisphere [7]. Subject to the temperature history of the cell, the resistivity of the amorphous PC material decreases due to the increase in crystal fraction. The resistivity of the amorphous material of

the active region changes including the effect the crystal fraction is given by [8]:

$$\rho_{active} = \rho_{amo} + C_f \cdot (\rho_{cry} - \rho_{amo}) \qquad (4)$$

where ρ_{amo} is the resistivity of the amorphous material [9] and C_f is the crystal fraction determined by JMA equation [10].

The resistance of the active region can be calculated using conformal mapping again to transform the normalized hemisphere top electrode and planar bottom electrode into two parallel plates with different lengths as shown in Fig. 4. Then, the same conformal mapping method as the SET state resistance model can

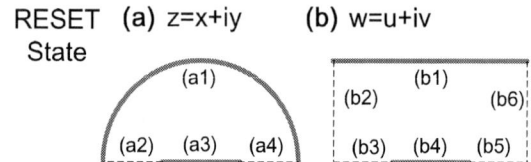

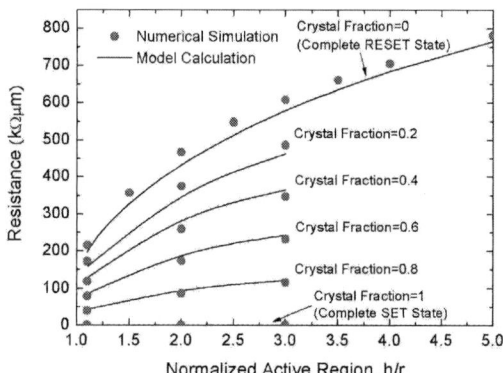

Figure 4. (a) RESET state PCM before conformal mapping (b) RESET state PCM after conformal mapping (c) A list of dimension and boundary condition of different side in Fig. 4(a) and Fig. 4(b)

Figure 5. Comparison between model calculation and numerical simulation of RESET state PCM resistance with different normalized active region

be applied to calculate the cell resistance by further transform the structure in Fig. 4b in to the parallel plates with same width. The final transformed dimension and resistance of the cell can be expressed as:

$$l_{RESET}' = \int_0^{\frac{\pi}{2}} \frac{d\theta}{\sqrt{\cosh^{-2}(c) \cdot \sin^2 \theta}} \qquad (5)$$

$$t_{RESET}' = \int_0^{\frac{\pi}{2}} \frac{d\theta}{\sqrt{1 - \cosh^{-2}(c) \cdot \sin^2 \theta}} \qquad (6)$$

$$R_{active} = \frac{1}{2} \cdot \rho_{active} \cdot \frac{t_{RESET}'}{l_{RESET}'} \qquad (7)$$

where l_{RESET}' and t_{RESET}' are transformed half width and thickness of the PCM cell respectively. The resistance of the crystalline material in series with the active region can be calculated using equation (3) and (7) as it is only geometry dependent:

$$R_{RESET} = \frac{1}{2} \cdot \rho_{active} \cdot \frac{t_{RESET}'}{l_{RESET}'} + \frac{1}{2} \cdot \rho_{cry} \cdot \left(\frac{t_{SET}'}{l_{SET}'} - \frac{t_{RESET}'}{l_{RESET}'} \right) \qquad (8)$$

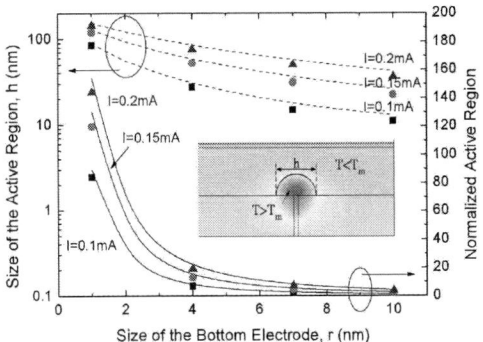

Figure 6. Model calculation and numerical simulation of the size of the active region, h, during RESET with different BEC, r, together with thermal simulation of the PCM cell

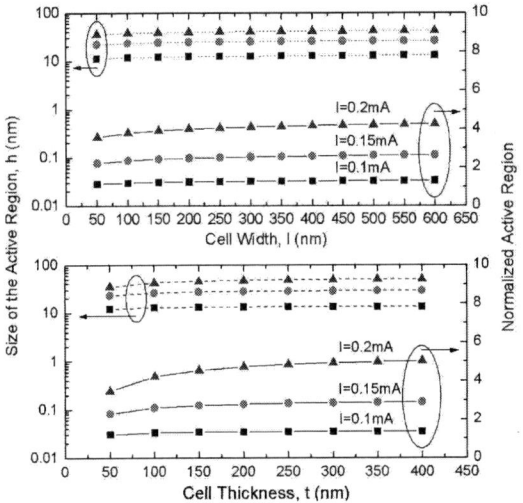

Figure 7. (a) Model calculation of the size of the active region, h, during RESET with different TEC, l, (b) Model calculation of the size of the active region, h, during RESET with different cell thickness, t

The model is compared with numerical simulation and the results are shown in Fig. 5. Besides the crystal fraction, the RESET state resistance also depends on the ratio of the diameter of the active region and the width of the bottom electrode. The RESET resistance increases gradually with larger normalized active region. When the normalized active region is small, most of the current flows from two sides of the hemisphere to the edge of the bottom electrode due to the shorter current path. With a higher crystal fraction, the active region resistivity decreases accordingly. When crystal fraction equals to 0, it is in complete RESET state which has a resistance dominated by the active region. As the crystal fraction increases and eventually converge back to the complete SET state when crystal fraction reached 1.

IV. THERMAL MODEL FOR RESET OPERATION

During RESET, a high current is applied to the cell to cause a temperature increase beyond the melting point of the material. With a fast cutoff of the applied current, the abrupt cooling or quenching forms the active region. The diameter of the active

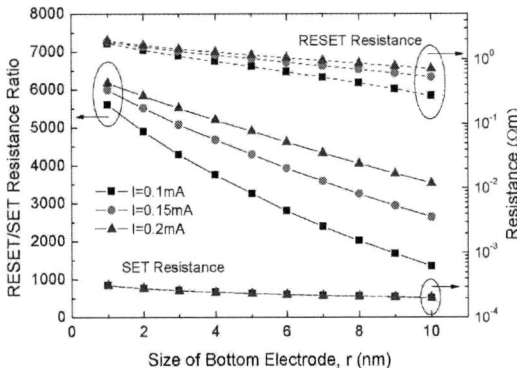

Figure 8. Model calculation of SET, RESET state resistance and corresponding RESET/SET resistance ratio with different size of bottom electrode, r, and Current (I=0.1mA, 0.15mA and 0.2mA)

region (h) has been formulated in [11] as an implicit function given by:

$$h = \sqrt{\frac{I^2 \rho_m}{2\pi^2 \kappa_s (T_m - T_r) f}} \quad (9)$$

where I is the amplitude of the programming current, ρ_m is the resistivity of the melting PC material, κ_s is the thermal conductivity of the PC material in crystalline state, T_m and T_r are the melting point of the PC material and the surrounding temperature respectively. f which is a parameter which is a weak function determined by the physical properties of the top and bottom electrode. A value of 0.6 is used in our work according to the formulation given in [11].

Fig. 6 shows the thermal simulation result of a PCM cell with the structure as given in Fig. 1. Due to the high current density at the bottom electrode, a high temperature is induced at the bottom of the structure by joule heating. The active region can be inferred by the region where the temperature is higher than the melting point of the PC material in the simulation. The size of the active region predicted by equation (9) as compared with the numerical thermal simulation is given in Fig. 6 which shows a reasonable agreement. The comparison is repeated by using different cell widths in Fig. 7a and cell thicknesses Fig. 7b.

V. DESIGN OPTIMIZATION AND SIMULATION RESULT

To have sufficient read margin, especially to allow intermediate states in multi-bit storage, it is desirable to maximize the fully RESET resistance and minimize the fully SET resistance. The RESET/SET resistance ratio can be calculated by:

$$Q = \frac{R_{RESET}}{R_{SET}} = \frac{\rho_{active}(\frac{t_{RESET}'}{l_{RESET}'}) + \rho_{cry}(\frac{t_{SET}'}{l_{SET}'} - \frac{t_{RESET}'}{l_{RESET}'})}{\rho_{cry}(\frac{t_{SET}'}{l_{SET}'})} \quad (10)$$

High read margin can be achieved by using a type of PC material which has a large difference in crystalline and amorphous state resistivity. However, changing the material is in generally difficult in the device fabrication. Another way to improve the RESET/SET resistance ratio is to optimize the cell geometry.

978-1-4673-1707-8/12 $31.00 © 2012 IEEE   103

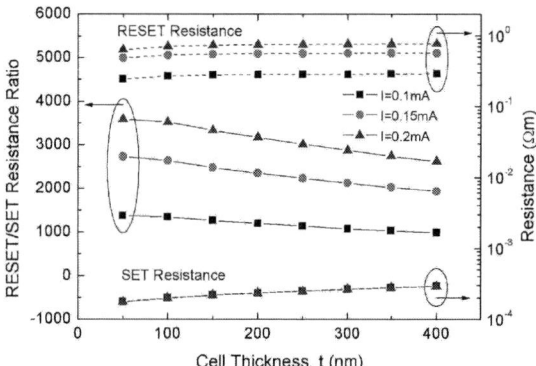

Figure 9. Model calculation of SET, RESET state resistance and corresponding RESET/SET resistance ratio with different cell thickness, t, and current (I=0.1mA, 0.15mA and 0.2mA)

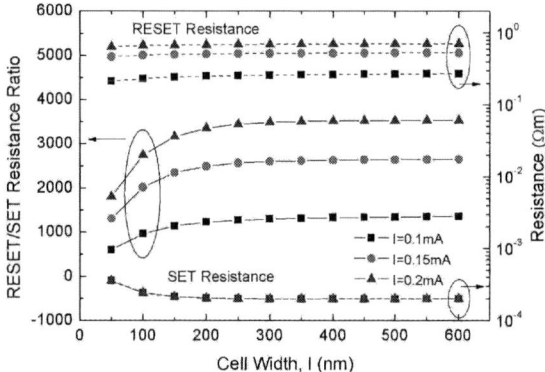

Figure 10. Model calculation of SET, RESET state resistance and corresponding RESET/SET resistance ratio with different cell width, l, and Current (I=0.1mA, 0.15mA and 0.2mA)

Fig. 8 shows the variation of the RESET/SET resistance ratio by changing the size of the bottom electrode. Reducing the bottom electrode increases the resistance of both SET and RESET state due to geometry effect. The increase in resistance, however, not only affects the read resistance but the thermal heating. During RESET, the use of smaller bottom electrode leads to more concentrated heating and larger active region as shown in Fig. 6. As a result, it has much stronger impact to increase the RESET resistance and leading to a higher RESET/SET resistance ratio. Therefore, it is desirable to have a small bottom electrode to improve the read margin, which agrees with experimental finding [12]. Fig. 9 shows that changing the cell thickness has a little change on the RESET/SET resistance ratio. Increase of the cell thickness increases the SET state resistance by introducing a longer current path and induces a larger active region due to slower heat loss rate. However, the increase in the normalized active region size is very small according to Fig. 7a for thick PC material and the RESET resistance remains more or less constant. As a result, increasing the cell thickness induces a slightly decrease of the RESET/SET resistance ratio by increasing the SET state resistance. Fig. 10 shows that the RESET/SET resistance ratio increases with cell width. Increasing the cell width

reduces the SET resistance by introducing more current paths and induces a larger active region due to slower heat loss rate. A wide cell results in only small increase in the RESET resistance. Also, as shown in Fig. 3, SET resistance saturates for large cell width. As a result, RESET/SET resistance ratio increases with cell width and approaches a constant value for large cell width.

VI. CONCLUSION

Using conformal mapping, PCM resistance models of SET state and RESET state including intermediate state for reading operation are developed and verified by numerical simulation. Based on the resistance and thermal models, analytical solution of the RESET/SET resistance ratio accounting for the physical properties of the material, PCM cell dimension and programming current, is obtained and used for simulation and study. Using this model, according to a desired RESET/SET resistance ratio, the dimension of the PCM cell and programming current amplitude can be determined accordingly which can facilitate the design of the PCM cell.

ACKNOWLEDGMENT

This work is supported by Hong Kong University Grant Council under the Area of Excellence Scheme with contract number AoE/P-04/08.

REFERENCES

[1] Stefan Lai, "Current status of the phase change memory and its future", IEEE, IEDM 2003, pp. 10.1.1-10.1.4

[2] S. Tyson, G. Wicker, T. Lowrey, S. Hudgens and K. Hunt, "Nonvolatile, high density, high performance, PC memory", Proc. Aero Space Conf., vol. 5, 2000, pp. 385-390

[3] S. R. Ovshinsky, "Electrically erasable, directly overwritable, multibit single cell memory elements and arrays fabricated thereform", U. S., Patent Editor, 1995

[4] F. Bedeschi, E. Bonizzoni, O. Khouri, C. Resta and G. Torelli, "A fully symmetrical sense amplifier for non-volatile memories", IEEE, ISCAS, 2004, II-625-8 vol. 2

[5] Yi Zhang, Jie Fengm Hao Wang, Bingchu Cai and Bomy Chen, "Modeling of two different operation modes of phase change material for phase-change random-access memory", Japanese Journal of Applied Physics, vol. 44, no. 4a, 2005, pp. 1687-1692

[6] P. Kumar, K. S. Bindra, N, Suri and R. Thangaraj, "Transport properties of a-Sn$_x$Sb$_{20}$Se$_{80-x}$ (8<x<18) chalcogenide glass", J. Phys. D: Appl. Phys. 39 (2006) pp.642-646

[7] Daewon Ha and Kinam Kim, "Recent advances in high density phase change memory (PRAM)", IEEE, Interational Symposium on VLSI-TSA 2007, pp. 1-4

[8] Y. –B. Liao, J. –T. Lin, and M. –H. Chiang, "Temperature-based phase change memory model for pulsing scheme assessment," in Proc. IEEE ICICDT, 2008, pp. 199-202

[9] Daniele Ielmini and Yuegang Zhang, "Analytical model for subthreshold conduction and threshold switching in chalcogenide-based memory devices", Journal of Applied Physics 102, 054517, 2007

[10] J. Malek, "The applicability of Johnson- Mehl- Avrami model in the thermal analysis of the crystallization kinetics of glasses", Thermochimica Acta, vol. 267, pp. 61-73

[11] B. Rajendran, J. Karidis, M-H. Lee, M. Breitwisch, G. W. Burr, Y-H. Shih, R. Cheek, A. Schrott, H-L. Lung and C. Lam, "Analytical model for RESET operation of phase change memory", IEEE, IEDM 2008, pp. 1-4

[12] A. L. Lacaita, "Physics and performance of phase change memories", IEEE, SISPAD 2005, pp. 267-270

978-1-4673-1707-8/12 $31.00 © 2012 IEEE

Drain-conductance optimization in nanowire TFETs

E. Gnani, S. Reggiani, A. Gnudi and G. Baccarani

ARCES and DEIS, University of Bologna, Viale Risorgimento 2, 40136 Bologna, Italy
Phone: +39 051 209 3773, Fax: +39 051 209 3779, E-mail: egnani@arces.unibo.it

Abstract— In this work we examine the problem of the non-linear output characteristics of tunnel FETs, and the related small drain conductance at low drain voltage, which prevents rail-to-rail logic operation and severely degrades the device dynamic properties compared with standard CMOS FETs. The problem is investigated with the help of an analytical model which highlights the constraints of the device design by splitting the effects of the tunneling probability from the density of states in the source, channel and drain, and makes it possible to design a nanowire TFET by an appropriate selection of the material, nanowire size and degeneracy levels in the source and drain regions. So doing, we remove the above characteristics' feature and recover a large drain conductance without degrading the subthreshold slope. The optimized device is numerically simulated using the k·p model, whose results are in fair agreement with the analytical one.

I. INTRODUCTION

Tunnel FETs (TFETs) have been the subject of intense investigation in the last decade, with the aim to achieve inverse subthreshold slopes significantly smaller that 60 mV/dec, and to open the avenue to low supply voltages and low-power consumption in logic circuits [1]–[7]. However, experimental results have been so far rather disappointing due to a number of deficiencies which are still largely unsolved: (i) the on-state currents are typically much smaller than those provided by standard CMOS FETs [1]–[4]; (ii) inverse subthreshold slopes smaller than 60 mV/dec are not sustained across several decades of current [1]–[3]; (iii) the output characteristics at low drain voltage exhibit a diode-like behavior, with a very small drain conductance which prevents rail-to-rail logic switching under dynamic conditions [8]; (iv) low-bandgap material TFETs exhibit an ambipolar effect, which degrades the subthreshold slope and the I_{ON}/I_{OFF} current ratio [3].

In a recent simulation work [8] it has been pointed out that the small drain conductance at low drain voltage significantly degrades the dynamic TFET performance by drastically reducing the rise and fall times of a TFET-based inverter. Also, the high saturation voltage compared with that of CMOS FETs reduces the inverter gain and lowers the static noise margins. In this work, we investigate the TFET drain conductance at low V_{DS} working out a simplified model suitable for surrounding-gate nanowire (NW) TFETs, identify the motivation for the diode-like characteristics and suggest design criteria meant to overcome the above effect. The optimized device is an InAs nanowire (NW) TFET having a 6×6 nm square cross-section, which exhibits an effective bandgap $E_G \simeq 0.6$ eV.

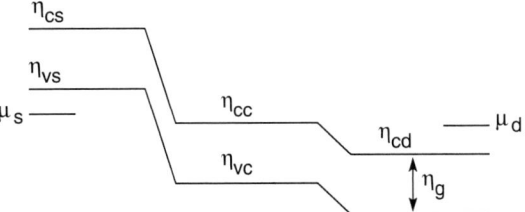

Fig. 1. Schematic of the band structure in an n-type TFET and definition of the band edges in the source, channel and drain.

II. FUNDAMENTAL EQUATIONS

The current flow through a surrounding-gate n-type NW-TFET under ballistic conditions can be represented by the Landauer equation, namely

$$I_D = \frac{2qk_BT}{h} \int_{\eta_{cd}}^{\eta_{vs}} T_t(\varepsilon) \left[f(\varepsilon - \mu_s) - f(\varepsilon - \mu_d) \right] d\varepsilon \quad (1)$$

where $T_t(\varepsilon)$ is the electron tunneling probability at the normalized energy ε, μ_s and μ_d are the Fermi levels in the source and drain regions, respectively, η_{vs} and η_{cd} are the normalized source valence-band and drain conduction-band edges, and $f(\varepsilon - \mu)$ is the Fermi function. We now consider the band structure schematically represented in fig. 1, and define the valence- and conduction-band edges in the source, channel and drain as indicated in the same figure. Finally, we assume for simplicity that only the lowest conduction and the upper valence subbands contribute to the current, which can be approximately true for fairly-narrow nanowires.

Under the assumption that $T_t(\varepsilon) \simeq 0$ in the energy interval $\eta_{vc} < \varepsilon < \eta_{cc}$, i.e. zero source-to-drain tunneling probability, the integral in (1) leads to the following expression

$$
\begin{aligned}
I_D = \frac{2qk_BT}{h} \Bigg\{ &T_{sc} \left[\ln \left(\frac{1 + \exp(\mu_s - \eta_{max})}{1 + \exp(\mu_s - \eta_{vs})} \right) \right. \\
&\left. - \ln \left(\frac{1 + \exp(\mu_d - \eta_{max})}{1 + \exp(\mu_d - \eta_{vs})} \right) \right] \theta(\eta_{vs} - \eta_{max}) \\
&+ T_{cd} \left[\ln \left(\frac{1 + \exp(\mu_s - \eta_{cd})}{1 + \exp(\mu_s - \eta_{vc})} \right) \right. \\
&\left. - \ln \left(\frac{1 + \exp(\mu_d - \eta_{cd})}{1 + \exp(\mu_d - \eta_{vc})} \right) \right] \theta(\eta_{vc} - \eta_{cd}) \Bigg\} \quad (2)
\end{aligned}
$$

where $\theta(\eta)$ is the step function, $\eta_{max} = \max\{\eta_{cc}, \eta_{cd}\}$ and $T_{sc}(\eta_{vs} - \eta_{vc})$, $T_{cd}(\eta_{cc} - \eta_{cd})$ are suitable averages of the tunneling probabilities across the source-channel and the

978-1-4673-1707-8/12 $31.00 © 2012 IEEE   105

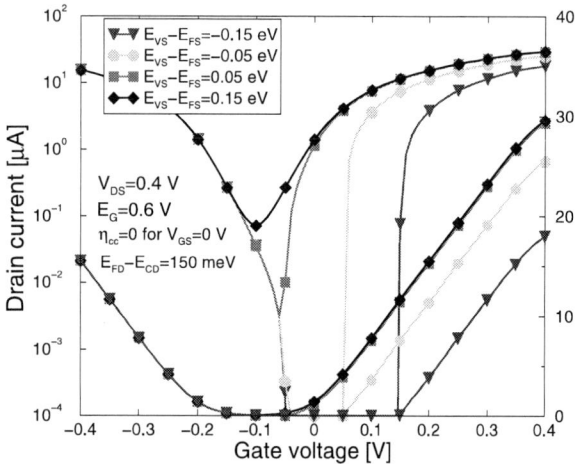

Fig. 2. Log and linear plot of the turn-on characteristics at $V_{\mathrm{DS}} = 0.4$ V for TFETs with degenerate drain ($E_{\mathrm{FD}} - E_{\mathrm{CD}} = 150$ meV) and three different levels of degeneracy in the source, namely: $E_{\mathrm{VS}} - E_{\mathrm{FS}} = 150, 50, -50, -150$ meV, represented by diamond, squares, dots and triangles, respectively. In this computation $T_{\mathrm{sc}} = T_{\mathrm{cd}} = 1$ and $E_G = 0.6$ eV.

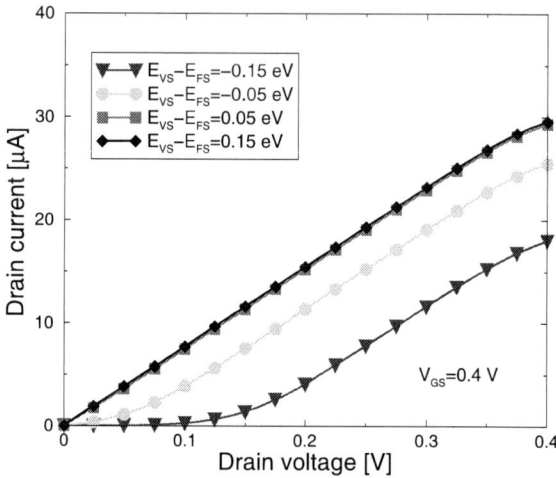

Fig. 3. Output characteristics at $V_{\mathrm{GS}} = 0.4$ V for TFETs with degenerate drain ($E_{\mathrm{FD}} - E_{\mathrm{CD}} = 150$ meV) and three different levels of degeneracy in the source, namely: $E_{\mathrm{VS}} - E_{\mathrm{FS}} = 150, 50, -50, -150$ meV, represented by diamonds, squares, dots and traingles, respectively. No current saturation occurs in the present voltage range. Here, $T_{\mathrm{sc}} = T_{\mathrm{cd}} = 1$ and $E_G = 0.6$ eV.

channel-drain tunneling windows. Their values can be evaluated from the WKB tunneling probability by integration over the bandgap of $\kappa(\varepsilon)$ from Flietner's expression and assuming a constant electric field. Such an integration leads to the following result

$$T_{\mathrm{sc}}(\eta_{\mathrm{vs}} - \eta_{\mathrm{vc}}) = \exp\left(-\frac{\pi \Lambda_{\mathrm{sc}}\sqrt{2m^* k_B T}\,\eta_g^{3/2}}{4\hbar(\eta_{\mathrm{vs}} - \eta_{\mathrm{vc}})}\right) \quad (3)$$

where $(\eta_{\mathrm{vs}} - \eta_{\mathrm{vc}})/\Lambda_{\mathrm{sc}}$ represents the normalized electric field at the source-channel junction, and Λ_{sc} is a suitable distance. A similar expression holds for T_{cd} against $(\eta_{\mathrm{cc}} - \eta_{\mathrm{cd}})$.

In (2) the first term in brackets accounts for the net electron flow from the source valence-band into the channel conduction-band, while the second term in brackets accounts for the net electron flow from the channel valence-band into the drain conduction-band. This latter term represents the ambipolar effect. The condition for the simultaneous occurrence of both contributions at some gate voltages is that $\eta_{\mathrm{vs}} - \eta_{\mathrm{cd}} > \eta_g$ and, in order to prevent it to occur, the inequality $v_{\mathrm{dd}} < \eta_g - (\eta_{\mathrm{vs}} - \mu_s) - (\mu_d - \eta_{\mathrm{cd}})$ must be fulfilled, with v_{dd} the normalized supply voltage.

A similar approach was proposed in [5] for the description of p-type graphene nanoribbon TFETs. However, besides neglecting the ambipolar effect, the integration of eq. (1) was carried out from the Fermi level, rather than the conduction-band edge, in the n-type source, up to the edge of the valence band in the channel. So doing, the degeneracy level in the source becomes totally uneffective on the device properties.

In order to split out the effect of the tunneling probability from the integral of the Fermi functions, we first assume that $T_{\mathrm{sc}} = T_{\mathrm{cd}} = 1$, and show in fig. 2 the turn-on characteristics of four n-type TFETs with different impurity concentrations in the source, such that $E_{\mathrm{VS}} - E_{\mathrm{FS}} = 150, 50, -50$ and -150 meV,

computed at $V_{\mathrm{DS}} = 0.4$ V. The bandgap is $E_G = 0.6$ eV and the drain degeneracy is $E_{\mathrm{FD}} - E_{\mathrm{CD}} = 150$ meV. The ambipolar effect is clearly visible in the figure for negative gate voltages. The gate-semiconductor work-function difference has been chosen such that $\eta_{\mathrm{cc}} = 0$ for $V_{\mathrm{GS}} = 0$ V. This implies that the TFETs with degenerate source (diamonds and squares) are conducting at $V_{\mathrm{GS}} = 0$ V, while the TFETs with non-degenerate source (dots and triangles) are still off at the same gate voltage.

It may be interesting to notice that, for the two degenerate source conditions, the overlap of the two tunneling currents originating from the source and the drain prevents the current falloff, thus negatively affecting the $I_{\mathrm{ON}}/I_{\mathrm{OFF}}$ current ratio. Instead, a non-degenerate source allows for a sudden transition from the off to the on state, with the elimination of the subthreshold region.

The output characteristics of the same TFETs are plotted in fig. 3 at $V_{\mathrm{GS}} = 0.4$ V. The drain conductance at low V_{DS} equals the limiting value $2q^2/h$ for the degenerate-source conditions, but drops down very significantly when the source is not degenerate. Another major problem is that no saturation of the output characteristics occurs for $V_{\mathrm{DS}} < V_{\mathrm{DD}}$. This lack of saturation would strongly reduce the voltage gain of a TFET-based inverter, with severely-degraded noise-immunity margins. If V_{DS} is vanishingly small and V_{GS} is large the ambipolar effect cannot take place, and the condition $\eta_{\mathrm{cd}} > \eta_{\mathrm{cc}}$ may be fulfilled, especially if the doping density in the drain is not too large. In this case the drain current becomes

$$I_D = \frac{2q k_B T}{h} T_{\mathrm{sc}} \left\{ \ln\left(\frac{1 + \exp\left(\mu_s - \eta_{\mathrm{cd}}\right)}{1 + \exp\left(\mu_s - \eta_{\mathrm{vs}}\right)}\right) - \ln\left(\frac{1 + \exp\left(\mu_d - \eta_{\mathrm{cd}}\right)}{1 + \exp\left(\mu_d - \eta_{\mathrm{vs}}\right)}\right) \right\} \theta(\eta_{\mathrm{vs}} - \eta_{\mathrm{cd}}) \quad (4)$$

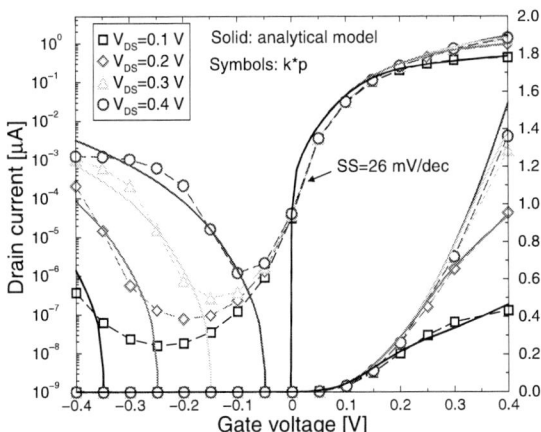

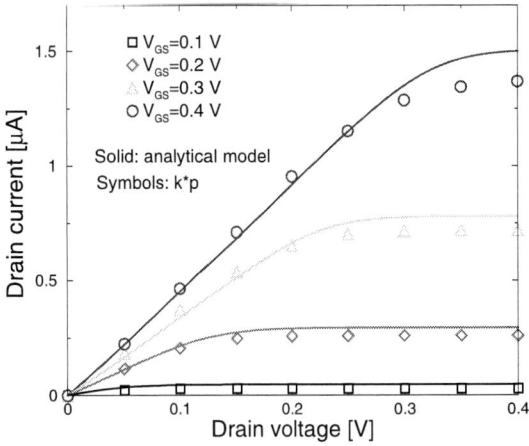

Fig. 4. Turn-on characteristics of the optimized InAs NW-TFET, with the gate voltage ranging between 0.1 and 0.4 V. The side of the assumed square cross-section is 6 nm, which provides a bandgap $E_G = 0.614\,\text{eV}$. The source degeneracy $E_{\text{VS}} - E_{\text{FS}} = 80\,\text{meV}$, while the drain degeneracy $E_{\text{FD}} - E_{\text{CD}} = 90\,\text{meV}$. Numerical simulations are carried out using the **k·p** model.

Fig. 5. Output characteristics of the optimized InAs NW-TFET as in fig. 4, for gate voltages ranging between 0.1 and 0.4 V. This plot shows that $I_{\text{ON}} \simeq 1.4\,\mu\text{A}$, which translates into a normalized on-state current of $0.23\,\text{mA}/\mu\text{m}$, and that the drain conductance at low V_{DS} is no longer independent of the gate voltage.

We may now observe that $\mu_d = \mu_s - v_{\text{ds}}$; $\eta_{\text{cd}} = \eta_{\text{cd}}^{(0)} - v_{\text{ds}}$ with v_{ds} the normalized drain-source voltage and $\eta_{\text{cd}}^{(0)}$ the drain conduction-band edge in equilibrium. The drain conductance g_d can now be evaluated at $v_{\text{ds}} = 0$ by differentiang either eq. (1) or eq. (4). We find

$$g_d = \frac{2q^2}{h} T_{\text{sc}} \left[f(\eta_{\text{cd}}^{(0)} - \mu_s) - f(\eta_{\text{vs}} - \mu_s) \right] \theta(\eta_{\text{vs}} - \eta_{\text{cd}}) \quad (5)$$

Eq. (5) tells us that the drain conductance can approach the prefactor $(2q^2/h)\,T_{\text{sc}}$ if $\mu_s - \eta_{\text{cd}}^{(0)} \gg 1$ and if $\eta_{\text{vs}} - \mu_s \gg 1$. If this is indeed the case, the first term in brackets nearly equals one, while the second term is much smaller than one. These conditions require that both the source and drain regions be heavily degenerate. If, on the other hand, either the source or the drain are not degenerate, the drain conductance drops down and may even fall close to zero if the two Fermi functions cancel out. Real devices typically exhibit thermal and tunneling currents not accounted for in this treatment, so that the actual drain conductance will never be exactly zero. On the other hand, the contribution of the tunneling currents to the drain conductance becomes negligibly small if the energy window between η_{vs} and η_{cd} totally disappears.

With reference to eq. (2), the device transconductance g_m can be expressed as

$$g_m = \frac{2q^2}{h} \Big\{ T_{\text{sc}} \left[f(\eta_{\text{cc}} - \mu_s) - f(\eta_{\text{cc}} - \mu_d) \right] \theta(\eta_{\text{vs}} - \eta_{\text{cc}}) \\ + T_{\text{cd}} \left[f(\eta_{\text{vc}} - \mu_s) - f(\eta_{\text{vc}} - \mu_d) \right] \theta(\eta_{\text{vc}} - \eta_{\text{cd}}) \Big\} \\ \left(-d\eta_{\text{cc}}/dv_{\text{gs}} \right) \quad (6)$$

Eq. (6) shows that, at low v_{gs} and high v_{ds} where the last term $(-d\eta_{\text{cc}}/dv_{\text{gs}}) \simeq 1$, the transconductance g_m is dominated by the first term in brackets $f(\eta_{\text{cc}} - \mu_s) \simeq \exp(\mu_s - \eta_{\text{cc}})$, which represents the exponential tail of the Fermi function.

The knowledge of the TFET current and transconductance makes it possible to compute the inverse subthreshold slope.

III. DEVICE OPTIMIZATION

One key aspect of TFET design is preventing the occurrence of the ambipolar effect, i.e. tunneling at the channel-drain junction, for positive gate voltages. By keeping the two conducting regimes wide apart enables a sudden falloff of the tunneling current, with clear benefits for the TFET leakage. According to the model, the abruptness of the transition between the off- and the on-state is much improved if the source is not degenerate. However, a moderate doping density in the source makes the drain conductance vanishingly small at low drain voltage, preventing rail to rail switching in logic operation. The device optimization can thus be achieved by selecting the semiconductor material, supply voltage and doping density within the source and drain regions. With $V_{\text{DD}} = 0.4\,\text{V}$, the smallest affordable bandgap is $E_G = 0.6\,\text{eV}$ so as to prevent the ambipolar effect. Also, a moderate degeneracy is required both in the source and drain regions, namely: $\eta_{\text{vs}} - \mu_s \simeq 3$ and $\mu_d - \eta_{\text{cd}} \simeq 3$. So doing, we accept a 10% degradation of the drain conductance but enforce the current falloff at $V_{\text{GS}} = V_T - 3k_B T/q$.

In order to validate the optimization methodology, we simulated an InAs NW-TFET with a 6×6 nm square cross-section, which provides an effective bandgap $E_G = 0.614\,\text{eV}$. The gate length was assumed to be $L_g = 20\,\text{nm}$, and so were the source and drain regions. The simulation was carried out using an in-house developed **k·p** model [9] implementing the complete eight-band Kane's Hamiltonian, which is expected to accurately describe the direct-gap semiconductor bands around the Γ point of the first Brillouin zone. The choice of InAs was suggested by the small electron effective mass and by the small bandgap which, on the other hand, can be tuned

978-1-4673-1707-8/12 $31.00 © 2012 IEEE

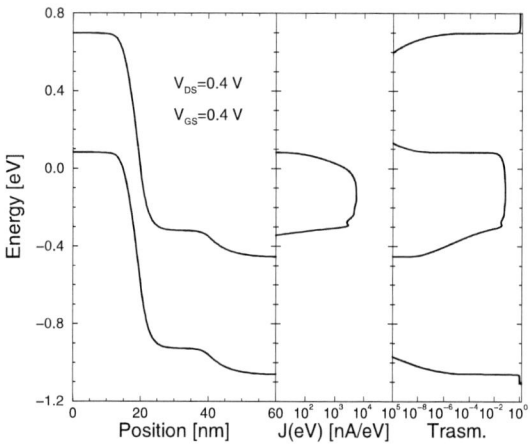

Fig. 6. Left: Band structure of the simulated InAs NW-TFET at $V_{GS} = V_{DS} = 0.4$ V. Center: Current spectrum within the tunneling window. Right: Tunneling probability. The mamximum value of the tunneling probability is 0.066, i.e. only 10% higher than the average value of 0.059 estimated from the ratio of the currents given by the simple analytical model and the **k·p** numerical simulation.

to the desired value by a suitable choice of the NW size. Besides, we fixed the impurity concentration in the source and drain regions in order to approach the optimal degeneracy conditions. The effective values of $E_{VS}-E_{FS}$ and $E_{FD}-E_{CD}$ turned out to be about 80 and 90 meV, respectively.

The turn-on characteristics of the simulated InAs TFET are compared in fig. 4 with the analytical model for V_{GS} ranging between -0.4 and 0.4 V and V_{DS} between 0.1 and 0.4 V. The I_{ON}/I_{OFF} ratio is well in excess of 10^4; the lowest SS is 26 mV/dec and $I_{ON} = 1.4\,\mu A$ at $V_{DD} = 0.4$ V, which translates into an on-state current density of 0.23 mA/μm. The output characteristics of the same device are plotted in fig. 5, showing that the device design provides a linear growth of the drain current at low V_{DS} and that the drain conductance is acceptably large. It may be worth mentioning that $g_d = 4.65\,\mu A/V$ at $V_{GS} = 0.4$ V, which scales with the theoretical value of $2q^2/h$ by a factor of 17, as for the drain current. The simplified model turns out to be in fair agreement with numerical simulation results, with a small deviation occurring near the current saturation. The change of the drain conductance with V_{GS} is due to the tunneling probability, which scales inversely with $(\eta_{cs} - \eta_{cc})$. Fig. 6 shows the subband profile within the simulated device (left), the current spectral density (center), and the tunneling probability vs. energy (right). The latter is modeled using a unique fitting parameter, namely Λ_{sc}, for all bias conditions.

IV. CONCLUSIONS

The problem of the small drain conductance in NW-TFETs is addressed with the help of an analytical model which makes it possible to gain a deeper insight into the device properties and to work out an optimal device design. The primary objective of this design is to maximize the drain conductance at low V_{DS} while, at the same time, preventing the ambipolar

effect, maximizing the on-state current and the steepness of the turn-on characteristics. In order to ensure a large tunneling probability we select InAs as the semiconductor material due to its small effective mass; next, we define the nanowire size in order to tune the bandgap so as to prevent the ambipolar effect with $V_{DD} = 0.4$ V; finally, we set the source and drain doping density such that $E_{VS} - E_{FS} = E_{FD} - E_{CS} \simeq 3k_B T$. With this choice, we achieve a drain conductance as large as 90% of its maximum theoretical value $(2q^2/h)\,T_{sc}$, a subthreshold slope SS $= 26$ mV/dec and an I_{ON}/I_{OFF} ratio much larger than 10^4, with $I_{ON} = 0.23$ mA/μm.

The present study shows that the performance penalty of TFETs is entirely due to the small tunneling probability, which turns out to be equal to 0.059 on average in our design. The small density of states is in fact compensated for by the high electron group velocity, leading to the pre-exponential factor $2qk_B T/h$ in the Landauer equation. An improvement in the tunneling probability can possibly be achieved by a suitably-designed heterostructure TFET architecture.

ACKNOWLEDGMENTS

This work has been partially supported by the EU Grant No. 257267 (STEEPER) via the IUNET Consortium, and by the Italian Ministry of University and Research (MIUR) via the FIRB-2010 Program "Futuro in Ricerca". Helpful discussions with Prof. Luca Selmi (Univ. of Udine) and Luca De Michielis (EPFL) are gratefully acknowledged.

REFERENCES

[1] W. Y. Choi, B.-G. Park, J. D. Lee, and T.-Y. K. Liu, "Tunneling Field Effect Transistors (TFETs) With Subthreshold Swing (SS) Less Than 60 mV/dec," *IEEE Electron Device Letters*, vol. 28, no. 8, pp. 743–745, August 2007.

[2] F. Mayer, C. Le Royer, J.-F. Damlencourt, K. Romanjek, F. Andrieu, C. Tabone, B. Previtali, and S. Deleonibus, "Impact of SOI and Si-GeOI and GeOI substrates on CMOS compatible Tunnel FET performance," in *International Electron Devices Meeting, 2008 – Technical Digest*. IEEE, 15-17 Dec. 2008, pp. 1–5.

[3] T. Krishnamohan, D. Kim, S. Raghunathan, and K. Saraswat, "Double-Gate Strained Heterostructure Tunneling FET (TFET) With Record High Drive Currents and less than 60 mV/dec Subthreshold Slope," in *International Electron Devices Meeting, 2008 – Technical Digest*. IEEE, 15-17 Dec. 2008, pp. 947–949.

[4] G. Dewey, B. Chu-Kung, J. Boardman, J. M. Fastenau, J. Kavalieros, R. Kotlyar, W. K. Liu, D. Lubyshev, M. Metz, N. Mukherjee, P. Oakey, R. Pillarisetty, M. Radosavljevic, H. W. Then, and R. Chau, "Fabrication, Characterization, and Physics of III-V Heterojunction Tunneling Field Effect Transistors (H-TFET) for Steep Sub-Threshold Swing," in *International Electron Devices Meeting, 2011 – Technical Digest*. IEEE, 15-17 Dec. 2011, pp. 785–788.

[5] Q. Zhang, T. Fang, H. Xing, A. Seabaugh, and D. Jena, "Graphene Nanoribbon Tunnel Transistors," *IEEE Electron Device Letters*, vol. 29, no. 12, pp. 1344–1346, December 2008.

[6] J. Knoch, S. Mantl, and J. Appenzeller, "Impact of the dimensionality on the performance of tunneling FETs: Bulk versus one-dimensional devices," *Solid-State Electronics*, vol. 51, no. 4, pp. 572–578, April 2007.

[7] M. Luisier and G. Klimeck, "Atomistic Full-Band Design Study of InAs Band-to-Band Tunneling Field-Effect Transistors," *IEEE Electron Device Letters*, vol. 31, no. 6, pp. 602–604, June 2009.

[8] A. Pal, A. B. Sachid, H. Gossner, and V. R. Rao, "Insights Into the Design and Optimization of Tunnel-FET Devices and Circuits," *IEEE Trans. Electron Devices*, vol. 58, no. 4, pp. 1045–1053, April 2011.

[9] M. Shin, "Full quantum simulation of hole transport and band-to-band tunneling in nanowires using the k·p method," *J. of Applied Physics*, vol. 106, p. 054505, 2009.

Comprehensive Statistical Comparison of RTN and BTI in Deeply Scaled MOSFETs by means of 3D 'Atomistic' Simulation

Salvatore M. Amoroso, Louis Gerrer,
Stanislav Markov, Fikru Adamu-Lema
Device Modeling Group
University of Glasgow
G12 8LT, Glasgow, UK

Asen Asenov
also with Gold Standard Simulations Ltd
G12 8QQ, Glasgow, UK

Abstract—We present a thorough statistical investigation of random telegraph noise (RTN) and bias temperature instabilities (BTI) in nanoscale MOSFETs. By means of 3D TCAD 'atomistic' simulations, we evaluate the statistical distribution in capture/emission time constants and in threshold voltage shift (ΔV_T) amplitudes due to single trapped charge, comparing its impact on RTN and BTI. Our analysis shows that, neglecting any impact of charge trapping on trans-characteristic degradation, the individual BTI ΔV_T steps are distributed identically as the RTN ΔV_T steps. However, the individual traps in a device cannot be considered as uncorrelated sources of noise because their mutual interaction is fundamental in determining the dispersion of capture/emission time constants in BTI simulation. These results are of utmost importance for profoundly understanding the differences and similarities in the statistical behavior of RTN and BTI phenomena.

I. INTRODUCTION

Due to the aggressive CMOS transistor scaling, RTN and BTI have emerged as tough challenges in the transistors design of contemporary and future CMOS technologies for both analog and digital applications [1,2]. The introduction of a Reliability-Aware design is today accepted as mandatory in order to maintain manufacturing yields [3]. An important paradigm shift [4] has recently identified the charge trapping in the oxide defects as uniquely responsible for both RTN and BTI instabilities, the latter being attributed in the past to reaction-diffusion phenomena [5]. Moreover, it has been experimentally confirmed that, similarly to ΔV_T amplitudes in RTN signal, the individual ΔV_T shifts observed in BTI measurements are exponentially distributed [6,7], corroborating the idea of a trapping phenomenon in presence of percolative source-to-drain conduction due to the discrete dopants in the channel. In this paper using 3D 'atomistic' simulations we shine light on the interplay between RTN and BTI degradation in contemporary and future CMOS transistors.

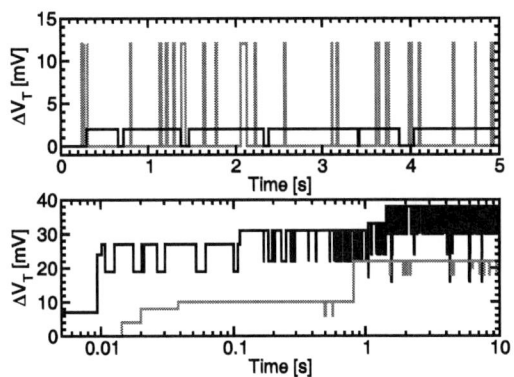

Fig. 1 RTN signal at $V_G = V_T$ for two different devices with one trap (up); BTI trace at $V_G = 1V$ for two different devices with different trap configurations.

II. SIMULATION METHODOLOGY

We have performed 3D simulations of a well-scaled 25 nm n-MOSFET device (featuring a 1.2 nm oxinitride and metal gate) using the GSS 'atomistic' simulator GARAND [8]. The oxide traps leading to RTN and BTI are modeled by means of three positional coordinates (x_T, y_T, z_T), one energy level (E_T) and a capture cross-section (σ). In the present study we considered x_T, y_T, z_T randomly distributed within the oxide boundaries, E_T randomly distributed within the range [$<E_F>$-0.15eV, $<E_F>$+0.15eV] (E_F being the quasi Fermi level at channel interface), and fixed cross-section $\sigma = 10^{-14}$ cm². The dynamic simulation of RTN and BTI signals is achieved within a Kinetic Monte Carlo (KMC) loop, as explained in [9]. After solving the 3D electrostatics and current continuity equations (at $V_G = 0.3V, V_D = 0.05V$) to evaluate the threshold voltage V_T, the average capture time $<\tau_c>$ for each single trap is computed integrating the tunneling gate current density that reaches the trap (WKB approximation) over an area equal to the trap cross-section σ [9]. The average emission time $<\tau_e>$ is calculated according to SRH statistics. Then the actual

This work has received funding from the *MORDRED* project (EU Project 261868)

978-1-4673-1707-8/12 $31.00 © 2012 IEEE

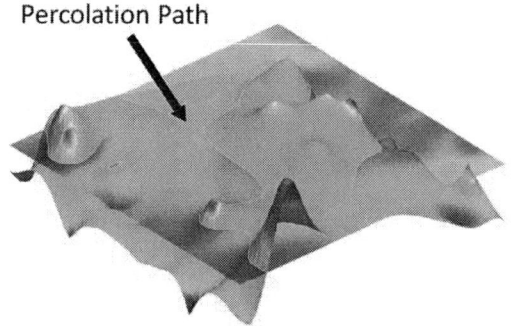

Fig. 2 3D distribution of the electrostatic potential at the channel interface crossed by a plan colored by current density illustrating the percolative conduction regime in nano FETs.

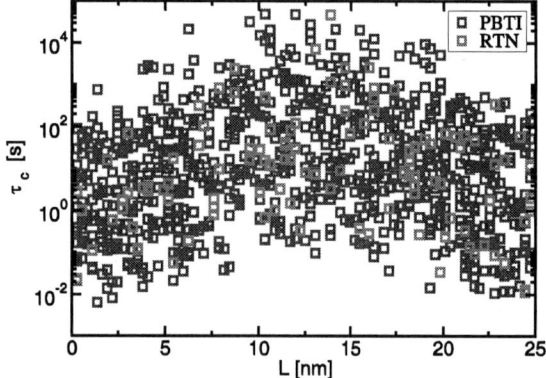

Fig. 3 Capture time distribution at $V_G=V_T$ for 1000 different devices featuring one single random trap.

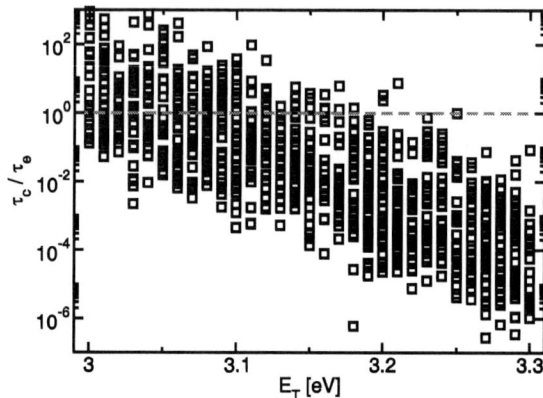

Fig. 4 Distribution of the ratio between capture and emission times versus trap energy level. A ratio of ~1 identifies the traps responsible for RTN.

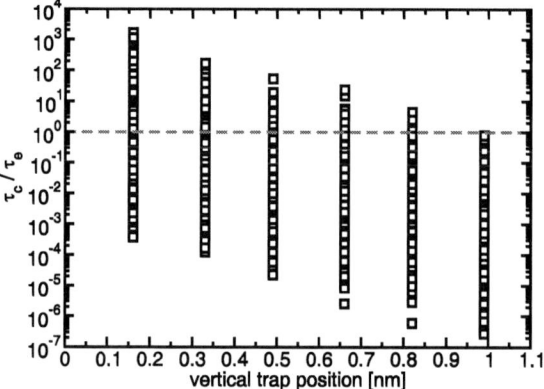

Fig. 5 Distribution of the ratio between capture and emission times versus trap distance from the Si/oxide interface. RTN traps are selected by ratio ~1.

capture and emission constants are randomly extracted from exponential distributions of average value $<\tau_c>$ and $<\tau_e>$.

Accordingly to these constants, a KMC-engine choose the next event (capture or emission) and the dynamic simulation time is increased by the extracted τ_c or τ_e. At this point the 3D electrostatics and current continuity equations are solved again to determine the threshold voltage shift (ΔV_T). The loop is repeated until the target time is reached.

It is important to mention that the device V_T is always read by means of a current criterion, considering the gate voltage allowing a drain current equal to 100nA. The impact of trapped charge on mobility or sub-threshold slope degradation is not taken into account in this work.

An example of simulated dynamic RTN and BTI is presented in Fig.1: both RTN and BTI traces show large device-to-device variability. This behavior in nanoscale MOSFETs is due to the percolative conduction regime in presence of random dopants (Fig.2): different devices have different dopants and traps configurations, so that both the capture/emission time constants and the ΔV_T amplitudes are strongly dispersed from device to device.

In our study we have considered a statistical ensemble of 1000 microscopically different devices with a maximum of 8 traps (corresponding to a density $\sim 1.2 \times 10^{12}$ cm^{-2}) for each device. In our simulation all the random traps can lead to BTI, while only the traps having $<\tau_c>/<\tau_e> \sim 1$ can lead to RTN.

III. RESULTS AND DISCUSSION

A. Capture/emission time constants variability

Fig. 3 shows the capture time constants along the channel length, obtained simulating 1000 different device **having one single trap** randomly positioned in the oxide. Owing to the discrete dopants and non-uniform electrostatics in the channel (Fig. 2), the capture time constants are strongly dispersed. BTI traps show larger dispersion compared to RTN traps. This is due to the broader underlying dispersion in energy level and vertical position of BTI traps. Indeed the RTN traps are characterized by a ratio $<\tau_c>/<\tau_e> \sim 1$, which selects traps close to the Fermi level (Fig. 4) and traps close to the channel interface (Fig. 5). However, BTI is a multi-traps phenomenon, whereas single-trap RTN is typically considered in the context of fresh devices. It is therefore mandatory to show how the capture time constants of a given trap are modified by the presence of **other filled traps in the device**. Fig. 6 shows the cumulative probability for each single trap, for one of the 1000 simulated devices, when we consider the presence of 8 traps in the oxide. It is clear that the mutual interaction of filled traps has a paramount impact on the time constant variability. The multiple charge-trapping shifts the time constants towards high values. The magnitude of the effect depends on the reciprocal position of the trap and can reach several decades (refer to Fig. 12).

978-1-4673-1707-8/12 $31.00 © 2012 IEEE

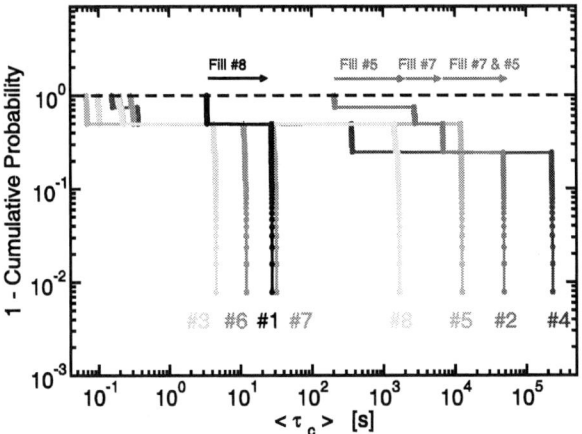

Fig. 6 Cumulative probability distribution of capture times for 1 device with 8 traps at $V_G=V_T$ illustrating their sensitivity on other traps occupancy status.

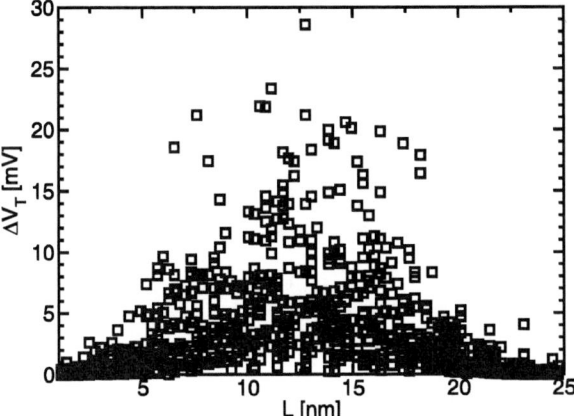

Fig. 7 Threshold voltage shift distribution along the channel length for 1000 devices with one single random trap.

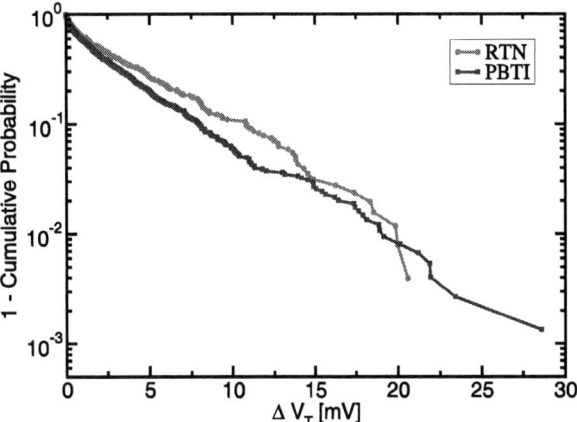

Fig. 8 Cumulative probability distribution of BTI/RTN ΔV_T for 1000 devices with 1single trap.

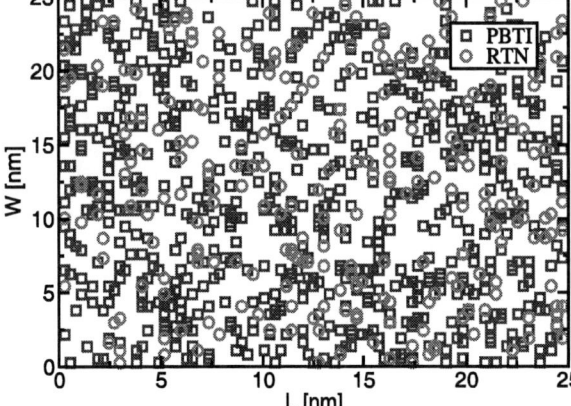

Fig. 9 Positions of simulated traps: both RTN and BTI traps are uniformly distributed over the active region.

Therefore, the broader dispersion in trap-characteristics increases the dispersion of the BTI time constants in the upper part of the scatter plot in Fig. 3. A very important conclusion is that individual traps cannot be modeled as uncorrelated sources of noise because their mutual interaction is fundamental in determining the dispersion of capture/emission time constants in BTI simulations.

B. ΔV_T amplitude variability

Fig. 7 shows the ΔV_T amplitudes along the channel length, obtained from 1000 different devices with **a** randomly positioned, **single trap** in the oxide. The atomistic dopants in the channel and the non-uniform electrostatics (Fig. 2) are the main contributor to the statistical dispersion in ΔV_T. Fig. 8 shows the cumulative distributions of ΔV_T for BTI and RTN. Both of them are characterized by an exponential behavior with a nearly identical slope (9.4 mV/dec for BTI and 10.5 mV/dec for RTN). The slightly higher slope for RTN is due to the fact that the criterion $<\tau_c>/<\tau_e>\sim1$ selects traps close to the channel interface (Fig. 5) that have a larger impact on ΔV_T. Fig. 9 shows that the selection criterion for RTN traps does not entail lateral localization, hence both RTN and BTI traps are uniformly distributed over the channel.

This explains the similarity in the cumulative distributions in Fig. 8. Fig. 10 shows the cumulative distributions of the *individual* (single charge-induced) ΔV_T for BTI and RTN when **several traps** (eight) are present in each device. The distributions of single charge-induced ΔV_T are barely affected by the presence of other filled traps (slopes 9.7 mV/dec for BTI and 10.5 mV/dec for RTN). In order to understand this, we report in Fig. 11 the cumulative probability of ΔV_T for each single trap, for one of the 1000 simulated devices, when we consider the presence of other traps in the oxide. Unlike the distributions of $<\tau_c>$ reported in Fig. 6, the distributions of ΔV_T are smooth and narrow, indicating a low mutual interactions between traps. This is because $<\tau_c>$ is calculated during the stress operation (that is at fixed gate bias, so that a filled trap has an effective impact on channel electrostatics), while ΔV_T is calculated during the reading operation (that is at fixed drain current, so that the conduction profile is re-organized at each trapping event in order to sustain the reading current). Such re-organization of the percolative current density profile after each trapping event is very clear in Fig. 12. It is also the reason why the maximum impact of each trap on ΔV_T depends on the overall configuration of filled and empty traps, as reported in Tab. 1.

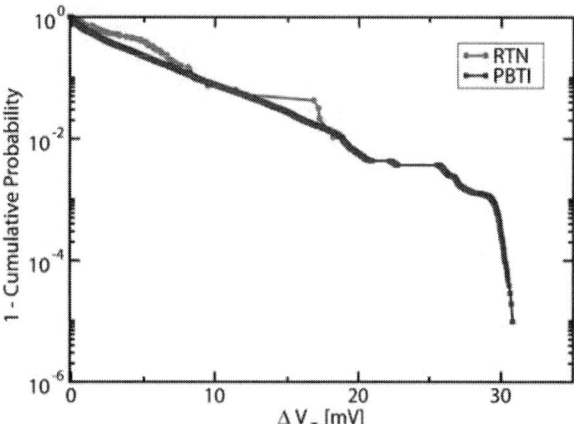

Fig. 10 Cumulative probability distribution of BTI/RTN single trapped charge-induced ΔV_T for 100 devices with 8 traps.

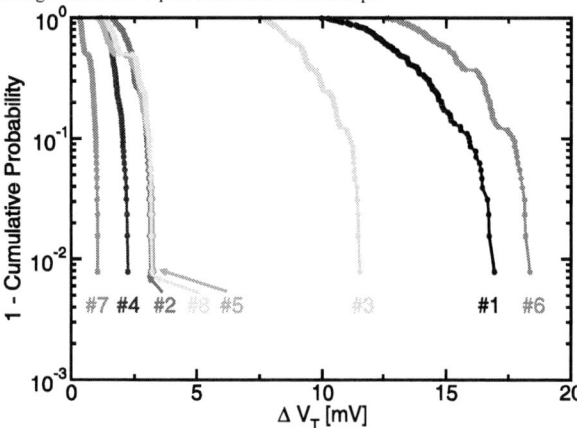

Fig. 11 Cumulative probability distribution of single charge trapped-induced ΔV_T for the same device in Fig.6 illustrating the sensitivity of each single trap on other traps occupancy status.

It is worth mentioning that experimental results in [7] show a BTI single charge-induced ΔV_T distribution larger than that of RTN. A model based on reaction-diffusion trap formation has been proposed in [7] to interpret the data. However we would like to highlight that this model forces a direct relation between ΔV_T and the channel carrier density and, in turn, with $<\tau_c>$. These relations have been recently demonstrated, both experimentally [6] and theoretically [10], to be erroneous. We suggest that a possible explanation of the larger BTI single charge-induced ΔV_T distribution may reside on the sub-threshold slope degradation following each trapping event in BTI experiments [11]. This effect is not captured in our simulation and will be object of further investigation.

IV. CONCLUSIONS

We have presented a comprehensive statistical comparison of RTN and BTI behavior in nanoscale MOSFETs, showing the importance of 3D physics-based simulations for an in-depth analysis of reliability in presence of variability effects.

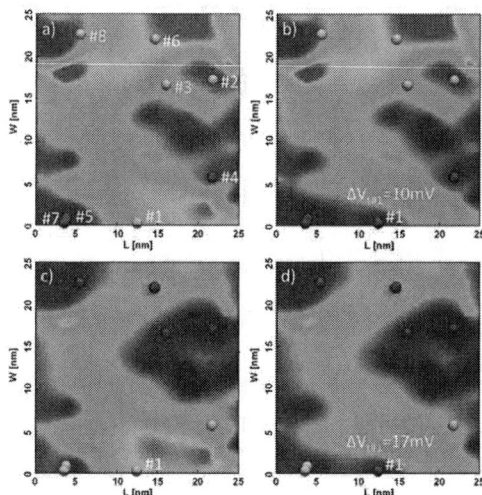

Fig. 12 Current density maps over the active region at $V_G=V_T$, showing the minimum (a-b) and the maximum (c-d) impact of trap #1 on ΔV_T. Filled traps are indicated in red, empty traps in yellow.

TRAP	Configuration MAX ΔV_T	MAX ΔV_T [mV]
#1	10100111	17.22
#2	00011011	3.14
#3	01111101	11.59
#4	10101110	2.27
#5	10110110	3.29
#6	01111101	18.46
#7	11100110	1.06
#8	11011001	3.20

Tab. 1 Overall trap configuration giving the maximum shift for each considered trap (same device of Fig.11). '1' indicate filled trap, '0' empty trap.

REFERENCES

[1] T. Nagumo et al, "Statistical characterization of trap position, energy, amplitude and time constants by RTN measurement of multiple individual traps", IEDM, pp. 628-631, 2010.

[2] S. V. Kumar et al. "Impact of NBTI on SRAM read stability and design for reliability", Int. Symp. on Quality Elec. Design, pp. 27-29, 2006.

[3] K. Takeuchi et al., "Comprehensive SRAM design methodology for RTN reliability", VLSI, pp. 130-131, 2011.

[4] T. Grasser et al, "The paradigm shift in understanding the bias temperature instability: from reaction–diffusion to switching oxide traps", IEEE TED, pp. 3652-3666, 2011.

[5] M. A. Alam et al., "A comprehensive model of PMOS NBTI degradation", Micr. Elec. Reliab., pp. 71-81, 2005.

[6] B. Kaczer et al., "Origin of NBTI variability in deeply scaled pFETs", IEEE IRPS, pp.26-32, 2010.

[7] J. P. Chiu et al., "A comparative study of NBTI and RTN amplitude distributions in high-κ gate dielectric pMOSFETs", IEEE EDL, pp.176-178, 2012.

[8] http://www.goldstandardsimulations.com

[9] S. M. Amoroso et al., "Three-dimensional simulation of charge-trap memory programming—part I: average behavior", IEEE TED, pp. 1864-1871, 2011.

[10] A. Mauri et al., "Impact of atomistic doping and 3D electrostatics on the variability of RTN time constants in Flash memories", IEDM, pp. 405-408.

[11] A. Shappir et al., "Subthreshold slope degradation model for localized-charge-trapping based non-volatile memory devices", Solid State Elect., pp. 937-941, 2003.

978-1-4673-1707-8/12 $31.00 © 2012 IEEE

Statistical Variability in 14-nm node SOI FinFETs and its Impact on Corresponding 6T-SRAM Cell Design

Xingsheng Wang[1]*, Binjie Cheng[1], Andrew R. Brown[2], Campbell Millar[1,2], Asen Asenov[1,2]

[1] Device Modelling Group
School of Engineering, University of Glasgow
Glasgow, Scotland U.K.
* Xingsheng.Wang@glasgow.ac.uk

[2] Gold Standard Simulations Ltd.
Rankine Building, Oakfield Avenue
Glasgow, Scotland U.K.

Abstract—**This paper presents a comprehensive statistical variability study of 14-nm technology node SOI FinFET which is optimized based on extensive exploration of TCAD design space. The variability sources, including random discrete dopants, gate and fin edge roughness, and possible metal gate granularity, are simulated and examined in term of their impacts on device parameters. The impact of intrinsic parameter fluctuations on a high density SOI FinFET 6T-SRAM cell is also investigated.**

I. INTRODUCTION

The increasing statistical MOSFET variability, dominated by random discrete dopants, critically affects the SRAM yield at the 32/28nm technology generation [1] and has made questionable the future of the conventional bulk CMOS scaling. FinFETs that tolerate low channel doping have been introduced at the 22nm node [2] in part to reduce the statistical variability and to enable the continuation of the CMOS scaling. However there are new important sources of statistical variability associated with FinFET fabrication among which the fin edge roughness formed by fin-body patterning becomes very important [3]. The high-k/metal gate is another important potential source of FinFET variability. The metal polycrystalline grains particularly formed during the high-temperature source/drain activation in gate-first technology leads to gate work-function variation depending on metal grain orientations, introducing metal gate granularity (MGG) as an important variability source [4]. The FinFET current is discretized by individual fins, which is different to planar transistors and affects the FinFET based SRAM and other circuits design. This paper presents a comprehensive statistical variability study of a FinFET on SOI substrate designed for the 14nm node low-power operation. The study includes all relevant variability sources. The impact of the statistical variability on SOI FinFET 6T-SRAM cells is also studied in terms of static noise margin variation.

II. DEVICE AND SIMULATION METHOD

The double-gate SOI FinFETs subject to this study are designed using the GSS 'atomistic' TCAD simulator GARAND [5]. The equivalent oxide thickness of the hafnium-based high-k insulator is 0.8nm, and mid-gap workfunction TiN is used for the metal gate. The fin height/width is 25nm/10nm and gate-length is 20nm. The low channel doping is $1\times10^{15}cm^{-3}$ and the maximum of the conformal source/drain doping is assumed to be $3\times10^{20}cm^{-3}$. It is reported that (110) sidewall orientation is not bad for electron mobility [6]. Stress engineering is also assumed. The drift diffusion device simulations are calibrated in respect of comprehensive 3D ensemble Monte-Carlo simulations performance study. Density-gradient quantum corrections are included and are mandatory for such small-scale thin-body transistors. The nominal n/p-channel FinFETs operating at 85°C achieve drive currents of ~0.9/0.8 mA/μm with off-leakage 10nA/μm at supply voltage 0.9V. The drain-induced barrier lowering (DIBL) is maintained at 56mV/V and 65mV/V respectively for n-channel and p-channel FinFET, indicating extremely good short-channel effects control due to double-gate action. The systematic changes of figures of merits against device geometry parameters are taken into account in order to explore design space, which facilitates the global compact model extraction. It is treated in the device and circuit co-design methodology.

Figure 1. The 3D view of SOI n-channel FinFET showing electron density inside the fin and potential on gate-oxide subject to random dopants, gate edge roughness, fin edge roughness, and metal gate granularity.

This work is supported in part by EU ENIAC joint undertaking project MODERN, and the Scottish Funding Council through the StatDes project.

978-1-4673-1707-8/12 $31.00 © 2012 IEEE

The 'atomistic' simulations of the template device are carried out with the GSS GARAND TCAD simulator. Random discrete dopants (RDD), gate- and fin- line edge roughness (GER and FER) with 30-nm correlation length, and metal gate granularity (MGG) are simulated individually and in combination with varying parameters. TiN metal grains features two work-functions with 0.2V difference and 40% and 60% occurrence probability [7]. Each individual-source ensemble simulates 1000 microscopically different samples, and combined-source ensembles simulate 200 samples.

III. SOI FinFET Statistical Variability

A. Threshold-voltage fluctuation

The threshold-voltage (V_T) is one of the most important device parameters with regard to the transistor performance. Threshold-voltage fluctuations have been extensively studied in the past, and in bulk planar MOSFETs are mainly due to random dopant fluctuations [4][8]. Figure 2 shows the threshold-voltage distributions due to distinct variability sources. In the undoped channel FinFETs, the random-dopant-induced variability is significantly reduced. Compared to planar devices, the width of the patterned fin-body may fluctuate, which introduces a new variability source of fin edge roughness. From the comparison, GER, FER and MGG contribute considerable variability to threshold voltage.

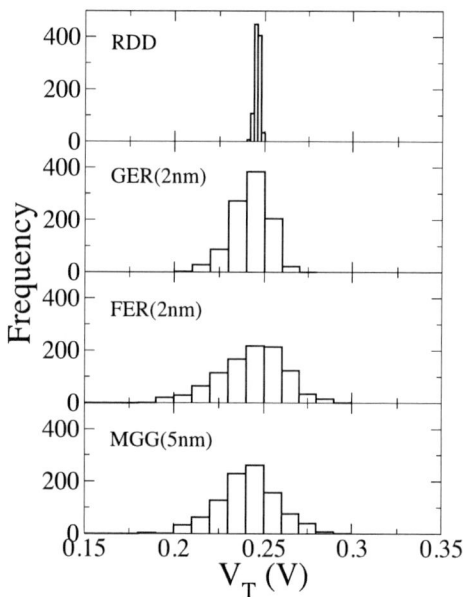

Figure 2. The distributions of threshold-voltages due to different variability sources. It is seen FER and MGG induce major variability. N-channel FinFET at V_D=0.9V.

In Figure 3 the normal Q-Q plots examine the distribution shape. RDD-induced V_T variability is close to a Gaussian distribution, but an accidental single dopant inside the channel can twist the distribution tail [3]. The V_T distribution due to GER with 3 times root mean square (3Δ) of 2nm is skewed with the prolonged tail towards smaller values. MGG with an

average metal-grain diameter of 5nm produces a V_T distribution that is close to Gaussian. V_T distribution due to FER with 3Δ of 2nm is also almost Gaussian. For this device structure, with the same LER magnitude, FER induced more variation than GER.

The threshold-voltage variability depends on the parameters of the variability sources. For example, with increasing LER magnitude, the standard deviation of threshold-voltage (σV_T) proportionally increases as shown in Figure 4(a), while large metal grain size also increases the variability demonstrated in Figure 4(b).

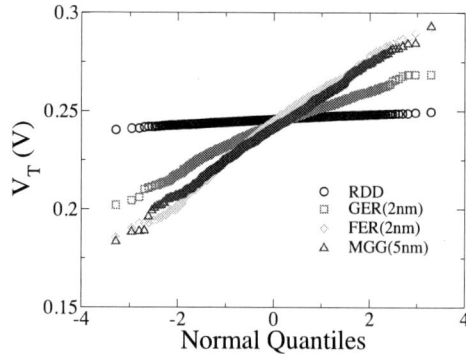

Figure 3. The normal Q-Q plot of threshold-voltage distributions due to different variability sources. V_T distribution due to GER is skewed with a prolonged tail towards to small values. All other distributions are close to Gaussian distributions. N-channel FinFET at V_D=0.9V.

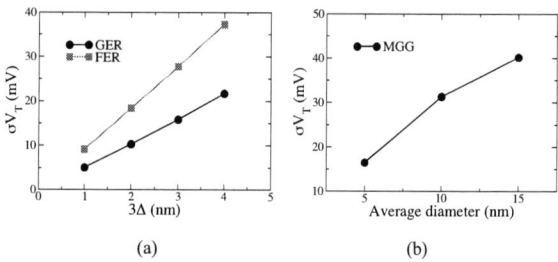

Figure 4. σV_T dependence on (a) LER magnitude and (b) metal-grain size. N-channel FinFET at V_D=0.9V.

The threshold-voltage variability due to particular combination of variability sources is examined in Figure 5 for gate-last and gate-first processes respectively. The standard deviation of the threshold-voltage of combined variability sources is 19mV and 26mV respectively for without MGG and with MGG included for n-channel transistors at V_D=0.9V. Due to additional MGG in gate-first process the distributions are more flat than those without MGG. With the increased drain-bias the distributions spread more, indicating that DIBL worsens the statistical variability. This is mainly due to enhanced LER impact at high drain-bias [1]. From the point of view of reduced variability, better short-channel effect in SOI FinFETs than bulk planar transistors can reduce statistical variability. These distributions are examined in Figure 6 showing close to Gaussian distribution.

978-1-4673-1707-8/12 $31.00 © 2012 IEEE

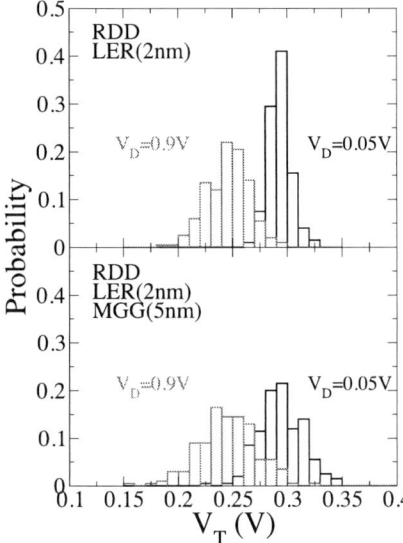

Figure 5. The distribution of threshold-voltages subject to combined major variability sources, at different drain bias. With MGG in gate-first process the variability increases, and with large drain-bias the variability increases. N-channel FinFET.

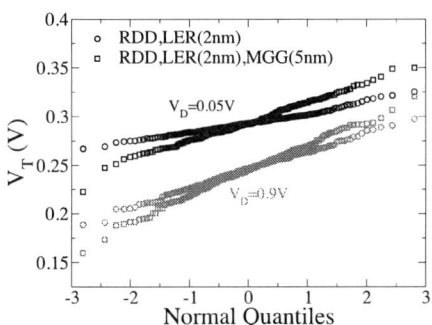

Figure 6. The normal Q-Q test on V_T distributions. All distributions are close to normal distribution. N-channel FinFET.

B. On-current and DIBL

The correlation between on-current (I_{ON}) and V_T is highly consistent when varying variability parameter values for a particular variability source. From Figure 7 the devices with RDD have the lowest correlation between V_T and I_{ON}, while devices with MGG have the strongest correlation. Simply modelling the threshold-voltage fluctuation is inadequate for completely capturing the statistical variability effect on transfer characteristics. In addition, the I_{ON} dependence on threshold-voltage is interestingly different for different variability sources, which is manifested by the different fitting line slopes. RDD introduces the largest dependence, and GER and FER have the least dependence. This explains why I_{ON} in the case of RDD has more variation than the proportion of V_T variation. This dependence also bridges I_{ON} variation and V_T variation given that I_{ON} is highly correlated with V_T. Depending on the variability sources, I_{ON} variation magnitude

can be different even if the V_T variation magnitude is the same. Therefore, except V_T variation, the on-current variation magnitude is also dependent on the variability source combination.

Figure 8 shows the DIBL variability. With MGG, DIBL shows more variation, and it also shows much less correlation with threshold voltage. Drain-bias affects the position of the potential barrier, which is subject to MGG-induced surface potential variation; therefore, DIBL shows more variability with MGG.

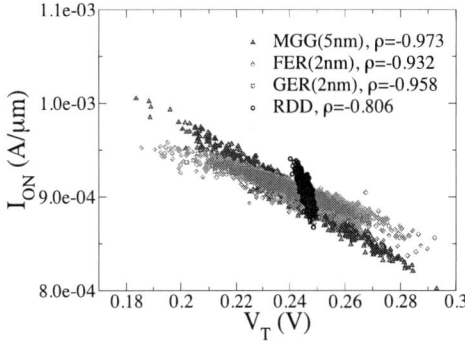

Figure 7. On-current variation and its correlation with threshold-voltage, and the dependence on variability sources. N-channel FinFET at $V_D=0.9V$.

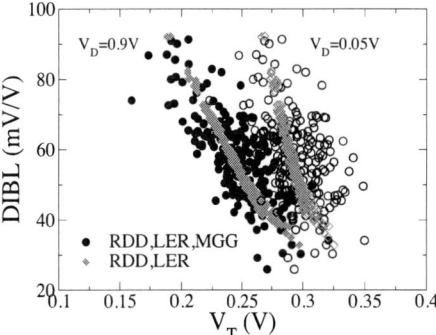

Figure 8. DIBL variation and its correlation with threshold-voltage for n-channel FinFET.

IV. IMPACT ON SOI FinFET SRAM

The GSS compact model extractor MYSTIC [5] is used to extract a statistical model library. A two stage statistical compact modelling approach is employed here [9]. At the first stage, the global exaction of BSIM-CMG model [10] based on nominal simulations is carefully performed. Secondly, the statistical extraction for ensemble 'atomistic' devices is carried out using a few well-identified model parameters, which closely represent statistical-variability effects. A high density 6T-SRAM cell with single-fin transistors is assumed, and the random cells are built using the pool of statistical model cards. Figure 9 demonstrates the simulation configuration, of which M1 to M6 are microscopically different single-fin SOI FinFETs.

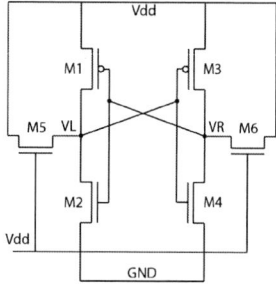

Figure 9.   The schematic SOI FinFET SRAM and its connection conditions.

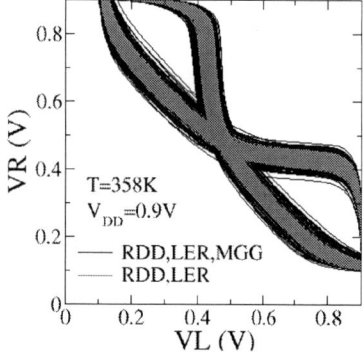

Figure 10. The SRAM butterfly characteristics subject to different combined variability sources.

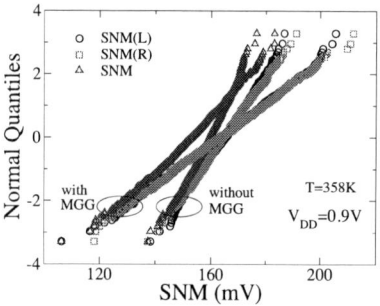

Figure 11. The Q-Q plots of SRAM static noise margin due to combined statistical variability sources of RDD and LER (2nm) with/without MGG (with average grain diameter of 5nm). It is seen that SNM shrinks with increased statistical variability.

The SRAM butterfly characteristics are plotted in Figure 10. It is seen that the butterfly curves fluctuate, subject to statistical variability sources. The static noise margin (SNM) is squeezed by statistical variability. With additional MGG variability in gate-first process, the curves show more variation with reduced SNM value. The SNM for each SRAM cell is extracted based on butterfly characteristics. SNM(L) and SNM(R) are the left- and right-margins in Figure 10. SNM is the minimum of these two. In Figure 11, SNM(L) and SNM(R) are close to Gaussian distribution, but SNM is skewed with the prolonged tail towards to small values. This is consistent with experimental measurements [11]. For yield

issues, the left side of the SNM distribution is more meaningful. From the figure, with increased variability from MGG, the SNM window significantly shrinks.

V. CONCLUSIONS

This paper presents a comprehensive simulation study of the statistical variability in generic 14nm technology node SOI FinFETs. The statistical variability is dominated by GER, FER and possible MGG, since RDD effects are significantly reduced by low-channel doping. The threshold-voltage fluctuation and on-current variability depart from the strong correlation due to RDD, while they have different average behaviour depending on the statistical variability sources. The DIBL is also affected by statistical variability. The SOI FinFET 6T-SRAM is simulated subject to statistical variability. The static noise margin is reduced with increased statistical variability.

ACKNOWLEDGMENT

The authors thank J. B. Kuang and Sani Nassif from IBM Research Division, Austin Research Laboratory, for helpful discussions.

REFERENCES

[1]   X. Wang, G. Roy, O. Saxod, A. Bajolet, A. Juge, and A. Asenov, "Simulation study of dominant statistical variability sources in 32-nm high-k/metal gate CMOS," *IEEE Electron Device Letters*, vol.33 no.5, pp.643-645, May 2012.

[2]   C. Auth, C. Allen, A. Blattner, *et al.*, "A 22nm High Performance and Low-Power CMOS Technology Featuring Fully-Depleted Tri-Gate Transistors, Self- Aligned Contacts and High Density MIM Capacitors" in *VLSI Tech. Sym.*, 2012.

[3]   X. Wang, A. R. Brown, B. Cheng, and A. Asenov, "Statistical variability and reliability in nanoscale FinFETs," in *IEDM Tech. Dig.*, pp.103-106, 2011.

[4]   X. Wang, A.R. Brown, N. Idris, S. Markov, G. Roy and A. Asenov, "Statistical Threshold-Voltage Variability in Scaled Decananometer Bulk HKMG MOSFETs: A Full-Scale 3-D Simulation Scaling Study," *IEEE Transactions on Electron Devices*, Vol. 58, No. 8, pp. 2293–2301, Aug. 2011.

[5]   GARAND, MYSTIC, http://www.goldstandardsimulations.com/

[6]   C.D. Young, M.O. Baykan, A. Agrawal, et al., "Critical discussion on (100) and (110) orientation dependent transport: nMOS planar and FinFET," in *VLSI Tech. Sym.*, pp.18-19, 2011.

[7]   A. R. Brown, N. M. Idris, J. R. Watling, and A. Asenov, "Impact of metal gate granularity on threshold voltage variability: a full-scale three-dimensional statistical simulation study," *IEEE Electron Device Letters*, vol. 31, no. 11, pp. 1199–1201, Nov. 2010.

[8]   A. Cathignol, B. Cheng, D. Chanemougame, A. R. Brown, K. Rochereau, G. Ghibaudo, and A. Asenov, "Quantitative evaluation of statistical variability sources in a 45-nm technological node LP N-MOSFET," *IEEE Electron Device Letters*, vol. 29, no. 6, pp. 609–611, June 2008.

[9]   B. Cheng, D. Dideban, N. Moezi, et al., "Statistical-Variability Compact-Modeling Strategies for BSIM4 and PSP," *IEEE Design & Test of Computers*, vol.27 no.2, pp.26-35, March/April 2010.

[10] BSIM-CMG, http://www-device.eecs.berkeley.edu/bsim/

[11] T. Hiramoto, M. Suzuki, X. Song, *et al.*, "Direct measurement of correlation between SRAM noise margin and individual cell transistor variability by using device matrix array," *IEEE Transactions on Electron Devices*, Vol. 58, No. 8, pp. 2249–2256, Aug. 2011.

Sensitivity-based Investigation of Threshold Voltage Variability in 32-nm Flash Memory Cells

Valentina Bonfiglio[1], Giuseppe Iannaccone[1,2]

[1]Dipartimento di Ingegneria dell'Informazione and [2]SEED Center, PUSL, Università di Pisa.
Email: {valentina.bonfiglio, g.iannaccone}@iet.unipi.it

Abstract— **We investigate variability of a 32 nm flash memory cell with a methodology based on sensitivity analysis performed with a limited number of TCAD simulations. We show that - as far as the standard deviation of the threshold voltage is concerned - our method provides results in very good agreement with those from three-dimensional atomistic statistical simulations, with a computational burden that is orders of magnitude smaller. We show that the proposed approach is a powerful tool to understand the role of the main variability sources and to explore the device design parameter space.**

Introduction

Non-volatile memory fabrication processes undergo even more aggressive scaling than CMOS technology for logic applications, as a means to increase bit density in response to the evolving demands of multimedia applications and mass storage. This exacerbates the device variability issue, which is especially acute in the case of multi-bit cells, where only few tens of electrons in the floating gate can separate two different logic levels [1].

The problem is particularly severe because floating gate cells must be designed and characterized for more than eight standard deviations, and therefore the second order moment of the probability distribution is hardly sufficient. [2]

In this paper we show that a recently proposed TCAD-based sensitivity analysis [3], can provide very interesting results at a small computational cost, at least for the calculation of the standard deviation of the threshold voltage. In the framework of the ENIAC Joint Undertaking MODERN project [4], we have considered a template device structure for a 32 nm CMOS flash memory cell, for which variability assessments based on three-dimensional atomistic statistical simulations and the impedance field method have been published [5]. We analyze the impact of variability sources such as random dopant distribution (RDD) [6], line-edge roughness (LER), line-width roughness (LWR), [7-8] interface trapped charge (ITC) [9], oxide thickness fluctuations (OTF) [10].

The template device structure is illustrated in Figure 1. It is a simplified polisilicon floating gate device with dimensions typical of a 32 nm technology (indicated in the table in Figure 1), generated at the crossing point of two orthogonal lines of width 32 nm. Control gate and floating gate consist of polysilicon and are separated by an ONO (oxide-nitride-onide) layer of 4-3-5 nm. The tunnel oxide thickness is 8 nm. Substrate is boron doped (2×10^{18} cm^{-3}), and arsenic doping of source and drain is symmetric with a maximum of 10^{20} cm^{-3}, Gaussian shape, and junction depth of 25 nm. Additional details are available in Ref. [5].

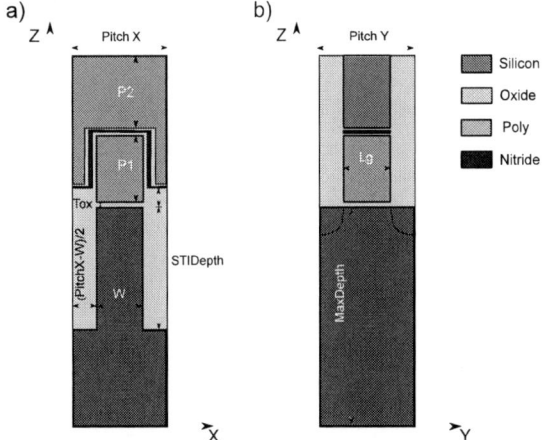

Geometrical parameters	Symbols and dimensions of layers	
	Symbol	Value
Cell X dimension	Pitch X	64 nm
Cell Y dimension	Pitch Y	64 nm
Active Area Width	W	32 nm
Gate Length	Lg	32 nm
Silicon substrate thickness	MaxDepth	0.5 um
Isoltation Depth	STIDepth	0.2 um
Tunnel oxide thickness	Tox	8 nm
	StepH	20 nm
Poly1 thickness	P1	70 nm
Poly2 thickness	P2	100 nm
ONO bottom oxide	ONO_bot	4 nm
ONO nitride	ONO_nit	3 nm
ONO top oxide	ONO_top	5 nm
Junction depth	xj	25 nm

Figure 1: Device structure and geometrical parameters of the template 32 nm flash memory under investigation.

METHODOLOGY

The approach proposed is described in detail in [3]. First, all process and geometry variability causes are expressed in terms of a set of synthetic independent variability sources. Then, TCAD-based sensitivity analysis is used to evaluate the contribution to the dispersion of electrical parameters (e.g. the threshold voltage V_{th}) of each independent source. This step is based on the assumption that the effect of each source is sufficiently small that first-order linearization is applicable. Also in the case of the 32 nm Flash memory [5], the variance of the threshold voltage due to combined effect computed with 3D atomistic statistical is shown to be very close to the sum of the variances due to individual effects, giving us confidence in the linear approximation.

As an example, let us consider the case of LER, considering the illustration in Fig. 2, where the 32 nm device is shown with the y axis running along the channel length direction, the x axis perpendicular to the device plane and the z axis running along the channel width.

We can translate line edge roughness in terms of the dispersion of the average position of both gate edges along the y axis (y_1 and y_2, where $\langle y_1 \rangle = 0$ and $\langle y_2 \rangle = L$). This in turn translates into gate length dispersion. We assume that parameters y_1, y_2 are only affected by LER and are physically independent. The average edge position is a random function $g(z)$ with zero mean value and Gaussian autocorrelation $r(d) \equiv \langle g(z)g(z+d) \rangle$ characterized by correlation length Λ_L and mean square amplitude Δ_L, i.e.:

$$r(d) = \Delta_L^2 e^{\frac{d^2}{2\Lambda_L^2}} \qquad (1)$$

from which we can write the variance of g as

$$\sigma_g^2 \equiv \langle g^2 \rangle = \frac{1}{W^2}\left\langle \int_0^W g(z_1)dz_1 \cdot \int_0^W g(z_2)dz_2 \right\rangle. \qquad (2)$$

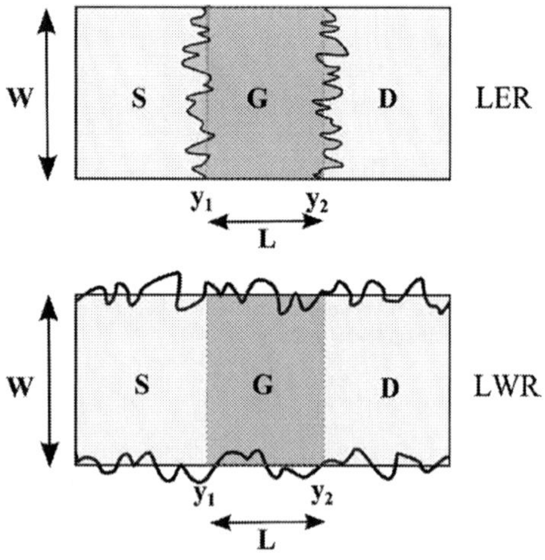

Figure 2: Illustration of the approach to the evaluation of line edge roughness (above) and line-width roughness (below).

If we compute (2) considering (1) we find:

$$\sigma_{LER}^2 = \frac{2\Delta_L^2 \Lambda_L}{W^2}\left[\Lambda_L\left(e^{-\frac{W^2}{2\Lambda_L^2}}-1\right)+\sqrt{\frac{\pi}{2}}W\mathrm{erf}\left(\frac{W}{\sqrt{2}\Lambda_L}\right)\right] \qquad (3)$$

The variance of V_{th} due to line edge roughness is:

$$\sigma_{V_{th}LER}^2 = \left(\frac{\partial V_{th}}{\partial y_1}\right)^2\sigma_{y_1}^2 + \left(\frac{\partial V_{th}}{\partial y_2}\right)^2\sigma_{y_2}^2 = 2\left(\frac{\partial V_{th}}{\partial L}\right)^2\sigma_{LER}^2, \quad (4)$$

where y_1, y_2 in (4) are the average gate edges indicated in Fig. 2. All required derivatives can be computed with TCAD sensitivity analysis as illustrated in Fig. 3 (left). The very same approach can be used for LWR.

In the case of OTF we must consider surface roughness with a two dimensional Gaussian autocorrelation

$$r(x_a,y_a,x_b,y_b) = \Delta_S^2 \exp\left(-\frac{(x_b-x_a)^2+(y_b-y_a)^2}{2\Lambda_S^2}\right), \quad (5)$$

characterized by correlation length Λ_S and mean square amplitude Δ_S, which corresponds to a variance of the average position of the interface:

$$\sigma_{SR}^2 = \frac{2\pi\Lambda_S^2\Delta_S^2}{L^2W^2}\left[L\cdot\mathrm{erf}\left(\frac{L}{\sqrt{2}\Lambda_S}\right)+\sqrt{\frac{2}{\pi}}\Lambda_S\left(e^{-\frac{L^2}{2\Lambda_S^2}}-1\right)\right]$$
$$\times\left[W\,\mathrm{erf}\left(\frac{W}{\sqrt{2}\Lambda_S}\right)+\sqrt{\frac{2}{\pi}}\Lambda_S\left(e^{-\frac{W^2}{2\Lambda_S^2}}-1\right)\right], \qquad (6)$$

The variance of the threshold voltage due to OTF is therefore

$$\sigma_{V_{th}SR}^2 = \sum_m \left(\frac{\partial V_{th}}{\partial s_m}\right)^2\sigma_{OTF}^2, \qquad (7)$$

where s_m are all positions of the interfaces between dielectric layers and between dielectric and conducting or semiconducting layers. Also in this case, all derivatives can be computed with TCAD simulations following the example of Figure 3 (right).

For LER and LWR, we consider a Gaussian autocorrelation with mean square amplitude $\Delta_L = 1.5$ nm and correlation length $\Lambda_L = 20$ nm. For OTF, we consider a Gaussian autocorrelation with with mean square amplitude $\Delta_S = 0.2$ nm and correlation length $\Lambda_S = 18$ nm.

Results are compared in Table I with those obtained from 3D atomistic simulations on 1000 samples performed with GARAND [5], in which the same statistical properties have been considered for LER, LWR, and OTF. The obtained standard deviation are practically identical. As in [5], the threshold voltage is defined with a current criterion of 100 nA for a drain-to-source voltage of 100 mV.

978-1-4673-1707-8/12 $31.00 © 2012 IEEE

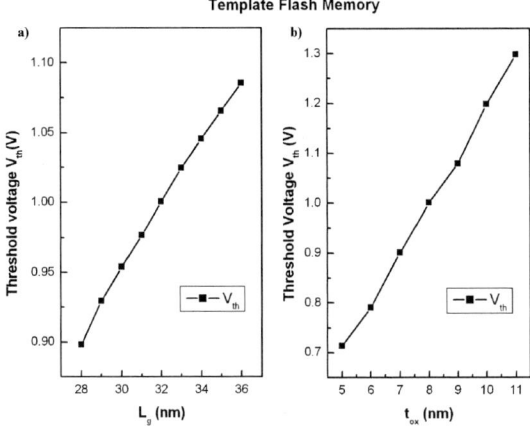

Figure 3 a) Threshold voltage as a function of gatelength Lg and b) threshold voltage as a function of tunnel oxide thickness t_{ox} for the template Flash Memory as computed from TCAD simulations.

For random discrete dopants (RDD) [6] and interface trapped charge (ITC) [10], we adopt an approach based on a propagator with a very coarse granularity, which is in principle very close to the concept of impedance field method [11]. As a difference with respect to the situation already described in [4], we here have to perform 3D simulations, since the Flash memory cells cannot be reduced to 2D structures.

For a given variation of doping concentration $\Delta N_A(x,y,z)$ with respect to the nominal value we can write the following expression for the variation of V_{th}:

$$\Delta V_{th} = \int K(x,y,z)\Delta N_A(x,y,z)dxdydz \qquad (8)$$

where $K(x,y,z)$ has the role of a propagator. The expression requires the linearity assumption to hold.

To conveniently compute the propagator K, we can assume that K is a smooth function of x, y, and z, and move from the continuum to a discrete space, partitioning the active area in small boxes. Now we can write:

$$\Delta V_{th} = \sum_i \Delta V_{th_i} = \sum_i K_i \Delta N_i \qquad (9)$$

The sum runs over all boxes, ΔN_i is the variation of the number of dopants in box i, and ΔV_{th_i} is the threshold voltage variation if only dopants in box i are varied.

In practice, we multiply doping in box i by a factor $(1+\alpha)$ and compute ΔV_{thi} with TCAD simulations. Therefore we have

$$\begin{aligned} \Delta N_i &= \alpha N_i \\ \Delta V_{th_i} &= \alpha K_i N_i \end{aligned} \qquad (10)$$

so that (9) becomes,

$$\Delta V_{th} = \sum_i \left(\frac{\Delta V_{th_i}}{\alpha}\right)\alpha = \sum_i \left(\frac{\Delta V_{th_i}}{\alpha}\right)\frac{\Delta N_i}{N_i} \qquad (11)$$

If we finally assume that doping variations in different boxes are independent Poisson processes, we can write

$$\sigma_{V_{thRDD}}^2 = \sum_i \left(\frac{\Delta V_{th_i}}{\alpha}\right)^2 \frac{1}{N_i} = \sum_i \sigma_{V_{thRDD}}^{2\ [i]}, \qquad (12)$$

The threshold voltage dispersion due to RDD only requires a single TCAD simulation for each box, and an integral of the doping profile in each box. To evaluate the most convenient level of granularity in device partitioning, we have made tests with different box sizes, as reported in the table in Figure 4.

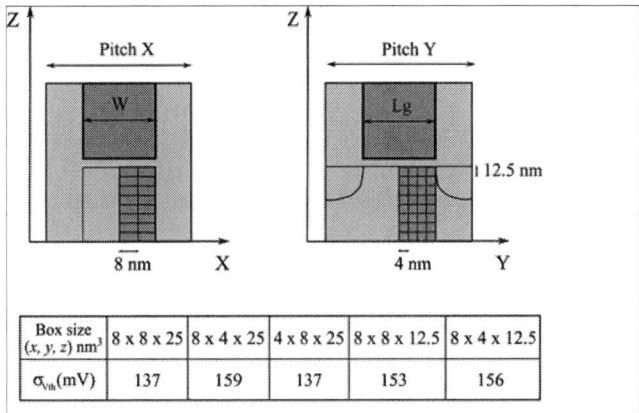

Box size (x, y, z) nm^3	8 x 8 x 25	8 x 4 x 25	4 x 8 x 25	8 x 8 x 12.5	8 x 4 x 12.5
$\sigma_{V_{th}}$(mV)	137	159	137	153	156

Figure 4 above: transversal (left) and longitudinal (right) device cross sections for the assessment of the proper box partitioning. Below: computed standard deviation of the threshold voltage as a function of the box size for different choices of the partition.

We have evaluated that a partition of the three dimensional silicon body in 64 boxes of size $8 \times 4 \times 12.5$ nm^3 represents a good trade-off between computing time and accuracy. Considering that we can exploit the symmetry of the structure also along the transport direction at very low drain-to-source voltage, only sensitivities corresponding to 32 boxes must be computed with TCAD simulations.

For ITC, the situation is similar: we assume an average trap density of 5×10^{11} cm^{-2} and partition the tunnel oxide in tales of $100 \times 8 \times 64$ nm^3, for a total of only four simulations, if the symmetry of the nominal structure is exploited. As can be seen in Figure 5, finer partitions do not lead to a different estimation of the threshold voltage dispersion.

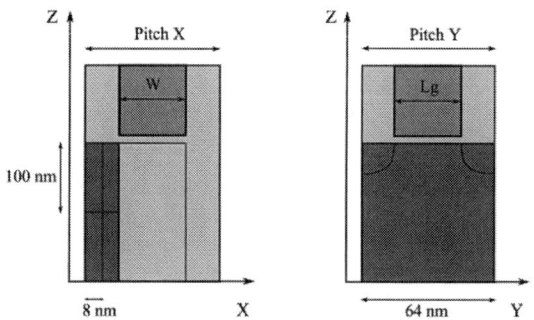

Box size (x, y, z) nm^3	16 x 64 x 200	8 x 64 x 100	16 x 64 x 50	8 x 32 x 50	4 x 32 x 50
σ_{vth} (mV)	27	59	56	59	59

Figure 5: Region partitioning in boxes $100 \times 8 \times 64$ nm^3 for the evaluation of propagators due to interface trapped charge. Left: transversal cross section. Right: longitudinal cross section.

The effect of RDD and ITC on the threshold voltage have been compared in Table 1 with direct simulation of a statistical ensemble done at the University of Glasgow through GARAND [5] obtained simulating samples of 1000 microscopically different devices. Considering that statistical simulations have been performed on ensembles of N=1000 devices, the mean square relative error on the estimated standard deviation of the threshold voltage is $(2N)^{-0.5}$, i.e., 2.2%: all terms lie within or very close to the error bars of statistical simulations.

TABLE 1 STANDARD DEVIATION OF THE THRESHOLD VOLTAGE DUE TO LER AND LWR OBTAINED WITH THE METHOD PROPOSED IN [3] AND WITH STATISTICAL SIMULATION IN [5].

σ_{Vth} (mV)	Our method [3]	Atomistic Sim. [5]
LER	46	48
LWR	28	26
OTF	14	14
RDD	156	144
ITC	59	67

CONCLUSION

We have proposed a methodology for the quantitative evaluation of the effects of the main mechanisms affecting threshold voltage variability, based on the careful identification of the main independent and relevant physical quantities. Our approach requires the calculation of partial derivatives of V_{th} with respect to device structure parameters, that can be obtained with a very limited number of TCAD simulations. We have shown that in all cases we are able to obtain results in good agreement with 3D atomistic statistical simulations [5] at a much smaller computational cost. We qualify this statement to the second order moment of the threshold voltage distribution,

because the proposed approach does not provide information on the far tails of the distribution, which are important for large Flash memory arrays, and would require extension of the method to higher order terms.

Our approach has some advantages over statistical modeling, not only because is orders of magnitude faster, but also because it represents a powerful tool for understanding the impact of individual factors and to efficiently explore the design space using tools already available and routinely used by technology developers .

ACKNOWLEDGMENTS

This work has been supported by the ENIAC project 12003 MODERN awarded to IUNET.

REFERENCES

[1] A. Calderoni, P. Fantini, A. Ghetti, A. Marmiroli, "Vth fluctuations in nanoscale floating gate memories, Proc. SISPAD, Sept. 9-11, 2008, pp. 49-52.

[2] A. Spessot, A. Calderoni, P. Fantini, A. S. Spinelli, C. Monzio Compagnoni, F. Farina, A. L. Lacaita, and A. Marmiroli, "Variability effects on the VT distribution of nanoscale NAND Flash memories," in Proc. IRPS, 2010, pp. 970–974.

[3] V. Bonfiglio and G. Iannaccone, "An Approach Based on Sensitivity Analysis for the Evaluation of Process Variability in Nanoscale MOSFETs," IEEE Transactions on Electron Devices, vol. 58, no. 8, pp. 2266-2273, 2011.

[4] Deliverable D.2.2.3 of the ENIAC MODERN Project, 2011.(Project website: www.eniac-modern.org).

[5] G. Roy, A. Ghetti, A. Benvenuti, A. Erlebach, and A. Asenov, "Comparative Simulation Study of the Different Sources of Statistical Variability in Contemporary Floating-Gate Nonvolatile Memory," IEEE Transactions on Electron Devices, vol. 58, no. 12, pp. 4155-4163, Dec. 2011.

[6] H.-S. Wong, Y. Taur, "Three-dimensional "atomistic" simulation of discrete random dopant distribution effects in sub-0.1 mm MOSFETs", Tech. Dig. IEDM 1993, pp. 705-708, 1993.

[7] A. Asenov, S. Kaya, and A. R. Brown, "Intrinsic parameter fluctuations in decananometre MOSFETs introduced by gate line edge roughness," IEEE Trans. Electron Devices, vol. 50, no. 5, pp. 1254–1260, May 2003.

[8] Ji-Young Lee, Jangho Shin, Hyun-Woo Kim, Sang-Gyun Woo, Han-Ku Cho, Woo-Sung Han, and Joo-Tae Moon, "Effect of line-edge roughness (LER) and line-width roughness (LWR) on sub-100-nm device performance", Proc. SPIE 5376, 426 (2004).

[9] A. Asenov, S. Kaya and J. H. Davies, "Intrinsic threshold voltage fluctuations in decanano MOSFETs due to local oxide thickness variations," IEEE Transactions on Electron Devices, vol. 49, pp. 112–119, 2002.

[10] C. L. Alexander, A. R. Brown, J. R. Watling, and A. Asenov, "Impact of single charge trapping in nano-MOSFETs— Electrostatics versus transport effects," IEEE Trans. Nanotechnol., vol. 4, no. 3, pp. 339–344, May 2005.

[11] W. Shockley, J. A. Copeland, R. P. James, "The impedance field method of noise calculation in active semiconductor devices", in Quantum Theory of Atoms, Molecules, and the Solid State, A tribute to John C. Slater. Edited by Per-Olov Loewdin. New York: Academic Press, 1966, p.537.

978-1-4673-1707-8/12 $31.00 © 2012 IEEE

Scaling of Trigate Nanowire (NW) MOSFETs Down to 5 nm Width: 300 K Transition to Single Electron Transistor, Challenges and Opportunities

V. Deshpande*, S. Barraud*, X. Jehl*, R. Wacquez*, M. Vinet*, R. Coquand*, B. Roche*, B. Voisin*, F. Triozon*,
C. Vizioz*, L. Tosti*, B. Previtali*, P. Perreau*, T. Poiroux*, M. Sanquer* and O. Faynot*
*CEA-LETI, Minatec campus and CEA-INAC, 17 rue des Martyrs, F- 38054 Grenoble.
Email: veeresh.deshpande@cea.fr

Abstract— For the first time we evidence the transition from a MOSFET operation to Single Electron Transistor (SET) behavior at 300 K in scaled nanowires (down to 5 nm width). In this paper we show that on scaling nanowire width from 20 nm down to 5 nm regime, together with achieving excellent short channel effect control (DIBL=12 mV/V for L_G=20 nm), we hit a dramatic transition in transport mechanism from monotonously increasing to periodically peaked I_D-V_G's. This transition is brought about by process induced channel potential variability (due to disorder) in nanowires and poses a challenge to further scaling. However, we show that it provides an exciting opportunity to cointegrate Single Electron Transistors with high-k/metal gate operating at room temperature (at V_D=±0.9 V!) with the state-of-the-art nanowire MOSFETs enabling large scale manufacturing of beyond Moore devices.

I. INTRODUCTION

Multigate FETs (MuGFET) owing to their improved short channel effect (SCE) control are considered prime choice for next nodes [1], [2]. Trigate NW-MOSFETs naturally form the successors to the current MuGFETs for extremely scaled gate lengths. Previous simulation works [3] have predicted a scaling law of L_G~3R (R - radius of NW) for reliable electrostatic integrity. So there is need for NW with channel diameter (or width for trigate), W=3-5 nm at the end-of-roadmap. The main challenge for scaling to sub-10 nm channel dimensions was predicted to be threshold voltage (V_T) increase due to quantum confinement [3].

In this paper we first demonstrate performance of Trigate NW MOSFETs with width scaling down to 5 nm and then show, for the first time, that on scaling width from 20 nm to about 5 nm the transport properties can be completely modified: we evidence a transition from normal FET operation to SET operation at room temperature. In depth analysis with low temperature measurements is done to shed light on the physical phenomena at play. Besides this, it is also first time demonstration of cointegration of room temperature operating SET with NW-MOSFET on state-of-the-art CMOS technology with high-k/metal gate.

II. DEVICE FABRICATION

Main steps of the process integration scheme used are given in Fig. 1. Starting from a SOI substrate (T_{Si}=12 nm),

- SOI Wafer (Mesa isolation)
- Active patterning (resist trimming)
- HfSiON/ALD TiN gate stack
- Gate patterning
- Spacer1 formation (10 nm or 25 nm)
- S/D Epitaxy (for RSD)
- LDD implantation
- Spacer2 formation
- HDD implantation
- Silicidation
- Back-End

Fig. 1: Flow chart of the process scheme used for trigate NW-MOSFETs on SOI.

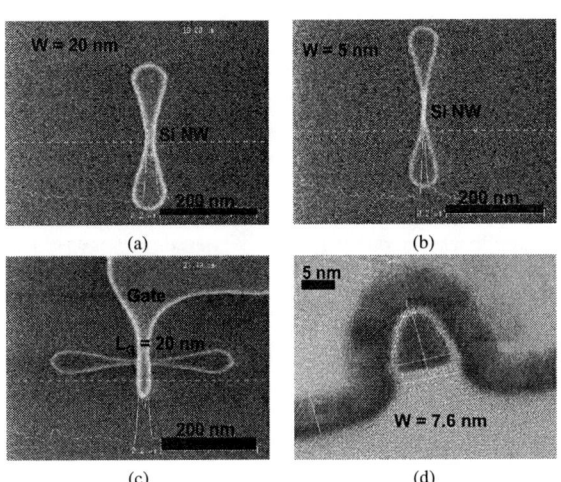

Fig. 2: SEM images of NWs with W=20 nm (a) and W=5 nm (b) after etching and after gate etching with L_G=20 nm (c). (d) TEM of a 7.6 nm wide nanowire.

NWs are patterned using DUV lithography. Resist trimming is performed during etching to reach NW widths down to 5 nm. High-k/Metal gate stack comprising of 2.3 nm HfSiON, 5 nm

TiN and 50 nm poly Si is used and gate down to 20 nm is patterned. Nitride spacers are then formed (CD Spacer=25 nm for 5-7 nm width NW and CD Spacer=10 nm for 20 nm wide NW). Standard SOI process flow is followed thereafter. Fig. 2 shows SEM image of NWs with two different widths after active area etching and the gate after etching. It can be seen that smallest width is around 5 nm.

III. RESULTS AND DISCUSSION

Fig. 3a and 3b show SCE control down to L_G=20 nm for NWs with different widths. For L_G=20 nm, DIBL and SS are 30 mV/V and 72 mV/dec respectively for NW width=20 nm. For the same L_G on reducing width to 7 nm, DIBL and SS are reduced to 12mV/V and 62 mV/V respectively showing excellent SCE control with W reduction. I_D-V_G plots for L_G=20 nm at NW width=20 nm and 7 nm are shown in Fig. 3c and Fig. 3d respectively. Despite excellent short channel effect control in NWs with W=5-7 nm, we have observed a variability leading to peculiar characteristics in some devices. All the subsequent measurements are performed at 300 K (Fig. 4-7) unless specified.

Fig. 4a shows the I_D-V_G characteristics of a NMOS with W=5 nm (device A) with classical FET behavior. However, device B with same nominal dimensions shows weak oscillations in the drain current at 300 K (Fig. 4b). These oscillations are due to single electron charging or 'Coulomb blockade' phenomenon. Disorder potential (Fig. 4b schematic) in the NW introduced during fabrication creates confinement of electrons leading to a small island in the NW [4] and the transport in such NWs becomes markedly different from the classical MOSFET case. Therefore, the reason for this transition and corresponding variability is due to channel potential profile variability introduced by disorder, with FET case having smooth channel potential and SET case having disorder potential (for instance due to LER [4], [5]).

Fig. 5 shows another device (device C), a PMOS exhibiting strong Coulomb oscillations in the drain current. The device is in fact operating as a very good Single Hole Transistor. Good peak to valley ratio is observed with quasi periodic oscillations. Fig. 6 shows the 'Coulomb Diamond' [6] (2D color plot of I_D with V_G and V_D) for device C. Charging energy calculated from this plot is estimated to be ~85 meV, which yields a total capacitance of the island ~2 aF. Such a small capacitance resulting in high charging energy is coherent with room temperature operation (25 meV). Fig. 7 shows another PMOS (device D) showing strong oscillations for very high V_D (=-0.9 V). It is an indication of a very small island with very high charging energy (> eV_D/2) and demonstrates operation of SET at 300 K, at supply voltages of next generation CMOS nodes.

In order to gain deeper understanding of the dramatically different regimes of transport with reduced thermal excitations, low temperature measurements were performed (differential conductance with lock-in setup is measured for various temperatures). Fig. 8a shows the evolution of G (=I_D/V_D)-V_G with temperature for NW-MOSFET (device E) with W=7 nm.

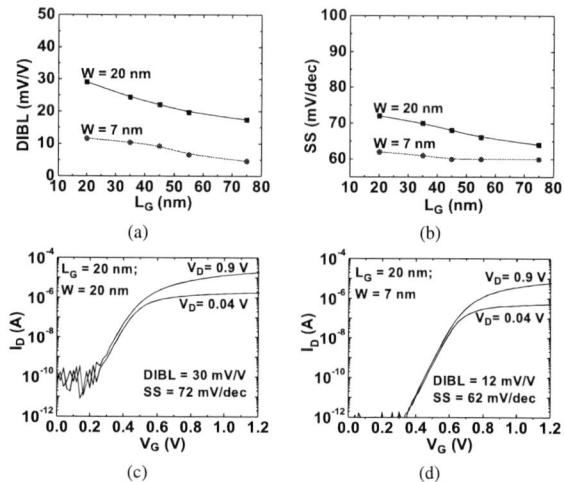

Fig. 3: SCE control by width scaling in Trigate NW-NMOSFETs. (a) DIBL vs. L_G for different width NWs down to L_G=20 nm. Drastic improvement in DIBL for L_G=20 nm from 30 mV/V (W=20 nm) to 12 mV/V (W=7 nm). (b) SS vs. L_G for different width NWs. Improvement for L_G=20 nm, from 72 mV/dec for W=20 nm to 62 mV/dec for W=7 nm. (c) and (d) I_D-V_G at V_D=0.04 V and 0.9 V for L_G=20 nm with W = 20 nm and W=7 nm respectively.

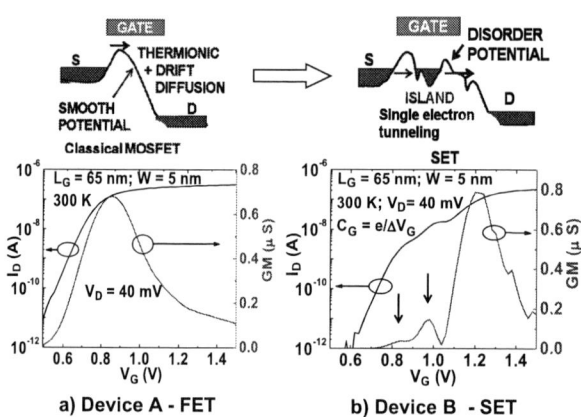

Fig. 4: I_D-V_G and transconductance (GM)-V_G plot for two NMOS (device A and B) with same dimensions. Transition from MOSFET to SET is observed due to channel potential variation (schematic above graphs) (a) Device A works as classical MOSFET. (b) Oscillations observed in I_D and GM of device B. Peaks marked by arrows (separation = 160 mV). Device B behaves as SET.

At 4.2 K clear Coulomb oscillations are observed. For a MOSFET, the channel potential is smooth, so confinement of electrons in the channel is caused by potential barriers formed below the spacers [7] and only occurs at lower temperature.

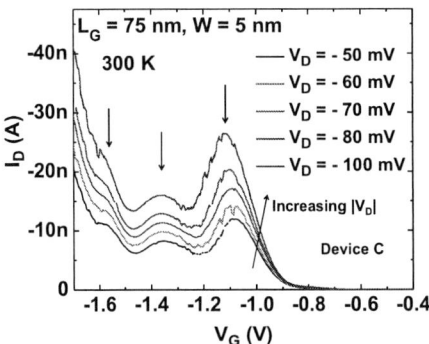

Fig. 5: I_D-V_G characteristics of a PMOS SHT (Single Hole Transistor) at 300 K - device C. Three peaks corresponding to single electron charging are observed. V_G period of the oscillations is about 280 mV giving gate capacitance C_G=0.57 aF.

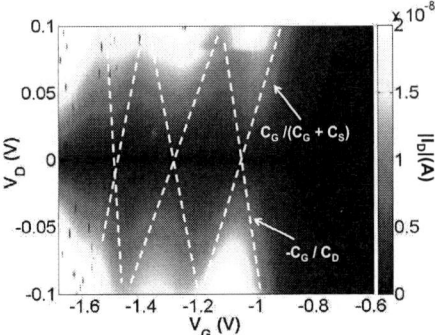

Fig. 6: 'Coulomb Diamond' plot for the device C at 300 K. Dashed lines show the diamonds. Slopes of the diamond (shown in fig) give C_S and C_D, which are 0.85 aF and 0.48 aF respectively. The charging energy: $[e^2/(C_S+C_D+C_G)]$ ~85 meV.

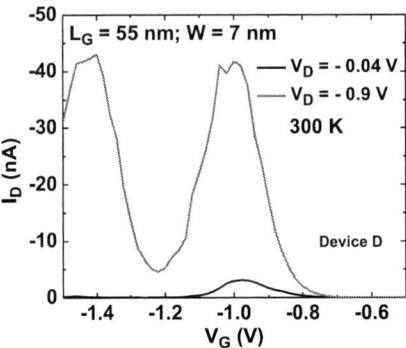

Fig. 7: I_D-V_G for a PMOS SHT with L_G = 55 nm showing very sharp oscillations at high V_D (= - 0.9 V) at 300 K.

The period of oscillations (e/C_G=ΔV_G ~6 mV) is consistent with the geometrical gate capacitance of the device (~28 aF for L_G=75 nm) showing island is defined by the whole

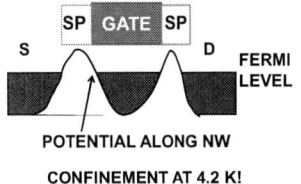

(a)

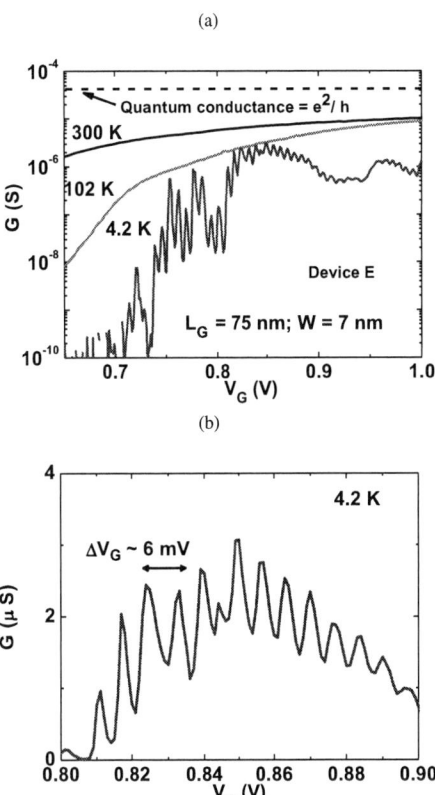

(b)

(c)

Fig. 8: (b) Conductance (G=I_D/V_D)-V_G evolution with temperature for a NMOS with classical MOSFET characteristics at 300 K. Since channel potential is smooth (FET case), confinement occurs only at 4.2 K due to potential barriers below spacers (see schematic Fig. 8a). Notice that the conductance is lower than quantum conductance (dashed line). (c) G-V_G at 4.2 K in linear scale showing period of oscillations ~6 mV.

channel. Where as in Fig. 9 notice the evolution of G-V_G with temperature for a SET type device (device F, oscillations at 300 K similar to B, C and D) with strong Coulomb oscillations visible at 115 K. Unlike device E, C_G for device F does not correspond to geometrical L_G, indicating smaller island in channel created by disorder and not by spacers. At lower temperatures the peaks split into multiple peaks either due to multiple islands or increased spectral resolution of quantum confined levels. In this purview an important point

978-1-4673-1707-8/12 $31.00 © 2012 IEEE   123

is to be noted in all these observations: oscillations (or peaks and valleys) in drain current in ultra-scaled NWs are often interpreted as result of diffusive transport through multiple 1D sub-bands [8]. It is clearly seen that conductance of the NW-MOSFETs is always much below quantum conductance ($G_Q \sim 4.10^{-5}$ S). But, for multiple sub-band population, it should be higher than G_Q. Thus, our results strongly indicate that this interpretation may be erroneous and disorder in sub-10 nm NWs enhances island formation and results in oscillations due to single electron charging (even at 300 K). So we consider MOSFET transition to SET transport to be one of the major challenges for scaling NWs to 5-7 nm widths (or diameters). But it also provides an exciting opportunity to co-integrate SET operating at 300 K with advanced CMOS NW-MOSFET to realize hybrid circuits enabling multivalued logic and neural network functions [9], [10].

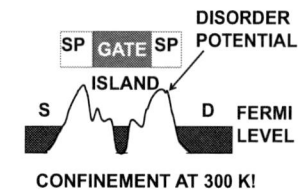

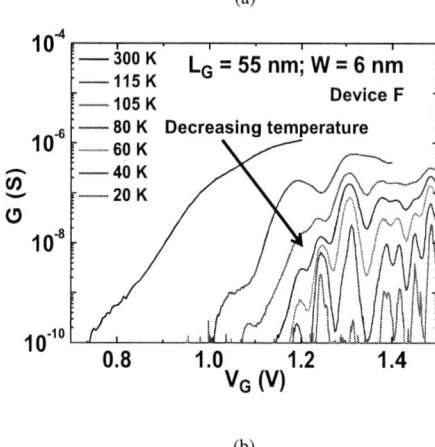

(b)

Fig. 9: G-V_G evolution with temperature for a NMOS showing Coulomb oscillations at 300 K. Higher V_G period due to high charging energy is observed. As shown in schematic (a), channel potential is disordered and leads to a small island and hence confinement occurs at 300 K! Note that unlike in schematic of Fig. 8a island is not due to spacers.

IV. CONCLUSION

In this work we demonstrate scaling of Trigate nanowire (NW)-MOSFETs (widths down to 5 nm) with excellent short channel effect control (DIBL=12 mV/V for L_G=20 nm). On reducing the width down to 5 nm we observe transition of MOSFET to single electron transistor. This is induced by disordered potential confinement creating islands in the nanowire. This challenge to the scaling of NW widths can also be seen as an opportunity to realize cointegration of SET and NWMOSFET in high-k/metal gate CMOS technology working at 300 K with V_D=±0.9 V.

ACKNOWLEDGMENT

Devices used in this study have been fabricated in the framework of the ST/IBM/LETI joint program. Authors acknowledge physical characterization team from STMicroelectronics-Crolles for TEM.

REFERENCES

[1] K. Tachi, M. Casse and, S. Barraud, C. Dupre and, A. Hubert, N. Vulliet, M. Faivre, C. Vizioz, C. Carabasse, V. Delaye, J. Hartmann, H. Iwai, S. Cristoloveanu, O. Faynot, and T. Ernst, "Experimental study on carrier transport limiting phenomena in 10 nm width nanowire cmos transistors," in *Electron Devices Meeting (IEDM), 2010 IEEE International*, dec. 2010, pp. 34.4.1 –34.4.4.

[2] S. Bangsaruntip, G. Cohen, A. Majumdar, Y. Zhang, S. Engelmann, N. Fuller, L. Gignac, S. Mittal, J. Newbury, M. Guillorn, T. Barwicz, L. Sekaric, M. Frank, and J. Sleight, "High performance and highly uniform gate-all-around silicon nanowire mosfets with wire size dependent scaling," in *Electron Devices Meeting (IEDM), 2009 IEEE International*, dec. 2009, pp. 1 –4.

[3] B. Yu, L. Wang, Y. Yuan, P. Asbeck, and Y. Taur, "Scaling of nanowire transistors," *Electron Devices, IEEE Transactions on*, vol. 55, no. 11, pp. 2846 –2858, nov. 2008.

[4] A. Lherbier, M. P. Persson, Y.-M. Niquet, F. m. c. Triozon, and S. Roche, "Quantum transport length scales in silicon-based semiconducting nanowires: Surface roughness effects," *Phys. Rev. B*, vol. 77, p. 085301, Feb 2008.

[5] M. Saitoh and T. Hiramoto, "Room-temperature operation of highly functional single-electron transistor logic based on quantum mechanical effect in ultra-small silicon dot," in *Electron Devices Meeting, 2003. IEDM '03 Technical Digest. IEEE International*, dec. 2003, pp. 31.5.1 – 31.5.4.

[6] K. Likharev, "Single-electron devices and their applications," *Proceedings of the IEEE*, vol. 87, no. 4, pp. 606 –632, apr 1999.

[7] M. Hofheinz, X. Jehl, M. Sanquer, G. Molas, M. Vinet, and S. Deleonibus, "Simple and controlled single electron transistor based on doping modulation in silicon nanowires," *Applied Physics Letters*, vol. 89, no. 14, p. 143504, 2006.

[8] N. Singh, F. Lim, W. Fang, S. Rustagi, L. Bera, A. Agarwal, C. Tung, K. Hoe, S. Omampuliyur, D. Tripathi, A. Adeyeye, G. Lo, N. Balasubramanian, and D. Kwong, "Ultra-narrow silicon nanowire gate-all-around cmos devices: Impact of diameter, channel-orientation and low temperature on device performance," in *Electron Devices Meeting, 2006. IEDM '06. International*, dec. 2006, pp. 1 –4.

[9] S. Mahapatra, V. Pott, S. Ecoffey, A. Schmid, C. Wasshuber, J. Tringe, Y. Leblebici, M. Declercq, K. Banerjee, and A. Ionescu, "Setmos: a novel true hybrid set-cmos high current coulomb blockade oscillation cell for future nano-scale analog ics," in *Electron Devices Meeting, 2003. IEDM '03 Technical Digest. IEEE International*, dec. 2003, pp. 29.7.1 – 29.7.4.

[10] S. J. Shin, C. S. Jung, B. J. Park, T. K. Yoon, J. J. Lee, S. J. Kim, J. B. Choi, Y. Takahashi, and D. G. Hasko, "Si-based ultrasmall multiswitching single-electron transistor operating at room-temperature," *Applied Physics Letters*, vol. 97, no. 10, p. 103101, 2010.

Active Strain Modulation in Field Effect Devices

Tom van Hemert and Raymond J.E. Hueting

MESA$^+$ Institute for Nanotechnology, University of Twente, Enschede, The Netherlands, email: t.vanhemert@utwente.nl

Abstract—**In this work we propose a novel feature for the transistor: a piezo-electric layer for strain modulation of the channel. The strain is formed at strong inversion only, to obtain a lower threshold voltage, but will be absent in the off-state to preserve the unstrained leakage current. Our results, obtained by combining electrical and mechanical finite element method simulation, demonstrate a seven mV/dec steeper subthreshold swing for a classical SOI transistor and ten mV/dec improvement for a silicon tunnel field effect transistor.**

I. INTRODUCTION

BESIDES conventional CMOS miniaturization novel devices are being researched to extend the performance. Major requirements are: higher switching frequency, increased functionality and reduced power consumption. The latter can be achieved with a smaller subthreshold swing (SS) yielding a reduced static power loss [1]. Currently, many disruptive device concepts are being investigated. Examples are: (a) the Impact Ionization field effect transistor (FET) [2] where the current is formed by avalanche, (b) the band to band tunnelling (B2B) FET [3], [4] where the current originates from a strongly field dependent band-to-band tunneling process, (c) the suspended gate transistor [5], in which the gate is physically put close to the silicon (Si) in on-state, however further away in off-state and (d) the incorporation of ferroelectric materials as a gate dielectric [6], [7], which could yield a negative differential capacitance.

For more than a decade strain has been applied in devices to boost the carrier mobility [8]–[11]. This enhancement is attributed to the offset of the energy bands due to strain. The offset also narrows the bandgap which in it turn increases the subthreshold current [11]. Consequently it is favorable to relax the semiconductor in the off-state, resulting in a low the leakage current (I_{off}), and to strain the semiconductor in the on-state, yielding an enhanced on-state current (I_{on}). Therefore we propose modulation of the strain with the the gate source voltage (V_{GS}), which we call strain modulation.

Common techniques to induce strain, such as employing materials with a built in stress or lattice mismatch, result in a constant strain. They cannot modulate the strain with V_{GS}. However, by employing a piezo-electric (PE) layer this may become possible. In earlier work [12] a complete substrate was strain modulated with a PE layer. We propose to attach the PE layer to an individual transistor and bias the PE layer with the V_{GS} of that transistor. To do this effectively we need to incorporate the PE layer into the transistor design.

A straightforward method to induce strain modulation is replacing the gate oxide in a bulk transistor with a PE layer. However, for bulk transistors the depletion charge should be small to obtain a steep SS, which yields a zero electric field in the PE layer in the subthreshold region, and hence no strain modulation to alternate the subthreshold current.

Therefore we propose an alternative solution. To displace, and hence generate strain, in a fully depleted (FD) SOI layer by compressing it with a V_{GS} biased PE layer. In this paper the silicon layer is part of either a classical transistor (an n-type diffusion FD silicon on insulator (SOI) FET) or a tunnel transistor (FD SOI B2B FET). However, any device with suitable strain dependent characteristics could be used.

We start with a brief discussion on the subthreshold current in strained Si in section II. In section III we discuss how the FET can be strained using one-dimensional analytical mechanical calculations and compare these with FEM simulations. In section IV we demonstrate the influence of strain modulation in the subthreshold characteristics for various devices and estimate the effect of strain modulation on the power consumption of a transistor. Finally the conclusions are drawn in section V.

II. THE SUBTHRESHOLD CURRENT

Strain in a crystalline material is a change of the interatomic distances and thus the lattice periodicity. This pertubs the band structure and related parameters such as the band alignment, the quantization effective masses, carrier mobility, and saturation velocity. The subthreshold current depends exponentially on the band alignment, hence this effect will be strong compared to the other strain related parameters and to obtain a good nevertheless simple estimation we can focus on the band alignment only.

In subthreshold the carriers in a classical fully depleted transistor move by diffusion. Hence the cross sectional area through which the current flows multiplied with the charge gradient and mobility gives the diffusion current [13];

$$
\begin{aligned}
I_D = \frac{t_S w_S \mu q u_t N_C}{L} &\left[1 - \exp \frac{-V_{DS}}{u_t} \right] \\
&\times \exp \left[\frac{\chi_S - \Delta E_C(\epsilon_{kk}) - \phi_m + V_{GS}}{u_t} \right],
\end{aligned}
\tag{1}
$$

where w_S, t_S and L are the SOI dimensions, μ the electron mobility, χ_S the semiconductor electron affinity, ϕ_m the metal work function, and ΔE_C is the change of the effective conduction band, given by the lowest of the different subbands, and ϵ_{kk} is the strain tensor. It can be readily observed that a negative ΔE_C increases the subthreshold current exponentially. The same equation can be derived for p-type transistors in which the highest valence band, scaling with ΔE_V, determines the current.

In a tunnel transistor the current is proportional to the tunneling probability, this can be described with various models [14]–[16], which all show an $E_G^{3/2}$ dependence. Hence the tunneling current scales with both ΔE_C and ΔE_V. This is in contrast to the classical transistor where the current scales with either one of them. Therefore the tunnel transistor is expected to be more responsive to strain modulation.

978-1-4673-1707-8/12 $31.00 © 2012 IEEE

Generally strain is either uniaxial or biaxial. This means that a tensile or compressive stress is applied along one or two of the coordinate axes. Along the non externally stressed axis the material will be more free to move. To preserve the volume change the strain will be opposite along the non externally stressed axis. Thus, as long as the applied stress is not triaxial, then the strain along at least one of the axes will be compressive.

The band alignment dependence on the strain is well known for various semiconductor materials [17]. If the PE layer deforms along the principle axis shear strain components can be neglected and we obtain,

$$\Delta E_{C,kk} = \Xi_d^\Delta (\epsilon_{11} + \epsilon_{22} + \epsilon_{33}) + \Xi_u^\Delta \epsilon_{kk}, \qquad (2)$$

here $k = 1, 2, 3$ are the three crystal coordinate axes, Ξ_d^Δ and Ξ_u^Δ are the deformation potentials [18], shown in table I. For Si $\Xi_u^\Delta > \Xi_d^\Delta$, hence the band edge shift is mainly determined by the term $\Xi_u^\Delta \epsilon_{kk}$. We stated that at least one of the strain components is likely to be negative, therefore at least one of the conduction band valley pairs will shift down to a lower energy level. Leaving out the shear strain terms out of [17] we find the valence band offset;

$$\Delta E_{V,kk} = -a(\epsilon_{11} + \epsilon_{22} + \epsilon_{33}) \pm \delta E,$$
$$\delta E = \frac{b}{\sqrt{2}}\sqrt{(\epsilon_{11}-\epsilon_{22})^2 + (\epsilon_{22}-\epsilon_{33})^2 + (\epsilon_{11}-\epsilon_{33})^2}, \qquad (3)$$

where a and b are the deformation potentials for holes, the $\pm$ sign separates the heavy and light holes. As long as $(\epsilon_{11}+\epsilon_{22}+\epsilon_{33})$ is negative the bands will move upward. Furthermore the δE term will always be positive. Therefore we can conclude that if one strain component is negative and relatively large, then the effective valence band will move upward, the effective conduction band (CB) downward, and bandgap narrowing can be observed.

III. APPLYING STRAIN

The strain induced in a PE needs to be transferred to the silicon layer. How this is done will have a major effect on the resulting transistor performance. To limit the amount of possibilities we simplify by modelling a silicon layer and a PE layer and neglect all other necessary layers to make a transistor. Also the concept should be viable to strain individual transistors. A further simplification iis to leave out shear deformations. When the PE is subject to an electric field the layer deforms in directions both perpendicular and parallel to the electric field. Both can be exploited to deform the silicon. We propose four different configurations to do such, these are shown in Fig. 1.

To estimate the strain obtainable using this four distinct configurations we derived a model from first order 1D equations;

$$\epsilon_{11} = \mathcal{E} \frac{-e_{31}w_\pi}{c_s w_\pi + c_{\pi 11} w_S}, \qquad (4a)$$

$$\epsilon_{33} = \mathcal{E} \frac{e_{33}w_\pi}{c_s w_S + c_{\pi 33} w_\pi}, \qquad (4b)$$

$$\epsilon_{33} = \mathcal{E} \frac{-e_{33}t_\pi}{c_s t_\pi + c_{\pi 33} t_S}, \qquad (4c)$$

$$\epsilon_{11} = \mathcal{E} \frac{e_{31}t_\pi}{c_s t_S + c_{\pi 11} t_\pi}, \qquad (4d)$$

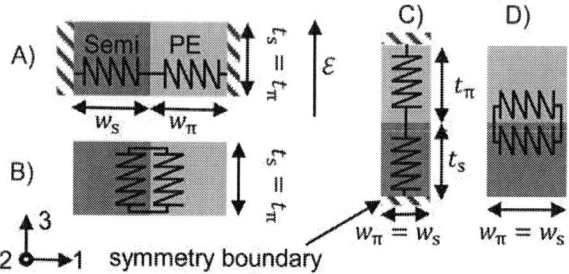

Fig. 1. Four basic configurations to induce strain into a silicon with a PE layer. the PE is subject to an electrical field, shown by $\mathcal{E}$ in the figure, however the required electrical contacts are not shown. At the highlighted boundaries of configurations A & C the displacement perpendicular to the boundary is zero.

TABLE I
MATERIAL PROPERTIES OF THE MATERIALS USED IN THIS WORK. THE PE MATERIALS PZT-5H, A SPECIFIC COMPOSITION OF PB(ZR,TI)O$_3$ (PZT), AND ALUMINUM-NITRIDE (ALN) ARE INDICATED WITH SUBSCRIPT π. SI HAS A SUBSCRIPT S. $\mathcal{E}_{cr}$ IS THE BREAKDOWN FIELD.

	$c_{\pi 11}$ [GPa]	$c_{\pi 33}$ [GPa]	e_{33} [C/m^2]	e_{31} [C/m^2]	$\mathcal{E}_{cr}$ MV/cm
PZT-5H [19]	127	117	23.3	-6.55	1
AlN [20]	345	395	1.55	-0.48	2
	Ξ_u^Δ [eV]	Ξ_d^Δ [eV]	a [eV]	b [eV]	c_s [GPa]
Si [21]	10.5	1.1	2.1	-2.33	166

where the subscripts indicate the direction and material, e is the piezoelectric charge constant, and c the stiffness.

In configurations A & C the strain in the PE is opposite to strain in the silicon, and the strain becomes independent of the PE stiffness c_π when a large t_π respectively w_π is used. Contrarily scaling in configurations B & D results in a c_s independent strain. For an as large as possible deformation an electric field close to breakdown is required, hence we choose $t_\pi = V_{DD}/\mathcal{E}_{cr}$. Furthermore, for configuration A & B we choose $w_S = 10$ nm. The material parameters are summarized in table I.

The calculated strain values for the different configurations and material combinations are shown in table II, including the optimum value for the scalable parameter. Clearly configurations B & C require a thin t_S.

Because PZT-5H is a ferroelectric material a large electric field can switch its polarity. Therefore the strain along the c-axis for fields in the range of $\mathcal{E}_{cr}$ is $\epsilon_{33} > 0$ and $\epsilon_{11} < 0$ & $\epsilon_{22} < 0$ [22]. As a result configurations C and D give a large compressive compressive strain in the silicon. In the remainder of this work we use configuration C with PZT-5H because it achieves a largest negative strain.

We simulated the strain using a 3D FEM package [23]. The results are shown in Fig. 2. The strain ϵ_{33} is large and negative, while the other components are much smaller, hence this will result in narrowing of the bandgap.

TABLE II
THE MAXIMUM OBTAINABLE STRAIN AND OPTIMIZED DIMENSIONS CALCULATED USING EQN. [4A-4D]

	A [%]	w_π [nm]	B [%]	w_π [nm]	C [%]	t_S [nm]	D [%]	t_S [nm]
Si&AlN	0.05	205	0.07	10	-0.17	0.2	-0.03	1
Si&PZT-5H	0.35	76	1.81	10	-1.26	1.4	-0.47	1

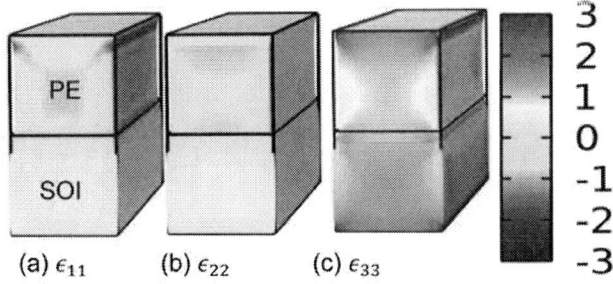

Fig. 2. 3D FEM simulation results for the strain obtainable with a 3D alternative of configuration C. $t_s = 10$ nm.

In Fig. 3 the 1D model, 2D, and 3D FEM results show how the strain depends on the SOI layer thickness. The results are all more or less in the same range, confirming our model and demonstrating that a thin SOI is required to obtain large strain.

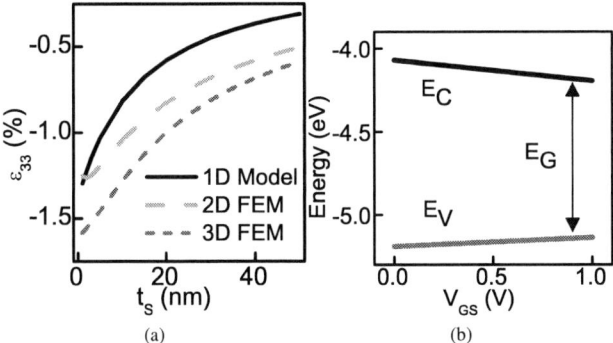

Fig. 3. (a) Modeled Eqn. [4c], 2D and 3D FEM results for the average Si strain in configuration C as a function of the SOI thickness. To obtain a single number the strain ϵ_{33} was averaged over the Si volume. (b) Simulated bandgap dependence on the applied bias, resulting from the biasing of the PE layer en hence strain in the silicon.

To calculate the strain vector a 3D FEM [23] simulation was performed. To simplify matters we simulated the PE and silicon layers only and neglected all other parts of the structure. From this simulation we obtain the strain vector, which is used as an input for the device simulator. The 3D strain vector is assumed uniform throughout the silicon layer in the device simulator. To include the bias dependency of the strain we assume $\epsilon = 0$ at $V_{GS} = 0$ V because zero electric fields results in zero strain, and use the simulated values of Fig. 2 for $V_{GS} = 1$ V, for all other V_{GS} we interpolate linearly. Then technology computer aided design (TCAD) input files have been built for each bias point and each of them has been simulated separately. This method is required because the TCAD platform [24] cannot sweep both strain and voltage. Note that this combination of mechanical and electrical simulation is not self-consistent, which is also not required because the strain depends on the applied voltage, and not on the electrical transistor architecture. The transfer characteristics, both with, and without strain modulation were extracted from the output files. Fig. 3b shows the simulated bandgap dependence on V_{GS}, which decreases significantly with the applied bias. To practically exploit the strain modulation effect we propose the 3D structure as schematically shown in Fig. 4. By contacting the PE gate to the source the bias V_{GS} is applied to the PE layer.

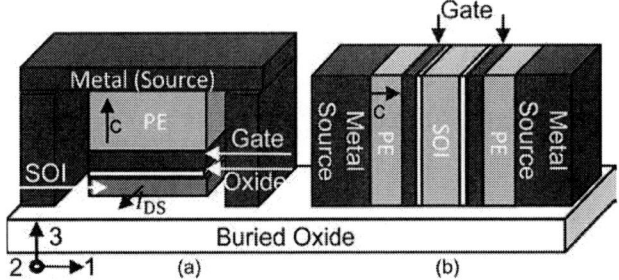

Fig. 4. 3D schematic of the proposed SOI transistor with strain induced by a PE layer. (a) compressing an SOI FET, requiring a stiff source metal, such as titanium-nitride to mimic the fixed boundary condition. (b) compressing a finFET using a PE with a 90 degrees rotated polarization (or c) axis.

TABLE III
THE DIMENSIONS AND PARAMETERS AS USED IN THE SIMULATION.

width	$w_S = w_\pi$	10	[nm]
length	L	50	[nm]
SOI thickness	t_S	10	[nm]
PE thickness	t_π	t_S	[-]
oxide thickness	t_{ox}	1	[nm]
drain bias	V_{DS}	1	[V]
FET			
gate work function	ϕ_m	5.1	[eV]
channel doping	N_A	10^{15}	[cm^{-3}]
drain/source doping	N_D	10^{20}	[cm^{-3}]
TFET			
gate work function	ϕ_m	3.9	[eV]
channel doping	N_A	10^{17}	[cm^{-3}]
drain doping	N_A	10^{20}	[cm^{-3}]
source doping	N_D	10^{18}	[cm^{-3}]

IV. ELECTRICAL RESULTS

In this section we investigate the effect of strain modulation on the transfer characteristics of the classical and the tunnel transistor in configuration C. In the simulations we neglected the quantum confinement effect, as it will result only in an horizontal offset of the curves. Fig. 5a shows the simulated transfer characteristics of the classical transistor without strain ($\epsilon = 0$), with a constant strain equal to the maximum strain when strain modulation is applied (-1.26 %), and with strain modulation. Without strain the transistor has a low leakage current, with strain it has a reduced threshold voltage and thus a higher higher on-current. With strain modulation we combine the best of both worlds. Therefore the strain modulation transistor shows a SS of 53 mV/dec, which is below the thermal limit of 60 mV/dec.

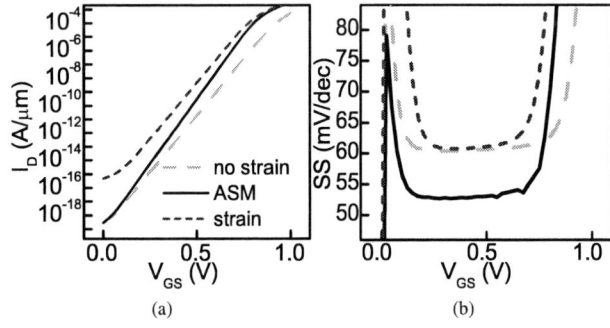

Fig. 5. The simulated effect of strain modulation on the transfer characteristics and SS of a classical transistor.

978-1-4673-1707-8/12 $31.00 © 2012 IEEE

We employ a nonlocal tunneling model [24] to calculate the tunneling current in the B2B transistor. Parameters have been obtained from [25]. The tunneling current is calculated using a nonlocal mesh, which extends 20 nm into both the source and channel region from the junction. Fig. 6a shows the transfer characteristics of the tunneling transistor. Compared to the classical transistor the tunneling transistor is more sensitive to strain modulation. This was explained in section II, where we stated that the tunneling depends on both the change in the conduction band and valence band.

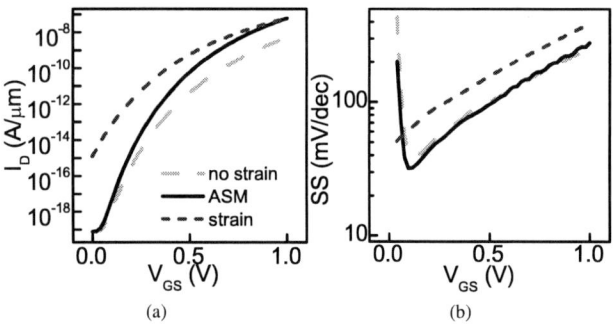

Fig. 6. The simulated effect of strain modulation on the transfer characteristics and SS of a tunnel transistor.

V. CONCLUSION

Strain modulation is a novel concept for combining the higher on-current of a strained device with the lower off-current of a device without strain. Hence it has the potential to improve the SS for all types of devices where the subthreshold current depends on the band alignment. With modeling we came up with a suitable geometry and demonstrated its tunability. With TCAD simulations we demonstrated that this technique has the potential to lower the SS of FETs. We found -1.26 % strain for a 10 nm wide and high SOI layer and 10 nm long PE layer on top. This was used as an input for the device simulator, which showed a 7 mV/dec better SS and 4 decades lower leakage current for the classical transistor and a 10 mV/dec improvement of SS and 5 decades lower leakage current for the tunnel transistor.

VI. ACKNOWLEDGEMENT

The authors would like to thank Alexandre Paternoster and prof. dr. ir. André de Boer of the applied mechanics department, University of Twente, for fruitful discussions and support on the structural engineering in this work. This work is supported by NanoNextNL, a micro and nanotechnology programme of the Dutch ministry of economic affairs, agriculture and innovation (EL&I) and 130 partners.

REFERENCES

[1] A. Chandrakasan and R. Brodersen, "Minimizing power consumption in digital CMOS circuits," Proceedings of the IEEE, vol. 83, no. 4, pp. 498–523, Apr. 1995.

[2] K. Gopalakrishnan et al., "I-MOS: a novel semiconductor device with a subthreshold slope lower than kT/q," IEEE IEDM, pp. 289–292, 2002.

[3] S. Banerjee et al., "A new three-terminal tunnel device," IEEE El. Dev. Lett., vol. 8, no. 8, pp. 347–349, Aug. 1987.

[4] A. C. Seabaugh and Q. Zhang, "Low-voltage tunnel transistors for beyond CMOS logic," Proc. IEEE, vol. 98, no. 12, pp. 2095–2110, Dec. 2010.

[5] N. Abelé et al., "Suspended-gate MOSFET: bringing new MEMS functionality into solid-state MOS transistor," in IEEE IEDM, Dec. 2005, pp. 479–481.

[6] S. Salahuddin and S. Datta, "Can the subthreshold swing in a classical FET be lowered below 60 mV/decade?" in IEEE IEDM, Dec. 2008, pp. 1–4.

[7] A. Rusu et al., "Metal-ferroelectric-metal-oxide-semiconductor field effect transistor with sub-60mV/decade subthreshold swing and internal voltage amplification," in IEEE IEDM, Dec. 2010, pp. 16.3.1 –16.3.4.

[8] S. Takagi et al., "Carrier-transport-enhanced channel CMOS for improved power consumption and performance," IEEE Trans. El. Dev., vol. 55, no. 1, pp. 21–39, Jan. 2008.

[9] J. L. Hoyt et al., "Strained silicon MOSFET technology," in IEDM Tech. Digest., Dec. 2002, pp. 23–26.

[10] S. E. Thompson et al., "Uniaxial-process-induced strained-Si: extending the CMOS roadmap," IEEE Trans. El. Dev., vol. 53, no. 5, pp. 1010–1020, May 2006.

[11] K. Rim et al., "Strained Si NMOSFETs for high performance CMOS technology," in VLSI, 2001, pp. 59–60.

[12] M. Shayegan et al., "Low-temperature, in situ tunable, uniaxial stress measurements in semiconductors using a piezoelectric actuator," Appl. Phys. Lett., vol. 83, no. 25, pp. 5235–5237, 2003.

[13] Y. Taur, "An analytical solution to a double-gate MOSFET with undoped body," IEEE El. Dev. Lett., vol. 21, no. 2, pp. 245–247, May 2000.

[14] E. O. Kane, "Zener tunneling in semiconductors," Journal of Physics and Chemistry of Solids, vol. 12, no. 2, pp. 181–188, 1960.

[15] G. A. M. Hurkx et al., "A new recombination model for device simulation including tunneling," IEEE Trans. El. Dev., vol. 39, no. 2, pp. 331–338, Feb. 1992.

[16] S. M. Sze and K. G. Ng, Physics of Semiconductor Devices. Wiley & Sons Inc., 2007.

[17] G. L. Bir and G. E. Pikus, Symmetry and strain-induced effects in semiconductors. Israel Program Sci. Translations, 1974.

[18] E. Ungersboeck et al., "The effect of general strain on the band structure and electron mobility of silicon," IEEE Trans. El. Dev., vol. 54, no. 9, pp. 2183–2190, Sep. 2007.

[19] M. A. Matin et al., "FE modeling of stress and deflection of PZT actuated micro-mirror: Effect of crystal anisotropy," Computational Materials Science, vol. 48, no. 2, pp. 349–359, 2010.

[20] K. Tsubouchi et al., "AlN material constants evaluation and SAW properties on AlN/Al₂O₃ and AlN/Si," IEEE Int. Ultrasonics Symp., pp. 375–380, 1981.

[21] M. V. Fischetti and S. E. Laux, "Band structure, deformation potentials, and carrier mobility in strained Si, Ge, and SiGe alloys," J. Appl. Phys., vol. 80, no. 4, pp. 2234–2252, Aug. 1996.

[22] M. Nguyen, "Ferroelectric and piezoelectric properties of epitaxial PZT films and devices on silicon," Ph.D. dissertation, University of Twente, June 2010.

[23] Comsol MultiPhysics, 3rd ed., COMSOL, Inc., Palo Alto, USA.

[24] Synopsys, Sentaurus Device User Guide. Synopsys, Inc, 2007.

[25] H. Virani et al., "Impact of electron velocity on the ION of n-TFETs," in IEEE ESSDERC, Sep. 2010, pp. 349–352.

Static and low frequency noise characterization of densely packed CNT-TFTs

Min-Kyu Joo[1, 2], Un Jeong Kim[3], Dae-Young Jeon[1, 2], So Jeong Park[1, 2],*
Mireille Mouis[1], Gyu-Tae Kim[2] and Gérard Ghibaudo[1]

1. IMEP-LAHC, Grenoble INP-MINATEC, 3 Parvis Louis Néel, 38016 Grenoble, France
2. School of Electrical Engineering, Korea University, Seoul 136-701, Republic of Korea
3. Frontier Research Laboratory, Samsung Advanced Institute of Technology, Suwon 440-600, Republic of Korea

Abstract— **Static and low frequency noise (LFN) character-izations in densely packed single-walled carbon nanotube thin film transistors (CNT-TFTs) are presented. To this end, the Y function method (YFM) is employed for parameter extraction in order to alleviate the influence of the channel access resistance. The low field mobility (μ_0), threshold voltage (V_{th}), mobility attenuation factor (θ) and on/off current ratio have been evaluated with respect to gate mask length (L_{mask}). The $1/f$ behavior of LFN has been interpreted with the carrier number and correlated mobility fluctuation model (CNF-CMF). A detailed analysis of the defect density surrounding the surface of carbon nanotube and the Coulomb scattering parameter has also been performed.**

I. INTRODUCTION

CNT-TFTs have been proposed as promising candidates of nano-scaled devices for large-scale, inexpensive and flexible electronics [1, 2]. Many research groups have tried to enhance their electrical performances and carrier transport properties based on various chemical and physical methods [2, 3]. In spite of the progress in device fabrication and carbon nanotubes (CNT) synthesis methods, there is still a lack of understanding in both static and LFN properties of CNT-TFTs as compared to standard silicon based devices. For example, reported mobility values for two dimensional CNT-TFTs are in the range $\sim 10^{-2}$ to $\sim 10^3$ cm^{2}V^{-1}s^{-1} [1, 4]. This huge variation might result from device fabrication methods, average nanotube density (ρ_{CNT}), portion of metallic path, mean length of CNT (L_{tube}), device geometry, channel access resistance (R_{sd}) and degree of CNT alignment as well as bad gate capacitance evaluation.

In this work, we fabricated several CNT-TFTs with high ρ_{CNT} with respect to L_{mask} (2, 3, 5, 7, 10 µm) to reduce the percolation-dominated transport influences. And we have employed, for the first time, the Y-function method (YFM) currently used in silicon devices, for the parameter extraction in CNT-TFTs. This technique enables to extract electrical parameters without R_{sd} influence. Moreover, we have carried out a detailed analysis of LF noise in these CNT-TFTs and provided a diagnostic of their LF noise sources.

II. DEVICE FABRICATION AND MEASUREMENTS

Single-walled carbon nanotubes (SWCNTs) were synthesized at 450°C by water-assisted CH$_4$ plasma enhanced chemical vapor deposition (PECVD) system on highly doped silicon substrates with 400 nm thick silicon dioxide (SiO$_2$). The desired channel locations have been pre-patterned by conventional photolithography. The average diameter (d_{CNT}) and L_{tube} of SWCNTs ranges from 0.8 to 1.3nm and ~1µm respectively. Average ρ_{CNT} is counted to be ~25 up to ~30 tubes/µm^2 by direct measurement from scanning electron microscope (SEM). To make a back-gate field-effect transistor (FET) configuration, Ti (10 nm)/Au (50 nm) were deposited by a standard electron-beam evaporation system for making source (S) and drain (D) electrodes on the SWCNTs layers. And then for a top-gate FET configuration, alumina (Al$_2$O$_3$) dielectric layer (t_{ox} = 50 nm) was deposited by atomic layer deposition (ALD) on the SWCNTs at 350°C. Afterwards, conducting metal Ti (10 nm)/Au (50 nm) was sputtered for the top-gate electrode. The cross section of the final device configuration is displayed in Fig. 1. The detailed fabrication and the quality of SWCNTs has been published elsewhere [2].

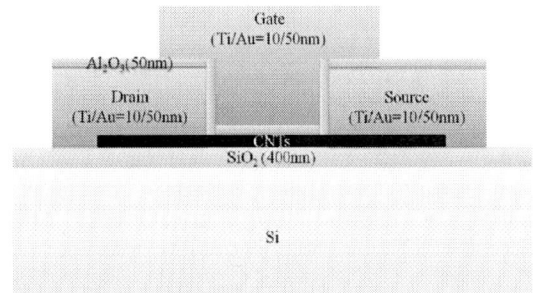

Fig. 1. Schematic cross section view of a typical top-gate CNT-TFT configuration.

Electronic mail: joom@minatec.inpg.fr (Min-Kyu JOO)

III. RESULTS AND DISCUSSION

A. Static parameter extraction

The $I_{ds}(V_{gs})$ transfer characteristics have been measured by an HP 4155B. Typical transfer curves, $I_{ds}(V_{gs})$, obtained for various L_{mask} of top-gate CNT-TFTs with fixed gate mask width W_{mask} (40 µm) biased at source-drain voltage ($V_{ds} = 50$ mV) are represented in Fig. 2. This figure confirms the overall good quality of these SWCNTs with low off-current, high on/off current ratio (~10^5 to ~10^6) and small subthreshold swing (~ 90 mV/decade) except for the case of $L_{mask} = 2$ µm. As in conventional Si-MOSFET (silicon based metal-oxide semiconductor field-effect-transistor), the on-current is inversely proportional to L_{mask}. These findings clearly indicate that the SWCNTs in CNT-TFTs are mostly behaving as semiconductor and that the channel resistance is proportional to L_{mask}. For the case of $L_{mask} = 2$ µm, there might be few metallic paths in the channel that could enhance the off-current between source and drain below threshold.

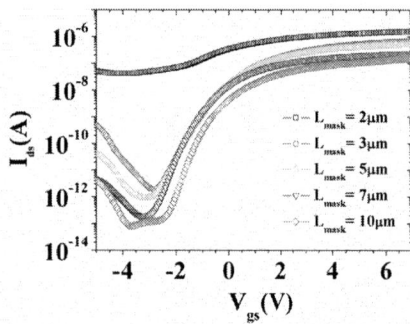

Fig. 2. Typical $I_{ds}(V_{gs})$ transfer curve for CNT-TFTs

For further electrical parameter evaluation in strong accumulation without R_{sd} influence, we employed the $YFM(V_{gs})$ defined as,

$$YFM(V_{gs}) = I_{ds}/\sqrt{g_m} = \sqrt{G_m \cdot V_{ds}}(V_{gs} - V_{th}) \qquad (1),$$

where $G_m = (W_{mask}C_{ox}\mu_0)/(L_{mask})$ is the transconductance parameter and C_{ox} is the top-gate oxide capacitance per unit area (F/cm^2) [5]. Once we obtain the YFM in accumulation region as shown in Fig. 3, we can easily extract V_{th} by linear extrapolation of YFM with respect to V_{gs} and the low-field mobility μ_0 from the slope of YFM. Then, θ and R_{sd} can easily be deduced by simple computation from $I_{ds}(V_{gs})$, V_{th} and G_m data [5]. The $YFM(V_{gs})$ characteristics show good linearity at high gate bias region and are well fitted by Eq. (1). The variation of V_{th}, μ_0 and θ with L_{mask} is shown in Fig. 4 to Fig. 5. The extracted θ values are almost constant with L_{mask} (≈ 0.05 V^{-1}). This means that there are negligible R_{sd} access resistance effects in source and drain electrodes. Therefore, the overall electrical performances of such CNT-TFTs are actually limited by the transport in the CNTs placed underneath the gate.

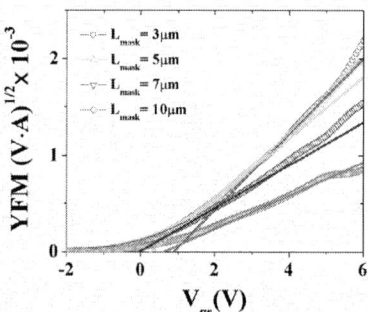

Fig. 3. The $YFM(V_{gs})$ and corresponding fitted lines

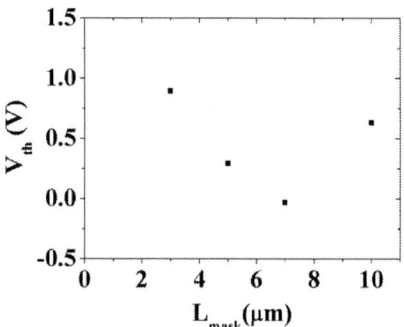

Fig. 4. Variation of V_{th} extracted from the YFM

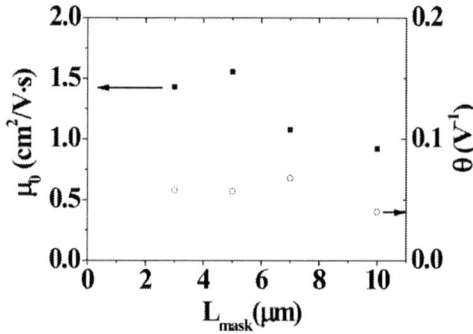

Fig. 5. The low field mobility μ_0 extracted from G_m and variation of mobility attenuation factor θ respect to L_{mask}

For the mobility calculation, we used a rigorous gate capacitance model for aligned multiple CNTs, the gate coupling capacitance being almost same that of the planar capacitance model ($C_{ox} = \varepsilon/t$) due to the high ρ_{CNT}, where ε ant t is the permittivity and the thickness of top-gate dielectric respectively [1]. Figure 5 shows the variation of μ_0 with gate mask length L_{mask}. Usually, the μ_0 is considered as an intrinsic mobility in MOS transistors and corresponds to the maximum mobility in the devices channel. Note that, by definition, it is not affected by θ and independent of V_{gs}. As can be seen from Fig. 5, μ_0 takes values in the range 1-1.7 cm^{2}V^{-1}s^{-1}, which are in the median range for such CNT-TFT devices in [1-4] and it increases slightly as L_{mask} is reduced. This dependence could be mainly attributed to the junctions between the tubes.

Hence, it could be enhanced by synthesizing longer CNTs to decrease the number of junctions in the channel.

B. Low frequency noise parameter extraction

LFN measurements have been carried out as a function of V_{gs} and L_{mask} from the subthreshold region to the strong accumulation one at low frequency ranging from $1\ Hz$ to $10\ kHz$. As seen in Fig. 6, the typical current noise power spectral density shows a $1/f$ dependence in all regimes ($V_{ds} = 50$ mV and $L_{mask} = 10$ μm). Due to the high ρ_{CNT}, the total LFN contribution could stem from ensembles of Lorentzian noise sources located in individual CNTs and/or junctions between CNTs. In fact, Appenzeller et al. have emphasized that the contact noise from the Schottky barriers in an individual CNT transistor could play a significant role [6]. However, in our case, due to the small R_{sd} value extracted from YFM and the L_{mask} dependency on channel resistance, we believe that this effect could be insignificant in our LFN measurements.

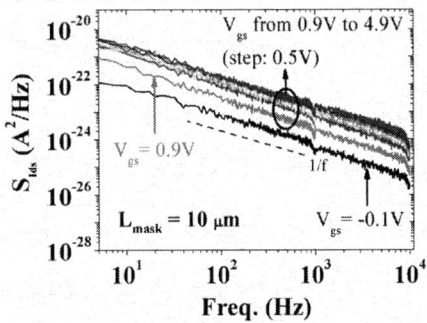

Fig. 6. The low frequency noise spectrum density (S_{Ids}) for different gate biases ($L_{mask} = 10$ μm).

The normalized drain current noise spectral density (PSD) at $f = 10\ Hz$ is displayed as a function of drain current in Fig. 7. The noise data have been well fitted with the conventional carrier number with correlated mobility fluctuations model (CNF-CMF) used in Si MOSFETs and given as [7],

$$\frac{S_{I_{ds}}}{I_{ds}^2} = \left(1 + \alpha \cdot \mu_{eff} \cdot C_{ox} \frac{I_{ds}}{g_m}\right)^2 S_{Vfb} \left(\frac{g_m}{I_{ds}}\right)^2 \quad (2),$$

where S_{Ids} is the drain current PSD (A²/Hz), μ_{eff} is the effective mobility (cm²V⁻¹s⁻¹), α is the Coulomb scattering coefficient. The flat band voltage spectral density associated with the interface charge fluctuations S_{Vfb} is given by

$$S_{Vfb} = \frac{q^2 k_B T N_{st}}{f W_{mask} L_{mask} C_{ox}^2} \quad (3),$$

where q is the electric charge, $k_B T$ is the thermal energy, N_{st} is the slow oxide trap surface density (eV⁻¹cm⁻²) and f is the frequency. The very good agreement between the CNF-CMF LF fluctuations model and the noise data clearly indicates that the LF noise sources in such CNT-TFT devices do stem from the trapping-detrapping of carriers into traps located in the

gate dielectric surrounding the semiconducting CNTs. Besides, it also shows that the Hooge mobility fluctuations (HMF) model is not adequate for the LF noise interpretation, especially at low drain current where the normalized drain current noise is flattening instead of increasing as $1/I_{ds}$ in HMF.

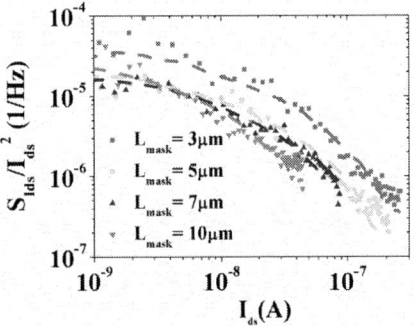

Fig. 7. Variation of the drain current normalized noise with drain current I_{ds} and corresponding fitted lines from the CNF-CMF model for various L_{mask} ($f = 10\ Hz$ at $V_{ds} = 50$ mV).

If we assume the planar capacitance value C_{ox} in Eq. (3), a slow oxide trap density $N_{st} \approx 10^{15}$ (eV⁻¹cm⁻²) would be deduced. This quantity appears unrealistic as regard to the surface atom density. This might be due to not a realistic capacitance model (more than 10 times overestimated). Even if the total CNTs placed underneath of gate dielectric are activated in gate potential, few of them and/or paths in CNT-TFTs can be participated in carrier transportation. In fact, in random network system, the main percolation paths play a significant role for overall device characterization. To get a better picture of the defect density around the 1D CNTs, we are considering that the CNTs are well aligned in the channel and that their capacitance per unit length is $C_{tube} = 2\pi\varepsilon/\ln(4t/d_{CNT})$ (F/cm). This yields a theoretical maximum number of parallel conduction channels/paths, $N_{ch} = W_{mask} C_{ox}/C_{tube}$, across the source and drain, which is for sure a strong overestimation of the real value. Then, the LF noise per each channel can be obtained by Eq. (4).

$$\frac{q^2 k_B T N_{st}}{f W_{mask} L_{mask} C_{ox}^2} = \frac{1}{N_{ch}} \frac{q^2 k_B T N_{tube}}{f L_{mask} C_{tube}^2} \quad (4),$$

where N_{tube} is the slow trap density surrounding the CNTs per unit length (eV⁻¹cm⁻¹). This relation relies on the area scaling property in LF noise analysis. Figure 8 shows the slow trap density N_{tube} derived using Eq. (4) as a function of L_{mask}. The obtained values for N_{tube} lie in the range from 0.6×10^{10} to 2.3×10^{10} eV⁻¹cm⁻¹, which is quite large compared to silicon nanowire MOSFETs. In addition, based on LF noise amplitude data from previous woks on individual CNT-FETs [3, 6, 8], we calculated that $N_{tube} \approx 0.1 - 1 \times 10^8$ eV⁻¹cm⁻¹. Our results are typically about 2 decades higher than in individual CNTs, which could be due to an overestimation of N_{ch} and to a slightly larger defect density in such densely packed CNT-TFTs with Al_2O_3 gate dielectric.

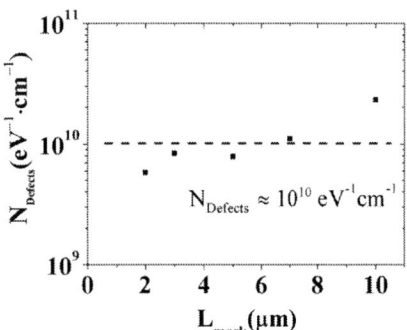

Fig. 8. Variation of the slow defect density N_{tube} (eV^{-1}cm^{-1})

For the α evaluation, the squared root of the input referred noise, $S_{Vgs}=S_{Ids}/g_m^2$, normalized by S_{Vfb}, has been used as,

$$\sqrt{\frac{S_{V_{gs}}}{S_{Vfb}}} = 1 + \alpha \cdot C_{ox} \cdot \mu_{eff} \cdot \frac{I_{ds}}{g_m} \qquad (6).$$

By this way, as displayed in Fig. 9, a straight line is obtained as a function of I_{ds}/g_m, allowing α to be extracted from the slope for each L_{mask}. As can be seen from Fig. 10, an average value of α lies in the range ~2×10^6 V.s/C. This value of α, extracted for the first time in CNTs, is much larger than in silicon based MOSFET devices ($\approx 10^4$ V.s/C), which might reflect a stronger Coulomb interaction in such almost 1D nano-structures.

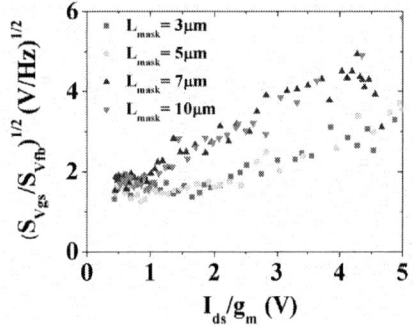

Fig. 9. Normalized input gate voltage noise $\sqrt{S_{V_{gs}}/S_{Vfb}}$ as a function of I_{ds}/g_m.

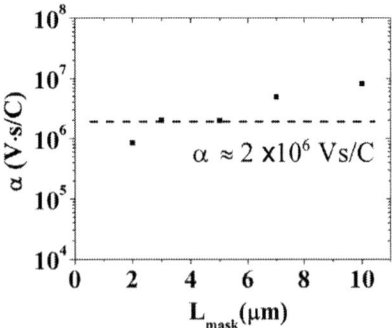

Fig. 10. Deviations of the Coulomb scattering coefficients

IV. CONCLUSION

A static and LFN characterization of CNT-TFTs has been carried out. The YFM has been employed for the electrical parameter extraction in CNT-TFTs to eliminate contact resistance effects. The various electrical parameters such as μ_0, V_{th}, θ, on/off current ratio and subthreshold swing have been evaluated and extracted versus gate mask length. The LF noise has been clearly attributed for the first time to carrier number with correlated mobility fluctuations. The trap linear density surrounding the CNTs in such densely packed TFTs deduced from LF noise is in the range of a few ~10^9 [eV^{-1}cm^{-1}], which is slightly higher than in individual CNT-FETs with top gate configuration.

ACKNOWLEDGMENT

This research was financially supported by the National Research Foundation of Korea (NRF) funded by the Ministry of Education, Science and Technology (2011K000623, R32-2011-000-10082-0(WCU), 2011-0031638). It has received partial support from the European Commission FP7/ICT Program, within the framework of the NanoFunction Network of Excellence, under grant agreement n° 257573.

REFERENCES

1. Kang, S.J., Kocabas, C., Ozel, T., Shim, M., Pimparkar, N., Alam, M.A., Rotkin, S.V., and Rogers, J.A.: 'High-performance electronics using dense, perfectly aligned arrays of single-walled carbon nanotubes', Nat Nano, 2007, 2, (4), pp. 230-236
2. Kim, U.J., Lee, E.H., Kim, J.M., Min, Y.-S., Kim, E., and Park, W.: 'Thin film transistors using preferentially grown semiconducting single-walled carbon nanotube networks by water-assisted plasma-enhanced chemical vapor deposition', Nanotechnology, 2009, 20, (29), pp. 295201
3. Kim, S., Kim, S., Janes, D.B., Mohammadi, S., Back, J., and Shim, M.: 'DC modeling and the source of flicker noise in passivated carbon nanotube transistors', Nanotechnology, 2010, 21, (38), pp. 385203
4. Zhiying, L., Zhi-Jun, Q., Zhi-Bin, Z., Li-Rong, Z., and Shi-Li, Z.: 'Mobility Extraction for Nanotube TFTs', Electron Device Letters, IEEE, 2011, 32, (7), pp. 913-915
5. Ghibaudo, G.: 'New method for the extraction of MOSFET parameters', Electronics Letters, 1988, 24, (9), pp. 543-545
6. Appenzeller, J., Yu-Ming, L., Knoch, J., Zhihong, C., and Avouris, P.: '1/f Noise in Carbon Nanotube Devices—On the Impact of Contacts and Device Geometry', Nanotechnology, IEEE Transactions on, 2007, 6, (3), pp. 368-373
7. Ghibaudo, G., Roux, O., Nguyen-Duc, C., Balestra, F., and Brini, J.: 'Improved Analysis of Low Frequency Noise in Field-Effect MOS Transistors', physica status solidi (a), 1991, 124, (2), pp. 571-581
8. Xu, G., Liu, F., Han, S., Ryu, K., Badmaev, A., Lei, B., Zhou, C., and Wang, K.L.: 'Low-frequency noise in top-gated ambipolar carbon nanotube field effect transistors', Applied Physics Letters, 2008, 92, (22), pp. 223114

Mechanically flexible double gate a-IGZO TFTs

Niko Münzenrieder, Christoph Zysset, Thomas Kinkeldei, Luisa Petti, Giovanni A. Salvatore and Gerhard Tröster

Institute for Electronics
Swiss Federal Institute of Technology Zurich
Zurich, 8092, Switzerland
muenzenrieder@ife.ee.ethz.ch

Abstract—In this paper, the concept of double gate transistors is applied to mechanically flexible amorphous Indium-Gallium-Zinc-Oxide (a-IGZO) thin film transistors (TFTs) fabricated on free standing plastic foils. Due to the temperature sensitivity of the plastic substrate, a-IGZO is a suitable semiconductor since it provides carrier mobilities of $\approx$ 10 cm^2/Vs when deposited at room temperature. Double gate TFTs with connected bottom and top gate are compared to bottom gate reference TFTs fabricated on the same substrate. Double gate a-IGZO TFTs exhibit a by 74% increased gate capacitance, a by 0,7 V higher threshold voltage, and therefore an up to 51% increased transconductance. The subthreshold swing and the on/off current ratios are improved as well, and reach excellent values of 69 mV/dec and $2x10^9$, respectively. The mechanical flexibility is demonstrated by showing device operation while the TFT is exposed to tensile strain of 0.55%, induced by bending to a radius of 5 mm.

I. INTRODUCTION

Electronic devices, especially thin-film transistors (TFTs) fabricated directly on flexible plastic substrates are a key requirement for a number of new large-area applications such as rollable displays, electronic skins or woven electronics for smart textiles [1]. While organic [2] and a-Si [3] based TFTs fabricated on flexible and temperature sensitive substrates in general suffer from low mobilities around 1 cm^2/Vs, a-IGZO TFTs are nearly unaffected by the choice of the substrate (rigid or flexible), and offer mobilities above 10 cm^2/Vs even when deposited at room temperature [4]. Besides the mobility, there are other important parameters to take into account for device performance, such as threshold voltage V_{th}, on-off current ratio I_{on}/I_{off}, and subthreshold swing SS. TFT performance parameters aim at a large I_{on}/I_{off}, a small SS, and a V_{th} which allows TFT operation between 0 V and 5 V (essential for the fabrication of digital circuits). In the past, several groups used different double gate structures [5], [6] to increase the coupling between the gate and the channel, and thereby improve the performance of a-IGZO TFTs, fabricated on rigid glass or Si substrates. Additionally, other approaches to increase the gate capacitance are the use of high k materials [7], or thinner gate oxide layers.

In this work the double gate concept was combined with a 10 nm thin gate oxide and applied to a-IGZO TFTs fabricated

This work has been scientifically evaluated by the Swiss National Scientific Foundation (SNSF), and financed by the Swiss Confederation and Nano-Tera.ch.

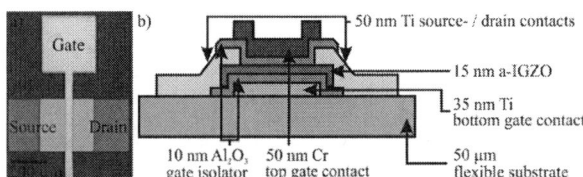

Figure 1. a) Micrograph of a fully processed flexible a-IGZO double gate TFT (W/L = 280 μm/ 10 μm), b) Double gate a-IGZO TFT schematic.

on flexible plastic foils. The bottom and the top gate were electrically connected to form an a-IGZO TFT controllable with a single gate voltage. The resulting n-type a-IGZO TFTs showed improved performance, and remained fully operational while subjected to mechanically induced strain of 0.55%. A-IGZO double gate TFTs while flat, as well as under mechanical strain, exhibit subthreshold slopes of 69 mV/dec and on-off current ratios > 10^9. In addition, the transconductance g_m of double gate TFTs was increased by 51% when compared to single bottom gate reference TFTs, fabricated on the same substrate.

II. FABRICATION

A micrograph of a fully processed flexible a-IGZO double gate TFT and the corresponding device cross section are shown in Figure 1. To ensure the successful fabrication of the presented a-IGZO double gate TFTs on flexible substrates, the following points had to be considered during the design:

- Atomic layer deposition (ALD) enables the deposition of thin (10 nm) and pinhole free Al_2O_3 isolation layers with good sidewall coverage.
- Evaporated Ti and Cr provide a sufficient adhesion on polyimide and a-IGZO, suitable for the fabrication of flexible TFTs. Additionally the work functions of these two materials are comparable (Φ_{Ti} = 4.33 eV, Φ_{Cr} = 4.44 eV) [8]. Therefore the work functions influence on the threshold voltage can be neglected if Ti and Cr are substituted as gate material.
- Ti, in contrast to other metals e.g. Au, Cu or Al, has a high resistivity against all wet etchants used to structure Al_2O_3, a-IGZO, and Cr. Hence, Ti is a suitable material for the bottom gate, which can be structured in a lift-off process, but should not be damaged by the chemicals used during further processing.

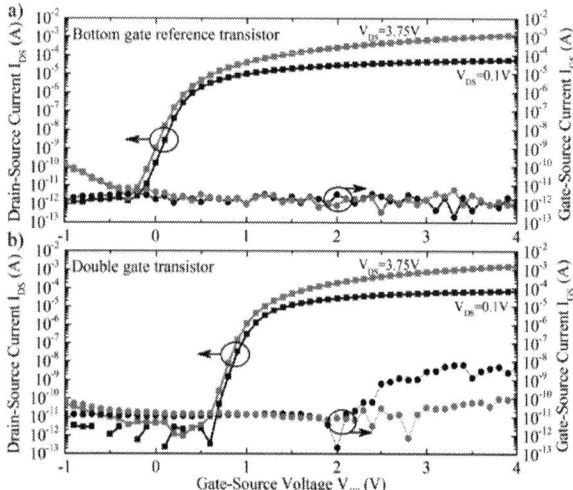

Figure 2. Typical a-IGZO TFT (W/L = 280 μm/10 μm) transfer characteristics measured in the linear and the saturation regime; a) bottom gate reference transistor, and b) double gate transitor manufactured on the same substrate.

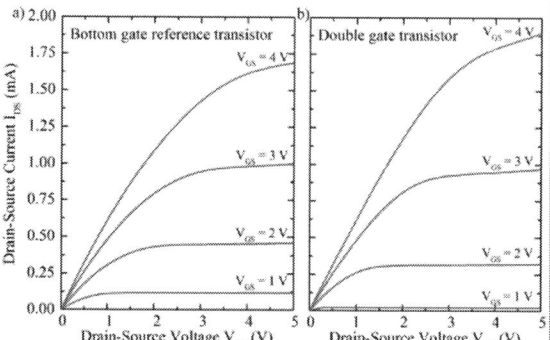

Figure 3. Typical a-IGZO TFT (W/L = 280 μm/10 μm) output characteristics; a) bottom gate reference transistor, and b) double gate transitor manufactured on the same substrate.

- Since wet etching is applicable to Cr, it is an appropriate material for the top gate contact.

To determine the influence of the double gate structure, flexible a-IGZO double gate TFTs, as well as standard single bottom gate reference devices were fabricated on the same free standing 50 μm-thick Kapton®E polyimide foil from DuPont. The total substrate size was 7.6 cm x 7.6 cm.

The manufacturing process for flexible a-IGZO TFTs on plastic substrates was as follows:

A. Single gate reference TFTs

Prior to fabrication, the substrate was cleaned by sonication in acetone and isopropanol for 5 minutes each, and was then pre-shrunk in a vacuum oven at 200 °C for 24 h. To increase the adhesion of the successive material layers, the top surface was treated with ozone for 60 minutes, using a UVOCS ultra violet ozone cleaning system. Next, negative MAN1420 photoresist and a Plassys MEB550SL e-beam evaporation system were used to deposit 35 nm Ti, and structure gate contacts in a lift-off process (photolithography mask 1). Resist leftovers were removed by an additional 60 min ozone treatment. A Picosun Sunale R-150B was used to deposit 10 nm Al_2O_3 as gate isolator by atomic layer deposition at 150 °C. Following, we deposited the 15 nm thick a-IGZO semiconducting layer using room temperature RF magnetron sputtering in a pure Ar atmosphere and a ceramic $InGaZnO_4$ target. The semiconductor was patterned by standard photolithography (mask 2) and diluted hydrochloric acid [9] ($HCl : H_2O = 1 : 120$). The Al_2O_3 gate isolator was structured into islands 20 μm wider than the semiconductor islands by photolithography mask 3 and AL-11 aluminium etchant from Cyantek heated to 50 °C [10]. We deposited and structured (mask 4) source and drain contacts (50 nm Ti) similar to the gate with another e-beam evaporation and lift-off step. A

second layer of Al_2O_3 was deposited and structured identical to the gate isolation layer. This concludes the fabrication process of the standard bottom gate TFTs, which served as reference for the fabricated double gate TFTs. In this case the second Al_2O_3 layer worked as device passivation [11].

B. Double gate TFTs

Double gate TFTs with an additional top gate connected with the bottom gate were fabricated on the same substrate. Therefore, 50% of the completed bottom gate TFTs were covered with 50 nm thick evaporated Cr. The Cr was then structured by standard photolithography and wet etching using again mask 1. The top Al_2O_3 layer served as second gate oxide in this case. Thereby the structuring of the Al_2O_3 into small islands ensured the electrical contact of the bottom and top gate on both sides of the channel region, whereas the 50 nm thick Cr is thick enough to establish a contact across the sidewalls of all previously structured layers (2x 10 nm Al_2O_3 + 15 nm a-IGZO).

III. RESULTS AND DISCUSSION

TFTs were characterized under ambient conditions using an Agilent technologies B1500A parameter analyzer with current-voltage, and capacitance-voltage measurement capabilities. Performance parameters were extrapolated from the transfer characteristics measured in the saturation regime using standard MOSFET equations to model the transistor current [8].

A. TFT characteristics

Figures 2a and 3a show transfer and corresponding output characteristics of a reference bottom gate a-IGZO TFT, and Figures 2b and 3b the equivalent measurements of an a-IGZO double gate TFT. Figure 4 compares the total measured gate capacitance of a bottom gate reference TFT and a double gate TFT. The W/L ratio is 280 μm/ 10 μm.

Compared to the reference single bottom gate TFT, the maximum drain-source current I_{DS} inside the double gate transistor is increased from 1.2 mA to 1.6 mA (V_{GS} =4 V). This increase is related to the stronger coupling between the gate and the channel. The stronger coupling is represented by the increased absolute gate capacitance of the double gate TFT (Figure 4). Due to the additional top gate, the area which

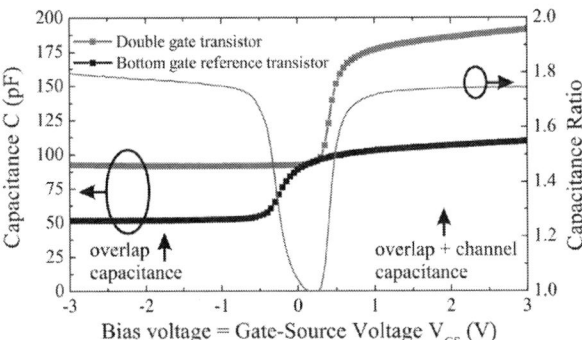

Figure 4. Measured gate capacitance and ratio between double gate TFT, and corresponding bottom gate reference TFT capacities (drain and souce contacts are grounded). The W/L ratio is 280 μm/10 μm

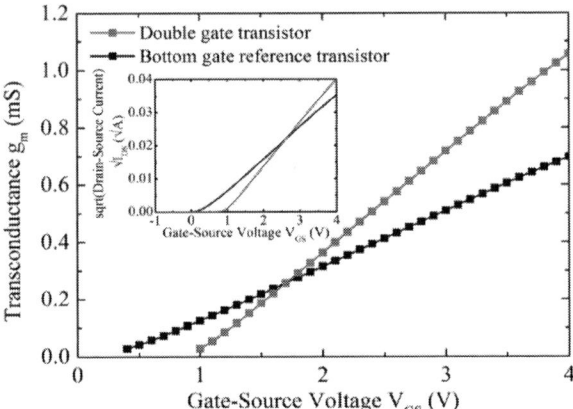

Figure 5. Calculated transconductance of a a-IGZO doube gate TFT, and coresponding bottom gate reference TFT for different values of V_{GS} at V_{DS} = 3.75 V. The W/L ratio is 280 μm/10 μm. The inset shows the square root of I_{DS} used for the calculation.

defines the absolute capacitance is increased. The increased gate area of the double gate TFT results in a gate capacitance of 191 pF in the on regime (V_{GS} =3 V). In contrast, the gate capacitance of the corresponding bottom gate reference TFT is 109 pF under equal measurement conditions. This corresponds to an increase of 74%.

The effective field effect mobility is 8.3 cm²/Vs for the bottom gate reference TFT and 8.5 cm²/Vs for the double gate TFT (calculated using a by 74% increases capacitance). The small increase of 2.5% is within the process variations of the presented flexible a-IGZO TFT.

The transconductance $g_m = \partial I_{DS}/\partial V_{GS}$ for the reference single bottom gate TFT and the double gate TFT is shown in Figure 5. Due to the higher gate capacitance of the double gate TFTs, g_m exhibits a steeper slope. At the same time, the increased threshold voltage of the double gate TFTs shifts the onset of the "on" region to higher V_{GS} values. Therefore the two g_m curves intercept at V_{GS} = 1.8 V. The absolute value of g_m at V_{GS} = 4 V is increased from 0.7 mS to 1.06 mS due to the additional top gate.

The threshold voltage of the double gate transistor is shifted by + 0.7 V, when compared to the reference single

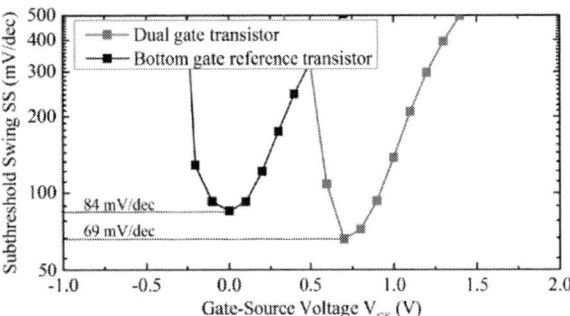

Figure 6. Subthreshold swing $(\partial \log I_{DS}/\partial V_{GS})^{-1}$ of a bottom gate reference TFT and a dual gate TFT for different gate voltages.

TABLE I. PERFORMANCE PARAMETERS OF A-IGZO DOUBLE GATE TFTS AND BOTTOM GATE REFERENCE TFTS.

Parameter	Transistor		Relative change
	Bottom gate reference	Double gate	
Gate capacitance (V_{GS} =3 V)	109 pF	191 pF	+ 74%
Effective field effect mobility	8.3 cm²/Vs	8.5 cm²/Vs	+ 2.5%
Transconductance (V_{GS} =4 V)	0.7 mS	1.06 mS	+ 51%
Threshold voltage	250 mV	950 mV	+ 700 mV
On-off ratio	9 x 10⁸	2 x 10⁹	x 2.2
Subthreshold swing	84 mV/dec	69 mV/dec	- 18%

bottom gate TFT. This shift is caused by the changed geometry [12] and in good agreement with previously published double gate TFTs [13]. The higher V_{th} value of 0.95 V ensures that the double gate TFT is totally turned off at V_{GS} = 0 V. This results in a by more than one order of magnitude decreased off-current.

The on-off current ratio is mainly improved due to the increased gate capacitance and therefore maximum drain-source current. The double gate TFT reaches a value of 2 x 10⁹

In Figure 6, the subthreshold swings (inverse of subthreshold slope), for the double gate a-IGZO TFT and the reference single bottom gate TFT, are calculated using the measurements shown in Figure 2. In [14] bottom gate a-IGZO TFTs with a similar geometry but 25 nm thick Al₂O₃ gate oxide exhibited a SS of 180 mV/dec. Due to the reduction of the gate oxide thickness [8] to 10 nm, the reference bottom gate TFT in this paper showed a decreased SS of 84 mV/dec. Furthermore, the additional top gate of the double gate a-IGZO TFT further improves the control of the channel potential [12] and reduces SS to 69 mV/dec.

The performance parameters of double gate TFTs and corresponding reference bottom gate TFTs are summarized in Table 1.

However, due to the thin gate oxide and the added top gate additional effects were observed. First, tunneling of carriers through the 10 nm thin gate oxide (single and double gate TFT) increases the gate leakage current I_{GS} in the "off" state

978-1-4673-1707-8/12 $31.00 © 2012 IEEE

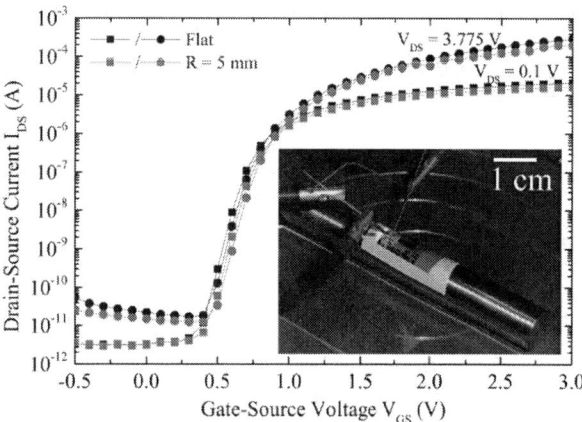

Figure 7. Transfer characteristic of the same a-IGZO double gate TFT (W/L = 280 μm/35 μm) measured while flat, and bent to a tensile radius of 5 mm. The inset shows the bent substrate during the measuremnt.

(Figure 2). The tunnel current increases with increasing V_{DS}, but does not impair devices operated with a supply voltage of 5 V. Second, the larger interface area of the double gate TFT induces a by at least one order of magnitude higher gate leakage current I_{GS} (Figure 2b).

B. Bendability

To investigate the flexibility of the fabricated a-IGZO double gate TFTs, bending tests were performed as follows: TFTs were attached to double sided tape and wound around a rod, in the way that tensile strain was applied parallel to the TFT channel. The radius of the employed rod was 5 mm, which corresponds to a tensile mechanical strain of 0.55%, calculated using the strain theory developed in [15]. The bent transistors were contacted with probe needles as usual. Figure 7 shows the transfer characteristic of an a-IGZO double gate TFT before bending, and while bent to a radius of 5 mm, as well as a photograph of the bent and contacted TFT.

The measurement demonstrated that the transistor remained fully operational when bent to a radius of 5 mm. The applied tensile strain induced a positive threshold voltage shift of 25 mV and a reduction of the effective field effect mobility by 7%, while the subthreshold swing stayed constant within the measurement inaccuracies. Compared to previous bending experiments with bottom gate a-IGZO TFTs [14], the observed shifts correspond to compressive strain in the TFT channel. This indicates different mechanical properties caused by the changed geometry of the double gate devices. Bending to even smaller radii is not possible because of the formation of cracks starting in the brittle Cr top gate contact. We believe that the use of more ductile metals like Cu would enable bending radii between 1 mm and 2 mm without the need of modifying the device structure.

IV. CONCLUSION

A double gate structure was combined with 10 nm thick Al_2O_3 gate oxide layers to fabricate TFTs on free standing flexible plastic foils. Double gate a-IGZO TFTs yield improved performance parameters compared to single bottom reference TFTs fabricated on the same substrate. The by 74% increased absolute gate capacitance increased the transconductance up to 51% (V_{GS} = 4 V). On-off current ratio increased by more than a factor of 2, while the subthreshold swing reached a value of 69 mV/dec. This is to our knowledge the smallest value ever reported on flexible a-IGZO TFTs.

Tensile mechanical strain of 0.55%, induced by bending the flexible a-IGZO double gate TFTs to a radius of 5 mm did not impair the device functionality significantly. In particular, the subthreshold swing remained unchanged while μ_{FE} decreased by 7% and V_{th} increased by 25 mV.

REFERENCES

[1] K. Cherenack, C. Zysset, T. Kinkeldei, N. Münzenrieder, and G. Tröster "Woven Electronic Fibers with Sensing and Display Functions for Smart Textiles", Adv. Mater vol 22, issue 45, pp. 5178-5182, Dec. 2010

[2] W. Li, H.E. Katz, A.J. Lovinger, and J.G. Laquindanum, "Field-Effect Transistors Based on Thiophene Hexamer Analogues with Diminished Electron Donor Strength," Chemistry of Materials, vol. 11, no. 2, pp. 458-465, Jan. 1999.

[3] K. H. Cherenack, A. Z. Kattamis, B. Hekmatshoar, J. C. Sturm, and S. Wagner, "Amorphous silicon thin-film transistors fabricated at 300ºC on a clear plastic substrate foil," IEEE El. Device Letters, vol. 28, no. 11, pp. 1004-1006, Oct. 2007.

[4] K. Nomura, et al., "Room-temperature fabrication of transparent flexible thin-film transistors using amorphous oxide semiconductors," Nature, vol. 432, pp. 488-492, Nov. 2004.

[5] H. Lim et al.," Double gate GaInZnO thin film transistors," Appl. Phys. Lett., vol. 93, no. 6, pp. 063505--063505, Aug. 2008.

[6] H.W. Zan et al.," Dual gate indium-gallium-zinc-oxide thin film transistor with an unisolated floating metal gate for threshold voltage modulation and mobility enhancement," Appl. Phys. Lett., vol. 98, pp. 153506- 153506-3, April 2011.

[7] . N.C. Su, S.J. Wang, and A. Chin, "High-Performance InGaZnO Thin-Film Transistors Using HfLaO Gate Dielectric," IEEE El. Device Letters, vol. 30, no. 12, pp. 1317-1319, Dec. 2009.

[8] S. M. Sze, and K. K. Ng "Physics of Semiconductor Devices," 3rd ed., John Wiley & Sons, Hoboken, 2007.

[9] J. B. Kim, C. Fuentes-Hernandez, W. J. Potscavage, Jr., X.-H. Zhang, and B. Kippelen," Low-voltage InGaZnO thin-film transistors with Al2O3 gate insulator grown by atomic layer deposition," Appl. Phys. Lett., vol. 94, no. 14, pp. 142107 - 142107-3, April 2009.

[10] B. Zhou, and W. F. Ramirez, "Kinetics and Modeling of wet etching of Aluminum Oxide by Warm Phosphoric Acid," J. Electrochem.Soc., vol. 143, no. 2, pp. 619-623, February 1996.

[11] I.-T. Cho, J.-M. Lee, J.-H. Lee and H.-I. Kwon, "Charge trapping and detrapping characteristics in amorphous InGaZnO TFTs under static and dynamic stresses", Semicond. Sci. Technol. 24, January 2009

[12] S. Zhang, R. Han, J.K.O. Sin, and M. Chan, "A novel self-aligned double-gate TFT technology," IEEE El. Device Letters, vol. 22, no. 11, pp. 530-532, Nov. 2001.

[13] K.S. Son et al., "Characteristics of double-gate Ga--In--Zn--O thin-film transistor," IEEE El. Device Letters, vol. 31, no. 3, pp. 219-22119, Mar. 2010.

[14] N. Münzenrieder, K. Cherenack, and G. Tröster "The effects of mechanical bending and illumination on the performance of flexible IGZO TFTs," IEEE Trans. El.. Devices, vol 58, issue 7, pp. 2041-2048, Jul. 2011

[15] H. Gleskova, S. Wagner, and Z. Suo,'a-Si:H thin film transistors after very high strain', JNCS, vol. 266-269, pp 1320-1324, May 2000.

Top-Down Fabricated ZnO Nanowire Transistors for Application in Biosensors

S.M. Sultan, K.Sun, M. R. R. de Planque, P. Ashburn, H.M.H. Chong

Electronics and Computer Science
University of Southampton
Southampton, SO17 1BJ United Kingdom
Email: sms08r@ecs.soton.ac.uk

Abstract— **Top-down ZnO nanowire FETs have been fabricated using mature photolithography, ZnO atomic layer deposition (ALD) and plasma etching. This paper investigates the effects of oxygen adsorption by measuring FET characteristics at different gate bias sweep rates and by characterizing hysteresis effects. Unpassivated devices exhibit a low threshold voltage shift of 5.4 V when the gate bias sweep rate is varied from 2500 V/s to 1.2 V/s and a low hysteresis width of less than 1.5 V. These results are considerably better than the state of the art for bottom-up as-fabricated ZnO nanowire FETs and demonstrate the suitability of this top-down technology for biosensor applications.**

I. INTRODUCTION

Over the last decade, nanowire devices have emerged as important candidates for biosensors [1]-[2]. There are many reasons why silicon nanowires are of interest, including high surface-to-volume ratio, high sensitivity, and real-time, label-free detection without expensive optical components. Most of this research has concentrated on silicon nanowires because of their compatibility with CMOS technology.

ZnO is also an attractive material for biosensor applications because the low growth temperature [3] is compatible with biosensor fabrication on cheap polymer substrates. Fabrication of ZnO nanowire FETs by bottom-up self-assembly has received considerable attention [4] and transistors with remarkable values of mobility have been reported. However, the transistors show undesirable hysteresis behaviour and the nanowires need to be covered with a passivation layer (typically a polymer) to improve the characteristics [5]. For example, Maeng *et al* [6] measured 11.6 V of threshold voltage shift on unpassivated bottom-up ZnO nanowire FETs, but this reduced to 3.3 V on PMMA passivated devices. Similarly, Hong *et al* showed a hysteresis width of 35 V for their unpassivated bottom-up nanowire devices [5]. Unfortunately, the use of a passivation layer is not compatible with biosensor applications, since the nanowire surface needs to be functionalized for target molecule capture. So far, there have been no reports on the

hysteresis behaviour of top-down fabricated ZnO nanowire FETs.

Recently the authors have developed a top-down approach to the fabrication of ZnO nanowire transistors and the resulting devices showed well behaved electrical characteristics with excellent values of breakdown voltage [7]. In this paper, we study the effects of hysteresis and gate bias sweep rate on top-down fabricated ZnO nanowire FETs. It is shown that a low threshold voltage shift of 5.4 V can be achieved for gate bias sweep rates in the range 2500 to 1.2 V/s, together with a low hysteresis width of <1.5 V without the use of a passivating layer. These results are significantly better than equivalent results for bottom-up fabricated ZnO nanowire FETs [5]-[6] and demonstrate that this top-down technology is well suited for biosensor applications.

I. EXPERIMENT

Fig. 1 shows a schematic of the top-down fabrication process. A p-type Si substrate was used as the transistor back-gate. A 200 nm SiO_2 layer was thermally grown and anisotropically reactive ion etched to form 100 nm pillars. A 34 nm (measured by ellipsometer) layer of ZnO was then deposited at $100^{\circ}C$ using remote plasma atomic layer deposition (ALD) in an Oxford Instruments Plasma Technology (OIPT) FlexAl system using diethyl zinc (DEZ)

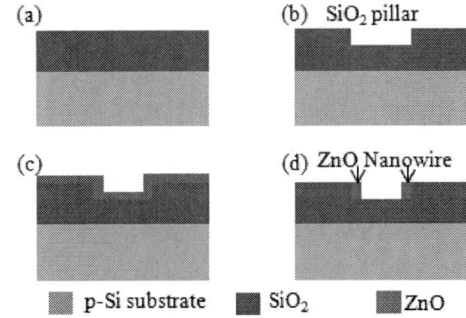

Fig. 1. Top-down fabrication process of ZnO nanowire FET (a) a 200 nm SiO_2 thermally grown through wet oxidation (b) SiO_2 dry etched to form ~100 nm pillars (c) ZnO thin film deposited by ALD (d) anisotropic ICP etch to obtain nanowires at the side of the oxide pillars.

This research is supported by the Malaysian Ministry of Higher Education through a doctoral scholarship.

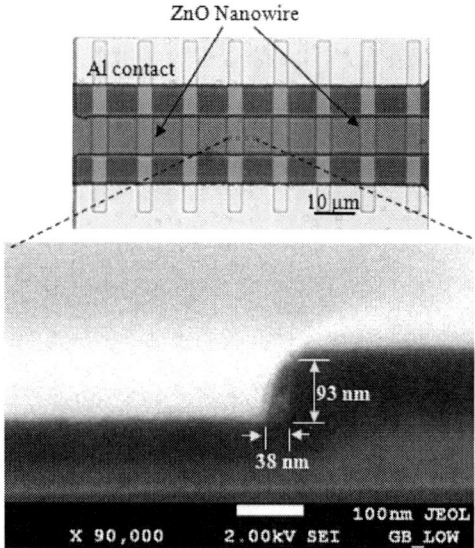

ZnO Nanowire

Al contact

10 μm

93 nm

38 nm

X 90,000   2.00kV SEI   GB_LOW   100nm JEOL

Fig. 2.   (top) Optical image of the ZnO nanowire array after metallization and (bottom) SEM cross-section of ZnO nanowire formed at the side of a SiO_2 pillar.

as the precursor, an RF power of 100 W, a pressure of 15 mTorr and an O_2 flow of 60 sccm. ZnO nanowires were fabricated using an anisotropic Inductively Coupled Plasma (ICP) CHF_3 etch to form nanowires at the sides of the SiO_2 pillars as shown in Fig. 1d. Aluminum source and drain electrodes were deposited by e-beam evaporation and patterned by lift-off. The Al contacts were then annealed in a Rapid Thermal Annealer at $350^{\circ}C$ for 2 mins.

Fig. 2 shows an optical image of the device taken after the metallization process. The light area is the Al contact pads, the red area is the ZnO source/drain and the nanowires are formed at each side of the green oxide pillars. An SEM cross-section was taken across the red dotted line and the nanowire dimensions were measured as shown in Fig. 2. Note that all measured results were tilt corrected as the sample was tilted during the measurement. The nanowire height and width were measured at 93 nm and 38 nm, respectively. The nanowire height is determined by the height of the SiO_2 pillar and the amount of over-etch of the ZnO layer (9% in this case). The nanowire width is 38 nm at the base, which is the same as the thickness of ZnO film after deposition.

The current-voltage characteristics of the devices were measured with an Agilent Technologies B1500 Semiconductor parameter analyzer. All electrical measurements were done in the dark at room temperature.

II.   RESULTS

Fig. 3a shows the transfer characteristics plotted on a semi-logarithmic scale measured at different back gate bias sweep rates (2500, 6 and 1.2 V/s) at a fixed drain voltage (V_D) of 1 V. The ZnO NWFET operates in n-type enhancement mode with a large modulation of channel conductance of 5 orders of magnitude.

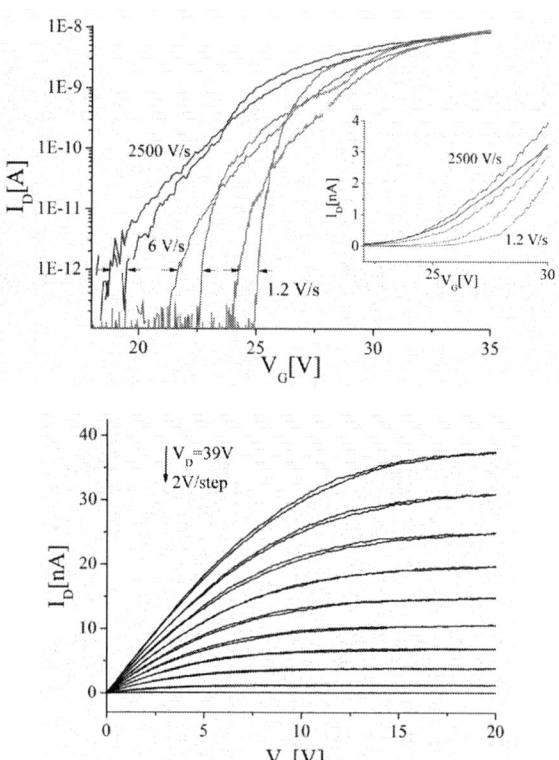

Fig. 3.   (a) Measured I_D-V_G curves for a ZnO dual nanowire FET with L= 10 μm for gate bias sweep rates of 2500, 6 and 1.2 V/s, measured at V_D=1V in ambient air. (inset) Linear I_D-V_G plots for gate bias sweep rates of 2500, 250, 100, 6 and 1.2 V/s. (b) I_D-V_D output characteristic of ZnO dual nanowire FET with L=10μm. The sweep rate for this measurement was 2 V/s.

Hysteresis is important for biosensor applications and the hysteresis width was defined as the difference in gate voltage during forward and reverse sweeps at a drain current, I_D of 1 pA. Hysteresis widths were found to be at 0.70 V, 1 V and 0.88 V at gate voltage (V_G) sweep rates of 2500, 6 and 1.2 V/s, respectively. When the gate voltage sweep rate decreased, the transfer curves were shifted in a positive direction as clearly shown in Fig. 3a. Single sweep I_D-V_G characteristics plotted on a linear scale are shown in the inset of Fig. 3a. As the gate bias sweep rate was decreased from 2500 to 1.2 V/s, the threshold voltage shifted in the positive gate bias direction from 24.5 V to 27.5 V. The threshold voltage was obtained by linear extrapolation of the I_D-V_G plot. This effect (V_T shift) has also been observed in ZnO nanowire FETs fabricated by bottom-up self assembly [6].

The field-effect mobility is calculated from the transconductance (g_m) using $\mu_e = g_m L^2/CV_D$, where L is the nanowire channel length and C is the capacitance between the nanowire and the back-gate, calculated following the method in [8]. Using the measured 93 nm x 38 nm nanowire dimensions (Fig. 2) and the 100 nm SiO_2 thickness gives C=4.14 x 10^{-16} F for a 10 μm long nanowire. The field effect

978-1-4673-1707-8/12 $31.00 © 2012 IEEE   138

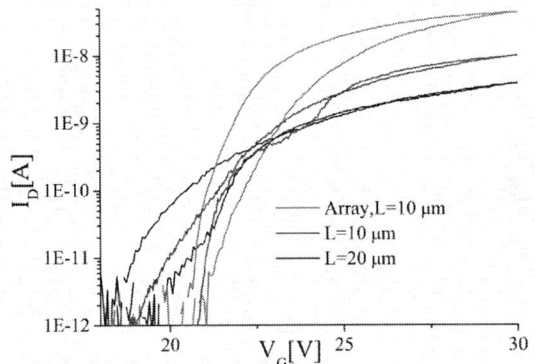

Fig.4 Transfer characteristics of three different ZnO nanowire FETs measured with a gate bias sweep rate of 5 V/s. The array device is comprised of 16 parallel nanowires.

mobility μ_e is found to vary with gate bias sweep rate, increasing from 1.05 cm^2/Vs to 5 cm^2/Vs as the gate bias sweep rate decreases from 2500 to 1.2 V/s at $V_D = 1$ V.

Fig. 3b displays the output characteristic of a 10 μm ZnO dual nanowire FET and shows clear pinch-off and saturation. The sweep rate for this measurement was 2 V/s. For nanowire biosensors, the doping concentration in the nanowire determines the biosensor sensitivity [2] and hence it is important to characterize this parameter. The doping concentration can be calculated from the measured resistivity (ρ) of the nanowires, which was found to be 57 Ω.cm at V_G-$V_T = 2$V and $V_D = 1$V. The electron carrier concentration is estimated from the relation $n_e = 1/\rho q\mu_e$ for different V_G values. The carrier concentration decreases from 1.0×10^{17} to 2.2×10^{16} cm^{-3} as the gate bias sweep rate decreases from 2500 to 1.2 V/s at V_G - $V_T = 2$ V and $V_D = 1$ V. This value is about the same as the typical carrier concentration of as-grown bottom-up ZnO nanowires [9]. A carrier concentration in this range is likely to deliver a high sensitivity in a ZnO nanowire biosensor.

To test the reproducibility of the hysteresis, Fig. 4 shows the transfer characteristics of 3 different devices: dual nanowire FETs with channel lengths of 10 μm and 20 μm and a 10 μm nanowire FET comprising 16 parallel nanowires. Hysteresis widths of 1.1 V, 1.5 and 0.6 V were obtained, respectively. The 20 μm nanowire FET shows the largest hysteresis width and the nanowire FET comprising 16 parallel nanowires shows the smallest hysteresis width. The better hysteresis width on the latter device may be because the nanowires with the smallest hysteresis tend to dominate the conduction.

III. DISCUSSION

Hysteresis in ZnO nanowire FETs is due to the depletion of electrons from the nanowires during the measurements, which has a strong dependence on the adsorbed oxygen in the atmosphere as well as ZnO nanowire/insulator interface quality. The positive shift in threshold voltage as we decrease the V_G sweep rate from 2500 to 1.2 V/s is caused by the depletion of carriers by the adsorption of oxygen during the

measurements. The positive gate bias induces more adsorption of oxygen ions, so when a slower gate bias sweep rate is applied, more electrons are trapped on the ZnO nanowire surface by the adsorbed oxygen ions [6]. Consequently, more voltage is required to turn on the device (high threshold voltage, V_T) and a larger hysteresis observed.

Slower gate bias sweep rates will strongly bond the captured oxygen ions which are adsorbed on the ZnO surface. Thus, slow gate bias sweep rates will cause more oxygen adsorption than fast gate bias sweep rate [6]. As a result, the nanowire conduction channel is pinched due to surface-depletion-induced channel narrowing. This pinching moves the channel away from the nanowire surface and leads to a transconductance increase due to a reduction in surface scattering. This effect explains the five-fold increase of carrier mobility obtained at slow gate bias sweep rates.

To better assess the performance of our ZnO nanowire FETs, Table I shows a comparison of values of threshold voltage shift obtained for our top-down ZnO nanowire devices with those of bottom-up fabricated devices reported by Maeng et al [6]. The results are presented for the same measurement conditions in which the gate bias sweep rate decreased from 2500 V/s to 1.2 V/s. In Maeng *et al*'s work [6], the ZnO nanowire FETs were measured under three different conditions; ambient air, nitrogen and air after PMMA passivation. For Maeng et al's measurements in air, the threshold voltage shift increases positively by 11.6 V as the gate bias sweep voltage rate decreases from 2500 V/s to 1.2 V/s. However, after PMMA passivation, the threshold voltage shift decreases to 3.3 V. In a nitrogen environment, the nanowire device was not influenced by the gate bias sweep rate, indicating that no oxygen was being absorbed on the nanowire surface. In our top-down ZnO device, the threshold voltage shift measured in air is only 5.4 V. This value is nearly as good as the passivated value of Maeng et al [6], but was obtained without the use of any passivation.

Fig. 5 displays a comparison of the curves in subthreshold region from [6] (black curves) and our work (red curves) at the same gate bias sweep rates of 2500, 250, 100, 6 and 1.2 V/s and for measurements in ambient air. The arrows indicate the range of the threshold voltage shift measured from both sets of data. The threshold voltage shift from our results is 53% lower than that reported by Maeng et al [6].

TABLE I.

THRESHOLD VOLTAGE SHIFT COMPARISON WITH BOTTOM-UP ZnO NANOWIRE DEVICES [6]

Reference	*Fabrication Method*	*Measurement Conditions*	V_T *shift [V]*
[6]	Bottom-up	air	11.6
		nitrogen	~0
		Air after passivation	3.3
This work	Top-Down	air	5.4

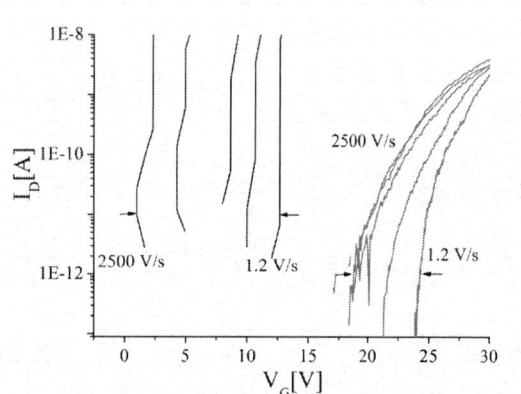

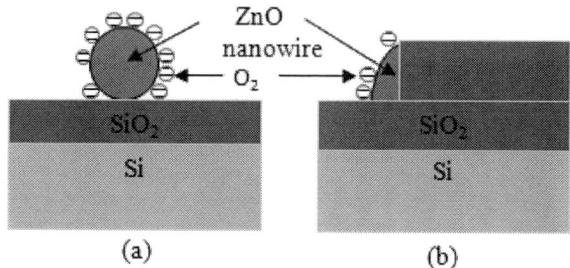

Fig. 6.   Cross-sectional schematics of ZnO nanowires (a) bottom-up ZnO nanowire (b) top-down fabricated ZnO nanowire

Fig.5   Comparison of I_D-V_G curves for our ZnO nanowire FETs (red) with those of Maeng et al [7] (black). Results are presented for gate bias sweep rates of 2500, 250, 100, 6 and 1.2 V/s.

Our top-down unpassivated ZnO nanowire devices also exhibit reproducible values of hysteresis width of less than 1.5 V which is significantly better than bottom-up unpassivated devices and comparable to passivated devices. Hong *et al* [5] showed a value of hysteresis width of 35 V on unpassivated rough nanowire devices. This value reduced to 5 V for unpassivated smooth nanowire devices. The excellent values of hysteresis obtained on our top-down ZnO nanowire devices is an important result for biosensor applications, because low values of hysteresis are needed when sensing measurements are made in body fluids.

Fig. 6 shows the schematic cross-section diagram of a typical bottom-up ZnO nanowire and our top-down ZnO nanowire. One of the surfaces of our nanowire is positioned against the SiO$_2$ wall, which means that less surface area is exposed to the air, resulting in less influence from the adsorption of oxygen ions due to the applied positive gate bias. This may be the reason for the 53% lower threshold voltage shift than that reported by Maeng *et al*. A smoother nanowire surface may also contribute to the better values of threshold voltage shift seen in our work. In addition, our ZnO film is grown by ALD, which may give better control over the film quality than typically obtained with bottom-up grown nanowires.

IV.   CONCLUSIONS

In conclusion, we have developed a simple top-down fabrication process based on a ZnO film deposited by remote-plasma ALD at low temperature of 100°C. The deposition process did not involve any post-deposition annealing and the ZnO nanowires were not passivated during the electrical measurements. A positive threshold voltage shift of 5.4 V obtained from the subthreshold plots as the gate bias sweep rate decreases from 2500 to 1.2 V/s which is 53% less than seen in bottom-up devices reported in the literature. A low value of hysteresis width of less than 1.5 V is achieved for all

devices, which is again better than comparable results for bottom-up fabricated devices reported in the literature. This technology provides ZnO nanowire FETs with stable transistor characteristics and opens up opportunities for the low-cost mass manufacturing of ZnO nanowire FETs for biosensor applications.

ACKNOWLEDGMENT

S.M.Sultan would like to thank the Faculty of Electrical Engineering, Universiti Teknologi Malaysia, Johor Bahru, for supporting her doctoral studies and the Southampton Nanofabrication Centre for the experimental work.

REFERENCES

[1] K.-I. Chen, B.-R. Li, and Y.-T. Chen, "Silicon nanowire field-effect transistor-based biosensors for biomedical diagnosis and cellular recording investigation," *Nano Today*, vol. 6, no. 2, pp. 131-154, 2011.

[2] M.M.A.Hakim, M.Lombardini K.Sun, F.Giustiniano, P.L.Roach, D.E.Davies, P.H.Howarth, M.R.R.de Planque, H.Morgan and P.Ashburn "Thin Film Polycrystalline Silicon Nanowire Biosensors," *Nano Letts.*, vol. 12, no. 4, p. 1868-1872, Mar. 2012.

[3] S. Jeon, S.Bang, S.Lee, S.Kwon, W.Jeong, H.Jeon, H.J.Chang and H.-H.Park, "Structural and Electrical Properties of ZnO Thin Films Deposited by Atomic Layer Deposition at Low Temperatures," *J. Electrochem. Soc.*, vol. 155, no. 10, p. H738-H743, 2008.

[4] S. J. Pearton, D. P. Norton, and F. Ren, "The promise and perils of wide-bandgap semiconductor nanowires for sensing, electronic, and photonic applications," *Small* , vol. 3, no. 7, pp. 1144-1150, 2007.

[5] W.-K. Hong, G. Jo, S.-S. Kwon, S. Song, and T. Lee, "Electrical Properties of Surface-Tailored ZnO Nanowire Field-Effect Transistors," *IEEE Tran. Electron Devices*, vol.55, no.11, p.3020-3029, 2008.

[6] J. Maeng, G.Jo, S.-S.Kwon, S.Song, J.Seo, S.-J.Kang, D.-Y.Kim, and T.Lee, "Effect of gate bias sweep rate on the electronic properties of ZnO nanowire field-effect transistors under different environments," *Appl. Phys. Lett.*, vol. 92, no. 23, p. 233120, 2008.

[7] S. M. Sultan, K. Sun, O.D.Clark, T.B.Masaud, Q.Fang, R.Gunn, J. Partridge, M. W. Allen, P. Ashburn, and H. M. H. Chong, "Electrical Characteristics of Top-Down ZnO Nanowire Transistors Using Remote Plasma ALD," *IEEE Electron Device Lett.*, vol.33, no.2, pp. 203-5, 2012.

[8] O. Wunnicke, "Gate capacitance of back-gated nanowire field-effect transistors," *Appl. Phys. Lett.*, vol. 89, no. 8, p. 4-6, 2006.

[9] J. Goldberger, D. J. Sirbuly, M. Law, and P. Yang, "ZnO nanowire transistors.," *J. Phys. Chem. B*, vol. 109, no. 1, pp.9-14, 2005.

Manufacturing Aspects of an Ultra-Thin Chip Technology

Evangelos A. Angelopoulos, Muhammad S. Al-Shahed, Wolfgang Appel, Stefan Endler, Saleh Ferwana,
Christine Harendt, Mahadi-Ul Hassan, Horst Rempp, Martin Zimmermann and Joachim N. Burghartz

Institute for Microelectronics Stuttgart (IMS CHIPS)
Stuttgart, Germany
Email: angelopoulos@ims-chips.de

Abstract—Ultra-thin silicon (Si) chips fabricated using the recently developed Chipfilm™ technology feature three distinct manufacturing issues, which are discussed in this paper. In Chipfilm™ technology a thin Si membrane is firmly attached to a conventional bulk Si wafer by vertical Si micro-anchors which, however, in the end are controllably fractured to make the thin chips detachable. The associated mechanical stability window is widened by adjusting the arrangement of the micro-anchors, so that chip detachment yields exceeding 99% are achieved. Another not yet reported issue relates to the process temperature and temperature uniformity within the membrane areas, which are thermally connected to the bulk substrate only by the anchors and at the chip edges during processing. Using rapid thermal oxidation and the local oxide thickness as a temperature monitor an on-chip oxide thickness variation of ±3 % is determined. Finally, the inherent deformation (warpage) of free-standing ultra-thin chips due to internal stresses is analyzed and their surface height variation is reduced to only ±4 μm using stress compensation techniques.

I. INTRODUCTION

Ultra-thin crystalline-Si chip technologies promote the development of novel concepts such as three-dimensional chip integration (3D ICs) and hybrid systems-in-foil (SiF) [1]. Conventional thin chip fabrication is based on wafer back-thinning, which is a subtractive approach. Wafer back-thinning suffers from poor thickness control and uniformity with decreasing target thickness and is, thus, best suitable for thickness down to 50 μm [2]. Silicon-on-insulator (SOI) wafers can overcome this problem and lead to thicknesses down to 10 μm [3], however, process cost is increased considerably. Moreover, when thinning down whole wafers costly carrier substrates are needed for handling purposes. The removal of damage induced by the thinning process from the chip backside and the final chip detachment from the carrier substrate without fracture remain additional challenges of the subtractive techniques [4].

In contrast, the recently introduced Chipfilm™ technology features an additive approach for the fabrication of ultra-thin chips and avoids these issues [5]. Its key features are the use of standard bulk Si wafers, excellent thickness control and on-wafer uniformity of the target chip thickness and the absence of any carrier substrate for handling. By forming cavities underneath the wafer surface and by using bulk Si micro anchors to firmly connect the thin Si membrane to the bulk wafer, conventional integration processes can be carried

out. Finally, a systematic fracture of these anchors allows for chip detachment without any damage to the chips.

Section II provides a short overview of the process steps required for fabricating ultra-thin chips using Chipfilm™ technology. In Section III the aspect of the mechanical stability of the membrane wafers is addressed and a solution towards a wider process window is proposed and experimentally verified. Section IV deals with the effect of thermal processes typically used for device integration on such wafers. Finally, Section V presents measurements of the inherent deformation of ultra-thin chips (warpage) and methods for its control. Conclusions follow in Section VI.

II. CHIPFILM™ TECHNOLOGY

The two basic processing blocks of Chipfilm™ technology include the pre- and post-process modules and the circuit integration in between [Fig. 1].

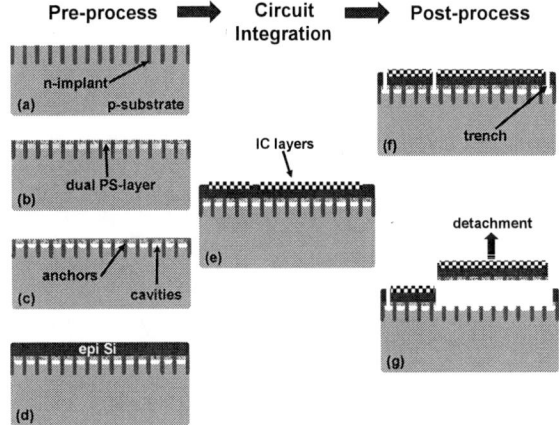

Figure 1. Complete process flow of Chipfilm technology used for the fabrication of ultra-thin chips.

The outcome of the pre-process module [Fig. 1a-d] is the fabrication of the Si membrane wafer that consists of an extensive ultra-thin crystalline Si membrane, which is kept in attachment to a Si wafer by bulk Si vertical micro anchors [Fig. 1d]. Optionally, smaller chip-sized membranes can be defined and separated by a bulk Si grid. The membrane wafer is fabricated by electrochemically forming a dual porous silicon (PS) layer-stack at the surface of a conventional p-type Si wafer [Fig. 1b] and by thermal annealing of the wafer in a hydrogen atmosphere to transform the bottom PS layer

This work was partly funded by the German BMBF projects ProMikron (FKZ 13N9746) and Ultimum (FKZ 16SV5136).

978-1-4673-1707-8/12 $31.00 © 2012 IEEE

into a buried cavity and the wafer frontside into a planar monocrystaline surface [Fig. 1c]; consecutively epitaxial Si growth leads to the required chip thickness [Fig. 1d]. Bulk Si vertical anchors are incorporated into the structure by masking the anchor points against PS formation by means of n-type implantation [Fig. 1a].

These Si membrane wafers can be used as substrates for standard circuit integration [Fig. 1e]. Finally, in the post-process module [Fig. 1f-g], the thin chips are singulated and detached from the wafer by using the Pick, Crack and Place™ technique. This step requires the etching of deep trenches alongside the chip perimeter down into the buried cavity, so that the chips are laterally separated and only held on the substrate by the vertical micro anchors [Fig. 1f]. By ensuring a sufficiently weak attachment, i.e. choosing the suitable anchor size and arrangement, the chips can be detached from the wafer using a conventional vacuum pick-tool [Fig. 1g]. Further details on Chipfilm™ technology can be found in [5], [6].

III. MECHANICAL STABILITY WINDOW

At all steps of the IC fabrication process the thin Si membrane and the buried anchors are exposed to the so-called Process Induced Stress (PIS), [6]. PIS originates from: (i) the residual stresses present in the deposited layers [Table I], (ii) possible temperature gradients in the wafer during thermal processes and (iii) wafer bending during processing, e.g. on vacuum chucks. Obviously, for a smooth IC integration process the mechanical stability of the membrane wafers must be ensured. On the other hand, the elementary requirement of Chipfilm™ technology is the controlled final detachment of all chips, which assumes a certain fragility of the underlying anchors. Thus a sufficiently wide process window should prevent the fracture or loss of chips due to PIS during processing while guaranteeing their retrieval in the end.

TABLE I. EXPERIMENTALLY DETERMINED STRESS VALUES IN THE VARIOUS IC LAYERS OF THE 0.5μM CMOS PROCESS

Layer	Thickness (μm)	Stress (MPa)	Bending Force (Nm)
Sintered PS*	1.5	+17	+26 (*backside)
Field oxide	0.58	-300	-174
IL dielectric	0.95	-30	-29
IM dielectric	1.3	-260	-338
Metal 1 / 2	0.56 / 0.85	+230	+129 / +196
Passivation	0.8	-440	-352

During circuit integration the anchors are shielded from the PIS by the enclosing bulk substrate and the epitaxial layer [Fig. 1e]. However, the stress on the anchors dramatically increases when trenches are etched along the chip periphery [Fig. 2a]. The free space created allows a bending moment to appear at the free edges of the chip, thus, forcing it to deform. Because the anchors are connected to the substrate, they resist this deformation and consequently become considerably more stressed. If their stress level remains below fracture limit the anchors remain intact and the chip stays firmly attached to the substrate. If an additional Externally Induced Stress (EIS), in

form of a moderate forced bowing of the wafer, is applied, the stress level of all anchors will be raised to a level at which the most stressed anchor breaks first.

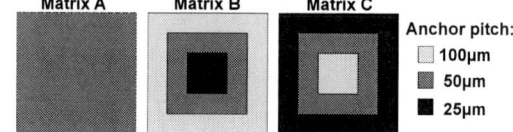

Figure 2. The 3 different arrangements of anchors realized for holding the rectangular chips on the substrate during wafer processing.

Finite Element Method (FEM) simulations using *ANSYS Workbench v.11* demonstrated that for a uniform anchor matrix having a constant anchor pitch the most stressed anchor is situated near the edge of the chip [Fig. 3: Matrix A – after trench]. After fracture of the first anchor a zipper-like consecutive breakage of anchors is initiated as the maximum stress propagates from one anchor to the next [Fig. 3: Matrix A – during zip-effect]. At the end of this zipping, only the outer most group of anchors are left intact to keep the chip weakly attached to the substrate. In contrast, in a non-uniform anchor matrix having gradually decreasing anchor pitch towards the chip center an additional group of central anchors finally remains intact to provide additional stability [Fig. 3: Matrix B – during zip-effect].

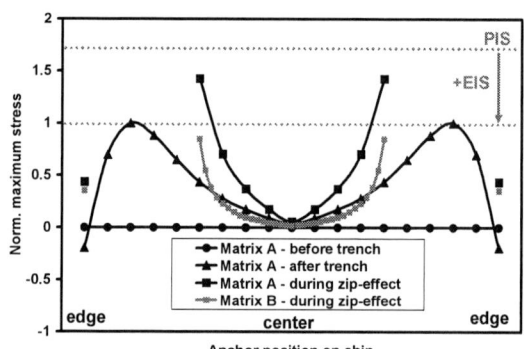

Figure 3. ANSYS mechanical simulations show the increase of stress on the anchors after the trench process (before – after), the effect of an externally applied stress (+EIS) to initiate the zipping of anchors and the stoppage of the zipping process when the anchor pitch is smaller at the center (Matrix B).

Previous experiments have shown that Si chips having a 16 μm epitaxial layer and lithographically defined n-type regions with diameters of 1 μm, 1.4 μm and 2 μm with a constant pitch of 100μm lead to unstable anchor matrices and chip detachment during the trench etching process [6]. Nevertheless, as summarized in Table II, if anchors with 1 μm lithographic diameter and a constant 50 μm pitch (Matrix A) are used, for low values of PIS these chip losses were avoided and after the additional EIS pick-yields > 99% were achieved. Experimentally the PIS value was varied according to the thickness of a thermal oxide grown on bare Si membrane wafers, using oxide thickness of 210nm (low PIS) and 640nm (high PIS). Also, the effective stress of the CMOS process was evaluated. For such high PIS this anchor matrix proved insufficiently stable and led to chip losses during trenching. A uniform design with 25 µm pitch prevented such losses, but proved too stable as very few chips could be detached. Most suitable for higher PIS values was found to be the non-uniform design of Matrix B which provided a 100% pick-yield. However, the non-uniformly distributed stress in the CMOS chips, which has both tensile and compressive components, resulted in a somewhat lower pick-yield of 69% and 93% for both Matrix A and Matrix B, respectively.

TABLE II.   PICK-YIELD FIGURES FOR DIFFERENT ANCHOR MATRICES AND STRESS CONDITIONS

		Layering	PIS	Chip losses after trench	Pick-yield (before EIS)	Pick-yield (after EIS)
Matrix A (50µm)		bare Si	min	0%	2%	100%
		+ SiO₂	low	1%	2%	99%
		+ SiO₂	high	7%	1%	93%
		+ CMOS	high	31%	< 5%	69%
Matrix A (25µm)		bare Si	min	0%	0%	16%
		+ CMOS	high	0%	0%	4%
Matrix B (25/50/100µm)		bare Si	min	0%	0%	91%
		+ SiO₂	low	0%	7%	100%
		+ SiO₂	high	0%	6%	100%
		+ CMOS	high	0%	< 5%	93%

IV. THERMAL PROCESSES

The experimental data of the mechanical stability window are based on a 0.5 µm CMOS technology which relies on furnace processes for thermal oxide growths and dopant diffusions. In such processes wafer heating occurs mainly by convection and temperature changes are relatively slow and uniform within the furnace. Thus, no difference in oxide thicknesses or diffusion depths between the membrane wafers and bulk wafers should be expected, which has been confirmed by their virtually identical measured device characteristics [6]. Nevertheless, in order to substantially reduce the thermal budget, more advanced integration schemes employ rapid thermal processes (RTP). In that case, a different thermal behavior is expected for Si membrane wafers. Similarly to the buried oxide in silicon-on-insulator wafers, the buried cavity acts as a thermal insulator that can lead to temperature discrepancies between wafer surface and the bulk wafer. On the other hand, the bulk Si grid at the periphery of each chip and the bulk Si anchors disrupting the cavities serve as efficient heat conductors. Moreover, a large vertical temperature gradient within the membrane wafers can lead to fracturing due the different thermal expansion of the membrane and the bulk part of the wafer.

For the investigation of the thermal behavior of the Chipfilm™ wafers, which is presented here for the first time, a rapid thermal oxidation (RTO) process was performed having a 35 °C / s ramp-up and -down and a 60 s soak step at 1100 °C using a *Centura RTP XE+* from *Applied Materials, Inc.* The thickness of the oxide grown was measured using a conventional laser ellipsometer and depends on the surface temperature during the process. The different membrane wafers processed were based on p⁺ bulk wafers having a 16 µm p⁻ epitaxial layer and cylindrical anchors with a diameter of ~1.7 µm arranged according to the three different matrices of Fig. 2. Lightly p-doped bulk monitor wafers were also processed as references.

Prior to the main experiment, the temperature sensitivity of the oxide thickness, i.e. the thickness – temperature curve, was determined by a series of RTO processes performed on bulk Si wafers using different soak temperatures ranging from 1080 °C to 1120 °C. As seen from Fig. 4, the linear fit of the oxide thickness measurements rendered a sensitivity of 0.083 nm / °C, which is in good agreement with the 0.089 nm / °C determined by TCAD simulation of the RTO process.

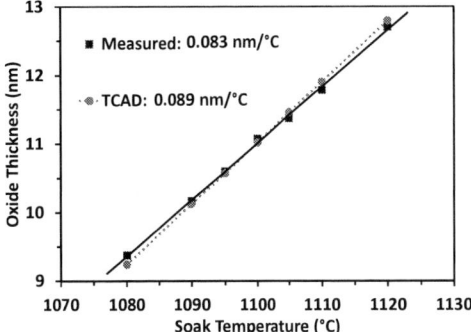

Figure 4.   Oxide growth sensitivity extracted from RTO experiments at different soak temperatures and compared to simulation results.

Next, the RTO process at 1100 °C was performed on Chipfilm™ wafers and no membrane damage was observed, indicating that the thermal stresses were not overly significant. Using a 60 µm step an oxide thickness map of 10,000 measurement points was produced covering a 6 x 6 mm² area at the center of each wafer. The thickness values were converted to temperature by the previously determined thickness sensitivity factor.

The average temperature on the surface of the membrane wafers was found to be +6 °C higher compared to the bulk reference wafers where the oxide thickness was 10.72 nm. The on-chip oxide thickness variation was measured between ±3% and ±4%. More specifically, the uniform design (Matrix A) resulted in the most uniform temperature distribution throughout the chip membrane [Fig. 5].

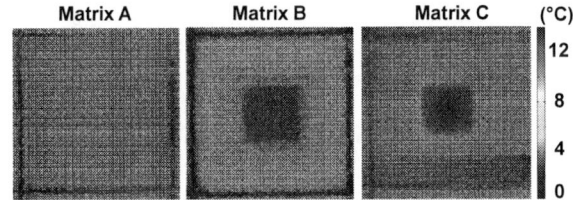

Figure 5.   Temperature difference on the surface of chips with the different anchor matrices (see Fig. 2) during RTO compared to a bulk wafer showing the heat dissipation through the bulk Si grid at their periphery and regions with varying anchor pitch.

However, the membrane temperature between the anchors was found to reach up to +7.9 °C compared to the colder grid region. On the other hand, in the non-uniform design of Matrix B this maximum difference was +7.1 °C near the edge where the anchors are furthest apart (100µm), while the 25 µm anchor pitch at the center provided a good thermal contact to the substrate and the temperature rose just 1-2 °C above the reference value. Finally, Matrix C delivered the worst performance in terms of on-chip temperature uniformity with the temperature steadily rising towards the center of the chip where the anchors are furthest apart (100μm) and also furthest from the bulk grid, reaching +9.2 °C. Notably, only Matrix B design resulted in a grid temperature equal to the reference wafer temperature. This indicates that all heat generated in the membrane could be dissipated through the vertical anchors and thus no dissipation through the grid was necessary, which makes the experimental results valid also for membrane wafers without any bulk grid.

V. INHERENT DEFORMATION OF ULTRA-THIN CHIPS

In contrast to standard chips with thicknesses in the range of several hundred micrometers, free-standing ultra-thin chips can appear severely deformed due to the inherent stress present in the IC layers, such as the various oxides, the metallization and the passivation layer. This deformation can have significant effect on the handling and packaging technologies used for ultra-thin chips and is therefore of increased importance.

The three main factors that determine the total internal stress distribution and hence the inherent chip deformation are: (i) the thickness of the chip's substrate (ii) the integration technology and the chip layout. For a given chip thickness, however, changing process parameters in well-established integration technologies is undesired. Therefore, only the adjustment of the chip layout can be a viable solution to control the stress distribution.

From Table I, it is evident that mainly the metal layers induce tensile stress and, thus, counteract the convex chip bow produced by all other compressively stressed layers. In addition, patterning the topmost metal layer can induce topography to the deposited passivation layer and reduces its plane stress by stress relief occurring at the vertical planes of the metal patterns. The effect of the metallization layout on the chip deformation was studied using rectangular Chipfilm™ chips with an edge of 4.6 mm, having a 16 μm epitaxial Si, followed by a 200 nm deposited oxide, a 1 μm patterned metal layer and an 850 nm passivation layer.

Studying the effect of metal layer on chip deformation has revealed that metal patterned in stripes, i.e. with axial symmetry, result in ellipsoidal chip deformation as the amount of stress in the passivation layer is partially relieved in the direction normal to the metal lines [Fig. 6-a, b]. In addition, the increase of metal coverage from 17% to 50% by using wider metal striped reduces the chip deformation [Fig. 6-b]. A significant change in deformation was observed in the case of chips with narrower and denser metal strips (from 83 to 250 strips / mm). As seen in Fig. 6-c, surface curvature perpendicular to the metal lines switched sign and became concave.

In contrast to the previous examples, a biaxially symmetric metal pattern, like the metal grid on the chip of Fig. 6-d, yields a very low chip deformation. Since the metal pattern is absent at the outermost part of the chip, which consists only

of the compressively stressed planar oxide and passivation layer, a convex bow is observed at its edges. The planarity of this chip is high with a surface height variation of just ±4 μm.

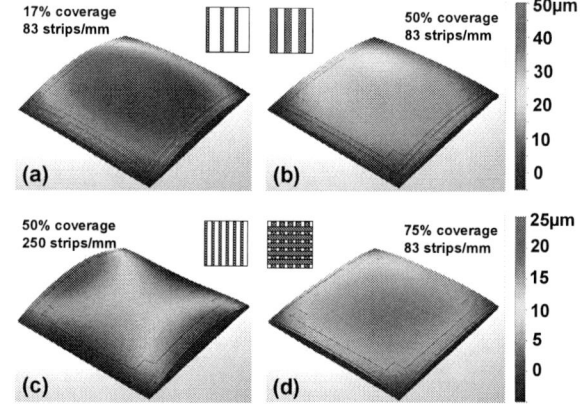

Figure 6. Optical profilometer images of the surface of ultra-thin chips with different metal paterns (inset illustrations) depicting the severe effect of the chip layout on its inherent deformation.

VI. CONCLUSION

The fabrication of ultra-thin chips using Chipfilm™ technology is a viable manufacturing process if certain aspects are observed. The bulk Si anchors supporting the thin membranes must have small diameters (< 3 μm) and be arranged in a non-uniform matrix which is denser at the chip center (Matrix B). While the anchors act as a heat sink during RTP, this matrix is also favorable for such processes as it effectively minimizes on-chip temperature variations. The planarity of the ultra-thin chips can be achieved by adjusting the metal interconnect layout and introducing biaxially symmetric dummy structures that result in high metal coverage (~ 50-75%) or increased topography.

ACKNOWLEDGMENT

The authors would like to thank all staff at the IMS CHIPS clean-room for their contributions to the development of Chipfilm™ technology and particularly E. Penteker, A. Albu, F. Letzkus and L. Diesing.

REFERENCES

[1] J. N. Burghartz, "Ultra-Thin Chip Technology and Applications", ISBN 978-1-4419-7275-0, Springer, New York, 2010.

[2] S. Takyu et al., "A Study on Chip Thinning Process for Ultra Thin Memory Devices", Proc. ECTC 2008, pp. 1511-1516, 2008

[3] R. Dekker et al., "A 10 μm Thick RF-ID Tag for Chip-in-Paper Applications", Proc. BCTM 2005, pp. 18-21, 2005.

[4] Eun-Kyung Kim, "Assessment of ultra-thin Si wafer thickness in 3D wafer stacking", Microelectronics Reliability, vol. 50, pp. 195-198, 2010.

[5] J. N. Burghartz et al., "A New Fabrication and Assembply Process for Ultrathin Chips", IEEE Trans. On El. Devices, vol. 56, pp. 321-327, 2009.

[6] E. A. Angelopoulos et al., "Ultra-Thin Chip Technology for System-in-Foil Applications", in Proc. of IEEE 2010 International Electron Devices Meeting, pp. 2.5.1-2.5.4, (2010).

Epitaxial Growth of Large-Area p^+n Diodes at 400°C by Aluminum-Induced Crystallization

Agata Sakic, Lin Qi, Tom L.M. Scholtes, Johan van der Cingel, Lis K. Nanver
Department of Microelectronics, Delft University of Technology
Dimes-TC, Feldmannweg 17, 2628 CT, Delft, The Netherlands
A.Sakic@tudelft.nl

Abstract—Aluminum-Induced Crystallization is applied on crystalline Si substrates. By this method a physical-vapor-deposited amorphous Si layer is successfully transformed into a monocrystalline solid-phase epitaxy (SPE) p-doped layer at an anneal temperature of 400°C. The as-grown epitaxial layer takes on the orientation and the lattice constant of the substrate. It is shown that a complete coverage over large areas is possible if the c-Si interface is free of nucleation centers. This can be achieved by the proper oxide-patterning and/or chemical treatments of the substrate surface before deposition of the Al mediator layer. High-quality p^+n diodes have been fabricated with areas up to 1×1 cm², having ideality factors down to 1.02 and low leakage currents in the 2-3 nA/cm² range. The full coverage by p^+ SPE-Si is confirmed by material analysis.

I. INTRODUCTION

Aluminum-Induced Crystallization (AIC) of amorphous Si has been receiving increasing attention from researchers in fields such as solar-cells, displays, sensors, thin films, and nanowires due to the capability to turn a thin-film of amorphous silicon on foreign substrates, for example glass or ceramic, into polycrystalline Si at processing temperatures below the Al-Si eutectic point of 577°C [1-5]. It is a potent alternative to laser-induced crystallization by delivering several tens of micron large grains that can be further enlarged by annealing steps. The same technology is recently applied for crystallization of amorphous Ge layers for large-area electronics and solar energy conversion [6]. So far, many aspects of the process have been investigated: the effects of the Al and Si layer thickness, deposition conditions, layer interfaces, and Al structure, all of which determine the Si nucleation rate, thickness, and crystalline quality of the as-crystallized Si film [1, 2, 4, 5].

In addition to these developments where AIC is used on low-cost substrates, it can also be used on c-Si wafers or seeded substrates where it results in a low-temperature epitaxy of Al-doped layers of Si, also known as Al-mediated Solid Phase Epitaxy (SPE) [7]. The applications of the as-grown p^+-doped layers include ultra-abrupt p^+-silicon elevated contacts and diodes [8], PNP bipolar transistor emitters [8], nanodevice integration [9], lateral SPE silicon-on-insulator growth [10], physical cryptography [11], solar cells [12], and counter-doping of Si thin films [13]. Since the Al-mediated SPE process can be performed at 400°C it is attractive as a low-cost add-on to fully-processed CMOS wafers.

In earlier work on Al-mediated SPE p^+n diodes on c-Si [7-9], ideal junction diode characteristics were only reached for small areas of a few microns large. In the present paper it is shown that the density and vicinity of nucleation sites on the substrate is decisive for the maximal area size that can be filled with Al-mediated SPE. In particular, the influence of the c-Si interface quality is studied with the aim of creating arbitrarily large defect-free Al-doped p^+n diodes at maximum temperature of 400°C. The characterization is carried out using scanning electron microscopy (SEM), high resolution transmission electron microscopy (HRTEM), nanobeam-diffraction (NBD), and atomic force microscopy (AFM) analysis. Finally, for the optimized surface treatment conditions, good quality diodes of up to 1×1 cm² in area are fabricated and characterized.

II. DEVICE FABRICATION

The fabrication process is schematically shown in Fig. 1. The starting substrates are n-type 2-5 Ωcm (100) wafers. First, thermal SiO_2 is grown and windows to the Si are patterned by either plasma etching or wet etching in buffered HF 1/7 (BHF 1/7) (Fig. 1a). Some of the wafers are then dip-etched for 4 min in a 0.55% HF solution and dried using the Marangoni effect of isopropanol alcohol (IPA) to ensure a native-oxide-free hydrogen-terminated surface. Then, a layer-stack of Al and α-Si is sequentially deposited by physical vapor deposition (PVD) at room temperature without vacuum break between the depositions (Fig. 1b), in order to prevent oxidation of the Al surface which may otherwise slow down the migration of Si and deteriorate the final crystallized Si quality [4]. The thickness of the deposited Al is either 100 nm or 200 nm, and that of the α-Si either 20 nm or 50 nm. The Si target is sputtered at 0.5 kW in Ar-atmosphere. After deposition, the stack is patterned around the oxide openings, and annealed for 40 min at 400°C in forming gas. During the annealing, the layer exchange takes place and the resulting Al-doped c-Si layer is deposited preferentially in the oxide openings (Fig. 1c). To form the anode contacts, PECVD oxide is deposited and windows to the SPE islands are patterned. A second layer of metal (PVD Al/1%Si) is then deposited for

TABLE 1 DESCRIPTION OF SAMPLE PROCESS VARIATIONS

Sample name	Oxide patterning	Post-etching treatment	Stack thickness [Al, Si] in nm
WetHF	wet BHF 1/7	HF 0.55% Marangoni-IPA	[200, 20] [100, 50]
WetBHF	wet BHF 1/7	×	[200, 20]
DryHF	dry	HF 0.55% Marangoni-IPA	[200, 20]

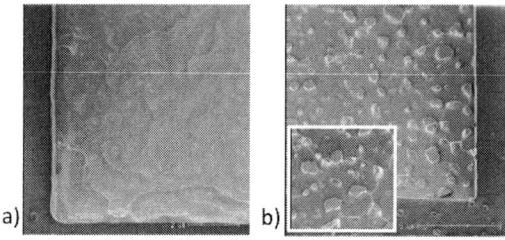

Fig. 2. SEM micrographs showing the SPE-Si structure after the Al/α-Si layer exchange at 400 °C and selective Al removal, for a) WetHF[200, 20], and b) DryHF[200, 20] samples. The inset clearly shows remnants of the original Al grain boundaries.

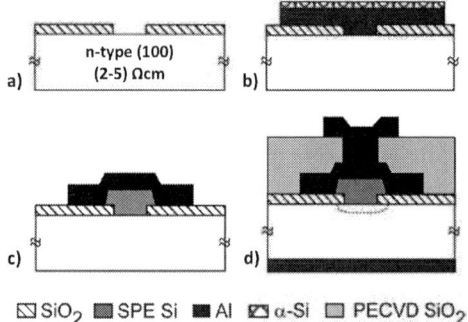

⬚ SiO₂ ■ SPE Si ■ Al ▨ α-Si ⬚ PECVD SiO₂

Fig. 1. Schematic process flow for fabricating Al-mediated SPE pⁿ diodes.

interconnect patterning. Finally, the back of the wafer is metalized and the wafers are alloyed for 40 min at 400°C in forming gas (Fig 1.d). An overview of the different process variations are given in Table I.

III. EXPERIMENTAL RESULTS

A. Influence of the substrate interface on AIC

Two distinctly different SPE patterns are observed for the different samples listed in Table 1. Examples are shown in Figs. 2a and 2b where the Al remaining after SPE is selectively etched away from the Si surface. The Si epi-layer seen in Fig. 2a displays a wavy pattern covering the whole surface as is representative for the WetHF samples, while for all the other samples isolated well-defined crystals of Si are scattered randomly across the surface as illustrated in Fig. 2b. These different results can be related to the perfection of the Si surface that the Al/α-Si stack is deposited on. The WetHF samples have a smooth native-oxide-free Si surface whereas the WetBHF samples are not well hydrogen-passivated against native oxide formation and the DryHF samples will have some degree of plasma-induced damage of the Si surface. The appearance of well-defined crystals in the latter two cases suggests that a significant number of nucleation centers for silicon epitaxy have been generated by this type of surface treatment, in contrast to the WetHF samples.

Additional tests were conducted on bare Si wafers, some of which were immersed only in BHF (WetBHF), while others were further dipped in HF followed by Marangoni drying (WetHF). As expected, SPE on the WetBHF set resulted in a pattern of isolated Si crystals while the WetHF samples

provided a continuous layer of thin Si across the whole wafer surface.

B. Growth kinetics and layer properties

The close-up image shown in the inset of Fig. 2b reveals thin tracks decorated by crystal sprouts that are a witness of the former grain boundaries in the original Al layer. This shows that the Si atoms that travel through grain boundaries during the annealing step coalesce soon after reaching the substrate, rather than diffusing along the Al/c-Si interface, and the preferential growth direction of the crystal is vertical. The maximum observed crystal size is ~ 1-2 μm² and, in the case shown in Fig. 2b, crystals are too far apart to merge and form larger islands. This is the limiting factor in filling large areas with this type of Al-mediated SPE process. In contrast, Fig. 2a displays an example of how a defect-free Al/c-Si interface results in a long lateral diffusion length of the Si atoms supplied from the α-Si layer through the Al grain boundaries. Thus, the whole available open area can be covered with SPE Si. The remnants of Al grain boundaries are still visible, but the substrate surface is fully covered with the Al-doped epi-Si layer. This conclusion is substantiated by the electrical characterization discussed in the next section.

Microscopy analysis (TEM, NDA, AFM) was performed on 60×60 μm² large windows on the WetHF[100,50] samples. After the layer exchange of Al and Si at 400°C, Al is selectively removed in HF and the samples are cleaned in a HNO₃ solution. As can be seen in Fig. 3, the TEM analysis substantiates that the growth is epitaxial. On lower-magnification TEMs some facet twinning is observed where the individual Si islands join together, mainly at the former Al grain boundaries, and the layer varies in height from a few atom layers up to 200 nm, which is the thickness of the original Al layer-exchange mediator layer.

In Fig. 4 the NBD pattern of the SPE Si is compared to that of the underlying c-Si substrate. The interatomic spacing (d-spacing) is determined and the lattice matching is within 0.01Å, which is within the measurement error of the NBD technique. The crystal orientation follows the substrate (100) orientation having only a slight tilt of 0.05 degrees observed around the <110> zone axis.

The AFM analysis shown in Fig. 5 supports the TEM picture of large height variations with peaks up to 100 nm high. It also makes clear that there is a significantly higher average thickness at the edge of the window near the oxide.

978-1-4673-1707-8/12 $31.00 © 2012 IEEE

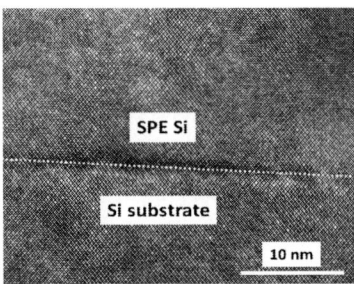

Fig. 3. HRTEM of SPE-Si on the c-Si substrate showing epitaxial growth. The sample is prepared according to the WetHF[200, 20] procedure.

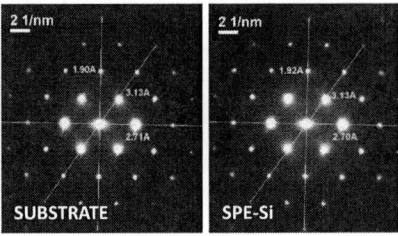

Fig. 4 Nanobeam-diffraction pattern of substrate Si and SPE-Si showing matching of the d-spacing for (111), (002), and (022) directions.

The thickening comes either from Si atoms deposited on the oxide layer that have sufficient diffusion lengths to reach the oxide-opening to silicon, or from defect-enhanced crystallization at the SiO_2 interface which has previously been documented for this type of Al-mediated SPE process [7].

C. Electrical characterization of Al-doped SPE p^+n diodes

The I-V characteristics of diodes fabricated in the WetBHF and DryHF procedures only show ideal p^+n behavior for the smallest window areas while for the larger sizes Schottky-like characteristics are observed. This is in accordance both with earlier results that were based on this type of processing [9] and with the SEM analysis that shows that coalescence of the SPE silicon into crystals is limited to islands of 2 to 3 microns in size. All in all, it can be concluded that large areas of the c-Si substrate are not covered with SPE Si in these cases.

In contrast, the WetHF SPE diodes are ideal for all sizes. Representative I-V characteristics are shown in Fig. 6 for the smallest and largest fabricated areas of 1×1 μm^2 and 1×1 cm^2, respectively, and compared to Shottky diodes of the same sizes. It is remarkable that the large 1×1 cm^2 SPE diode has a saturation current per unit area that is of the same low level as the small diode, and also that an ideality factor as low as 1.02 is recorded. This indicates a good SPE Si coverage over the whole 1×1 cm^2 surface as already suggested by the SEM/TEM analysis. Moreover, it can be seen in Fig. 7 that both the forward and reverse currents scale as expected with the diode area. The low leakage current corroborates that there is a very low defect-density at the interface of the SPE Si and the c-Si substrate as observed in the HRTEM (Fig. 4). To determine the mechanism behind the leakage current, the temperature behavior was monitored using Arrhenius-like plots [14]. Diode areas of 1×1 μm^2, 40×40 μm^2, and 60×60 μm^2 were measured from 25 to 190°C, and the extracted activation

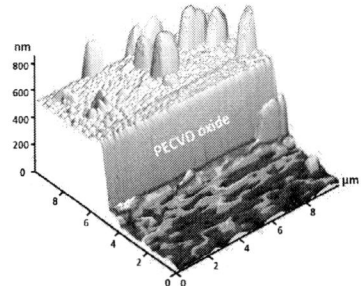

Fig. 5. AFM analysis of an SPE-Si surface near the oxide at the window perimeter where the SPE-Si layer is thicker (red color); for the WetHF[100, 50] sample.

energy E_a was 1.05 eV, 0.97 eV, and 0.94 eV, respectively, at 2 V reverse bias. These values are all close to the Si bandgap value E_g, indicating that the dominant origin of the dark current is from ideal diffusion over the depletion regions and not from defect-related generation-recombination currents.

According to earlier SIMS analysis [15], the Al-mediated SPE-Si layer contains a high concentration of Al which p-dopes the layer up to the limits set by the solubility and diffusivity of Al in Si at the given annealing temperature, in this case at 400°C. Hole injection from this p-region can be verified by the method proposed in [16], where adjacent SPE islands are connected in a lateral pnp configuration. Measurements of this type confirmed that for the given layer-stacks and anneal cycle there is a high level of p-doping in the SPE islands, which is in accordance with the low saturation current and ideality of the diode characteristics. Due to the high anode doping, the diode breakdown voltage is set by the lower substrate doping, in this case the Si wafer with 2-5 Ωcm resistivity. The recorded 65 V breakdown for 1×1 μm^2 diodes matches well with previously reported 70 V [17], while for the larger diodes of 1×1 cm^2 the average measured breakdown occurs at ~ 85 V. This suggests that the bulk breakdown is

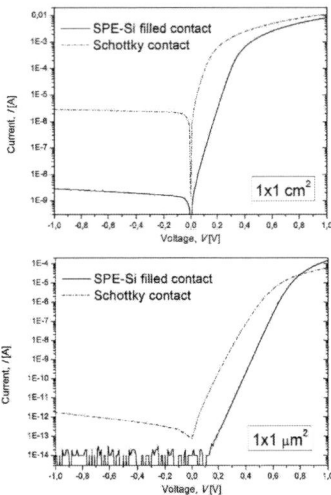

Fig. 6. Measured I-V characteristics of 1×1 μm^2 and 1×1 cm^2 WetHF SPE-Si diodes compared to those of Schottky diodes of the same size.

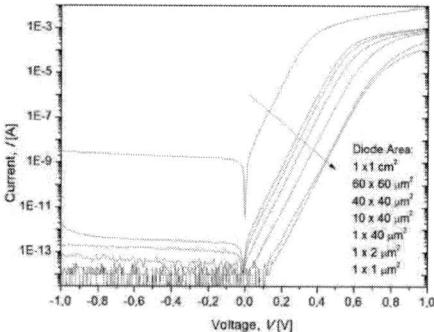

Fig. 7. Measured I-V characteristics for the set of WetHF[200, 20] SPE-Si diodes with anode areas: 1×1 cm^2, 60×60 μm^2, 40×40 μm^2, 40×10 μm^2, 40×1 μm^2, 1×2 μm^2, and 1×1 μm^2.

higher than the perimeter breakdown which is to be expected since no guard rings are implemented in these structures.

The Al layer used in the AIC process should not be less than about 50 nm thick, otherwise the resulting Si crystalline quality is compromised [18]. In this work an Al thicknesses of 100 nm or 200 nm is used and in this range no significant influence on diode properties is observed provided the same α-Si thickness is applied. With respect to the thickness of α-Si, samples were compared with either a 20 nm or 50 nm α-Si layer. While the saturation current and the ideality factor of the diodes remained unaffected by the available amount of Si, the series resistance did slightly increase as the Si thickness was increased. This is observed over several batches and for different diode sizes, and cannot be explained by the spread in the substrate wafer resistivity. It is assumed that the increase in resistance is caused by the limited annealing time and temperature made available for the layer exchange process: a layer of non-crystallized α-Si may remain on the surface, thus increasing the resistance of the layer. For example, 10 devices with an area of 40×40 μm^2 were measured to have a series resistance of 315 ± 20 for the 20-nm-Si device, and 485 ± 65 Ω for the 50-nm-Si device. These are very high series resistances but previous work on Al-mediated SPE islands has shown that low-ohmic resistivity values less than 10^{-7} Ωcm can be achieved [9]. In this flow the corrupted AIC-Al layer is removed before the final metallization.

IV. CONCLUSIONS

Amorphous Si thin films have been crystallized over arbitrarily large c-Si surfaces at 400°C using an Aluminum-Induced Crystallization process. An epitaxial layer coverage that is continuous, monocrystalline, and highly p-doped with Al, is achieved by minimizing the density of nucleation centers on the wafer surface, achieved here by wet-patterning of the isolation oxide layer in BHF 1/7 and H-passivating the c-Si interface with a HF 0.55% dip-etch followed by IPA-Marangoni drying. The epitaxial growth of Si takes on the orientation and the lattice constant of the underlying c-Si substrate. A defect-free interface is verified by HRTEM imaging and by the low leakage of p$^+$n diodes of areas ranging from 1×1 μm^2 up to 1×1 cm^2. The very large

diodes have ideality factors as low as 1.02, and leakage current levels of only 2-3 nA at 1 V reverse bias. The fact that such arbitrarily large high-quality diodes can be made at sub-metallization temperatures make this process very attractive as an add-on to fully-processed CMOS or for diode processing in large-area electronics and c-Si solar cells.

REFERENCES

[1] S. Gall, M. Muske, I. Sieber, O. Nast, W. Fuhs, "Aluminum-induced crystallization of amorphous silicon," Journal of Non-Crystalline Solids, vol. 299–302, Part 2, pp. 741–745, 2002.

[2] O. Nast, S. Brehme, S. Pritchard, A. G Aberle, S. R Wenham, "Al-induced crystallisation of Si on glass for thin-film solar cells," Solar Energy Mat. and Solar Cells, vol. 65, Issues 1–4, pp. 385-392, 2001.

[3] Y. Wang, V. Schmidt, S. Senz, U. Gösele, "Epitaxial growth of Si nanowires using Al catalyst", Nature Nanotech. 1, pp. 186-189, 2006.

[4] O. Nast and A. J. Hartmann, "Influence of interface and Al structure on layer exchange during aluminum-induced crystallization of amorphous silicon", Journal of Applied Physics, vol. 88, pp. 716-724, 2000.

[5] O. Nast, T. Puzzer, L. M. Koschier, A. B. Sproul, and S. R. Wenham, "Aluminum-induced crystallization of amorphous silicon on glass substrates above and below the eutectic temperature," Applied Physics Letters, vol. 73(22), pp. 3214-3216, 1998.

[6] S. Hu and P. C. McIntyre, "Nucleation and growth kinetics during metal-induced layer exchange crystallization of Ge thin films at low temperatures", J. Applied Physics, vol. 111 (4), nr. 044908, 2012.

[7] Y. Civale, G. Vastola, L.K. Nanver, R. Mary-Joy, and J.-R. Kim, "On themechanisms governing the aluminum-mediated solid-phase epitaxy of silicon," J. Electronic Materials, vol. 38 (10), pp.2052-2062, 2008.

[8] Y. Civale, L.K. Nanver, P. Hadley, E.J.G. Goudena, and H. Schellevis, "Sub-500°C solid-phase epitaxy of ultra-abrupt p$^+$-Si elevated contacts and diodes," IEEE Electron Device Lett., Vol. 27, pp. 341–343, 2006.

[9] Y. Civale, L.K. Nanver, and H. Schellevis, "Selective solid-phase silicon epitaxy of p$^+$ aluminum-doped contacts for nanoscale devices," in IEEE Trans. on Nanotechnology, Vol. 6 (2), pp. 196–200, 2007.

[10] A. Sakic, Y. Civale, L. K. Nanver, C. Biasotto, V. Jovanovic, "Al-mediated Solid-Phase Epitaxy of Silicon-On-Insulator", MRS Spring Meeteing Symposium A, Proc. Vol. 1245, pp. 1245-A20-03, 2010.

[11] C. Jaeger, M. Algasinger, U. Rührmair, G. Csaba, and M. Stutzmann, "Random pn-junctions for physical cryptography", Applied Physics Letters, vol. 96 (17), pp. 172103-1-3, 2010.

[12] K. Sharif, H. H. Abu-Safe, H. Naseem, W. Brown, M. Al-Jassim, H. M. Meyer, "Epitaxial Si thin films by low-temperature Al induced crystallization of amorphous Si for solar cell applications", IEEE 4th World Conf. Photovoltaic Energy Conversion, pp. 1676 – 1679, 2006.

[13] M. S. Haque, H. A. Naseem, and W. D. Brown, "AIC and counter-doping of phosphorous-doped hydrogenated amorphous Si at low temperatures", J. Applied Physics, vol. 79, pp. 7529-7536, 1996.

[14] H. Lee, "Characterization of shallow silicided junctions for sub-quarter micron ULSI technology. Extraction of silicidation induced Schottky contact area", Trans. Electron Devices, vol.47 (4), pp.762-767, 2000.

[15] Y. Civale, L. K. Nanver, S. G. Alberici, A. Gammon, and I. Kelly, "Accurate SIMS Doping Profiling of Al-Doped Solid-Phase Epitaxy Silicon Islands' in Electrochem. Solid-State Lett. 11, H74, 2008.

[16] G. Lorito, L. Qi, L. K. Nanver, "Lateral bipolar structures for evaluating the effectiveness of surface doping techniques," 2011 IEEE Conference on Microelectronic Test Structures, pp. 108-113, 2011.

[17] Y. Civale, R. Mary-Joy, L. K. Nanver, "Electrical Characterization of Layer-Exchange Solid-Phase Epitaxy Si Diode Junctions", Proc. 10th Annual Workshop SAFE, pp. 408-411, 2007.

[18] G.J. Qia, S. Zhangb, T.T. Tangb, J.F. Lib, X.W. Sunb, X.T. Zenga, Experimental study of Al-induced crystallization of amorphous silicon thin films", Surface & Coatings Tech., vol. 198, pp. 300– 303, 2005.

978-1-4673-1707-8/12 $31.00 © 2012 IEEE

Current-Voltage Characteristics of Vertical Diodes for Next Generation Memories

Hokyun An, Kong-Soo Lee, Yoongoo Kang,
Seonghoon Jeong, Wonseok Yoo, Jae-Jong Han,
Bonghyun Kim, Hanjin Lim, Seokwoo Nam,
Gi-Tae Jeong, Ho-Kyu Kang, Chilhee Chung
Process Development Team, Semiconductor R&D Center
Samsung Electronics Co., Ltd.
Hwasung, Korea
hkano.an@samsung.com

Byoungdeog Choi
School of Communication and Information Engineering
Sungkyunkwan Univerisity
Suwon, Korea
bdchoi@skku.edu

Abstract—In this paper, current-voltage-temperature (I-V-T) characteristics of vertical diodes realized by different selective epitaxial growth techniques have been investigated. Diodes by the batch-type cyclic SEG process at low temperature have shown eligible performances for vertical switches, including ideality factor of 1.08, off-current of 1.0×10^{-12} A and on/off-ratio of 2.4×10^{8}. The optimization of crystallographic defects and series resistance is expected to be the most critical for the performances of vertical diodes for next generation memories.

I. INTRODUCTION

As charge-based memory devices shrink rapidly, conventional memories such as dynamic random access memory (DRAM) and Flash memory are expected to suffer from serious problems including device integration, electrical performance, as well as reliability. Immature performance of extreme ultraviolet (EUV) photolithography hinders further scale-down beyond 20 nm and the increase in the density of memories. Apart from the device integration, the electrical performance and reliability will be of greater concern. When the design rule is scale-down to sub-20nm, both memories will face the serious reduction of storage area in capacitors and floating poly-silicons (F-Polys). In addition, charge loss by leakage current in capacitors and F-Polys will lead to the potential failure in electrical performance and reliability. It is natural that new approaches are worthy of world-wide attention to develop, so called, next generation memories with different storage media such as magnetic RAMs (MRAMs) and phase change memories (PCMs), which are not charge-based but resistance-based devices [1~5]. A cross-point structure is rigidly recommended to reduce the cell area and to improve cost-effectiveness for next generation memories. The scheme of vertical switches is the pre-requisite for the integration of the cross-point structure, which replaces conventional horizontal MOSFET switches.

We have also investigated the properties of epitaxial growth processes to develop vertical diodes. The roles of process conditions in solid-phase epitaxy (SPE) and selective epitaxial growth (SEG) on the physical and electrical characteristics of vertical switches have been explored [6-8]. In this study, current-voltage (I-V) characteristics of vertical diodes made by different SEG processes are analyzed. In forward bias condition, diffusion and recombination components are separated and we derive ideality factors of various diodes. Recombination current is compared to conventional Shockley diode model, and series resistance is defined from I-V curves. Reverse bias condition is also investigated.

II. EXPERIMENTAL PROCEDURE

The process sequence and SEG parameters of preparing vertical diodes are published elsewhere [6-8]. The process conditions of co-flow and cyclic SEG are summarized in Table I. Fig. 1 shows the structure of test element module for measuring electrical characteristics of vertical PN diodes. P and As diffuse out from heavily doped Si active (N+Base) to SEG-Si during SEG process to form N- regions. B was implanted to SEG pillars to form P+ regions. Both ends of a diode were connected to metal lines via metal contacts for the evaluation of electrical properties. Current-voltage (I-V) characteristic was measured by Precision Semiconductor parameter analyzer, Agilent 4156C. Voltage was swept from -2.0 V to +2.0 V and measurement temperature was varied from -20 ℃ to +85 ℃. The other processes and structures including dimensions for the diode configuration were identical except for the method to form diode pillars. A single

TABLE I. Growth Conditions for Different SEG Processes

SEG	Temperature (℃)	Gases	Pressure
Co-flow	≥ 800	$H_2/DCS/HCl$	Tens of Torr
Cyclic	≤ 700	$H_2/SiH_4/Cl_2$	Full pumping

978-1-4673-1707-8/12 $31.00 © 2012 IEEE

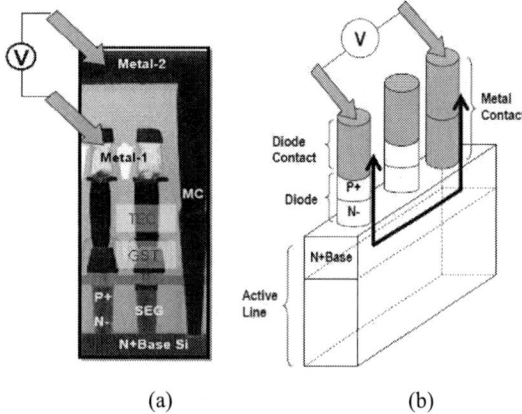

Fig. 1. Electrical measurement structure for vertical diodes; (a) TEM of full structure, and (b) schematic drawing.

mold structure was adopted for all diodes. The point in the stable selectivity window for the cyclic SEG was chosen to compare its ability to co-flow SEG-Si diodes.

III. RESULTS AND DISCUSSION

$$I = \frac{qn_i WA}{\tau_{SCR}} \exp(\frac{qV_A}{n_1 kT}) + qn_i^2 FA(\frac{\sqrt{D_N}}{\sqrt{\tau_N \cdot N_A}} + \frac{\sqrt{D_P}}{\sqrt{\tau_P \cdot N_D}}) \exp(\frac{qV_A}{n_2 kT}) \quad (1)$$

$$I = I_{01} \exp(\frac{qV_A}{n_1 kT}) + I_{02} \exp(\frac{qV_A}{n_2 kT}) \quad (2)$$

Classical Shockley diode equation is summarized as (1) and (2) excluding high field effects such as high level injection and series resistance. Total diode current is the sum of minority carrier diffusion current in the quasi-neutral region (QNR), recombination-generation current in the space-charge region (SCR), and surface generation-recombination current [9]. In most cases, the third term is neglected and total diode current is expressed in the equations mentioned above, where A is the area of diode, and n_i is the concentration of intrinsic carriers, τ_n, τ_p are the minority carrier lifetimes of electrons and holes , and τ_{SCR} is the lifetime in the space-charge-region, and D_N, D_P are the diffusion coefficients of electrons and holes, and L_N, L_P are the diffusion lengths of electrons and holes, and N_A, N_D are the concentrations of acceptors and donors, and W is the width of depletion layer, and V_A is the applied voltage, and k is the Boltzmann constant, and T is the temperature of measurement, and n_1, n_2 are the ideality factors in SCR and QNR, respectively. F is a correction factor that depends on the sample geometry, e.g., denuded zones on defective substrates, epitaxial layers on heavily or lightly doped substrates, silicon-on-insulator (SOI), etc. [9]

Current-voltage-temperature curves were measured, as shown in Fig.2 and Fig. 3. Ideality factors in the diffusion current component were 1.36, 1.08 for diodes built by co-flow, and cyclic SEG, respectively. The n values were extracted from semi-log (I) vs V plot with the coefficients of correlation

(R^2) bigger than 0.995. The values are in a good agreement with the density of crystallographic defects in diodes according to the pillar processes [6-8]. Although quantitative evaluation of the defects is not possible at this point, qualitative comparison of the crystalline defects between diodes with different methods reveals that the ideality factor increases with the density of defects in diode pillars. The ideality factors from 1.08 to 1.36 imply that vertical silicon diodes are successfully functioning in accordance with the Shockley-Read-Hall (SRH) theory, where recombination centers or traps within silicon bandgap in the space-charge region impede the current flow through diodes, and induce the deviation from ideal diodes. Crystallographic defects including stacking faults, micro-twins and grain boundaries are known to induce trap levels within the bandgap and act as R-G centers. The n value of 1.08 for cyclic SEG diodes means that the diodes are almost ideal to the literatures and the density of defects are quite rare in SEG-Si region. The temperature dependence of the ideality factors was also investigated. We could find no obvious dependency of the n values on temperature, as shown in Fig. 4.

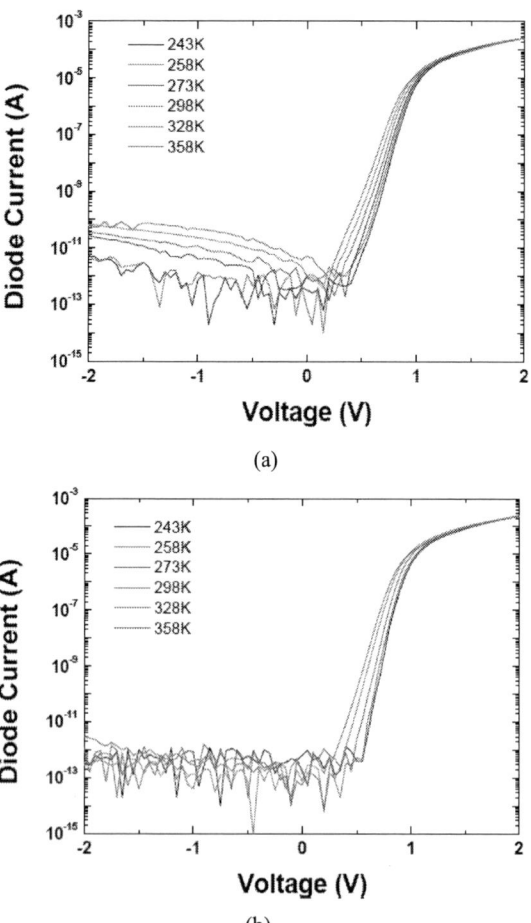

Fig. 3. I-V-T curves of vertical diodes; (a) co-flow SEG-Si, and (b) cyclic SEG-Si diodes

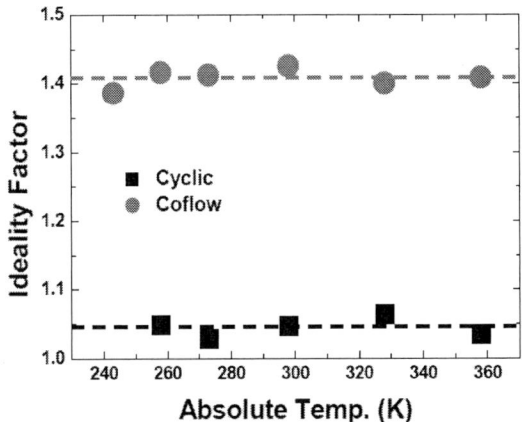

Fig. 4. Temperature dependency of the n values in SEG-Si diodes.

$$I = I_S(e^{q(V-Ir_s)/nkT} - 1) \qquad (3)$$

$$I\frac{dV}{dI} = IR_S + \frac{nkT}{q} \qquad (4)$$

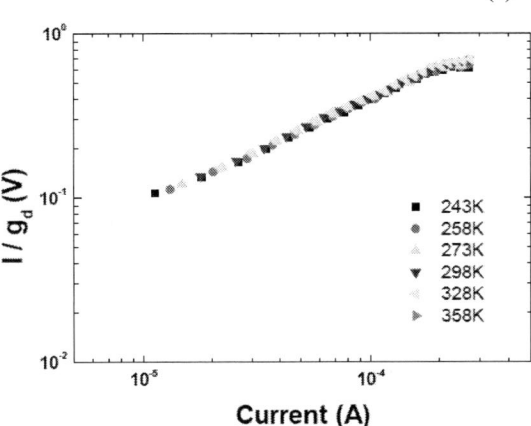

Fig. 5. I/g_d-I curve for co-flow SEG-Si diode

The recombination component at low forward bias regions ($V_A \le 0.5$ V) showed the ideality factors around 3.2 regardless of the formation processes, which implies that the difference in crystallographic defects doesn't play a significant role on the ideality factor. The n value exceeding 2 in the low field regime cannot be explained by the theoretical SRH model. Such non-ideality has been observed in solar cell applications, where the dark I-V characteristics of most silicon solar cells deviate from that expected by classical diode theory, and the recombination current at biases smaller than 0.5V is orders of magnitude higher than expected from the carrier lifetime [10],[11]. Steingrube et al. demonstrated that higher ideality factor than 2 and high recombination current are observed in heavily and artificially damaged pn-junctions by cleaving along (100), by laser cutting, and by diamond scratching on the surface of solar cells [11]. By virtue of such artificial damages, the local density of recombination centers in the pn junction is higher than in the bulk by orders of magnitude. The authors suggested three possible mechanisms; 1) coupled defect level model, 2) deep donor-acceptor-pair recombination, and 3) the local extension of the recombination region across the edge of the cell (due to electrostatic charging). All the three models can clarify the fact that favorable recombination takes place easily via inhomogeneous, continuous distribution of high density trap centers by damages. In this experiment, silicon pillars are surrounded by mold oxide layers deposited by CVD process. Unlike the pn junction of MOSFETs whose sidewall faces STI after thin thermal oxide has been formed to cure the dry etch damage, the cylindrical surface of vertical diodes are facing CVD oxide layers for this experiment. Additional surface states and fixed charge at the perimeter interface are abundant along the silicon pillar/mold oxide interface. The high density of inhomogeneous and continuous trap sites along the diode interface provides convenient recombination paths by the models mentioned above. Although absolute level of recombination current is also affected by the number of crystallographic defects, the convergence of the ideality factors to 3.2 regardless of the defect density strongly imply that the conduction through the inhomogeneous and continuous traps at diode/mold interface is a crucial mechanism in sub-100 nm vertical diodes.

At high bias regime, the forward I-V curve faces additional disturbance including high-level injection and series resistance. Voltage drop by series resistance becomes considerably high in large diode current regime. When series resistance is taken into account, (2) must be modified as (3) where, r_S is series resistance. In order to extract r_s, I/g_d versus I curve is necessary, where g_d represents diode transconductance (dI/dV). The slope of I/g_d-I curve shows the value of r_S, as shown in (4). Fig. 5 shows I/gd-I curves, measured at various temperatures, and the extracted series resistance values of SEG-Si diodes are displayed in Fig. 6. The series resistance of cyclic SEG-Si diodes is ~3300 ohm, which is larger than co-flow SEG by ~300 ohm. No obvious temperature dependence is observed for series resistance. Since forward I-V properties in the high current region generally involves two factors, high-level injection (HLI) and series resistance, the portion of HLI should be considered. However, it is believed that the HLI phenomena in our experiment can be of little effect by the consideration provided below. Sheet resistance (R_{SH}) values of active N+Base region are intimately correlated to the series resistance in Fig. 6. The values of cyclic SEG diodes were larger than that of co-flow SEG by approximately 15 % (Fig. 7). Since the patterns for real I-V contained a number of sheets of N+Base active, total difference in resistance between co-flow and cyclic SEG diodes is 288 ohm, which is consistent with the result in Fig. 6. Higher temperature of co-flow SEG process can activate large amount of dopants in N+Base region.

Reverse I-V properties of SEG-Si diodes measured at various temperatures also show considerable difference according to SEG processes. It is evident that cyclic SEG diodes show lower leakage current than co-flow diodes.

978-1-4673-1707-8/12 $31.00 © 2012 IEEE

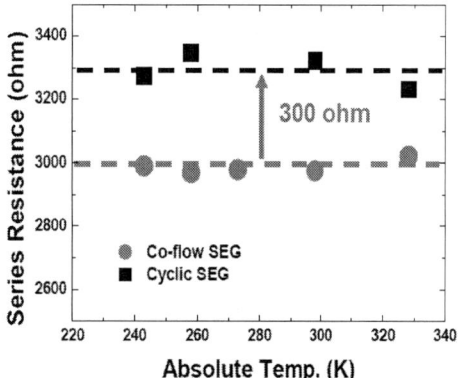

Fig. 6. Series resistance of vertical diodes by SEG process extracted from I/g$_d$-I curve.

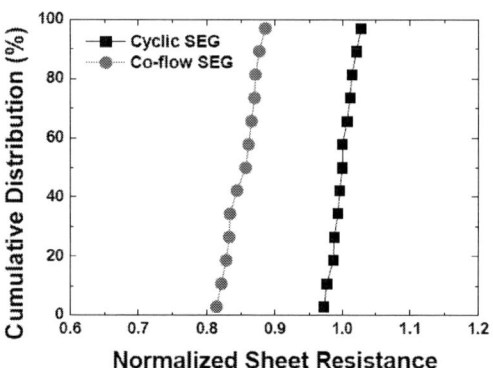

Fig. 7. The distribution of normalized R_{SH} for SEG-Si diodes

Furthermore, the leakage current of co-flow diodes increases much more rapidly than that of cyclic diodes as the temperature increases. The off-current of different diode processes, which was measured at 298 K, is in the same context with the order of the crystallographic defects (cyclic SEG < co-flow SEG) such as staking faults, twins, and perimeter interface according to the diode processes. Since the diameter of diodes pillars are less than 100 nm, we could not find an appropriate electrical and analytical tools or methods to determine the actual concentration of activated dopants in vertical diodes yet. Dopant profiling will be studied in near future.

IV. CONCLUSION

The electrical characteristics of vertical diodes have been studied to compare the performance of different processes. The ideality factors in the diffusion current regime clarify that vertical silicon diodes are still behaving in the similar manners as classical diode theories. In addition, trap levels within the silicon bandgap created by crystallographic imperfections are explicitly related to on- and off-current. The n values exceeding 2 (~ 3.2) in the low field regime ($V_A \leq 0.5V$) is attributed to perimeter interface formed between silicon pillars and surrounding mold layers, which is deduced to be in charge of such serious non-ideality.

REFERENCES

[1] S.C. Oh, J.H. Jeong , W.C. Lim, W.J. Kim, Y.H. Kim, H.J. Shin, J.E. Lee, Y.G. Shin, S. Choi and C. Chung, "On-axis scheme and Novel MTJ structure for sub-30nm Gb density STT-MRAM", *IEDM Tech. Dig.*, pp. 12.6.1 (2010)

[2] Y. Kim, S.C. Oh, W.C. Lim, J.H. Kim, W.J. Kim, J.H. Jeong , H.J. Shin, K.W. Kim, K.S. Kim, J.H. Park, S.H. Park, H. Kwon, K.H. Ah, J.E. Lee, S.O. Park, S. Choi, H.K. Kang, C. Chung, "Integration of 28nm MJT for 8~16Gb level MRAM with full investigation of thermal stability", *Symp. On VLSI Technology*, pp. 210, 2011.

[3] J.H. Oh, J.H. Park, Y.S. Lim, H.S. Lim, Y.T. Oh, J.S. Kim, J.M. Shin, J.H. Park, Y.J. Song, K.C. Ryoo, D.W. Lim, S.S. Park, J.I. Kim, J. Yu, F. Yeung, C.W.Jeong, J.H. Kong, D.H. Kang, G.H. Koh, G.T. Jeong, H.S. Jeong, and Kinam Kim, "Full Integration of Highly Manufacturable 512Mb PRAM based on 90nm Technology", *IEDM Tech. Dig.*, pp. 2.6.1 (2006)..

[4] D.H. Kang, J.S. Kim, Y.R. Kim, Y.T. Kim, M.K. Lee, Y.J. Jun, J.H. Park, F. Yeung, C.W. Jeong, J. Yu, J.H. Kong, D.W. Ha, S.A. Song, J. Park, Y. Park, Y.J. Song, C.Y. Eum, K.C. Ryoo, J.M. Shin, D.W. Lim, S.S. Park, W.I. Park, K.R. Sim, J.H. Cheong, J.H. Oh, J.I. Kim, Y.T. Oh, K.W. Lee, S.P. Koh, S.H. Eun, N.B. Kim, G.H. Koh, G.T. Jeong, H.S. Jeong, and Kiman Kim, "Novel heat dissipating cell scheme for improving a reset distribution in a 512M phase-change random access memory (PRAM)", *Symp. on VLSI Technology*, pp. 96-97, 2007.

[5] K. J. Lee, B. H. Cho, W. Y. Cho, S. B. Kang, B. G. Choi, H. R. Oh, C. S. Lee, H. J. Kim, J. M. Park, Q. Wang, M. H. Park, Y. H. Ro, J. Y. Choi, K. S. Kim, Y. R. Kim, I. C. Shin, K. W. Lim, H. K. Cho, C. H. Choi, W. R. Chung, D. E. Kim, Y. J. Yoon, K. S. Yu, G. T. Jeong, H. S. Jeong, C. K. Kwak, C. H. Kim, and Kinam Kim, "A 90 nm 1.8V 512 Mb Diode-Switch PRAM With 266 MB/s Read Throughput", *IEEE J. Solid-State Circuits*, **43** (2008), pp. 150.

[6] K.S. Lee, D.H. Yoo, J.J Han, Y.W. Hyung, S.S. Kim, C.J. Kang, H.S. Jeong, J.T. Moon, H. Park, H. Jeong, K.R. Kim, and B. Choi, "Selective Epitaxial Growth of Silicon for Vertical Diode Application", *Japanese Journal of Applied Physics*, **49** (2010), pp. 08JF03.

[7] K.S. Lee, J.J. Han, B.H. Kim, H.J. Lim, S.W. Nam, H.K. Kang, C.H. Chung, H.S. Jeong, H. Park, H. Jeong, K.R. Kim and B. Choi, "Highly manufacturable silicon vertical diode switches for new memories using selective epitaxial growth with batch-type equipment", *Semiconductor Science and Technologies*, **26** (2011), pp. 055022.

[8] K.S. Lee, J.J. Han, H.J. Lim, S.W. Nam, C. Chung, H.S. Jeong, H. Park, H. Jeong, and B. Choi, "Cost-Effective Silicon Vertical Diode Switch for Next Generation Memory Devices", *IEEE Electon Device Letters*, **33** (2012), pp. 242.

[9] D.K. Schroder, Semiconductor Material and Device Characterization, 3rd Edition, Wiley Inter-science, 2006.

[10] A. Schenk, and U. Krumbein, "Coupled defect-level recombination: Theory and application to anomalous diode characteristics", *J. Appl. Phys.* **78** (1995), pp. 3185.

[11] S. Steingrube, O. Breitenstein, K. Ramspeck, S. Glunz, A. Schenk, and P.P. Altermatt, "Explanation of commonly observed shunt currents in c-Si solar cells by means of recombination statistics beyond the Shockley-Read-Hall approximation", *J. Appl. Phys.* **110** (2011), pp. 014515.

Si Tunneling Transistors with High On-Currents and Slopes of 50 mV/dec Using Segregation Doped NiSi$_2$ Tunnel Junctions

L. Knoll, Q.T. Zhao, S. Trellenkamp, A. Schäfer, K. K. Bourdelle* and S. Mantl

Peter Grünberg Institut 9 (PGI-9/IT), JARA-FIT, Forschungszentrum Jülich, 52425 Jülich, Germany
* SOITEC, Parc Technologique des Fontaines, 38190 Bernin, France.

l.knoll@fz-juelich.de

Abstract—**Planar and nanowire (NW) tunneling field effect transistors (TFETs) have been fabricated on ultra thin strained and unstrained SOI with shallow doped Nickel disilicide (NiSi$_2$) source and drain (S/D) contacts. We developed a novel, self-aligned process to form the p-i-n TFETs which greatly easies their fabrication by tilted dopant implantation using the high-k/metal gate as a shadow mask and dopant segregation. Two methods of dopant segregation are compared: Dopant segregation based on the "snow-plough" effect of dopants during silicidation and implantation into the silicide (IIS) followed by thermal outdiffusion. High drive currents of up to 60 µA/µm of planar p-TFETs were achieved indicating good silicide/silicon tunneling junctions. The non linear temperature dependence of the inverse subthreshold slope S indicates typical TFET behavior. Strained Si NW array n-TFETs with omega shaped HfO$_2$/TiN gates showed high drive currents of 7 µA/µm @ 1V V$_{dd}$ and steep inverse subthreshold slopes with minimum values of 50mV/dec due to the smaller band gap of strained Si and optimized electrostatics.**

I. INTRODUCTION

Band-to-band tunneling field effect transistors (TFET) are most promising for ultralow power applications. Due to the enhanced switching performance and low power consumption TFETs show distinct advantages as compared to standard CMOS [1]. However, silicon TFETs suffer from very low on-currents, mainly due to the relatively large band gap of

~1.1 eV. A lower band-gap material like strained Si or SiGe enhances the band-to-band tunneling probability and thus the on-currents. Also the doping profile of the tunnel junction affects sensitively the tunneling probability. The advantage of silicon based materials is the availability of the highly advanced Si technology which helps to meet the critical requirements for high-k/metal gate stack and S/D junction formation.

In this paper we make use of this by studying TFETs with silicon and strained Si, which offers a lower band gap than unstrained Si. Along this line we investigated two dopant segregation methods to form the silicide tunnel junctions with abrupt doping profiles at S/D at low temperatures. We investigated planar and nanowire array TFETs on ultrathin SOI and strained SOI.

II. FABRICATION

Planar TFETs were fabricated on SOI substrates with top silicon thicknesses of 15 nm and 6 nm, respectively. A simple process without additional implantation masks was developed as sketched in Fig.1. Using the TiN/HfO$_2$ gate stack as a shadow mask tilted B$^+$ and As$^+$ implantations at an angle of 45° and 135° allow the formation of aligned n$^+$ and p$^+$ regions self-aligned at the gate edges. For B an implantation energy of 1 keV and a dose of 2×10^{15}/cm^2 and for As 5 keV and a dose of 2×10^{15}/cm^2 were used. Silicidation was performed at 700°C to form epitaxial NiSi$_2$ layers for S/D. Simultaneously dopant segregation (DS) occured and formed shallow highly doped pockets at the edge of the silicide, producing p$^+$-pocket on the right side and n$^+$-pocket on the right side of the gate. We assume that only the dopants of the "shadow" regions form the junctions, while the other dopants are incorporated in the metallic silicide and have no electrical effects. This process is labeled "DS" in Fig. 1. Alternatively, epitaxial NiSi$_2$ was formed first, than implanted at tilt angles as stated above and RTP annealed at 700°C to drive-out and activate the implanted dopants and recover the crystal structure. This process is known as "Implantation Into Silicide" (IIS). The silicide is epitaxial NiSi$_2$ formed with only 3 nm Ni [2]. All the processes were performed at fairly low temperatures with T$_{max}$ = 700°C. Planar and nanowire devices were fabricated. Fig. 2 shows a TEM image of the source part of a planar device on 15 nm SOI (left), and an SEM image of an Ω-gated nanowire array device (right). These Si nanowires have a width of 30 nm and a thickness of ~6 nm made from ultrathin

---

This work was partially supported by the EU project STEEPER.

978-1-4673-1707-8/12 $31.00 © 2012 IEEE

SOI with electron beam lithography and dry etching. The HfO$_2$ with a thickness of 3 nm was deposited conformally by atomic layer deposition and TiN by vapor deposition.

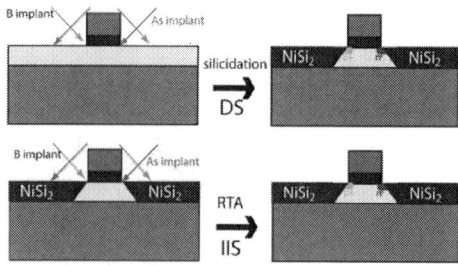

Fig. 1: TFET fabrication process using tilted B$^+$ and As$^+$ ion implants into Si (DS process) or into epitaxial NiSi$_2$ (IIS process) as S/D contacts. Highly doped pockets at the silicide edges are formed by dopant segregation.

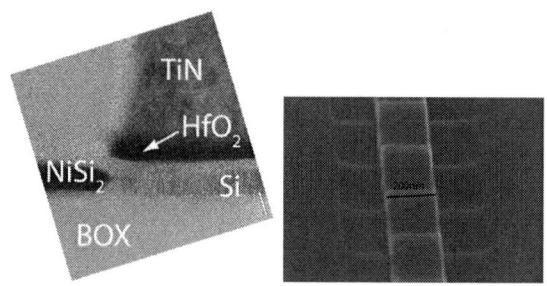

Fig. 2: Cross sectional TEM image of a planar device with epitaxial NiSi$_2$ source after IIS perfectly aligned with the HfO$_2$/TiN gate (left) and an SEM image of a NW-array TFET with a 200 nm HfO$_2$/TiN gate (right)

III.   RESULTS AND DISCUSSION

A TFET consists of a gated p-i-n diode which allows under appropriate reverse bias conditions band to band tunneling. A schematic band structure of a TFET with silicide Schottky contacts and highly doped pockets under reverse bias is shown in Fig. 3. Dopant segregation produces highly doped regions close to the NiSi$_2$ contacts leading to strong band bending. As a consequence, the effective Schottky barrier is lowered to very small values (< 0.2 eV). With other words, the ultrathin barrier becomes transparent via carrier tunneling. By applying this method we have fabricated Schottky-Barrier MOSFETs with output currents exceeding 1mA/µm and remarkable RF performance [3]. Here, we make use of this method for TFETs. As compared to a standard TFET, a Schottky barrier controls the carrier injection from the silicide into the (electron) reservoir formed behind the Schottky barrier. Band to band tunneling is expected between the source reservoir and the channel when an energy window between the source Fermi energy and the valence band edge of the channels opens at appropriate gate and drain bias. For efficient BTBT the tunneling junction should be aligned with the fringing field of the gate edge to obtain a large electric field and the conduction and valence bands should come sufficently close to minimize the tunneling path (see arrow in Fig 3).

Abrupt dopant profiles are needed for this purpose. A great advantage of DS is that very steep dopant profiles can be achieved at low T process [3,4]. In a p-type TFET minority carriers, holes, tunnel into the channel and electrons from the channel into the reservoir. Only at appropriate gate and drain bias a narrow energy window opens to enable BTBT as in Fig. 3.

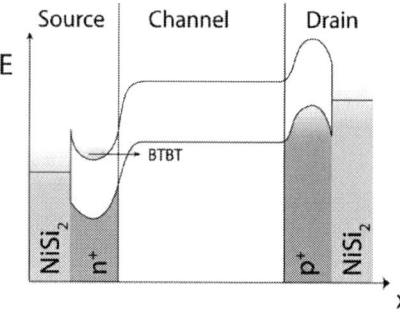

Fig. 3: Schematic band diagram of a TFET with dopant segregation at the NiSi$_2$ Schottky source/drain contacts under reverse bias. Band to band tunneling is indicated beween the n+ Source region and the intrinsic

Fig.4 shows the output characteristics measured at 300 K of a planar p-TFET, which was fabricated with the DS process. The 400nm gate length device shows perfect saturation. The corresponding transfer characteristics are presented in Fig.5. It has to be mentioned that the noise in the off-state stems from the measurement setup and not from gate leakage. The subthreshold regime shows two different slopes labeled as regions I and II with inverse subthreshold slopes (SS) of 94 mV/dec and 190 mV/dec, respectively. Region II indicates a voltage dependent slope, typical for TFETs. Fig. 6 shows the Id-Vg characteristics measured from 100 K to 350 K at V$_d$ = -0.1 V. Region I extends from about 200 K - 350 K. We assume that trap assisted tunneling (TAT) occurs in this region as reported in [5] and [6], while in region II BTBT may dominate. This is further substantiated, by the results of Fig. 7, showing SS(T) of region II. For T > 200 K, SS(T) decreases rapidly, when, as we assume, TAT and BTBT occur simultaneously. At T < 200 K, SS(T) changes less rapidly and becomes even constant at T < 100 K which indicates BTBT. For comparison, the linear SS(T) dependence of a conventional MOSFET fabricated with the same process is presented. We conclude that remaining end of range defects, steming from the ion implantation before the silicidation, cause TAT in the tunneling junction which degrades the TFET performance drastically.

Therefore, we investigated also IIS for tunnel junction formation. The IIS process has the advantage that the dopant implantation occurs into the metallic silicide and not into the Si channel material and the dopants are driven out by thermal annealing. Residual defects in the metallic contacts will have no effect on the tunnel junction behavior. In addition, we further improved the electrostatic gate control by reducing the silicon thickness to 6nm and the formation of nanowire array transistors. These measures improve significantly the natural

978-1-4673-1707-8/12 $31.00 © 2012 IEEE

length λ, a measure for good gate control [7]. The IIS process is also advantageous for doping of nanowires which is anyway a complex issue due to dopant deactivation [8].

devices of Fig. 4 the improvement may come partly from a higher dopant concentration in the junction since we used the same implant parameters as for the somewhat thicker films

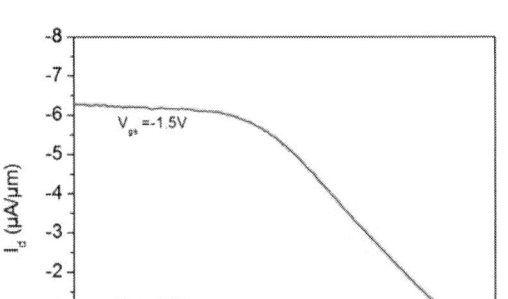

Fig. 4: Output characteristics of a 400 nm planar p-TFET at 300 K

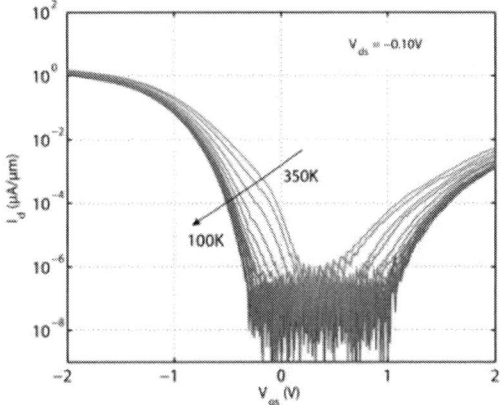

Fig. 6: Id-Vgs characteristics of a planar p-TFET measured between 350 K and 100 K

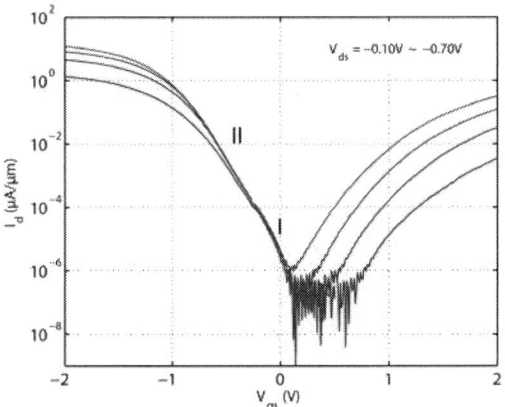

Fig. 5: Transfer characteristics of a p-TFET at 300 K

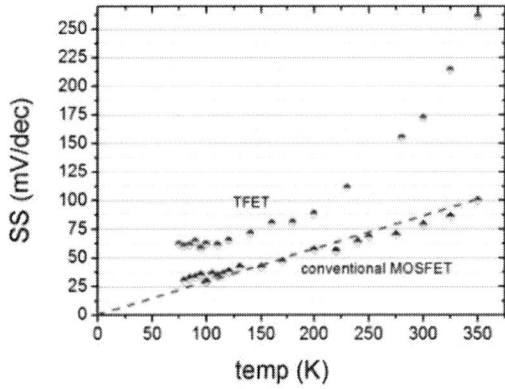

Fig. 7: Temperature dependence of SS extracted from region II of a planar TFET in comparision with aconventional p-MOSFET.

The TFET performance was further improved by using nanowire structures fabricated with 6nm SOI substrates. Fig.8 shows the transfer characteristics of a nanowire array p-TFET with a 200 nm HfO_2/TiN gate at 300 K. The nanowire has a cross section of 30×6 nm^2. In this case the IIS process was employed with an activation temperature of 700°C after tilted ion implantations into the single crystalline $NiSi_2$ layers. The average slope in the Id range, 10^{-7} to 10^{-4} µA/µm, amounts to ~90 mV/dec. In contrast to Fig. 6, no kink of the transfer curve appeared, which indicates a more perfect tunnel junction due to the IIS process. Compared to DS devices of Fig. 4 and Fig. 5, IIS generates obviously less or no damage to the silicon. We assume that the absence of implantation induced damage in the silicon led also to a better silicide-gate alignment which helps to enhance the tunneling current as explained above. The on-current reaches a record value of 60 µA/µm at $V_d = -0.7$ V and $V_g = -1.5$ V for Si NW TFETs. As compared to the

Even better performance was achieved by using strained SOI (SSOI). The strained Si of SSOI wafers, supplied by SOITEC has a tensile strain of 1%, corresponding to a stress of 1.4 GPa. Previously, we have shown that patterning of the layer into nanowires transforms the biaxial strain into uniaxial strain since the strain across the narrow wire relaxes while the strain along the wire fully maintains. This leads to an enhancement of the electron mobility and the output current of about a factor of two [8]. Since the BTBT transmission probability in a TFET decreases exponentially with band gap of the junction material, a small reduction of the band gap as in the strained Si nanowires will lead to a further improvement of the TFET. Due to the high resistance of the tunnel junction the enhancement of the mobility will have less effect than in a normal MOSFET.

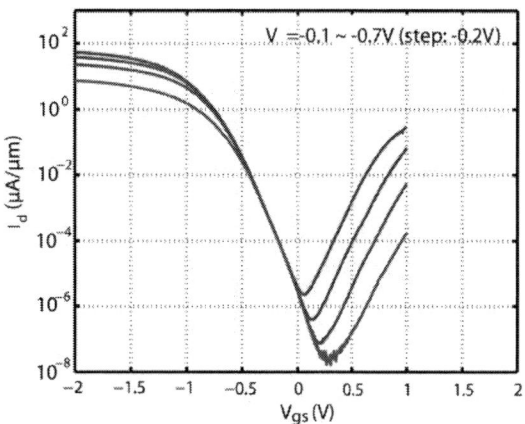

Fig. 8: Id- Id-Vgs characteristics of a 200 nm gate length nanowire array p-TFET measured at 300K

The uniaxial strained Si (sSi) nanowires have a smaller band gap, around 0.8 eV as compared to 1.1 eV of Si, which is beneficial for BTBT. Fig. 9 shows the transfer characteristics of sSi nanowire array n-TFETs with a nanowire dimensions of 30×10 nm^2. In addition, the gate leakage current I_g is also shown. The minimum slope measured at $V_d = 0.18$ V amounts to 50 mV/dec, clearly below values attainable in a normal MOSFET. The on-current at $V_d = 0.5$ V reaches 7 µA/µm which is also high compared with published values. Interestingly the n-type TFET, shown in Fig. 9, improved significantly. The out-diffusion was performed at 450°C instead of 700°C to minimize Boron diffusion. Thus, we assume that the slower Boron diffusion in strained silicon is beneficial for n-type TFETs. The IIS process may also be more suited to maintain the strain in the wires, since no ion implantation occurs in the Si channel region of the device.

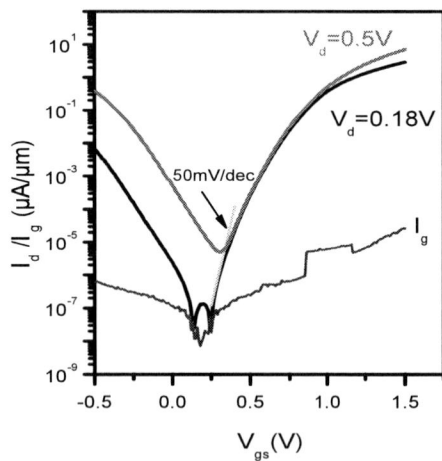

Fig. 9: Id- Id-Vgs characteristics of strained Si nanowire array n-TFET with a gate length of 200 nm measured at 300K.

IV. CONCLUSIONS

A simple, low temperature process for TFET fabrication has been developed using tilted implantation to from the junctions at S/D. Since no implantation masks are needed further scaling should be easily possible. While the DS process gives rise to TAT, IIS seems much more appropriate to form silicided tunnel junctions providing with very large on currents for Si TFETs. For both, DS and IIS, a maximum T of only 700°C is sufficient to activate the dopants. As compared with planar devices Si nanowire array TFETs showed better slopes and very high on-currents up to 60 µA/µm at V_{dd}=-0.7 V. A remarkable improvement was achieved with 30×10 nm^2 NW array TFETs with uniaxial strained Si nanowires delivering a minimum slope of 50 mV/dec over 3 orders of magnitude of I_d at 300K.

Acknowledgement

This work is supported by European project STEEPER.

REFERENCES

[1] Adrian M. Ionescu and Heike Riel, "Tunnel field-effect transistors as energy-efficient electronic switches", Nature 16;479(7373):329-37, 2011

[2] L. Knoll, Q.T. Zhao, S. Habicht, C. Urban, B. Ghyselen and S. Mantl, "Ultrathin Ni Silicides With Low Contact Resistance on Strained and Unstrained Silicon", *IEEE Electron Device Lett., 31(2010)350*

[3] C. Urban, C. Sandow, Q.T. Zhao, J. Knoch , S. Lenk and S. Mantl, "Systematic study of Schottky barrier MOSFETs with dopant segregation on thin-body SOI", Solid-State Electronics 54 (2010) 185–190, 2010

[4] S. F. Feste *et al*," Formation of steep, low Schottky-barrier contacts by dopant segregation during nickel silicidation", J. Appl. Phys. 107, 044510, 2010

[5] D. Leonelli et al, "Silicide Engineering to Boost Si Tunnel Transistor Drive Current", *J. J. Appl. Phys. 50, 04DC05, 2011*

[6] E. Simoen, F. De Stefano, G. Eneman, B. De Jaeger, C. Claeys and F. Crupi, "On the Temperature and Field Dependence of Trap-Assisted Tunneling Current in Ge p+n Junctions"., *IEEE Electron Device Lett., 30(2009)562*

[7] J. Knoch, "Optimizing tunnel FET performance – Impact of device structure, transistor dimensions and choice of material", VLSI-TSA, 2009

[8] Z. Zhang, Z. Qiu, R. Liu, M. Östling and S.-L. Zhang, "Schottky-Barrier Height Tuning by Means of Ion Implantation Into Preformed Silicide Films Followed by Drive-In Anneal", *IEEE Electron Device Lett., 28, no. 7, 2007*

[9] S. Habicht, S. Feste, Q.-T. Zhao, Dan buca and S. Mantl, „ Electrical characterization of Ω-gated uniaxial tensile strained Si nanowire-array metal-oxide-semiconductor field effect transistors with <100>- and <110> channel orientations", *Thin Solid Films Vol. 520, Issue 8, 2012*

A Comparative Analysis of Tunneling FET Circuit Switching Characteristics and SRAM Stability and Performance

Yin-Nien Chen, Ming-Long Fan, Pi-Ho Hu, Ming-Fu Tsai, Chia-Hao Pao, Pin Su and Ching-Te Chuang

Department of Electronics Engineering & Institute of Electronics, National Chiao Tung University, Hsinchu, Taiwan

E-mail:Chingte.chuang@gmail.com

Abstract—**With steep sub-threshold slope, tunneling FETs (TFETs) are promising candidates for ultra-low voltage operation, achieving low leakage current and superior performance compared with the conventional MOSFETs. However, the broad soft transition region in the Id-Vgs characteristics, where Id increases slowly to reach saturation following the steep slope region, results in large cross-over region/current in an inverter, thus degrading the Hold/Read Static Noise Margin (H/RSNM) of TFET SRAMs. The Write-ability and Write Static Noise Margin (WSNM) of TFET SRAMs are constrained by the uni-directional conduction characteristics caused by the asymmetric source-drain structure and large cross-over contention of the Write access transistor and the holding transistor. In this paper, we present a detailed analysis of TFET circuit switching characteristics/performance and compare the stability/performance of several TFET SRAM cells using atomistic TCAD mixed-mode simulations. A robust 7T Driver-Less (DL) TFET SRAM cell is proposed. The proposed 7T DL TFET SRAM cell, with decoupled Read current path from cell storage node and push-pull Write action with asymmetrical raised-cell-virtual-ground Write-assist, provides significant improvement in Read/Write stability and performance.**

I. INTRODUCTION

Voltage scaling is an efficient way to achieve ultra-low power consumption of circuits. However, continual reduction of supply voltage drives MOSFET digital circuits into sub-threshold operation, causing exponential increase in the delay. The fundamental limit of the sub-threshold swing of MOSFET device is 60mV/dec at 300K. Reduction of V_T (threshold voltage) of MOSFET device to maintain a reasonable drive current at low supply voltage inevitably causes dramatic increase of leakage current. Recently, Tunneling FET (TFET) which utilizes the band-to-band tunneling as the conduction mechanism have emerged as one of the most promising candidates for ultra-low voltage/power operation [1-5]. Several studies of TFET device demonstrating the progress of fabrication and experimental results have been reported [6-9]. The results provide promising demonstration of steep sub-threshold slope TFET devices. One of the major application constraint of TFET is the uni-directional current-conduction characteristic. For logic circuits, the uni-directional characteristics does not present a serious problem, except for pass-transistor based circuits for which an TFET with opposite current-conducting direction can be added at the expense of area/performance if bi-directional characteristics is required. For SRAMs, however, the uni-directional characteristics severely impact the robustness due to conflicting Read/Write requirements and stringent cell area constraint [10-13]. In this work, we present a detailed analysis of TFET circuit switching characteristics/performance and compare the stability/performance of several TFET SRAM cells using atomistic TCAD mixed-mode simulations. A 7T Driver-Less (DL) TFET SRAM cell is proposed. The robustness and performance advantages of the proposed 7T DL TFET SRAM cell are demonstrated.

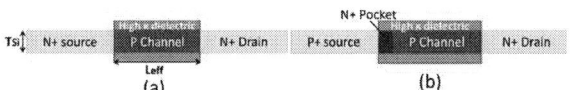

Fig. 1. Structures of (a) n-type MOSFET and (b) n-type PNPN TFET

TABLE I.

	MOS	TFET(PNPN)
L_{eff}	25nm	25nm
T_{Si}	6nm	6nm
EOT	0.6nm	0.6nm
Source Doping	2E20 cm^{-3}	2E20 cm^{-3}
Drain Doping	2E20 cm^{-3}	2E20 cm^{-3}
Pocket Doping	--	1.2E20 cm^{-3}

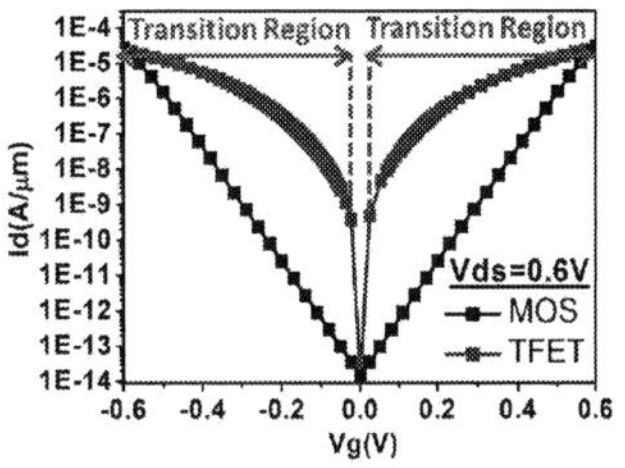

Fig. 2. Ids-Vgs characteristics at Vds = 0.6V of n-/p-type DG MOSFETs and DG PNPN/NPNP TFETs with the same Ioff=1.4E-14 A/μm and Table-I: Parameters used in the study for MOSFET and PNPN TFET

II. DEVICE STRUCTURES AND CHARACTERISTICS AND TCAD SIMULATION METHODOLOGY

Device Structure and TCAD Simulation Methodology

In this work, we consider the PNPN type TFET [5] for its capability to achieve sub-threshold swing well below 60mV/dec at 300K, and hence a superior I_{on}/I_{off} ratio at ultra-low voltage. The conventional MOSFET is also considered for comparison. The n-type device structures are shown in Fig. 1. Double-Gate (DG) structures are used, and the device parameters are listed in Table-I. Fig. 2 shows the Ids-Vgs characteristics of the TFET (PNPN) and MOSFET (MOS) at Vds = 0.6V. The TFET (PNPN) and MOSFET (MOS) devices/circuits are analyzed using atomistic TCAD mixed-mode simulations [14]. The nonlocal band-to-band tunneling model which is applicable to arbitrary tunneling barrier with non-uniform electric field is used for TFET simulations. The tunneling paths are dynamically determined according to the gradient of the band energy. The TFET (PNPN) device I_{on}/I_{off} ratio is calibrated with [5] and the off-state current is calibrated with available Si TFET experimental data [6, 7]. The DG MOSFET device with same off-sate current is used for fair ultra-low power operation comparison.

Device and Circuit Switching Characteristics

Fig. 3(a) shows the I_{on}-I_{off} comparison between TFET and MOSFET with different supply voltages from 0.6V to 0.2V. As can be seen, the on current degradation of MOSFET is much more severe than TFET as Vdd scales down, since MOSFET would operate in sub-threshold regime. On the other hand, the reduction of off current

978-1-4673-1707-8/12 $31.00 © 2012 IEEE

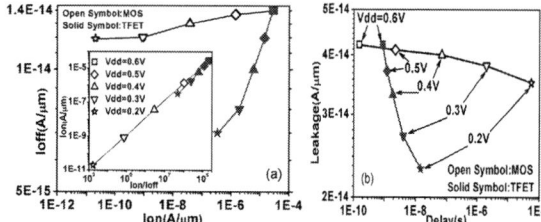

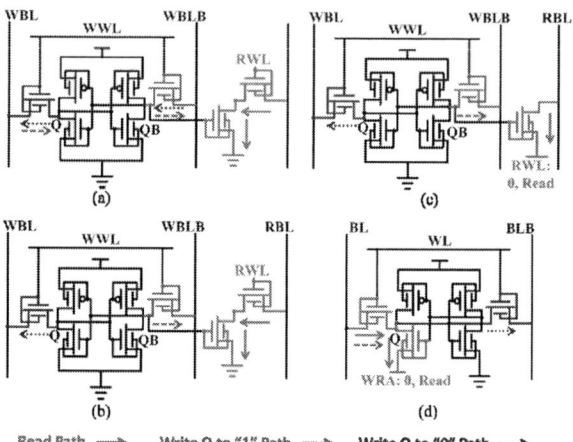

Fig. 3 (a) Ion-Ioff comparisons of DG MOSFET and DG PNPN TFET at various supply voltages. Inset shows Ion-Ion/Ioff ratio. (b) Leakage-delay comparisons of five-stage inverter chain (FI = FO = 1) with MOSFET devices and TFET devices at various supply voltages.

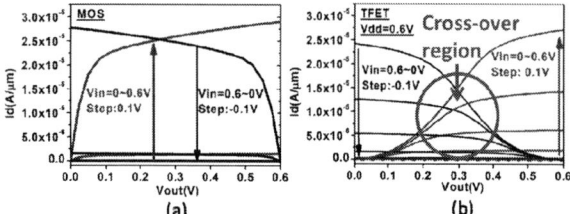

Fig. 4 Device switching characteristics of (a) DG MOSFET and (b) DG PNPN in an inverter.

of TFET exhibits significantly stronger/better dependence on Vdd compared with the MOSFET counterpart, since the off current of TFET is dominated by band-to-band tunneling leakage with stronger Vdd sensitivity than the off current of MOSFET dominated by the sub-threshold leakage. Thus, TFET provides superior I_{on}/I_{off} ratio of about eight orders of magnitude even at Vdd=0.2V (inset of Fig. 3(a)), validating its capability to operate at ultra-low voltage with low leakage while maintaining performance.

Fig. 3(b) shows the leakage-delay comparison of five-stage inverter chain (FI=FO=1) with TFET and MOSFET devices. As can be seen, at Vdd=0.6V, the leakage current of TFET inverter is the same as the MOSFET inverter, while the delay of TFET inverter is larger than that of MOSFET inverter due to its smaller on current at Vdd=0.6V. As Vdd scales down, TFET inverter exhibits superior leakage-delay performance compared with MOSFET inverter due to its inherently better I_{on}/I_{off} characteristics at low supply voltages.

Fig. 4(a) and 4(b) show the switching Id-Vds characteristics of TFET and MOSFET devices in an inverter. There are two distinct features for the TFET case. First, at low Vds, the channel potential pinning effect locates the TFET Id-Vds in the tunnel-junction-limited region. With enough Vds, a small increase in Vds causes the current to increase rapidly resulting in upward concaved shape for the switching Id-Vds characteristics in Fig. 4(b). Second, due to the broad soft transition region following the initial steep slope region (the region in the Id-Vgs characteristics in Fig. 2 where Id increases slowly with Vgs), the cross-over region (where both n-type and p-type TFET conduct) for the TFET inverter is significantly larger/broader than that for the MOSFET inverter. These two features, together with the characteristics of uni-directional current-conduction, play important roles in determining the stability and performance of TFET SRAMs.

III. TUNNELING FET BASED SRAMs

Several prior studies on TFET SRAMs used artificially-built Verilog-A look-up tables [10-13] to model the uni-directional current-conduction to bypass the physics based TCAD modeling/simulations. In this work, we conduct the investigation on the stability and performance of TFET SRAMs using direct atomistic TCAD mixed-mode simulations to capture the underlying physics

Fig. 5 Cell structures and corresponding Read/Write paths for: (a) conventional 8T MOSFET SRAM, (b)8T TFET SRAM, (c)7T TFET SRAM [10] and (d) 6T TFET SRAM [11].

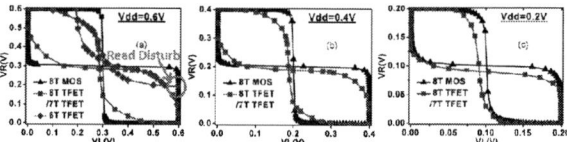

Fig. 6 Comparisons of butterfly curves of 8T MOSFET SRAM cell and 7T/8T TFET SRAM cells in Hold/Read modes. Also shown in (a) is the butterfly curve for 6T TFET SRAM cell.

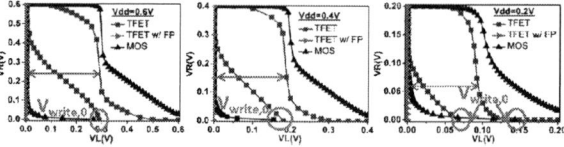

Fig. 7 Comparisons of butterfly curves of 8T MOSFET SRAM cell and 7T/8T TFET SRAM cells in Write mode.

and detailed device operation/switching characteristics. In the following sections, we examine the stability of published TFET SRAM cells and the conventional 8T TFET SRAM cell. The conventional 8T cell [15] is considered due to its technical viability with uni-directional TFET, as opposed to the conventional 6T SRAM cell which does not function with uni-directional TFET. The effectiveness of Write-assist circuit techniques for TFET SRAM is also explored.

Fig. 5 shows the cell structures and corresponding paths of Read/Write operation of published TFET SRAM cells (Fig. 5 (c)(d)), the conventional 8T MOSFET SRAM cell (Fig. 5 (a)) and 8T TFET SRAM cell (Fig. 5 (b)).

Read/Hold Modes:

As can be seen, the 6T TFET SRAM cell (Fig. 5(d)) utilizes inward access transistor for Read, hence suffers from large Read-disturb and shows worse Read Static Noise Margin (RSNM) (smaller than 50mV at Vdd=0.6V (Fig. 6(a)). Both the 7T TFET SRAM cell and the 8T TFET SRAM cell utilize the Read buffer to decouple the Read current path from the cell storage node to eliminate the Read-disturb, except that the 7T TFET SRAM cell uses one transistor instead of two stacked transistors in the 8T TFET SRAM cell for Read. Notice that one transistor Read in the 7T TFET cell is possible due to the uni-directional current flow characteristics of TFET that prohibits the

978-1-4673-1707-8/12 $31.00 © 2012 IEEE

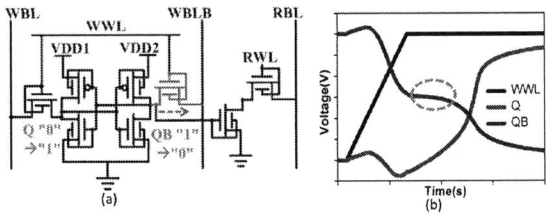

Fig. 8 (a) Write operation of 8T TFET SRAM cell and (b) the corresponding Write transient waveforms.

current from the Read transistors of unselected cells on the selected bit-line from flowing into the bit-line (The forward-bias current is negligible). Fig. 6 compares butterfly curves of the 8T MOSFET SRAM cell and 7T/8T TFET SRAM cells in Hold and Read modes at Vdd=0.6V, 0.4V and 0.2V, respectively. As can be seen, the transition of the Voltage Transfer Characteristics (VTC) of 7T/8T TFET SRAM cells is not as sharp as that for the 8T MOSFET SRAM cell. This is due to the broader cross-over region in TFET inverter discussed earlier, which degrades the Hold/Read SNM. However, as Vdd scales down, this effect becomes less significant as the soft transition region in the TFET Id-Vgs characteristics is reduced, and the Hold/Read SNM approaches that for the 8T MOSFET SRAM cell.

Write Mode:

As shown in Fig. 5, all TFET SRAM cells use one transistor during "Write" operation, since unlike MOSFET SRAM cell, TFET transistor can only conduct current uni-directionally. As such, the Write Static Noise Margin (WSNM) of TFET SRAM cells are significantly worse than that of the 8T MOSFET SRAM cell (with push-pull action during Write) as show in Fig. 7. There are three major obstacles faced by the TFET 7T/8T SRAM cells during Write. First, since the access transistor is faced-outward (pull-down of the storage node), there is no "push" action hence no "disturb" in the Write "1" VTC, resulting in degraded WSNM. Second, while the Write "0" disturb is comparable among 8T MOSFET SRAM cell and 7T/8T TFET SRAM cells, due to the broader cross-over region in TFET inverter, the Write "0" VTC of 7T/8T TFET SRAM cells stay considerably higher than the Write "0" VTC of the 8T MOSFET SRAM cells, thus further degrading the WSNM. Similar to the case for Hold/Read SNM, this effect becomes less significant as Vdd scales down. Third, the use of only one (pull) transistor during "Write" operation and the broad cross-over region between competing currents from the access n-type TFET and the holding p-type TFET cause a plateau in the low-going cell storage node voltage (dashed circle region in Fig. 8(b)), thus retarding the Write-ability and Write Performance. At beginning of the Write operation, the outward-faced access n-type TFET pulls QB down. As QB goes "low", the holding current from the holding p-type TFET rapidly increases, forming a plateau region with QB decreasing slowly due to the large cross-over region until QB and Q cross each other and the latch feedback effect kicks in to complete the Write operation.

Write-Assist:

In order to improve the Write-ability and Write performance of TFET SRAM cells without adding cell transistors (e.g. using full transmission gate instead of single pass-transistor improves the Write operation at the cost of adding two p-type access transistors), Write-assist circuit techniques such as footed virtual ground (as that used by the 6T TFET SRAM cell in Fig. 5(d)) and Floating- Power (FP) line Write [16] can be used. In Fig. 7, we show that the use of

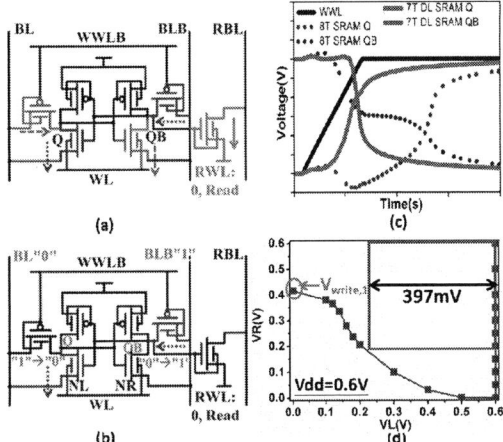

Fig. 9 (a) Cell structure and corresponding Read/Write paths of the proposed 7T Driver-Less (DL) TFET SRAM cell, (b) schematics of Write operation, (c) Write transient waveforms for the proposed 7T DL TFET SRAM cell and 8T TFET SRAM cell, and (d) WSNM of 7T DL TFET SRAM cell.

Floating Power (FP) Write-assist improves the WSNM significantly for the 7T/8T TFET SRAM cells.

IV. 7T DRIVE-LESS TFET SRAM

Based on the analysis in Section III, the use of extra reading transistor to decouple the Read current path from the storage node eliminates Read-disturb and enhances RSNM; while the uni-directional characteristics and broad soft Id-Vgs transition region of TFET device significantly reduce the WSNM. Thus there is a need for a TFET SRAM with better Write-ability and Write performance.

Operation Principles of 7T Drive-Less TFET SRAM:

Fig. 9(a) shows the proposed 7T Driver-Less (DL) TFET SRAM cell structure and the corresponding Read/Write paths. The 4T DL FinFET SRAM cell introduced in the literature [17] exploits the capability of independent-gate control, using a single double-gate FinFET device to replace the function of two conventional transistors. The main idea is to combine the access transistor and pull-down transistor into one independently-controlled-gate pull-down device (NL/NR) with its back-gate serving as the access transistor. The source node of NL/NR is connected to bit-line BL/BLB, which are pre-discharged to "Low" state. We adopt the driver-less scheme and add two p-type TFET as Write access transistors on each side with their source nodes connected to BL/BLB. During Write "0" operation, BLB, i.e. the source node of the NR, is charged "High", thus facilitating the pull-up of the cell storage node QB (Fig. 9(b)). As such, $V_{write,1}$ becomes very close to Vdd as shown by the vertical straight line near Vdd (0.6V) in Fig. 9(d), resulting in significantly improved WSNM compared with the 8T TFET SRAM. The added p-type access TFET (pass "1" better than n-type TFET) helps pull QB to "High" state while NL pulls Q to "Low" state, thus eliminating the plateau region for the low-going cell storage node QB and significantly improving the Write performance compared with the 8T TFET SRAM (Fig. 9(c)). In essence, the proposed 7T DL TFET SRAM cell provides push (p-type Write access TFET) and pull (e.g. NL) Write action, as well as asymmetrical raised-cell-virtual-ground Write-assist (e.g. raised NR source node). Notice that due to the uni-directional characteristic of TFET device, the Write-disturb in the half-selected cells on the selected bit-line through the half-on NR devices [18] is eliminated as shown in Fig. 10.

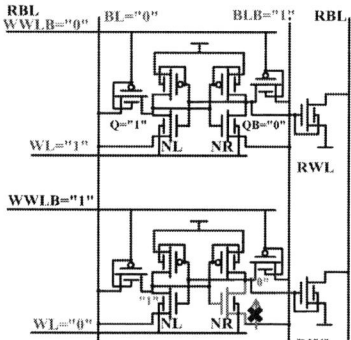

Fig. 10 Illustration of eliminated Write-disturb in half-selected cells during Write operation in 7T DL TFET SRAM cell.

Comparisons of Stability and Performance of TFET SRAMs:

We now compare the stability and performance of 8T MOSFET SRAM cell, 8T TFET SRAM cells w/ and w/o Floating-Power (FP) Write-assist, and the proposed 7T DL TFET SRAM cell.

Stability:

Fig. 11(a) compares the Hold/Read Static Noise Margin (H/RSNM) of 8T MOSFET SRAM cell, 8T TFET SRAM cell and 7T DL TFET SRAM cell. As can be seen, the proposed 7T DL TFET SRAM cell has slightly better Hold/Read SNM than the 8T TFET SRAM cell, and both approach that of the 8T MOSFET SRAM cell at Vdd=0.2V. Fig. 11(b) compares the Write Static Noise Margin (WSNM) of the 8T MOSFET SRAM cell, 8T TFET SRAM cell w/o and w/ Floating-Power (FP) Write-assist, and the proposed 7T DL TFET SRAM cell. The proposed 7T DL TFET SRAM cell can be seen to offer the best WSNM among all cells.

Performance:

We consider SRAM array with 16 cells per bit-line. Fig. 11(c) shows the Read access time versus Vdd for the various cells considered. The Read access time is defined as the time from when the selected Word-Line (WL) reaches half-Vdd to when the Read Bit-Line (RBL) is pulled down to half-Vdd. The worst–case bit-line data pattern is considered. As can be seen, the Read performance of TFET SRAMs outperforms MOSFET SRAM by about one and three orders at Vdd=0.4V and Vdd=0.2V, respectively. Fig. 11(d) compares the Time-to-Write, defined as the time from the 50% activation of the word-line to the time when the voltages of two cell storage nodes cross each other. The Write performance of TFET SRAMs, can be seen to also outperforms MOSFET SRAM by about one and three orders at Vdd=0.4V and Vdd=0.2V, respectively. Furthermore, the proposed 7T DL TFET cell shows over 30% Write performance improvement compared with the 8T TFET SRAM using Floating-Power Write-assist.

V. CONCLUSIONS

We have presented a detailed comparative analysis of TFET circuit switching characteristics/performance and stability/performance of several TFET SRAM cells using atomistic TCAD mixed-mode simulations. Our results indicated that TFET SRAM cell without separating Read current path from the storage node had degraded RSNM due to large Read disturb and broad soft transition region following the initial steep slope region in the Id-Vgs characteristics which resulted in large (inverter) cross-over region/current. TFET SRAMs with single-type access transistor for Write suffered from lack of push-pull action due to TFET's uni-directional conduction characteristics and large cross-over contention from the holding transistor, resulting in insufficient WSNM

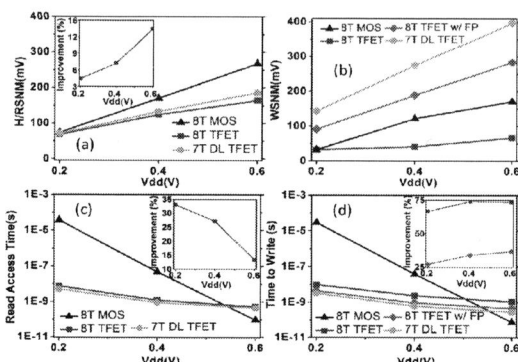

Fig. 11 Comparisons of stability and Performance of 8T MOSFET SRAM cell, 8T TFET SRAM cell w/ and w/o floating-power Write-assist, and the proposed 7T DL TFET SRAM cell in (a) Hold/Read Static Noise Margin, and (b) Write Static Noise Margin. (c) Read Access Time considering worst-case bit-line pattern, and (d) Time-to-Write. The inset of (a) shows the margin improvement of 7T DL TFET SRAM cell with respect to the 8T TFET SRAM cell. The inset of (c) shows the Read performance improvement of 7T DL TFET SRAM cell with respect to the 8T TFET SRAM cell, and the inset of (d) shows the Write performance improvement of 7T DL TFET SRAM cell with respect to 8T TFET SRAM cell w/ and w/o FP Write-assist

and necessitating Write-assist. A robust 7T DL TFET SRAM cell was proposed. The proposed 7T DL TFET SRAM cell, with decoupled Read current path from cell storage node and push-pull Write action with asymmetrical raised-cell-virtual-ground Write-assist, provided significant improvement in Read/Write stability and performance.

ACKNOWLEDGMENT

This work was supported in part by the Ministry of Education in Taiwan under the ATU Program, by the Ministry of Economic Affairs in Taiwan under Contract 100-EC-17-A-01-S1-124, and by National Science Council of Taiwan. The authors are grateful to National Center for High-Performance Computing in Taiwan for computational facilities and software supports.

REFERENCES

[1] K. K. Bhuwalka et al., *IEEE Trans. Electron Devices*, vol. 51, no. 2, pp. 279–282, Feb. 2004.

[2] C. Aydin et al., *Appl. Phys. Lett.*, vol. 83, no. 8, pp. 1653–1655, Aug. 2003.

[3] Q. Zhang et al., *IEEE Electron Device Lett.*, vol.27, no.4, pp. 297–300, 2006.

[4] K. Boucart et al., *IEEE Trans. Electron Devices*, vol. 54, no. 7, pp. 1725–1733, Jul. 2007.

[5] A. Tura et al., *Trans. Electron Devices*, vol. 57, no. 6, pp. 1362–1368, 2010.

[6] F. Mayer et al., *IEDM. Tech. Dig.*163, 2008.

[7] R. Gandhi et al., *Electron Device Lett.*, vol. 32, no. 4, pp. 437-439, 2011.

[8] A. C. Ford et al., *Applied Physics Letters*, vol. 98, no. 11, pp. 113105, 2011.

[9] S. H. Kim et al., *VLSI Symp. Tech. Dig.*, 2009, pp. 178–179.

[10] D. Kim et al., *Symp. Low Power Electronics and Design*, pp. 219–224, 2009.

[11] J. Singh et al., *Proc. ASP-DAC*, pp.181-186, 2010.

[12] X. Yang et al., *Proc. Design Automation and Test in Europe*, pp. 1-6, 2011.

[13] V. Saripalli et al., *Symp. Nanoscale Architectures*, pp. 45-52, 2011.

[14] *Sentaurus User's Manual*, 2011.

[15] L. Chang, et al., *Symp. VLSI Tech. Dig.*, pp. 128-129, 2005.

[16] M. Yamaoka, et al., *ISSCC Digest of Tech. Papers*, pp. 480-481, Feb. 2005.

[17] B. Giraud et al., *Proc. Int. Conf. Circuits Syst.*, pp. 3022–3025, 2007.

[18] M.-L. Fan et al., *Trans. Electron Devices*, vol. 58, no. 3, pp. 609–616, 2011

978-1-4673-1707-8/12 $31.00 © 2012 IEEE

Tunnel FET with Non-Uniform Gate Capacitance for Improved Device and Circuit Level Performance

Cem Alper, Luca De Michielis, Nilay Dagtekin, Livio Lattanzio and Adrian M. Ionescu

Ecole Polytechnique Fédérale de Lausanne, Switzerland

Abstract—We propose and report the significant improvement obtained by a non-uniform gate capacitance made by appropriate combination of high-k and low-k regions over the tunneling and the channel regions of a heterostucture TFET (called HKLKT-FET). In addition to significantly enhanced I_{ON} and subthreshold swing, we find that this structure offers great improvements for the dynamic switching energy (66% saving) and propagation delay (∼3X fast operation) compared to a heterostructure TFET (HeTFET) due to the reduction of the Miller effect. We compare and benchmark the proposed device against a 65nm low stand-by power (LSP) CMOS technology, and we show that at a supply voltage of $V_{DD} = 0.4V$, TFETs can have smaller propagation delays compared to CMOS operating in the subthreshold region.

I. INTRODUCTION

In order to address the increasing demand for low-power digital applications, CMOS circuits operating in the subthreshold region have been proposed as a compromise between dynamic power and performance. Subthreshold logic that uses MOSFETs operating at subthreshold supply voltages have already been demonstrated [1]. Although significant power reduction due to supply voltage scaling can be attained by this approach, the performance of such a technology is limited due to the thermionic emission that dictates a minimum subthreshold swing (SS) of 60mV/dec at room temperature.

Tunnel FETs (TFETs) have been introduced as one of the major candidate overcoming the supply voltage scaling issues of MOSFET devices. Unlike MOSFETs, TFETs are not limited by the thermionic emission constraint, therefore they can offer steeper switching, becoming very attractive especially for low supply voltage applications. Indeed, sub-60 mV/dec operation has been experimentally achieved for various material configurations such as carbon nanotube TFETs [2], or Germanium source devices [3]. At the same time, the obvious choice of all-Silicon based implementations are not suitable for optimum TFET performance indirect and large bandgap of Silicon, which results in extremely low ON currents (I_{ON}). Therefore, heterojunction TFET has emerged as a promising candidate, which aims to benefit from the narrower bandgap of materials such as Germanium or Indium Arsenide [4]. In parallel, the choice of high-k materials as gate oxides is a very common solution to enhance the I_{ON} of TFETs [5], [6]. These materials enable to achieve, for a given oxide thickness, higher I_{ON} due to increased band bending at a given voltage, as the electrostatic coupling from the gate is improved.

However, the increased electric field caused by the high-k gate oxide can also result in significant band-to-band tunneling at the drain-channel interface and may serve as an ambipolar leakage mechanism [7]. Furthermore, the high-k oxide also increases the total gate capacitance of the device and in consequence, energy per switching rises since it is proportional to the load capacitance ($E_{sw} \sim C_L V_{DD}^2$).

Being the tunneling process highly localized around the source-channel interface (see figure 1), this paper will show how to drastically suppress these disadvantages of high-k gate oxide without degrading the device performance.

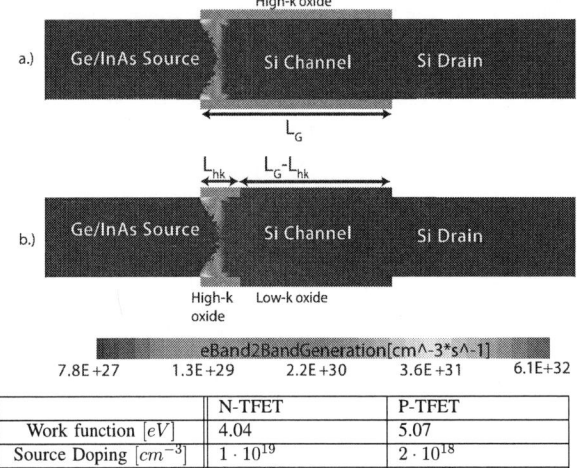

Fig. 1. Device Structures for (a) HeTFET (b) HKLKTFET with band-to-band tunneling generation for electrons denoted. Besides the parameters reported in the table, the simulated devices feature an oxide thickness $t_{ox} = 2.5nm$, gate length $L_G = 50nm$ and silicon body thickness $t_{Si} = 20nm$.

	N-TFET	P-TFET
Work function $[eV]$	4.04	5.07
Source Doping $[cm^{-3}]$	$1 \cdot 10^{19}$	$2 \cdot 10^{18}$
Drain Doping $[cm^{-3}]$	$1 \cdot 10^{19}$	$1 \cdot 10^{19}$
Channel Doping $[cm^{-3}]$	$5 \cdot 10^{15}$	$5 \cdot 10^{15}$

II. DEVICE-LEVEL CONSIDERATIONS

In this paper we have compared a conventional double gate p-i-n/n-i-p heterojunction TFET with high-k gate oxide (HeTFET, Figure 1a), with an optimized TFET design (HK-LKTFET, Figure 1b) that features a tandem of high-k/low-k oxides. In particular, HfO_2 ($\epsilon_r = 22$) has been used as high-k material while common SiO_2 ($\epsilon_r = 3.9$) has been used as the low-k oxide. The source materials for n-type and p-type TFETs are Ge and InAs respectively. The intrinsic region is p-doped (Boron). Abrupt junctions have been assumed. It should be noted that the device poses some challenges in terms of

978-1-4673-1707-8/12 $31.00 © 2012 IEEE

fabrication, due to the non-uniform gate oxide structure. One possible solution to this, however, would be the adoption of the technique proposed by Long et al. in 2001 [8], employing multiple material depositions and photolithography steps in order to achieve non-uniform gate dielectrics along the channel direction.

The device-level simulations were performed using Sentaurus Device 2010.12 with a non-local band-to-band tunneling model, which dynamically determines the tunneling path starting from the beginning of the tunneling path with a direction of the valence band gradient [9]. Also, Shockley-Read-Hall recombination model has been activated to account for the trap-assisted tunneling under high electric field, present in the tunnel junction during device operation.

Figure 2 reflects the progressive reduction of the small signal total gate (C_{gg}) and gate-to-drain (C_{gd}) HKLKTFET capacitances obtained by reducing the high-k oxide length L_{hk}. While in MOSFET, operating in the linear region, the total gate capacitance is known to be equally distributed between the source and drain terminals [10], the reverse biased junction at the source channel interface of TFET devices makes the entire gate capacitance to to fall upon the drain terminal, causing much higher C_{gd} ($\simeq C_{gg}$) components.

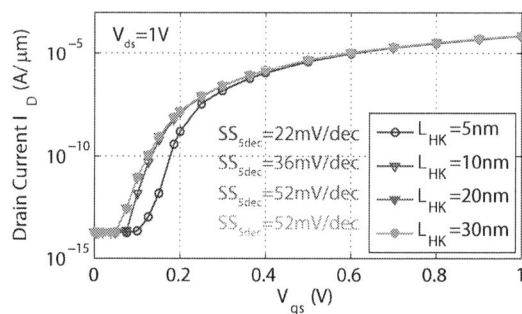

Fig. 3. Transfer characteristics for various high-k oxide lengths. Gate workfunction shifted to 4.14eV

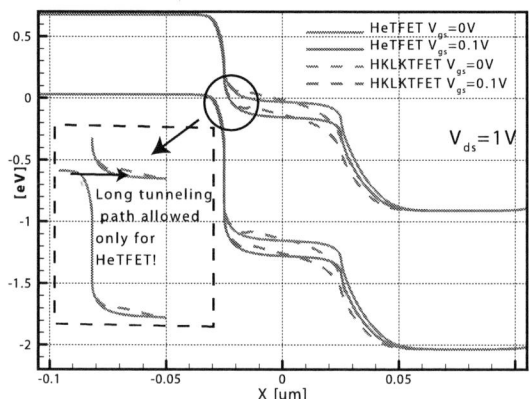

Fig. 4. The bandstructure during the switch ON behavior for HeTFET (solid) and HKLKTFET (dashed).

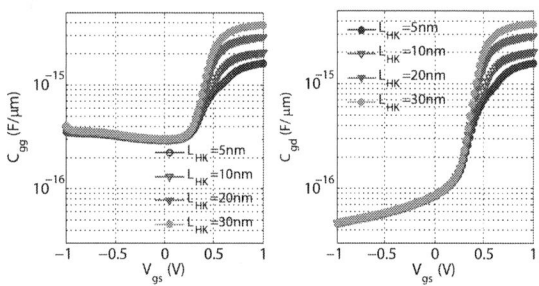

Fig. 2. Total gate capacitcance (left) gate-to-drain capacitance (right) with respect to the gate voltage for different high-k oxide lengths.

In line with the results of Choi et. al. [7], we observe an optimum L_{hk} for which the subshreshold swing is minimized (Figure 3). This is caused by the modification of the electrostatics of the device (Figure 4), which alters the band structure in the channel near the source and results in a more sensitive modulation of tunneling distance with respect to the gate voltage, due to the alignment of the source valence band and channel conduction band minimum [7]. This can be interpreted as HKLKTFET being equivalent to a TFET with low-k gate oxide for low gate voltages and once a certain *onset voltage* V_{onset} is reached, it abruptly switches to a TFET with high-k gate oxide. For this reason, lower OFF currents and steeper subthreshold swing values can be achieved by the HKLKTFET structure. By optimizing the device design to keep the onset voltage at the desired voltage level, more abrupt switching with negligible ON current loss is possible (see figure 3).

Figure 5 shows the $I_D - V_{DS}$ of the two TFETs for different V_{GS}. For gate voltages higher than the onset voltage, the dif-

ference between the output characteristics of the two cases is negligible. This can be explained by noting that, the modification of the gate oxide mainly changes the characteristics in the low voltage region when the local minimum of the conduction have not yet contributed to the current. But as the channel gets strongly inverted, we regain the similar output characteristic since it is the local minimum that actually governs the band-to-band generation characteristics. Furthermore, it has to be noted that both TFETs show considerably higher r_{ds} compared to MOSFET of equal channel length, indicating a superior immunity to short channel effects (SCE) [11]. Although initial qualitative explanations have been provided in the literature which argue the difference in charge concentration in the channel as the reason of the device output characteristics [12], the exact quantitative analysis regarding the output resistance still needs to be done. Superlinear behavior for low V_{ds} can be observed, which is typical of TFETs due to drain voltage effecting the bias on the tunnel junction in the source-channel interface.

III. CIRCUIT-LEVEL SIMULATIONS

In order to assess the circuit-level performance, three stage inverter chain and ring oscillator circuits have been simulated.

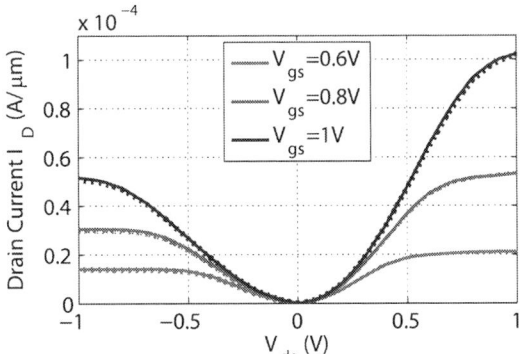

Fig. 5. Output characteristics for (Solid) HeTFET (Dashed) HKLKTFET with $L_{hk} = 10nm$ for various gate voltages. For gate voltages significantly higher than V_{onset}, output characteristics change is negligible.

Gate length L_G and high-k region length are 50nm and 10nm, respectively. For this study, lookup table based approach has been used, as it is the prevailing solution in the literature [13]. The model assumes a first-order expansion with respect to terminal voltages for each terminal current. The current through any terminals at a given time t can then be written as:

$$I_d(t) = I_{DC}(V_{gs}, V_{ds}) + \frac{\partial Q_d}{\partial v_d}\frac{\partial v_d}{\partial t} + \frac{\partial Q_d}{\partial v_g}\frac{\partial v_g}{\partial t} + \frac{\partial Q_d}{\partial v_s}\frac{\partial v_s}{\partial t}$$

$$= I_{DC}(V_{gs}, V_{ds}) + C_{dd}\frac{\partial v_d}{\partial t} + C_{dg}\frac{\partial v_g}{\partial t} + C_{ds}\frac{\partial v_s}{\partial t}$$

where Q_i is the charge associated to each terminal, I_{DC} is the DC current and the $C_{ij} = \frac{\partial Q_i}{\partial v_j}$ are defined as the small-signal capacitances that contribute in the transient switching. Circuit-level simulations were conducted through Verilog-A models that make use of the lookup tables explained.

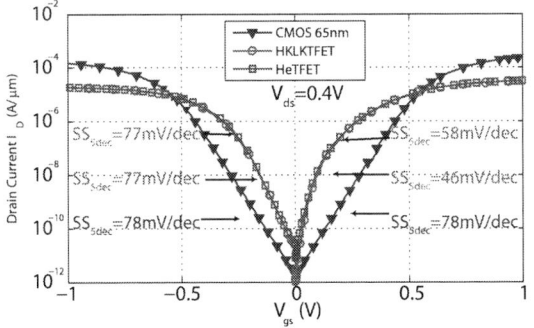

Fig. 6. Transfer characteristics for HeTFET, HKLKTFET and LSP CMOS 65nm technology node.

As it can be noted easily through the theoretical analysis of the subthreshold behavior, TFETs are intrinsically suitable

for low supply voltage operations [4]. Indeed it can be seen in figure 6 that, owing to their superior subthreshold characteristics, a supply voltage window exists where TFETs can deliver higher current than the MOSFET operated in subthreshold regime. Note that for p-type TFETs, HKLK modification does not result in sub-60mV/dec switching as for the n-type case, due to the use of InAs as the source material, whose band-gap is relatively small, a relatively high reverse-bias leakage at the source-channel junction is observed.

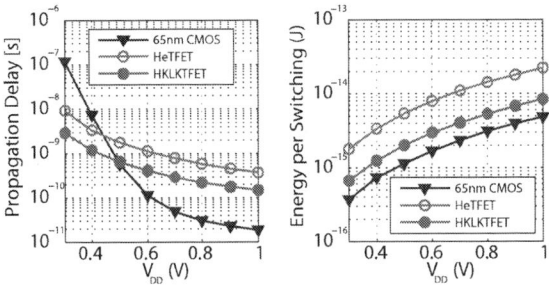

Fig. 7. Average propagation delay (left) and Energy per switching (right) as a function of supply voltage.

To assess the switching performances with respect to different supply voltages, the circuit-level transient simulations were conducted in a range of V_{DD} as shown in figure 7. Comparison of timing and energy dissipation performance of HeTFET, HKLKTFET and a 65-nm CMOS technology complying to ITRS low standby power requirements [14] is given. Owing to its superior subthreshold swing and reduced capacitance,, HKLKTFET is able to outperform CMOS for supply voltages lower than $\sim 0.5V$ in terms of propagation delay and oscillation frequency.

Another striking observation concerning figure 7 is that HKLKTFET being almost three times faster than HeTFET, simply due to the drastic reduction in the oxide capacitance without at the same ON current. Furthermore, this reduction also affects the switching energy dissipation, where a notable decrease for HKLKTFET is evident. Major point to underline here as a serious setback for both TFET structures is that, due to higher oxide capacitance, energy per switching values for both HeTFET and HKLKTFET is always higher than CMOS.

For switching analysis, the components of the load capacitance are shown for a fan-out-1 inverter chain in figure 8. The effective load capacitance ($C_{L,eff}$) for such a structure is given as:

$$C_{L,eff} = 2C_{gd1} + 2C_{gd2} + C_{sd1} + C_{sd2} + C_{gg3} + C_{gg4}$$

where C_{gd} and C_{sd} terms are intrinsic capacitances of the driving stage and C_{gg} terms are the gate capacitances of the following inverter stage being driven. The factors 2 in front of C_{gd} components is due to the *Miller effect*, that effectively doubles the capacitance seen on the output node. Combined with the highly asymmetric distribution of the gate capacitance that lies almost entirely on the gate-drain side

[15], this effective load is actually boosted for both TFET structures, meaning that for common source configuration, which is the case for the vast majority of digital applications, TFET has a disadvantage compared to MOSFET with an equal oxide capacitance. Furthermore, increased C_{gd} results in larger overshoots for TFET devices [15], as can be seen from the transient plots in figure 9. It is seen that HKLKTFET combines the advantages of CMOS and HeTFET as it has much higher ON current than MOSFET (13x more) and lower device capacitance than HeTFET ($\sim 66\%$ lower).

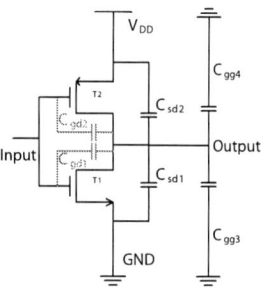

Fig. 8. The circuit-level schematic indicating the capacitive load components of a fan-out-1 inverter.

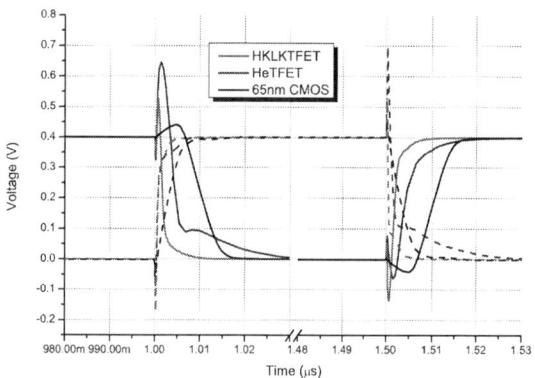

Fig. 9. (a) Transient simulations for $V_{DD} = 0.4V$. Input signals are denoted as dashed. High overshoot and undershoot amplitudes can be noted due to increased Miller capacitance for TFETs.

IV. CONCLUSIONS

Device-level and circuit-level simulations were analyzed to compare the performance of two different TFET structures and a 65nm CMOS technology. Compared to CMOS operating in the subthreshold region, heterostructure TFET structures are shown to be the preferred choice for low supply voltage applications due to their steeper subthreshold swings which enables them to achieve smaller delays for a given leakage power requirement. It has been shown that high oxide capacitance for HeTFET results in much slower switching speed (around 66% slower in the V_{DD} range of investigation) as well as extremely high switching energies ($\sim$ 2x more) compared

to HKLKTFET. This study is thus aiming to emphasize that, together with an effort for the I_{ON} boosting, TFETs need improvements focusing on the device capacitance optimization as well. In the overall, HKLKTFET stands out as an extremely attractive modification to high-k gate oxide TFETs, which can offer steeper SS, faster operation and higher energy efficiency.

ACKNOWLEDGMENT

This work was supported by the European Communitys Seventh Framework Program (FP7/2007-2013) under grant agreement no. 257267.

REFERENCES

[1] A. Wang and A. Chandrakasan, "A 180-mv subthreshold fft processor using a minimum energy design methodology," *IEEE J. Solid-State Circuits*, vol. 40, pp. 310–319, 2005.

[2] J. Knoch and J. Appenzeller, "A novel concept for field-effect transistors - the tunneling carbon nanotube FET," in *63rd Device Research Conference Digest, 2005. DRC '05.*, vol. 1, pp. 153–156, IEEE.

[3] S. H. Kim, H. Kam, C. Hu, and T.-j. K. Liu, "Germanium-Source Tunnel Field Effect Transistors with Record High I ON / I OFF," *Technology*, pp. 178–179, 2009.

[4] A. M. Ionescu and H. Riel, "Tunnel field-effect transistors as energy-efficient electronic switches," *Nature*, vol. 479, pp. 329–337, Nov. 2011.

[5] K. Boucart and A. Ionescu, "Double-gate tunnel fet with high-k gate dielectric," *Electron Devices, IEEE Transactions on*, vol. 54, pp. 1725 –1733, july 2007.

[6] R. Asra, M. Shrivastava, K. Murali, R. Pandey, H. Gossner, and V. Rao, "A tunnel fet for vdd scaling below 0.6 v with a cmos-comparable performance," *Electron Devices, IEEE Transactions on*, vol. 58, pp. 1855 –1863, july 2011.

[7] W. Choi, "Hetero-gate-dielectric tunneling field-effect transistors," *Electron Devices, IEEE Transactions on*, vol. 57, no. 9, pp. 2317–2319, 2010.

[8] W. Long, Y. W. Liu, and D. Wollesen, "Non-uniform gate/dielectric field effect transistor."

[9] Synopsys, *Sentaurus Device User Guide*, 2010.

[10] S. Mookerjea, R. Krishnan, S. Datta, and V. Narayanan, "Effective Capacitance and Drive Current for Tunnel FET (TFET) CV/I Estimation," *Electron Devices, IEEE Transactions on*, vol. 56, pp. 2092–2098, Sept. 2009.

[11] T. Nirschl, S. Henzler, J. Fischer, M. Fulde, A. Bargagli-Stoffi, M. Sterkel, J. Sedlmeir, C. Weber, R. Heinrich, U. Schaper, J. Einfeld, R. Neubert, U. Feldmann, K. Stahrenberg, E. Ruderer, G. Georgakos, A. Huber, R. Kakoschke, W. Hansch, and D. Schmitt-Landsiedel, "Scaling properties of the tunneling field effect transistor (tfet): Device and circuit," *Solid-State Electronics*, vol. 50, no. 1, pp. 44 – 51, 2006. ¡ce:title¿Special Issue: Papers selected from the 2005 ULIS Conference¡/ce:title¿.

[12] A. Mallik and A. Chattopadhyay, "Drain-dependence of tunnel field-effect transistor characteristics: The role of the channel," *Electron Devices, IEEE Transactions on*, vol. 58, pp. 4250 –4257, dec. 2011.

[13] J. Zhuge, A. S. Verhulst, W. G. Vandenberghe, W. Dehaene, R. Huang, Y. Wang, and G. Groeseneken, "Digital-circuit analysis of short-gate tunnel FETs for low-voltage applications," *Semiconductor Science and Technology*, vol. 26, p. 085001, Aug. 2011.

[14] "Itrs roadmap [online]. available: http://public.itrs.net."

[15] S. Mookerjea, R. Krishnan, S. Datta, and V. Narayanan, "On enhanced miller capacitance effect in interband tunnel transistors," *Electron Device Letters, IEEE*, vol. 30, pp. 1102 –1104, oct. 2009.

From FinFET to Nanowire ISFET

Michal Zaborowski, Daniel Tomaszewski, Piotr Dumania, Piotr Grabiec

Division of Silicon Microsystem and Nanostructure Technology
Institute of Electron Technology ITE
Warsaw, Poland
e-mail: mzab@ite.waw.pl

Abstract— **A p-type FinFET manufacturing process has been presented. It is a starting point for development of H^+ ion-sensitive n-type nanowire FETs (ISFETs). In the paper, new process steps are pointed out together with SEM examination. Characteristics of the n-type junctionless FETs have been measured in buffer solutions in a beaker and in contact with a single drop of the liquid. ISFET current versus pH and voltage versus pH curves have been presented and discussed. Relatively small hysteresis and drifts in pH measurements have been found.**

I. INTRODUCTION

Multi-gate MOSFETs exhibit a number of advantages as compared to single-gate ones [1]. Because of a good gate control over channel conduction they are predestined for low-voltage applications. FinFETs are good candidates for 22nm technology and beyond [2]. Obviously, manufacturing of such devices is based on sophisticated photolithography and plasma patterning processes. Such techniques allowing for large volume fabrication of ultra-low size devices are available only for large companies. In the small laboratory case an electronolithography-based manufacturing of FinFETs is possible, but it allows in reality only for a very small scale processing. In ITE clean-room a smart technique for fabrication of near 100 nm silicon Fins has been developed, which in some sense remains within the gap between the two approaches mentioned above, and expressed in terms of productivity. The resulting FinFET devices are briefly described and characterized in the next chapter.

Semiconductor sensors seem to be among the most interesting areas of application of the Fin transistors [3]. The Si-Fin that is fixed only at the bottom to BOX layer can be considered as a nanowire (NW). Its conductance can be controlled by potential of a surrounding liquid analyte. Development of the fabrication process toward H^+ ion-sensitive FET (ISFET) has been the aim of the work.

II. FinFET DEVICE FABRICATION

P-type FinFETs have been fabricated in p SOI wafers, which have been doped with phosphorus ions to change the wafers into n type. Two hundred nanometer wide and narrower Si Fins have been prepared using a PaDEOx technique [4,5]. In this technique, the SOI Fins are plasma etched in Bosch process with use of SiO_2 hard mask. Width of the mask is process controlled. The mask originates from short range lateral thermal oxidation of silicon along an edge of nitride pattern. Length of the Fins is controlled by standard 365 nm lithography. Next a gate thermal oxide and poly-Si gate electrode have been grown over the Fins. This part of the flow chart is presented in the left column of Table 1. The FinFETs have been covered with PSG/SiO_2 passivation layer. Then contact holes have been patterned and Ti:W/Al metallization has been done by magnetron sputtering step. The details of the Fin and poly-Si gate are shown in Fig.1.

In the experiment described above the double-Fin p-channel FinFETs with different channel drawn lengths have been manufactured, namely 2.5, 3.0 and 6.0 μm. They have been used as a reference for further experiments. Their electrical characterization has been done using a set-up consisting of Keithley 2600A series SMUs,.

Within the characterization task the measurements of output and input I-V characteristics have been done. In Fig. 2a a family of I_{DS}-V_{DS} curves of the FinFET with the channel length L=6.0 μm is shown. The device exhibits a correct behavior typical for long-channel devices. In Fig. 2b a set of I_{DS}-V_{GS} curves of the FinFETs (L=6.0 μm) is shown.

TABLE I.   COMPARISON OF THE FINFET AND ISFET PROCESSES

	FinFET Processes		**ISFET Processes**
A	Gate oxidation	A	Gate oxidation
B	Poly-Si deposition	B	Gate nitride LPCVD
C	Gate photolithography	C	Contact photolithography
D	Poly-Si plasma etching	D	Nitride plasma etching
E	Passivation PSG/oxide CVD	E	Oxide wet etching
F	Contact photolithography	F	Metal PVD
G	Contact etching plasma/wet	G	Metal photolithography and etching
H	Metal PVD	H	Passivation oxide CVD
I	Metal photolithography and etching	I	Active and bond areas photolithography
		J	Oxide wet etching

The work was partially supported by the Ministry of Science and High Education in Poland under grant NRO2 0010 06/2009.

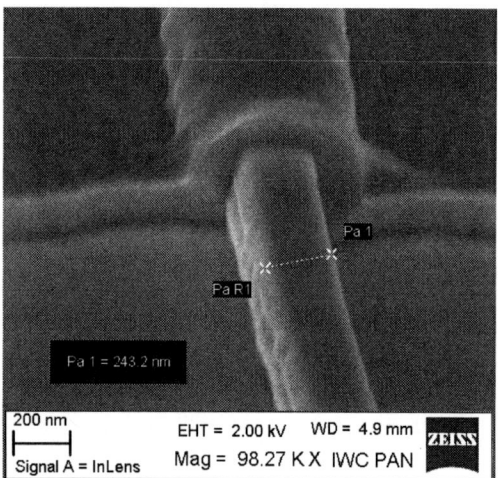

Figure 1. SEM image of an oxidized SOI Fin and poly-Si gate.

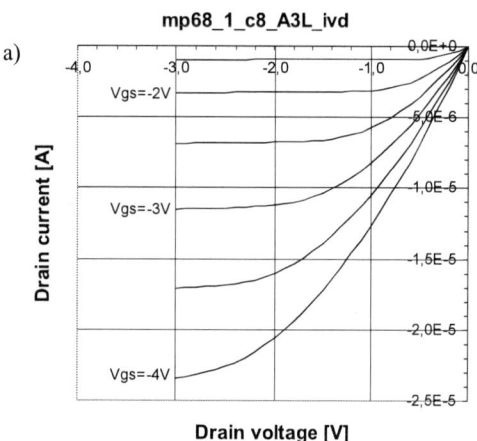

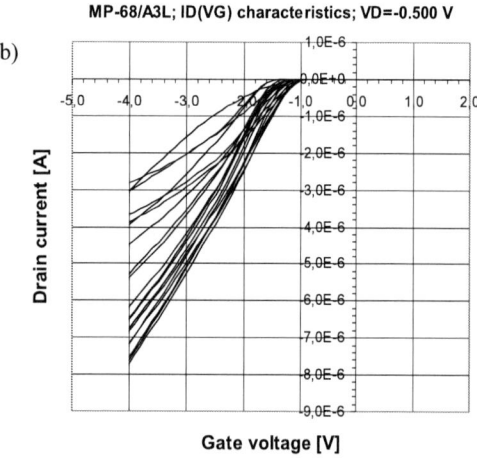

Figure 2. I-V characteristics of the 6 μm long channel dual-Fin p-MOSFETs: a) output characteristics of a single device; V_{GS}=-1.5, ..., -4V; b) input characteristics of 20 devices from a wafer; V_{DS}=-0.5V

Based on this chart it may be stated, that most of the devices exhibit a threshold voltage of the order of -1.0..-1.4 V. Thus they operate as normally-off MOSFETs. A large spread of the I-V curves may be observed. The most probable reason for this effect is a large, not fully controlled value of the series resistance of the FinFETs. Minimization and better control of series resistance seem to be the most important issues for this technology. Nevertheless we have assumed, that the FinFET process described above may be used as a basis for development of NW-based detectors of ions in liquid analytes. This work is described in the next chapter.

III. NANOWIRE ISFET PROCESS

A stoichiometric silicon nitride Si_3N_4 reveals a good sensitivity to H^+ ions. This layer is more stable as compared to a thermal oxide [6]. Differences between the Fin MOSFET and newly developed NW ISFET processes are listed in the Table. 1. First, a 14 nm-thick silicon oxide layer has been grown by thermal oxidation of the n-type doped silicon Fins (nanowires). Next, a 12 nm-thick silicon nitride layer has been formed in a LPCVD process. Since there are no gate electrodes in basic ISFET devices, the contact patterning is the next operation. Selective etching of the nitride and oxide layers has been followed by Ti:W and Al metal magnetron sputter deposition. Patterned metallization paths and a remainder of the nitride areas are then covered with a pure CVD 750 nm-thick oxide layer. This coating has been chosen in order to protect terminals against an influence of surrounding body potential. Finally, openings in the oxide coating have been done in ion-sensitive and bonding areas. The final dual-nanowire ISFET is shown in Fig.3. Parts of the nanowires covered with the dielectric passivation layer are responsible for the series resistance of the NWs. These devices have always even numbers of the NW channels connected in parallel. Multiplication of the NWs allows for significant enhancement of device current efficiency and sensor ability.

IV. CHARACTERIZATION OF ISFET SENSORS

Chips containing several NW ISFET devices have been assembled in gold covered PCB foil. Ultracompressed bonds and all edges of the chip have been electrically insulated with epoxy resin (Fig. 4). Electrical measurements of the assembled NWs remaining in contact with the drops of the liquid analyte of the given pH level have been done. The output characteristics have been measured for a number of reference electrode – source voltages. The compact reference electrode necessary in ISFET measurements plays a role of the gate contact in standard MOSFETs. The electrode has been dipped in the analyte drop. Based on the dense grid of the output I_{DS}-V_{DS} curves extraction of the input I_{DS}-V_{GS} curves is possible (subscript G denotes the reference electrode). In Fig. 5 the resulting curves are shown, which have been obtained for V_{DS}=100mV and for solution pH=7. It may be stated, that the devices under test reveal a relatively low transconductance equel approximately $2 \cdot 10^{-7}$ S/fin.

978-1-4673-1707-8/12 $31.00 © 2012 IEEE

a)

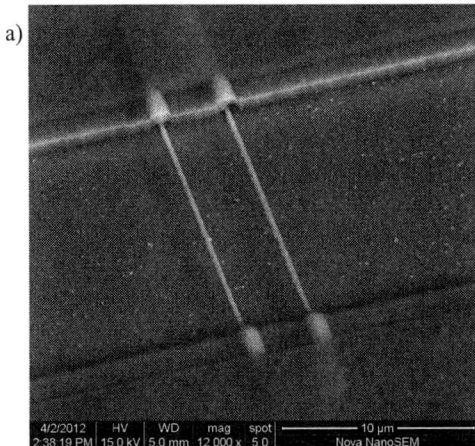

b)

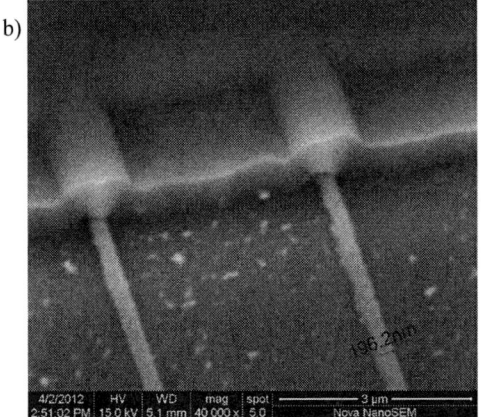

Figure 3.   SEM images of an active area of the dual-NW ISFET.

Figure 4.   Optical image of the chip with pH sensors, mounted and edge-insulated; a drop of analyte is visible in the middle.

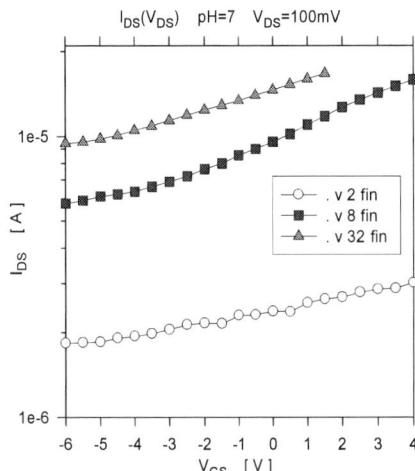

Figure 5.   Reference electrode voltage control of 2, 8 and 32-NW ISFETs (L=10μm) in pH=7 buffer solution in a beaker.

Two reasons may be mentioned to explain the low transconductance of the n-type NWs under test. Firstly, the channel n-type doping might be too high, which would lead to too high channel conductivity, and thus to a weak response to the electric field originating from the Fin surrounding. Secondly, in the case of n-type channel due to the positive net charges at the Si-SiO_2 interfaces, the NWs under the test very probably reveal an enhanced n-type conductivity. It is worthwhile to mention, that the output I-V characteristics, from which the presented in Fig. 5 input ones have been extracted, exhibit rather resistive behavior. These observations remain in agreement with results presented in [7] for the NW FETs with metal gates. In those devices only slightly higher transconductances have been obtained. In the ISFETs, considered here a potential drop in the liquid solution should be taken into account. It decreases the effective NW channel control by the reference electrode. Hence lower transconductances have been measured.

Furthermore, it may be noticed, that the curves shown in Fig.5 are not scalable with respect to a number of fins. The spread of NW conductance may be a reason for this effect. But due to a large discrepancy from an expected linear scaling rule there should be another factor, which at the moment is not known.

In Fig. 6 the output I_{DS}-V_{DS} curves of 32-NW ISFET sensor are shown. The sensor has been exposed to drops of pH=5, 6 and 7 buffer solutions. For the given pH the reference electrode voltage V_{ref} has been varied from -0.5V to 0.5V. It may be stated, that pH value has a significant impact on the ISFET behavior, whereas in accordance with previous remarks the reference electrode potential effect is weak.

978-1-4673-1707-8/12 $31.00 © 2012 IEEE   167

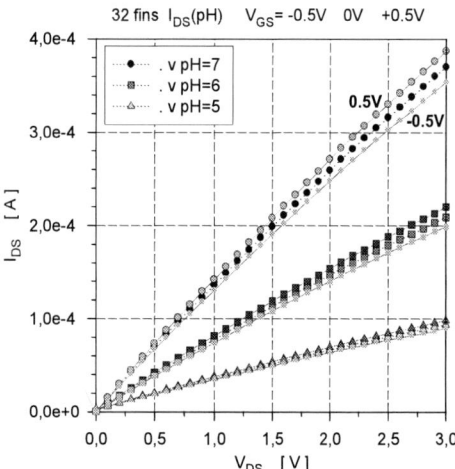

Figure 6.   Output characteristics of 32-NW ISFET sensor contacted with drops of pH=5, 6 and 7 buffers for V_{ref} varying from -0.5V to 0.5V.

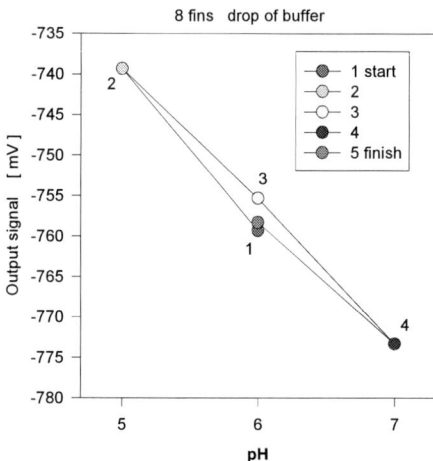

Figure 8.   Dependence of the 8-NW ISFET sensor output voltage on pH of a drop of buffer solution (extracted from Fig. 7).

Results of sensor stability measurements are shown in Fig.7. In this experiment the 8-NW ISFET has been assembled in a source follower circuit. The ISFET has been periodically exposed to drops of buffer solutions of different pH levels. The output voltage of the sensor has been measured in time during the whole experiment. A short-time drift of the sensor response may be stated for each pH level. Also sensor responses to different liquid drops for the same pH exhibit a small long-time drift illustrated in Fig. 7 by the regression lines. A voltage versus pH curve extracted from these data gives evidence of relatively small hysteresis of the output signal – Fig. 8.

SUMMARY

In the presented work selected aspects of n-type nanowire manufacturing, assembling and electrical characterization have been described. Experience acquired in experiments with p-type FinFETs has allowed for better n-type ISFET development. The new device is still CMOS-compatible. It has been positively evaluated as H+ ion concentration sensor. It exhibits rather low transconductance, which although not well understood, does not hinder its use in pH measurement application. The output signal hysteresis and short time stability remain in a range typical for Si ISFETs with the nitride dielectric.

The developed NW ISFET-type devices may be used in sensor applications, where a small amount of analyte is required. Moreover their size is compatible with the biological object scale, so it seems to be promising from biological sample investigation perspective.

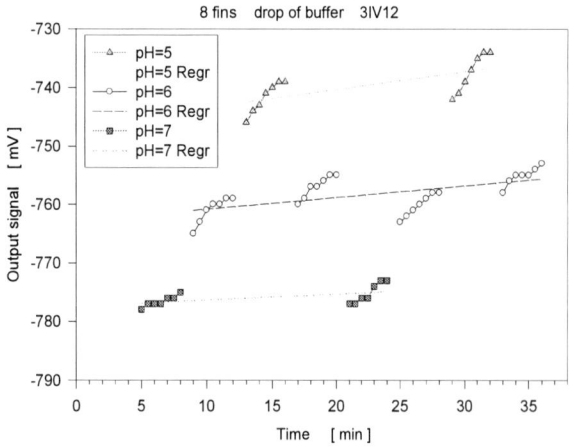

Figure 7.   Time-dependence of output voltage measured in the source follower circuit with 8-NW sensor exposed to drops of three buffer solutions (without exposition of the active area to the air) .

REFERENCES

[1] J. P. Colinge, "FinFETs and Other Multi-Gate Transistors", Springer, New York, 2008.

[2] Intel 22nm 3-D Tri-Gate Transistor Technology, 2011-05-02 in http://newsroom.intel.com/docs/DOC-2032

[3] R. Yan et al., "Performance Analysis of SOI Junctionless Nanowire Transistors", ELTE2010 Conf. Wroclaw, 22-25 Sept. 2010.

[4] M. Zaborowski et al., "Nanoscale pattern definition by edge oxidation of silicon under the Si_3N_4 mask – PaDEOx" Acta Physica Polonica A, Vol.116 (2009), pp.139-141.

[5] M. Zaborowski, D. Tomaszewski, A. Panas, P. Grabiec, "Double-fin FETs based on standard CMOS approach" Microelectronic Engineering Vol. 87, Issue: 5-8, May - August, 2010, pp. 1396-9.

[6] P. Bergveld, "Thirty years of ISFETOLOGY", Sensors and Actuators B, 88 (2003), pp 1-20.

[7] M. Zaborowski, D. Tomaszewski, A. Panas, P. Grabiec, "Characterization of Test Devices for Development of Nanowire Sensor FETs", to be presented at 19th Int. Conf. Mixed Design of Integrated Circuits and Systems, MIXDES, Warsaw, 24-26th May, 2012.

Micro- and Nano-link Ultra-Low Power Heaters for Sensors

A. W. Groenland, E. Vereshchagina, A. Y. Kovalgin, R. A. M. Wolters, J.G.E. Gardeniers, and J. Schmitz
MESA+ Institute for Nanotechnology, University of Twente, Enschede, The Netherlands
email: a.y.kovalgin@utwente.nl

Abstract—A new microfabricated device for heating and sensing in gases is presented. It is based on the resistive heating of a micro- or nano-metric hollow cylinder of titanium nitride, and measurement of its (temperature-dependent) resistance. This article presents the fabrication and temperature calibration of the device, and illustrates its function as flow meter and thermal conductivity meter. A temperature of 280 °C is achieved at a power consumption of only 5.5 μW, orders of magnitude less than existing commercial hotplate devices. The thermal time constant can be as low as 60-120 microseconds.

I. INTRODUCTION

Contemporary micro hotplates can be electrically heated to 100-600 °C with small time constants (~100 μs). Such micro hotplates are commonly used in chemical sensors [1] and mass flow meters [2] based on temperature changes. Because of the elevated operating temperatures, the devices can also behave as chemical actuators, i.e., in micro-reactors providing energy in the form of heat to initiate thermo-activated chemical reactions [3]. When coated with catalytically active surface coatings, hotplates can be used as catalytic microreactors or catalytic combustion sensors such as Pellistors [1, 4] for the detection of hydrocarbons.

One significant drawback of conventional hotplates is their relatively high power consumption. Commercially available platinum wire based ('classic') devices require a power around 100 mW [4]. Other types of hotplates, that are fabricated in the last few decades using micro-technology, are based on a suspended membrane with metal thin film resistors and still require 10-40 mW [5, 6].

As the heat losses scale with the hot plate size, a smaller resistive heater can yield a better power-temperature balance. A new generation of micro hotplate devices was recently reported, with a power consumption of only a few milliwatts [7, 8]. In that approach, the heat was generated by a conductive nano-link, sandwiched between two polysilicon electrodes. The link was formed by antifusing (breakdown) a thin silicon dioxide (SiO_2) dielectric between the electrodes, followed by controllable electrical programming. While these devices proved the viability of power reduction by downsizing, the approach is difficult to industrialize due to the

antifusing / programming sequence as well as limited predictability and reliability of these hot plates.

In this work, we present a new fabrication approach to realize the link, aiming to overcome these limitations. The conductive link is formed directly by microfabrication, by first etching a hole in a SiO_2 layer on top of the first electrode, and filling this hole in with the second electrode material (TiN) via Atomic Layer Deposition (ALD). The device design and the fabrication procedure are detailed in Section II. In Section III, we show the electrical and thermal characterization of these hotplates with an outer link diameter of 2-6 μm ('micro-link') and 100 nm ('nano-link'). The work is summarized in the Conclusion.

II. DESIGN AND FABRICATION

A cross-section and a three-dimensional schematic of the device design are shown in Figure 1. The width W of the electrodes is 10 μm. The link is realized in the center of a suspended membrane of 40×40 μm.

For the device fabrication, standard 4" boron-doped (10^{15} atoms/cm³) <100> silicon wafers were processed as described in ref. [9]. Briefly, this included

1) deposition of a 100 nm low stress silicon nitride (SiRN) layer,

2) sputtering and further patterning a layer of 100-nm-thick TiN to make the bottom electrode (see Fig. 1),

3) deposition of a 100-nm-thick SiO_2 by plasma enhanced chemical vapor deposition (PECVD),

4) sputtering and patterning a 100-nm-thick TiN layer to form the top electrode,

5) etching a hole in the SiO_2 layer for the micro- or nano-link by UV or e-beam lithography, respectively, and

6) ALD of 7 nm (micro-link) or 15 nm (nano-link) of TiN to form the link between the electrodes.

The fabrication of 1-μm-thick aluminum (Al) contact pads and release of the SiRN membrane in a KOH solution finalized the structure. A cross-section (made by the focused ion beam (FIB) in combination with high-resolution scanning electron microscopy – HRSEM) of the fabricated nano-link device is shown in Fig. 2.

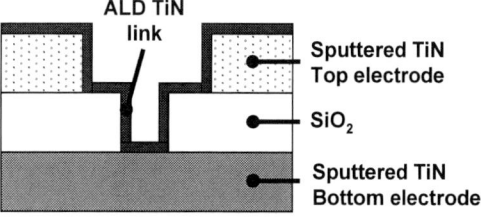

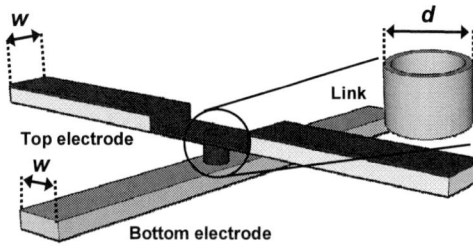

Figure 1.   Cross-section (top) and 3D schematic (bottom) of micro- and nano-link-based devices.

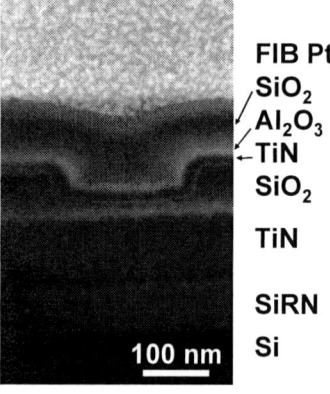

Figure 2.   HRSEM image showing a cross-section through the nano-link. The upper platinum layer is added for image contrast; it is not present in the actual devices.

III.   EXPERIMENTAL

For the electrical characterization, IV-measurements were carried out using a HP4156B parameter analyzer and a Keithley 4200 semiconductor characterization system in combination with a Cascade Microtech or Karl Süss PM8 probe station. For the temperature calibration measurements, the temperature controlled chuck of the probe station was used. The infrared images were obtained using a XenIC (XEVA) camera equipped with an InGaAs detector (λ=0.9 – 1.7 µm) and mounted on the probe station. The infrared measurements were calibrated to the device temperature using

a large-area meander-type heater composed of the same layer stack as the link-based devices.

The fabricated devices are characterized using the four-point method, before and after the membrane release step. A voltage (V_{src}) is forced between two adjacent electrodes, leading to a vertical current through the link (I_{src}). The link resistance is measured excluding the electrode resistance, i.e. by measuring the voltage drop (V_{diff}) over the two opposite electrodes. The four-point link resistance (R_{link}) is calculated as $R_{link} = V_{diff} / I_{src}$. The power consumed in the link is calculated as $P_{diff} = V_{diff} \times I_{src}$.

A.   Micro-link devices

The temperature coefficient of resistance (TCR) of TiN links was obtained in the temperature range of 25-300 °C using the thermo-chuck of the probe station. A linear IV-behavior was observed for all the temperatures. The TCR is calculated as 3.0×10^{-4} /°C for a 7-nm-thick TiN layer. For a comparison, TCR values of a flat ALD TiN layer, measured with a Greek Cross (GC) test structure, were found to be 3.3×10^{-4} /°C [10]. In this respect, the hollow cylinder-shaped ALD TiN layer behaves similarly to a flat ALD TiN layer.

Next, the membrane release etch was carried out and the IV-characteristics were measured again. A linear IV-behavior was observed in the corresponding voltage range. From the change of R_{link} versus P_{diff}, the link temperature can be calculated at each P_{diff} provided the known TCR. To verify the electrically induced temperatures, infrared emission measurements were performed. They indicated a temperature of 175 °C and were in a good agreement with the electrically measured link temperature of 171 °C [9].

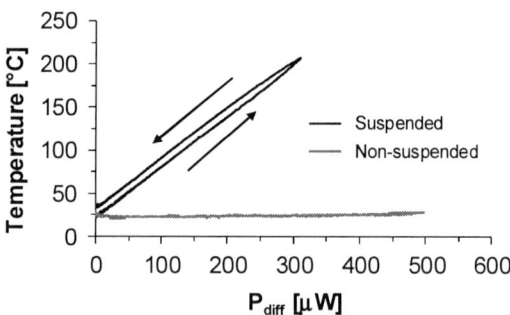

Figure 3.   Comparison of link temperature versus applied power for a micro-link device before ("non-suspended") and after ("suspended") the membrane release etch. Temperature is calculated from the resistance change and TCR.

As can be seen in Fig. 3, the micro-link devices require a power of about 0.3 mW to achieve a link temperature of 200 °C. This gives a thermal efficiency exceeding 650 °C/mW, which is much higher compared to state-of-the-art hot surface devices [11]. A little hysteresis is observed, this is attributed to the thermal inertia of the surrounding media (e.g. SiRN membrane). Importantly, the devices on non-suspended membranes indicated no heating.

When operating conditions are changed, a definite time interval (thermal response) is required to reach thermal equilibrium. To evaluate the thermal link response, we measured the time needed for a heater to change its resistance from the "off" state to 63% of the operating (stable steady-state) value. This time is required to be in the millisecond range for rapid sensing and heating applications. The measurements were carried out in open air. Figure 4 shows the dynamic response of a micro-link obtained as a result of a square voltage pulse of 1.2 V. As can bee seen, the resistance rise time is in the range of 60-120 microseconds, i.e. much shorted than required for applications.

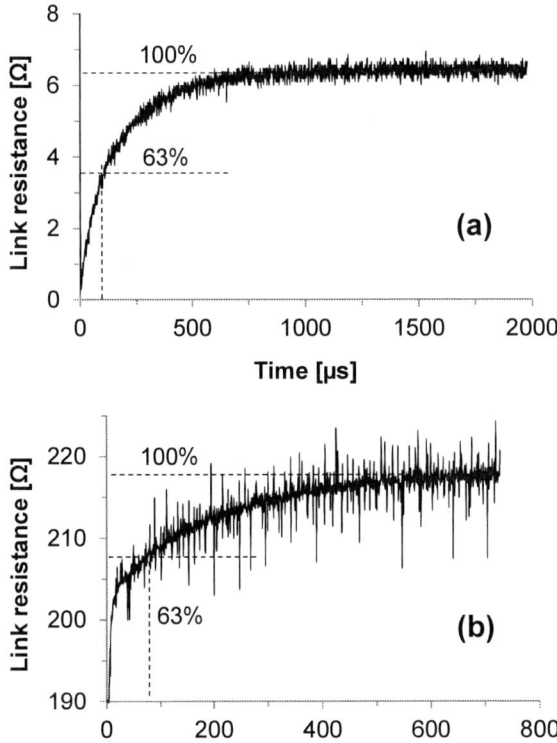

Figure 4. Real-time response of the link resistance in open air at an applied voltage of 1.2 V for low- (a) and high-ohmic (b) micro-link devices.

When a heating element is used to monitor the temperature change, the sensitivity is determined by the dependence of the resistance on temperature. For thermal conductivity or gas flow detectors, the sensing principle is based on measuring the heat removal by the gas from the heated element. In Figure 5, a response of the micro-link heater to switching between nitrogen and hydrogen flow is shown. The clear response is an indication of this device's ability to behave as a thermal conductivity sensor for gases. It further provides additional evidence that the surface of the heating element can indeed be maintained at an elevated temperature.

B. Nano-link devices

Devices with a nano-link were characterized before and after the membrane release step, using the same procedure as for the micro-link devices. The TCR was extracted in the same manner, i.e. on a heated chuck. Again a linear IV-behavior is observed for all the temperatures. The TCR is calculated as -3.3×10^{-3} /°C. The TCR is negative and higher in absolute terms compared to that of the micro-link device and of the flat ALD TiN layer. This is most likely a result of the dominating high contact resistance (R_c) due to the small contact area, which is in series with the nano-link resistance. The corresponding specific contact resistance is calculated to be $5.5 \times 10^{-4} \Omega cm^2$, which is a high value [12] and most likely the result of an interfacial layer between the nano-link and the bottom electrode (see Fig. 2). Contact resistances are known to exhibit a negative TCR as result of a decreasing barrier at elevated temperatures [13].

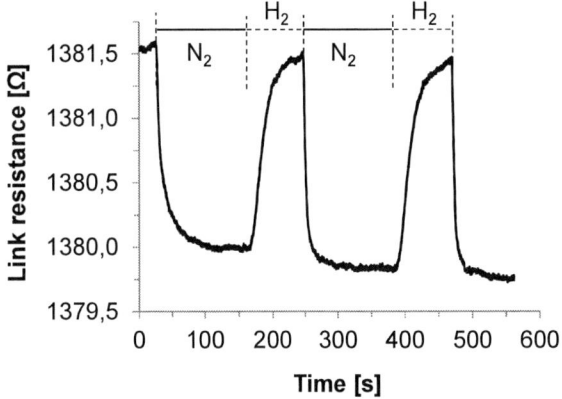

Figure 5. Measured resistance of the micro-link as a function of time upon exposure to pulses of hydrogen (H_2) and nitrogen (N_2). The measurements were performed at power levels below 1 mW. Switching between N_2- and H_2-pulses is indicated. The delayed sensor response (compared to that in open air shown in Fig. 4) is related to the gas diffusion to the device surface.

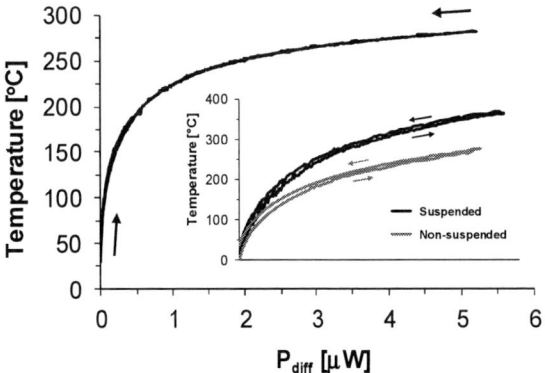

Figure 6. Link temperature versus applied power for a nano-link device after the membrane release step. The inset shows a comparison before ("non-suspended") and after ("suspended") the membrane release. The temperature is calculated from the resistance change and TCR.

Based on the TCR, the temperature-versus-power characteristics were measured. A nonlinear behavior is observed, as can be seen in Figure 6. Contrary to the micro-link devices, heating the link is localized, giving the parabolic-shape curve. In this case, the link area of 100 nm is too small to verify the temperature using the infrared setup.

The effect of the membrane release step is not very significant for nano-link devices (see the inset in Fig. 6). Similar parabolic-shape curves are observed for devices on suspended and non-suspended membranes. This indicates the local heating of the link, regardless the state of the membrane. For the same power dissipation in the link, the link temperature is on average less than 100 °C lower for non-suspended devices compared to the ones on suspended membranes. This indicates that release of the membrane has much less effect on the temperature of a nano-link device. The latter is probably due to the fact that the very local and small-in-volume nano-link heaters are encapsulated into SiO_2. From Fig. 6, for a suspended nano-link, a link temperature of 250 °C can be obtained by applying only 2 μW of power.

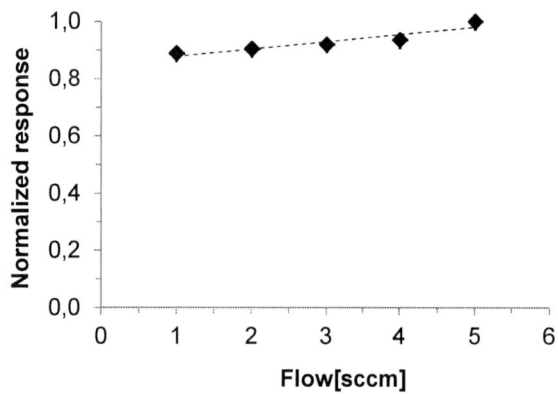

Figure 7.   Normalized response of the nano-link device to flow of nitrogen in the range of 0-5 sccm.

In Figure 7, a normalized output of the nano-link device upon exposure to nitrogen flow in the range from 0 to 5 sccm is shown. As can be seen, for the flow between 1 and 5 sccm, the curve is more or less linear and has a low slope indicating low sensitivity to the flow. However, when flow/no-flow conditions are compared, one can logically expect a gradual decrease of the sensor response to flow in the range between 0 and 1 sccm. On one hand, this can be utilized for "digital" sensing (i.e. flow conditions correspond to 1 and no-flow to 0) of gas flows at low powers. On the other hand, this opens the possibility to utilize nano-links in sensors designed to measure very small flows of gases, i.e. less than 1 sccm. The accuracy of the flow control in the experimental setup was unfortunately limited to 1 sccm. More experiments are therefore required to confirm the latter.

IV.   CONCLUSION

We presented a new approach to miniaturize micro hotplates by utilizing conformal atomic layer depositions in a small contact hole between electrodes. By ohmic heating of the small conductive link, a well-defined temperature can be generated. In combination with a resistance measurement, such devices can be used as gas flow sensors and thermal conductivity sensors, as conceptually demonstrated in this paper; and perhaps as well as chemical reaction actuators. By using TiN as the conductive link material, a reasonably high temperature coefficient of resistance is obtained, leading to reasonably high temperature sensitivity.

The miniaturization resulted in very low power consumption for the demonstrated devices. The devices with micro-links reached a link temperature of 200 °C at a power consumption of 0.3 mW, whereas the nano-link devices exhibited a link temperature of 250 °C while applying 2 μW only, raising to 280 °C at 5.5 μW.

ACKNOWLEDGMENT

The authors thank A. K. van Langen-Suurling and A. J. van Run of Kavli Institute of Nanoscience (Delft, NL) for their help with e-beam lithography. M. A. Smithers and V. J. Gadgil of the MESA+ Institute are acknowledged for the HRSEM and FIB work, respectively. R. P. G. Sanders (TST, University of Twente) is gratefully acknowledged for the assistance in the measurements of the heater-time constants. This work was supported by the Dutch Technology Foundation STW under project 07682.

REFERENCES

[1] M. Gall, *Sensors and Actuators B: Chemical*, 16, 260-264 (1993).

[2] H. E. de Bree, *Acta Acustica United with Acustica*, 89, 163-172 (2003).

[3] R. M. Tiggelaar, P. v. Male, J. W. Berenschot, J. G. E. Gardeniers, R. E. Oosterbroek, M. H. J. M. d. Croon, J. C. Schouten, A. v. d. Berg, and M. C. Elwenspoek, *Sensors and Actuators A: Physical*, 119, 196-205 (2005).

[4] P. T. Moseley and B. C. Tofield, Solid State Gas Sensors. Bristol: IOP Publishing Ltd, Technohouse, Redcliffe Way, Bristol BS1 6NX, England, 1987.

[5] C. Ducso, M. Adam, P. Furjes, M. Hirschfelder, S. Kulinyi, and I. Barsony, *Sensors and Actuators, B: Chemical*, 95, 189-194 (2003).

[6] S. Z. Ali, F. Udrea, W. I. Milne, and J. W. Gardner, *Journal of Microelectromechanical Systems*, 17, 1408-1417 (2008).

[7] A. Y. Kovalgin, J. Holleman, G. Iordache, T. Jenneboer, F. Falke, V. Zieren, and M. J. Goossens, *Journal of the Electrochemical Society*, 153, H181-H188 (2006).

[8] A. Y. Kovalgin, J. Holleman, and G. Iordache, *IEEE Sensors Journal*, 7, 18-27 (2007).

[9] A. W. Groenland, R. A. M. Wolters, A. Y. Kovalgin, and J. Schmitz, *ECS Trans.* 35 (30), 25-34, 2011.

[10] A.W. Groenland, R.A.M. Wolters, A.Y. Kovalgin, J. Schmitz, *IEEE transactions on semiconductor manufacturing*, 25 (2). pp. 1-24, 2012.

[11] H.-Y. Lee, S. Moon, S.J. Park, J. Lee, K.-H. Park and J. Kim, Electronics Letters, 44 (25) (2008).

[12] S.R. Wilson, C.J. Tracy, J.L. Freeman Jr, *Handbook of multilevel metallization for integrated circuits: material, technology and applications*. New Jersey: Noyes publications, 1993.

[13] T. Schwamb, B.R. Burg, N.C. Schirmer, D. Poulikakos, *Appl. Phys. Lett.*, 92, 243106-3 (2008).

High performance printed N and P-type OTFTs for complementary circuits on plastic substrate

S. Jacob[1], M. Benwadih[1], J. Bablet[1], I. Chartier[1], R. Gwoziecki[1], S. Abdinia[2], E. Cantatore[2], L. Maddiona[3], F. Tramontana[3], G. Maiellaro[4], L. Mariucci[5], G. Palmisano[6], R. Coppard[1]

[1]CEA-LITEN, Grenoble, France; [2]Eindhoven University of Technology, Department of Electrical Engineering, MSM, Eindhoven, The Netherlands; [3]STMicroelectronics, Catania, Italy; [4]University of Catania, DIEEI, Italy now with STMicroelectronics, Catania, Italy, [5]CNR-IMM, Rome, Italy; [6]University of Catania, DIEEI, Italy
stephanie.jacob@cea.fr

Abstract—**This paper presents a printed organic complementary technology on flexible plastic substrate with high performance N and P-type Organic Thin Film Transistors (OTFTs), based on small-molecule organic semiconductors in solution. Challenges related to the integration of both OTFT types in a common complementary flow are addressed, showing the importance of surface treatments. Data on single devices and elementary complementary digital circuits (inverters and ring oscillators) are presented, demonstrating that a robust and reliable flow with high electrical performances can be established for printed organic devices.**

I. INTRODUCTION

Organic electronics has attracted significant interest as possible inexpensive and flexible alternatives to inorganic devices. Indeed, tremendous progresses have been made during the last decade, opening promising perspectives for applications such as flexible displays, logic circuits for radio-frequency identification (RFID) tags and sensing devices [1]. Processing materials in solution makes possible to use printing processes that are well appropriate for large area flexible substrates and thus for low cost production.

Recently, significant advances have been made on the development of solution processable organic semiconductors (OSC) and both P- and N-type materials that present high performance and good air-stability are now available [2, 3].

The integration of both P- and N-type OSC enables the fabrication of complementary circuits which increase the speed and reliability of organic-based electronics [4, 5].

Several organic complementary circuits have already been reported with solution deposited materials [3, 4, 6, 7], but there is still no work presenting at the same time a full printed complementary technology on flexible substrate with high mobility P- and N-type semiconductors.

In a previous work [8], we already presented full printed organic complementary circuits on plastic substrates with a first generation of P- and N-type semiconductors with lower mobilities ($\mu_N < 0.06 \text{cm}^2.\text{V}^{-1}.\text{s}^{-1}/\mu_P < 0.04 \text{cm}^2.\text{V}^{-1}.\text{s}^{-1}$). In this work, we present a printed organic complementary technology made on flexible foils (sheet-to-sheet technology) with N- and P-type small-molecule semiconductors, allowing higher mobilities. Several mask sets have been designed and processed with this technology, containing digital and analog circuits which are presented more into details in [9].

II. COMPLEMENTARY PROCESS FLOW

The organic complementary circuits are fabricated on 11cm x 11cm flexible foils using a top-gate bottom-contact structure for both N and P-type devices (Fig. 1). The process starts on a 125µm-thick polyethylene-naphtalate (PEN) foil. Gold is sputtered to a thickness of 30 nm and then patterned either by photolithography or directly by laser ablation, forming the source and drain electrodes as well as the 1st level of interconnection lines between gates. Then, a Self-Assembled Monolayer (SAM) is deposited to optimize electron injection in the Lowest Unoccupied Molecular Orbital (LUMO) of the N-type organic semiconductor [10]. The N-type OSC (Polyera ActivInk®) is first patterned by printing methods, in order to form the individual devices, leading to a final thickness in the range of 50-200nm. Then, the source/drain electrodes and the PEN in P-type areas are cleaned with an O_2 UV-free plasma during 180s to prepare the surface for the SAM deposition and P-type OSC (TIPS-pentacene) printing. The thickness of P-type OSC patterns is also in the 50-200nm range. The common fluoropolymer dielectric (CYTOP®) is screen-printed on top of both semiconductors and then annealed, leaving open areas for via holes, with a final thickness of 750nm. Finally, a silver-ink conductor is screen-printed on the top of the dielectric and annealed at 100°C, forming in the same step the gate electrodes for devices and the 2nd level for interconnection.

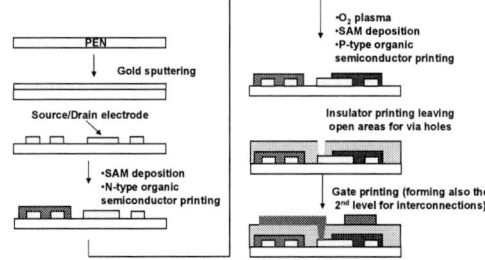

Figure 1. Schematic view of the complementary process flow.

978-1-4673-1707-8/12 $31.00 © 2012 IEEE

Transfer curves of N- and P-type devices with L=100μm and W=2000μm processed with our complementary technology are plotted on Fig. 2, exhibiting mobilities of 1.5 cm².V⁻¹.s⁻¹ for P-OTFTs and 0.55 cm².V⁻¹.s⁻¹ for N-OTFTs. The typical electrical characteristics of transistors are listed in table I. In addition to high mobility, both P and N-type transistors present high ratio between the On and Off currents as well as steep subthreshold slopes.

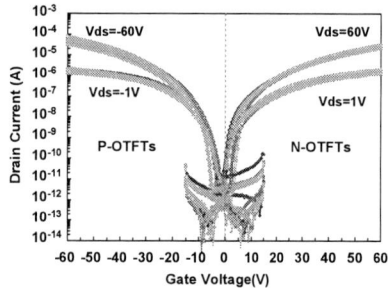

Figure 2.   Transfert characterics in linear (|Vd|=1V) and saturated (|Vd|=60V) regime of 8 P-OTFTs and 8 N-OTFTs of the complementary process flow. L=100μm W=2000μm.

TABLE I.   ELECTRICAL CHARACTERISTICS OF L=100μM/W=2000μM TRANSISTORS PROCESSED WITH THE COMPLEMENTARY FLOW

	P-OTFT	N-OTFT
Vt_{sat} (V)	-20	18
V_{onset} (V)	-0.2	1
μ_{sat} (cm².V⁻¹.s⁻¹)	1.5	0.55
Ion_{sat} L/W (A)	~2x10⁻⁶	>10⁻⁶
$Ioff_{sat}$ L/W (A)	~2x10⁻¹³	~ 5x10⁻¹⁴
Ion/Ioff	~10⁷	> 2x10⁷
Subthreshold Slope (V/dec)	2.4	1.2

III.   PROCESS OPTIMIZATIONS

As described in the previous section, a plasma surface treatment of PEN and gold electrodes is required before SAM and P-OSC deposition. First, it is necessary to get a sufficient wettability of the P-OSC on PEN. Indeed, an oxygen plasma treatment enables to decrease the contact angle with water from 70° (without treatment) to 30°. The plasma treatment is also critical for improving the injection of holes between the source/drain electrodes and the P-OSC, by cleaning the surface of gold. One of the main issues of the integration of both OTFT types in a common complementary flow is the introduction of this plasma treatment between the N-OSC deposition and the P-OSC one. The N-type semiconductor is deposited first and thus it is exposed to the plasma. To study the effects of a plasma treatment exposure and especially the effect of UV rays on the N-OSC, several foils have been processed, with different treatments: O₂ Reactive Ion Etching (RIE) plasma with 2 exposure times (60s and 100s), UV ozone exposure of 5 min and no treatment. The electrical performances of N-type transistors obtained in the different

cases are compared on Fig. 3(a), showing that all the treatments induce a decrease of the mobility. In particular, the UV ozone treatment leads to a strong deterioration of the N-OTFT electrical characteristics, demonstrating that the N-OSC is damaged under UV rays exposure. Fig. 3(b) also shows that the onset voltage is shifted towards negative voltages when the exposure time of RIE plasma increases from 60s to 100s, which confirms that RIE plasma treatments must be avoided to obtain the best performance from the N-OSC.

Further to previous conclusions, another plasma treatment with an UV-free option has been tested, with several exposure times from 150s to 240s. Firstly, we have studied the capability of this treatment to clean sufficiently the PEN/gold electrodes surface of P-type OTFTs. For this, the ON resistance (R_{ON}) of P-type devices has been calculated from the output characteristics Id-Vd according to (1).

$$R_{ON}(Vgs) = \frac{\partial Vds}{\partial Ids}\bigg|_{Vds=0V} \qquad (1)$$

The tested transistors have identical geometries (especially same channel length) and identical intrinsic mobilities, which means their channel resistance is also comparable. As R_{ON} is the sum of channel and contact resistances, comparing it is equivalent to compare the contact resistance of the different P-type transistors, which is a good indicator of the gold electrodes cleaning. Fig. 4(a) shows that without any plasma treatment, R_{ON} is 1.5 to 6 times higher, demonstrating the necessity of a surface treatment for P-OTFTs. Moreover, we see that there is an optimum point for 180s exposure time.

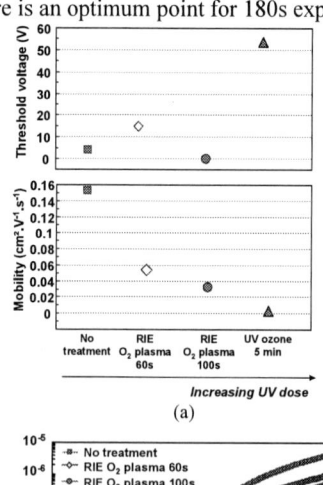

(a)

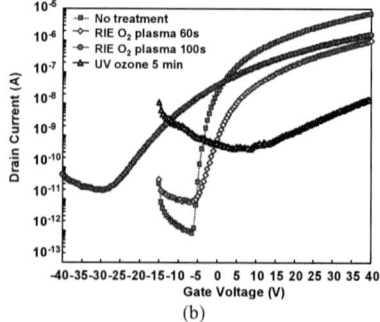

(b)

Figure 3.   Effect of plasma treatments on threshold voltage and mobility (a) and on transfert characteristics of N-OTFTs (b) in saturated regime (Vds=40V). L=50μm W=2000μm.

In addition, Fig. 4(b) and (c) show that the threshold voltage and mobility of N-type devices exposed to UV-free plasma are the same than without plasma treatment whatever the exposure time, validating that the N-OSC is not damaged by the UV-free plasma treatment.

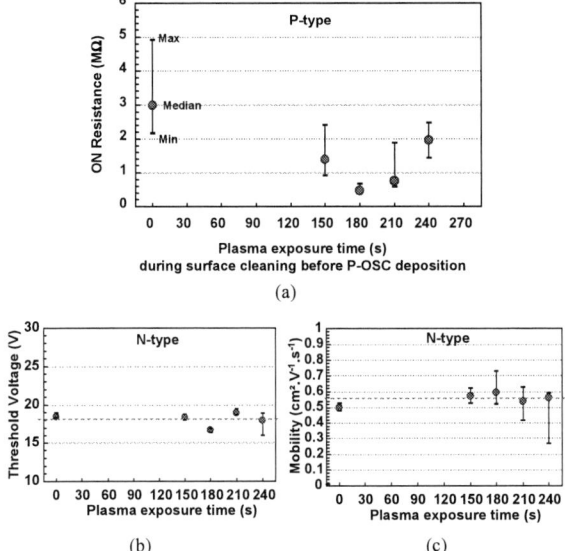

Figure 4. Effect of UV-free plasma treatement done on P-type areas after N-OSC deposition on the cleaning of P-OTFT contacts (a) and also on the saturation threshold voltage (b) and mobility (c) of N-OTFTs. L=100μm W=2000μm. |Vgs|=60V.

IV. STABILITY OF THE COMPLEMENTARY TECHNOLOGY

A. Single devices

After optimizing the process flow, the question of stability and repeatability of our complementary technology has been addressed. Fig. 5 represents the threshold voltage and mobility of several P- and N-OTFTs on 9 sheets processed between October 2011 and March 2012 with our complementary flow (only minor process changes between the different sheets). We observe that the threshold voltage is very robust for N-type devices but there is some dispersion for P-type OTFTs from sheet to sheet. The mobility of P-OTFTs from a same sheet presents more dispersion than the N-OTFTs one, which is characterized by tight distributions, while from sheet to sheet, the mobility is quite repeatable for both P- and N-type transistors.

The yield has also been studied, considering the ratio between Ion and Ioff currents as criterion. A transistor is considered functional if the Ion/Ioff ratio is higher than the set criterion. The yield of P- and N-type transistors has been computed on 5 sheets for different criteria (from Ion/Ioff >10^1 to Ion/Ioff >10^6) and for a long (100μm) and a shorter (20μm) channel length (Fig. 6). Transistors exhibit high yield rates. With a criterion of Ion/Ioff > 10^3 (which guarantees functional circuits), the yield is higher than 98% for both N- and P-OTFTs and is similar for L=100μm and L=20μm. Moreover, the yield remains still high for more severe criteria until Ion/Ioff =10^5.

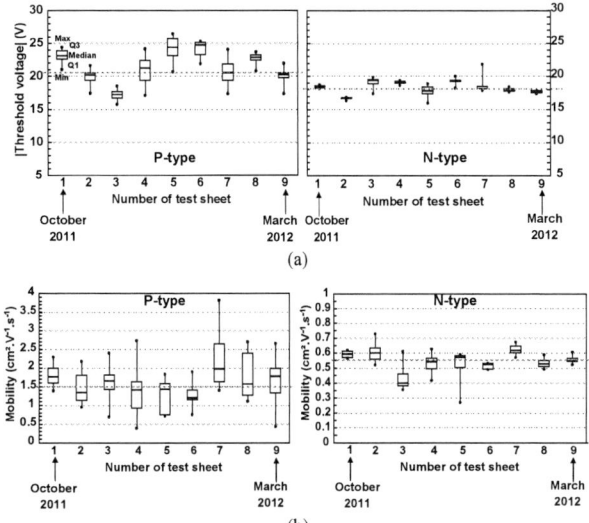

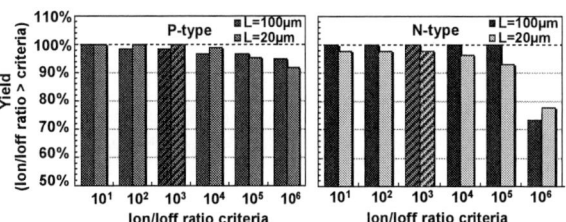

Figure 5. Repeatability of threshold voltage (a) and mobility (b) in saturation regime (|Vds|=60V) of P and N-type transistors on 9 sheets processed between October 2011 and March 2012. For each sheet, 8 P-OTFTs and 8 N-OTFTs with L=100μm W=2000μm have been tested.

Figure 6. Yield of P- (a) and N-type (b) OTFTs for 2 channel lengths (L=100μm and L=20μm / W=2000μm) with Ion/Ioff ratio as criterion for functional circuits. Median values are calculated on 60 transistors for L=100μm and 85 transistors for L=20μm, processed on 5 foils.

B. Complementary circuits

Several mask sets have been designed based on our Design Tool Kit. Different digital and analog circuits, such as logic gates, flip-flops, envelope detectors and comparators have been processed on 11cm x 11cm foils (Fig. 7). Results on simple digital circuits like inverters and a ring oscillator are presented, the measurements of more complex circuits being detailed in [9].

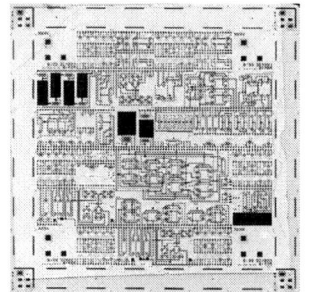
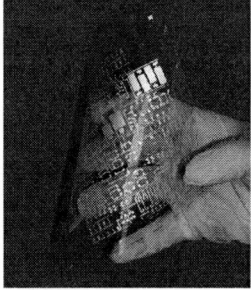

Figure 7. Pictures of a 11cm x 11cm plastic foil with printed single devices and complementary digital and analog circuits.

Fig. 8 shows the static output characteristics and corresponding gain of complementary inverters for 2 supply voltages (40V and 20V) and with 3 ratios of the P- and N-type transistors channel widths (W_P and W_N), the channel length L staying constant. As the drain current of the P-OTFTs is higher than the N-OTFTs one (μ_P=1.5 cm².V^{-1}.s^{-1} and μ_N=0.55 cm².V^{-1}.s^{-1}), with equally sized P- and N-type transistors, the transition between high and low levels is shifted towards higher voltages than VDD/2. An optimized design enables to balance the inverter. Indeed, by increasing W_N, the drain current of the N-OTFT is also increased close to the P-OTFT one and the transition is brought back in the region of VDD/2, as shown on Fig.8. We can also remark that our complementary inverters present relatively high gains, with a maximum of 20.

The 7-stage ring oscillator reaches high oscillation frequencies varying from 1.2kHz @ 40V to 200Hz @ 20V, corresponding to delay/gate values of 60µs and 350µs, respectively (Fig. 9). Moreover, as shown on Fig.10, after 6 months in ambient air, the oscillation frequency remains basically unchanged (1.4kHz@40V). We observe a modification of the pull down behaviour, which may derive for instance from the fact that our plastic sheets are not passivated.

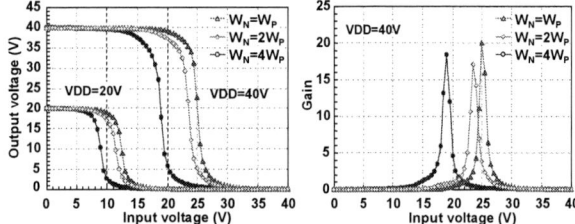

Figure 8. Static output characteristic and corresponding gain of 3 complementary inverters with different ratios between P- and N-type OTFTs channel widths, W_P and W_N. L=20µm W_P=1000µm and W_N=1000µm/2000µm/4000µm.

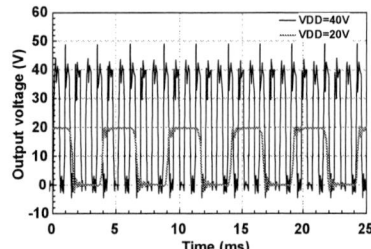

Figure 9. Transient response of a 7-stage complementary ring oscillator with equally sized P-and N-type OTFTs for VDD=40V and 20V. L=20µm W=1000µm.

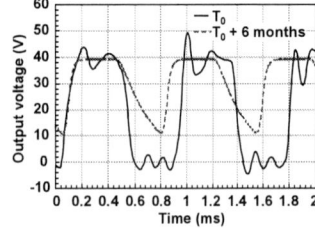

Figure 10. Shelf-lifetime of the ring oscillator of Fig. 9 kept for 6 months in ambient air without passivation. VDD=40V.

V. CONCLUSION

We have developed a printed organic complementary technology on flexible plastic substrate, compatible with large-area printing processes. Both P- and N-OTFTs are processed with small-molecule organic semiconductor solutions. Several challenges related to the integration of both P- and N-type OTFTs in a common flow have been addressed. In particular, the importance of the surface treatment between the N- and P-OSC deposition has been demonstrated. Our organic complementary technology exhibits high electrical performances and stable electrical characteristics, with a yield going from 98% to 100% for both P- and N-OTFTs, depending on the chosen criterion. Digital and analog circuits have been processed with this technology on several plastic sheets. Measurements of complementary inverters and the associated ring oscillators have been presented, showing that high gains as well as high frequencies over the kHz could be achieved. Shelf-lifetime results measured on an oscillator without passivation kept for 6 months in ambient air demonstrates that reaching long term stability with organic electronics is realistic. Moreover, the more complex digital and analog circuits as flip-flops, envelope detectors and comparators presented in [9] are functional, demonstrating the potential of this technology for higher-complexity organic complementary circuits.

ACKNOWLEDGMENTS

This work was funded in the frame of the European FP7 project COSMIC (grant agreement n° 247681). The authors want to thanks Polyera Corp. for supplying N-type semiconductor ActivInk® material.

REFERENCES

[1] A. C. Arias, J. D. MacKenzie, I. McCulloch, J. Rivnay and A. Salleo, "Materials and Applications for Large Area Electronics: Solution-Based Approaches", Chem. Rev., vol. 110, pp. 3-24, 2010.

[2] I. McCulloch et al., "Liquid-crystalline semiconducting polymers with high charge-carrier mobility", Nat. Mater., vol: 5, pp. 328, 2006.

[3] H. Yan et al., "A high-mobility electron-transporting polymer for printed transistors", Nature, vol. 457, p. 679, 2009.

[4] T. N. Ng et al., "Electrical stability of inkjet-patterned organic complementary inverters measured in ambient conditions", Appl. Phys. Lett., vol. 94, issue 23, 2009.

[5] H. Klauk, U. Zschieschang; J. Pflaum and M. Halik, "Ultralow-power organic complementary circuits", *Nature, vol. 445, p. 745,* **2007.**

[6] H. Yan et al, "Solution Processed Top-Gate n-Channel Transistors and Complementary Circuits on Plastics Operating in Ambient Conditions", Adv. Mater., vol. 20, issue 18, p. 3393, 2008.

[7] R. Blache, J. Krumm and W. Fix, " Organic CMOS Circuits for RFID Applications", ISSCC Dig. of Tech. Papers, pp. 208, 2009.

[8] A. Daami et al., "Fully printed organic CMOS technology on plastic substrates for digital and analog applications", ISSCC Dig. of Tech. Papers, p. 328, 2011.

[9] S. Abdinia et al., "Design of analog and digital building blocks in a fully printed complementary organic technology", submitted to ESSCIRC, 2012.

[10] D. Boudinet et al., "Modification of gold source and drain electrodes by self-assembled monolayer in staggered n- and p-channel organic thin film transistors", Organic Electronics, vol. 11, p. 227, 2010.

A Gate-Last $In_{0.53}Ga_{0.47}As$ Channel FinFET with Molybdenum Source/Drain Contacts

Xingui Zhang, Hua Xin Guo, Xiao Gong, and Yee-Chia Yeo.

Dept. of Electrical and Computer Engineering and NUS Graduate School of Integrative Sciences and Engineering,
National University of Singapore (NUS), Singapore 117576.
Phone: +65-6516-2298, Fax: +65-6779-1103, Email: yeo@ieee.org

Abstract — We demonstrated $In_{0.53}Ga_{0.47}As$ channel FinFETs with self-aligned Molybdenum (Mo) contacts for the first time. By using self-aligned Mo contacts formed on *in-situ* doped n^{++} $In_{0.53}Ga_{0.47}As$ source and drain, series resistance of ~250 $\Omega\cdot\mu m$ was achieved, which is the lowest value ever reported for $In_{0.53}Ga_{0.47}As$ non-planar devices. A gate-last process was used. FinFET with channel length of 500 nm and EOT of 3 nm has an I_{ON}/I_{OFF} of over 10^5 and peak G_m of 255 $\mu S/\mu m$ at $V_D = 0.5$ V.

INTRODUCTION

New materials and novel device structures are needed to overcome the challenges faced by silicon (Si) complementary metal-oxide-semiconductor (CMOS) technology. III-V compound semiconductors such as Indium Gallium Arsenide (InGaAs) are attractive alternative channel materials for sub-14 nm technology in metal-oxide-semiconductor field-effect transistors (MOSFETs) due to their higher electron mobility than that of Si and potential integration on Si substrates [1]-[7].

In addition, FinFETs are more scalable than planar MOSFETs. Promising results on InGaAs channel FinFETs have been achieved recently [8]-[12]. However, aggressive scaling of fin width in FinFET for control of short-channel effects may lead to high series resistance (R_{SD}). R_{SD} includes the resistance of the doped source/drain (S/D), the contact resistance (R_C) between metal and semiconductor, and the metal resistance. High R_{SD} could limit the drive current performance of InGaAs FinFETs.

Materials such as NiAuGe have been employed as S/D contacts for InGaAs FinFETs and were formed by a non-self-aligned method. The reported values of R_{SD} in those FinFETs are higher than 1 $k\Omega\cdot\mu m$ [8],[9],[11]. Development of a reliable, CMOS compatible, and low-resistance S/D ohmic contact for InGaAs FinFET is needed to realize high drive current performance. In addition, self-alignment of the S/D contacts to the channel or gate electrode is desirable for reduced S/D access resistance and for continual scaling of transistor footprint [13]-[18]. Recently, we demonstrated the integration of self-aligned Ni-InGaAs S/D contacts for InGaAs multiple-gate FETs [12]. The devices have Ni-InGaAs S/D contacts formed self-aligned to the gate electrode and show a low R_{SD} of 364 $\Omega\cdot\mu m$. It was found that the R_C (79 $\Omega\cdot\mu m$) between Ni-InGaAs and n^{++} $In_{0.53}Ga_{0.47}As$ could contribute a significant portion of R_{SD} [12]. In

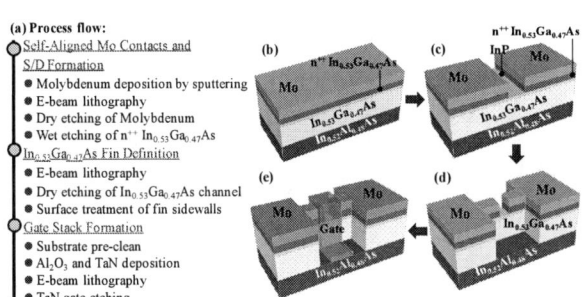

(a) Process flow:
- ○ Self-Aligned Mo Contacts and S/D Formation
  - ● Molybdenum deposition by sputtering
  - ● E-beam lithography
  - ● Dry etching of Molybdenum
  - ● Wet etching of n^{++} $In_{0.53}Ga_{0.47}As$
- ○ $In_{0.53}Ga_{0.47}As$ Fin Definition
  - ● E-beam lithography
  - ● Dry etching of $In_{0.53}Ga_{0.47}As$ channel
  - ● Surface treatment of fin sidewalls
- ○ Gate Stack Formation
  - ● Substrate pre-clean
  - ● Al_2O_3 and TaN deposition
  - ● E-beam lithography
  - ● TaN gate etching

Fig. 1. (a) Process flow for fabrication of a novel $In_{0.53}Ga_{0.47}As$ FinFET with self-aligned Mo contacts. (b) Starting $In_{0.53}Ga_{0.47}As$ substrate with blanket Mo deposition. (c) Device after self-aligned Mo contacted S/D formation. (d) Schematic of the device after $In_{0.53}Ga_{0.47}As$ fin formation. (e) Final structure of the FinFET with Mo contacts self-aligned to the $In_{0.53}Ga_{0.47}As$ channel.

Ref.[12], Ni-InGaAs S/D contacts were self-aligned to gate electrode but not to the $In_{0.53}Ga_{0.47}As$ channel.

Non-alloyed Mo has been reported to form ohmic contact on n^{++} $In_{0.53}Ga_{0.47}As$ with very low R_C [2],[19]. In this work, we realized a novel $In_{0.53}Ga_{0.47}As$ channel FinFET with self-aligned Mo contacts for the first time. By realizing low-resistance Mo S/D contacts self-aligned to $In_{0.53}Ga_{0.47}As$ channel, the FinFETs show low R_{SD} of 250 $\Omega\cdot\mu m$ and good drive current performance.

DEVICE FABRICATION

A. Self-Aligned Molybdenum (Mo) Contacts

The process flow for the FinFET fabrication is summarized in Fig. 1 (a). InP wafers served as the starting substrates for sequential growth of 300 nm of undoped $In_{0.52}Al_{0.48}As$, 50 nm of undoped $In_{0.53}Ga_{0.47}As$ channel, 2 nm of undoped InP capping, and 30 nm of Si-doped n^{++} $In_{0.53}Ga_{0.47}As$ (~5×10^{19} cm^{-3}). After native oxide removal, the sample was loaded into a sputtering chamber for Mo (~32 nm) deposition [Fig. 1. (b)]. Mo-contacted S/D was patterned by electron beam lithography (EBL) and formed by dry etching of Mo and citric-acid based wet etching of n^{++} $In_{0.53}Ga_{0.47}As$ [Fig. 1. (c)], using positive resist as an etch mask. InP served as an etch stop layer and a capping layer for the top surface of $In_{0.53}Ga_{0.47}As$ fin, which will be formed next.

978-1-4673-1707-8/12 $31.00 © 2012 IEEE

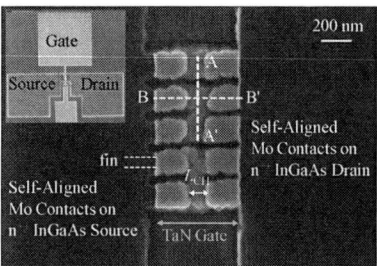

Fig. 2. SEM shows top view of a 5-fin $In_{0.53}Ga_{0.47}As$ FinFET with Mo contacts on n^{++} $In_{0.53}Ga_{0.47}As$ S/D. The recess of Mo and n^{++} $In_{0.53}Ga_{0.47}As$ defines the device channel. The inset shows the FinFET layout.

B. FinFET Fabrication

$In_{0.53}Ga_{0.47}As$ fin was patterned by EBL and formed by Cl_2-based plasma etching using negative resist as mask [Fig. 1. (d)]. Etch damage on $In_{0.53}Ga_{0.47}As$ fin sidewalls was removed by wet etching in H_2SO_4: H_2O_2: H_2O = 1: 1: 120 solution. After pre-gate clean, the sample was loaded into the atomic layer deposition (ALD) tool for deposition of ~6.7 nm Al_2O_3. 70 nm TaN metal layer was deposited by sputtering. The gate electrode was then patterned and etched using Cl_2-based plasma. Device fabrication was completed after the gate stack formation. The final structure is shown in Fig. 1. (e).

RESULTS AND DISCUSSION

Scanning electron microscopy (SEM) in Fig. 2 shows the top view of a 5-fin $In_{0.53}Ga_{0.47}As$ FinFET with Mo contacts formed on top of n^{++} $In_{0.53}Ga_{0.47}As$ S/D. The recess of Mo and n^{++} $In_{0.53}Ga_{0.47}As$ defines the device channel. The $In_{0.53}Ga_{0.47}As$ fins are oriented in the horizontal direction. Device layout is shown in the inset.

Transmission electron microscopy (TEM) image in Fig. 3 (a) shows a device cross-section along the dashed line A-A' in Fig. 2. The $In_{0.53}Ga_{0.47}As$ fin has a top width of ~80 nm, bottom width of ~125 nm, and height of 52 nm [Fig. 3 (b)]. A zoomed-in view of the rectangular region shows the 2 nm InP capping layer on top of the fin and that TaN/Al_2O_3 was conformally formed on the fin top and sidewall surfaces [Fig. 3 (c)]. Fig. 4 shows a device cross-section along the dashed line B - B' in Fig. 2. The Mo contacts appear as a darker layer on the surface of n^{++} $In_{0.53}Ga_{0.47}As$, lying adjacent to the $In_{0.53}Ga_{0.47}As$ channel. Mo layer has a thickness of about 32 nm as shown in the inset. In our previous work [12], S/D Ni-InGaAs contacts were formed self-aligned to the gate electrode. There is still a separation between the channel and S/D contacts due to the gate-to-S/D overlapping. The channel and S/D contacts were connected by an n^{++} $In_{0.53}Ga_{0.47}As$ capping layer with thickness of 30 nm and sheet resistance of ~55 Ω/square. In this work, the non-alloyed Mo contacts were formed self-aligned to the $In_{0.53}Ga_{0.47}As$ channel. The zoomed-in view (the inset) of the S/D regions in Fig. 4 confirms the self-alignment of Mo contacts to the channel.

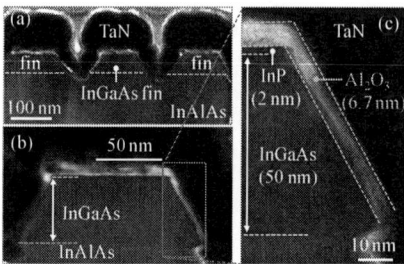

Fig. 3. (a) Cross-section TEM along A - A' in Fig. 2 shows $In_{0.53}Ga_{0.47}As$ fin structure. (b) The $In_{0.53}Ga_{0.47}As$ fin has top width of ~80 nm, bottom width of ~125 nm, and height of 52 nm. (c) A zoomed-in view of the rectangular region indicates the conformally formed gate stack on top and sidewalls of the $In_{0.53}Ga_{0.47}As$ fin.

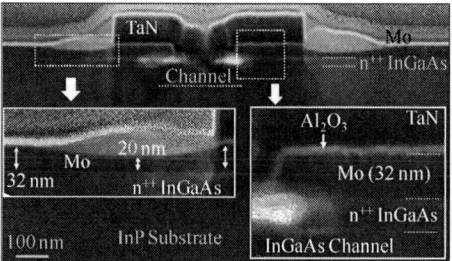

Fig. 4. TEM images show the device cross-section along B - B' (indicated in Fig. 2). Mo contacts, appearing as a darker layer, were on the surface of n^{++} $In_{0.53}Ga_{0.47}As$ S/D. Zoomed-in view of the S/D regions shows the self-alignment of Mo with the channel.

Transfer Length Method (TLM) test structures were also fabricated on Si-doped n^{++} $In_{0.53}Ga_{0.47}As$ (~5×10^{19} cm^{-3}) for the extraction of contact resistance R_C. Fig. 5 (a) plots total resistance R_T (logarithm scale) between two Mo contacts as a function of the contact spacing d from a TLM structure. The circles are experimental data and the solid lines are fitted curve by using linear fitting. The vertical axis was plotted as logarithm scale so as to be able to see the intercept which is ~50 $\Omega \cdot \mu m$. The inset shows plot of total resistance versus d in linear scale, indicating good linear relation between total resistance and contact spacing d. Contact resistance R_C is extracted from the intercept to be ~24 $\Omega \cdot \mu m$, which is comparable to the value reported in Ref. [2]. Statistical plot in Fig. 5 (b) shows the distribution of specific contact resistivity, which is in the level of $1 \times 10^{-7} \sim 1 \times 10^{-6}$ $\Omega \cdot cm^2$ and ~10 times lower than that of Ni-InGaAs [12],[17].

Fig. 6 (a) shows I_D-V_G and G_m-V_G plots of a single-fin $In_{0.53}Ga_{0.47}As$ FinFET with channel length L_{CH} of 500 nm and fin width W_{fin} of 90 nm. Drain voltage V_D of 0.05 and 0.5 V were applied. The I_D and G_m values were normalized by the width of top channel (W_{top}) and two sidewalls ($2W_{side}$), as obtained from the TEM. Good transfer characteristics with I_{ON}/I_{OFF} of over 10^5 and peak G_m of 255 $\mu S/\mu m$ at $V_D = 0.5$ V were observed. The extrinsic G_m is reasonable, taking the large channel length L_{CH} of 500 nm and large equivalent oxide thickness (EOT) of ~3 nm into consideration. The device has a subthreshold swing (SS) of 286 mV/V and drain

978-1-4673-1707-8/12 $31.00 © 2012 IEEE   178

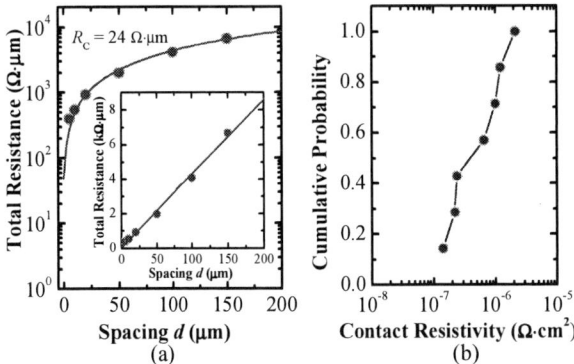

Fig. 5 (a) Total resistance versus contact spacing d (logarithm scale) of a TLM shows low contact resistance of ~24 $\Omega\cdot\mu$m. The inset shows a linear plot of total resistance versus d. (b) Statistical plot shows the distribution of specific contact resistivity measured from a number of TLM structures.

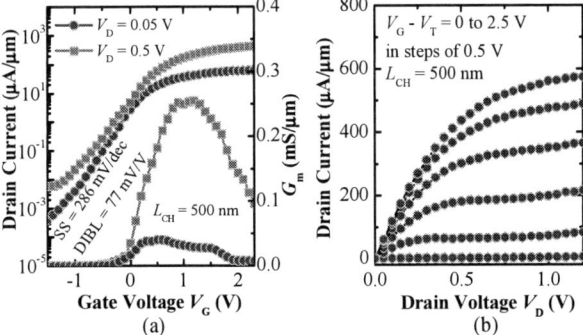

Fig. 6 (a). I_D–V_G and G_m–V_G of a single-fin In$_{0.53}$Ga$_{0.47}$As FinFET with L_{CH} = 500 nm and W_{fin} = 90 nm, showing I_{ON}/I_{OFF} of over 10^5. Drain current and transconductance were normalized by $W_{top} + 2W_{side}$. The device shows DIBL of 77 mV/V and peak G_m of 255 μS/μm at V_D = 0.5 V. (b) I_D–V_D characteristics of the same device in Fig. 8, showing good saturation and pinch-off characteristics.

induced barrier lowering (DIBL) of 77 mV/V. Although a better swing could be achieved by reducing the fin width and EOT, the large SS here is probably also due to the etch damage of InGaAs fin. Besides wet etching used in this work, a better technique is still needed to efficiently remove or repair the etch damage. I_D-V_D [Fig. 6 (b)] of the same device demonstrates good saturation and pinch-off characteristics. The gate over-drive (V_G −V_T) was varied from 0 to 2.5 V in steps of 0.5 V.

Fig. 7 (a) shows low gate dielectric leakage current density (J_G) which was normalized by gate and fin overlapped area. Fig. 7 (b) plots the extrinsic peak G_m of devices for different L_{CH} at V_D = 0.5 V, showing increased G_m with reduction of L_{CH}. Fig. 8 (a) plots the on-state resistance (R_{ON}) in the linear regime (V_D = 0.05 V) as a function of as-printed channel length $L_{As-printed}$ at three specified gate overdrives ($V_G − V_T$) of 1, 1.5, 2 V. The equation for the solid curve is given by $R_{ON} = R_{SD} + (L_{As-printed} - \triangle L)[W\mu C_{ox}(V_G - V_T)]^{-1}$, and was used to fit the data points (solid symbols). The three

fitted curves were extrapolated to show an intersection point. The intercept gives the R_{SD} of ~250 $\Omega\cdot\mu$m, which is the lowest value ever reported for InGaAs non-planar devices.

The intercept also gives a negative value of $\triangle L$ = -100 nm, which means the actual L_{CH} is 100 nm larger than the as printed channel length $L_{As-printed}$. This was also confirmed by cross-sectional TEM.

The fact that L_{CH} is larger than $L_{As-printed}$ is due to the lateral etch in the step of removal of n^{++} In$_{0.53}$Ga$_{0.47}$As by wet etching, which defined the device channel length. All the L_{CH} values mentioned in this paper is the actual L_{CH} after correction and $L_{As-printed}$ mentioned in Fig. 8 (a) is the as-printed channel length.

R_{SD} represents the sum of series resistance component in the source R_S and in the drain R_D, i.e. $R_{SD} = 2R_S = 2R_D$. The plot in Fig. 8 (b) shows the estimated component elements of the source resistance R_S. Mo resistance (R_{metal}) is calculated to be about 36 $\Omega\cdot\mu$m based on the Mo sheet resistance and the device source geometry as shown in Fig. 2. The Mo layer here has a high sheet resistance of ~40 Ω/square. The calculated metal resistivity is ~128 $\mu\Omega\cdot$cm which is 25 times higher than the reported value of 5~6 $\mu\Omega\cdot$cm in Ref. [20]. This could be due to an un-optimized metal sputtering process [21]. Mo contact resistance is ~24 $\Omega\cdot\mu$m as obtained

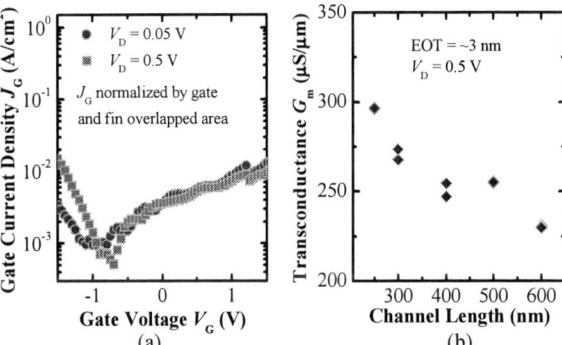

Fig. 7 (a) J_G as function of V_G shows low gate leakage current density. (b) Extrinsic peak G_m of devices with different L_{CH} (W_{fin} = 90 nm). The applied drain voltage is 0.5 V and EOT is about 3 nm.

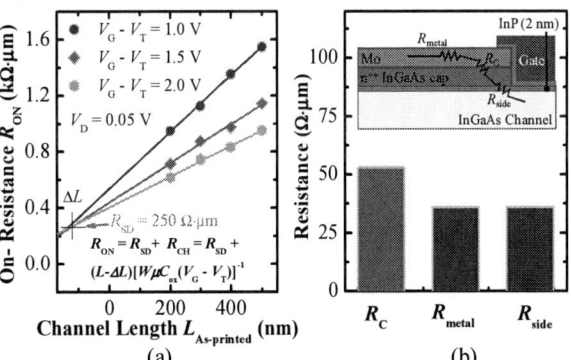

Fig. 8 (a) R_{ON}–$L_{As-printed}$ of FinFETs with W_{fin} = 90 nm. Lowest R_{SD} of 250 $\Omega\cdot\mu$m for In$_{0.53}$Ga$_{0.47}$As FinFETs was achieved. (b) The plot indicated the estimated component elements of the source resistance ($R_{SD} = 2R_S = 2R_D$).

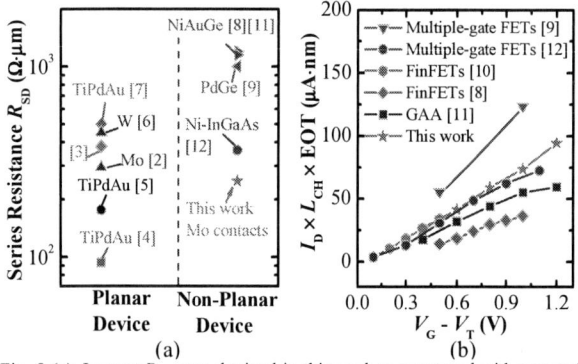

Fig. 9 (a). Lowest R_{SD} was obtained in this work as compared with reported InGaAs non-planar devices with non-self-aligned or self-aligned contacts. (b) $I_D \times L_{CH} \times$ EOT versus $(V_G - V_T)$ of InGaAs non-planar devices reported in the literature as well as the devices in this work.

from TLM structures and it contributes ~53 $\Omega \cdot \mu m$ to the source resistance, taking the contact width into consideration (Mo only contacts the top of fin but not the sidewalls of the fin). The remaining resistance (denoted as R_{side}) is ~36 $\Omega \cdot \mu m$, which includes InP (~2 nm) barrier resistance and spreading resistance in source. From the above calculation, we may conclude that R_C, R_{metal}, and R_{side} are quite comparable and have almost equal contribution to the total source resistance R_S in this self-aligned FinFET structure.

Fig. 9 (a) benchmarks R_{SD} of this work with reported values for InGaAs planar and non-planar devices. Ref. [4] reported the lowest R_{SD} of 93 $\Omega \cdot \mu m$ for $In_{0.53}Ga_{0.47}As$ planar MOSFETs. International Technology Roadmap for Semiconductors (ITRS) requires R_{SD} below 131 $\Omega \cdot \mu m$ for III-V multiple-gate FETs. However, the reported R_{SD} values for InGaAs non-planar devices in Ref. [8], [9], and [11] are over 1 k$\Omega \cdot \mu m$ and well above the ITRS requirement. Much lower R_{SD} of 364 $\Omega \cdot \mu m$ for multiple-gate $In_{0.53}Ga_{0.47}As$ device was achieved by forming Ni-InGaAs S/D contacts self-aligned to gate [12]. Lowest R_{SD} of 250 $\Omega \cdot \mu m$ for non-planar device was achieved in this work, with contribution from self-alignment of the Mo contact to $In_{0.53}Ga_{0.47}As$ channel and its low R_C on in-situ doped n^{++} $In_{0.53}Ga_{0.47}As$ S/D. Further reduction of R_{SD} is possible by optimizing the device structure and fabrication process. Fig. 9 (b) shows $I_D \times L_{CH} \times$ EOT versus overdrives $(V_G - V_T)$ of $In_{0.53}Ga_{0.47}As$ non-planar devices reported in the literature and the devices in this work. Drive current performance in this work is comparable to that of the best reported $In_{0.53}Ga_{0.47}As$ non-planar devices taking the large EOT of ~3 nm into consideration.

CONCLUSION

$In_{0.53}Ga_{0.47}As$ FinFETs with Mo contacts self-aligned to channel were realized for the first time. The devices achieved lowest R_{SD} of 250 $\Omega \cdot \mu m$ due to the self-alignment of Mo contacts as well as its low R_C. FinFET with self-aligned Mo contacts demonstrates good transfer characteristics with high I_{ON}/I_{OFF} ratio and good performance.

ACKNOWLEDGEMENT

Research grant from the National Research Foundation (NRF) of Singapore (Award number NRF-RF2008-09) is acknowledged.

REFERENCE

[1] M. Radosavljevic et al., "Advanced high-k gate dielectric for high-performance short-channel $In_{0.7}Ga_{0.3}As$ quantum well field effect transistors on silicon substrate for low power logic applications" Tech. Dig. – Int. Electron Devices Meet., 2009, 319.

[2] T.-W. Kim et al., "60 nm self-aligned-gate InGaAs HEMTs with record high-frequency characteristics" Tech. Dig. – Int. Electron Devices Meet., 2010, 696.

[3] Y. Sun et al., "Scaling of $In_{0.7}Ga_{0.3}As$ buried-channel MOSFETs" Tech. Dig. – Int. Electron Devices Meet., 2008, 367.

[4] Y. Yonai et al., "High drain current (>2 A/mm) InGaAs channel MOSFET at V_D = 0.5 V with shrinkage of channel length by InP anisotropic etching" Tech. Dig. – Int. Electron Devices Meet., 2011, 307.

[5] M. Egard et al., "High transconductance self-aligned gate-last surface channel $In_{0.53}Ga_{0.47}As$ MOSFET" Tech. Dig. – Int. Electron Devices Meet., 2011, 303.

[6] N. Waldron et al., "90 nm self-aligned enhancement-mode InGaAs HEMT for logic applications" Tech. Dig. – Int. Electron Devices Meet., 2007, 633.

[7] R. Terao et al., "InP/InGaAs composite metal-oxide-semiconductor field-effect transistors with regrown source and Al_2O_3 gate dielectric exhibiting maximum drain current exceeding 1.3 mA/μm" Appl. Phys. Express. 4, 054201 (2011).

[8] Y. Q. Wu et al., "First experimental demonstration of 100 nm inversion-mode InGaAs FinFET through damage-free sidewall etching" Tech. Dig. – Int. Electron Devices Meet., 2009, 331.

[9] H.-C. Chin et al., "III-V multiple-gate field-effect transistors with high-mobility $In_{0.7}Ga_{0.3}As$ channel and epi-controlled retrograde-doped fin" IEEE Electron Device Lett., 32, 146 (2011).

[10] M. Radosavljevic et al., "Electrostatics improvement in 3-D tri-gate over ultra-thin body planar InGaAs quantum well field effect transistors with high-k gate dielectric and scaled gate-to-drain/gate-to-source separation" Tech. Dig. – Int. Electron Devices Meet., 2011, 765.

[11] J. J. Gu et al., "First experimental demonstration of gate-all-around III-V MOSFETs by top-down approach" Tech. Dig. – Int. Electron Devices Meet., 2011, 769.

[12] X. Zhang et al., "Multiple-gate $In_{0.53}Ga_{0.47}As$ channel n-MOSFETs with self-aligned Ni-InGaAs contacts" 221st Electrochem. Soc. Meeting 2012, (to appear).

[13] X. Zhang et al., "III-V MOSFETs with a new self-aligned contact" Dig. Tech. Pap.- Symp. VLSI Technol., 2010, 233.

[14] X. Zhang et al., "Self-aligned contact metallization technology for III-V metal-oxide-semiconductor field effect transistors" J. Vac. Sci. Technol. B, 29(3), 032209-1 (2011).

[15] X. Zhang et al., "$In_{0.7}Ga_{0.3}As$ channel n-MOSFET with self-aligned Ni–InGaAs source and drain" Electrochem. Solid-state Lett., 14 , H60 (2011).

[16] X. Zhang et al., "A self-aligned Ni-InGaAs contact technology for InGaAs channel n-MOSFETs" J. Electrochem. Soc., 159(5), H511 (2012).

[17] L. Czornomaz et al., "Self-aligned S/D regions for InGaAs MOSFETs" European Solid-State Device Research Conference (ESSDERC), 2011, 219.

[18] S. H. Kim et al, "Self-aligned metal source/drain $In_xGa_{1-x}As$ n-MOSFETs using Ni-InGaAs alloy" Tech. Dig. – Int. Electron Devices Meet., 2010, 596.

[19] A. K. Baraskar et al., "Ultralow resistance, nonalloyed ohmic contacts to n-InGaAs" J. Vac. Sci. Technol. B, 27(4), 2036 (2009).

[20] Bass J., 1.2.1 Pure metal resistivities at T= 273.2 K. in: K.-H. Hellwege, J. L. Olsen (ed.), Springer Materials. The Landolt-Börnstein Database. (ISBN 978-3-540-11082-8)

[21] Z.-H. Li et al., "Molybdenum thin film deposited by in-line DC magnetron sputtering as a back contact for Cu(In,Ga)Se2 solar cells", Appl. Surf. Sci., 257, 9682 (2011).

Complementary RF-LDMOS Transistors Realized with Standard CMOS Implantations

Andreas Mai and Holger Rücker

IHP, Im Technolgiepark 25, 15236 Frankfurt (Oder), Germany

Email: mai@ihp-microelectronics.com

Abstract— Complementary lateral-drain-extended MOS transistors (CLDMOS) were integrated in a 0.13 μm SiGe BiCMOS technology. The LDMOS devices were realized in the dual-gate-oxide CMOS process without additional process steps. Drift regions were formed by the lightly-doped drain (LDD) implantations of 3.3V NMOS and PMOS transistors of the baseline process. The NLDMOS transistors use a combination of n-LDD and p-LDD to form the low-doped drift region whereas the PLDMOS drift region consists of the p-LDD implantation. Stable operation with less than 10% parameter variation in 10 years was achieved up to voltages of 6 V and 10 V for complementary LDMOS devices with different layouts. Peak transit frequencies f_T of 14.5 GHz and 9 GHz were demonstrated for PLDMOS transistors with $V_{DD,max}$ of -6 V and -10 V, respectively.

I. INTRODUCTION

The design of single chip communication systems requires advanced CMOS technologies with integrated high-voltage transistors for analog, radio frequency, and power management functionalities. MOS transistors with lightly-doped lateral extensions of the drain regions (LDMOS) are widely used for this purpose because they can be integrated in a cost effective manner by adding a small number of masked ion implantation steps [1]–[4]. Complementary LDMOS transistors are required for several analog and power management circuits such as power amplifiers in high-performance DC/DC converters [5].

The realization of n-type LDMOS devices without any additional processing effort was demonstrated in [6] for a 0.13 μm SiGe BiCMOS technology with dual gate oxides for 1.2 V and 3.3 V CMOS. In the present paper, this concept is extended to complementary RF-LDMOS transistors. NLDMOS and PLDMOS devices for operating voltages of 6 V and 10 V were realized for different gate lengths and drift region layouts using the gate oxide of the 3.3 V CMOS transistors. NLDMOS devices for operating voltages $V_{DD,max}$ of 10 V and 6 V showed peak transit frequencies f_T of 32 GHz and 14 GHz, respectively [6]. The PLDMOS devices presented here exhibit $f_T = 14.5$ GHz at $V_{DD,max} = -6$ V and $f_T = 9$ GHz at $V_{DD,max} = -10$ V and breakdown voltages BV_{DSS} of -14 V.

II. DESIGN DESCRIPTION

The complementary LDMOS (CLDMOS) devices were integrated in the 0.13 μm SiGe BiCMOS process of IHP [7]. The technology offers high-speed SiGe heterojunction bipolar transistors ($f_T = 240$ GHz, $f_{max} = 330$ GHz) together with 1.2 V and 3.3 V CMOS transistors, a set of passive RF components, and seven layers of aluminum interconnects

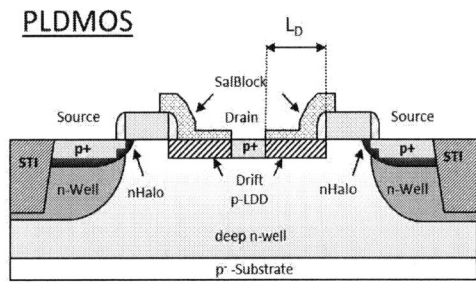

Fig. 1. Cross sections of the CLDMOS transistors. The drift region of the NLDMOS is formed by a partly superposition of the n-LDD and p-LDD implantations of the 3.3 V NMOS and PMOS devices. The n-doped drift region near the gate edge (width L_c) is formed by the n-LDD implantation only. The PLDMOS drift region consists of the p-LDD implantation.

to address RF and mm-wave applications with a high level of integration. The CMOS process features two gate oxide thicknesses of 2 nm and 7 nm for 1.2 V and 3.3 V CMOS transistors, respectively. The integration of the complementary LDMOS devices uses only process steps of the core CMOS flow and does not affect the HBT process.

Figure 1 shows schematic cross sections of the CLDMOS devices. Both types of LDMOS devices share the 7 nm gate oxide of the 3.3 V MOS transistors. The doping profile at the source side is defined by the standard p- or n-well, the heavily-doped drain (HDD) and the halo of the 1.2V-CMOS transistors (n-HDD and p-halo for NLDMOS and vice versa). The strongly asymmetric channel profiles for both device types are beneficial for DC and RF performance allowing short channel lengths and support high transconductance g_m

978-1-4673-1707-8/12 $31.00 © 2012 IEEE

and high transit frequencies f_T. The well implantations are restricted to the source regions in order to reduces the lateral electrical field at the drain side of the gate. In conjunction with the 1.2 V CMOS halo implantations at the source side this enables low standby currents and improves the breakdown voltages BV_{DSS}. The sensitivity of V_T to alignment variations of the CMOS well mask edges for both transistor types are effectively suppressed by the halo implantations due to the higher doping concentration near the surface compared to the CMOS wells [4], [8].

The lightly-doped drift region of the NLDMOS is realized by a superposition of the n-LDD and p-LDD implantations of the 3.3 V NMOS and PMOS devices. The n-LDD implantation is self-aligned to the gate edge whereas the mask edge for the p-LDD implantation is placed over the drift region with a distance of $L_C = 0.1\,\mu m$ to the gate edge. The partly compensation of the drift region results in a higher net doping of the drift region near the gate edge. This improves the trade-off between on-resistance and breakdown voltage for the NLDMOS [6]. The drift region of the PLDMOS is formed by the p-LDD implantation of the 3.3V PMOS. The drain of the PLDMOS is isolated from the p-substrate by the deep n-well. Cobalt salicide formation is blocked in the drift regions by a nitride layer which is used in the baseline process for the formation of unsalicided poly-silicon resistors.

III. RESULTS AND DISCUSSION

In the present approach, the same n-LDD and p-LDD implantations are used for 3.3 V CMOS transistors and for the formation of the drain extensions of NLDMOS and PLDMOS transistors. The implant conditions have to be chosen with respect to the performance of the four types of devices. In the following, device characteristics are discussed as a function of n-LDD dose ranging from $6 \cdot 10^{12}\,cm^{-2}$ to $1.2 \cdot 10^{13}\,cm^{-2}$ and p-LDD dose ranging from $5 \cdot 10^{12}\,cm^{-2}$ to $9 \cdot 10^{12}\,cm^{-2}$.

A. DC characteristics of CLDMOS and 3.3 V CMOS

Figure 2 shows the saturation drain currents I_{DS} and on-resistances R_{ON} of 3.3 V CMOS transistors as a function of LDD implantation doses. Error bars indicate the standard deviation across the eight inch wafer. Reducing the p-LDD dose from $9 \cdot 10^{12}$ to $5 \cdot 10^{12}\,cm^{-2}$ results in an increase of R_{ON} by $\approx 0.2\,\Omega mm$ and a decrease of I_{DS} by 4% for the 3.3V PMOS. The variation of the p-LDD dose has a much stronger impact on I_{DS} and R_{ON} of the PLDMOS transistor as shown in Fig. 3. This is due to the increase of the resistance of the drift region with decreasing p-LDD dose. The breakdown voltages BV_{DSS} of he PLDMOS device was found to be -14 V for all investigated p-LDD dose values. This value is determined by the breakdown voltage of the pn junction between drain and deep n-well.

For the NLDMOS, we found that BV_{DSS}, R_{ON}, and I_{DS} are determined mainly by the net implantation dose in the compensated part of the drift region. Fig. 4 shows the on-resistances R_{ON} of an NLDMOS with gate length $L_G = 0.5\,\mu m$ and drift length $L_D = 1.4\,\mu m$ as a function of

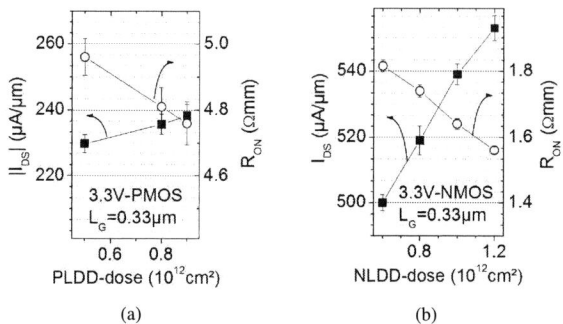

Fig. 2. Drain currents I_{DS} and on-resistances R_{ON} of 3.3 V PMOS (a) and 3.3 V NMOS (b) transistors for different n- and p-LDD doses.

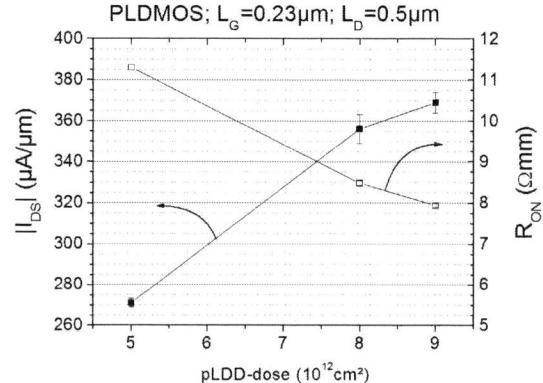

Fig. 3. Drain current I_{DS} and on-resistance R_{ON} of PLDMOS transistors vs. p-LDD dose.

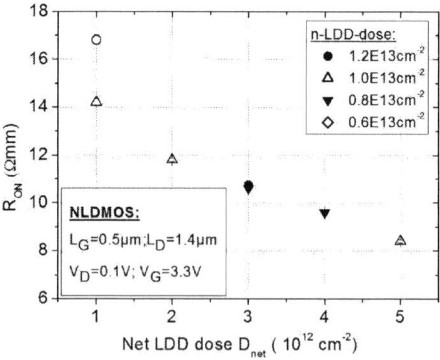

Fig. 4. On-resistances R_{ON} of NLDMOS transistors with different net LDD doses in the compensated part of the drift region part and various n-LDD dose values.

the net LDD dose of the compensated drift region for various combinations of n-LDD and p-LDD doses. The two points for the net dose $D_{net} = 3 \cdot 10^{12}\,cm^{-2}$ and n-LDD doses of $8 \cdot 10^{12}\,cm^{-2}$ and $1.2 \cdot 10^{13}\,cm^{-2}$ coincide. Only for the

978-1-4673-1707-8/12 $31.00 © 2012 IEEE

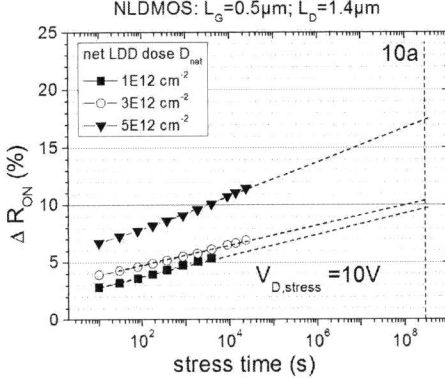

Fig. 5. R_{ON} degradation vs. stress time for NLDMOS devices with L_G=0.5 μm and L_D=1.4 μm for a stress voltage of $V_{D,stress}$=10 V depending on different net LDD doses.

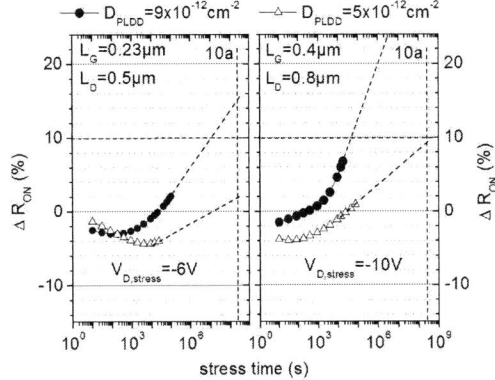

Fig. 6. R_{ON} degradation vs. stress time for PLDMOS devices with two drift and gate lengths and for a p-LDD dose of $5 \cdot 10^{12}$ cm^{-2} at stress voltage of $V_{D,stress}$ of -6 V and -10 V.

lowest net dose D_{net} of $1 \cdot 10^{12}$ cm^{-2} with almost complete compensation of n-LDD and p-LDD, we see a difference for two n-LDD dose variants.

For the final choice of appropriate LDD doses, device characteristics such as BV_{DSS}, R_{ON}, and I_{DS} have to be traded off against hot-carrier injection (HCI) immunity. Device degradation due to HCI for various LDD doses is addressed next.

B. Hot-carrier injection

The impact of net LDD dose variations on the high-voltage capability of the NLDMOS devices were investigated by voltage stress measurements. The NLDMOS devices have a gate length of 0.5 μm and a drift length of 1.4 μm and were stressed at drain voltages of 10 V for time intervals up to $2 \cdot 10^4$ s. The gates were biased at ≈1.7 V during stress corresponding to the maximum substrate current which is the worst case for HCI stress degradation. The on-resistance turned out to be the most sensitive parameter against HCI stress. Figure 5 shows the relative change of R_{ON} for devices with different net LDD dose variants. The device with a net LDD dose of $5 \cdot 10^{12}$ cm^{-2} is most vulnerable to high-voltage stress. The extrapolated degradation of R_{ON} for 10 V operation for 10 years is more than 15 %. The device degradation due to HCI is reduced for smaller D_{net} and is less than 10 % for $D_{net} \leq 3 \cdot 10^{12}$ cm^{-2}. This is due to a reduced electrical field strength at the drain-sided gate edge for the lower doping of the drift regions. The drift regions are depleted at $V_D = 10$ V for $D_{net} \leq 3 \cdot 10^{12}$ cm^{-2}.

PLDMOS transistors with different p-LDD doses were stressed at gate voltages of -1.1 V corresponding to the maximum gate current (worst case condition). Fig. 6 shows the degradation of R_{ON} which is also for PLDMOS devices the most sensitive parameter with respect to HCI degradation. We observed for all devices an initial decrease followed by an increase of R_{ON}. For PLDMOS transistors with $L_G = 0.23$ μm, $L_D = 0.5$ μm and a p-LDD dose of $9 \cdot 10^{12}$ cm^{-2} an

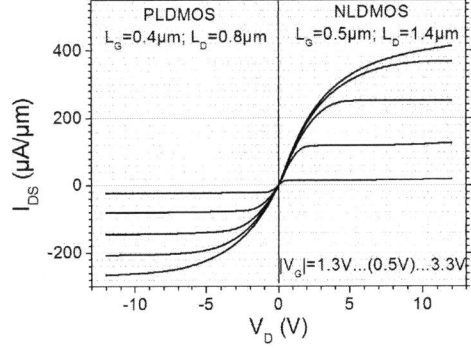

Fig. 7. Output characteristics for the preferred complementary LDMOS devices capable for $|V_{DD,max}|$=10V operation.

R_{ON} degradation of 15 % after 10 years was extrapolated for $V_D = -6$ V. For the smaller p-LDD dose of $5 \cdot 10^{12}$ cm^{-2} the degradation of R_{ON} was much less. Larger PLDMOS devices with $L_G = 0.4$ μm, $L_D = 0.8$ μm were investigated for maximum operating voltages $V_{DD,max}$ of -10 V. These devices show an extrapolated degradation of R_{ON} of 10 % within 10 years for the p-LDD dose of $5 \cdot 10^{12}$ cm^{-2} (Fig. 6).

Based on the described HCI degradation of NLDMOS and PLDMOS devices, we have chosen a p-LDD dose of $5 \cdot 10^{12}$ cm^{-2} and a n-LDD dose of $8 \cdot 10^{12}$ cm^{-2} corresponding to a net dose $D_{net} = 3 \cdot 10^{12}$ cm^{-2} in the compensated part of the NLDMOS drift region. These dose values enable stable operation of NLDMOS and PLDMOS devices up to 10V for appropriate gate lengths and drift lengths. Output characteristics of the complementary 10 V devices are plotted in Fig. 7.

C. RF-LDMOS characteristics

S-parameter measurements were used to study RF characteristics of the devices. Transit frequencies f_T and max-

TABLE I

DC AND RF PARAMETERS OF CLDMOS TRANSISTORS WITH DIFFERENT DRIFT LENGTHS AND GATE LENGTHS AND 3.3 V CMOS DEVICES. $V_{DD,max}$

RELATES TO AN ON-RESISTANCE DEGRADATION OF 10% IN 10 YEARS UNDER WORST BIAS CONDITIONS. *(RESULTS OF NLDMOS6 FROM [6])

Type	L_G [μm]	L_D [μm]	I_{Off} [$pA/\mu m$]	I_{ON} [$\mu A/\mu m$]	R_{ON} [Ωmm]	BV_{DSS} [V]	$V_{DD,max}$ [V]	$f_{T,max}$ [GHz]
			$V_{DD,max}$ V_G=0V	$V_{DD,max}$ V_G=3.3V	V_D=0.1V V_G=3.3V	$V_G = 0V$		V_D=5V
PLDMOS6	0.23	0.5	<1	290	7.6	-14	-6	14.5
PLDMOS10	0.4	0.8	<1	230	12.4	-14	-10	9
NLDMOS6*	0.2	0.8	<1	430	4.9	25	6	32
NLDMOS10	0.5	1.4	<1	420	8.4	27.5	10	14
PMOS	0.33	no	<1	230	4.7	-6.5	-3.3	
NMOS	0.33	no	<1	520	1.75	7.5	3.3	

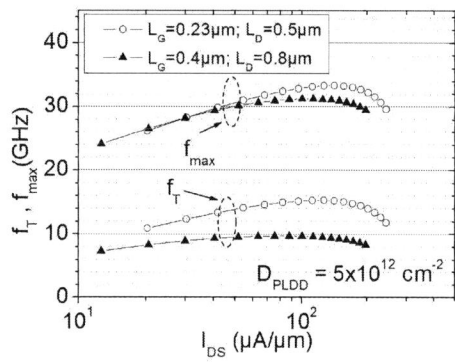

Fig. 8. Cutoff frequency f_T and maximum oscillation frequency f_{max} vs. drain current for PLDMOS transistors with two device geometries and with a P-LDD dose of $5 \cdot 10^{12}$ cm^{-2}.

imum oscillation frequencies f_{max} as a function of drain current are plotted in Fig. 8 for the two PLDMOS devices. The larger device with $L_G = 0.4\,\mu$m, $L_D = 0.8\,\mu$m, and $V_{DD,max} = -10$ V shows a peak f_T of 9 GHz while the smaller device with $L_G = 0.23\,\mu$m, $L_D = 0.5\,\mu$m and $V_{DD,max} = -6$ V shows a peak f_T of 14.5 GHz. Basic DC and RF parameters are summarized in Table I including data of NLDMOS devices discussed in detail in [6]. A comparison of the demonstrated PLDMOS figures of merit with previously reported data is given in Table II indicating a RF performance close to that of state-of-the-art PLDMOS devices with dedicated mask steps for drift region formation.

IV. SUMMARY

Complementary RF-LDMOS transistors were realized in a $0.13\,\mu$m SiGe-BiCMOS technology without additional process effort. Drift regions of NLDMOS and PLDMOS devices are formed by combinations of n-LDD and p-LDD implantations of the 3.3V CMOS. The channel doping is defined by CMOS wells and halo implants of the 1.2V CMOS at the source side. The resulting self alignment of the drift regions and of the major channel doing to the gate edges facilitates short

TABLE II

DC AND RF-PARAMETERS OF RECENTLY PUBLISHED RF-PLDMOS

TRANSISTORS

	BV_{DSS} [V]	$V_{DD,max}$ [V]	f_T [GHz]
Ref. [3]	-13	-5.5	16.8
Ref. [5]	-19	-10	11
this work	-14	-6	14.5
	-14	-10	9.5

minimum gate lengths and excellent RF parameters.

NLDMOS and PLDMOS transistors for maximum operation voltages of 6 V and 10 V with less than 10 % parameter drift within ten years were realized for different device layouts. Peak cutoff frequencies of 14.5 GHz and 32 GHz were demonstrated for the 6V PLDMOS and 6V NLDMOS, respectively.

ACKNOWLEDGMENT

The authors would like to thank the IHP clean room staff for the excellent support and D. Schmidt and C. Wipf for measurements.

REFERENCES

[1] K. Ehwald et al., "A Two Mask Complementary LDMOS Module Integrated in a 0.25 μm Sige:C BiCMOS Platform", Proc. of ESSDERC, pp. 121-124, 2001.

[2] G.. Baldwin et al., "90 nm CMOS RF Technology with 9.0 V I/O Capability for Single-Chip Radio", Digest of Symposium VLSI Technology, pp. 62-63, 2003.

[3] R.A. Bianchi et al., "High voltage devices in advanced CMOS technologies", Proc. of CICC, pp. 363-366, 2009.

[4] Z. Lee et al., "A modular 0.18 μm Analog RFCMOS Technology Comprising 32 GHz F_T RF-LDMOS and 40 V Complementary MOSFET devices", Proc. of IEEE BCTM, pp. 126-129, 2006.

[5] R. Sorge et al., "Complementary 0.25 μm RF LDMOS Module for 12 V DC/DC Converter and 6GHz Power Applications", Proc. of IEEE SiRF, pp. 57-60, 2011.

[6] A. Mai et al., "Drain-Extended MOS Transistors Capable for Operation at 10 V and at Radio Frequencies", Proc. of ESSDERC, pp. 110-113, 2010.

[7] H. Rücker et al., "A 0.13μm SiGe BiCMOS Technology Featuring fT/fMAX of 240/330 GHz and Gate Delays Below 3 ps", Proc. of IEEE BCTM, pp. 166-169, 2009.

[8] A. Mai et al., "Cost-Effective Integration of RF-LDMOS Transistors in 0.13 μm CMOS Technology", Proc. of IEEE SiRF, pp. 124-128, 2009.

TCAD degradation modeling for LDMOS transistors

S. Reggiani, G. Barone, E. Gnani, A. Gnudi

ARCES and DEIS, University of Bologna
Bologna, Italy, email: sreggiani@arces.unibo.it

S. Poli, M.-Y. Chuang, W. Tian, R. Wise

Texas Instruments, Inc.
Dallas, Texas

Abstract—**Physically-based models of hot-carrier stress and dielectric field-enhanced thermal damage have been incorporated in the framework of a TCAD tool with the aim of investigating the electrical stress degradation in integrated power devices over an extended range of stress biases and ambient temperatures. An analytical formulation of the distribution function accounting for the effects of the full band structure has been employed for the hot-carrier modeling. A quantitative understanding of the kinetics and local distribution of degradation are achieved, and the drift of the most relevant parameters is nicely predicted on an extended range of stress times and biases.**

I. INTRODUCTION

High voltage lateral double-diffused DMOS (LDMOS) transistors are of great interest as they are needed for a variety of analog applications and can be easily implemented in mixed-signal technologies. Previous studies of rugged LDMOS devices with shallow-trench isolation (STI) mainly focused on the device characterization and optimization [1, 2], whereas few works addressed also the hot-carrier stress (HCS) analysis [3-6]. In [7], the HCS experimental characterization of the STI-LDMOS device has been extended to very high V_{GS} and operating temperatures to the purpose of gaining an insight on the physical mechanisms which cause SiO_2 interface damage. A TCAD tool was used to correlate the electrical parameters drift to local temperatures and electric fields [8], but it could not be applied to predict degradation effects due to the limitations of the incorporated degradation model in presence of non-uniform lattice heating.

In this work, new physically-based models for hot-carrier stress and dielectric field-enhanced thermal damage have been incorporated in the framework of the TCAD tool. The experimental results of the linear drain-current drifts ($\Delta I_{d,lin}$) and of the threshold-voltage shifts (ΔV_t) have been nicely predicted over an extended range of stress times and biases, revealing the role played by the different physical mechanisms in different stress conditions.

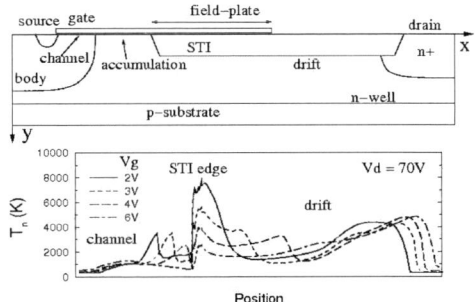

Figure 1. Top: 2D schematic view of the STI-based LDMOS device. Bottom: Electron temperature at the Si/SiO_2 interface at different V_{GS} in a 70V-LDMOS. The hot spot of electrons at the STI edge plays the most relevant role in degrading the Si/SiO_2 interface.

II. EXPERIMENTAL RESULTS

In Fig. 1, a 2D schematic view of the device under investigation is presented. An n-type LDMOS with maximum V_{DS} from 60 to 70 V and $V_{GS} = 5$ V is investigated. The device has been stressed by applying $V_{DS} = 70V$ at different V_{GS} biases. The high electric field in the proximity of the source-side STI corner is the main cause of HCS effects: the presence of a high density of "hot" electrons close to the Si/SiO_2 interface (Fig. 1, bottom) leads to a fast trap generation. The consequent trapping of charges in the on-current condition causes large shifts of the drain current itself. This is confirmed by the analysis of $\Delta I_{d,lin}$ as a function of V_{GS} reported in Fig. 2 (top). A first relative maximum of the current drift takes place at $V_{GS} \approx 2$ V, when the highest electron temperature T_n is expected. When further increasing V_{GS}, $\Delta I_{d,lin}$ is strongly reduced due to the reduction of T_n, reaching a minimum at $V_{GS} \approx 3\text{-}4$ V. At even higher gate biases $\Delta I_{d,lin}$ increases to large degradation drifts reaching its maximum value at the highest V_{GS}. By observing the maximum lattice temperature $T_{L,max}$ as a function of V_{GS} (Fig. 2, bottom), the strong increase of $\Delta I_{d,lin}$ may be ascribed to a thermally-activated degradation mechanism taking place in the high-V_{GS} stress regime and compensating for the T_n reduction at the STI corner.

Work supported by the SRC Research Contract No. 2011-VJ-2161

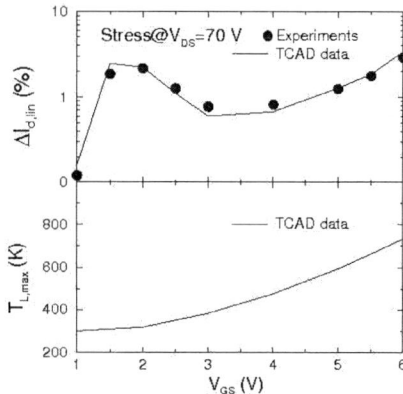

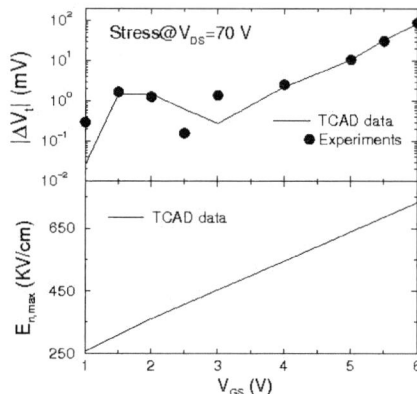

Figure 2. Top: Relative linear current degradation as a function of the stress gate voltage. The shifts are measured at long stress times. TCAD simulations have been performed with the new degradation model. Bottom: Maximum lattice temperature at the Si/SiO$_2$ interface as a function of the stress V$_{GS}$. The LDMOS experiences a strong self-heating at high V$_{GS}$.

Figure 3. Top: Absolute value of the threshold voltage shift as a function of the stress gate voltage. The shifts are measured at long stress times. TCAD simulations have been performed with the new degradation model. Bottom: Maximum normal elecric field at the Si/SiO$_2$ interface as a function of the stress V$_{GS}$. The peak is localized within the channel at the source side.

A different trend is observed for the ΔV_t curves (Fig. 3, top), being V$_t$ extracted as the intercept of the tangent in the inflexion point (maximum transconductance) with the V$_{GS}$ axis. First, negligible negative shifts are measured up to V$_{GS}$ = 3 V, as an effect of the $\Delta Id_{,lin}$ degradation itself. On the contrary, the sharp increase of ΔV_t at higher V$_{GS}$ clearly reflects a strong degradation localized within the channel of the device, enhanced by the peak of the electric field at the Si/SiO$_2$ interface, as shown by the correlation with the trend of the maximum normal electric field E$_{n,max}$ at the source side of the channel (Fig. 3, bottom). In the latter regime, the longitudinal electric field in the channel is low, thus preventing any hot-carrier effect (Fig. 1, bottom).

I. TCAD APPROACH

The simulation flowchart is shown in Fig. 4 (left). First, the carrier transport is solved at the stress bias. Two-dimensional TCAD simulations have been carried out using [8]. Drift-diffusion equations have been solved coupled with the heat transfer equation. The electron temperature is calculated as a post-processed solution of the energy-balance equation. A nice calibration of the simulation deck was carried out in previous works [2, 5].

The HCS module calculates the distribution function and solves the rate equations for the interface-trap concentration N$_{it}$. The electron distribution function f is determined in each position (x,y) along the Si/SiO$_2$ interface by means of a non-Maxwellian formulation which reproduces the electron density and T$_n$ calculated by the TCAD solution accounting for the non-parabolicity effects of the numerical full-band structure, as available in [8]. In Fig. 4 (right), the analytical f as a function of energy has been compared with the numerical solution of the Boltzmann transport equation based on the Spherical-Harmonics-Expansion (SHE) approach [8]. Four different positions along the Si/SiO$_2$ interface have been analyzed, corresponding to different electric-field regimes. The high-energy tail of the distribution function observed at the STI corner is nicely predicted up to 10 eV. The

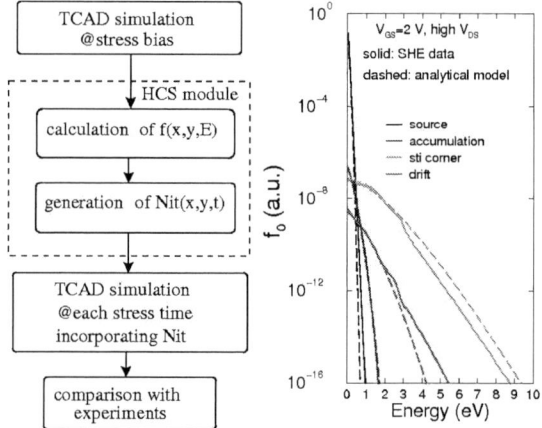

Figure 4. Left: Simulation flow-chart. The HCS module generates N$_{it}$ by using the TCAD results at the stress bias. Right: comparison between the electron distribution function obtained by means of the SHE-BTE solver and from our analytical model. Four different positions along the Si/SiO$_2$ interface are reported, showing a nice agreement up to very high energies.

distribution function is used to calculate the scattering rate integrals which model the interaction between hot electrons and the Si-H bond. Two different hot-carrier degradation mechanisms have been incorporated, namely, the single-electron (SE) interaction and the multiple-vibrational-excitation (MVE) model [9, 10]. They are controlled by integrals with the same functional structure. For the SE model, the scattering rate integral directly gives the forward reaction rate k_{SE}:

$$k_{SE} = \int_{E_A}^{\infty} f(E)g(E)v(E)\sigma(E)dE,$$

where E_A is the activation energy, g and v are the density of states and group velocity taken from [8], and σ is the reaction cross-section, given by:

978-1-4673-1707-8/12 $31.00 © 2012 IEEE

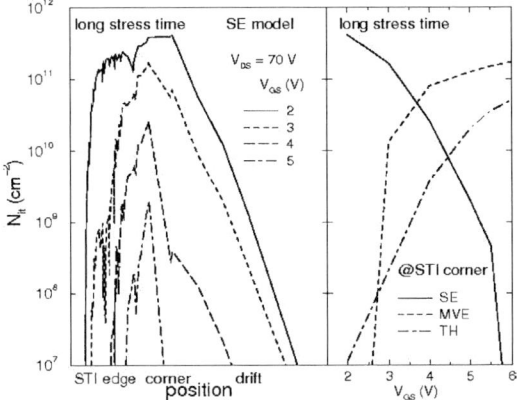

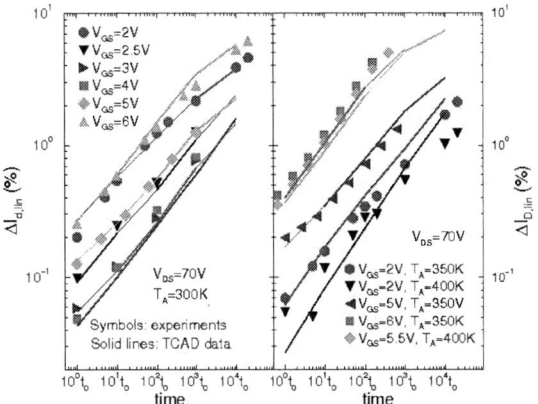

Figure 5. Left: Interface trap concentration due to SE processes vs. position along the Si/SiO$_2$ interface at a fixed stress time for different stress biases. A significant peak is obtained, localized at the STI corner. Right: Interface trap concentration at the STI corner given by the SE, MVE and TH processes at a fixed stress time for different V$_{GS}$ biases.

Figure 6. Measured (symbols) and simulated (solid lines) Id,lin shifts vs. stress time for (left) different V$_{GS}$ at T$_A$= 300K and (right) different V$_{GS}$ and T$_A$ up to 400K. A reference t$_0$ is fixed for the stress time range. More than four decades are covered by the measurement.

$$\sigma(E) = \sigma_0 \left(\frac{E - E_A}{K_B T_L} \right)^p,$$

being p = 11 [9] and σ_0 the model parameters. The MVE model has been implemented following [10].

In addition to the hot-carrier contributions, the role played by the local electric field and lattice temperature has been modeled by using the theoretical formulation in [11]. The thermally-activated (TH) dielectric degradation is predicted in the presence of large electric fields and high lattice temperatures even if no current flows at the Si/SiO$_2$ interface. The reaction rate reads:

$$k_{TH} = \upsilon_0 \exp\left[-\frac{E_A - p \cdot E_{ox}}{K_B T_L} \right],$$

where υ_0 is the interaction frequency with the lattice, E_A is the energy required for activating the bond breakage in the absence of field, p is the effective dipole moment, and E_{ox} is the normal electric field at the oxide.

The N$_{it}$ rate equations have been solved by assuming that the effect of the trap generation is negligible on the device electrostatics and transport, thus keeping the temperature and field distributions of the fresh condition. In addition, the models have been modified in order to take into account the finite distribution of the activation energies for the Si/SiO$_2$ bonds in a disordered medium like the amorphous SiO$_2$ [12], eliminating the assumption of discrete E$_A$ values in previous equations. Therefore, E$_A$ is substituted by a bond-dispersion energy distribution and the reaction rates become functions of the bond energy. Following [13], the reaction rates have been modeled as:

$$k_i(E) = k_i(E_A) \exp\left[-\frac{E - E_A}{\lambda K_B T_n} \right],$$

where i = SE, MVE, TH, and λ is a model parameter accounting for the effect of the tail of the distribution function.

The workflow ends with the simulation of the turn-on characteristic of the device in stressed conditions. In order to account for the degradation effects, the density of charge trapped at the Si/SiO$_2$ interface is needed at each node along the interface and at each stress time. It has been checked that for V$_{GS}$ > 1 V the electron Fermi level is close to the minimum of the conduction band at any position along the Si/SiO$_2$ interface. This implies that the interface-trap concentration would lead to an equivalent negative trapped-charge density. Thus, the generated N$_{it}$ distribution has been incorporated in the simulation set-up by assuming an acceptor trap density with a single energy level at the mid-gap, leading to fully occupied interface states.

The linear turn-on characteristics of the stressed device have been simulated by considering also the effect of the charged trap states on the carrier mobility by using the model available in [8] and calibrating the parameters on experiments.

II. RESULTS AND DISCUSSION

The $\Delta I_{d,lin}$ drift predicted by the new TCAD approach are reported in Fig. 2 for different V$_{GS}$, nicely comparing with experiments. For low V$_{GS}$, the major role is played by the SE hot-carrier trap formation, which gives a peak of N$_{it}$ localized at the STI corner strongly correlated to the high-energy tail of the electron distribution function. This is confirmed by the N$_{it}$ profiles shown in Fig. 5 (left) calculated by accounting only for the SE model at different V$_{GS}$ biases. At V$_{GS}$ ≈ 3-4 V, a minimum of the $\Delta I_{d,lin}$ drifts is reached, which depends on the different roles played by the hot-carrier SE and MVE processes giving N$_{it}$ distributed at the STI edge and along the drift region. In addition, the role played by N$_{it}$ localized along the drift region clearly showed the need of using a different set of parameters in the planar interface with respect to the non-planar one. It is worth observing that a different orientation is experienced at the STI edge, leading to different physical and chemical features. The SE and MVE contributions at the STI corner are reported in Fig. 5 (right) as functions of V$_{GS}$, along with the contribution given by the thermally-activated dielectric degradation, showing that the latter becomes relevant at the highest V$_{GS}$.

978-1-4673-1707-8/12 $31.00 © 2012 IEEE

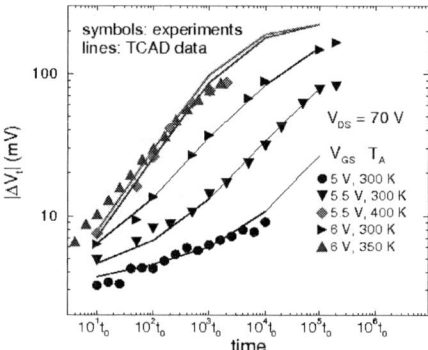

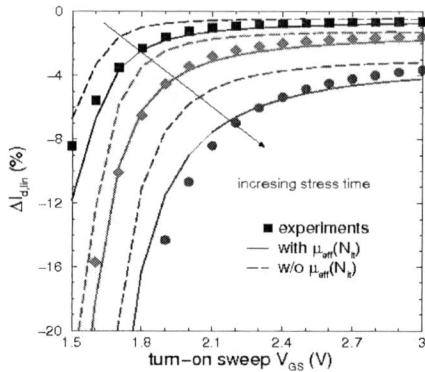

Figure 7. Threshold voltage shift vs. stress time. Symbols: experiments for different V_{GS} and T_A, ranging from a few mV to more than 100 mV. Solid lines: TCAD predictions.

Figure 8. $\Delta I_{d,lin}$ drift as a function of V_{GS} for increasing stress times. Symbols: experiments. Dashed lines: TCAD data without mobility degradation. Solid lines: TCAD data with mobility degradation.

In Fig. 6, the predicted $\Delta I_{d,lin}$ drifts are reported as functions of different stress times, showing the nice agreement of the TCAD results with experiments. The SE model has been calibrated at biases where its contribution on the $\Delta I_{d,lin}$ drifts is dominant, i.e., $V_{GS} < 3$ V, and accounting for its temperature dependence ($V_{GS} = 2$ V and $T_A = 350$ and 400 K reported in Fig. 6, right). The MVE and TH models have been mainly calibrated at higher V_{GS} both at room temperature (Fig. 6, left) and at higher T_A (Fig. 6, right). The TH contribution becomes the major term for the highest V_{GS} and T_A, while it is the most relevant one in the channel region at almost any V_{GS} for long stress times due to the limited contribution of hot carriers.

The ΔV_t shifts reported in Fig. 3, top, show that the TCAD results nicely predict both the negligible negative shifts measured up to $V_{GS} = 3$ V, and the larger positive shifts up to $V_{GS} = 6$ V. This nice result has been obtained through the MVE and TH models. A more accurate analysis has been carried out by considering the ΔV_t shift as a function of the stress time (Fig. 7). At $V_{GS} = 5$ V, the curve reflects the presence of both MVE and TH contributions. The MVE degradation is relevant at short stress times and gives an initial faster degradation but limited to a lower shift level, and it is mainly due to the presence of the high current density within the channel region. At long stress times a sharp increase of ΔV_t is predicted, which is due to the onset of TH contribution. The different trend at short and long stress times is also visible at $V_{GS} = 5.5$ and 6 V at room temperature. At higher T_A, the TH degradation becomes the most relevant one and the curves are mainly characterized by the higher slope associated with the field-enhanced degradation effect.

It is worth mentioning the role played by the effective mobility degradation. Following the analysis reported in [6], the mobility degradation due to the interface-trap generation needs to be accounted for. The latter can be incorporated in the TCAD simulations as a Coulomb-scattering term which can be added to the Matthiessen formulation [8]. The mobility model has been calibrated on different sets of stressed turn-on curves. In Fig. 8, the experimental $\Delta I_{d,lin}$ drifts extracted for a fixed stressed condition at different V_{GS} biases are reported along with the simulation results obtained with and without mobility degradation. The effect of the mobility degradation on the $\Delta I_{d,lin}$ drifts becomes significant when investigating the curves for V_{GS} values up to about 3 V.

III. CONCLUSION

A fast numerical degradation approach suited for STI-LDMOS devices has been presented and verified against experiments. Physically-based models of hot-carrier stress and dielectric field-enhanced thermal damage have been incorporated. By using an analytical formulation for the electron distribution function, the HCS worst-case conditions of a STI-based LDMOS device are nicely predicted on an extended range of drift variations. A quantitative understanding of the degradation contributions on the most relevant parameter drifts is achieved.

REFERENCES

[1] P. Hower et al., "A Rugged LDMOS for LBC5 Technology" Proc. ISPSD 2005, pp.327-330.

[2] S. Reggiani et al., "Explanation of the Rugged LDMOS Behavior by Means of Numerical Analysis", IEEE Trans. on ED 56, p. 2811, 2009.

[3] J. Chen et al., "On-Resistance Degradation Induced by Hot-Carrier Injection in LDMOS Transistors With STI in the Drift Region", IEEE El. Dev. Lett. 29, p. 1071-1073,2008.

[4] J. Chen et al., "An Investigation on Anomalous Hot-Carrier-Induced On-Resistance Reduction in n-type LDMOS Transistor", IEEE Trans. Dev. Mat. Rel. 9, p. 459-464, 2009.

[5] S. Poli et al., "Investigation on the temperature dependence of the HCI effects in the rugged STI based LDMOS transistor," Proc.ISPSD 2010, pp. 311 – 314.

[6] S. Reggiani et al., "Physics-Based Analytical Model for HCS Degradation in STI-LDMOS Transistors", IEEE Trans. on ED 58, pp. 3072-3080, 2011.

[7] S. Poli et al., "Temperature dependence of the threshold voltage shift induced by carrier injection in integrated STI-Based LDMOS transistors", IEEE El. Dev. Lett. 32, pp. 791-793, 2011.

[8] Synopsys Inc., "Sentaurus device simulator (release D-2010.03)," 2010.

[9] S. Tyaginov et al., "Interface traps density-of-states as a vital component for hot-carrier degradation modeling", Microelectronics Reliability 50, pp. 1267–1272, 2010.

[10] I. Starkov et al., "Hot-carrier degradation caused interface state profile - simulations vs. experiment", Journal of Vacuum Science and Technology – B 29, pp. 01AB09-1-01AB09-8, 2011.

[11] J. W. McPherson et al., "Complementary model for intrinsic time-dependent dielectric breakdown in SiO2 dielectrics", J. Appl. Phys. 88, pp. 5351-5359, 2000.

[12] D. Varghese et al., "OFF-State Degradation in Drain-Extended NMOS Transistors: Interface Damage and Correlation to Dielectric Breakdown", IEEE Trans. on ED 54, pp. 2669-2678, 2007.

[13] K. Hess et al., "Simulation of Si-SiO2 Defect Generation in CMOS Chips: From Atomistic Structure to Chip Failure Rates", IEDM Tech. Dig. 2000, pp. 93-96.

Pulsed I(V) - pulsed RF measurement system for microwave device characterization with 80ns/45GHz

Mario Weiß, Sébastien Fregonese, Marco Santorelli, Amit Kumar Sahoo, Cristell Maneux and Thomas Zimmer
Université Bordeaux 1, Laboratoire IMS, CNRS - UMR 5218
Cours de la Libération - 33405 Talence Cedex, France
Email : mario.weisz@ims-bordeaux.fr

Abstract— **This paper presents a combined pulsed I(V) - pulsed RF state-of-the-art measurement system. Isothermal DC and AC measurement data can be achieved allowing a complete characterization and exploration of the save operating area (SOA) of advanced SiGe:C HBTs. System configuration, measurements and accuracy issues are presented.**

I. INTRODUCTION

The first stride towards encroaching the THz-gap with silicon-based technology was made within the DOTFIVE project [1] in demonstrating the first Silicon Germanium (SiGe) Heterojunction Bipolar Transistor (HBT) with 500 GHz operating frequency at room temperature [2] and in establishing the international leadership of the European mm-wave community. The follow-up project DOTSEVEN is targeting the development of SiGe:C HBTs technologies with cut-off frequencies f_{max} of around 700 GHz.

The higher the speed of advanced HBTs, the higher are their current densities and internal electric fields. This leads to increased self-heating and potential thermal instabilities. The measured DC and AC characteristics become intrinsically temperature dependent: That means, for two different bias points the internal device temperature is different. This makes compact model parameter extraction very cumbersome because these parameters are very often temperature dependant e.g. access (base, emitter, collector) resistances, transit times, saturation currents.

In order to characterize the temperature dependence isothermal data is necessary. This can be achieved by performing DC and S-parameter measurements in pulsed condition. Even if pulsed measurements may not completely eliminate self-heating, they do reduce the device temperature significantly and thus allow (i) achieving quasi-isothermal measurement data, (ii) extracting thermal resistance R_{th} and capacitance C_{th} [3], (iii) extending the measured bias range towards higher voltages and current densities. The latter is necessary, in order to verify the compact model and its extracted parameters under realistic conditions: the device characterization needs to go beyond the standard measurements since most high frequency applications require large-signal operation, in which currents and charges have to be modeled accurately over a sufficiently wide bias range.

This paper presents an automatic measurement setup for heterojunction bipolar transistors (HBTs). It can be also applied to advanced MOSFET devices, but here we focus on HBTs. It enables simultaneous measurements of pulsed DC and pulsed S-parameter characteristics. Measurements with pulse widths down to 80 ns are shown for DC and 150 ns for RF in a broad frequency range (500 MHz–45 GHz) with current resolution in the µA range. These performances are beyond those of the measurement systems in [4–6]. In Section II, a detailed description of the pulse system is given including the general principle, the experimental setup and the related measurement procedure. In Section III, pulsed I(V) and pulsed S-parameter measurements for state of the art high speed HBTs are presented.

II. PULSE MEASUREMENTS

A. Principle

Pulse measurements are based on recording the transistor response to short pulsed stimuli applied at its terminals. The main idea is that if the pulse width is sufficiently short, there is not enough time for the internal temperature of the device to change significantly.

A pulse generator produces short pulses starting from a defined DC quiescent condition at low power dissipation. The applicability for different device technologies and geometries is limited by the internal resistance and the I(V) compliance values of the generator. In order to avoid high-frequency oscillations, the pulses should pass through a bias tee with a $50\,\Omega$ RF load on its RF port. A calibration that takes into account the resistance of bias tees and cables is required. The pulse width T_w and the pulse period T_p must be adapted to define the duty cycle D for the device under test (DUT), as $D = Tp/Tw \cdot 100\%$.

The pulse duration must be large enough to achieve sufficient measurement precision, but much smaller than the thermal time constant to ensure isothermal behavior of the device. Besides, the duty cycle has to be large enough for the device temperature to be defined solely by the DC quiescent point. In dependence of the DUT, a duty cycle of 0.1%–5% is generally adequate. Since the pulses are applied for a very small time compared with the total period, pulsed measurements cause little stress to the devices, so they allow measurements outside the SOA without damaging the

978-1-4673-1707-8/12 $31.00 © 2012 IEEE

transistors and study phenomena such as the electrical breakdown.

Moreover, it is possible to perform pulsed RF measurements. An RF and a DC pulse can be superimposed using DC and RF port of a bias tee (see Fig. 1(a)). The pulse form strongly depends on the input resistance of the pulse generator and of the DUT as well as lumped parasitics of cables and bias tees. Therefore, the pulse form should be verified with an oscilloscope in order to define the measurement window to be in a steady region of the pulse (see Fig. 1(b)).

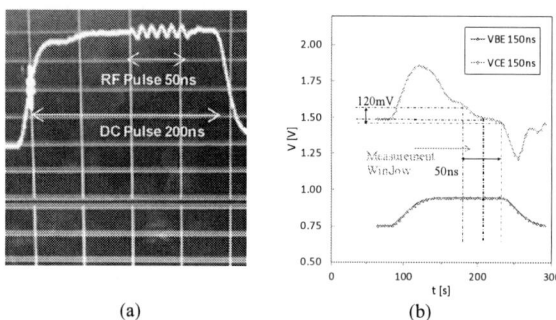

(a)   (b)

Fig. 1: (a) Oscilloscope measurement of synchronized DC and RF pulse. (b) Measured base pulse $V_{BE}(t)$ and collector voltage $V_{CE}(t)$ with indicated measurement window for a pulse width T_w of 150 ns.

B. Experimental Setup

Pulse measurements were carried out with the pulsed DC analyzer Keithley 4225-SCS in combination with the VNA Rohde & Schwarz ZVA67 (with ZVAX extension unit) as shown in Fig. 2. The 4200-SCS consists of two pulse measurement units (PMUs) that generate the pulse and measure the pulse response. The pulses are applied directly to the gate/base of the DUT, while the collector/drain voltage is held constant and the emitter/source is connected to ground. The PMU allows to create pulses larger than 80 ns with minimum rise and fall times of 20 ns. Accurate measurements were obtained in a time window between 60% and 90% of the DC pulse duration as shown in Fig. 1(b).

In order to synchronize the DC and RF pulses, the 4200-SCS sends a trigger signal generated with the pulse generator unit (PGU) to the ZVA67. The synchronization has been adapted with delay parameters and verified with the oscilloscope LeCroy Wavepro960. The input resistance of channel 1 (base monitor) and channel 2 (collector monitor) of the oscilloscope are set to $50\,\Omega$ and $1\,M\Omega$, respectively, to prevent the system from oscillations. A reasonable trade-off between accuracy and measurement time has been reached using an averaging factor between 5 and 10.

In this setting, a duty-cycle of up to 50% is possible, but good results were achieved using values between 1% and 10%. Below 1%, the RF pulsed measurements exhibit insufficient accuracy, while above 10%, thermal dispersion can occur in the DUT. RF pulsed measurements could be performed up to 50 GHz, but cables and connectors set the limit to 45 GHz. The maximum measurable current for the system is 200 mA, while the maximum admissible voltage is

10 V. The measured RF data is transferred from the ZVA67 to the 4200-SCS over a GPIB interface.

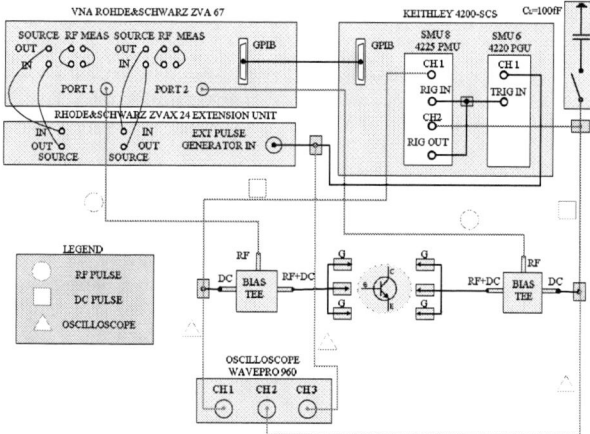

Fig. 2: Experimental set-up of the pulse measurement system, where DC and RF pulses are applied through the DC and RF ports of the bias tees at the base and the collector of the DUT.

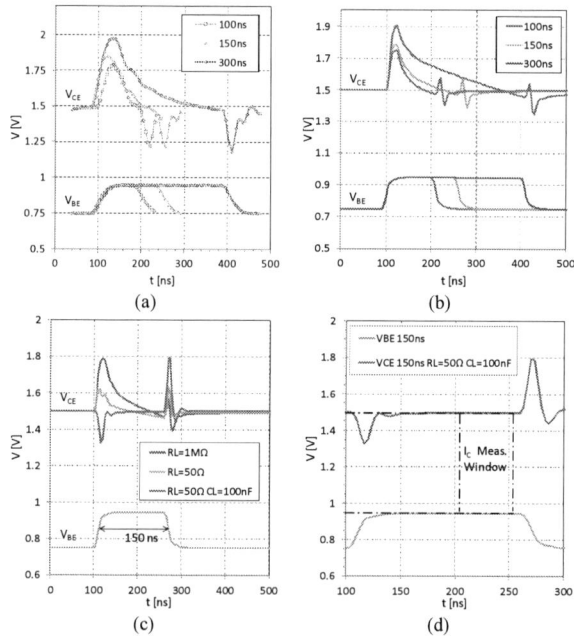

(a)   (b)

(c)   (d)

Fig. 3: Base voltage pulses $V_{BE}(t)$ from 0.75 V to 0.95 V with a pulse widths T_w of 100 ns, 150 ns and 300 ns and a constant collector voltage V_{CE} of 1.5 V are (a) applied to the HBT and (b) simulated with an HBT model. (c) V_{CE} stability optimization by applying (i) a load resistance R_L of $50\,\Omega$ or (ii) R_L of $50\,\Omega$ and a load capacitance C_L of 100nF at the collector node of the HBT model. (d) Enlarged image section of pulse simulation with stabilized condition (ii) with indicated measurement window for collector current $I_C(t)$.

C. V_{CE} Stability Optimization

Measurements for base voltage pulses $V_{BE}(t)$ from 0.75 V to 0.95 V with a pulse width T_w of 100 ns, 150 ns and 300 ns and a constant collector voltage V_{CE} of 1.5 V applied at the base of a high speed HBT are shown in Fig 3(a). During the pulse, a V_{CE} overshoot is observed that results in a maximum

978-1-4673-1707-8/12 $31.00 © 2012 IEEE

deviation of 15% in the measurement window for a base pulse of 100 ns. Similar overshoot effects have been reproduced with circuit simulations in Fig. 3(b). The schematic of the simulation represents an identical set-up connecting a pulse and a standard voltage source through a network of passive elements to the base and the collector node of the transistor model. The stability of V_{CE} during the pulse can be improved by applying (i) a load resistance R_L of $50\,\Omega$ or (ii) a load resistance R_L of $50\,\Omega$ and a load capacitance C_L of 100nF at the collector node as shown in Fig. 3(c). Best results are achieved with the latter configuration (see Fig. 3(d)). However, the current flowing through the capacitance modifies the measured collector current I_C. In order to assure correct I_C values, normal pulsed I(V) characteristics are measured in a first run without connecting the capacitance. In a second step, the capacitance is connected and pulsed RF measurements are carried out.

D. Pulsed I(V)

Accurate device characterization with pulse measurements requires efficient external software. For automatic measurements a special library was implemented in the 4200-SCS. It is very complex to take into account the impact of various device technologies (input and output impedances) on the measurement system. A reliable and generalized procedure is mandatory. Forced bias conditions must be reached with a corrective algorithm. For example, due to load impedance at the collector, the I_C response produces a negative V_{CE} pulse that has to be compensated. An error value for the V_{CE} pulse correction using the load impedance, the initial voltage and the measured current response is calculated. A compensation pulse is generated using this error value. The preceding steps are repeated until the error value is below a desired limit.

E. Pulsed S-parameter

To characterize the small-signal behavior of active devices, low RF power has to be used. The source level has been chosen to be -28 dBm in order to ensure the linear condition and at the same time have a sufficient accuracy level. Since the VNA measures during the full pulse period, a small duty cycle brings the RF signal level close to the VNA noise level [5]. In this setting, pulsed S-parameters with a duty cycle below 1% exhibit insufficient accuracy (see in Fig. 4(a) measured open standard).

First of all, a power calibration with a RF power of -20 dBm is performed that adapts the source level for each frequency to achieve a constant RF power level. After that, a standard SOLT calibration with a power level of -28 dBm is carried out. The first step is a normal pulsed I(V) measurement without connecting the capacitance. Then, pulsed S-parameters are measured using the capacitance in order to stabilize the DC pulse. Biasing level and pulse synchronization are checked by monitoring port 1, port 2 and the trigger trace on the oscilloscope (see Fig. 2). Pulsed S-parameters have been compared with standard S-parameter measurements, as shown for a measured open (Fig. 4(a)(b)) and thru (Fig. 4(c)(d)). Very good agreement was achieved between conventional and pulsed RF measurements with a duty cycle D of 5%.

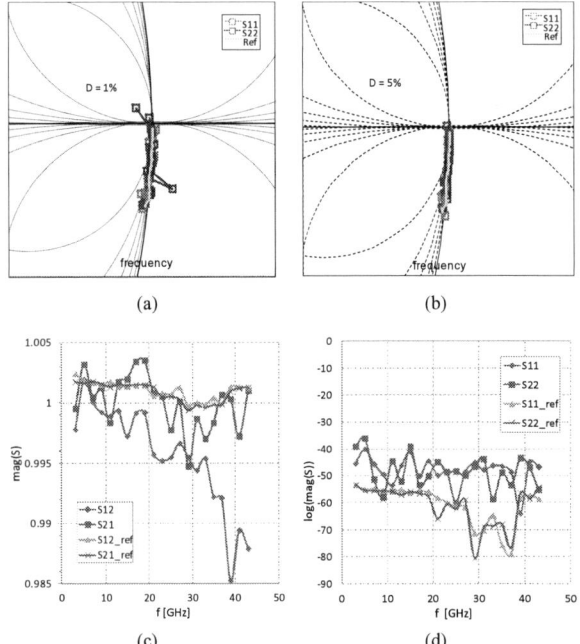

Fig. 4: Standard (as reference) and pulsed S_{11}, S_{22} of an open (Smith chart) with a duty cycle D of (a) 1% and (b) 5%. Standard and pulsed (c) S_{12}, S_{21} and (d) S_{11}, S_{22} of a thru with duty cycle D of 5%.

III. MEASUREMENT RESULTS

In this paragraph DC and RF device characteristics as well as their bias dependence are presented. On wafer measurements have been carried out on high speed NPN SiGe HBTs [7] in GSG configuration at room temperature. The measured devices were fabricated within the BiCMOS9MW process of STMicroelectronics. This quasi self-aligned trench isolated technology has breakdown voltages, $BV_{CEO} = 1.6\,V$, $BV_{CBO} = 5.5\,V$, transition frequency $f_T = 230\,GHz$ and maximum oscillation frequency $f_{max} = 290\,GHz$.

A. DC Results

Fig. 5(a) and 5(c) show pulsed measured output curves obtained for an HBT with a drawn emitter window $A_E = 0.27x5\,\mu m^2$ and $A_E = 0.84x5\,\mu m^2$. Measurements were carried out with pulse widths from 80 ns up to $50\,\mu s$. Good agreement between long pulse widths and standard DC characteristics validates the accuracy of the system. A realistic behavior of I_C is obtained that increases with longer pulse widths. The longer the pulse width, the more DC power is dissipated in the device and, therefore, the higher is the internal device temperature. Fig. 5(b) and 5(d) show I_C increase for a given V_{CE} as a function of the pulse width. These characteristics have been extracted from Fig. 5(a) and 5(c), respectively, by making a vertical section at a given V_{CE} and collecting the measured current values for each pulse width. It can be observed that isothermal measurements are achieved for the larger device when the pulse width is below 100ns (Fig. 5 (b)). For the smaller device, only quasi-isothermal measurements are achieved, even for the narrowest pulse width of 80ns (Fig. 5 (d)).

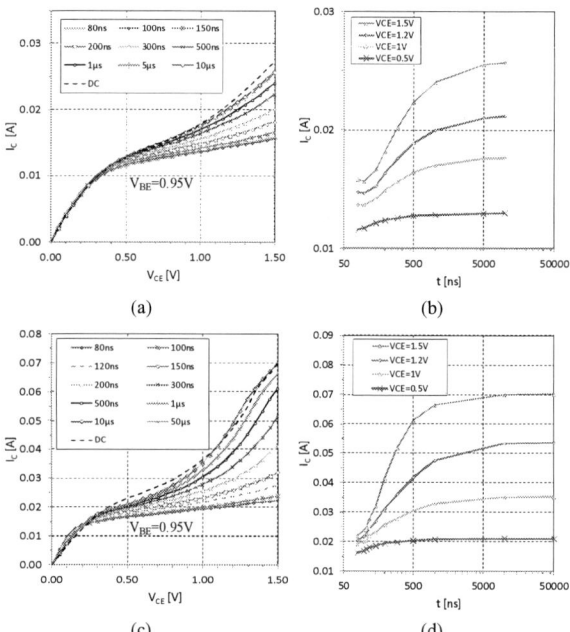

(a)　　　　　　　　　　(b)

(c)　　　　　　　　　　(d)

Fig. 5: Output characteristics for V_{BE}=0.95 V measured with various pulse widths from 80 ns up to 50 µs for an HBT with a drawn emitter area (a) A_E=0.27x5 µm² and (c) A_E=0.84x5 µm². Collector current I_C versus pulse width T_w for different collector emitter voltages V_{CE} for an HBT with a drawn emitter area (b) A_E=0.27x5 µm² and (d) A_E=0.84x5 µm².

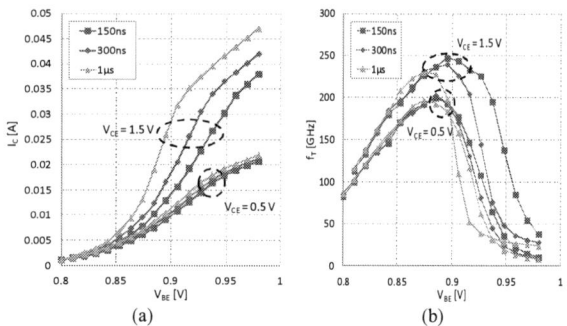

(a)　　　　　　　　　　(b)

Fig. 6: (a) Forward Gummel plot and (b) transit frequency f_T versus base emitter voltage V_{BE} measured with 150 ns, 300 ns and 1 µs pulse widths at constant collector emitter voltages V_{CE}=[0.5,1.5] V for an HBT with a drawn emitter area A_E=0.5x5 µm².

B. RF Results

Fig. 6 shows the Forward Gummel and the current gain cut-off frequency f_T versus V_{BE} for an HBT with a drawn emitter area of A_E=0.5x5 µm². S-parameters are obtained up to 45 GHz as a function of V_{BE} from 0.8 V to 0.98 V for V_{CE}=[0.5, 1.5] V. S-parameters were deembedded with open structure and the f_T is extracted at 15 GHz. For V_{CE}=0.5 V internal self-heating can be neglected. Therefore, measurements for various pulse widths should be close, as shown in Fig. 6. On the other hand, for V_{CE}=1.5 V, the transistor is significantly heated resulting in higher I_C and lower peak f_T for longer pulse widths.

IV. CONCLUSION

An advanced pulsed I(V) and pulsed RF measurement system has been presented. It allows to adapt the duty cycle, the pulse width as well as the synchronization of DC pulse and RF pulse. Pulsed I(V) measurements for an advanced SiGe:C HBT technology have been presented with pulse widths between 80 ns and 50 µs. Collector bias voltage has been stabilized to enable more accurate pulsed S-parameter measurements. Good accuracy has been achieved with a duty cycle D of 5%.

Self-heating inside the device has been significantly decreased using small pulse widths. This enables new possibilities for electrical and thermal device characterization especially for high nonlinear and high frequency active devices. Moreover, compact model parameter extraction methodology becomes more accurate and the safe-operating-area of the device can be widened.

ACKNOWLEDGEMENT

This work is part of the RF2THZ SiSoC project supported by the European EUREKA Program CATRENE and the French Ministry for Economics and Industry and of the Dotseven project supported by the European Commission through the Seventh Framework Program for Research and Technological Development.

REFERENCE

[1] DOTFIVE, "Towards 0.5 THz Silicon/Germanium Heterojunction Bipolar Technology." [Online]. Available: EUFP7 funded IP, number 216110, http://www.dotfive.eu/.

[2] B. Heinemann, R. Barth, D. Bolze, J. Drews, G. G. Fischer, A. Fox, O. Fursenko, T. Grabolla, U. Haak, D. Knoll, R. Kurps, M. Lisker, S. Marschmeyer, H. Rucker, D. Schmidt, J. Schmidt, M. A. Schubert, B. Tillack, C. Wipf, D. Wolansky, and Y. Yamamoto, "SiGe HBT technology with fT/fmax of 300GHz/500GHz and 2.0 ps CML gate delay," in *Electron Devices Meeting (IEDM), 2010 IEEE International*, 2010, pp. 30.5.1–30.5.4.

[3] A. K. Sahoo, S. Fregonese, M. Weiss, N. Malbert, and T. Zimmer, "Electro-thermal characterization of Si-Ge HBTs with pulse measurement and transient simulation," in *Solid-State Device Research Conference (ESSDERC), 2011 Proceedings of the European*, 2011, pp. 239 –242.

[4] B. Schaefer and M. Dunn, "Pulsed measurements and modeling for electro-thermal effects," in *Bipolar/BiCMOS Circuits and Technology Meeting, 1996., Proceedings of the 1996*, 1996, pp. 110 –117.

[5] J.-P. Teyssier, P. Bouysse, Z. Ouarch, D. Barataud, T. Peyretaillade, and R. Quere, "40-GHz/150-ns versatile pulsed measurement system for microwave transistor isothermal characterization," *IEEE Trans. Microwave Theory Techn.*, vol. 46, no. 12, pp. 2043–2052, Dec. 1998.

[6] A. Saleh, M. A. Chahine, T. Reveyrand, G. Neveux, D. Barataud, J. M. Nebus, R. Quere, Y. Bouvier, J. Godin, and M. Riet, "40 ns pulsed I/V set-up and measurement method applied to InP HBT characterization," in *2009 IEEE Radio Frequency Integrated Circuits Symposium*, Boston, MA, USA, 2009, pp. 401–404.

[7] G. Avenier, M. Diop, P. Chevalier, G. Troillard, N. Loubet, J. Bouvier, L. Depoyan, N. Derrier, M. Buczko, C. Leyris, S. Boret, S. Montusclat, A. Margain, S. Pruvost, S. T. Nicolson, K. H. . Yau, N. Revil, D. Gloria, D. Dutartre, S. P. Voinigescu, and A. Chantre, "0.13µm SiGe BiCMOS Technology Fully Dedicated to mm-Wave Applications," *IEEE Journal of Solid-State Circuits*, vol. 44, no. 9, pp. 2312–2321, Sep. 2009.

978-1-4673-1707-8/12 $31.00 © 2012 IEEE

Novel Deep Trench Buried-Body-Contact (DBBC) of 4F² Cell for Sub 30nm DRAM technology

Youngseung Cho, Yoosang Hwang, Huijung Kim, Eunok Lee*, Soojin Hong*, Hyunwoo Chung, Daeik Kim,
Jiyoung Kim, Yongchul Oh, , Hyeongsun Hong, Gyo-Young Jin, Chilhee Chung
DRAM TD Team, Process Development P/J 1* Semiconductor R&D Center, Samsung Electronics Co.
San#16 Banwol-Dong, Hwasung-City, Gyeonggi-Do, Korea, 445-701
Email: ys0.cho@samsung.com

Abstract — Novel Deep Trench Buried-Body-Contac (DBBC) has been successfully developed for 4F² DRAM cells on sub-30nm technology node. The critical requirements of thermal stability, shallow junction depth, and conformal source-drain doping profile for the contact are achieved by using an ultra thin Ti silicide ohmic layer and PLAD technique, which also show excellent electrical performance and process feasibility for the development of 4F² DRAM cell on the 30nm node and beyond.

I. INTRODUCTION

For the last decade, DRAM technology has drastically evolved towards reducing chip size and solving the problems incurred from the cell size scaling [1]. The transition from 8F² to 6F² in the array configuration has been a key driver to reduce the cell size with the same lithography tool. Moreover, innovative cell structure of RCAT or BCAT has played a prominent role to enhance the performance and reduce power consumption of the scaled devices [2,3]. As the feature size of DRAMs is now approaching the 20 nm technology node, another big transition from 6F² to 4F² cell is required to continue the cell size scaling without introducing a nano-lithography technology like EUV tool [4]. Compared with 6F² cell, the 4F² cell reduces the cell size by 33% with the same lithography tool or the patterning cost by extending present ArF technology for one more generation.

The key features of 4F² cell are a Buried-BL(bit line) and vertical channel compared with a Stacked-BL and U-shaped channel in the 6F² cell [5]. The vertical channel enables unit cell to cut down 1F size along the x-axis and eventually cut down 2F² area because an unit cell consists of 1F active and 1F field region, while the Buried-BL enables to maximize the contact area of storage node without shadowing active area by Stacked-BL (Figure 1).

However, forming a Buried-BL into a deep and narrow Si trench has some critical challenges. First of all, the Buried-BL should be more thermally stable than Stacked-BL because the Buried-BL is built up before forming the gate stack which has a high temperature oxidation process. Another big challenge is to make a conformal source-drain doping profile along the surface of Si, while maintaining the shallow junction depth to assist hole-draining from a channel to a Si substrate.

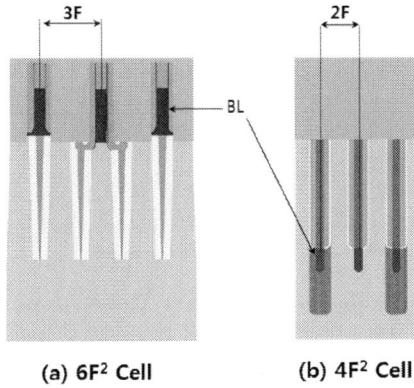

Figure 1.   Schematic structure of (a) 6F² and (b) 4F² in Bit Line direction

This paper proposes a novel Deep-Buried-Body-Contact (DBBC) connecting a source-drain node on sidewall of Si pillar to a Buried-BL. The DBBC has demonstrated low contact resistance, excellent thermal stability and shallow junction profile through Plasma Doping (PLAD) technology and ultra thin silicide ohmic layer.

II. EXPERIMENT & DISCUSSION

A. Process flow

The process sequence for fabricating the DBBC is outlined in Figure 2. After ion implantation for adjusting threshold voltage of a channel, the pillar-shaped active region is built up by crossing line-shaped patterns through DPT and dry etching process. An oxide and nitride spacer protects the sidewall of Si pillars from dopant penetration during a following ion implantation step, and isolates the Buried-BL from unrelated cells. After carrying out the 2nd trench etching,

978-1-4673-1707-8/12 $31.00 © 2012 IEEE

a PLAD process executes and forms a conformal and shallow arsenic doping profile along the unprotected Si surface. Finally, ultra-thin Ti film is deposited in consecutive nitrogen ambient followed by tungsten deposition and etching it back into the Si trench.

ISO. Patterning

Si Trench Etch

Spacer Deposition

Spacer Etch

Si Trench Etch

Bottom S/D Doping

Metal Deposition

Metal Etch Back

Figure 2.   Process flow of Deep Buried Body Contact

B. Critical Challenges

The 4F^2 cell typically features a Buried-BL structure which means the BLs should be formed before gate stacks, and thus BLs undergo a high temperature process of gate oxidation. Therefore, the thermal stability is clearly essential to the Buried-BL. Co/TiN/W layer is widely used for a plugging material of which silicide layer agglomerates together to form larger cluster during high temperature process and eventually deteriorate the characteristics of a contact. Figure 3 of EELS shows the cluster is an agglomerated layer of Co silicide. Figure 4 shows the agglomeration of Co silicide layer after 850C heat treatment and the cluster size becomes larger as the heat treatment time increases. As a result, the contact resistance rapidly increases as the total heat budget increases because the ohmic contact area decreases as the cluster's surface area decreases. Therefore, developing an ultra thin silicide layer and inhibiting the agglomeration is a straightforward approach to achieve better thermal stability. Also, even thinner ohmic layer is essential to have a shallower junction depth.

The floating body is more critical and fundamental problem in the 4F^2 cell to obtain sufficient retention time because depletion edge of the BL junction laterally expands to the end of Si pillar and blocks the discharge of holes generated in a channel. On the 30nm technology node, the Si body is fully depleted by BL bias and hence additional component is required to drain the holes to the Si substrate. And the shallow BL junction is still indispensible to effectively drain the holes and minimize the side-effects from the additional structure. Also, conformal N-type doping is crucial to overlap source-drain to the gate and reduce the series resistance. However, shallow and conformal doping profile in the 300 nm deep trench is beyond the limit of standard beamline implantation method. Therefore, we need a more advanced doping technology like PLAD showing a wide spectrum in the angular distribution of the implanted ions owing to the scattering in the plasma and thus hitting the sidewall and bottom of the trench with the same probability.

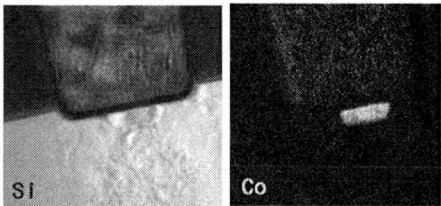

Figure 3.   EELS (Electron Energy Loss Spectroscopy) of Co/TiN/W with heat budget [850°C, 80 minutes]

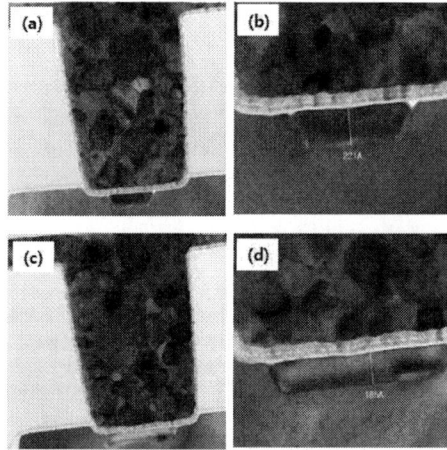

Figure 4.   TEM images of fabricated contact (a) Co silicide, heat budget [850°C, 80 minutes], (b) Agglomeration image of (a), (c) Co silicide, heat budget [850°C, 60 minutes], (d) Agglomeration image of (c) [Test conditions : same contact area , same thickness of Co/TiN/W]

C. Experimental results

Forming an ultra-thin silicide film for ohmic contact layer is a key approach to prevent the agglomeration and hence improve the thermal stability. The consecutive nitrogen ambient during a deposition of Ti remains just a few atomic-layer of Ti for ohmic sheet while converting the rest Ti to TiN for a barrier metal film.   While a conventional Co silicide ohmic sheet is required more than 100Å thick to have the conformal coverage along the bottom and sidewall of a Si

978-1-4673-1707-8/12 $31.00 © 2012 IEEE

trench, an ultra thin Ti silicide film by the novel process just shows less than 10Å thick (Figure 5).

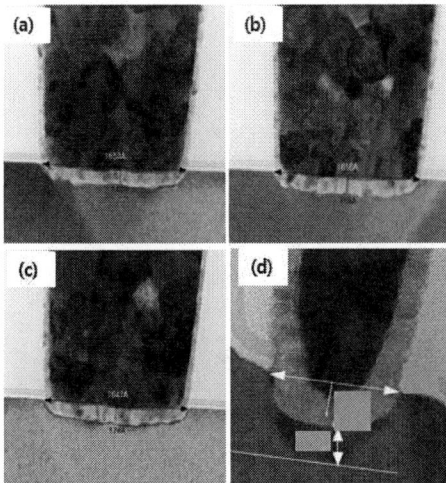

Figure 5. TEM images of fabricated contact (a) thin Ti_xSi_y with heat budget [850°C 40 minutes], (b) thin Ti_xSi_y with heat budget [850°C, 60 minutes], (c) thin Ti_xSi_y with heat budgets [850°C ,80 minutes] (d) Co_xSi_y with conventional heat budget [850°C, 7 minutes]

Physically, the ultra thin Ti silicide film shows the excellent thermal stability enduring a thermal treatment without any Ti silicide agglomeration up to 850°C 80minutes. Electrically, the contact resistance drastically decreases 55% and the cumulative distribution is significantly improved through the ultra-thin Ti silicide ohmic film (Figure 6 & 7).

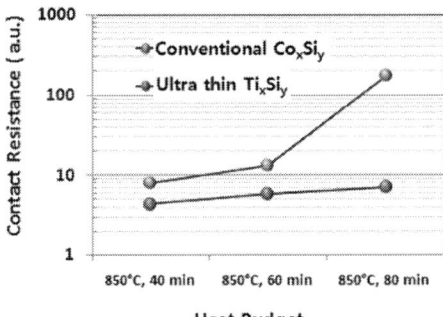

Figure 6. Contact resistance of conventional Co silicide and ultra thin Ti silicide with thermal treatment conditions

In order to perform the conformal and shallow junction in the deep Si trench, the PLAD technique has been introduced and successfully moved up the peak point of arsenic doping close to the Si surface. The cross section image of DBBC taken by HV-SSRM (High Vacuum Scanning Spread Resistance Microscopy) shows the shallow and conformal doping profile of arsenic along the Si surface (Figure 8). The SIMS profile of Figure 9 also shows that, through the PLAD technique, the projected range (Rp) becomes extremely shallow of 3.5nm deep from the Si surface and of 45nm deep

even after a heat treatment of 850°C 60min, while a standard beam-line technique shows 79nm deep Rp with the same heat treatment (Figure 9). With the help of lowering contact resistance by the ultra-thin Ti silicide ohmic layer and series resistance by PLAD technique, the current driving capability of $4F^2$ cell improves 28% with the same threshold voltage, which is a great breakthrough for developing a viable $4F^2$ cell (Figure 10).

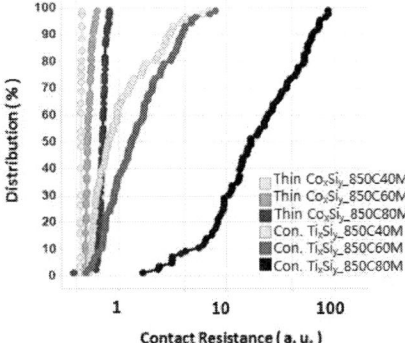

Figure 7. Cumulative distribution of contact resistance for conventional Co_xSi_y and ulta thin Ti_xSi_y with thermal treatment conditions

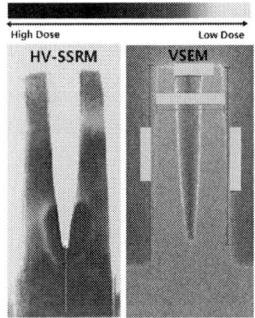

Figure 8. HV-SSRM (High Vacuum Scanning Spreading Resistance Microscopy) of arsenic distribution in buried body Area , VSEM image of buried body contact

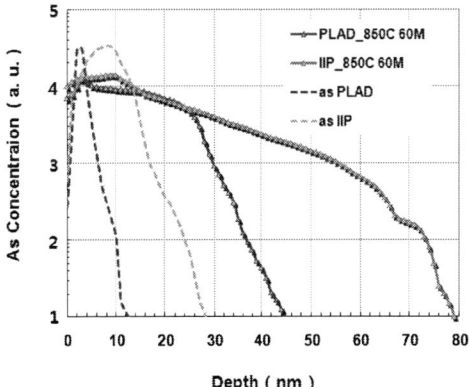

Figure 9. SIMS (Second Ion Mass Spectroscopy) for arsenic PLAD doping and arsenic conventional IIP doping with thermal treatment conditions.

978-1-4673-1707-8/12 $31.00 © 2012 IEEE

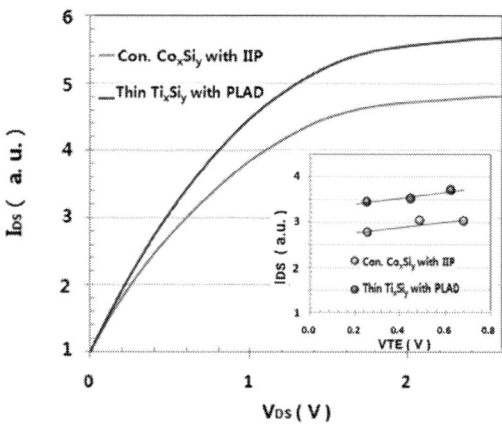

Figure 10. I_{DS}-V_{DS} curves of a fabricated conventional Co_xSi_y with IIP and ultra thin Ti_xSi_y with PLAD (Insert : I_{DS}-V_{TE} characteristic curves) in $4F^2$ cell [vertical channel transitor]

III. CONCLUSION

The novel DBBC structure solves one of the critical challenges of the $4F^2$ DRAM cell on 30nm feature size. Less than 10Å of Ti silicide ohmic layer and ultra shallow junction of 3nm deep has been achieved and shown excellent performances of 55% reduction of contact resistance and 28% increase of driving current. The DBBC also shows excellent thermal stability up to 850°C 80 minutes and integration flexibility to develop a $4F^2$ DRAM cell on 30nm node and beyond.

REFERENCES

[1] Ki-Whan Song, "A 31 ns Random Cycle VCAT-Based 4F2 DRAM With Manufacturability and Enhanced Cell Efficiency," *IEEE J. Solid-State Circuits*, pages 880–888, 2010.

[2] Ji-Young Kim, "The Breakthrough in Data Retention Time of DRAM Using Recess-Channel-Array Transistor (RCAT) for 88 nm Feature Size and Beyond," *Proc. Symp. VLSI Technol., Dig. Tech. Papers*, 2003.

[3] T. Schloesser, "A 6F2 Buried Wordline DRAM Cell for 40nm and Beyond," *IEDM Tech. Dig.*, 2008.

[4] Jae-Man Yoon, "A Novel Low Leakage Current VPT(vertical pillar transistor) Integration for 4F DRAM Cell Array with Sub 40 nm Technology," *DRC Tech. Dig.*, 2006.

[5] Kinam Kim, "From The Future Si Technology Perspective: Challenges and Opportunities," *IEDM Tech. Dig.*, 2010.

Z^2-FET Used as 1-Transistor High-Speed DRAM

Jing Wan,[1a)] Cyrille Le Royer,[2] Alexander Zaslavsky,[1,3] and Sorin Cristoloveanu[1]

[1]IMEP-INPG/Minatec, 3 Parvis Louis Néel, 38016 Grenoble Cedex 1, France
[2] CEA, LETI, MINATEC, F-38054 Grenoble, France
[3] School of Engineering, Brown University, Providence, Rhode Island 02912, USA
[a)] Corresponding author: wanj@minatec.inpg.fr

Abstract—We have recently demonstrated a new device named Z^2-FET (*zero* subthreshold swing and *zero* impact ionization) and proposed it as a 1-transistor DRAM. The device is built on an FD-SOI substrate and operates by feedback between carrier flows and injection barriers. We now present additional results obtained from extensive experiments and simulations. Experimentally, the I_{ON}/I_{OFF} ratio exceeds 10^9 and supply voltage (V_{DD}) scales down to 1.1 V with the DRAM retention time as high as 0.15 s at 75 °C. In simulation, the access time reaches below 1 ns and the Z^2-FET can be scaled down to 30 nm. We also discuss various operation modes.

I. INTRODUCTION

The conventional dynamic random access memory (DRAM), combining one transistor and one external capacitor (1T-1C DRAM), has shown good reliability and high integration density [1]. However, the external capacitor is not scalable, as it needs to store enough charge and maintain a long retention time. Thus it requires high aspect-ratio structure, which is challenging to fabricate [2]. The access speed of DRAM is also limited by the required minimum charge storage [1].

The single-transistor capacitor-less DRAM (1-T DRAM) is of great interest due to its compact size [3-4]. Most 1-T DRAMs use the floating body effect, where the stored majority carriers control the flow of minority carriers. Most floating body memories use impact ionization or band-to-band tunneling for writing, leading to slow write speeds and requiring relatively high V_{DD} [5]. These problems have been mitigated recently by using the bipolar writing mode [6]. Another interesting 1-T DRAM is thyristor-based: it shows high integration density and access speed [7-8], but requires precise doping control to obtain stable bipolar characteristics under various temperatures [9]. A field effect diode (FED) with two front gates was used for electrostatic discharge protection and proposed as a capacitor-less memory device [10-12]. Recently, we demonstrated the use of the Z^2-FET as capacitor-less and high speed DRAM using transient feedback [13]. The Z^2-FET is simpler and more compact, with a single front gate and an undoped channel.

Here we systematically study the Z^2-FET used as a 1-T DRAM. The dc (direct current) measurements show sharp switching and gate-controlled hysteresis, resulting from the feedback between carrier flows and their injection barriers, as confirmed by simulation. Unlike the thyristor, the Z^2-FET shows good temperature stability, and does not involve impact ionization or doping-sensitive bipolar action. The Z^2-FET is used as 1-T DRAM with the charge directly stored in the gate capacitor and non-destructively read out through internal feedback amplification. We demonstrate experimentally the scaling of V_{DD} down to 1.1 V. The retention time is studied in detail as the function of temperature, biasing and device size.

II. DEVICE STRUCTURE AND DC PERFORMANCE

A. Device structure and DC characteristics

The Z^2-FET is a forward-biased gated *pin* diode built on an FD-SOI substrate with channel partially covered by a gate (L_G) and the rest ungated (L_{IN}), schematically shown in Fig. 1(a). The device operates with either backgate voltage or surface charge Q_S. In a *p*-type device, the front gate is adjacent to the n^+ doped drain and negatively biased. Either a backgate voltage $V_{BG} > 0$ or a positive Q_S on the ungated region is required. Figure 1(b) shows the scanning electron microscope (SEM) image of the *n*-type device, where the gate is adjacent to the p^+ doped drain and positively biased, whereas $V_{BG} < 0$. The device is similar in layout to an asymmetric tunneling FET (TFET) and is fabricated in an advanced SOI process [14-15], featuring HfO_2 gate oxide and raised source/drain. Figures 1(c) and (d) show the measured I_D-V_D curves as a function of V_G in *p*-type and *n*-type devices, respectively. The device is initially in the OFF state at low $|V_D|$, and turns on sharply as $|V_D|$ increases beyond a turn-on voltage (V_{ON}). As $|V_D|$ sweeps back below 0.8 V, the device is turned off. The V_{ON} is linearly controlled by V_G and thus a large controlled hysteresis is obtained.

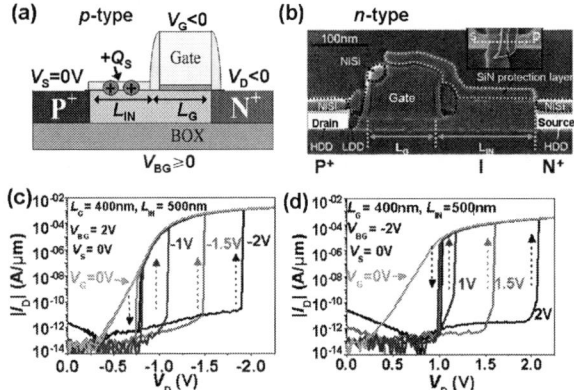

Fig. 1: (a) Schematic structure of the *p*-type Z^2-FET and (b) SEM image of the *n*-type device [14-15]. The device either operates with backgate voltage or surface charge Q_S. Experimental I_D-V_D curves in *p*-type (c) and *n*-type (d) V_{BG}-operated Z^2-FETs show sharp switching and gate-controlled hysteresis. The device parameters are $T_{ox} = 3$ nm HfO_2, $T_{Si} = 20$ nm, $T_{BOX} = 140$ nm, $L_G = 400$ nm and $L_{IN} = 500$ nm.

B. Operation principle

The operation principle is understood by TCAD simulation in Silvaco [16]. The simulated I_D-V_D curves reproduce the experimental results well, see Fig. 2(a). The dots in Fig. 2(a) correspond to the simulated results at $V_G = -2$ V including impact ionization. They show no difference, indicating that the impact ionization is not a factor in the Z^2-FET. The electron and hole injection barriers are formed in the L_G and

978-1-4673-1707-8/12 $31.00 © 2012 IEEE

L_{IN} regions by the $V_G < 0$ and $V_{BG} > 0$ (or positive Q_S), respectively, blocking carrier injection under low $|V_D|$, see Fig. 2(b). This biasing scheme emulates a virtual $p/n/p/n$ thyristor, but without recourse to any channel doping. As $|V_D|$ increases towards $|V_{ON}|$, the L_G region is depleted, and thus the electron barrier is reduced, causing electron injection from the drain into the channel. The electrons flow to the source and induce a potential drop that reduces the hole injection barrier. This permits hole injection and initiates positive feedback that completely eliminates the injection barriers, see the $V_D = -2$ V curve in Fig. 2(b). As a result, the device turns on sharply to a high current [13], similar to the feedback FET (FB-FET) [17].

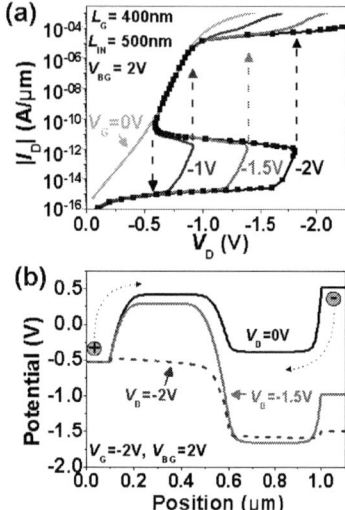

Fig. 2: (a) Simulated I_D-V_D curves of the p-type V_{BG}-operated Z^2-FETs reproducing the experimental results in Fig. 1(c). Including the impact ionization (dots) has no effect on the simulation results. (b) Surface potential profile for different V_D values, showing electron and hole injection barriers that are eliminated at $V_D = 2$V.

C. Reliability and scalability

In the absence of impact ionization or doping-related bipolar action, the characteristics of Z^2-FET are relatively insensitive to temperature (T) variation. Figure 3 shows that the $|V_{ON}|$ of the Q_S-operated Z^2-FET decreases by only ~0.12 V as T increases by 80 °C. Simulations show that the V_{BG}-operated device is scalable down to $L_G = L_{IN} \sim 30$ nm given an advanced SOI structure with ultra-thin $T_{Si} = 5$ nm, $T_{BOX} = 15$ nm and $T_{ox} = 1$ nm, which is helpful to enhance the controllability of front and back gates, see Fig. 4.

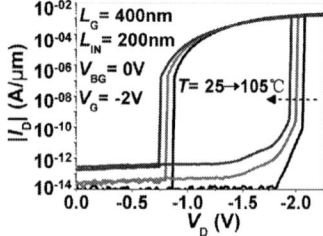

Fig. 3: Experimental I_D-V_D measurements on Q_S-operated Z^2-FET vs. T, showing small temperature sensitivity of the I_D-V_D hysteresis. The Q_S-operated device is similar to the V_{BG}-operated device in Fig. 1, except that $L_{IN} = 200$ nm, $T_{ox} = 6$ nm SiO$_2$, $V_{BG} = 0$ and $Q_S \sim 10^{12}$ cm^{-2} formed in the CVD-deposited SiO$_2$ layer on L_{IN} region [13].

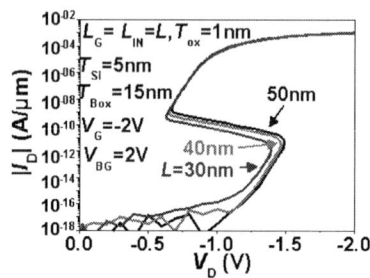

Fig. 4: Simulated scaling of the V_{BG}-operated Z^2-FET ($T_{Si} = 5$ nm, $T_{ox} = 1$ nm SiO$_2$ and $T_{BOX} = 15$ nm), showing adequate I_D-V_D hysteresis down to $L_G = L_{IN} = 30$ nm.

III. 1-T DRAM APPLICATION

A. 1-T DRAM operation using the Z^2-FET

Thanks to the V_G-controlled hysteresis and temperature insensitivity, the Z^2-FET is well-suited for 1T-DRAM memory application. Figures 5(a) and (b) show the experimental DRAM operation using Q_S-operated Z^2-FET for logic "0" and "1", respectively. The experimental rise/fall times are 15 ns, limited by our equipment. The two logic states are distinguished by the charge stored on the front-gate capacitance C_G, see Fig. 6. For writing "0", $V_G = V_D = 0$ is used to discharge C_G through the drain junction, as shown in Fig. 6(a). In contrast, writing "1" is achieved by turning on the Z^2-FET by applying $V_G = 0$ and $V_D = -1.3$ V. In the ON state, electrons and holes are injected into the channel and the holes are stored on C_G, as the device returns back to hold stage – see Fig. 6(b).

The logic states are read out by pulsing V_D from 0 to -1.3 V while keeping $V_G = -1.7$ V. For logic "1", the transient current due to the C_G discharging generates a voltage drop at the drain junction and causes electron injection from drain. This triggers feedback to turn on the device, with the current

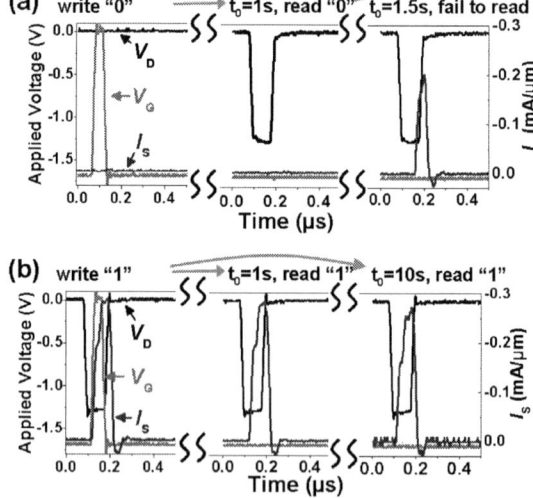

Fig. 5: Experimental results show the DRAM operation waveforms using the Q_S-operated Z^2-FET. (a) The logic "0" is written by V_G pulse and read out correctly by V_D pulse after a delay of $t_0 = 1$ s, but not after $t_0 = 1.5$ s, due to limited retention time t_{re}. (b) The logic "1" is written by simultaneous V_G and V_D pulses and read out correctly by V_D pulse after $t_0 = 1$ and 10 s (t_{re} is unlimited in logic "1").

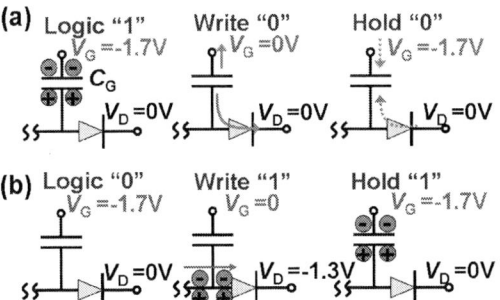

Fig. 6: Schematic writing of (a) "0" and (b) "1" logic states illustrated with an equivalent circuit including the gate capacitor (C_G) and channel-drain junction.

reaching 200 µA/µm, see Fig. 5(a). Conversely, for logic "0", no charge is stored on C_G. Since there is no discharging current, the device stays in the OFF state during the read pulse. Figure 5 (a) shows that the read of logic "0" fails only after a delay $t_0 > 1.5$ s due to the recharging of C_G by the reverse leakage current, indicated by the dashed arrow in the right-most panel of Fig. 6(a). Thus, while logic "0" requires periodic refreshing, "1" is stable and needs no refreshing.

B. Z^2-FET DRAM performance

The supply voltage of the Z^2-FET DRAM is scalable down to 1.1 V experimentally, which is lower than floating body memories and conventional 1T-1C DRAMs [1, 5], see Fig. 7(a) and (b). The retention time actually improves to 5.5 s due to lower leakage current, but the readout current of the "1" logic state is reduced to ~60 µA/µm.

The ultimate simulated write/read times of our device are very short, down to 1 ns, as shown in Ref. 13. Compared to the 1T-1C DRAM, where a large amount of charge is required to drive the external amplifier, the Z^2-FET DRAM needs less charge storage ΔQ_G because of its internal feedback amplification. Basically, the memory effect is triggered not by ΔQ_G, as in SOI 1T-DRAMs, but by the induced discharge current $\Delta Q_G/\Delta t$.

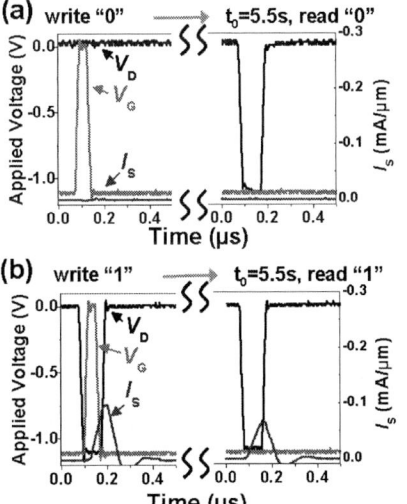

Fig. 7: Transient measurements showing that the Z^2-FET DRAM in Fig. 5 operates under $|V_{DD}| = 1.1$V with t_{re} increasing to 5.5 s, albeit with lower current for logic "1".

Unlike the standard 1T-1C DRAM, the reading of Z^2-FET DRAM is not only nondestructive but also helps to prolong the retention time (t_{re}) of the logic "0", as shown in Fig. 8. During the readout of the "0" level, the device stays in the OFF state and the residual charges accumulated in C_G are evacuated by the reading pulse. Conversely, the readout of logic "1" turns on the device and regenerates the stored charge in the channel.

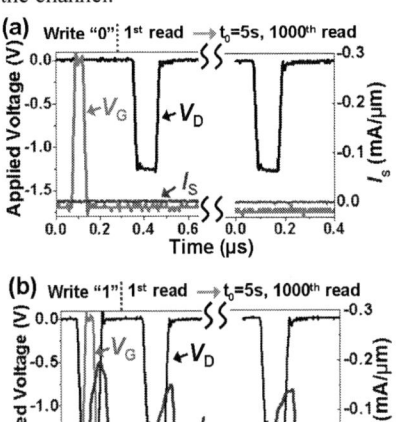

Fig. 8: Measurements on Q_S-operated Z^2-FET DRAM show non-destructive reading of logic (a) "0" and (b) "1" states. The reading pulse V_D is applied periodically every 5ms and outputs correctly after 5 s. The retention of the logic "0" is prolonged by the reading pulses, compared to Fig. 5.

C. Retention time dependence of temperature and scaling

The retention time t_{re} of "0" is determined by the leakage of the drain junction and the capacitance C_G, and thus depends on the device dimensions, biasing voltages, and temperature T, as shown in Fig. 9. We find that t_{re} decreases due to reverse drain junction leakage if either T, $|V_G|$ in the holding stage, or $|V_D|$ in the reading stage are increased. Also, downscaling L_G reduces C_G and thus reduces t_{re}, see Fig. 9(c).

D. Alternative modes of operation

The 1T-DRAM using the V_{BG}-operated Z^2-FET in Fig. 1(c) shows similar behavior, see Fig. 10. Using V_{BG} instead of Q_S is advantageous for controllability and reliability. No degradation is observed after cycling the write/read sequence 6×10^{10} times.

An alternative operation mode of the Z^2-FET uses the source-side MOSFET to write the C_G, as shown in Fig. 11(a). The C_G is charged through a MOSFET, as in a standard 1T-1C DRAM, but the stored charge is still read out through the internal feedback, ensuring less required charge and higher speed. This mode is suitable for a device with two independent gates. Here, we use the V_{BG}-operated Z^2-FET for demonstration, as shown in Fig. 11(b), where the C_G is initially discharged through the drain junction (write "0"), and then recharged by the V_{BG} pulse turning on the source-side MOSFET. The readout correctly outputs high current. This mode may be advantageous because of design rules analogous to the conventional 1T-1C DRAM.

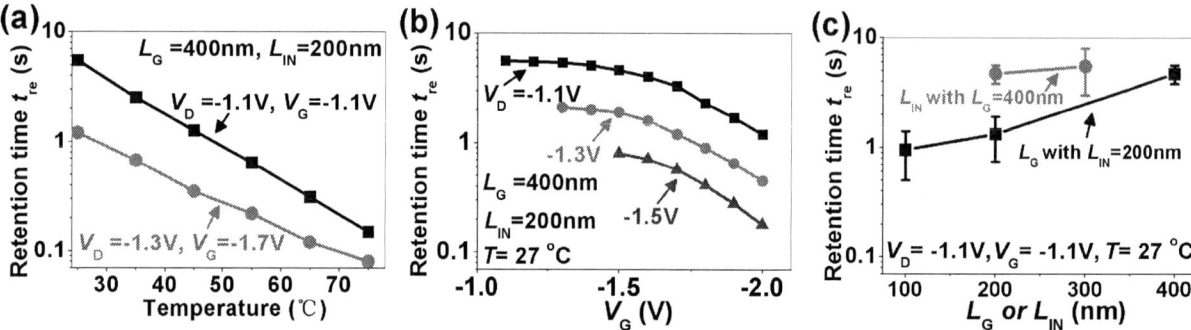

Fig. 9: Experimental results show the dependence of retention time t_{re} of logic "0" on the (a) temperature T; (b) applied V_G and V_D in holding and reading stages, respectively; and (c) dimensions (L_G and L_{IN}). The devices are Q_S-operated Z^2-FET DRAMs.

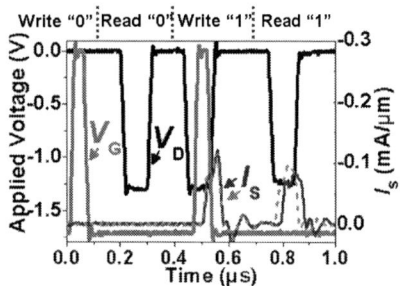

Fig. 10: DRAM operation waveform of the V_{BG}-operated Z^2-FET with $V_{BG} = 2$ V. The dashed curve shows the output current after 6×10^{10} cycles.

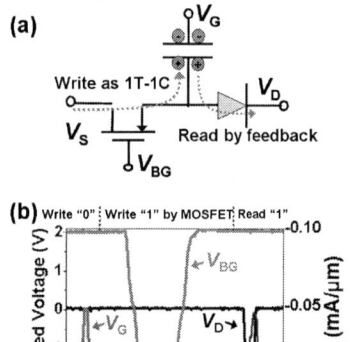

Fig. 11: (a) Schematic view and (b) experimental demonstration of the DRAM mode using the source MOSFET for writing and internal feedback for reading.

IV. CONCLUSION

We have systematically studied the use of Z^2-FET as a 1-T DRAM. The device possesses $I_{ON}/I_{OFF} > 10^9$, $|V_{DD}| \sim 1.1$ V, $t_{re} \sim 0.15$ s at 75 °C, simulated access time of ~1 ns and scalability down to 30 nm. The high performance and compact form are of interest for future memory generations.

ACKNOWLEDGMENTS

The work at Minatec is funded by the RTRA program of the Grenoble Nanosciences Foundation. A. Zaslavsky also acknowledges support by the U.S. National Science Foundation (award ECCS-0701635). Another author (CLR) also acknowledges support by the European STEEPER project (FP7/2007-2013, grant agreement n°257267).

REFERENCES

[1] K. W. Song, J. Y. Kim, H. Kim, H. W. Chung, K. Kim, H. W. Park, et al., "A 31ns random cycle VCAT-based 4F2 DRAM with enhanced cell efficiency," in *Proc. Symp. VLSI Circuits*, 2009, pp. 132-133.

[2] W. Mueller, G. Aichmayr, W. Bergner, E. Erben, T. Hecht, et al., "Challenges for the DRAM Cell Scaling to 40nm," 2005, pp. 4 pp.-339.

[3] M. Bawedin, S. Cristoloveanu, and D. Flandre, "A capacitorless 1T-DRAM on SOI based on dynamic coupling and double-gate operation," *Electron Device Letters, IEEE,* vol. 29, pp. 795-798, 2008.

[4] E. Yoshida and T. Tanaka, "A capacitorless 1T-DRAM technology using gate-induced drain-leakage (GIDL) current for low-power and high-speed embedded memory," *Electron Devices, IEEE Transactions on,* vol. 53, pp. 692-697, 2006.

[5] T. Hamamoto and T. Ohsawa, "Overview and future challenges of floating body RAM (FBRAM) technology for 32 nm technology node and beyond," *Solid-State Electronics,* vol. 53, pp. 676-683, 2009.

[6] S. Okhonin, M. Nagoga, E. Carman, R. Beffa, and E. Faraoni, "New generation of Z-RAM," in *Tech. Dig. -Int. Electron Devices Meet.*, 2007, pp. 925-928.

[7] H. J. Cho, F. Nemati, R. Roy, R. Gupta, K. Yang, et al., "A novel capacitor-less DRAM cell using thin capacitively-coupled thyristor (TCCT)," in *Tech. Dig. -Int. Electron Devices Meet.*, 2005, pp. 311-314.

[8] R. Gupta, F. Nemati, S. Robins, K. Yang, V. Gopalakrishnan, J. Sundarraj, et al., "32nm high-density high-speed T-RAM embedded memory technology," in *Tech. Dig. -Int. Electron Devices Meet.*, 2010, pp. 12.11.11-12.11.14.

[9] K. Yang, R. Gupta, S. Banna, F. Nemati, H. J. Cho, M. Ershov, et al., "Optimization of Nanoscale Thyristors on SOI for High-Performance High-Density Memories," in *Intern. SOI Conf.*, 2006, pp. 113-114.

[10] A. A. Salman, S. G. Beebe, M. Emam, M. M. Pelella, and D. E. Ioannou, "Field effect diode (FED): a novel device for ESD protection in deep sub-micron SOI technologies," in *Tech. Dig. -Int. Electron Devices Meet.*, 2006, pp. 107-111.

[11] Y. Yang, A. Gangopadhyay, Q. Li, and D. E. Ioannou, "Scaling of the SOI field effect diode (FED) for memory application," in *Intern. Semicond. Dev. Res. Symp.*, 2009, pp. 1-2.

[12] U. E. Avci, D. L. Kencke, and P. L. D. Chang, "Floating-Body Diode—A Novel DRAM Device," *Electron Device Letters, IEEE,* vol. 33, pp. 161-163, 2012.

[13] J. Wan, C. Le Royer, A. Zaslavsky, and S. Cristoloveanu, "A Compact Capacitor-Less High-Speed DRAM Using Field Effect-Controlled Charge Regeneration," *Electron Device Letters, IEEE,* vol. 33, pp. 179-181, 2012. See also the French patent no. FR11/03232, Oct. 21, 2011.

[14] F. Mayer, C. Le Royer, J. F. Damlencourt, K. Romanjek, F. Andrieu, C. Tabone, B. Previtali, and S. Deleonibus, "Impact of SOI, Si1-xGexOI and GeOI substrates on CMOS compatible Tunnel FET performance," in *Tech. Dig. -Int. Electron Devices Meet.*, 2008, pp. 163-167.

[15] J. Wan, C. Le Royer, A. Zaslavsky, and S. Cristoloveanu, "Tunneling FETs on SOI: Suppression of ambipolar leakage, low-frequency noise behavior, and modeling," *Solid-State Electronics,* vol. 65-66, pp. 226-233, 2011.

[16] "Silvaco (Atlas version 2. 10. 4. R) .".

[17] A. Padilla, C. W. Yeung, C. Shin, C. Hu, and T. J. K. Liu, "Feedback FET: A novel transistor exhibiting steep switching behavior at low bias voltages," in *Tech. Dig. -Int. Electron Devices Meet.*, 2008, pp. 171-174.

978-1-4673-1707-8/12 $31.00 © 2012 IEEE

A 5.61 pJ, 16 kb 9T SRAM with Single-ended Equalized Bitlines and Fast Local Write-back for Cell Stability Improvement

Qi Li, Bo Wang, and Tony T. Kim
VIRTUS, IC Design Centre of Excellence
Nanyang Technological University, Singapore 639798
Email: thkim@ntu.edu.sg

Abstract— **A 5.61 pJ, 16 kb 9T SRAM is implemented in 65nm CMOS technology. A single-ended equalized bitline scheme is proposed to improve both read bitline voltage swing and sensing timing window. A fast local write-back allows the half-select-free write operation without performance degradation. The test chip shows a minimum operating voltage of 0.24V and a minimum energy of 5.61 pJ at 0.3V.**

I. INTRODUCTION

Robust SRAMs with a wide operating supply voltage range are critical in many ultra-dynamic voltage scaling (UDVS) applications such as processors, implantable biomedical devices, wireless sensor nodes, and portable electronics. The primary goal of UDVS is to provide high performance during normal operation modes while significantly reducing power and energy consumption during ultra-low voltage operation modes [1], [2]. Conventional 6T SRAMs fail to deliver reliable UDVS operation due to the deteriorated Static Noise Margin (SNM), poor write margin, and reduced bitline sensing margin. In UDVS systems, decoupled SRAM cells with a dedicated read port have been employed to eliminate disturbing current from bitlines into data storage nodes and consequently to improve SNM at a cost of additional devices [3-7]. Poor write margin issues have been addressed by design techniques such as boosted wordline voltage and collapsed cell supply. Write-back schemes have been utilized in many low-voltage SRAMs for improving the stability of half-selected cells by using an additional clock cycle. The degradation in bitline sensing margin demands various read-assist circuit techniques like bitline leakage manipulation (reduction, compensation, and equalization) [5-7], a virtual-ground replica scheme [3], sense amplifier redundancy [5], etc. Although various ultra-low voltage SRAMs have been successfully demonstrated through the improved cell stability of the decoupled SRAM cells, the smaller read bitline sensing margin and the stability of half-selected cells are still remained as significant challenges to the reliable ultra-low voltage operation.

In this paper, we demonstrate a 0.3 V, 5.61 pJ, 16 kb SRAM with enhanced bitline sensing margin and improved cell stability. Circuit techniques such as a novel 9T SRAM cell, a single-ended equalized bitline scheme, and a fast local write-back scheme are proposed.

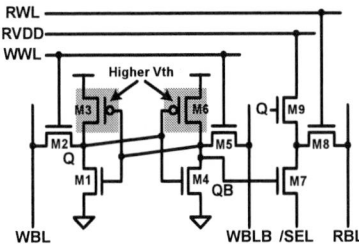

Figure 1. Proposed 9T SRAM cell and control.

II. PROPOSED 9T SRAM: CELL AND BITLINE STRUCTURE

The proposed 9T SRAM cell is illustrated in Fig. 1. It has a dedicated read port consisting of 3 NMOS transistors (M7, M8, and M9). Compared to the conventional decoupled 8T SRAM cell, M9 is added for equalizing the read bitline leakage in unselected rows and providing a pull-up current path in a selected row to improve bitline sensing margin in case of reading '1'. During read (RWL=RVDD='1' and /SEL='0'), the pre-charged read bitline (RBL) is conditionally discharged depending on 'Q' and 'QB'. When reading data '1' (Q='1'), a pull-up path is formed from RWL to RBL through M9, which is different from the conventional 8T SRAM cell where both cell read current and bitline leakage current pull down RBL. The pull-up path slows down RBL discharging speed of data '1' and improves RBL sensing margin. In stand-by modes, RVDD and /SEL are held to VDD to save the leakage current in the read port. The write operation of the proposed SRAM cell is the same as the conventional 8T SRAM. To implement a good write margin using minimally sized devices, high-V_{th} devices are employed in the PMOS loads.

The 16kb SRAM array has 4 sub-arrays of 256 rows by 16 columns and 4 IOs. Hierarchical bitlines are employed for reducing delay and power consumption. While the optimal number of cells per local bitline is 32, we selected 64 cells per

---

This work was supported by MediaTek Inc. and Academic Research Fund Tier1 Grant by Ministry of Education, Singapore.

978-1-4673-1707-8/12 $31.00 © 2012 IEEE

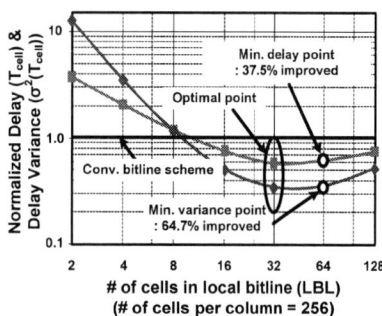

Figure 2. Optimal hierarchical bitline for minimizing read delay and read delay variation.

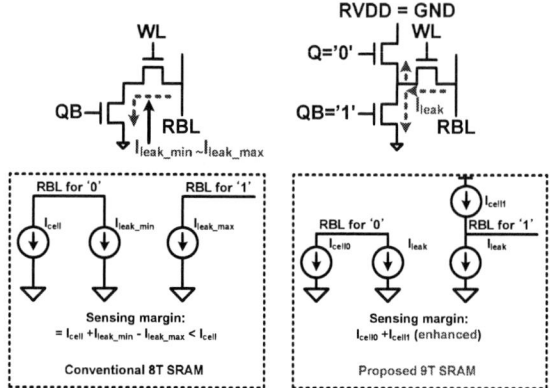

Figure 3. Principle of the equalized bitline.

local bitline in consideration of the area overhead (Fig. 3). The chosen hierarchical bitline structure improves the read delay and delay variation by 37.5% and 64.7% respectively.

III. SINGLE-ENDED EQUALIZED BITLINES

Bitline leakage is a primary concern in reliable bitline sensing since it becomes significant compared to the SRAM cell read current, particularly at ultra-low supply voltages. This limits the number of cells in a bitline small. An equalization scheme for differential bitlines was proposed to enhance the bitline sensing ability without changing the number of cells per bitline [7]. However, the previous equalization scheme used a 6T SRAM cell and is not applicable to single-ended bitlines. Fig. 3 compares the principle of the proposed single-ended equalized bitlines with the conventional 8T SRAM bitlines. In the conventional 8T SRAM, the amount of the bitline leakage is determined by the data pattern stored in a column. For example, the worst case read '0' happens when QB of the selected cell is '1' and QB of the unselected cells is '0'. This produces the overall pull-down current of '$I_{cell} + I_{leak\_min}$'. Similarly, the worst case data '1' happens when QB of the selected cell is '0' and QB of the unselected cells is '1', which generates the pull-down current of 'I_{leak_max}'. For correct RBL sensing, '$I_{cell} + I_{leak\_min}$' should be always larger than 'I_{leak_max}'. This constraint limits the

maximum number of cells per bitline. However, in the proposed 9T SRAM cell, the leakage from the unselected rows is constant since both RVDD and /SEL are GND, and one of M7 and M8 (Fig. 1) is turned on while the other is turned off by 'Q' and 'QB'. This generates the current margin of '$I_{cell0} + I_{cell1}$' for RBL sensing. The detailed operation of the single-ended equalized bitlines is illustrated in Fig. 4. During read operation, RVDD in unselected rows is held to GND, which forces the equalized bitline leakage to flow from RBL to GND. Compared to the conventional read port without M9, the RBL levels in the proposed scheme are not affected by the column data. When reading data '0', the RBL discharging speed is determined by the summation of the pull-down read current and the constant bitline leakage. Similarly, when reading data '1', the summation of the pull-up current through M8 and M9 (Fig. 1) and the constant bitline leakage controls the discharging speed. Consequently, the proposed single-ended equalized bitline provides better sensing margin in terms of voltage swing and sensing timing window (Fig. 4 (right)), which further lowers the minimum operating voltage.

IV. FAST LOCAL WRITE-BACK FOR STABILITY IMPROVEMENT

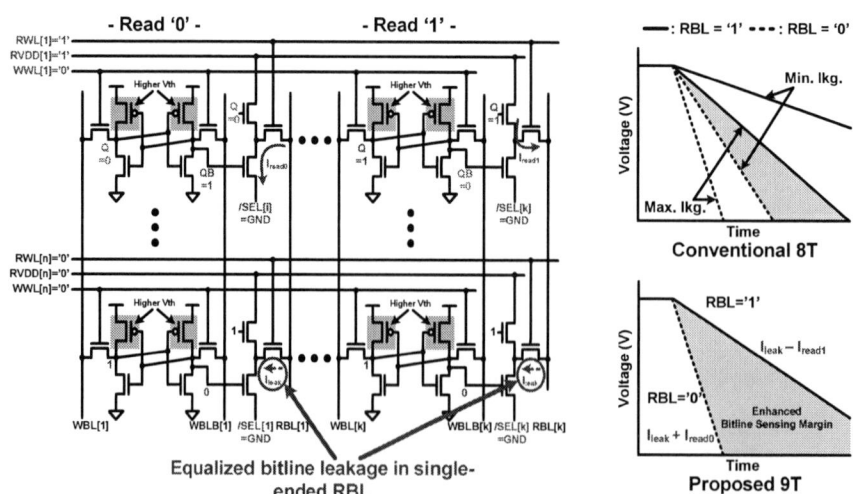

Figure 4. Single-ended equalized read bitline for improved sensing margin.

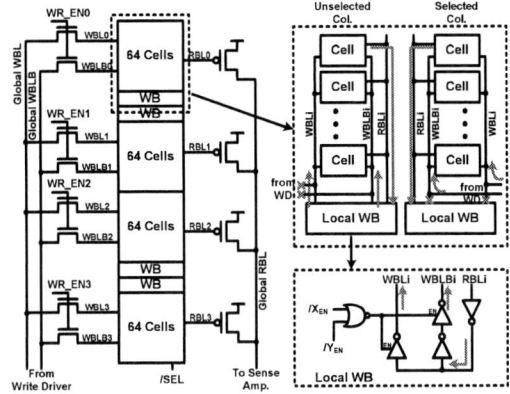

Figure 5. Fast local write-back for improving cell stability.

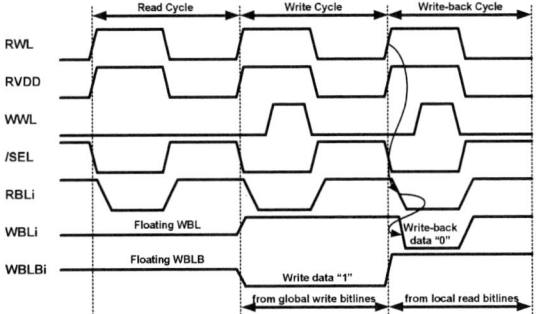

Figure 6. Timing diagram of the proposed fast local write-back scheme.

Decoupled SRAM cells improve cell stability significantly in read operation [3-7]. Thus, the worst case cell stability occurs in the half-selected cells during write operation. The half-selected cells have the same SNM as that of the 6T SRAM cell. Pulsed wordline improves the dynamic cell stability by reducing the time period of cell node disturbance by the pre-charged write bitlines [8]. However, the pulsed wordline cannot remove the half-selection issue completely and accurate pulse width control for the reliable write operation and the minimal disturbance is remained as a challenging task. Write-back schemes can eliminate the half-selection problem by firstly reading data from unselected columns and writing the read data into write bitlines in the same columns [3]. However, previous write-back schemes require additional time for reading data in each write operation, which limits the performance. In this work, a fast local write-back scheme without performance degradation is proposed. Fig. 6 explains the simplified bitline architecture with the proposed fast local write-back scheme. Four sets of the local bitlines are connected to the global bitlines through switches and PMOS sensing devices. A write cycle starts by asserting the read bias condition (Fig. 6). The inserted read operation quickly generates read data in the local read bitlines (RBLi) and the addressing signals (WR_EN, $/X_{EN}$, and $/Y_{EN}$) determine the actual write path. In the selected column, the data in the global write bitlines is selected for normal write operation while the data from the inserted read operation is written back into the local write bitlines (WBLi and WBLBi) of the unselected columns. After the inserted read operation, a

delayed write wordline (WWL) is enabled for writing target data and read data into the enabled cells. Due to the fast read operation through the local read bitlines, the additional delay for the inserted read operation is small enough to make the write-back speed similar to the read speed through the global bitlines. Consequently, the proposed fast local write-back eliminates the half-selection problem without significant performance degradation. Fig. 6 illustrates a read/write timing diagram including the proposed write-back operation.

V.   MEASUREMENT RESULTS

A 16 kb SRAM with the proposed techniques was fabricated in a 65 nm CMOS technology. The 16 kb array is configured with 256 rows and 64 columns. Each column is divided into four sub-blocks for realizing the hierarchical bitline structure and the fast local write-back operation. Fig. 7 summarizes the measured read access time over different supply voltage levels. The read access time is 4.88 μs at the minimum operating supply voltage of 0.24 V. Increasing supply voltage from the minimum operating voltage reduces the read access time exponentially while the supply voltage is in the sub-threshold region. When the supply voltage is in the super-threshold region, the reduction in the read access time from the raised supply level is slowed down. Fig. 8 is a sample screenshot of the SRAM read operation. The test chip is fully functional down to 0.24V. Fig. 9 shows the measurement results of the read/write power, the leakage power, the average power, and the energy consumption from the test chip. The average write power is larger than the average read power by ~25% (Fig. 9(a)). This is primarily due to the inserted read operation required by the proposed fast local write-back operation. Note that the read/write power plots show two different slopes because the maximum operating frequency degrades exponentially at a lower supply

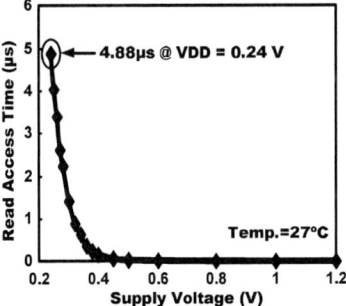

Figure 7. Measured read access time.

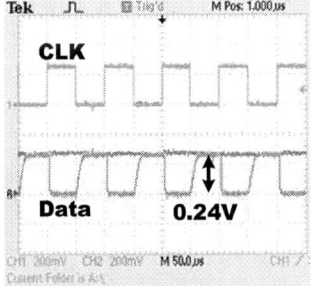

Figure 8. Sample measured waveforms.

978-1-4673-1707-8/12 $31.00 © 2012 IEEE

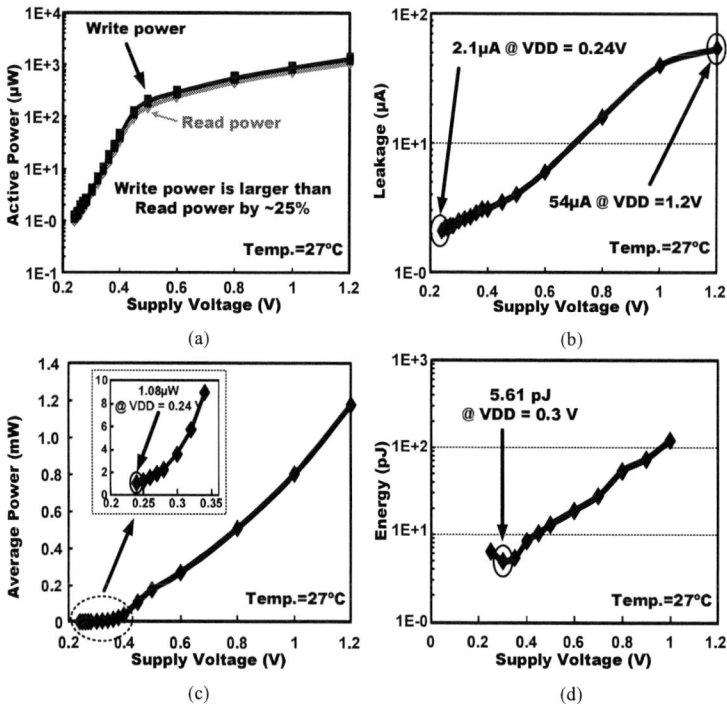

Figure 9. SRAM measurement results: (a) read/write power, (b) leakage current, (c) average power (50% read and 50% write), and (d) energy consumption.

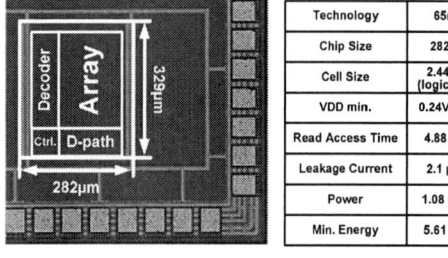

Technology	65nm CMOS
Chip Size	282 x 329 μm²
Cell Size	2.44 x 0.72 μm² (logic design rule)
VDD min.	0.24V @ 256 rows
Read Access Time	4.88 μs @ 0.24V
Leakage Current	2.1 μA @ 0.24V
Power	1.08 μW @ 0.24V
Min. Energy	5.61 pJ @ 0.30V

Figure 10. Chip micrograph and test chip summary.

voltage region. The leakage current changes from 54 μA to 2.1 μA by lowering the supply voltage from 1.2 V down to 0.24 V (Fig. 9(b)). The average power of the test chip is 1.08uW with the maximum operating frequency at 0.24V (Fig. 9(c)). The minimum energy of 5.61 pJ was achieved at 0.3 V (Fig. 9(d)). The summary of the SRAM test chip is given in Fig. 10.

VI. CONCLUSION

A 9T SRAM with circuit techniques for improving bitline sensing margin and removing half-selected cells is presented. The proposed single-ended bitline equalization scheme eliminates the data dependency of the read bitline leakage and always provides positive bitline sensing margin regardless of the amount of read bitline leakage. Consequently, it enhances the bitline voltage swing and sensing timing window. We also proposed a fast local write-back scheme to implement half-select-free write operation. The inserted read operation and the write-back operation through the local read bitlines facilitates the high write-back speed comparable to the normal read

speed. A test chip was fabricated in a 65nm CMOS technology. The hardware implementation demonstrates a minimum operating voltage of 0.24 V and a minimum energy of 5.61 pJ at 0.3 V. The proposed circuit techniques can be employed in SRAMs for UDVS systems where ultra-low voltage operation is strongly demanded.

REFERENCES

[1] A. Wang and A. Chandrakasan, "A 180-mV subthreshold FFT processor using a minimum energy design methodology," IEEE J. Solid-State Circuits, vol. 40, no. 1, pp. 310–319, Jan. 2005.

[2] S. Hanson et al., "A Low-Voltage Processor for Sensing Applications with Picowatt Standby Mode," IEEE J. Solid-State Circuits, vol. 44, no. 4, pp. 1145-1155, Apr. 2009.

[3] T. Kim, J. Liu, J. Keane, and C. Kim, "A High-Density Subthreshold SRAM with Data-Independent Bitline Leakage and Virtual-Ground Replica Scheme", IEEE International Solid-State Circuits Conference, pp. 330-331, Feb. 2007.

[4] J.-J. Wu et al., "A large σVTH/VDD tolerant zigzag 8T SRAM with area-efficient decoupled differential sensing and fast write-back scheme", IEEE Symposium on VLSI Circuits, pp. 103-104, June 2010.

[5] N. Verma and A. Chandrakasan, "A 65nm 8T Sub-Vₜ SRAM Employing Sense-Amplifier Redundancy", IEEE International Solid-State Circuits Conference, pp. 328-329, Feb. 2007.

[6] T. Kim, J. Liu, C. Kim, "A Voltage Scalable 0.26V, 64kb 8T SRAM with Vmin Lowering Techniques and Deep Sleep Mode," IEEE J. Solid-State Circuits, vol. 44, no. 6, pp. 1785-1795, June 2009.

[7] Y. Ishii et al., "A 28-nm dual-port SRAM macro with active bitline equalizing circuitry against write disturb issue", IEEE Symposium on VLSI Circuits, pp. 99-100, June 2010.

[8] M. Khellah et al., "Wordline & Bitline Pulsing Schemes for Improving SRAM Cell Stability in Low-Vcc 65nm CMOS Designs", IEEE Symposium on VLSI Circuits, pp. 9-10, June 2006.

An Advanced Statistical Compact Model Strategy for SRAM Simulation at Reduced V_{DD}

P. Asenov[1], D. Reid[2], S. Roy[1], C. Millar[2,1], A. Asenov[2,1]

[1] Device Modelling Group, School of Engineering, University of Glasgow
[2] Gold Standard Simulations Ltd
Email : 0405152a@student.gla.ac.uk

ABSTRACT

Accurate statistical compact model extraction and circuit simulation are key issues in contemporary SRAM design. The high statistical variability of the small SRAM cell transistors in combination with high density leads to yield problems determined by 5-6σ deviations from the mean. The compact modeling approach presented in this paper utilizes a firm understanding of the physical phenomenon underlying device variability. Its illustration is based on comprehensive 'atomistic' 3D device simulations. Extracted statistical models are then utilized in SRAM SNM simulation, and benchmarked against Gaussian V_T based simulation. The results show that aside from the increasing error in the yield estimate with the reduction of the supply voltage V_{DD}, Gaussian V_T simulations also fail to capture the decorrelation between nominal SNM and SNM.

1. INTRODUCTION

Statistical variability, which is exacerbated by aggressive transistor scaling is the major factor affecting SRAM yield and functionality. Beyond the 28nm CMOS technology generation it dictates the transition to new device architectures including the competing FD SOI and FinFET like solutions. Statistical variability steams from the discreteness of charge and the granularity of matter. In bulk CMOS, the dominant statistical variability sources includes random discrete dopants (RDD) [1], line edge roughness (LER) [2] and polysilicon/metal gate granularity (P/MGG) [3].

Typically the statistical variability (mismatch) in the SRAM analysis *is* captured in compact models using a single parameter, the threshold voltage, or few uncorrelated parameters modelling threshold voltage and on-current effects[4][5]. Failing to accurately model the impact of statistical variability leads to large errors when modelling the trade-off between SRAM performance, power, area and yield [6].

In this paper we present the extraction of an optimal set of statistical compact model parameters that accurately reproduce the distributions and the correlations between key transistor figures of merit obtained from comprehensive physical simulation or measurements. We then we compare results of statistical SRAM static noise margin (SNM) simulations at multiple V_{DD} levels, using the extracted statistical compact models with the results obtained using only Gaussian V_T distribution. The focus is on the previously reported decorrelation between SNM at high V_{DD} and low V_{DD} due to the effects of varying DIBL [7].

2. Physical simulation

The 25nm template transistor in this study was designed and supplied by GSS Ltd. [8] and is representative for the 22/20nm technology generation. The simulations are carried out using the atomistic 3D simulator Garand [8]. Extremely large statistical samples of 10,000 transistors were simulated [9] subject to RDD, LER and MGG. The spread of simulated device transfer characteristics at low drain voltage is shown in Figure 1. For this typical he range of I_{off} spans more than 5 orders of magnitude, with a wide range of threshold voltages, as well as a significant spread in I_{on} The simulations were performed at high and low drain bias to capture the performance of the devices under all typical operating conditions. It is important to note that although this paper focuses on device characteristics generated through simulation, it is equally applicable for data acquired through measurement.

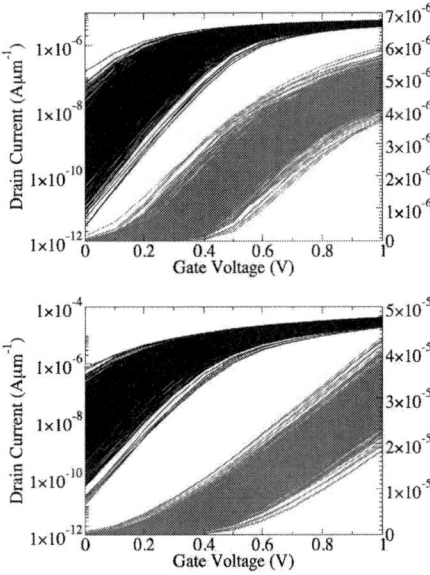

Figure 1: Simulated statistical I_D-V_G characteristics of the n-MOSFET subject to RDD, LER and MGG. Top V_{DD} =50 mV, bottom V_{DD} =1 V

3. SCM Extraction

The fundamental basis of the statistical extraction strategy has been previously described by Cheng *et al.* [10]. This is a two-stage process, and employs the statistical compact model extraction tool Mystic [8]. First a nominal compact model based on a 'uniform' TCAD simulations or device measurements containing no sources of variability is extracted. This model is calibrated to capture gate length, body bias and temperature dependence for the underlying technology. Sensitivity analysis is applied to select a subset of the standard BSIM4 parameters that are capable of capturing the variations in device performance due to atomistic effects. The analysis combines an in-depth knowledge of device physics, the effects of stochastic variability on device performance and an

978-1-4673-1707-8/12 $31.00 © 2012 IEEE

intimate knowledge of the BSIM4 compact model equations and parameters. Each physical effect of stochastic variability at high and low drain voltage (namely threshold voltage, on-current, off current, subthreshold slope, drain induced barrier lowering (DIBL), mobility, vertical field dependence) is modelled by a specifically selected parameter, which is directly extracted. In the case of the 10,000 25nm n-channel ensemble, it was found that 8 carefully chosen and controlled parameters can accurately and completely capture the performance of the device ensemble, this will be proved below.

In order to illustrate the difficulties in selecting the statistical compact model parameters Figure 2 compares the I_D-I_G characteristics of the 'uniform' transistor used to extract the nominal compact model and three 'atomistic' transistors with extreme behavior from the 10,000 statistical sample.

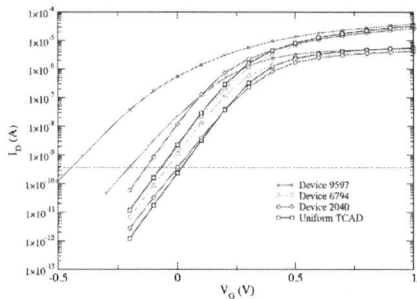

Figure 2: I_D-V_G characteristics of the 'uniform' transistor and three 'atomistic' transistors with extreme behavior.

For example device 6794, indicate very small (only 30mV) DIBL despite the fact that the transistor, have low drain threshold voltage below the threshold voltage of the uniform transistor resulting in an increase leakage.

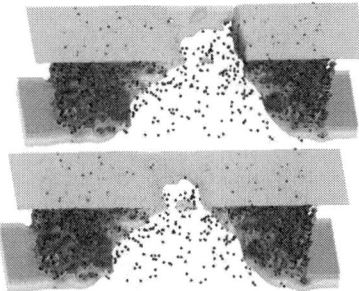

Figure 3: Electron concentration contours for Device 6794 at high drain (top) and low drain (bottom) at a constant current criterion. Acceptor dopants (blue) and Donor dopants (red) are also shown

Figure 3 shows that the region of the percolation path is an area where there are few dopants which leads to a reduction in the gate control of current flow in this region. A current percolation path forms at both high and low drain voltage. However at high drain voltage three acceptor dopants near the drain are exposed and dramatically reduce the impact of the drain voltage on the potential barrier along the current percolation path and which leads to the extremely low DIBL in this transistor.

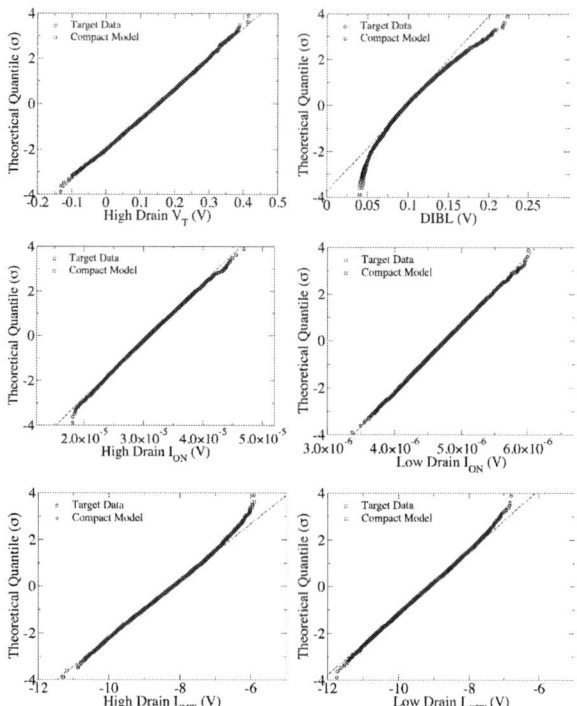

Figure 4 :Comparison between the distribution of key figures of merit of the nMOSFET extracted from the TCAD simulation and from the compact model.

Figure 4 provides a comparison between the device figures of merit of device simulated using the extracted compact models and those from atomistic simulation while Figure 9 illustrates their correlations.

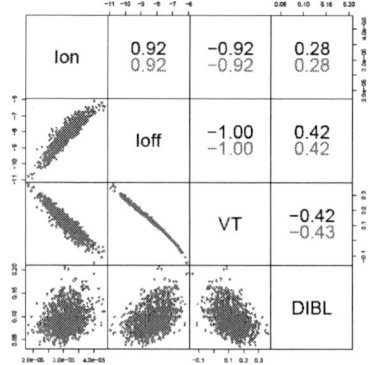

Figure 5: Comparison of correlations between extracted figures of merit for the nMOSFET from TCAD simulations and from the compact models at V_D=1.0V.

The extraction methodology is capable of very accurately capturing the effect of variability over the whole range of device operation and subject to non-Gaussian distribution of some of the figures of merit.

3.1 SRAM SNM Simulation

SNM simulations and calculations were performed using the standard methodology described in [11]. DC SPICE simulations produce a "butterfly curve". Static noise margins are calculated for both the left and right loop and the overall static noise margin is the minimum of these two values. We define a stability fail as a cell with SNM less than 4% nominal V_{DD} or 40mV. In order to investigate the effect of statistical variability alone, the SRAM simulations in this paper are performed at the typical/typical corner at room temperature.

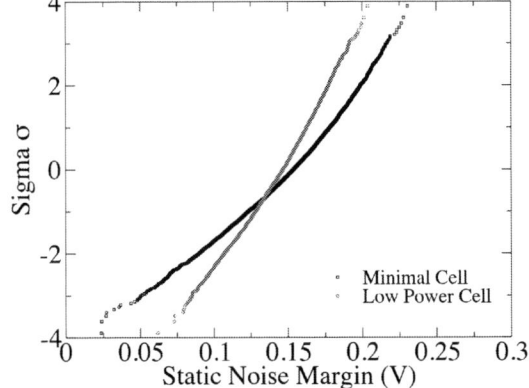

Figure 6: SNM of minimal cell and low power cell cell

Initial simulations were performed with a minimal sized cell (~0.09µm), scaled relative to Intel's 32nm cell (0.172µm), however, as shown in Figure 6 this cell was too sensitive to statistical variability, with failures seen within the first 10,000 simulations even at the typical/typical corner. In order to study SNM as a function of V_{DD} at the 22nm technology node, a larger, more variability stable, 'low power' cell was designed, with an increased cell area (~0.120µm).

SNM distributions for 10,000 cells at V_{DD} =1V to 0.5V are shown in Figure 7. As can be seen in the figure, even for as few as 10,000 samples Gaussian V_T simulations under estimate the effect of statistical variability on SNM. The discrepancy is worst and high supply voltage, due to maximum impact of DIBL, within the lower tail of the distribution, which defines the worst performance of the circuit.

Though the deviation looks relatively small on the plot, if we use these distributions for yield estimation, the discrepancy increases. We use Generalized Lambda Distribution (GLD) fits to the distributions of SNM at 1V to estimate failure rate of the cell, this is shown in Figure 8. As the graph shows full model simulations predict ~5,000 cells per billion, while the Gaussian V_T simulations predict ~400 parts per billion, at this failure criteria there is an order of magnitude difference in predicted yield.

As supply voltage is reduced, the correlation with SNM at V_{DD} = 1V is reduced. This reduction in correlation is due to the variability in the DIBL of the devices. The extracted compact models capture variability in DIBL and Gaussian V_T models have no way to capture this. This divergence is shown in Figure 9, firstly as the evolution of correlation between SNM at nominal supply voltage and reduced supply voltage, then a plot of SNM at V_{DD} = 1V against SNM at V_{DD} = 0.5V shows the increased spread of SNM at low V_{DD} using the accurate full model approach.

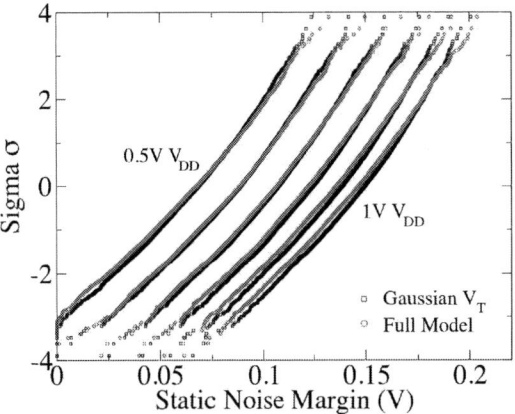

Figure 7: SNM at multiple V_{DD} levels, simulated with full models and Gaussian V_T models

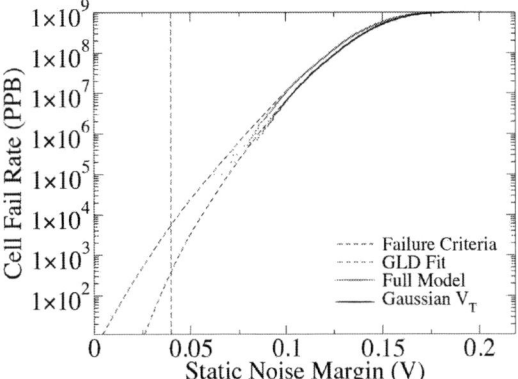

Figure 8: GLD based yield predictions

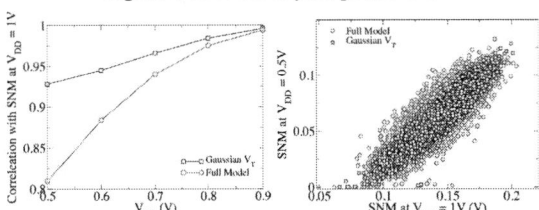

Figure 9: Correlation coefficients between SNM at V_{DD} = 1V and lower supply voltages (left) and SNM at V_{DD} = 1V against SNM at V_{DD} = 0.5V (right)

To emphasize the error introduced through Gaussian V_T simulation of SNM as a function of supply voltage, Figure 10 shows a set of cells, chosen as they have a V_{DD} = 1V SNM between the range of 120-125mV. As supply voltage to the cells are reduced, we see full model simulations predict a much larger spread in SNM than Gaussian V_T simulation. This result is most relevant when considering the calculation of minimum supply voltage for the SRAM block, where the cells reliably retain information V_{DDmin}. This is an important metric which defines minimum leakage, and supply voltage during "write assist" cycles. If an estimation of V_{DDmin} is calculated based on Gaussian

V_T based simulation SNM dependence on drain bias will be incorrectly calculated.

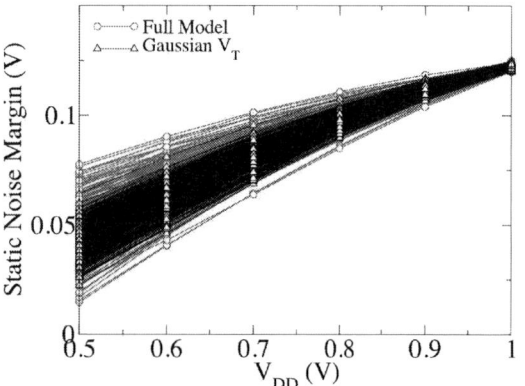

Figure 10: SNM as a function of V_{DD} for cells with SNM between 120mV and 125mV at V_{DD}=1V

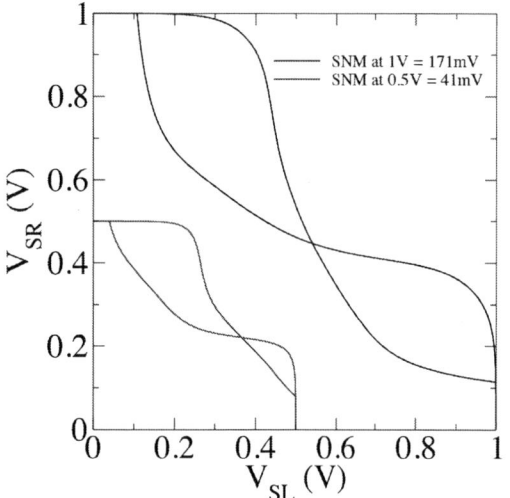

Figure 11: An instance of a cell cell with extreme shift in SNM

For illustrative purposes, Figure 11 shows an instance of a cell simulated with the full compact model approach. The cell has SNM of 171mV at 1V V_{DD}, for cells of this type Gaussian V_T simulations predict SNM of between 80mV and 110mV (see Figure 9, right plot) at 0.5V V_{DD}. The full model simulations of the circuit show that actual SNM for this cell is 41mV, moreover

full model simulations predict that for a cell of this type, SNM at 0.5V V_{DD} could be between 40mV and 120mV.

4. CONCLUSIONS

A methodology to propagate device level variability information to the circuit level through the use of Statistical Compact Modelling is described. A practical application for the models extracted is SRAM simulation. The paper shows that using full model simulation has multiple advantages of the basic Gaussian V_T simulation approach, especially when considering V_{DD} scaling and the calculation of V_{DDmin}. Full model simulations capture the decorrelation between high drain SNM and SNM at reduced supply voltage. This is due to the fact that the extracted compact models capture DIBL variability not present in the Gaussian V_T simulations.

5. REFERENCES

[1] A. Asenov, G. Slavcheva, A. R. Brown, J. H. Davies and S. Saini, *Increase in the random dopant induced threshold fluctuations and lowering in sub-100nm MOSFETs due to quantum effects: A 3-D density-gradient simulation study*, IEEE Trans. Electron Devices, vol. 48, no.4, 2001, pp. 271–350.

[2] A. Asenov, S. Kaya, A. R. Brown, "*Intrinsic parameter fluctuations in decananometer MOSFETs introduced by gate line edge roughness*" IEEE Trans. Electron Devices, vol. 50, no.5, 2003, pp. 1254–1260

[3] A. R. Brown, G. Roy and A. Asenov, "*Impact of Fermi level pinning at polysilicon gate grain boundaries on nano-MOSFET variability: A 3D simulation study*" in Proc. 36th ESSDERC, 2006, pp.451-454.

[4] K. Agarwal, S. Nassif, *The Impact of Random Device Variation on SRAM Cell Stability in Sub-90-nm CMOS Technologies*, IEEE Trans. On VLSI Systems, Vol. 16

[5] B. H. Calhoun, A. P. Chandrakasan, *Static Noise Margin Variation for Sub-threshold SRAM in 65-nm CMOS*, IEEE Journal of Solid-State Circuits, Vol. 41, No. 7, July 2006

[6] P. Asenov, F. Adamu-Lema, S. Roy, C. Millar, A. Asenov, G. Roy, U. Kovac, D. Reid *The Effect of Compact Modelling Strategy on SNM and Read Current variability in Modern SRAM* in proc. SISPAD 2011, p 283 - 286

[7] M. Suzuki, T. Saraya, K. Shimizu, A. Nishida, S. Kamohara, K. Takeuchi, S. Miyano, T. Sakurai, and T. Hiramoto, *Direct Measurements, Analysis, and Post-Fabrication Improvement of Noise Margins in SRAM Cells Utilizing DMA SRAM TEG* 2010 Symposium on VLSI Technology Digest of Technical Papers P191-19

[8] url=www.goldstandardsimulations.com

[9] A. Asenov, *Simulation of Statistical Variability in Nano MOSFETs* in 2007 Symposium on VLSI Technology, p. 86 – 87

[10] B. Cheng, D. Dideban, N. Moezi, C. Millar, G. Roy, X. Wang, S. Roy, A. Asenov, *Statistical-Variability Compact-Modeling Strategies for BSIM4 and PSP*, IEEE Design & Test of Computers, Vol. 27, Issue 2, March-April 2010

[11] R. Herald, P. Wang, *Variability in Sub-100nm SRAM Designs*, in Proc. Int. Conf. Comput.-Aided Des., 2004, pp. 347–352

Multibranch Mobility Characterization: Evidence of Carrier Mobility Enhancement by Back-Gate Biasing in FD-SOI MOSFET

C. Navarro[*¶], N. Rodriguez[*], L. Donetti[*], A. Ohata[†], F. Gamiz[*],
F. Andrieu[‡], O. Faynot[‡], C. Fenouillet-Berangerand[‡§], S. Cristoloveanu[¶]

[*]Nanoelectronics Research Group, CITIC-University of Granada, Granada, Spain
[†]Dept. Electronics, Osaka City University, Osaka, Japan
[‡]CEA-LETI-Minatec, Grenoble, France
[§]STMicroelectronics, Crolles, France
[¶]IMEP-Minatec, Grenoble, France

Abstract—The multibranch mobility analysis is used for the detailed characterization of the carrier mobility in SOI-MOSFETs. This technique shows the actual mobility dependence on the effective field in the device, allowing the separation of the contributions of the carriers located at the front- and back-interfaces. Measurements indicate that the mobility increases in thin and thick-BOX transistors when a back-gate bias is applied. The results demonstrate the impact of the distributions of carriers and electric field in the transistor body.

I. INTRODUCTION

In the last 10 years, the carrier transport in ultrathin Silicon-On-Insulator (SOI)-MOSFETs has been a subject of intensive study [1], [2], [3]. Mobility versus inversion charge or effective field plots have been constantly used to benchmark the experimental data for demonstrating the benefits of new technology modules or the need for further process optimization. However, the validity of the effective field representation to generate Universal Mobility Curves [4] in SOI devices is questionable if two or more channels contribute to the transport [5]. The reason is the lack of information to capture the particular mobility dependence for different carrier distributions in the body of the device when a back-gate bias is applied. In order to overcome this issue, the study of the mobility under different bias conditions is usually performed in terms of gate bias or inversion charge concentration [6]. The bias conditions must be analyzed carefully, since the coupling between the interfaces [7] may lead to either separation or intermixing of the channels. This methodology has two main disadvantages: (i) the comparison of mobility under different biases is not necessarily accurate, since the electric field profiles in the body of the device are substantially different when a back-gate bias is applied; (ii) the distinction between the conduction at the front- and back- interfaces becomes impossible when the film thickness is reduced to the decananometer range due to the supercoupling effect [8]. More elaborated characterization techniques such as the quantitative mobility spectrum analysis are able to distinguish and identify the different carrier species, but in contrast, they require the use of specialized facilities

able to generate very large magnetic fields ($> 10T$) [9].

In this work, we de-correlate the relationship between hole mobility and effective field by the combination of numerical simulations and experimental measurements in advanced SOI transistors with different Buried Oxide (BOX) thicknesses. A calibrated Poisson-Schroedinger solver has been used to evaluate the actual effective field value from its integral definition. Our approach results in useful multibranch mobility plots. As we will show, this representation contains more information about the underlying transport mechanisms in the device than conventional plots (mobility vs gate bias, inversion charge or experimental effective field determination). Preliminary results [10] are now substantiated by comparing the behavior of electrons and holes.

The paper is divided as follows: in Section II, we present experimental measurements of mobility in ultrathin SOI transistors. They show a remarkable mobility enhancement when a back-gate bias is applied. In Section III, the procedure to generate multibranch mobility plots is illustrated. Application to our devices is described and discussed in Section IV. Finally the main conclusions are drawn in Section V.

II. SPLIT-CV MOBILITY RESULTS

The SOI transistors were characterized by the split-C(V) technique [11] producing usual mobility vs. gate-voltage plots for different back-gate bias conditions. The samples under study were fully depleted (FD) MOSFETs on 300mm Unibond wafers with hafnium-based gate insulator/metal gate fabricated at LETI and STMicroelectronics. Two batches of samples were investigated: *i)* transistors with Buried Oxide (BOX) thickness of $T_{BOX} = 145nm$ [12] and *ii)* transistors with ultrathin BOX ($T_{BOX} = 10nm$) and implanted ground plane [13].

Experimental hole mobility results are shown in Figure 1. Increasing the back-gate bias produces overall benefit in the carrier mobility, whether the BOX is ultrathin, Figure 1.a, or thick, Figure 1.b. Similar curves are obtained for electrons (not shown). These results agree with previously published works [6], [14], [15] revealing the impact of the high-*k* gate-stack

978-1-4673-1707-8/12 $31.00 © 2012 IEEE

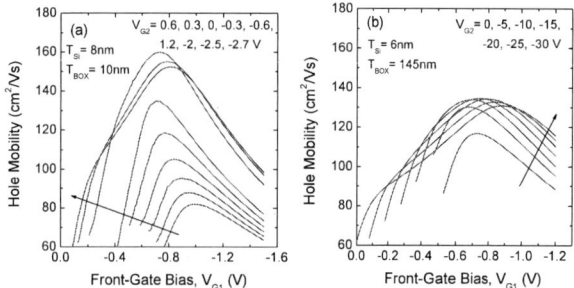

Fig. 1. Experimental hole mobility extracted by split-C(V) technique as a function of the front-gate voltage for different values of the back-gate bias. **(a)** $T_{Si} = 6nm$, $T_{BOX} = 145nm$, $L = W = 10\mu m$, EOT=1.3nm. **(b)** $T_{Si} = 8nm$, $T_{BOX} = 10nm$, $L = W = 10\mu m$, EOT=1.3nm.

technology and the benefit of volume conduction regime. The mobility gain, $(\mu_{max} - \mu_0)/\mu_0$, where μ_{max} is the maximum mobility at a given back-gate bias, and μ_0 the maximum mobility at $V_{G2} = 0V$, has been represented in Figure 2 as a function of the normalized back-gate bias (V_{G2}/T_{BOX}). The enhancement of both electron and hole mobilities reaches 70% in the $T_{BOX} = 10nm$ devices, whereas for the case of $T_{BOX} = 145nm$ the improvement is more modest: 40% for electrons and 15% for holes. The impact of the different body thickness of the *thin* and *thick*-BOX devices is expected to be minimal according to [16]. In the following sections, the multibranch analysis will clarify the reasons for these differences, highlighting also the role of each interface. Such detailed analysis is beyond the capability of conventional $\mu - V_G$ or $\mu - Q_{inv}$ representations.

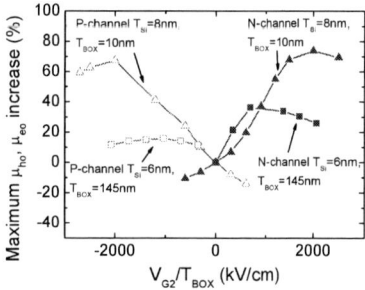

Fig. 2. Enhancement of the maximum of low-field carrier mobility, with respect to $V_{G2} = 0V$, by applying several back-gate biases (V_{G2}) in P-channel and N-channel devices with two BOX ($T_{BOX} = 145nm, 10nm$) thicknesses.

III. MULTIBRANCH REPRESENTATION

The basics of the multibranch mobility representation is the integral definition of the effective field. E_{eff} is defined as the *average* of the *local* transverse electric field, $E(z)$, pondered by the local inversion charge concentration $p(z)$ [17]:

$$E_{eff} = \frac{\int |E(z)| p(z)\, dz}{\int p(z)\, dz} \qquad (1)$$

where, z, is the transversal direction of the device (perpendicular to the gate).

The integration limits in Eq. 1 corresponds to the physical boundaries of the transistor body where the electric field is not negligible: from the interface to the border of the depletion region in bulk and Partially Depleted SOI-MOSFETs, or between the physical limits of the Si film in Fully Depleted Single-Gate or Double-Gate SOI devices.

The role of the simulations within this methodology is only to calculate the carrier and electric field distributions needed to perform the calculation of Eq. 1. All mobility data is purely experimental. As compared to the conventional procedure to generate $\mu - E_{eff}$ plots, we do not use the inversion charge from split-C(V) to compute the effective field with analytical approximations [3]. Such approximations are more or less valid and we prefer not to rely on them.

The methodology to elaborate *multibranch mobility* plots can be summarized in three steps:

(i) The effective mobility is experimentally determined by the front-gate split-C(V) method [3], leading to a $\mu_{eff} - V_{G1}$ plot (Fig. 1) for each V_{G2} value.

(ii) A Poisson-Schrödinger self-consistent solver is used to determine the actual profiles of $E(z)$ and $p(z)$ in Eq. 1 and perform the calculation of E_{eff}. The solver is calibrated to reproduce the $Q_{inv} - V_{G1}$ curves obtained from the front-gate split-C(V) measurements by tuning the work-function and doping (Fig 3.a). For the case of the conduction-band the solution is performed under the effective-mass approximation whereas for the case of the valence-band a six-band $k \cdot p$ model has been implemented [18] in the solver to calculate the energy dispersion relationship, $E(k)$.

(iii) Once the $\mu_{eff} - V_{G1}$ (Fig. 1) and $E_{eff} - V_{G1}$ (Fig. 3.b) tables are determined in (i) and (ii), they can be combined to complete the actual $\mu_{eff} - E_{eff}$ representation.

This procedure can be applied similarly to the back-channel by extracting the mobility from the back-gate split-C(V) [3]. Note that the $Q_{inv} - V_{G1}$ is the nexus between the simulations and the experimental results. The split-C(V) technique only provides the total inversion charge concentration, i.e., the integral of $p(z)$. For the evaluation of the actual effective electric field it is necessary to know the local distribution of the inversion charge and not only its integral.

Figure 3.a illustrates an example of the accuracy of the calibration of the simulator for the $T_{BOX} = 10nm$ device. The inversion charge obtained from the Split-C(V) measurements is divided by the transistor area ($L = W = 10\mu m$) and fitted by tuning the work-function and ground-plane doping. Typically two extreme values of the back-gate bias (i.e. $V_{G2} = 0V, -2V$ for the thin-BOX devices) are selected for the best fit in all the back-gate bias range.

In Figure 3.b, we show the results of the effective field calculated from Eq. 1. For the case of $V_{G2} = 0V$, the effective field increases monotonically with the front-gate bias (V_{G1}) as predicted by the analytical approximations. The most interesting case, leading to the multibranch plots, occurs when

978-1-4673-1707-8/12 $31.00 © 2012 IEEE

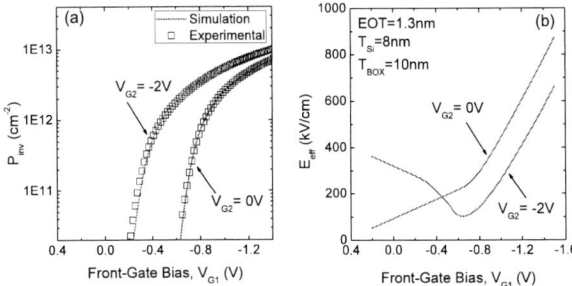

Fig. 3. (a) Comparison of the inversion charge concentration as a function of the front-gate bias between the experimental results, obtained by the split-C(V) technique, and the simulation results after calibration. Two values of back-gate bias have been considered. (b) Effective field evaluated with Eq. 1 as a function of the front-gate bias for two values of V_{G2}. When the negative back-gate bias is applied there is a range from $150kV/cm$ to $380kV/cm$ where two different front-gate biases can lead to the same value of effective field. Same device as in Figure 1.a.

a back-gate bias is applied (see $V_{G2} = -2V$ in Fig. 3.b). In this situation, since the potential at the BOX interface is relatively large, increasing the front-gate bias first decreases the effective field (the potential difference between the two interfaces is reduced). When the top-interface is inverted ($V_{G1} \simeq -0.65V$) the effective field increases again due to the rapid enhancement in the carrier concentration at the front-interface; obviously, the weight of the electric field in Eq. 1 is strongly modified. The immediate consequence is straightforward: there is a range of effective field values (from $150kV/cm$ to $380kV/cm$) where, for a given effective field value, two carrier distributions, and therefore two mobility values, are possible, leading to two different branches of mobility [5]. This result is particularly relevant in advanced SOI MOSFETs where the carrier mobilities near the front and back interfaces are rather contrasting.

IV. EXPERIMENTAL APPLICATION

The complete multibranch curves of our devices at different bias conditions are shown in Figure 4 (a & c: PMOS, b & d: NMOS). In all cases, for $V_{G2} = 0$, the curves show the typical phonon and surface roughness limited mobility evolution with the effective field as the inversion charge at the front channel increases. For non-negligible back-gate bias ($V_{G2} = +/-2V$ or $+/-30V$) the importance of the multibranch representation becomes evident. In Figure 4.a (PMOS device), for $V_{G2} = -2V$ increasing the front-gate bias, first enhances the back-channel (because V_{T2} decreases): the hole current only flows at the back-interface until the return point is achieved ($E_{eff} \simeq 80kV/cm$) when the front-interface becomes also enriched with holes. From this point, both channels are enhanced and the mobility decreases due to the larger impact of the electric field and the interface roughness scattering. The comparison with the multibranch plot for electrons (Fig. 4.b) clearly reveals the different impact of the mobility of the front and back-interfaces depending of the type of the carriers: for

electrons, the carrier mobility at the back-interface is larger than at the front-interface (note the loop of the mobility curve in Figure 4.b around the return point, $E_{eff} \simeq 100kV/cm$). By contrast, in the case of holes, the hole mobility branch of the back-interface always presents the lower mobility.

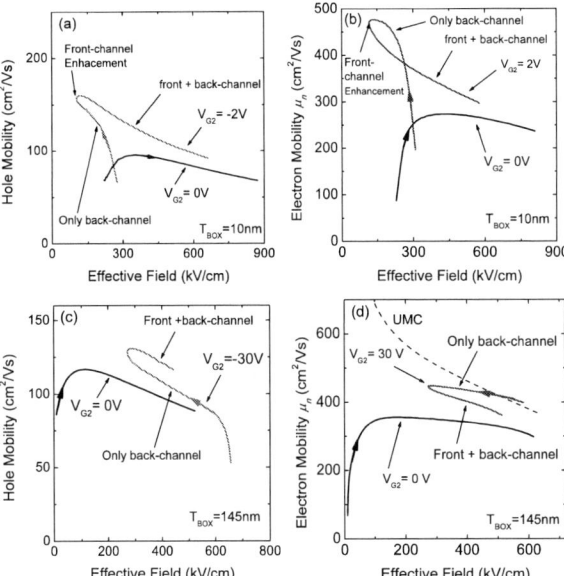

Fig. 4. Multibranch plots for hole and electron mobilities with different back-gate bias conditions: (a) PMOS, $T_{BOX} = 10nm$, $T_{Si} = 8nm$ (b) NMOS, $T_{BOX} = 10nm$, $T_{Si} = 8nm$ (c) PMOS, $T_{BOX} = 145nm$, $T_{Si} = 6nm$ (d) NMOS, $T_{BOX} = 145nm$, $T_{Si} = 6nm$. Universal mobility curve (UMC) is shown for comparison.

If we consider the 145nm-BOX devices (Figure 4. c & d), the back-channel is already strongly inverted at $V_{G2} = 30V$ (NMOS) or $V_{G2} = -30V$ (PMOS): for both electrons and holes, the multibranch mobility curves corresponding to the back channel or to the back+front channels have rather parallel paths (before the return points); this is due to the fixed potential at the back-interface when the back-channel is strongly inverted, leading to negligible changes in the electric field at the back-interface when V_{G1} is swept.

In order to investigate the physical origins of the higher carrier mobility at the front-interface when the current flow is driven by holes, we show in Figure 5 the carrier distribution for the $T_{BOX} = 10nm, T_{Si} = 8nm$ devices, at $E_{eff} = 200kV/cm$, where the multibranch behavior of the mobility is clearly manifested. By analyzing the profiles, it is remarkable that the volume conduction regime is more significative in the case of holes than for electrons due to the different curvature of the energy dispersion relationship of the valence and conduction bands ($\partial^2 E(k)/\partial k^2$). Therefore, electrons are going to be more affected by remote scattering mechanisms induced by the high-k dielectric and surface related scattering events when the front-channel is enriched. For the case of holes, (Figures 4.a & c), when only the back-channel is active,

978-1-4673-1707-8/12 $31.00 © 2012 IEEE

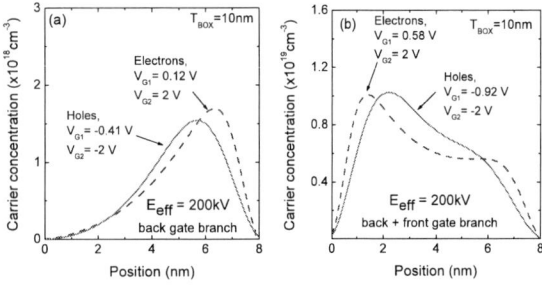

Fig. 5. Carrier concentration profiles for electrons and holes at $E_{eff} = 200kV/cm$ corresponding to the different branches of mobility in Figure 4.a & b. (a) back-gate branch (b) back+front-gate branch

carriers are mainly affected by the back-interface; increasing the front-gate bias induces a strong volume conduction regime leading to a reduction of the surface scattering events from both interfaces, and hence increasing the mobility. This benefit adds to the already reported increase of the population of the light-hole bands in DG mode [15].

Finally, from Figure 2, the mobility enhancement obtained by applying back-gate bias is much more pronounced in the thin-BOX than in the thick-BOX devices, especially when the current is driven by electrons. Nevertheless, if the comparison is now made considering the same values of the effective electric field (Figure 4), this outstanding enhancement partially vanishes. In the following table we summarize the maximum mobility enhancement due to back-gate bias when the front channel is strongly inverted in all devices (front+back gate branch and $E_{eff} > 350kV/cm$); the comparison is performed according to Figure 4.

TABLE I

Device	$T_{BOX} = 10nm$	$T_{BOX} = 145nm$
μ_h enhancement	< 26%	< 12%
μ_e enhancement	< 26%	< 23%

The carrier mobility gain is still larger in thin-BOX transistors, but the difference with thick-BOX devices is no longer as dramatic as in Figure 2. For the thin-BOX devices the enhancement that can be achieved in electron and hole mobility is the same, whereas an asymmetry is still manifested in the thick-BOX transistor between positive and negative carriers.

V. CONCLUSION

We have applied a novel multibranch mobility analysis for the characterization of carrier transport in advanced SOI-MOSFETs. The methodology is based on the accurate evaluation of the integral of the effective field. This technique reveals the actual mobility dependence with the effective field in the body of the devices. The results show that the larger mobility enhancement in thin-BOX transistors with respect to thick-BOX devices, when a back-gate bias is applied, is closely

related to an electric field reduction rather than technological factors.

ACKNOWLEDGMENT

This work was partially supported by projects TEC 2011-028660, TIC2010-6902, and KAKENHI (22560334).

REFERENCES

[1] T. Ernst, S. Cristoloveanu, G. Ghibaudo, T. Ouisse, S. Horiguchi, Y. Ono, Y. Takahashi, and K. Murase, "Ultimately thin double-gate SOI MOSFETs," *IEEE Trans. Electr. Dev.*, vol. 50, no. 3, pp. 830–838, 2003.

[2] K. Uchida, H. Watanabe, A. Kinoshita, J. Koga, T. Numata, and S. I. Takagi, "Experimental study on carrier transport mechanism in ultrathin-body soi n- and p- MOSFETs with SOI thickness less than 5nm," *Proceedings of the International Electron Devices Meeting (IEDM'02)*, 2002, pp. 47–50.

[3] A. Ohata, M. Casse, and S. Cristoloveanu, "Front- and back-channel mobility in ultrathin SOI-MOSFETs by front-gate split CV method," *Solid-State Electronics*, vol. 51, no. 2, pp. 245–251, 2007.

[4] S. Takagi, S. Iwase, A. Toriumi, and H. Tango, "On the universality of inversion-layer mobility in Si MOSFETs: part I. effects of substrate impurity concentration," *IEEE Trans. Elec. Dev.*, vol. 41, no. 12, pp. 2357–2362, Dec. 1994.

[5] S. Cristoloveanu, N. Rodriguez, and F. Gamiz, "Why the Universal Mobility is not," *IEEE Trans. Elec. Dev.*, vol. 57, no. 6, pp. 1327–1333, 2010.

[6] A. Ohata, Y. Bae, C. Fenouillet-Beranger, and S. Cristoloveanu, "Mobility Enhancement by Back-Gate Biasing in Ultrathin SOI MOSFETs with Thin BOX,," *Electron Device Letters*, vol. 33, no. 3, pp. 348–350, 2012.

[7] H. Lim and J. G. Fossum, "Current-voltage characteristics of thin-film SOI MOSFETs in strong inversion," *IEEE Trans. on Electr. Dev.*, vol. 31, no. 4, pp. 401–407, 1984.

[8] S. Eminente, S. Cristoloveanu, R. Clerk, A. Ohata, and G. Ghibaudo, "Ultra-thin fully-depleted SOI MOSFETs: special charge properties and coupling effects," *Solid-State Electronics*, vol. 51, no. 2, pp. 239–244, 2007.

[9] J. R. Meyer, C. A. Hoffman, J. Antoszewski, and L. Faraone, "Quantitative mobility spectrum analysis of multicarrier conduction in semiconductors," *Journal Appl. Phys.*, no. 81, p. 709, 1997.

[10] C. Navarro, N. Rodriguez, A. Ohata, F. Gamiz, F. Andrieu, O-Faynot, C. Fenouillet-Berangerand, and S. Cristoloveanu, "Multibranch Mobility Analysis for the Characterization of FD-SOI Transistors," *IEEE Elec. Dev. Lett., in press.*

[11] C. G. Sodini, T. W. Ekstedt, and J. L. Moll, "Charge accumulation and mobility in thin dielectric MOS transistors," *Solid State Eletron.*, vol. 25, no. 9, p. 833, Dec. 1982.

[12] F. Andrieu et al., "Low leakage and low variability ultra-thin body and buried oxide (UT2B) SOI technology for 20nm low power CMOS and beyond," in *Symposium on VLSI Technology,*, San Francisco, CA, 2010, pp. 57–58.

[13] C. Fenouillet-Beranger et al., "FDSOI devices with thin BOX and ground plane integration for 32 nm node and below," *Solid-State Electronics*, vol. 53, no. 7, pp. 730–734, 2009.

[14] N. Rodriguez, S. Cristoloveanu, and F. Gamiz, "Evidence for mobility enhancement in double-gate silicon-on-insulator metal-oxide-semiconductor field-effect transistors," *Journal of Applied Physics*, vol. 102, pp. 083 712–1–8, Oct. 2007.

[15] S. Kobayashi, M. Saitoh, and K. Uchida, "Hole mobility enhancement by double-gate mode in ultrathin-body silicon-on-insulator p-type metal-oxide-semiconductor field-effect transistors," *Jap. J. App. Phys.*, no. 106, pp. 024 511–6, 2009.

[16] F. J. Gamiz, J. B. Roldan, P. Cartujo-Cassinello, J. E. Carceller, J. A. Lopez-Villanueva, and S. Rodriguez, "Electron Mobility in Extremely Thin Single-Gate Silicon-on-Insulator Inversion Layers," *J. Appl. Phys.*, vol. 86, no. 11, pp. 6269–6275, 1999.

[17] M. Shoji and S. Horiguchi, "Electronic structures and phonon limited electron mobility of double-gate silicon-on-insulator Si inversion layers," *IEEE Trans. Elec. Dev.*, vol. 85, no. 5, pp. 2722–2731, 1999.

[18] L. Donetti, F. Gamiz, and N. Rodriguez, "Simulation of hole mobility in two-dimensional systems," *Semiconductor Science and Technology*, no. 24, pp. 035 016–035 023, 2009.

978-1-4673-1707-8/12 $31.00 © 2012 IEEE

The role of the temperature on the scattering mechanisms limiting the electron mobility in metal-oxide-semiconductor field-effect-transistors fabricated on (110) silicon-oriented wafers

Philippe Gaubert*‡, Akinobu Teramoto*, Shigetoshi Sugawa*†, and Tadahiro Ohmi*

*New Industry Creation Hatchery Center, Tohoku University
†Graduate school of Engineering, Tohoku University
Aza-Aoba 6-6-10, Aramaki, Aoba-ku, Sendai 980-8579, Japan
TEL: +81-22-795-3977, FAX: +81-22-795-39860
‡gaubert@fff.niche.tohoku.ac.jp

Abstract—The scattering mechanisms limiting the electron mobility in Si(110) MOSFETs have been studied in function of the temperature. They have been compared to the ones limiting the electron mobility in Si(100) MOSFETs. It appeared that the lower electron mobility encountered for the (110) orientation was coming from a stronger limitation due to the sole Coulomb and surface roughness scatterings. Indeed, the phonon-limited mobility have been found similar for both orientations. Furthermore, contrary to what it is commonly assumed, the surface roughness scattering mechanisms are not independent of the temperature. Like the Coulomb-limited mobility, the surface roughness-limited mobility will greatly vary at high temperature while they will reach a constant value when the temperature will be reduced.

I. Introduction

The mobility is one of the key parameter for high performance Metal-Oxide-Semiconductor Field-Effect-Transistor (MOSFET). Nowadays research is focusing on enhancing the mobility by inducing stress [1], changing material [2] or even changing the crystallographic orientation of the wafer [3]. The change from the conventional (100) silicon oriented wafers for the (110) ones is a very promising approach since the hole mobility is highly enhanced. Unfortunately, the electron mobility is at the same time degrading because of a strong contribution of the surface roughness scatterings [4]. However, the recovery and therefore the enhancement of the performance of the n-MOSFETs for that new orientation has been recently achieved by using new transistors which are working on accumulation mode rather than on inversion one [5]. Their drivability can become even greater than that of the conventional orientation since there is a bulk contribution adding to the conduction layer [6]. therefore, the study of the mobility in inversion mode Si(110) n-MOSFETs is mandatory since it is the first step toward the modeling of the accumulation mode Si(110) n-MOSFETs. In addition, the surface roughness-limited mobility is unusually strong for these devices and therefore their study is a very good chance to understand in more detail the surface roughness scatterings mechanism.

In this paper, the extraction of the electron mobility in Si(110) n-MOSFETs has been carried out for several temperatures. Its modeling has been done in order to separate each scattering mechanism limiting it. The Coulomb, phonon and surface roughness scattering mechanisms that are limiting the electron mobility in Si(110) n-MOSFETs have been compared with that of the conventional Si(100) n-MOSFETs. Moreover, their evolution with the temperature has been analyzed. Finally, the modeling of the drivability of the Si(110) n-MOSFETs has been achieved.

II. Experimental results

Inversion-mode fully depleted n-channel silicon-on-insulator (SOI) MOSFETs have been fabricated on bonded SOI (110) and (100) crystallographic silicon-oriented wafers. The doping concentration N_d, the silicon thickness of the SOI layer t_{SOI}, the thicknesses of the buried oxide layer and of the SiO_2 gate insulator have been respectively evaluated around 10^{16} cm^{-3}, 50 nm, 100 nm and 7 nm.

The measurement of the capacitance-gate voltage C-V_g and of the drain current-gate voltage I$_d$-V_g required for the evaluation of the effective mobility μ_{eff} have been carried out on L/W=100 μm/100 μm transistors at a constant drain voltage V_d=100 mV for various temperatures T ranging from 213° K to 473° K. Moreover, the I$_d$-V_g curves have been collected for each temperature on several others transistors featuring a gate width equal to 20 μm and a gate length varying from 20 μm down to the sub-micron scale. η, which is the averaging of the effective electric field E$_{eff}$ over the carrier distribution has been taken equal to 1/2 for the Si(100) wafers [7] and 1/3 for the Si(110) ones [8]. The S parameter and the threshold voltage V$_{th}$ have been evaluated afterwards for all geometry and for all temperature. In addition to notice a decrease of the S parameter and an increase of V$_{th}$, the reduction of the temperature resulted as well in a reduction of the dispersion of the S parameters and V$_{th}$. As it is possible to notice in Fig. 1 (a), while the gate voltage V$_g$ is increased,

978-1-4673-1707-8/12 $31.00 © 2012 IEEE

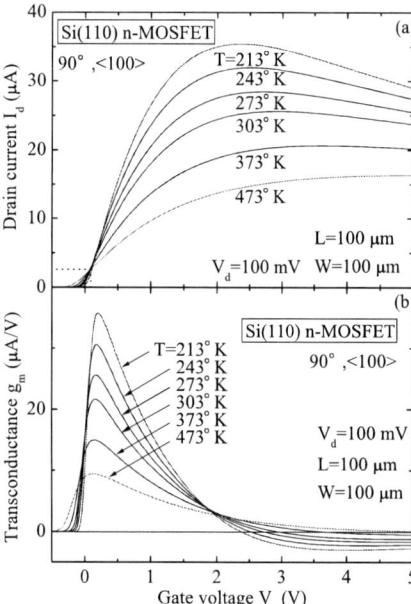

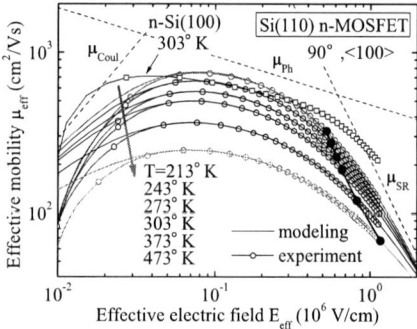

Fig. 1. Drain current I_d (a) and transconductance g_m (b) versus the gate voltage V_g at V_d=100 mV for a Si(110) n-MOSFET measured for several temperatures.

Fig. 2. Effective mobility μ_{eff} versus the effective electric field E_{eff} for a Si(110) n-MOSFET measured at several temperatures. The result for a si(100) n-MOSFET at ambient temperature has been reported with the dashed line. The symbols represent the experimental data while the full lines represent the modeling according to Eq. 1. The three main scattering mechanisms have reported with the dashed lines.

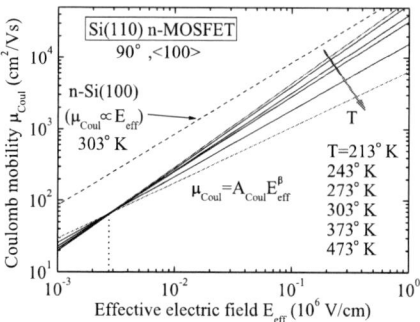

Fig. 3. Coulomb-limited mobility μ_{Coul} versus the effective electric field E_{eff} for a Si(110) n-MOSFET calculated at several temperatures. The result for a si(100) n-MOSFET at ambient temperature has been reported with the dashed line.

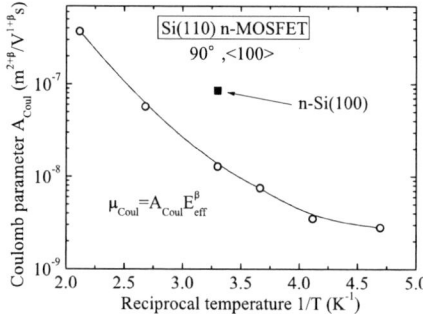

Fig. 4. Coulomb parameter A_{Coul} versus the reciprocal temperature for a Si(110) n-MOSFET. The result for a si(100) n-MOSFET at ambient temperature has been reported with the full symbol.

the drain current I_d will saturate to finally decrease. This trend, attributed to the unbalance between the rise of the carrier number and the fast degradation of the mobility [4] is not visible for the conventional Si(100) oriented wafers other than at low temperature. The consequence is a negative transconductance such as the ones showed in Fig. 1 (b). This trend is vanishing for temperature higher than 473° K while it is amplified with a reduction of the temperature, although the drain current is enhanced. μ_{eff} has been plotted in Fig. 2. The mobility of the Si(110) n-MOSFETs, which is decreasing while the temperature is increased, is also lower than that

of the Si(100) one. The scattering mechanisms have been studied by the means of the modeling of μ_{eff}. By taking into account the Coulomb, phonon and surface roughness scattering mechanisms, μ_{eff} has been written according to the Matthiessen rule:

$$\frac{1}{\mu_{eff}} = \frac{1}{\mu_{SR}} + \frac{1}{\mu_{Ph}} + \frac{1}{\mu_{Coul}} =$$
$$\frac{1}{A_{Ph}E_{eff}^{-0.3}} + \frac{1}{A_{Coul}E_{eff}^{\beta}} + \frac{1}{A_{SR}E_{eff}^{\gamma}}. \quad (1)$$

μ_{Coul} is the Coulomb-limited mobility, proportional to E_{eff}^{β} where β is a fitting parameter [9]. μ_{Ph} is the phonon-limited mobility, generally proportional to $E_{eff}^{-0.3}$ [8]. Finally, the surface roughness-limited mobility μ_{SR} is proportional to E_{eff}^{γ} [10] in which γ is usually found between -1 and -3. A_{Coul}, A_{Ph} and A_{SR} are finally fitting parameters respectively associated to the Coulomb, phonon and surface roughness scatterings.

The results concerning the Coulomb-limited mobility are presented in Fig. 3 and 4. μ_{Coul} is temperature dependent, with β varying between 0.8 for the highest temperature and 1.2 for the lowest. The variation of the scattering rate with E_{eff} is close to the unit for both wafers, indicating a good agreement with the literature. However, the limitation is stronger for the

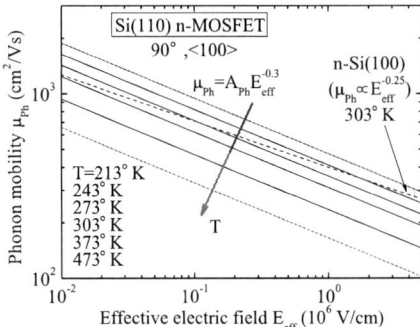

Fig. 5. Phonon-limited mobility μ_{Ph} versus the effective electric field E_{eff} for a Si(110) n-MOSFET calculated at several temperatures. The result for a si(100) n-MOSFET at ambient temperature has been reported with the dashed line.

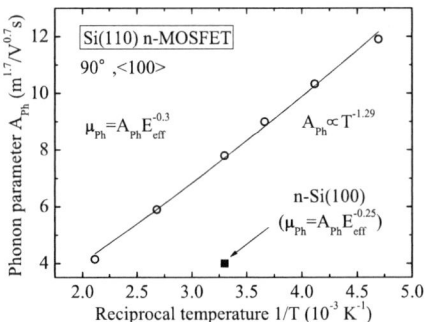

Fig. 6. Phonon parameter A_{Ph} versus the reciprocal temperature for a Si(110) n-MOSFET. The result for a si(100) n-MOSFET at ambient temperature has been reported with the full symbol.

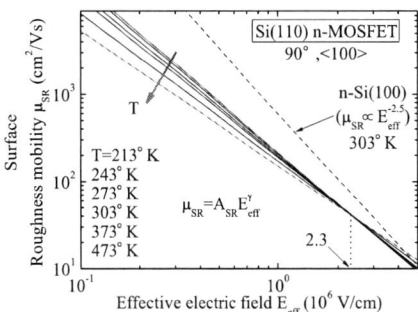

Fig. 7. Surface roughness-limited mobility μ_{SR} versus the effective electric field E_{eff} for a Si(110) n-MOSFET calculated at several temperatures. The result for a si(100) n-MOSFET at ambient temperature has been reported with the dashed line.

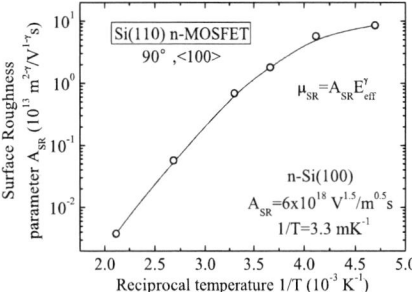

Fig. 8. Surface roughness parameter A_{SR} versus the reciprocal temperature for a Si(110) n-MOSFET.

Si(110) transistors. Fig. 3 revealed a fixed point at which μ_{Coul} is independent of the temperature. This fixed point is actually originating the crossing point seen in Fig. 1 (a) around V_g=100 mV and noticed by Mereu et al. [11]. Below this point, the electrons, that have their energy increased by the temperature, scatter less since the strength of the Coulomb interaction is decreased. To finish, Fig. 4 strongly suggests that μ_{Coul} might become independent of the temperature at low temperature.

Regarding the phonon-limited mobility, the results have been reported in Fig. 5 and 6. For both wafers, μ_{Ph} are relatively similar like stated by Takagi et al. [8]. However, the power law had to be slightly adjusted to smaller value ($\mu_{Ph} \propto E_{eff}^{-0.25}$) for the conventional Si(100) wafers. In addition, the phonon scatterings are temperature dependent and are therefore varying according to $T^{-1.29}$, agreeing with the literature.

The results regarding the surface-roughness-limited mobility μ_{SR} are presented in Fig. 7, 8 and 9. Contrary to the well accepted statement declaring that μ_{SR} is almost independent of the temperature, it clear from the data that this is not the case. Actually, in Fig. 8 and 9, γ and A_{SR} are clearly temperature dependent, however both quantities appear to converge toward a constant value for lower temperature. Below 200° K, μ_{SR} becomes regardless of the temperature and γ

reaches a constant value roughly equal to -2, corresponding to the reported data found in the literature. Another arousing observation in Fig. 7 is the presence of a fixed point in E_{eff} such as the one previously seen for μ_{Coul}. It is worth noticing that our results and the simulation carried out by Fischetti et al. [12] have a similar behavior even if this last work has been done for the Si(110) p-MOSFETs. This threshold electric field roughly corresponds to the breakdown of the gate oxide and is therefore very difficult to experimentally study. Above this threshold electric field, the increase of the temperature will reduce the surface roughness scattering rate. In Fig. 7, while the surface roughness scattering rate is lower for the Si(100) n-MOSFETs, it degradation with E_{eff} is the fastest. Since both wafers have an identical very low microroughness measured below 0.08 nm, it is possible to think that the Si(100) wafers have an intrinsic grainier surface. This might be explained by the lower density of atom at the Si/SiO$_2$ interface for the Si(100) wafers. At the same time, the higher surface density of atom for the Si(110) wafers might be also responsible for the shorter wave function at the interface causing a higher scattering rate and therefore a lower surface-roughness-limited mobility μ_{SR} for the Si(110) n-MOSFETs. μ_{SR} is obviously surface dependent and therefore contributes, with μ_{Coul} which is also surface dependent, to degrade the mobility of the Si(110) n-MOSFETs. Indeed, as summarized in Fig. 10, μ_{Ph} remains almost regardless of the surface.

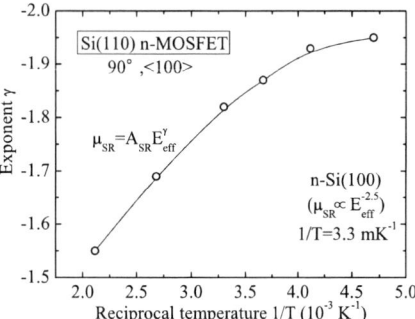

Fig. 9. Surface roughness exponent γ versus the reciprocal temperature for a Si(110) n-MOSFET.

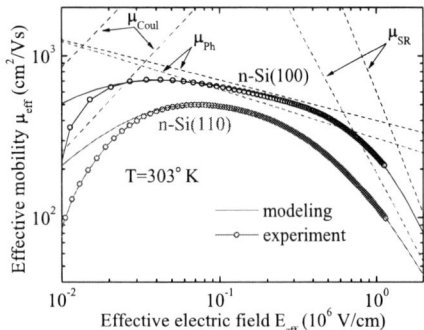

Fig. 10. Effective mobility μ_{eff} versus the effective electric field E_{eff} for a Si(110) and Si(100) n-MOSFETs measured at 303° K. The symbols represent the experimental data while the full lines represent the modeling according to Eq. 1. The three scattering mechanisms have reported as well with the other lines.

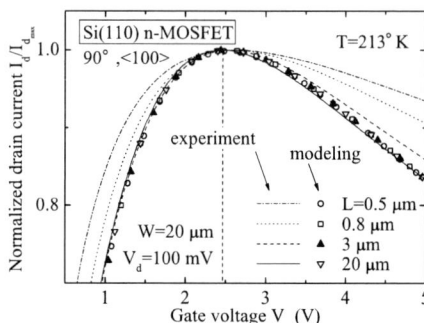

Fig. 11. Normalized drain current I_d calculated with the maximum drain current $I_{d_{max}}$ versus the gate voltage V_g for a Si(110) n-MOSFET at 213° K. The symbols represent the modeling while the lines refer to the experimental data.

seems to disappear when the temperature is lowered down. Finally, the physical nature of the Si/SiO$_2$ interface seems to define the intrinsic surface roughness scattering rate.

REFERENCES

[1] P. Packan, S. Cea, H. Deshpande, T. Ghani, M. Giles, O. Golonzka, M. Hattendorf, R. Kotlyar, K. Kuhn, A. Murthy, P. Ranade, L. Shifren, C. Weber and K. Zawadzki, High Performance Hi-K + Metal Gate Strain Enhanced Transistors on (110) Silicon, in *Int. Electron Device Meet. Tech. Dig.* (2008) 63-66.

[2] S. Beysserie, J. Branlard, S. Aboud, S. M. Goodnick and M. Saraniti, Comparative Analysis of SOI and GOI MOSFETs, *IEEE Trans. Electron Devices* **53** (2006) 2545-2550.

[3] P. Gaubert, A. Teramoto, W. Cheng, T. Hamada and T. Ohmi, Different mechanism to explain the 1/f noise in *n*- and *p*-SOI-MOS transistors fabricated on (110) and (100) silicon-oriented wafers, *J. Vac. Sci. Technol. B* **27** (2009) 394-401.

[4] P. Gaubert, A. Teramoto, W. Cheng and T. Ohmi, Relation Between the Mobility, 1/f Noise, and Channel Direction in MOSFETs Fabricated on (100) and (110) Silicon-Oriented Wafers, *IEEE Trans. Electron Devices* **57** (2010) 1597-1607.

[5] W. Cheng, A. Teramoto, M. Hirayama, S. Sugawa and T. Ohmi, Impact of Improved High-Performance Si(110)-Oriented MetalOxideSemiconductor Field-Effect Transistors Using Accumulation-Mode Fully Depleted Silicon-on-Insulator Devices, *Jpn. J. Appl. Phys.* **45** (2006) 3110-3116.

[6] P. Gaubert, A. Teramoto, S. Sugawa and T. Ohmi, Hole Mobility in Accumulation Mode Metal-Oxide-Semiconductor Field-Effect Transistors, *Jpn. J. Appl. Phys.* **51** (2012) 04DC07.

[7] S.-I. Takagi, A. Toriumi, M. Iwase and H. Tango, On the Universality of Inversion Layer Mobility in Si MOSFET's: Part I-Effects of Substrate Impurity Concentration, *IEEE Trans. Electron Devices* **41** (1994) 2357-2362.

[8] S.-I. Takagi, A. Toriumi, M. Iwase and H. Tango, On the Universality of Inversion Layer Mobility in Si MOSFETs: Part II-Effects of Surface Orientation, *IEEE Trans. Electron Devices* **41** (1994) 2363-2368.

[9] F. Stern and W. E. Howard, Properties of Semiconductor Surface Inversion Layers in the Electric Quantum Limit, *Phys. Rev.* **163** (1967) 816-835.

[10] G. Mazzoni, A. L. Lacaita, L. M. Perron and A. Pirovano, On Surface Roughness-Limited Mobility in Highly Doped n-MOSFETs, *IEEE Trans. Electron Devices* **46** (1999) 1423-1428.

[11] B. Mereu, C. Rossela, E. P. Gusev and M. Yang, The role of Si orientation and temperature on the carrier mobility in metal oxide semiconductor field-effect transistors with ultrathin HfO$_2$ gate dielectrics, *J. Appl. Phys.* **100** (2006) 014504.

[12] M. V. Fischetti, Z. Ren, P. M. Solomon, M. Yang and K. Rim, Six-band k-p calculation of the hole mobility in silicon inversion layers: Dependence on surface orientation, strain, and silicon thickness, *J. Appl. Phys.* **94** (2003) 1079-1095.

The turning point seen in Fig. 1 (a) which corresponds to the maximum drain current $I_{d_{max}}$ has been investigated. This inflection point is regardless of the gate size like testifies Fig. 11. Moreover, $I_{d_{max}}$ has been found to be proportional to the reciprocal of the temperature. In addition, the corresponding μ_{eff} plotted in Fig. 2 with the full symbols appeared to vary according to $E_{eff}^{-2.1}$.

To finish, the modeling of I_d has been carried out. A disagreement between the experiment and the modeling which was increasing while the gate length was reduced has been acknowledged like testifies Fig. 11. The short channel effect and especially the saturation of the velocity which is a stronger limiting factor in the case of the Si(110) n-MOSFETs has been first favored. However, further studies revealed that the surface roughness attenuation-like behavior was actually originating from the access series resistances.

III. CONCLUSION

Since the phonon-limited mobility is regardless of the wafer orientation, the higher surface roughness and Coulomb interactions rate are the reason for the reduction of the electron mobility in Si(110) MOSFETs when compared to that of the Si(100) ones. In addition, the surface roughness-limited mobility is clearly temperature dependent. However, this dependence

New parameter extraction method based on split C-V

for FDSOI MOSFETs

I. Ben Akkez[1,2], A. Cros[2], C.Fenouillet-Beranger[2,3], F. Boeuf[2], Q. Rafhay[1], F. Balestra[1], G. Ghibaudo[1],

1) IMEP-LAHC, MINATEC Campus, 3 Parvis Louis Néel, 38016 Grenoble, Cedex 1, France.
2) STMicroelectronics, 850, rue J. Monnet, BP. 16, 38921 Crolles, France.
3) CEA-LETI, MINATEC Campus, 17 rue des Martyrs, 38054 Grenoble, Cedex 9, France.
Email : benakkei@minatec.grenoble-inp.fr.

Abstract - **A new parameter extraction methodology based on split C-V is proposed for FDSOI MOS devices. To this end, a detailed capacitance theoretical analysis is first conducted emphasizing the usefulness of the Maserjian function. Split C-V measurements carried out on various FDSOI CMOS technologies show that the Maserjian function exhibits a power law dependence with inversion charge as $\propto Q_i^{-2}$ whatever the carrier type and gate oxide thickness. This feature enables to confirm the validity of a two-parameter simple capacitance model and allows reliable MOSFET parameter extraction.**

I. INTRODUCTION

Fully Depleted (FD) Silicon On Insulator (SOI) technology is a promising candidate for future sub 28nm CMOS generations. Indeed, the use of ultra thin body and buried oxide thickness (UTBB) enables better technology scalability, providing ideal subthreshold slope and low drain-induced barrier lowering (DIBL). Moreover, midgap/high-k metal gate stack with undoped SOI films allows for great improvement in variability as compared to bulk technology [1-3]. In this context, an accurate characterization and parameter extraction method from capacitance characteristics are mandatory for precise technology qualification and device modeling. In bulk MOS structures, this is realized by measuring the C-V curves in depletion to accumulation, and, extracting the flat band voltage and gate oxide thickness using the Maserjian function [4]. This is no longer feasible in FDSOI devices where full depletion occurs and where accumulation cannot be formed as in bulk structures

In this work, we propose a simple and efficient methodology for split C(V) analysis and related parameter extraction in advanced FDSOI MOSFETs accounting for their specificities as regard to bulk architectures.

II. THEORETICAL BACKGROUND

The charge conservation equation in FDSOI structures can be expressed, neglecting the depletion in the substrate under the BOX, as,

$$Q_i + Q_d = C_{ox1}\left(V_{g1} - V_{fb1} - V_{s1}\right) + C_{ox2}\left(V_{g2} - V_{fb2} - V_{s2}\right), \quad (1)$$

where Q_i is the inversion charge (>0), $Q_d = q.N_a.t_{si}$ with N_a being the acceptor channel impurity concentration and t_{si} the Si film thickness, $C_{ox1,2}$ is the front/back gate oxide capacitance, $V_{s1,2}$ is the front/back surface potential, $V_{g1,2}$ the front/back gate voltage and $V_{fb1,2}$ the front/back flat band voltage.

Based on charge conservation, the front gate-to-channel capacitance $C_{gc} = dQ_i/dV_{g1}$ can be derived as,

$$\frac{1}{C_{gc}} = \frac{1}{C_{ox1}} + \frac{1}{\partial Q_i/\partial V_{s1}} + \frac{C_{ox2}}{C_{ox1}}\frac{1}{\partial Q_i/\partial V_{s2}}. \quad (2)$$

For standard UTBB FDSOI, $C_{ox2} \ll C_{ox1}$ such that, in normal front gate operation with nearly grounded body, Eq (2) reduces to,

$$\frac{1}{C_{gc}} \approx \frac{1}{C_{ox1}} + \frac{1}{\partial Q_i/\partial V_{s1}}. \quad (3)$$

Note that, in contrast to bulk devices, there is no contribution of depletion term in FDSOI capacitance due to full depletion regime.

Under Boltzmann's statistics, Q_i is proportional to $\exp(\beta V_{s1})$ with $\beta = q/kT$, such that Eq. (3) yields,

$$C_{gc} = \frac{\beta Q_i C_{ox1}}{C_{ox1} + \beta Q_i}. \quad (4)$$

It should be noted such an inversion charge dependence for the capacitance results in a Lambert W function variation of Q_i with gate voltage [5].

Besides, a useful function in capacitance analysis is the so-called Maserjian function Y_m, which is independent of gate oxide capacitance defined as [4],

$$Y_m = \frac{1}{C_{gc}^3} \cdot \frac{dC_{gc}}{dV_{g1}} = -\frac{d\left(1/C_{gc}\right)}{dQ_i}, \quad (5)$$

reducing here to,

$$Y_m = \frac{1}{\beta Q_i^2}. \quad (6)$$

Note that, unlike in bulk structures, Y_m is also independent of channel doping concentration in FDSOI devices due to full depletion. So, it cannot be used as in bulk devices for doping level and V_{fb} extraction [4].

The derivative of C_{gc} with respect to V_{g1} can be obtained from the above equations as,

$$\frac{dC_{gc}}{dV_{g1}} = Y_m C_{gc}^3 = \frac{\beta^2 Q_i C_{ox1}^3}{\left(C_{ox1} + \beta Q_i\right)^3}. \quad (7)$$

Typical simulated variations of C_{gc} and $|dC_{gc}/dV_{g1}|$ versus Q_i are shown in Fig. 1 (a). dC_{gc}/dV_{g1} passes through a maximum for a charge threshold value $Q_{ith} = C_{ox1}/(2\beta)$. As can be seen from Fig. 1 (b), the $dC_{gc}/dV_{g1}(V_{g1})$ corresponding plot enables to locate the threshold voltage from the maximum peak position for both electron (N) and hole (P) channels.

It can be shown that, for Boltzmann's statistics, the N and P type threshold voltages can be well approximated as,

$$V_{tn} \approx V_{fb1} + kT \ln\left(\frac{Q_{ith}^{2}}{2q\varepsilon_{si}n_{i}kT}\right) + \frac{Q_{ith}+Q_{d}}{C_{ox1}}, \quad (8a)$$

$$V_{tp} \approx V_{fb1} - kT \ln\left(\frac{Q_{ith}^{2}}{2q\varepsilon_{si}n_{i}kT}\right) - \frac{Q_{ith}-Q_{d}}{C_{ox1}}. \quad (8b)$$

where n_i is the intrinsic carrier density, kT the thermal voltage and ε_{si} the silicon permittivity.

Therefore, in Fig. 1 (b), the distance between the N and P type peaks is mainly function of Q_{ith} as,

$$V_{tn} - V_{tp} = 2kT \ln\left(\frac{Q_{ith}^{2}}{2q\varepsilon_{si}n_{i}kT}\right) + 2\frac{Q_{ith}}{C_{ox1}}, \quad (9)$$

whereas the mid position between the peaks depends mostly of V_{fb1} as,

$$\frac{V_{tn} + V_{tp}}{2} = V_{fb1} + \frac{Q_{d}}{C_{ox1}}. \quad (10)$$

So, if $V_{fb1}=0$ and $N_a=n_i$, the N and P $dC_{gc}/dV_{g1}(V_{g1})$ peaks are centered just around zero. Any deviation from symmetry will reflect a non-zero V_{fb1} value and/or a channel doping level $N_a>n_i$.

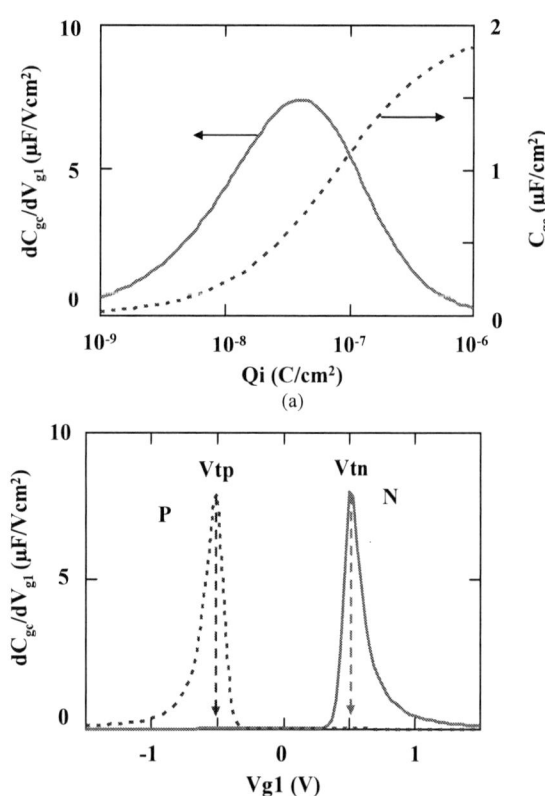

Fig.1. Variations of (a) C_{gc} and dC_{gc}/dV_{g1} vs Qi and (b) $|dC_{gc}/dV_{g1}|$ vs V_{g1} (Si film thickness t_{si}=8nm, C_{ox1}=2x10⁻⁷F/cm², C_{ox2}=1.38x10⁻⁸F/cm², $V_{fb1}=V_{fb2}$=0).

III. RESULTS AND DISCUSSION

Split C-V measurements have been carried out with an Agilent HP4284 on large area ($10\times10\mu m^2$) N and P channel MOSFETs as well as MOS gated diode structures (see Fig. 2) allowing both channel type formation on the same device issued from several FDSOI technologies having undoped 8nm Si film and 25nm BOX thicknesses. The AC signal of the impedance analyzer has amplitude of 25mV and frequency of 1MHz.

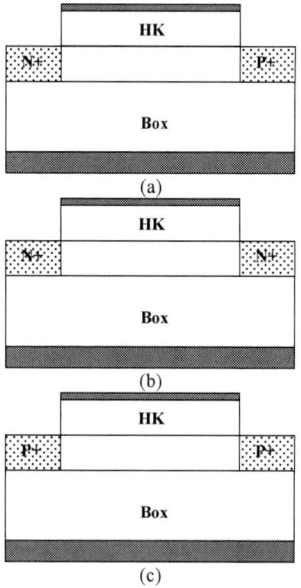

Fig.2. Schematics of FDSOI test structures used for split C-V measurements: a) MOS gated diode, b) NMOSFET and c) PMOSFET.

Typical $C_{gc}(V_{g1})$ obtained on MOS gated diode structures are shown in Fig. 3 (symbols) illustrating the onset of both P and N channel formation. These data have been well fitted using the capacitance model of Eq. (4) using an adjusted value of β i.e. β_1=25/V and C_{ox1}=1.9μF/cm² for both carrier types. As is usual in split C-V technique, the inversion charge has been calculated by integration of $C_{gc}(V_{g1})$ curves.

More $C_{gc}(V_{g1})$ characteristics have been measured on another FDSOI technology featuring 2 gate oxides, GO1 and GO2 with N and P channel CMOSFETs (see Fig. 4, symbols). Similarly, the data have been well reproduced by the model of Eq. (4) with given parameters (see caption).

Indeed, the adequacy of the capacitance model of Eq. (4) can be verified experimentaly by directly checking if the Maserjian function Y_m obeys the inversion charge dependence of Eq. (6).

This is clearly illustrated in Figs 5 and 6 where $Y_m(Q_i)$ characteristics have been obtained on MOS gated diodes and on N and P type MOSFETs having two gate oxides thicknesses (GO1 and GO2). Note the almost perfect superposition of all the data whatever the channel type and gate oxide variation (GO1 vs GO2). Note also the remarkable power law dependence of $Y_m(Q_i)$ as Q_i^{-2}, which confirms the validity of Eq. (6) over three decades of inversion charge, but with an appropriate value of β. This slight deviation of β could possibly be attributed to the

978-1-4673-1707-8/12 $31.00 © 2012 IEEE

inadequacy of Bolzmann's statistics and/or the onset of quantum confinement effects.

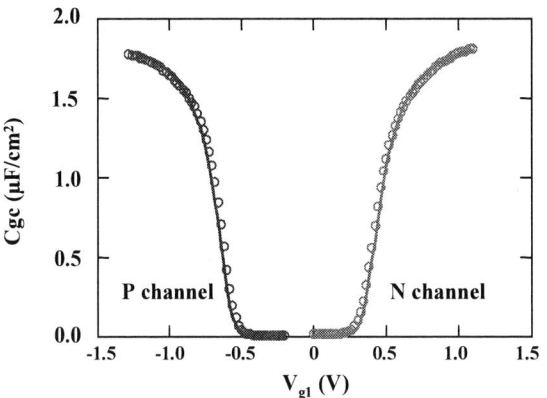

Fig.3. Experimental (symbols) and modeled (lines) $C_{gc}(V_{g1})$ characteristics for N and P channel (MOS gated diode 10x10μm², C_{ox1}=1.9μF/cm², β_1=25/V).

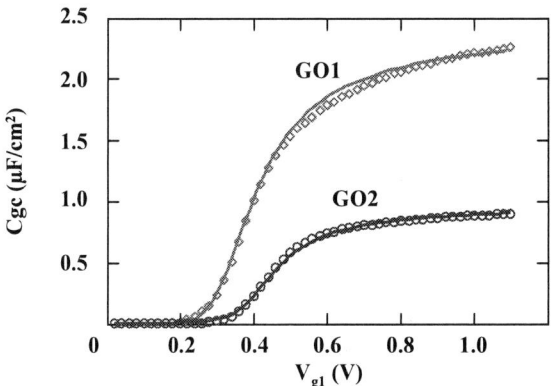

Fig.4. Experimental (symbols) and modeled (lines) $C_{gc}(V_{g1})$ characteristics for 2 gate oxides GO1 (C_{ox1}=2.4μF/cm², β_1=20/V) and GO2 (C_{ox1}=1.0μF/cm², β_1=20/V) (N channel MOSFET 10x10μm²).

It should also be mentioned that such plots of $Y_m(Q_i)$ for both carrier types obtained on the same MOS device are not possible on bulk MOSFETs. This is due to the presence of the depletion capacitance and to the transition between the flat band and depletion regimes in bulk structures, which does not permit to both carriers to behave symmetrically. It is also not adequate for bulk MOS devices with polysilicon gate where poly depletion effect might occur in strong inversion and perturbs the Y function [4]. This feature is a specificity of FDSOI MOS devices, enabling such a capacitance analysis.

As was indicated in section II and illustrated in Fig. 7, the derivative of $C_{gc}(V_{g1})$ can further be used for the extraction of the capacitive threshold voltages, V_{tn} and V_{tp}, for N and P channels from the peak voltage locations (Here, V_{tn}=0.4V and V_{tp}=-0.7V).

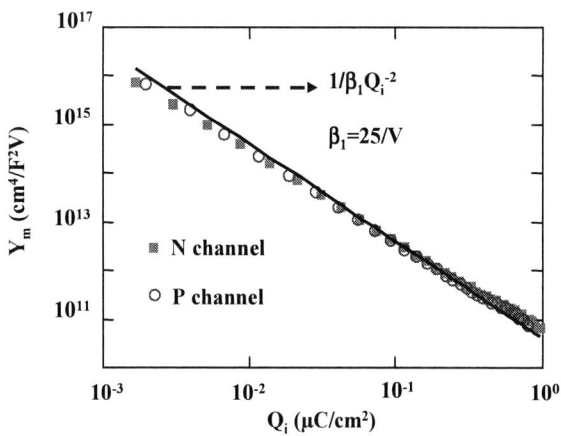

Fig.5. Experimental (symbols) $Y_m(Q_i)$ characteristics for N and P channel (MOS gated diode 10x10μm², C_{ox1}=1.9μF/cm²).

Note the very good agreement between the experimental data (symbols) and the modeling results (lines) obtained with Eqs (8). In particular, the difference between V_{tn} and V_{tp} is well accounted for by the simple model of Eq. (9), emphasizing the consistency of the capacitance analysis. Moreover, according to Eq. (10), the median value between V_{tn} and V_{tp} provides the flat band voltage V_{fb1} (Here, V_{fb1}=-0.15V). Note that the term Q_d/C_{ox1} in Eq. (10) can here be neglected due to the small values of t_{si} and N_a, and, the large value of C_{ox1}.

Instead of the fitting procedure based on Eq. (4) employed in Figs 3 and 4 to extract the front gate oxide capacitance, one may also use a practical method relying on Eq. (3). Indeed, the reciprocal capacitance can be plotted versus the reciprocal inversion charge such that a straight line could be obtained as,

$$\frac{1}{C_{gc}} = \frac{1}{C_{ox1}} + \frac{1}{\beta_1 Q_i} . \qquad (11)$$

Then, as shown in Fig. 8, a linear regression can be used to extract the reciprocal gate oxide capacitance, $1/C_{ox1}$, from the y-axis intercept and the value of β_1 from the inverse of the slope.

An alternative approach could also take advantage of the $Y_m(Q_i)$ property to construct a function which returns a quantity equivalent to the gate oxide thickness as,

$$Thickness(Q_i) = \left(\frac{1}{C_{gc}} - Y_m Q_i \right) \varepsilon_{ox} . \qquad (12)$$

An example of $Thickness(Q_i)$ variation illustrating the direct extraction of the equivalent oxide thickness from the plateau observed for large inversion charge is given in Fig.9, where $t_{ox1} \approx 1.75$ nm is extracted.

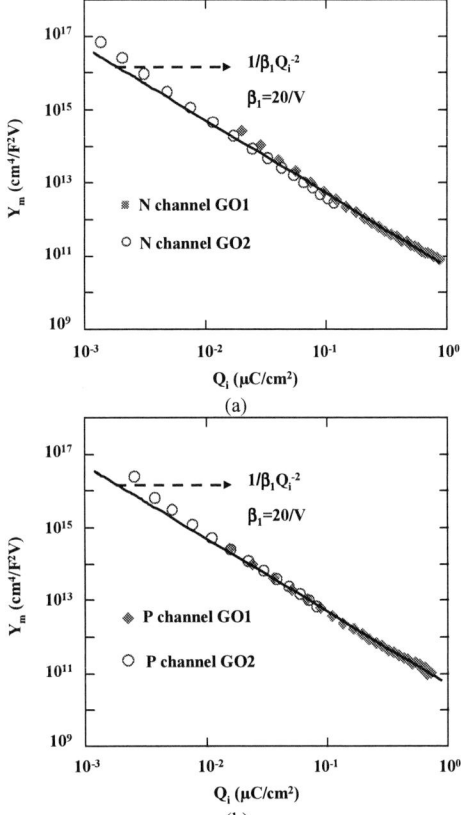

(a)

(b)

Fig.6. Experimental (symbols) $Y_m(Q_i)$ characteristics for N (a) and P (b) channels and for 2 gate oxides GO1 (C_{ox1}=2.4μF/cm²) and GO2 (C_{ox1}=1.0μF/cm²) (MOSFET 10x10μm²).

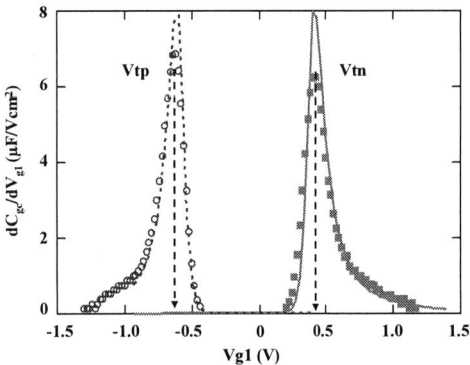

Fig.7. Experimental (symbols) and modelled (lines) variations of $|dC_{gc}/dV_{g1}|$ with gate voltage V_{g1} for N and P channels (MOS gated diode 10x10μm², C_{ox1}=1.9μF/cm²).

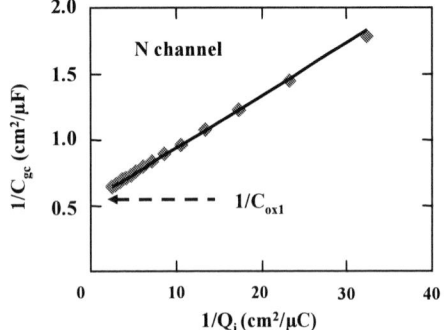

Fig.8. Linear plot of $1/C_{gc}$ versus $1/Q_i$ allowing extraction of gate oxide capacitance (MOS gated diode 10x10μm², C_{ox1}=1.9μF/cm²).

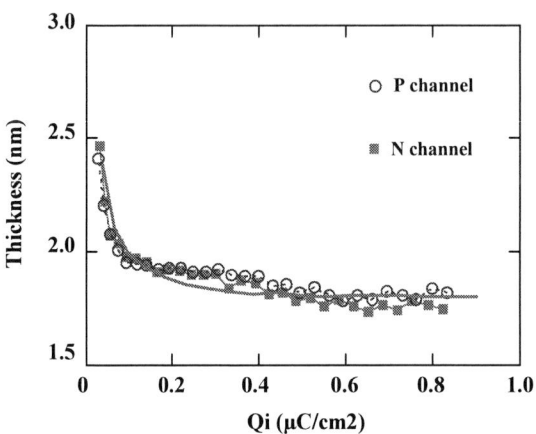

Fig.9. Experimental (symbols) and modeled (line) variations of the equivalent thickness function, *Thickness*, with inversion charge (N and P channel MOSFETs 10x10μm²).

IV. CONCLUSIONS

A new split C-V based parameter extraction methodology has been proposed for FDSOI MOS devices. First, a detailed theoretical analysis for the capacitance has been performed showing the utility of the Maserjian function. Split C-V measurements have been carried out on various FDSOI CMOS technologies and showed that the Maserjian function exhibits a power law dependence with inversion charge as $\propto Q_i^{-2}$, whatever the carrier type and gate oxide thickness. This feature confirms the validity of the established two-parameter capacitance model and allows for a reliable MOSFET parameter extraction in FDSOI devices.

ACKNOWLEDGEMENTS

This work has been partly supported by Reaching 22 Catrene project and Labex on nanoelectronics MINOS.

REFERENCES

[1] C. Fenouillet-Beranger et al, "Fully-depleted SOI technology using high-k and single-metal gate for 32 nm node LSTP applications featuring 0.179 μm² 6T-SRAM bitcell," Proc. IEDM'07, p. 267.

[2] O. Weber et al, "High immunity to threshold voltage variability in undoped ultra-thin FDSOI MOSFETs and its physical understanding," Proc. IEDM'08, p. 245.

[3] N. Sugii et al, "Comprehensive study on Vth variability in silicon on Thin BOX (SOTB) CMOS with small random-dopant fluctuation: Finding a way to further reduce variation," Proc. IEDM'08, p. 249.

[4] G. Ghibaudo, et al, "Improved method for the oxide thickness extraction in MOS structures with ultra thin gate dielectrics", IEEE Transaction on Semiconductor Manufacturing, 13, 152 (2000).

[5] A. Tsormpatzoglou et al, "Analytical modelling for the current–voltage characteristics of undoped or lightly-doped symmetric double-gate MOSFETs", Microelectronic Engineering, 87, 1764 (2010).

Methodology for Extracting the Characteristic Capacitances of a Power MOSFET Transistor, Using Conventional On-Wafer Testing Techniques

C. Kerner, I. Ciofi, T. Chiarella, S. Van Huylenbroeck

IMEC vzw, Kapeldreef 75, B-3001 Leuven, Belgium,

Tel;.: +32 16 28 8035, e-mail: kernerc@imec.be

Abstract — **A methodology for extracting the characteristic reverse transfer-, input- and output-capacitance on power MOSFET transistors is presented in this work. We show that by using standard CV setup and measurement techniques, these dynamic characteristics can be obtained from separate measurements of the three capacitance components: Gate-to-drain, gate-to-source and drain-to-source capacitances. Our method is validated against industry-like approaches, using dedicated complex circuits and procedures. The advantage of our approach lies in its simplicity, flexibility and applicability to common electrical testing equipment and holds both for wafer level and package level characterization.**

I. INTRODUCTION

The increasing requirement for electrical testing on many different-purpose MOSFET devices demands both flexibility and standardization of measurement equipment and techniques in order to collectively meet these demands. During the drive to finalized product phase, accurate and flexible electrical evaluation on wafer-level is required. In this work, usage of conventional CV extraction techniques is investigated for evaluating the reverse transfer-, input- and output capacitances of power MOSFET's. In comparison, industrial standards typically consider direct measurement of these capacitances using a dedicated and complex setup [1,2].

II. BACKGROUND AND METHODOLOGY

A. Background

A power MOSFET device, as depicted in Fig.1a [3,4], acts as a high speed switch, intrinsically limited only by its internal capacitances which get charged and discharged during device operation. Datasheets of regular powerFETs usually state the input-, output- and reverse-transfer capacitance [5], respectively Ciss, Coss and Crss, typically for f=1MHz, Vgs=0V (gate turned off) and Vds=25V (drain bias). While these reflect the device behavior, physically, they can be better understood by their capacitance components, the gate-to-drain (Cgd), gate-to-source (Cgs) and drain-to-source (Cds) capacitances, through the following relationships

$$Ciss = Cgd + Cgs,$$
$$Coss = Cgd + Cds,$$
$$Crss = Cgd \ (Miller \ \text{capacitance}).$$

These components, as illustrated in Fig.1b, depend on the specific design of the device and the biasing [3]. The Cgs component constitutes the overlap capacitance between the polysilicon gate and metal source electrodes and underlying highly doped N+ source and P-body regions. The capacitance arising from

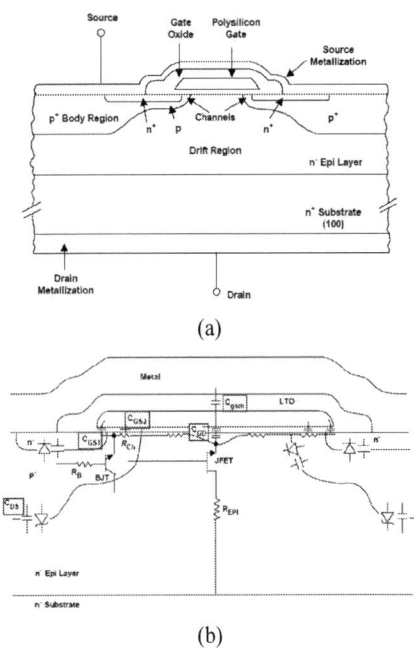

(a)

(b)

Fig. 1. Power MOSFET [4] (a) cross-section view, (b) parasitics.

the P-N junction between the source and drain electrodes is reflected by Cds. The *Miller* capacitance, Cgd, is the series combination of the gate-oxide capacitance and the depletion capacitance and influences all three CV circuits (Ciss/Coss/Crss). Varying the drain bias changes the depletion region and diode response while the gate bias controls the inversion mechanism. Cgd has the largest impact at low bias and becomes effectively zero at high drain bias when the device is in inversion, blocking the gate-to-drain path, thus leaving Cgs the main contributor to Ciss at Vds=25V. Here, the charging of these *cold* capacitances (Vgs remains 0V canceling out drain inversion impact on the circuit) through the drain bias, should be accurately reflected in partial or global measurement for all three circuits. Due to the thick gate-oxide here, gate leakage or breakdown issues are not of consideration for accurate measurements.

B. Methodology

Industry rated power MOSFETs have specific measurement procedures associated with their performance specs. Capacitances are for instance described by the JEDEC [1] and MIL [2] standards but lack detail on measurement circuit and configuration. These setups are not commercially available for R&D purposes or for on-wafer evaluation and not straight-forward to implement. Funaki et. al [6] present a circuit that allows the (LCR meter) measurement on three-terminal power devices but still induces a fair level of complexity. We approach this challenge from an electrical wafer testing point of view with standard (commercially available) equipment and circuitry. Setup calibration is relaxed and complexity minimized. Our setup is shown in Fig.2, involving a PA300 probestation, a HP4284 LCR meter

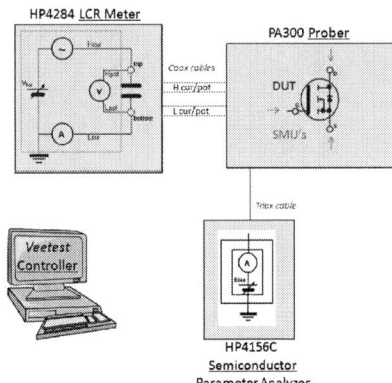

Fig. 2. Measurement setup comprising probestation, LCR meter and semiconductor parameter analyzer.

and HP4156C semiconductor parameter analyzer. A simple *Veetest* software program controls the LCR meter, setting the bias and frequency conditions, but is not crucial for effective implementation. Capacitance (C) and dissipation factor (D) values are extracted from impedance measurements via the LCR meter, run through the two-element equivalent circuit model of the tool [7]. For each measurement, two device terminals are connected to the Hi/Lo inputs of the LCR meter and the third terminal is either biased or grounded externally (to the LCR) by means of the semiconductor parameter analyzer in stand-by mode. This prevents the use of expensive high-voltage splitters or damaging the LCR meter.

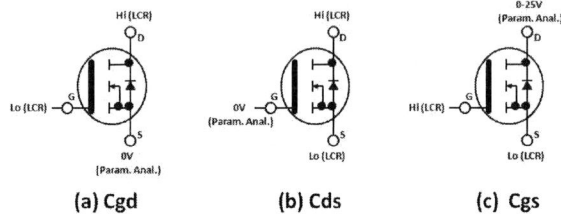

(a) Cgd   **(b) Cds**   **(c) Cgs**

Fig. 3. Implemented test circuits for (a) Cgd, (b) Cds and (c) Cgs.

Cabling between probestation and measurement equipment involves coax/triax connections. Consequently, we define the three capacitance circuits in Fig.3, namely the component CV's, between which we switch by simply changing the connections to the DUT. For (a) Cgd, the drain and gate are put on the Hi/Lo LCR meter input, and the source electrode is grounded via the parametric analyzer. Similarly, for (b) Cds, drain and source form the Hi/Lo of the LCR meter, and the gate terminal is set to ground. Finally, (c) Cgs is realized between gate and source (Hi/Lo) with the drain biased at 25V through the semiconductor parameter analyzer. While in principle, a one point measurement is sufficient for each circuit to obtain the capacitance at Vds=25V and Vgs=0V (spec condition) for quick extraction of Ciss/Coss/Crss, we are interested in the full CV response versus drain bias to test the validity of the measurement approach and device physics considerations. The quality of the measurement itself is assessed through monitoring the frequency and bias dependency of the C and D parameters together with the OPEN parasitic capacitance of each configuration [7]. Last, we put our procedure against and industrial setup.

III. MEASUREMENTS AND DATA DISCUSSION

A commercially available APT6060 BNR [8] N-Channel Enhancement Power MOSFET was used as DUT. Both packaged and bare-Si piece samples were used, but solely the

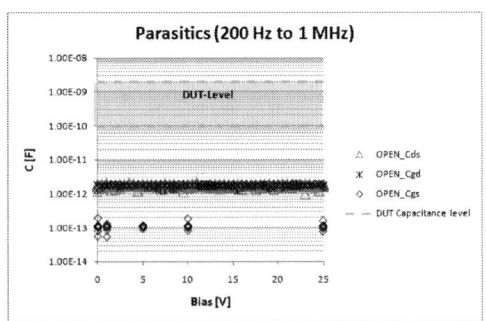

Fig. 4. OPEN parasitic capacitance vs. bias for f=200Hz to 1MHz, for each measurement circuit.

data extracted on Si-level is reported, since its behaviour was similar to the packaged device. For packaged device measurement, we replace the probestation by a test fixture box (HP16442 A).

978-1-4673-1707-8/12 $31.00 © 2012 IEEE   222

A. OPEN Measurements

As initial check of the setup influence and parasitics, the OPEN parasitic capacitance was extracted on each circuit. The measured capacitances are depicted in Fig.4, showing low and effectively constant values versus bias and frequency. The data over all frequencies was overlaid and DUT capacitance levels (much higher) indicated. For Cgd and Cds, the parasitic capacitance was below 1.9pF, whereas for Cgs, a value below 0.19pF was extracted.

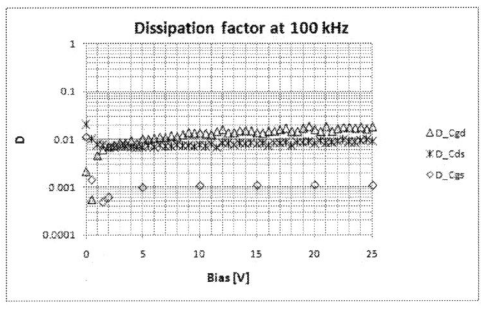

(a)

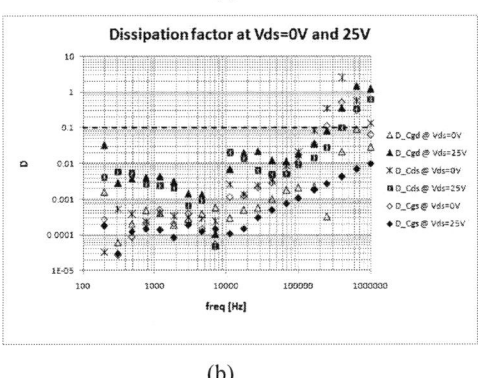

(b)

Fig. 5. Dissipation factor D (a) at f=100 kHz vs. Vds and (b) at Vds=0V, 25V vs. frequency.

B. Capacitance and Dissipation Measurement of CV component circuits (DUT)

We measured 20 frequencies (logarithmic step) from 200Hz to 1MHz versus drain bias and with zero applied gate bias. The *AC* bias component was kept at 30mV after checking that it was not impacting the measurement. The series capacitance and dissipation factor were recorded during the LCR meter sweep, using the Cs-D equivalent circuit model and the capacitances corrected for the corresponding OPEN parasitic capacitance values. A value of D, consistently below 0.1 and corresponding stable C value vs. frequency indicated accurate measurement [7]. We plot D for each circuit at f=100kHz versus drain bias in Fig.5a, showing a stable value with bias and below the limit of 0.1. The D values at Vds=0V and 25V versus frequency are shown, respectively, in Fig.5b, indicating accuracy for frequencies up to and beyond 100kHz overall but an increase in value, i.e. possible loss of measurement accuracy, when approaching the 1MHz domain. This is simply related to the response of the measurement circuit and does not reflect actual frequency-dependency of

the DUT. The increase in capacitance when approaching 1MHz may also simply be due to parasitic inductance. Its

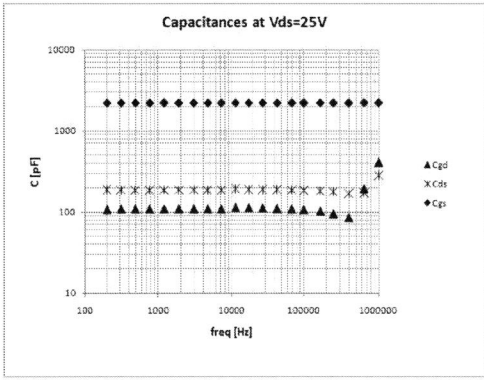

(a)

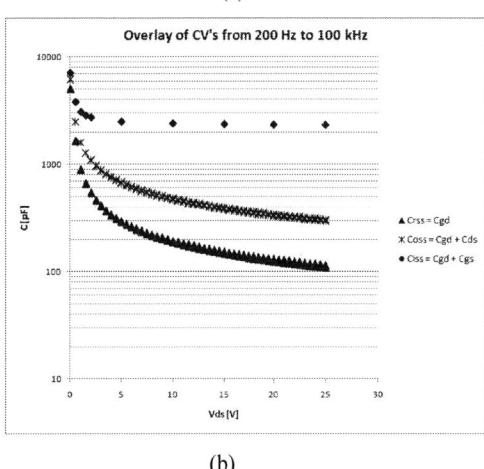

(b)

Fig. 6. DUT capacitance (a) at Vds=25V vs. frequency and (b) extracting the CV circuits, from 200Hz to 100kHz.

trend is shown in Fig.6a, indicating a stable value at Vds=25V vs. frequency until the vicinity of f=1MHz.

C. Extraction of Crss, Coss and Ciss

A frequency limit of 100kHz was chosen and the global circuit CV's, Crss, Coss and Ciss extracted, from the measured Cgd, Cds and Cgs circuits and overlaying all data sets between f=200Hz and 100kHz. Fig.6b shows that the extracted CV's overlay for each circuit and for the selected frequencies. Consistent trends versus drain bias and thus, accurate CV measurements were performed, optimal below f=1MHz for our setup.

D. Benchmark against industrial method

In order to validate our measurement approach and assumed device behaviour, we benchmarked the obtained data against a set of reference data obtained using a technique relying on voltage division at 10kHz [9]; low frequency used to avoid the effect of parasitic inductance in the setup. Both partial and global circuit data match the reference data, shown in Fig.7a and 7b, respectively. Data sets at f=100kHz were

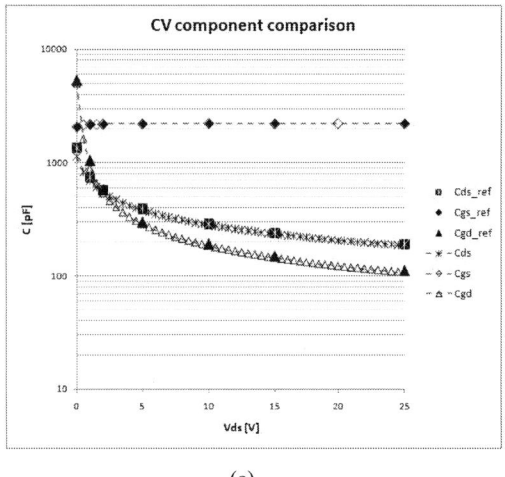

CV component comparison

(a)

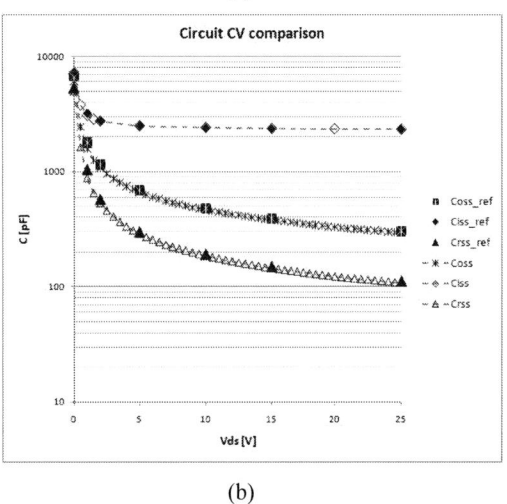

Circuit CV comparison

(b)

Fig. 7. Comparison of data vs. industrial reference for (a) the component and (b) the full circuit CV's.

selected here from our extractions whereas the reference was obtained at f=10kHz. The frequency choice, as usually stated in the datasheets of power MOSFETs, reflects a point for consistent/accurate measurement rather than specific device behaviour. This allows us to overlay the reference data at f=10kHz with our data taken at f=100kHz, where we obtain high measurement accuracy. Also, impact of device leakage currents or series resistance in general was negligible and thus not affecting the CV measurements. Further, Funaki et. al [6] also get highest measurement accuracy for frequencies up to 300 kHz with their circuit and find low impact of employed measurement circuit. Fig 8a shows that all compared data also match well in the low bias region, verifying both the involved procedures and related impact or assumptions on the device behavior. Partial measurement and global measurement pick up similar charging trends versus drain bias. The spec sheets of power MOSEFT devices typically state the Ciss, Coss and Crss values at Vds=25V, rather than full curves. The extracted values at this drain bias, plotted in fig. 8b are comparable for both procedures.

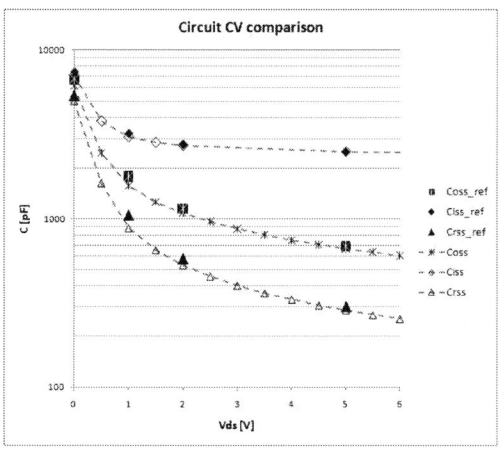

Circuit CV comparison

(a)

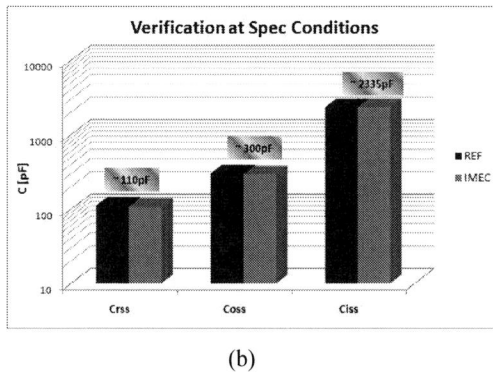

Verification at Spec Conditions

(b)

Fig. 8. Comparison of capacitance data with reference for (a) low Vds and (b) at spec, i.e. Vds=25V (Vgs=0V).

IV. CONCLUSIONS

In summary, an accurate method for extracting on-wafer the reverse-transfer-, output- and input-capacitance using conventional equipment with simplified procedure was presented. We validated our method versus an industrial procedure, finding both approaches to match well. This allows for implementation of our method for standard R&D type wafer testing during power MOSFET development. Future work is planned to widen the study by investigating different power MOSFET devices (in terms of architecture, switching behaviour and CV range) and to extend the benchmarking.

REFERENCES

[1] JDEC Standard JESD24;
http://www.jedec.org/sites/default/files/docs/jesd24.PDF .

[2] Military Standard MIL-STD-75-; http://snebulos.mit.edu/projects/reference/MIL-STD/MIL-STD-750D-Notice5.pdf .

[3] J. Dodge, Advanced Power Tech., Power MOSFET Tutorial, Appl. Note APT-0403 Rev B, http://www.microsemi.com/en/sites/default/files/micnotes/APT0403.pdf .

[4] V. Barkhordarian, Int. Rectifier, Power MOSFET Basics, http://www.irf.com/technical-info/appnotes/mosfet.pdf .

[5] Datasheet example for an Advanced Power technology (APT) device, http://www.datasheetcatalog.org/datasheet/

978-1-4673-1707-8/12 $31.00 © 2012 IEEE

AdvancedPowerTechnology/mXqyrrw.pdf

[6] Funaki et. al, IEEE Transac. On Power Electronics, Vol.25(6), 2009, http://ieeexplore.ieee.org/stamp/ stamp.jsp?arnumber=05062317 .

[7] I. Ciofi at. al, Microelectr. Eng., Vol. 87, 2010.

[8] Commercial device APT6060BNR, available at http://www.icelect.com/search2.php?BUY=APT6060BNR .

[9] Reference data courtesy of D. Van Zyl, AEI, Fort Collins, Colorado, CO 80525, USA.

A Gate Modulated Avalanche Bipolar Transistor in 130nm CMOS Technology

Robert K. Henderson, Eric A. G. Webster, Richard Walker
Institute for Integrated Micro and Nano Systems
School of Engineering, The University of Edinburgh,
King's Buildings, Mayfield Road, Edinburgh, EH9 3JL, UK

Abstract —A novel Geiger-mode avalanche bipolar transistor structure is realized in a 130nm, low-voltage CMOS technology. A MOS transistor formed within the base region of the device allows gate modulation of the output pulse rate. In bipolar operation, the device generates Poisson-distributed digital output pulses at rates from 1kHz to 20MHz, linearly related to emitter currents in the range 10nA to 1μA. In MOS operation, the mean pulse rate varies exponentially over 4-5 decades as the gate voltage changes by 300mV and consumes less than 180μA drain-source current. The gate input eliminates the input current of the avalanche bipolar transistor, enabling capacitive sensor interfaces and direct device-level, analogue-digital conversion. The device is fully compatible with low-voltage CMOS circuits and standard digital process steps.

I. INTRODUCTION

Avalanche bipolar transistors capable of counting single electrons were recently demonstrated in a 0.35μm high voltage CMOS process [1]. The so-called single electron bipolar avalanche transistor (SEBAT) is derived from the single-photon avalanche diode (SPAD) and early avalanche transistor implementations [2][3]. In SPADs, avalanche breakdown is triggered by photoelectrons entering a high field region [4]. Breakdown is rapidly halted by a quenching circuit, generating large Geiger-mode voltage pulses which can be processed by standard digital logic circuits. In SEBATs, single electrons are injected by an emitter-base junction into the collector-base junction, triggering avalanche pulses. Applications investigated include sigma-delta analogue-digital conversion and bio-impedance sensors [5].

In this paper, we propose a Gate Modulated Avalanche Transistor (GMAT) by constructing a MOS transistor within the base region of a SEBAT. The GMAT modulates the output pulse stream by field effects and dispenses with the requirement of a base-emitter input current. Based on the SPAD of [6] and bipolar [7], this is the first avalanche transistor implemented in low-voltage CMOS technology [8]. Resistive quench and capacitive level shift circuits render all input and outputs of the GMAT compatible with conventional supply voltages [9]. The GMAT offers a device-level analogue to digital interface for capacitive sensors, ion-sensitive field-effect transistors and switched-capacitor ADCs.

II. DEVICE DESIGN

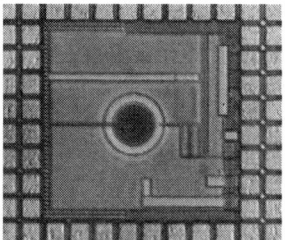

Figure 1. GMAT device micrograph

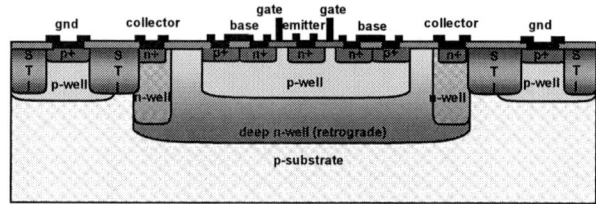

Figure 2. Cross-section of the GMAT

Fig. 1 shows a photomicrograph of the GMAT test structure implemented in a 130nm CMOS image sensor process. The device cross-section in Fig. 2 shows an 8.5μm p-well active diameter within a 19.5μm circular device. P+ contacts are arranged at the p-well periphery. The p-well is surrounded by a region of p-well block, creating a guard ring within the retrograde doping profile of the deep n-well, and constitutes a SPAD structure [6]. A circular, 3.3V-compatible, double-oxide MOS transistor is drawn at the centre of the p-well. This device has a 2.5μm gate diameter (~7.85μm width) and 0.5μm length. The central n+ source/drain diffusion also forms the emitter of a SEBAT, with base as the p-well and collector as the deep-n-well. P-well is also blocked from around the periphery of the n-well to prevent lateral p-well/n-well diode breakdown as the n-well is subjected to high voltages to activate the p-well/deep-n-well high field region as in [9]. STI is blocked from the active region of the device. A

978-1-4673-1707-8/12 $31.00 © 2012 IEEE

400kΩ unsalicided polysilicon resistor is used to quench avalanche breakdown.

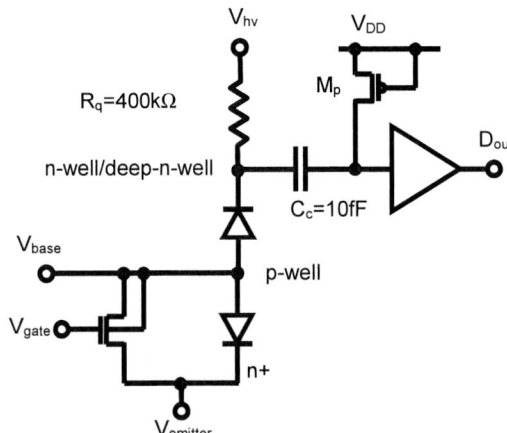

Figure 3.   GMAT device and interface circuitry

The GMAT is coupled to a digital buffer via a 10fF metal-oxide-metal capacitor. This component level shifts the high voltage avalanche pulses (around 15V) to the low voltage supply of the logic circuits (1.8V). An off-PMOS is used to provide a sub-threshold leakage current to initialize the buffer input to a high logic state [9]. Fig. 3 shows the equivalent circuit of the GMAT and its interface circuitry. The NMOS transistor is placed in parallel with the base-emitter junction of the SEBAT. The collector-base diode formed by the p-well/deep n-well undergoes avalanche breakdown induced by electrons injected from the emitter.

III.   AVALANCHE BIPOLAR OPERATION

The operation of the device as a SEBAT is described first. In this mode the gate voltage V_g=0V and the emitter is grounded. Onset of Geiger mode pulsing occurs at a collector voltage V_{hv}=14.35V. with a supply voltage V_{dd}=1.8V. The device is operated at V_{hv}=15.25V with a mean dead time of 20ns and a median count rate of around 1kHz. The test setup is shielded from light during all measurements. Fig. 4 shows the plot of average count rate with base voltage V_b at several excess bias voltages. The device shows a clear exponential increase of dark count over 3-4 decades to a maximum of around 20MHz. The noise floor of this device prevents observation of this trend below V_{be}=0.1V. In previous trials, the SPAD structure has yielded a median dark count rate of 9Hz, leading us to believe a significant increase in dynamic range is attainable [6].

Fig. 5 shows the flow of current in the SEBAT, measured as the device operates in Geiger mode. Superimposed is an estimate of the capacitive current due to the avalanche rate based on a 40fF junction capacitance and a 1.8V excess bias. Above Vbe=0.4V the count rate decreases rapidly as well as the emitter current. A constant base and quench resistor current of around 3μA continues to flow, reducing the collector voltage by 1.3V and causing the base-collector junction to operate in linear avalanche mode (or in Geiger mode with reduced voltage excursions which are unable to

trigger the CMOS inverter). The emitter injection efficiency of our device is very low, estimated at around 2e-4%. This is thought to be due to the relatively thick base region with high doping.

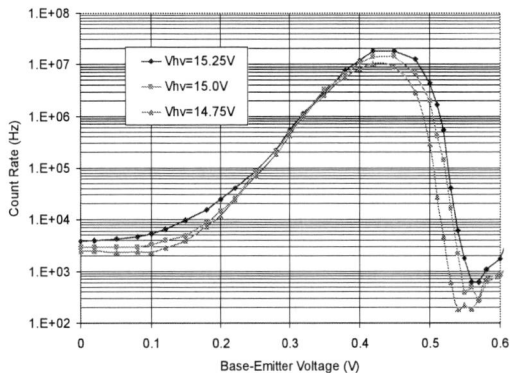

Figure 4.   Average pulse rate versus base emitter voltage

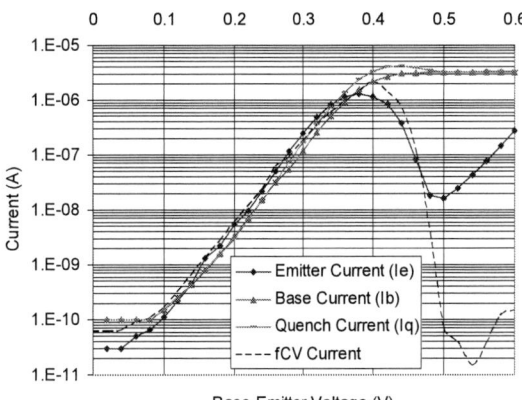

Figure 5.   Current flow in SEBAT mode

IV.   GATE MODULATED OPERATION

Gate modulated operation of the GMAT is illustrated in Fig. 6. The base-emitter junction is biased at a fixed potential obtaining the mean pulse rate at V_g=0V from Fig. 4. As the gate voltage is swept from 0V to 3V the count rate is decreased from the initial rate by around 2 decades commencing when the gate-base voltage exceeds 1.5V. Fig. 7 shows the emitter current variation with gate voltage at several base-emitter voltage conditions. When the gate-base voltage is below threshold (0.6V) the emitter current corresponds to the level measured in SEBAT operation in Fig. 5. As the MOS transistor is turned on, current starts to flow from base to emitter through the channel. The current attains a value of a few 100μA in the triode region at high gate-base voltages. Fields induced by the gate bias voltage divert carriers to flow at the surface of the device rather than at depth where may enter the high field region and cause avalanche breakdown.

978-1-4673-1707-8/12 $31.00 © 2012 IEEE   227

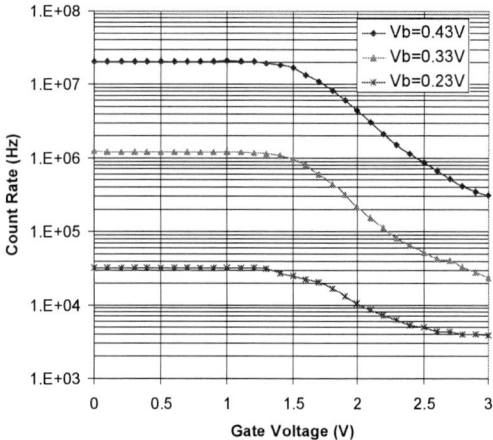

Figure 6. Gate modulated count rate at various base voltages

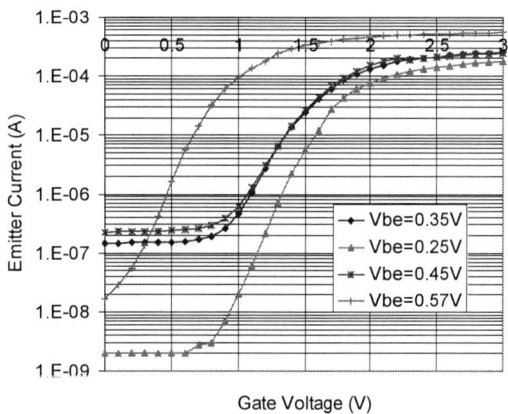

Figure 7. Emitter current variation in GMAT operation

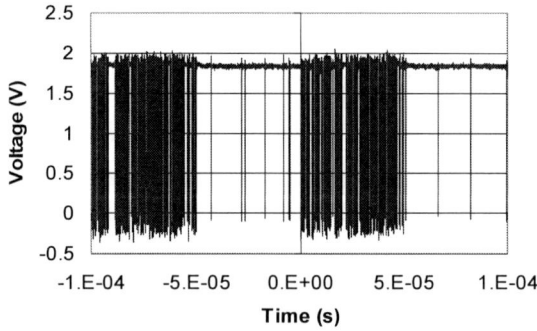

Figure 8. Modulated pulse stream for a 10kHz 1V to 3V square wave applied to the gate voltage and Vbe=0.35V

Fig. 8 shows the response of the GMAT to a square wave input. Two Poisson rates are induced by each gate voltage.

A more sensitive mode of operation is observed by varying the gate voltage when the base-emitter voltage is set at a value between 0.5V and 0.6V. In this condition, the Geiger-mode pulse rate has dropped to a rate below the noise count rate at V_{be}=0V. The device is in a condition where the excess bias voltage has dropped close to breakdown due to a continuous current flow of around 3μA across the 400kΩ quench resistor to the base. The emitter current has also dropped indicating the forward bias level of the emitter-base junction is reduced. This suggests a lower potential within the base region around the emitter. As the MOS is turned on by increasing the gate voltage, the current direction in the base reverses and the channel carries current from base to emitter. Fig. 9 shows that the pulse rate varies over around five decades from 100Hz to 10MHz as the gate voltage 1.0V to 1.3V. Increasing V_{hv} decreases the gate voltage required to cause the transition to count rates.

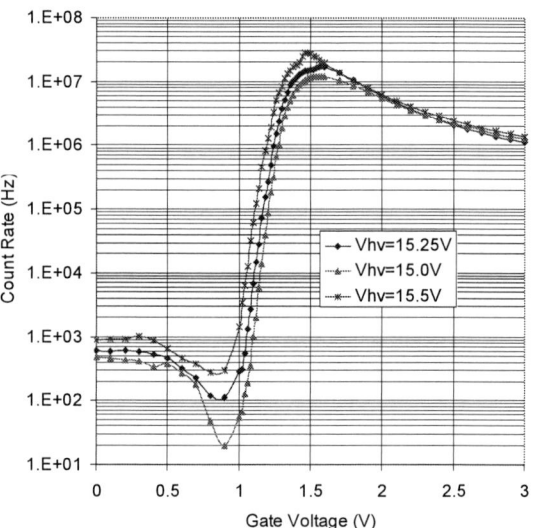

Figure 9. Gate modulated pulse rate for V_{be}=0.57V

At gate voltages above 1.5V, it can be seen from Fig. 9 that the count rate declines steadily with a slope similar to the plots in Fig. 6. Variation of V_{hv} causes little change in the count rates in this region of operation. The current consumption of the device is higher than in the around 180μA in the region of most rapid transition.

V. NOISE PERFORMANCE

By introducing a MOS gate into the SEBAT it is expected to introduce 1/f noise into the device rather than simply shot noise. Fig. 10 and Fig. 11 show histograms of pulse inter-arrival times in two different bias conditions. In Fig. 10 the mean arrival rate is around 500kHz and in Fig. 10 around 10MHz. The histograms for low count rates conform to the expected exponential form expected from the Poisson distribution. As the mean count rate approaches the dead time significant distortion is observed in the plot at short inter-arrival times. This leads to a higher standard deviation and lower signal to noise ratio than predicted by Poisson statistics.

978-1-4673-1707-8/12 $31.00 © 2012 IEEE

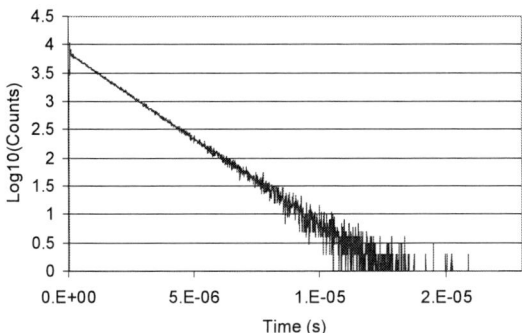

Figure 10. Pulse inter-arrival time histogram at V_{be}=0.57V and V_g=1.1V

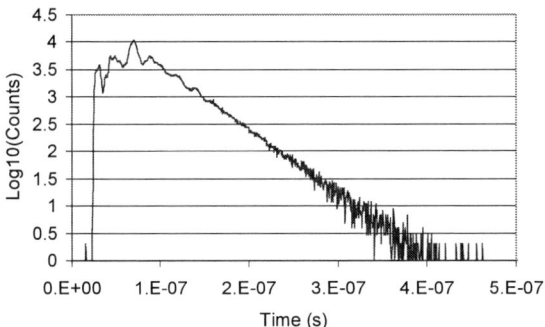

Figure 11. Pulse inter-arrival time histogram at V_{be}=0.57V and V_g=1.3V

We define an excess noise factor F in order to study the departure from ideal Poissonian statistics as:

$$F = \frac{\sqrt{\overline{N}}\sigma_N}{\overline{N}} \qquad (1)$$

where N is the pulse rate and σ_N is the measured standard deviation of the pulse rate.

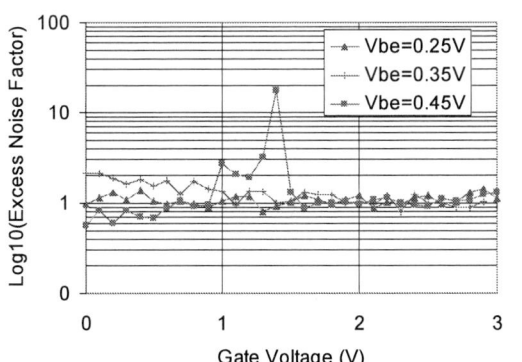

Figure 12. Excess noise factor of the GMAT

Fig. 12 shows the excess noise factor of the GMAT for three base-emitter operating voltages. The excess noise factor

is close to the ideal value of unity in the region of most interest above 1.5V where the gate voltage starts to modulate the output pulse rate. A peak in the excess noise factor is observed close to the onset of gate modulation for high pulse rates (V_{be}>0.45V).and is believed to be due to pulses merging as the dead time approaches the mean pulse rate. Two pulses which merge will trigger the output inverter only once departing significantly from the inter-arrival standard deviation predicted by Poissonian statistics.

VI. CONCLUSIONS

We have shown measured results of a new avalanche device compatible with digital CMOS operating voltages and manufacturing steps. Further TCAD device modeling and theoretical investigation are required. Optimization of the gate, emitter and base dimensions will be undertaken to reduce the current consumption and increase the emitter injection efficiency. Array implementation of GMATs is expected to offer compact, high rate analogue to digital conversion and capacitive sensing solutions for nano-scale CMOS technologies.

ACKNOWLEDGMENT

The authors would like to thank ST Microelectronics for chip fabrication and design support.

REFERENCES

[1] M. Lany, G. Boero, and R. Popovic, "Electron counting at room temperature in an avalanche bipolar transistor," Appl. Phys. Lett., col 92, No. 2, pp. 022111 – 022111-3, 2008.

[2] D. Hamilton, J. Gibbons, and W. Shockley, "Physical principles of avalanche transistor pulse circuits," Proc. IRE, 47, vol. 6, 1959, pp. 1102 – 1108.

[3] J. Carroll, and A. Winstanley, "Transistor improvements using an IMPATT collector," Electronics Letters, 10, (24), 1974, pp. 516 – 518.

[4] A. Rochas, M. Gani, B. Furrer, P. Besse, R. Popovic, G. Ribordy, and N. Gisin, "Single photon detector fabricated in a complementary metal–oxide–semiconductor high-voltage technology," Rev. Sci. Instrum., vol. 74, pp. 3263–3270, 2003.

[5] M. Lany and R. Popovic, "Current and voltage ADC using a differential pair of single-electron bipolar avalanche transistors", IEEE Bipolar/BiCMOS Circuits and Technology Meeting (BCTM), October 13–14, 2009, Capri, Italy.

[6] J. Richardson,.E. A. G. Webster, L. Grant, R. Henderson," Scaleable Single-Photon Avalanche Diode Structures in Nanometer CMOS Technology", IEEE Trans. Electr. Dev., vol. 58, 7, pp. 2028-2035, July 2011.

[7] E. Vittoz, "MOS transistors operated in the lateral bipolar mode and their application in CMOS technology", IEEE J. Solid-State Circuits, vol. 18, No. 3, pp. 273-279, June 1983.

[8] E. A. G. Webster; J. A. Richardson; L. A. Grant; R. K. Henderson, "A single-electron bipolar avalanche transistor implemented in 90nm CMOS", accepted for publication in J. Solid-State Electronics, 2012.

[9] E. A. G. Webster, J. Richardson; L. Grant, D. Renshaw, D.; R. Henderson, An infra-red sensitive, low noise, single-photon avalanche diode in 90nm CMOS, International Image Sensor Workshop (IISW), Hokkaido, Japan, 8-11 June 2011.

Low-noise and large-area CMOS SPADs with Timing Response free from Slow Tails

Danilo Bronzi, Federica Villa, Simone Bellisai, Bojan Markovic, Simone Tisa, Alberto Tosi, Franco Zappa

Dip. Elettronica e Informazione - Politecnico di Milano
Piazza Leonardo da Vinci 32, I-20133 Milano, Italy
franco.zappa@polimi.it

Sascha Weyers, Daniel Durini, Werner Brockherde, Uwe Paschen

Fraunhofer Institute for Microelectronic Circuits and Systems IMS
Finkentraße 61, D-47057 Duisburg, Germany
Werner.Brockherde@ims.fraunhofer.de

Abstract — **This paper reports the design and the characterization of Single-Photon Avalanche Diodes (SPADs) fabricated in a standard 0.35 μm CMOS technology aimed at very low noise and sharp timing response. We present the investigation on the breakdown voltage, photon detection efficiency (PDE), dark count rate (DCR) and timing response on devices with different dimensions and shapes of the active area. Results show uniform breakdown voltage among different structures, PDE above 50% at λ = 420 nm, DCR below 50 cps at room temperature and timing response with no exponential tail and typical full-width at half-maximum of 77 ps and 120 ps for 10 μm and 30 μm active areas, respectively. The fabricated devices enable the fabrication of imagers with CMOS SPAD arrays suitable for advanced applications demanding extremely low noise and picosecond timing accuracy.**

I. INTRODUCTION

Single-photon avalanche diodes (SPADs) [1] are semiconductor devices that have been known for decades and have been exploited in several fields where single optical photons are to be detected: chemistry, physics, biology; laser ranging; optical time-domain reflectometry; single molecule detection; astronomy; and photon correlation techniques.

Best performing SPADs are produced in custom technologies, which allow the complete tailoring of fabrication parameters and processing conditions in order to achieve optimized devices with state-of-the-art performances: low dark count rate (DCR), high photon detection efficiency (PDE) and picosecond timing jitter even with large-area devices [1]. Main drawbacks of custom processing are costs and mainly, the impossibility to monolithically integrate SPADs and their surrounding electronics on the same chip (substrate). Hence the difficulty to fabricate imagers with very large pixel counts without employing wafer-bonding approaches.

Recently, SPADs were successfully implemented also in standard CMOS technologies, thus opening the way to cost-effective production of complete SPAD-based imagers. Major concern with standard processing is the introduction of contaminants and "defects" that, even with no impact on the functionality of the electronics, could drastically impair detector performances, mainly the DCR noise. In fact, lattice dislocations and generation-recombination centers could increase thermal generation, band-to-band tunneling contributions and carrier trapping, thus affecting also afterpulsing and further increasing DCR [2]. Moreover, as the depleted region is usually accommodated in an n-well, the avalanche process is triggered by the photo-generated holes, which have low triggering probability compared to electrons; hence, also PDE and timing jitter result impaired [1].

Different research groups are pushing the development of SPAD imagers in CMOS scaled-technologies [3]–[8], aimed at reaching megapixel chips. Often these approaches resulted in drastic drawbacks in active area dimensions (of just a few micrometers), operating excess bias (of few Volts), DCR (higher than hundreds of counts per second, cps) and PDE (of just few tens % in the visible range).

In this paper, we present the design of novel CMOS SPADs fabricated in a 0.35 μm cost-effective CMOS automotive technology that sets the new state-of-the-art in terms of large active area and very low DCR and sharp timing precision. Such detectors will be the building block for SPAD smart pixels of advanced 2D and 3D imagers with single-photon sensitivity and in-pixel processing.

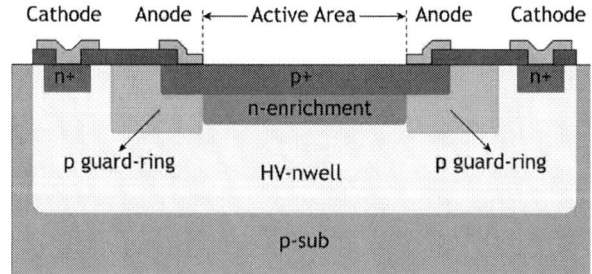

Fig. 1. Cross-section of the SPAD in 0.35 μm CMOS technology. Diameter of the active area varies between 10 μm and 30 μm diameter.

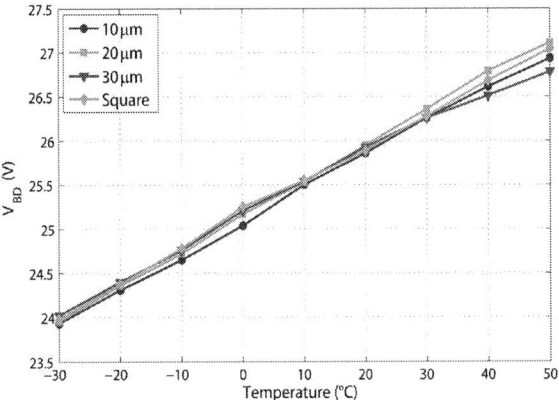

Fig. 2. Breakdown voltage (V_{BD}) as a function of the temperature for different SPAD structures.

II. DEVICE FABRICATION

The SPAD is a p-n junction, reverse-biased well above the breakdown voltage. At this bias, the detector works in the so-called Geiger-mode and a single photo carrier injected into the depletion layer can trigger a self-sustaining avalanche process. As a consequence, a single photon absorbed by the Silicon in which the SPAD is fabricated produces a standard, macroscopic (milliamps), and fast (sub-nanosecond) rising-edge current pulse, which marks the arrival time of the detected photon [9].

The device is monolithically integrated with an active quenching circuit. The structure of the device itself is shown in Fig. 1: a deep low-doped n-well implant and a p+ shallow implant form the p-n junction; the defined high-field region in the active area yields a breakdown voltage of $V_{BD} = 26$ V; a p-doped guard-ring smoothes down the peripheral electric field surrounding p+ defined active area of the SPAD, thus preventing edge breakdown.

III. EXPERIMENTAL CHARACTERIZATION

We performed a full characterization of the four different fabricated SPADs: three of them were circular with 10 μm, 20 μm and 30 μm diameter active area; and one was square-shaped with an active area equivalent to a 22.5 μm circular SPAD. All the measurements, except the ones for the breakdown voltage, were carried out using the active quenching circuit monolithically integrated with the photodetectors.

A. Breakdown Voltage

The breakdown voltage (V_{BD}) was measured as a function of the temperature, using a curve tracer and a climatic chamber. The results show a homogeneous breakdown voltage value among the four structures over the whole temperature range (Fig. 2). The breakdown voltage increases linearly with temperature, with an interpolated dependence given by:

$$V_{BD}(T) = 37.8 \text{ mV/K} \cdot T + 14.8 \text{ V} \qquad (1)$$

Fig. 3. Dark Count Rate of the CMOS SPADs, with different area and shapes, as a function of the temperature and at different applied overvoltage (V_{EX}) with 300 ns hold-off time.

with a $V_{BD} = 26.1$ V at room temperature (T = 300 K).

B. Dark Count Rate

The Dark Count Rate (DCR) was measured at different temperatures and different excess bias voltages ($V_{EX} = V_{POL} - V_{BD}$) with a fixed hold-off time ($T_{HOLD} = 300$ ns) to reduce the detrimental effects of afterpulses on measurements. Fig. 3 shows that dark counts decrease significantly from 50 °C to 0 °C because of thermal generation, whereas at low temperature the main contribution is tunneling (either trap-assisted or band-to-band).

It is possible to gain a deeper insight into the SPAD performances looking at Table I, which reports the most important DCR values and shows that, at room temperature with an overvoltage of 5 V, all the SPAD structures have extremely low dark count rates with a maximum of 43 cps for the biggest SPAD and a minimum of 6 cps for the smallest one. Even at the highest temperature and bias voltage (T = 50 °C, $V_{EX} = 6$ V) the DCR is still very low (<1 kcps) for the small and medium structures, and moderate for the 30 μm SPAD (~2 kcps); at the lowest temperature and bias voltage (T = 0 °C, $V_{EX} = 4$ V) the DCR is negligible. Because of the low DCR the fabricated SPADs can be used without any cooling.

978-1-4673-1707-8/12 $31.00 © 2012 IEEE

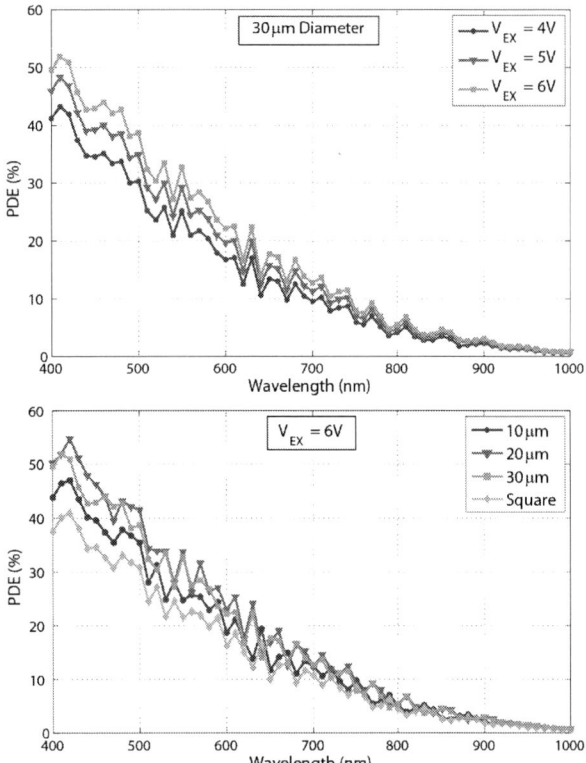

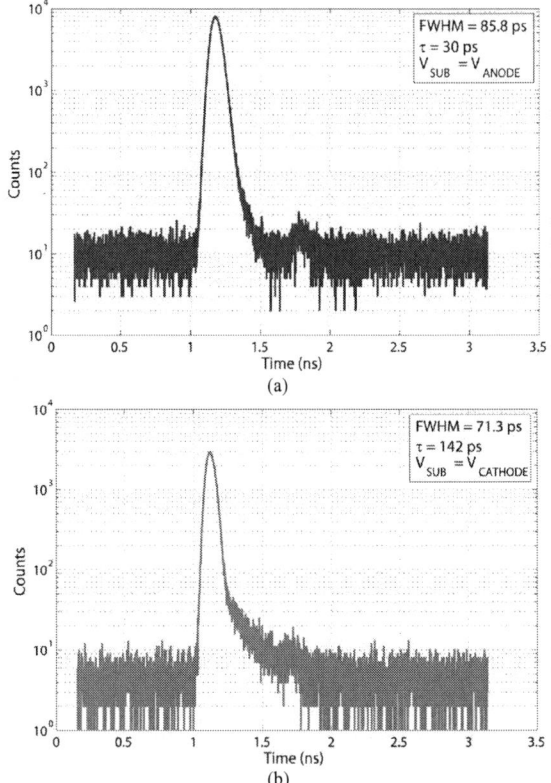

Fig. 4. Photon detection efficency as a function of the wavelength and the excess bias voltage (V_EX) for different structures. The PDE increases with the overvoltage because the triggering probability increases. The 10 μm and square-shaped SPADs have lower efficiency because the guard rings caused a reduction of the effective active area.

Fig. 5. Timing waveforms showing the effect of substrate biasing and wavelength on FWHM and exponential tail. At 390 nm wavelength, when the substrate is kept at the anode voltage, no tail is visible (a), whilst when the substrate is kept at the cathode voltage the exponential decay is visible (b). Both timing waveforms show a secondary peak caused by a reflection of the laser pulse; in the first figure, the faint peak is recognizable, whereas in the second one the peak is hidden by the diffusion tail.

C. Photon Detection Efficiency

The Photon Detection Efficiency (PDE) was measured over the visible range and at different excess bias, by means of a monochromator and an integrating sphere.

Fig. 4 shows the PDE curves of the four analyzed structures: the peak is around 420 nm with a maximum value of 40% for the square-shaped SPAD, of almost 50% for 10 μm diameter area and above 50% for bigger areas. The results show also an enhanced efficiency in the ultraviolet region, and still good values in the near-infrared range. The 10 μm and the square structures have lower efficiency because of a further diffusion of the guard rings into the high-field region that causes the effective active area to shrink. Nevertheless these results are much better than those reported so far [6].

D. Timing Resolution

Timing responses were characterized by Time-Correlated Single Photon Counting (TCSPC) method at three different wavelengths ($\lambda = 390$ nm, 520 nm and 780 nm) using high-repetition rate (80 MHz) mode-locked lasers (Menlo Systems, TC-1550) and a computer board module (Becker & Hickl, SPC-130). The whole system had an overall jitter of 19 ps.

Table II summarizes the measured full-widths at half-maximum (FWHM) as functions of excess bias voltage, wavelength and SPAD area: the timing jitter increases at lower bias voltage, as the triggering probability is reduced; increases on bigger area because the avalanche build-up has a wider statistical spread and moderately varies with wavelength.

TABLE I. DARK COUNT BENCHMARK VALUES OF DIFFERENT SPADS. V_EX = 5V AND ROOM TEMPERATURE IDENTIFY THE STANDARD CONDITION. MINIMUM VALUES AND MAXIMUM VALUES ARE REFERRED RESPECTIVELY TO LOWEST AND HIGHEST OVERVOLTAGE AND TEMPERATURE VALUES.

Diameter	Dark Count Rate (cps)		
	Standard Condition	Minimum Values	Maximum Values
10μm	6	9m	119
20μm	24	60m	590
30μm	43	500m	1920
Square	20	80m	496

978-1-4673-1707-8/12 $31.00 © 2012 IEEE

TABLE II. FULL-WIDTH AT HALF MAXIMUM MEASURED AS A FUNCION OF THE WAVELENGHT FOR DIFFERENT SPAD AREAS AND SHAPES, WITH 5 V OVERVOLTAGE. FOR THE 30 μm SPAD, ONE MEASURE WAS PERFORMED AT DIFFERENT EXCESS BIAS VOLTAGES (V_{EX} = 4 V, 5 V, 6 V).

Diameter	Full-Width at Half Maximum (ps)		
	390 nm	*520 nm*	*780 nm*
10μm	88.6	77.1	119
20μm	127.6	120.7	140.6
30μm	142.5 (V_{EX} = 4 V) 134 (V_{EX} = 5 V) 112.9 (V_{EX} = 6 V)	120.7	141.5
Square	123.7	116.8	155.2

Fig. 5 (a) shows a 10 μm SPAD timing response which does not display the typical exponential tail. Such feature is obtained thanks to the thick depleted region of reverse-biased cathode-to-substrate junction, which squeezes the neutral layer hence almost avoiding the diffusive effects of photo generated carriers [1], [2], [9].

In order to confirm that tail-less waveforms are caused by neutral layer squeezing, we performed further measurements with SPAD substrate equipotential to either anode or cathode voltage, by using an external quenching circuit. Fig. 5 (b) shows that the exponential tail is clearly recognizable when the substrate is short-circuited to the cathode, i.e. the parasitic diode is slightly reverse-biased and the depleted zone is small enough not to shrink the neutral region. A different FWHM is visible as well, because the squeezed neutral layer causes a higher resistance, which lowers the electric field as the current rises thus worsening the timing jitter.

The very low dark counting rate and the absence of the diffusion tail is extremely welcome when signals either with very fast decays (less than 100 ps) or with very faint peaks following the main one must be recorded, i.e. when very high dynamic range is needed. For instance, in photon migration measurements or time-resolved diffuse optical spectroscopy, detectors with no diffusion tail can boost the measurement dynamic range, hence allowing the detection of late diffusive photons, with very faint intensities compared to the main peak, e.g. 5 decades in 1 ns as in Ref. [10].

IV. CONCLUSION

In this paper we presented the characterization of four SPAD structures (three circular devices with 10 μm, 20 μm and 30 μm diameter active area and a square-shaped device with an active area equivalent to a 22.5 μm circular SPAD), fabricated in a high-voltage 0.35μm CMOS technology. The devices exhibit very low dark count rate at room temperature

and negligible DCR at lower temperature. Due to a shallower depleted zone within the n-well and the n-well p-substrate p-n-junction, the photon detection efficiency reaches the peak around 420 nm with a maximum value of 40% for the square-shaped SPAD, almost 50% for the 10 μm diameter SPAD, and above 50% for the bigger area SPADs, decreasing its value with longer wavelengths. The timing responses are characterized by the absence of slow tails with best timing performances of 77 ps for the 10 μm diameter SPAD and about 120 ps for the other structures. Because of the excellent performances, the fabricated devices are suitable to implement large SPAD-based arrays that could be used for low light level imaging or in those applications that require picoseconds timing resolution and detection of fast signal with slow components.

ACKNOWLEDGMENTS

This work was supported by the "MiSPiA" project, under the ICT theme of the EC 7th Framework Program (FP7, 2007-2013), grant agreement n. 257646.

REFERENCES

[1] S. Cova, M. Ghioni, A. Lacaita, C. Samori, and F. Zappa, "Avalanche photodiodes and quenching circuits for single-photon detection," *Appl. Opt.*, vol. 35, pp. 1956–1963, 1996.

[2] M. Ghioni, A. Gulinatti, I. Rech, F. Zappa, and S. Cova, "Progress in silicon single-photon avalanche diodes," *IEEE J. Sel. Topics Quantum Electron.*, vol. 13, no. 4, pp. 852–862, Jul./Aug. 2007.

[3] L. Pancheri and D. Stoppa, "A SPAD-Based pixel linear array for high speed time-gated • uorescence lifetime imaging," in *ESSCIRC, 2009. ESSCIRC '09*, pp. 428–431, 2009.

[4] C. Niclass, C. Favi, T. Kluter, F. Monnier, E. Charbon, "Single-Photon Synchronous Detection," *IEEE J. Solid-State Circuits*, vol.44, no.7, pp.1977-1989, July 2009.

[5] F. Guerrieri, S. Tisa, A. Tosi, F. Zappa, "Two-dimensional SPAD imaging camera for photon counting", *IEEE Photon. J.*, vol. 2, no 5, pp. 759-774, 2010.

[6] C. Niclass, M. Sergio, and E. Charbon, "A single-photon avalanche diode array fabricated in 0.35 μm CMOS and based on an event-driven readout for TCSPC experiments," *Proc. SPIE*, vol. 6372, p. 63720S, 2006.

[7] D. Stoppa, D. Mosconi, L. Pancheri, and L. Gonzo, "Single-photon avalanche diode CMOS sensor for time-resolved • uorescence measurements," *IEEE Sensors J.*, vol. 9, pp. 1084–1090, Sep. 2009.

[8] C. Niclass, M. Gersbach, R. Henderson, L. Grant and E. Charbon, "A 130nm CMOS single-photon avalanche diode," *Proc. SPIE*, vol. 6766, 2007.

[9] S. Tisa, F. Zappa, A. Tosi, and S. Cova, "Electronics for single photon avalanche diode arrays", *Sensors and Actuators A*, vol. 140, pp. 113–122, 2007.

[10] A. Tosi, A. Dalla Mora, F. Zappa, A. Gulinatti, D. Contini, A. Pifferi, L. Spinelli, A. Torricelli and R. Cubeddu, "Fast-gated single-photon counting technique widens dynamic range and speeds up acquisition time in time-resolved measurements," *Opt. Express*, vol. 19, no. 11, pp. 10735 - 10746, 18 May 2011.

Extreme Temperature 4H-SiC Metal-Semiconductor-Metal Ultraviolet Photodetectors

Wei-Cheng Lien, Albert P. Pisano
Applied Science and Technology&
Berkeley Sensor and Actuator Center,
University of California, Berkeley
Berkeley, CA 94709, USA
Email: wclien@berkeley.edu

Dung-Sheng Tsai, Jr-Hau He
Department of Electrical Engineering
& Institute of Photonics and
Optoelectronics, National Taiwan
University
Taipei 10617, TAIWAN, R. O. C.

Debbie G. Senesky
Department of Aeronautics &
Astronautics, Stanford University
Stanford, CA 94305, USA

Abstract—This work demonstrates high-temperature operation of metal-semiconductor-metal photodetectors (MSM PDs) using lightly Al-doped epitaxial 4H-SiC thin films. The responsivity of the PDs under 325 nm illumination is 0.116 A/W at a 20 V bias at room temperature. The photo-to-dark current ratio of SiC MSM PDs is as high as 1.3×10^5 at 25°C and is 22 at 400°C. The rise time of PDs is increased slightly from 594 μs to 684 μs from room temperature to 400°C. These results support the use of 4H-SiC thin films photodetectors in extreme high-temperature applications.

I. INTRODUCTION

Photodetectors (PDs), especially for ultraviolet (UV) detection, have drawn interest for use in chemical and biological analysis, combustion flame monitoring, and optical communication devices [1-3]. In the past few years, different types of PDs have been developed including photoconductor, Schottky barrier photodiodes, p-n and p-i-n photodiodes, avalanche photodiode, phototransistor, metal-insulator-semiconductor structures (MIS), and metal-semiconductor-metal (MSM) photodiodes [4].

Most operation environments under UV radiation require the PDs to work at elevated temperatures. However, conventional Si-based PDs, with a narrow bandgap of 1.12eV, are limited to low operation temperatures (below 125°C) due to generation of thermal carriers, significant shifts in the optical properties, and device ageing under UV radiation, leading to the deterioration in spectral response [3]. For example, the Si photodiodes exhibit a high dark current density of 10 mA/cm^2 and 10 A/cm^2 at 300°C and 500°C, respectively [5]. Wide bandgap materials, such as diamond, AlN, GaN and SiC, are potential candidates for the UV photodetection at high temperature. However, several obstacles need to be overcome before employing the wide bandgap semiconductor as UV photodetectors. The first limitation is the high dopant activation energy is required for

these materials making the heavily doped layers with different type of dopants are difficult to achieve [6]. The second limitation, which comes with the consequence of the first limitation, is that the lack of reliable Ohmic contact with wide bandgap semiconductor and metals at high temperature [2, 6]. These drawbacks hinder the development of p-n, p-i-n photodiodes and phototransistor by using the wide bandgap materials. On the other hand, MSM PDs are operated based on the two back to back Schottky contacts and thus do not require the Ohmic contact which is beneficial for wide bandgap semiconductors. Furthermore, compared to the p-i-n diode or Schottky barrier photodiode, MSM PDs offer high speed, low capacitance operation and can be readily integrated with field effect transistors (FETs) in a single chip without extra complicated fabrication steps [4, 7]. Therefore, MSM PDs using wide bandgap materials can be considered for UV detections at high temperature.

Among the variety of wide bandgap materials, 4H-SiC has high thermal conductivity (4.9 W cm^{-1} K^{-1}, three times larger than Si), strong chemical bond, high electron saturation velocity (2×10^{-7} cm s^{-1}, two times larger than Si) which enables 4H-SiC PDs to operate at high temperature, high power and high radiation environments with high operation speed [6]. In this paper, we characterize a 4H-SiC MSM PD and demonstrate a photodetection scheme with working temperatures as high as 450°C. This study paves the way for UV light detector applications in harsh, high-temperature conditions.

II. EXPERIMENT

Figure 1(a) is a schematic image of the 4H-SiC MSM PD structures used in this work. The PDs were fabricated on p-type 3-inch research grade 4H-SiC wafers (sheet resistivity < 2.5 Ωcm) with a 7 um lightly p-type Al-doped epitaxial layer with a carrier concentration of 2×10^{15} cm^{-3}. As shown in Figure 1(b), the planar MSM PDs were defined using photolithography, e-beam evaporation and metal lift-off process with active areas of 500×158 μm^2 and utilized 8-μm-wide, 150-μm-long interdigitated 20 nm Cr/150 nm Pd electrodes with 8-μm-wide spacing on the 4H-SiC substrates.

The crystal structure of the films was monitored with transmission electron microscopy, TEM (JEOL JEM-2100F, operating at 200 kV). The transmission spectra were measured with a JASCO V-670 UV-visible spectrometer in the spectral range from 250 nm to 800 nm. Photocurrent was generated under the illumination of a He–Cd laser at a wavelength of 325 nm with laser power density of 1.02×10^4 W/m^2. The Keithley 4200-SCS semiconductor characterization system with Tungsten probe tips were used to measure I-V characteristics of the fabricated PDs. The time-resolved measurements were measured by the data acquisition system (DAQ) with SRS low noise preamplifier and assisted by a mechanical chopper to switch on/off the UV light. For high-temperature characterization, the PDs are heated on hot plate and the device temperature was monitored with calibrated thermocouple.

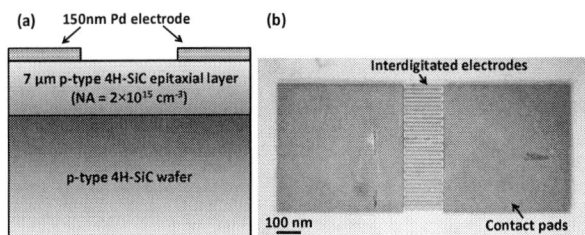

Figure 1. (a) Cross-sectional schematic diagram (b) Optical image of 4H-SiC MSM PD.

III. RESULTS AND DISCUSSION

The cross-sectional TEM image of a 7-um-thick, lightly p type Al-doped epitaxial 4H-SiC films and its corresponding electron diffraction pattern are shown in Figure 2. There is no obvious structural defect such as micropipes observed in the epitaxial thin films. The results suggest that the lightly Al-doped epitaxial 4H-SiC thin films exhibit the good crystallinity and are hexagonal structure as expected.

The transmission spectrum of 4H-SiC substrates is shown in Figure 3. The 4H-SiC substrate shows nearly no absorption of visible/IR signals implying that the as-fabricated 4H-SiC MSM PDs have intrinsic visible-blindness. The optical absorption coefficient (α) can be calculated as follows [8]:

$$\alpha = 1/d \, ln(I_0/I) \quad (1)$$

where d is the thickness of substrate, and I_0 and I are the intensities of the initial and transmitted light, respectively. The optical bandgap (E_g) of 4H-SiC can be calculated based on the Tauc relation [8, 9]:

$$\alpha h\upsilon = A(h\upsilon - E_g)^{1/2} \quad (2)$$

where h is Planck's constant, υ is photon frequency. By extrapolating the linear region (red line in Figure 3) of the

$(\alpha h\upsilon)^2$ versus energy ($h\upsilon$) plot as shown in Figure 3, the optical bandgap of 4H-SiC substrate is approximately 3.23 eV.

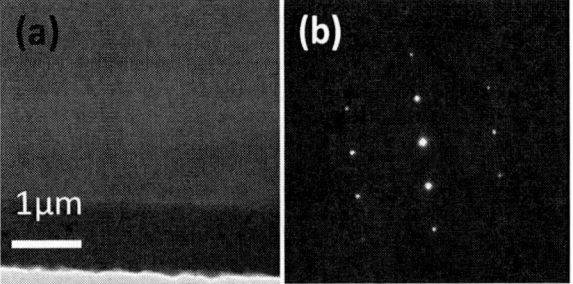

Figure 2. (a) Cross-sectional TEM image (b) Electron diffraction pattern of p-type Al-doped epitaxial 4H-SiC layer with $N_A = 2 \times 10^{15}$ cm^{-3}.

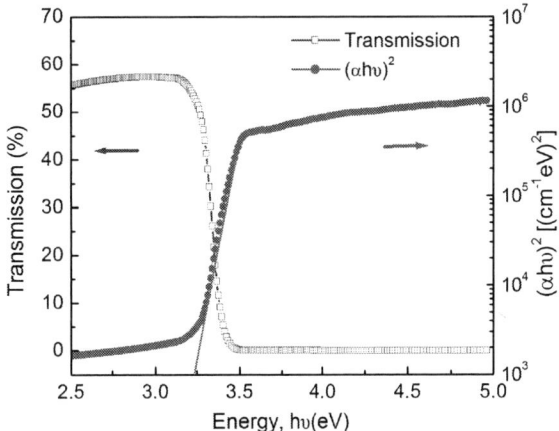

Figure 3. The transmission spectrum (left y axis) and $(\alpha h\upsilon)^2$ (right y axis) vs. $h\upsilon$ plot of 4H-SiC substrates.

The I-V curves of the 4H-SiC MSM PD in the dark and under a 325 nm illumination at room temperature are shown in Figure 4. The dark current of the 4H-SiC MSM PD is approximately 6×10^{-11} A at 5 V bias which corresponds to a leakage current density of 1.5×10^{-7} A/cm^2. The photocurrent of PD is approximately five orders of magnitude larger than the dark current at a 5 V bias. The responsivity (R_i) of PD can be obtained as follows [4]:

$$R_i = I_{photo}/P_{opt} \quad (3)$$

where I_{photo} is photocurrent and the P_{opt} is the optical power which is 4×10^{-4} W in our case. The calculated responsivities under the illumination of 325 nm He-Cd laser are 0.0167 A/W and 0.116 A/W at 5 V and 20 V bias, respectively. The sensitivity factor, photo-to-dark current ratio (PDCR), which is defined as follows [10]:

$$PDCR = (I_p - I_d)/I_d \quad (4)$$

978-1-4673-1707-8/12 $31.00 © 2012 IEEE   235

where I_d is the dark current and I_p is the photocurrent under illumination. The response curve in Figure 5 shows that the 4H-SiC MSM PDs are capable of significant UV light sensing up to 400°C. This is mainly due to the small levels of dark currents and high thermal stability of the 4H-SiC films at high temperatures. More specifically, a further increase in temperature lowers the sensitivity factor of SiC PDs due to an increase in dark current at a higher temperature by generating the thermal carriers which cannot be completely eliminated [10].

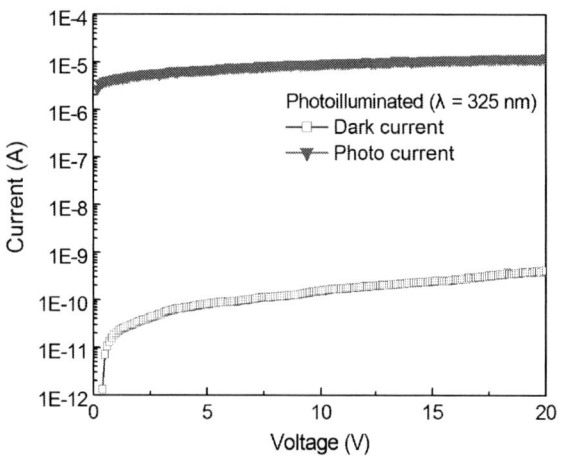

Figure 4. I-V curves of the 4H-SiC MSM PDs measured in the dark and under a 325 nm illumination at room temperature.

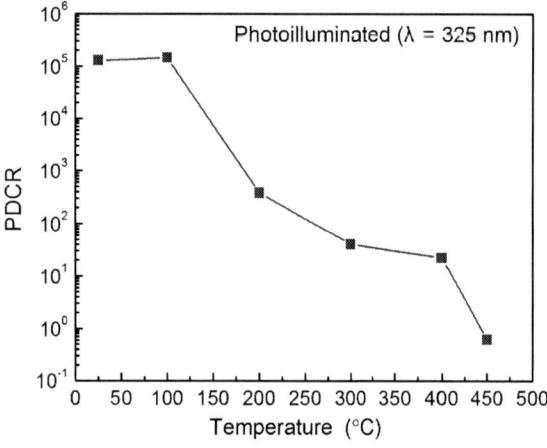

Figure 5. PDCR value as function of temperature under a 5 V bias.

The time dependence of the steady-state photo and dark current at 450°C is shown in Figure 6, one can see that the averaged photo current (2.5×10^{-5} A) is still higher than the averaged dark current (1.55×10^{-5} A) at 450°C but with the small PDCR value of 0.62. However, our setup of high temperature measurement cannot eliminate the thermal noise

generated from the Tungsten probe tips due to the increased contact resistance as temperature increases. Therefore, the contributions of the measured dark current are not only from the thermal carriers of MSM PDs but also from the thermal noise of the probe tips used in the characterization set up. Suitable high temperature ceramic packages [11] or a high temperature probe station might be able to reduce the amount of the thermal noise generated from the probe-testing. Based on the specification from one vendor of a thermal probing system, the thermal noise floor is only 2 fA at 300°C and increases to 10 nA at 600°C [12]. The working limitation of the operating temperature for 4H-SiC MSM PDs can be realized with proper high temperature package and measured under such thermal probing system.

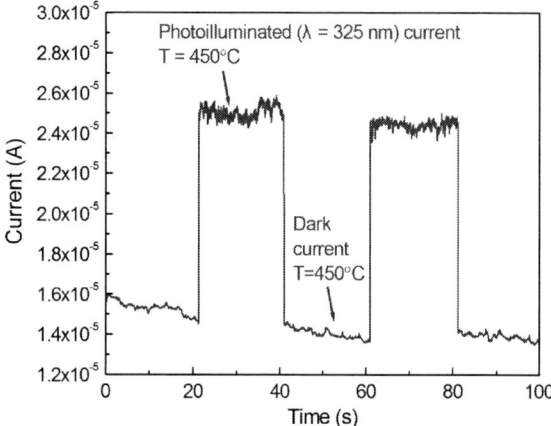

Figure 6. The time dependence of the steady-state photocurrent and dark current at 450°C.

The operation speed of the MSM PDs can be determined by performing the time-resolved measurement. The quantitative parameters of describing how fast the photodetector responds to external light illumination are rise time and fall time. The rise time is defined as the time required reaching the steady-state photocurrent and is the time difference between the 10% and 90% of photocurrent. Fall time means the time required for the decay when the optical excitation is interrupted and is the time difference between the 90% and 10% signals [13]. The time constant is extracted by fitting exponent decay of transient photocurrent as follows:

$$I_{photo}(t) = I_{ss} \, exp \, (-t/\tau_0) \qquad (5)$$

where $I_{photo}(t)$ is the transient photocurrent, I_{ss} is the steady-state current, and t is the time. Figure 7 shows the transient photocurrent of 4H-SiC MSM PDs at room temperature and 400°C. The extracted rise time, fall time and time constant are summarized and listed on Table 1. The rise time and fall time of PD at room temperature are 594 µs and 699 µs, respectively. The rise time and fall time are increased as temperature increases implying that the operation speed of PD is decreased at higher temperature. One possible reason is that the bandwidth of transit-time-limited is proportional to the

978-1-4673-1707-8/12 $31.00 © 2012 IEEE

saturation velocity at fixed spacing between fingers as follows [14]:

$$f_{tr} = (\ 0.441/\sqrt{2})(v_s/s)\qquad (6)$$

where f_{tr} is transit-time-limited 3-dB bandwidth, v_s is the saturation velocity, and s is spacing between electrode fingers. The hole and electron saturation velocity of 4H-SiC is decreased as temperature increases [15]; therefore, the bandwidth is decreased and the speed of the PD is decreased at higher temperature.

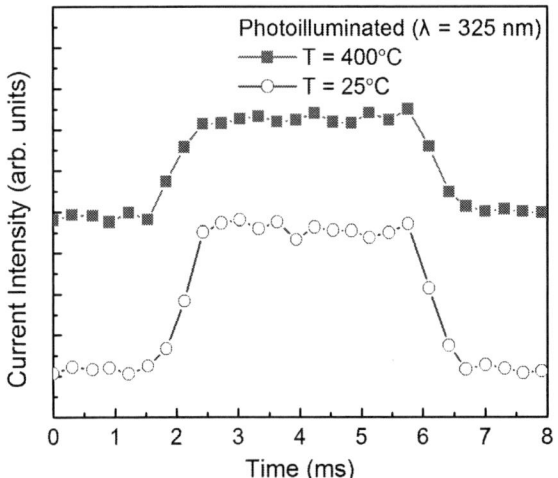

Figure 7. The transient photocurrent of 4H-SiC MSM PD at room temperature and 400°C.

TABLE I.   RESPONSE TIME OF FABRICATED 4H-SIC MSM PHOTODETECTORS

Temperature (°C)	Rise Time (μs)	Fall Time (μs)	Time Constant (μs)
25	594	699	422
400	684	786	441

IV.   CONCLUSION

In summary, the MSM PDs employing lightly Al-doped epitaxial 4H-SiC thin films were successfully fabricated and characterized with working temperatures up to 450°C. The PDCR value is 1.3×10^5 at 25°C and decreases to 22 at 400°C under a 325 nm illumination. However, the operation speed of PDs does not significantly decrease as temperature increases. The rise time of the PD is 594 μs and 684 μs at 25°C and 400°C, respectively. The fall time of PD is 699 μs and 786 μs at 25°C and 400°C, respectively. High temperature packaging engineering and high temperature probe stations with minimum thermal noise is required to fully characterize the working temperature limit of 4H-SiC MSM PDs. This work demonstrates that the 4H-SiC holds promise for the next-generation visible-blind UV PDs for the operation within high temperature conditions.

ACKNOWLEDGMENT

The authors would like to thank D.-T. Lien for his assistance in time-resolved measurements.

REFERENCES

[1] W.-R. Chang, Y.-K. Fang, S.-F. Ting, Y.-S. Tsair, C.-N. Chang, C.-Y. Lin, and S.-F. Chen "The hetero-epitaxial SiCN/Si MSM photodetector for high-temperature deep-UV detecting applications," IEEE Electron Device Lett., vol. 24, pp. 565-567, September 2003.

[2] E. Monroy, F. Omnes, and F. Calle, "Wide-bandgap semiconductor ultraviolet photodetectors," Semicond. Sci. and Tech., vol. 18, pp. R33-R51, April 2003.

[3] A. Vijayakumar, R. M. Todi, and K. B. Sundaram, "Amorphous-SiCBN-Based metal-semiconductor-metal photodetector for high-temperature applications," IEEE Electron Device Lett., vol. 28, pp. 713-715, August 2007.

[4] S. M. Sze and K. K. Ng, Physics of Semiconductor Devices, 3rd ed., Wiley-Interscience: Hoboken, 2007, pp. 663–741.

[5] D. M. Brown, E. T. Downey, M. Ghezzo, J. W. Kretchmer, R. J. Saia, Y. S. Liu, E. G. Gati, J. M. Pimbley, and W. E. schneider, "Silicon carbide UV photodiodes," IEEE Trans. on Electron Devices, vol. 40, pp. 325-333, Feburary 1993.

[6] Stephen E. Saddow, and A. Agarwal, Advances in Silicon Carbide Processing and Applications, 1st ed., Artech house, Inc.: Norwood, 2004, pp. 1-27, 109-153.

[7] S. Assefa, F. N. Xia, S. W. Bedell, Y. Zhang, T. Topuria, P. M. Rice, and Y. A. Vlasov, "CMOS-integrated high-speed MSM germanium waveguide photodetector," Optics Express, vol. 18, pp. 4986-4999, March 2010.

[8] A. A. Ogwu, E. Bouquerel, O. Ademosu, S. Moh, E. Crossan, and F. Placido, "The influence of rf power and oxygen flow rate during deposition on the optical transmittance of copper oxide thin films prepared by reactive magnetron sputtering," Journal of Phys. D-Applied Phys., vol. 38, pp. 266-271, January 2005.

[9] J. Tauc, "Optical properties and electronic structure of amorphous Ge and Si," Mater. Res. Bulletin, vol. 3, pp. 37-46, November 1968.

[10] W.-C. Lien, D.-S. Tsai, S.-H. Chiu, D. G. Senesky, R. Maboudian, A. P. Pisano, and J.-H. He, "Low-Temperature, ion Beam-assisted SiC thin films with antireflective ZnO nanorod arrays for high-temperature photodetection," IEEE Electron Device Lett., vol. 32, pp. 1564-1566, November 2011.

[11] P. G. Neudeck, S. L. Garverick, D. J. Spry, L.-Y. Chen, G. M. Beheim, M. J. Krasowski, and M. Mehregany, "Extreme temperature 6H-SiC JFET integrated circuit technology," Physica Status Solidi A-Applications and Mater. Sci., vol. 206, pp. 2329-2345, October 2009.

[12] "Signatone, Inc.," http://www.signatone.com.

[13] K. Wang, F. Chen, N. Allec, and K. S. Karim, "Fast lateral amorphous-selenium metal-semiconductor-metal photodetector with high blue-to-ultraviolet responsivity," IEEE Trans. on Electron Devices, vol. 57, pp. 1953-1958, August 2010.

[14] J. Kim, W. B. Johnson, S. Kanakaraju, K. S. Karim, "Improvement of dark current using InP/InGaAsP transition layer in large-area InGaAs MSM photodetectors," IEEE Trans. on Electron Devices, vol. 51, pp. 351-356, March 2004.

[15] S. Potbhare, N. Goldsman, A. Lelis, J. M. Mcgarrity, F. B. Mclean, and D. Habersat, "A physical model of high temperature 4H-SiC MOSFETs," IEEE Trans. on Electron Devices, vol. 55, pp. 2029-2040, August 2008.

A silicon photomultiplier with >30% detection efficiency from 450-750nm and 11.6μm pitch NMOS-only pixel with 21.6% fill factor in 130nm CMOS

Eric A. G. Webster, Richard J. Walker, Robert K. Henderson

Institute for Integrated Micro and Nano Systems
School of Engineering, The University of Edinburgh,
King's Buildings, Mayfield Road, Edinburgh, EH9 3JL, UK
e.webster@ed.ac.uk, robert.henderson@ed.ac.uk

Lindsay A. Grant

Imaging Division
STMicroelectronics (R&D) Ltd.
Edinburgh, UK, EH12 7BF
lindsay.grant@st.com

Abstract— A 16×16 Silicon Photomultiplier (SiPM) is reported in a 130nm CMOS imaging technology with a photon detection probability of >30% from 450-750nm. The SiPM demonstrates a 21.6% fill factor with an 11.6μm pitch and 8μm diameter Single-Photon Avalanche Diodes (SPADs). This is achieved using a new SPAD structure with integrated resistor and capacitor. NMOS-only pixel electronics are used to improve fill factor and to implement an addressable array of SPADs that are isolated from the array and column load. A 1T DRAM in each pixel is implemented to inhibit the output of high dark count rate (DCR) SPADs. The SiPM also achieves: a median DCR of ≈200Hz at 1.2V excess bias; low after pulsing; and a SPAD timing jitter of ≈95ps at 654nm with a column delay of ≈100-200ps.

I. Introduction

Silicon Photomultipliers (SiPMs) are arrays of common-cathode connected Single Photon Avalanche Diodes (SPADs) that are used for low light photon counting in nuclear science applications, such as Positron Emission Tomography (PET) and high energy particle physics. SiPMs were developed in the '90's [1] and have seen rapid development since [2]. SiPMs are typically manufactured in custom processes and have detection efficiencies around 30% which peak in the blue-green region [2]. Varying fill factors have been realized of up to ≈80% for large pixel pitches and small numbers of cells. Back-side illuminated (BSI) SiPMs have also been developed that can offer wider spectral responses [2]. Conventional SiPMs output an analogue waveform where the magnitude corresponds to the number of triggered devices.

The first SiPMs implemented in CMOS were recently reported by Phillips as the so-called digital-SiPM (dSiPM) [3]. Integrated electronics is leveraged to provide embedded processing such as defect correction, energy determination, time-of-arrival digitization and communications. The large tile-able sensors have very high (>50%) fill-factor (FF) and pixel size (≈60×30μm). For PET, the detector sensitivity peak of ≈450nm is well matched to the most common scintillator materials.

Smart CMOS SiPM architectures are also enabling on chip fluorescence lifetime calculation, however the full CMOS pixels with isolated SPADs offer 10% FF and 22.5μm pitch [4]. New circuit techniques such as the shared deep n-well approach [5], offer improved fill factors with NMOS only pixels of 20.8% FF at 25μm pitch [6]. Shared well approaches enable 70% FF at 25μm pitch with CMOS pixel electronics outside the array for time-of-flight applications [7]. However, fluorescence lifetime favors red wavelengths for low phototoxicity and time-of-flight sensors require near infrared light for invisibility. Conventional shallow junction SPADs offer only a few percent photon detection probabilities (PDP) at these wavelengths.

Long-wavelength sensitive CMOS compatible SPADs have recently been developed [8,9]. We present the first array implementation of these devices in the form of a SiPM with a broad spectral response and in-pixel memory. Small pitch and high fill factor are maintained through innovations in shared substrate, NMOS-only circuits, compact quench and level-shift implementations. The 11.6μm pitch is the smallest SPAD pixel with embedded intelligence and is compatible with future VGA resolution single photon image sensors.

II. SPAD Structure & CMOS SiPM Implementation

A 130nm CMOS imaging process was chosen for this work as it offered both small geometry MOSFETs for array integration as well as dielectric stack optimization for light transmission [10]. This process was also compatible with the long wavelength sensitive SPAD [8] where the multiplication junction is formed with the deep n-well (DNW) and the p-substrate. Similar to [8], the natural guard ring is formed by blocking p-well around the periphery and using the doping characteristics of the epitaxial layer on p-substrate and the implant spreading of the DNW. This device demonstrates 3-8× improvement in PDP at red and NIR wavelengths compared to conventional CMOS SPADs. The p-well in the device reported in [8] is replaced by n-well to recover the blue and green sensitivity.

This work was supported by The University of Edinburgh and STMicroelectronics (R&D) Ltd., Edinburgh, U.K.

978-1-4673-1707-8/12 $31.00 © 2012 IEEE

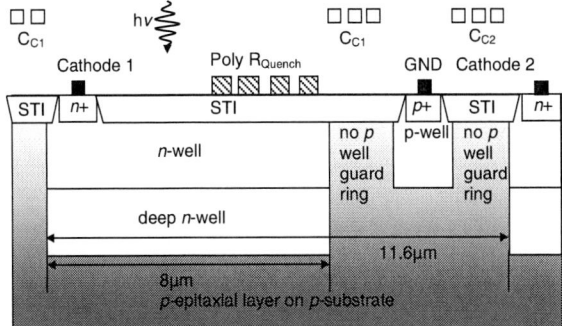

Fig. 1. **Simplified SPAD and pixel cross-section**

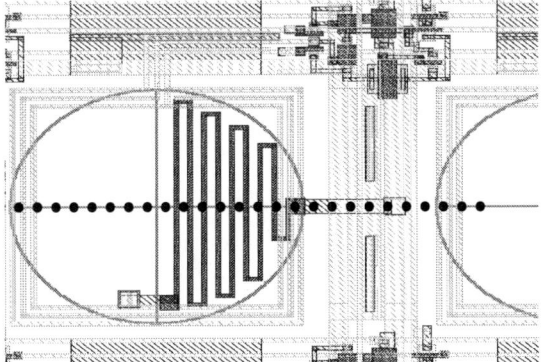

Fig. 2. **Pixel layout showing the placement of the poly resistor, coupling capacitor, and NMOS transistors. Cross-section (Fig. 1) marked as dots.**

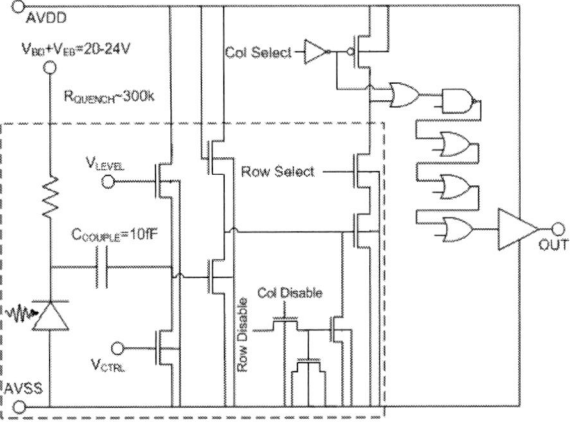

Fig. 3. **SiPM and pixel (enclosed) schematic.**

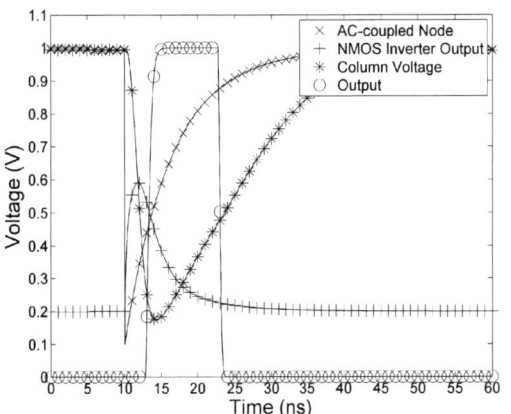

Fig. 4. **Voltages in pixel, column, and output on SPAD event.**

To achieve high fill factor/small pitch, a novel combined approach to the device and circuit design is taken that involves placing circuit elements inside the SPAD structure. The IR-sensitive SPAD requires a polysilicon resistor and metal-oxide-metal (MOM) coupling capacitor [8]. These are combined with the SPAD by placing the polysilicon resistor on top of some of the active region above shallow trench isolation (STI) oxide to prevent oxide breakdown. This approach was taken because unlike a conventional planar SPAD, this device is sensitive to long wavelength light and it is known these wavelengths pass through polysilicon. The guard ring would otherwise be wasted fill factor as it is not light sensitive, so to make the most of this, the MOM capacitor was formed in a ring above it. A cross-section of the device structure with integrated resistor and capacitor is illustrated in Fig 1 and Fig. 2.

An advantage of the new SPAD [8] structure is that it is possible to share the substrate as is done in conventional SiPMs [2]. The approach was taken to use NMOS only logic in the pixel because *p*-well is automatically placed next to the guard ring and NMOS transistors could be placed very close to the SPAD. The use of PMOSFETs would have reduced the fill factor due to well-spacing design rules. The pixel and SiPM schematic is illustrated in Fig. 3.

The SPAD produces negative pulses on the moving node which are coupled through by a ≈10fF MOM capacitor to an NMOS inverter. The V_{LEVEL} and V_{CTRL} NMOS transistors are used to maintain the bias voltage high on the coupled node and to prevent the node being pumped above AVDD by SPAD pulses, respectively. The NMOS inverter then drives a

column OR function where all the pixels in a column share the same pull-up PMOS which is located at the edge of the array. The advantage of this approach compared to conventional SiPMs is that the array capacitance is not seen by each SPAD, reducing the charge per pulse and hence after pulsing. The distributed column inverter can be designed to have sufficient strength to drive the column load.

The 16 rows and columns are addressed by two 4 to 16 decoders. Row and column addressing is implemented with an NMOS in the pixel and an inverter connected to the column PMOS. An OR gate is used to ensure the correct logic state is forced when the column is disabled. A simple four-level, two-input OR tree is used to produce a pulse train for each column, connected to an output buffer. In the future, any CMOS column logic could be implemented depending on the application.

The waveforms in the signal path are illustrated in Fig. 4 from a SPICE simulation showing the operation of the NMOS inverter and column pull down logic. The current consumption of the SiPM with no SPAD events is ≈1mA, indicating ≈4µA/pixel. The inverter design is a trade-off between current consumption and speed. It is not possible to have positive edges from the SPAD to simplify the electronics, as in [6].

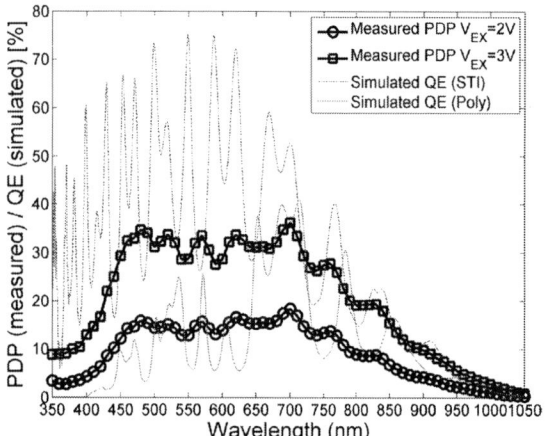

Fig. 5. **Measured PDP of the SiPM compared to the simulated QE**

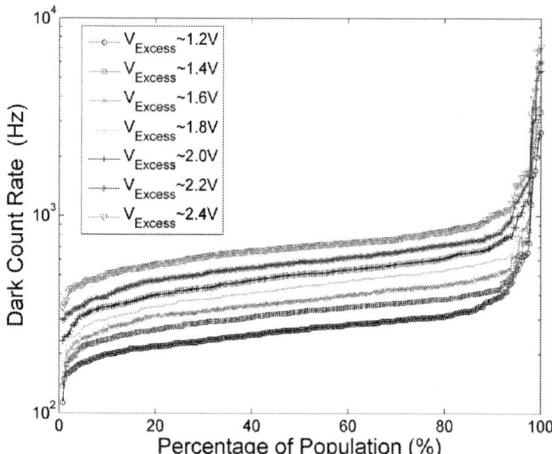

Fig. 7. **Dark Count Rate Distribution vs. Voltage**

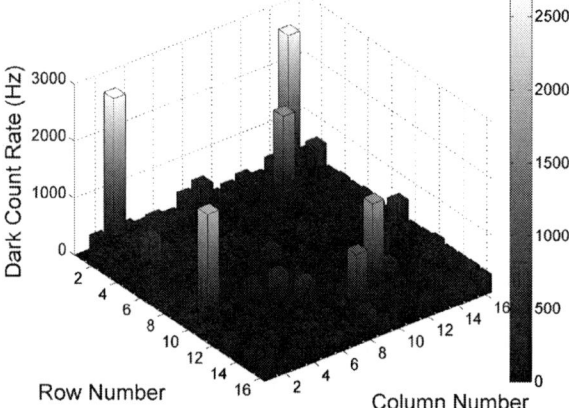

Fig. 6. **Array DCR map at 1.2V excess bias**

Additionally, a 1T DRAM cell is included in-pixel to inhibit the output of high DCR devices. This is similar to the approach used in [3] to disable noisy SPADs. However, in this case, the output is inhibited rather than disabled because of the high voltage. The use of a DRAM for this purpose is advantageous over SRAM because it is compact, improving fill factor. Moreover, the refresh disadvantage of DRAM is minimized because only a small fraction of noisy SPADs require turned off by storing a logic one and can therefore be refreshed rapidly. The DRAM, however, operates on a separate row and column addressing structure requiring two further 4 to 16 decoders and additional metal routing across the array. The DRAM was found to have a hold time of 4ms under illumination sufficient to saturate the SiPM. This was achieved by using a long and narrow thick oxide read/write transistor to minimize leakage, and a thin oxide MOS capacitor to maximize the capacitance that were both completely covered in a metal light shield.

III. RESULTS

The SiPM was found to operate with a breakdown voltage of around 20V, consistent with isolated SPAD test structures of the same design. This V_{BD} is higher than 14.9V in the 90nm process [8], consistent with dose-scaling trends in CMOS.

The PDP of a single SPAD in the array was measured with a calibrated monochromator and reference photodiode. The PDP of the whole SiPM was not measured as the output count rate would have saturated due to pile up, distorting the result. The results are illustrated in Fig. 5. It can be observed that the PDP is >30% from 450-720nm which is much wider than conventional SPAD arrays [3,4,5]. The results are consistent with the extended spectral response of the device used [8]. The simulated Quantum Efficiency (QE) is shown for the device through STI and through the polysilicon using the Transfer Matrix Method at normal incidence with Synopsys Sentaurus TCAD. The simulations show how the STI biases the spectral response towards blue by making the junction effectively closer to the surface. The polysilicon transmits light around 650-750nm and this corresponds to a peak in the measured PDP. The combined effect of the STI and the poly is to produce an interestingly flat spectral response believed to be the first of its kind. In the realised layout of the device (Fig. 2) the polysilicon takes up about half of the SPAD active area. If the SPAD's dimensions were increased then the relative contribution of the poly would reduce.

A DCR map of the array is illustrated in Fig. 6 at 1.2V excess bias. No positional-dependence of the DCR is evident, indicating that there is no column 'droop' due to the drive capacity of the pull down NMOSFETs. The DCR yield is illustrated in Fig. 7 at different bias voltages showing a median DCR of ≈200Hz at 1.2V excess bias (V_{EB}) and increasing regularly with voltage to 2.4 V_{EB}. The DCR is quite high for a SiPM but it is thought that this can be improved with process modification.

No obvious crosstalk was identified from analysis of the likelihood of SPADs having elevated DCR next to very high DCR devices. Unfortunately, it was not possible to perform a detailed characterization of crosstalk because it was only possible to inhibit the output of SPADs, not turn them off. However, an interesting crosstalk-like effect was observed at higher excess bias where the DCR increases exponentially while milliamperes of current starts to be drawn on the high voltage supply ($V_{BD}+V_{EX}$). This was attributed to emitter action by the source and drain of V_{LEVEL} and V_{CTRL} transistors,

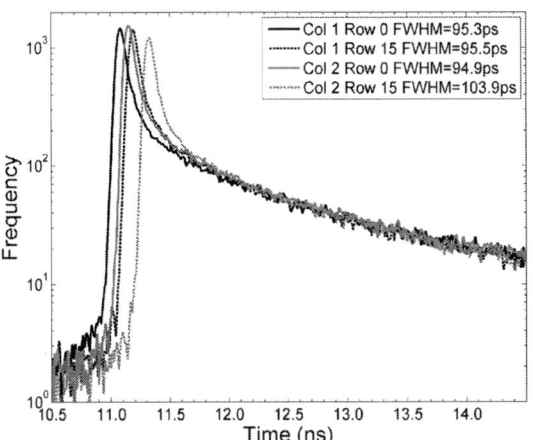

Fig. 8. **System jitter results illustrating column propagation delay.**

respectively. At high excess bias, the coupled node is forced negative with respect to the substrate at the bottom of the SPAD edge. This injects electrons into the substrate which then flow through the multiplication region, triggering avalanche breakdown and increased DCR while others flow through the guard ring and add to the current. This negative behavior can be mitigated by designing the charge division on the coupled node to attenuate the voltage pulse.

The after pulsing of selected devices was measured with the time window technique [8] and the results are displayed in Table I. The after pulsing is quite low and exhibits the expected correlation to DCR.

TABLE I.   AFTER PULSING OF SELECTED DEVICES

Device Address	Single Device Parameters		
	Voltage (V)	DCR (Hz)	After pulsing probability (%)
Column 1 Row 9	21.4	2975	0.572
Column 1 Row 9	22.4	7231	4.584
Column 1 Row 8	21.4	451	0.483
Column 1 Row 8	22.4	886	1.474
Column 6 Row 15	22.4	892	1.003

The system jitter results at 2V excess bias are displayed in Fig. 8. The jitter was found not to vary appreciably along the column from testing devices 1 and 16 in columns 1 & 2. The results show that there is a ≈100-200ps delay from SPADs at the top of the column but that the jitter of each SPAD is independent of the propagation delay of the signal along the column bus. The figures include the jitter of the SPAD (unknown), laser (49ps), output buffer (≈30ps), oscilloscope (25ps) the pixel and column OR-tree jitter (unknown).

IV. CONCLUSION

A hybrid conventional/digital SiPM with extended spectral response and 21.6% fill factor with basic in-pixel functionality was presented. Compared to existing SiPMs the realised fill factor is quite low; however, compared to intelligent CMOS SPAD arrays the pixel pitch is roughly half that of prior work with comparable fill factor. It is thought that process modification could be employed to further improve the PDP and fill factor with the same pitch. Alternatively, the pitch could be increased and with larger area SPADs, the fill factor boosted to comparable levels while maintaining the functionality offered by the NMOS pixel.

The SPAD is compatible with back-side illumination CMOS imaging processes [8]. Utilisation of the SiPM in BSI would eliminate the problem of optical transmission through the resistor. In this configuration, the spectral response would peak in the blue-green. The wafer thickness could be specifically designed to optimise sensitivity at a scintillator emission wavelength and this would not require fully depleting the substrate as done with conventional SiPMs [2].

ACKNOWLEDGMENT

The authors thank Sara Pellegrini and Brent Hearn at STMicroelectronics, Edinburgh, for measurement assistance.

REFERENCES

[1] G. Bondarenko et al "Limited Geiger-mode microcell silicon photodiode: new results," Nuclear Instruments and Methods in Physics Research A Vol 442 No 1-3 pp. 187-192, 2000

[2] N. Otte, "The Silicon Photomultiplier - A new device for High Energy Physics, Astroparticle Physics, Industrial and Medical Applications," presented at SNIC Symposium, Stanford, California, 2006.

[3] T. Frach et al "The Digital Silicon Photomultiplier - Principle of Operation and Intrinsic Detector Performance," presented at IEEE Nuclear Science Symposium, Orlando, FL 2009.

[4] D. Tyndall et al "A 100Mphoton/s Time-Resolved Mini-Silicon Photomultiplier With On-Chip Fluorescence Lifetime Estimation in 0.13μm CMOS Imaging Technology," presented at ISSCC, 2012

[5] L. Pancheri, D. Stoppa, "Low-Noise CMOS Single-Photon Avalanche Diodes with 32 ns Dead Time," Proc. 37th IEEE European Solid-State Device Research Conference (ESSDERC), Munich, Germany 11-13 Sept. 2007, pp. 362 - 365

[6] L. Pancheri, N. Massari, F. Borghetti, D. Stoppa, "A 32x32 SPAD Pixel Array with Nanosecond Gating and Analog Readout," presented at the International Image Sensor Workshop, Japan, 2011

[7] C. Niclass, M. Soga, H. Matsubara, S. Kato, "100m-range 10-frame/s 340x96-pixel time-of-flight depth sensor in 0.18μm CMOS", Proc. 36th IEEE European Solid-State Circuits Conference (ESSCIRC), Helsinki, Finland, 12-16 Sept. 2011, pp. 107-110.

[8] E. A. G. Webster, J. A. Richardson, L. A. Grant, D. Renshaw, R. K. Henderson, "A Single-Photon Avalanche Diode in 90nm CMOS imaging technology with 44% photon detection efficiency at 690nm", IEEE Electron Device Letters, Vol. 33, No. 5, May 2012, pp. 694-696.

[9] S. Mandai, M. W. Fishburn, Y. Maruyama, E. Charbon, "A wide spectral range single-photonavalanche diode fabricated in anadvanced 180 nm CMOS technology," Optics Express, Vol. 20, No. 6, pp. 5849-5877, March 2012.

[10] M. Cohen et al, "Fully Optimized Cu based process with dedicated cavity etch for 1.75μm and 1.45μm pixel pitch CMOS Image Sensors," IEDM, 2006.

978-1-4673-1707-8/12 $31.00 © 2012 IEEE

Low-power DRAM-compatible Replacement Gate High-k/Metal Gate Stacks

R. Ritzenthaler, T. Schram, E. Bury*, J. Mitard,
L.-Å. Ragnarsson, G. Groeseneken*,
N. Horiguchi, and A. Thean
imec, Kapeldreef 75, 3001 Leuven, Belgium
*also with KU Leuven
e-mail: romain.ritzenthaler@imec.be

A. Spessot, C. Caillat, V. Srividya, and P. Fazan
Micron Technology Belgium BVBA
imec Campus, Kapeldreef 75,
3001 Leuven, Belgium

Abstract— In this paper, the feasibility of High-k/Metal Gate (HKMG) Replacement Metal Gate (RMG) stacks for low power DRAM compatible transistors is assessed. It is shown that traditional RMG gate stacks cannot be used because of the additional anneal needed in a DRAM process. New solutions are developed, and a PMOS stack HfO_2/TiN with TiN deposited in three times combined with Work Function metal oxidations is demonstrated, featuring a Work Function of 4.95 eV. On the NMOS side, a new solution based on the use of oxidized Ta as a diffusion barrier is proposed, and a HfO_2/TiN/Ta/TiAl/TiN/TiN gate stack featuring an aggressive Work Function of 4.35 eV (allowing a Work Function separation of 600 mV between NMOS and PMOS) is demonstrated.

Keywords-component; DRAM periphery transistors, RMG (Replacement Metal Gate), Work Function Engineering.

I. INTRODUCTION

Dynamic random access memory (DRAM) technology is constantly working for ways to speed up the peripheral logic performance in order to reduce the gap with microprocessor performance and DRAM periphery transistors are pushed to adopt characteristics used by high performance logic devices [1]. Therefore, introduction of High-K, Metal Gate (HKMG) [2] is becoming necessary for next technology nodes. In order to target low power mobile applications, low leakages are also a necessity. All of these new modules must also stay compliant with process compatibility requirements: for DRAM compatible applications, the transistors used in the periphery of the memory cell have typically a thicker Electrical Oxide Thickness (EOT) compared to Logic applications (low-leakages requirement), but also have to sustain an extra anneal (so called "DRAM Anneal" (DA), typically several hours) [3]. Additionally, Replacement Metal Gate (RMG) [4] schemes are seriously considered for DRAM compatible transistors, since they can allow avoiding early aggressive thermal budget.

In this paper, the feasibility of gate-last DRAM compatible gate stacks for Work Function (WF) tuning is investigated using a capacitor flow. The crucial differences with stacks used for Logic applications are highlighted, and innovative solutions for both NMOS and PMOS Work Function tuning are proposed.

II. RMG DRAM-COMPATIBLE PROCESS SPECIFICITIES

A. RMG integration

In an RMG integration scheme, a poly-Si capped dummy gate stack is first deposited, featuring in a "high-k first" scheme (HKF, used in this work) a SiO_2 Interfacial Layer (IL)/high-k stack followed by an Etch Stop Layer (ESL); the ESL serves to protect the high-k during the dummy gate removal (Fig.1 a). In order to tune the Work Function, one or several metal layers are then deposited (Fig. 1 b), and the gate stack is completed by a fill metal.

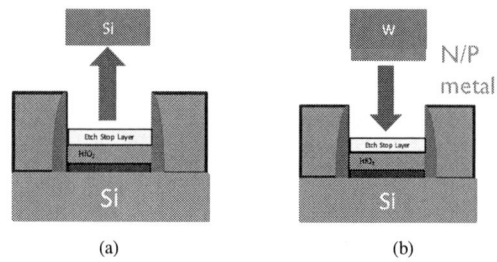

(a) (b)

Fig. 1: Schematics illustrating the RMG dummy gate removal (a) and metal gate deposition (b) steps.

B. PMOS stacks

For the stack targeting PMOS applications, RMG HKF gate stacks traditionally used for Logic applications [5] (here with a thicker SiO_2 interfacial layer to account for stronger gate leakage control requirements) have been submitted to the DRAM anneal (Fig. 2). These stacks feature a TiN ESL (using two deposition techniques, Fig. 2), followed by a second TiN layer deposition. It can be clearly seen that the implementation of the DRAM anneal (temperature above 600°C during several hours selected in this work) seriously reduces the Work Function, suggesting a move toward mid-gap for the TiN based stacks (Fig. 2). However, by adding a second thin TiN layer deposited on top of the ESL (i.e. all included three TiN depositions) and optimizing its thickness, a Work Function of 4.95 eV is demonstrated (Fig. 2). This Work Function increase is correlated with the TiN thickness increase and with the oxidation of the Work Function metals [5].

978-1-4673-1707-8/12 $31.00 © 2012 IEEE

It is noteworthy that in this case, a long anneal like the DRAM anneal is even beneficial (Fig. 2), since it improves the oxygen diffusion in the gate stack.

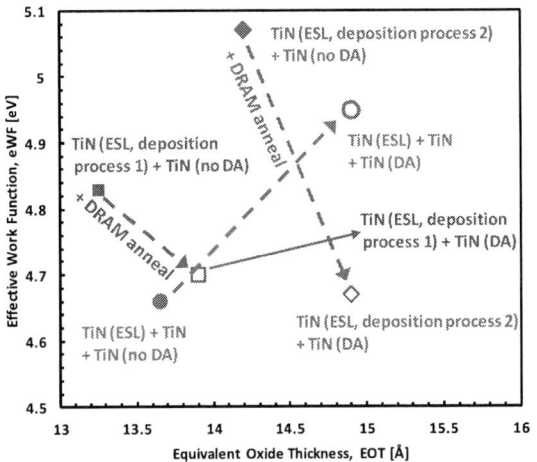

Fig. 2: Effective Work Function vs. Equivalent Oxide Thickness for stacks targeting PMOS applications without (closed symbols) and with (open symbols) DRAM anneal.

C. NMOS stacks

For gate stacks targeting NMOS applications, TiAl alloys have been reported as interesting candidates, with work functions down to 4.2 eV for TiN(ESL)/TiAl/TiN stacks [6]. However, Al being a fast diffuser, it is expected that the aggressive thermal budget imposed by DRAM compatible applications will considerably narrow the process window. Indeed, SIMS analysis carried out on a TiAl/TiN stack show that the DRAM anneal induces a significant diffusion of Al in TiN (Fig. 3). Besides, it is noteworthy that the Al profiles obtained after a 5 min anneal and after a several hours anneal are pretty similar, indicating that most of the diffusion occurs within a short time. This is coherent with the fact that the Work Function obtained with the "double DRAM anneal" is very similar to the one obtained with a single DA (diamonds, Fig. 4).

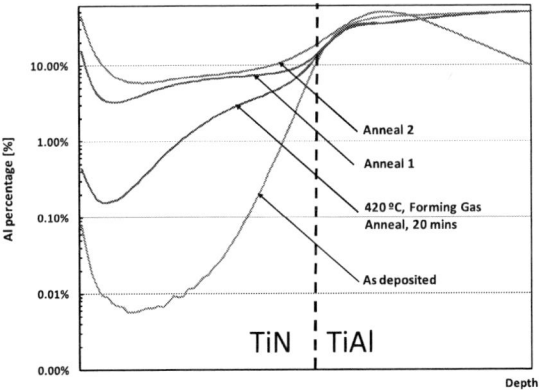

Fig. 3: Al concentration in a TiN/TiAl/HfO$_2$ stack obtained by SIMS analysis (blanket wafers) as deposited, after a 420°C/20 min Forming Gas Anneal, Anneal 1 (DRAM anneal, 5 min), and Anneal 2 (DRAM anneal, several hours).

Since the TiN ESL layer has to block an excessive Al diffusion, an increase of the ESL thickness might help circumventing the problem. However, an increase of the metal gate thickness comes at a price of a very significant Work Function increase, making this option highly unsuitable for NMOS applications (Fig. 4). Therefore, the combination of TiN(ESL) and TiAl Work Function tuning material in a DRAM compatible flow seems to have a fundamental issue.

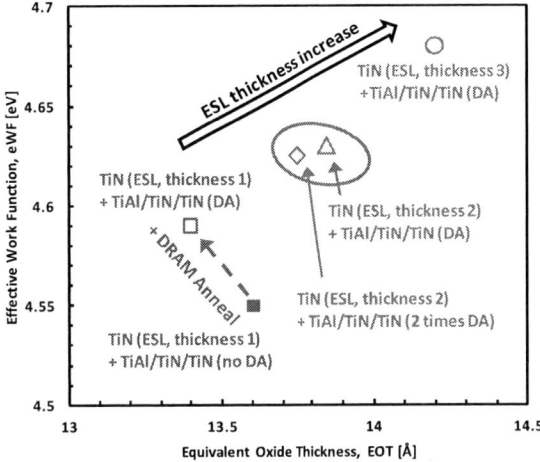

Fig. 4: Effective Work Function vs. Equivalent Oxide Thickness for stacks targeting NMOS application without (closed symbols) and with (open symbols) DRAM anneal.

Investigating other materials as diffusion barrier, we found that Ta-based stacks constitute excellent diffusion barriers for Al (Fig. 5, using a TaN layer). It can be seen that Al still diffuses in the TaN layer, but with low concentrations even after the DRAM anneal. The resistance to the Al diffusion is not the sole parameter for the choice of the layer deposited on top of the ESL; for CMOS process reasons related to the etch (not shown here), the best candidate for the NMOS stack integration is finally Ta combined with an oxidation.

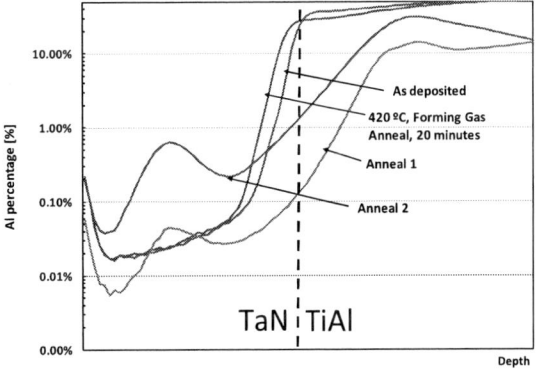

Fig. 5: Al concentration in a TaN/TiAl/HfO$_2$ stack obtained by SIMS analysis (blanket wafers) as deposited, after a 420°C/20 min Forming Gas Anneal, Anneal 1 (DRAM anneal, 5 min), and Anneal 2 (DRAM anneal, several hours).

III. DEVICE ELECTRICAL PERFORMANCE

A. Device Manufacturing.

Based on the above learning, the following stacks (Fig. 6) have been manufactured:

- For the PMOS stack, a SiO_2 Interfacial Layer / HfO_2/ TiN(ESL) layer is deposited before Poly-Si dummy gate removal. After the dummy gate removal, two additional TiN layers combined with oxidations and the fill metal are deposited to complete the gate stack.

- For the NMOS stack, the gate stack until dummy poly removal remains essentially the same. After dummy gate removal, an oxidized Ta layer is deposited as a diffusion barrier. The gate stack is followed by a Ti_XAl_{1-X} + TiN deposition, and completed by the last TiN and fill metal layers.

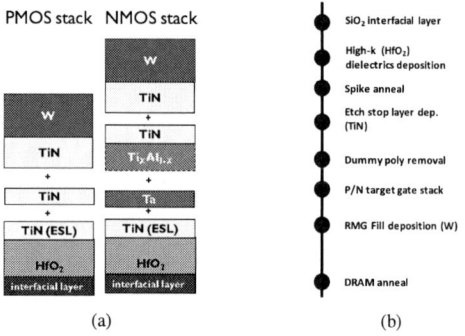

Fig. 6: Schematics of the fabricated gate stacks (NMOS/PMOS) (a), and related process flow (b).

B. Electrical Performance

The accumulation capacitances for the PMOS and NMOS stacks are shown in Fig. 7. The flat-band voltage modulation is clearly noticeable, and the interface state density (estimated using the conductance) features shows reasonable values (see inset, Fig. 7).

Using Hauser fitting approach [7], the Effective Work Function (EWF) and EOT are also extracted. EOT values are respectively around 11.5 Å for NMOS and 13 Å for PMOS (Fig. 8); it is slightly higher than commonly reported EOT used in Logic stacks, but necessary for DRAM compatible applications since the target here is more on low gate leakages than high drive currents. The gate current density (extracted at V_{FB} -0.6V, with V_{FB} the flat- band voltage) moves accordingly and exhibits values below 100 mA/cm^{-2}. On the other hand, EWF is measured around 4.95 eV for the PMOS stack, and feature a very attractive EWF around 4.35 eV for the NMOS stack, ensuring an EWF separation of 600 meV between NMOS and PMOS.

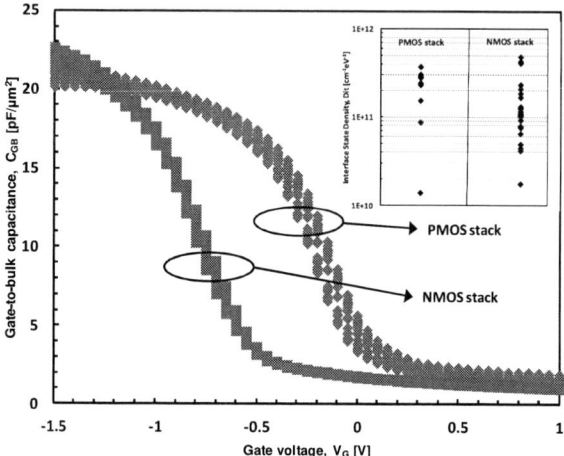

Fig. 7: Gate-to-bulk capacitance for the PMOS stack (SiO_2 IL/HfO_2/TiN-ESL/TiN/TiN combined with Oxidations & NMOS stack (SiO_2 IL/HfO_2/TiN-ESL/Ta/TiAl/TiN/TiN) vs. gate voltage V_G; Both capacitances are measured in a p-type well. Inset shows the interface state density D_{it}.

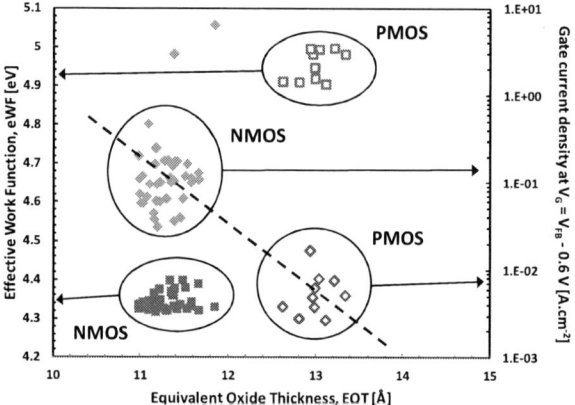

Fig. 8: Gate Effective Work Function (squares) and gate current density (at V_{FB} - 0.6 V, diamonds) vs. EOT for NMOS (closed symbols) and PMOS (open symbols).

C. Reliability

Bias temperature instabilities (BTI) and especially negative BTI are currently major concerns regarding device reliability. BTI manifests itself as an unwanted shift of the threshold voltage (ΔV_{TH}) during operation of the device and is caused by pre-existing or newly generated interface and bulk defects. An accurate evaluation of this BTI degradation, necessary for a correct lifetime extrapolation, can be performed by the extended measure-stress-measure (eMSM) technique [8]. With this technique, the ΔV_{TH} relaxation is typically observed over many decades in time. The ΔV_{TH} is estimated using the time zero drain current/gate voltage characteristic. Recently, it has been shown, in analogy with this I-V eMSM technique, that the same considerations could apply to capacitance measurements [9], where the ΔV_{TH} relaxation is then measured as a shift in gate capacitance at time zero V_{TH_0} or V_{FB_0} on transistor or capacitor structures, respectively (C-V eMSM).

978-1-4673-1707-8/12 $31.00 © 2012 IEEE

Reliability-wise, RMG stacks do not exhibit significant differences compared to Gate First stacks [9]. However, it is noteworthy that the NMOS stack does not behave as commonly expected (Fig. 9): instead of having a ΔV_{FB} (flat-band voltage shift) monotonically decreasing as a function of the relaxation time (due to hole detrapping at the Si/SiO_2 interface), it increases first, then reaches a *turning point* and decreases later on. This turnaround is observed at increasing relaxation times as a function of stress time (t_{stress}) and eventually drops out of the measurement window. Notable is that ΔV_{FB} never exceeds 20 mV, which is a much lower shift than generally observed. Both observations are an indication of two competing mechanisms.

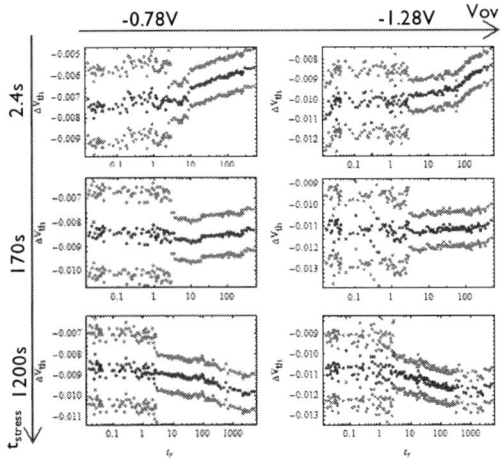

Fig. 9: Flat-band voltage shift ΔV_{FB} reliability vs. relaxation time after various stress times (3σ error bars are plotted in red). Gate voltage V_G is equal to flat-band voltage V_{FB} during relaxation, and V_{FB}-0.78V (left) and V_{FB}-1.28V (right) during stress.

This *anomalous* behaviour has been reported before on Dy based stacks [10], and can be related to the fact that electrons originating from the gate are trapped in the HfO_2 layer (Fig. 10.a). Subsequently, during relaxation (Fig. 10.b), the initial negative V_{FB} shift is then related to the electron detrapping dominating over the hole detrapping, followed by the hole detrapping eventually taking over (Fig. 9). Both mechanisms are concomitant over the whole observed time window since absolute V_{FB} shifts are much smaller than commonly observed (<20 mV). The most conceivable origin of these electron defect centers is the Al originating from the gate stack. Indeed, Al-diffusion-induced defects are known to exhibit a positive V_{FB} shift [11].

Extracting the predicted 10-year lifetime V_{ov} thus becomes irrelevant since standard technique is based on using the (back-extrapolated) V_{FB} shift after 1ms of relaxation for different V_{ov}. To assess the quality of the high-k stack in terms of defects, the two mechanisms need to be decoupled under the presumption they are uncorrelated. Hole detrapping commonly exhibits a power law relaxation with an exponent 0.10 ~ 0.13, a fully consistent model for the electron detrapping mechanism still has to be defined. Concluding, reliability-wise it can be stated

that this gate stack performs well in terms of absolute V_{FB} shift, albeit the electron detrapping remains to be fully understood.

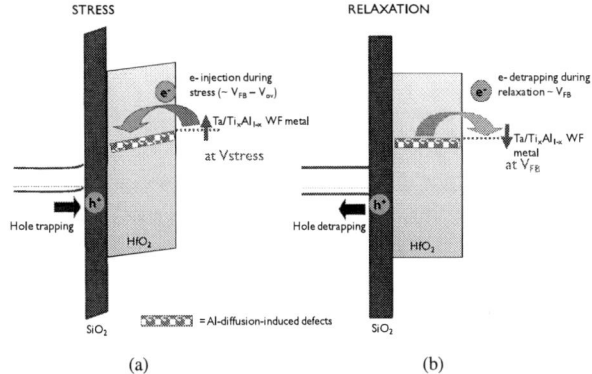

Fig. 10: band structure in a p-type Si substrate/SiO_2/HfO_2 gate stack under stress (negative applied voltage) (a) and during relaxation (at flat-band voltage) (b).

IV. CONCLUSIONS

In this work, we have successfully demonstrated a NMOS and a PMOS gate stack for DRAM compatible RMG applications. It is shown that RMG gate stacks used for Logic application are not suitable because of the high thermal budget applied. It is also shown that solutions based on oxidations and use of Ta on top of the etch stop layer allow to circumvent the high thermal budget imposed by DRAM process for NMOS. The EWF for the PMOS stack (resp. NMOS stack) has a Work Function of 4.95 eV (resp. 4.35 eV), featuring a very attractive EWF separation of 600 meV between NMOS and PMOS. Gate leakages, EOT, and interface state density are well in the range of acceptable values. Reliability is very encouraging, but remains to be fully understood for these gate stacks; as a next step, the dual NMOS/PMOS integration in a CMOS transistor flow has also to be assessed.

ACKNOWLEDGEMENTS

The imec sub-22 nm program members, the imec pilot line and amsimec electrical characterization facilities are greatly acknowledged for their support.

REFERENCES

[1] S. Y. Cha, *IEDM 2011 Short Course*, 2011.

[2] Martin M. Frank, proc. Of the *41th European Solid-State Device Research Conference (ESSDERC)*, pp. 25-33, 2011.

[3] C. Ortolland et al., *VLSI Symp 2010*, pp. 185-186, 2010.

[4] K. Mistry et al., *IEDM 2007 Tech. Dig.*, pp. 247-250, 2007.

[5] Z.Li et al., Microelectronic Engineering 87 (2010) 1805–1807.

[6] A. Veloso et al., *VLSI Symp 2011*, pp. 34-35, 2011.

[7] J. R. Hauser and K. Ahmed, *1998 international conference on characterization and metrology for ULSI technology*, pp. 235-239, 1998.

[8] B. Kaczer et al., *IRPS 2008*, pp. 20-27, 2008.

[9] E. Cartier et al., *IEDM 2011 Tech. Dig.*, pp. 441-444, 2011.

[10] R. O'Connor et al., *Appl. Phys. Lett. 93*, 053506 (2008).

[11] W. Wang et al., *J. Appl. Phys. 105*, 064108 (2009).

On The UTBB SOI MOSFET Performance Improvement In Quasi-Double-Gate Regime

V. Kilchytska, D. Flandre

ICTEAM / ELEN, Université catholique de Louvain (UCL),
1348 Louvain-la-Neuve, Belgium
e-mail address: valeriya.kilchytska@uclouvain.be

F. Andrieu

CEA-Leti, MINATEC Campus
17 rue des Martyrs, 38054 Grenoble Cedex 9, France
e-mail address: francois.andrieu@cea.fr

Abstract—**This work investigates the simultaneous electrostatic improvement and performance enhancement of UTBB SOI MOSFETs obtained in quasi-double-gate (QDG) regime (i.e. simultaneously biasing gate and substrate (or ground plane) as $V_{sub}=k*V_g$) as a strong function of k-multiplication factor, when compared to a standard single-gate mode. QDG mode is demonstrated to allow threshold voltage tuning and on-current enhancement without off-state current degradation, of interest for digital applications (e.g. switches). Improved performance in QDG mode combined with lowered DIBL and enhanced gain are of interest for high-precision low-frequency analog applications. The work finally quantifies the resulting gate area decrease in QDG mode, potentially exploitable in actual circuit implementations.**

I. INTRODUCTION

Ultra-thin body and buried oxide (UTBB) fully depleted (FD) SOI MOSFETs are widely recognized as a promising candidate for 20 nm-node and beyond, due to outstanding electrostatic control of short channel effects (SCE) [1,2]. Introduction of a highly-doped layer underneath thin buried oxide (BOX), so called ground-plane (GP), targets suppression of detrimental parasitic substrate coupling and opens multi-threshold voltage (V_{Th}) and dynamic-V_{Th} opportunities within the same process as well as the use of back-gate control schemes [1,2].

Emulation of double-gate (DG) regime (through the simultaneous gate and substrate control) was reported to result in a significant improvement of device behavior and analog figures of merit (FoM) in FD SOI MOSFETs [3-5]. In those works the simultaneous sweeping of the gate bias, V_g and substrate bias, V_{sub} was realized as $(V_{sub}-V_{Th2})= t_{BOX}/t_{gox}\cdot(V_g-V_{Th1})$ where t_{gox} and t_{BOX} are gate oxide and BOX thicknesses and V_{Th1} and V_{Th2} are front and back V_{Th}, respectively. Simplified realization of DG regime, or so called asymmetric DG (ADG) regime, consists in substrate (or GP contact) to front gate connection, i.e. $V_{sub}=V_g$. Such regime applied to UTBB devices was shown to improve their electrostatic behavior and to bring about 10-20% enhancement in analog FoM [2,6].

In this work we realize quasi-DG (QDG) operation by simultaneously sweeping the gate and substrate biases as $V_{sub}=k*V_g$ with k-factor varied from 1 (as in ADG regime [6]) to 25. Taking into account actual t_{gox} and t_{BOX} (see next

section), realization of ideal DG mode in strong inversion would require in our devices a k of about 22. Additionally, "front" and "back" V_{Th} should be taken into account. As they are strongly interdependent (particularly in UTBB devices) and moreover dependent on drain voltage, V_d and length, ideal DG regime realization becomes very complex in a practice.

In this work we investigate UTBB MOSFET performance improvement which one can expect from application of QDG mode with different "k-factors" as compared to a standard single gate (SG) operation mode. First, we assess effect of QDG mode on UTBB MOSFET electrostatic behavior, through key device parameters including threshold voltage, subthreshold slope S, maximum transconductance G_{mmax} and drain-induced barrier lowering DIBL. Next, we analyze the perspectives of such operation mode for performance enhancement through such figures of merit as drain current, I_d, transconductance over drain current ratio G_m/I_d, Early Voltage V_{EA} and Intrinsic gain A_v (for analog applications) as well as ability of simultaneous on-current I_{on} and off-current I_{off} control (for digital applications). The target applications are reconfiguration or sleep-mode switches in digital circuits and high-precision amplifiers in analog, which typically use relatively large-size transistors and low-frequency operation, so that the area overhead of the k-factor generation circuits (e.g. by switched capacitors) and the frequency bandwidth limitation of the GP contact can be neglected to the first order.

II. DEVICE DESCRIPTION

The devices are fabricated at CEA-Leti on 300 mm UNIBOND™ SOI wafers with 25 nm-thin BOX. The Si body is thinned down to approximately 7.5 nm and is left undoped in the channel. The Si substrate underneath the BOX is a standard p-type substrate with resistivity of 20 Ω.cm; additional As implantation is used to realize n-type GP, i.e. a localized highly-doped ($\sim 10^{18}$ cm^{-3}) layer just under the BOX. The gate stack consists of HfSiON (with equivalent t_{gox} of 1.2 nm) and a TiN gate electrode. More process details are presented in [1]. The studied devices are n-MOSFETs with a drawn gate length, L from 40 to 100 nm and channel width, W of 10 µm. Physical gate length was from 28 to 88 nm, respectively.

The work has been partly funded by the FNRS (Belgium), by the FP7 NoE "EuroSOI+" and by Catrene "Reaching 22" projects.

978-1-4673-1707-8/12 $31.00 © 2012 IEEE

III. EXPERIMENTAL RESULTS AND DISCUSSION

A. Electrostatic behavior

Fig.1 shows an example of transfer I_d-V_g characteristics in saturation and I_d-V_d output characteristics of UTBB SOI MOSFET operated in SG and QDG modes with different k-factors from 1 to 25. The device biased in QDG mode clearly exhibits improved characteristics and the improvement enhances with k-factor. Furthermore, as can be noticed from Fig. 1b, I_d enhancement is not purely related to the V_{Th} shift but also to the mobility improvement as discussed hereafter.

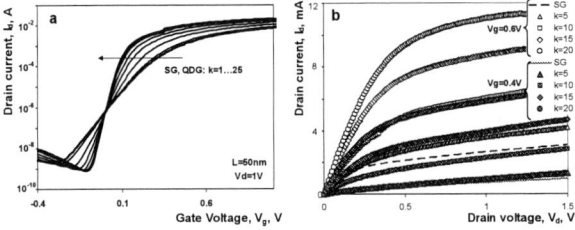

Figure 1. I_d-V_g curves in saturation (V_d=1V) (a) and I_d-V_d curves at different V_g (b) for 50 nm-long UTBB MOSFET biased in SG mode and in QDG mode with different k-factors. W=10μm.

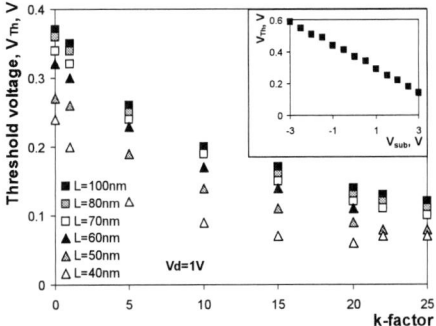

Figure 2. V_{Th} vs. k-factor in QDG mode (k=0 corresponds to SG mode) for UTBB MOSFETs with different lengths; inset gives V_{Th} as a function of a constant V_{sub} in a standard SG mode for UTBB MOSFET with L=100nm.

Fig.2 presents V_{Th} as a function of "k-factor" for UTBB devices with different lengths. QDG mode allows V_{Th} tuning from 0.35 to 0.1 V as a function of k. Application of a constant V_{sub} from -3 to +3V (i.e. nominal range for this technology) allows a bit larger V_{Th} modulation from 0.6 to 0.15 V (inset in Fig. 2). However, it is worth pointing out that in the case of QDG mode off-state current (i.e. I_d at V_g=0 V), I_{off} is evidently independent of k-factor and exactly the same as in SG mode. Therefore, contrarily to the constant V_{sub} bias mode, V_{Th} reduction in QDG mode is not accompanied by I_{off} increase.

Additionally, SCE measured as V_{Th} reduction with length shortening (or so called charge sharing (CS), assessed here as V_{Th}(L=100nm)-V_{Th}(L=40nm)) decreases with k increase: from 0.13 in SG mode, it drops to 0.08 and 0.05 for QDG mode with k of 20 and 25, respectively. Improved SCE in QDG mode are confirmed by strong DIBL reduction (Fig. 3), allowing the shorter devices which do not satisfy ITRS specifications in SG operation to fit in the required limits in QDG

mode. One can note that DIBL can also be reduced by the means of a constant negative V_{sub} application (inset in Fig. 3). However, firstly, a lesser improvement is observed, and secondly, negative V_{sub} pushes V_{Th} to higher values (inset in Fig. 2) and therefore can't be considered as a solution for LVT option. In the case of QDG mode, DIBL reduction can be achieved simultaneously with V_{Th} reduction and without I_{off} degradation.

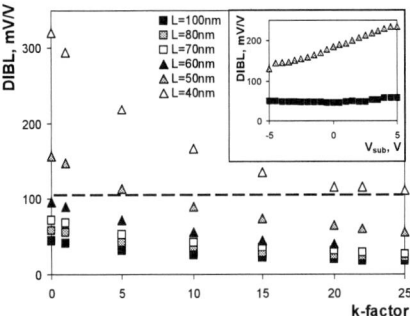

Figure 3. DIBL vs. k-factor in QDG mode (k=0 corresponds to SG mode) for UTBB MOSFETs with different L; inset gives DIBL vs. a constant V_{sub} in a standard SG mode for UTBB MOSFETs with L of 50 and 100nm.

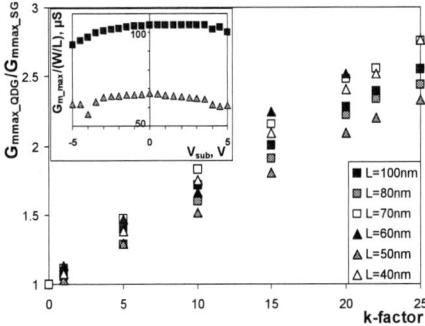

Figure 4. G_{mmax} improvement in QDG mode vs. k-factor (k=0 corresponds to SG mode) for devices with different L; inset gives G_{mamx}/(W/L) vs. a constant V_{sub} in a SG mode for devices with L of 50 and 100 nm. V_d=20mV.

Fig. 4 shows G_{mmax} improvement provided by QDG mode compared to SG; up to 3 times higher values can be obtained with k increase. Similar values were previously reported for a DG mode realized in FD SOI MOSFETs [3-5] and related to simultaneous activation of "top" and "bottom" channels accompanied by realization of volume inversion (VI) conditions and hence mobility, μ improvement [3,4]. Particular importance of VI-related enhancement is evident in the case of our UTBB devices with 7.5 nm Si film thickness. From one side, so thin Si film supposes quasi-VI operation regime even in SG mode, while from another side, impact of interface roughness and remote scattering on the carrier transport is rather strong. Indeed, G_{mmax} vs. V_{sub} (inset in Fig. 4) features G_{mmax} degradation at both positive and negative V_{sub}. Negative V_{sub} pushes channel to the top Si interface and hence enhances scattering from high-k oxide interface (and remote from the metal gate), while positive V_{sub} attracts channel to the bottom thus enhancing bottom interface scattering and possible μ decrease as a result of nGP implantation (e.g. higher doping at the Si film bottom, or charges in the BOX, at interface). Contrarily to the constant V_{sub} case, in QDG mode, "mean channel position" is pushed to the center of the Si film, which is profitable from the μ point of view. Moreover, QDG regime allows electric field lowering (due to potential flattering), thus also improving μ.

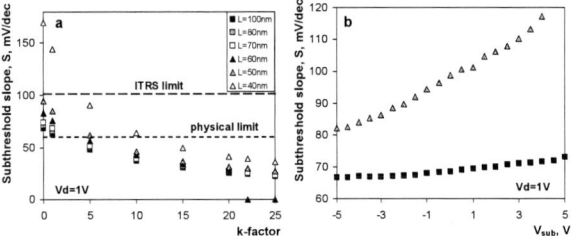

Figure 5. (a) S vs. k-factor in QDG mode (k=0 corresponds to SG mode) for UTBB MOSFETs with different lengths; (b) S vs. a constant V_{sub} in a standard SG mode for UTBB MOSFETs with L of 50 and 100nm. V_d=1V.

Fig. 5 shows strong S improvement in UTBB MOSFETs operated in QDG mode. The improvement is particularly strong for short devices. With k increase, S decreases below "theoretical limit" defined by $\ln(10) \cdot kT/q$ and approaches 20 mV/dec almost independently of L for high values of k. S reduction below its physical limit was previously reported for FD SOI MOSFETs biased in DG mode [5, 7]. [7] relates low S in a virtual DG mode to the fact that k-factor for pure DG realization differs from weak to strong inversion. In QDG mode, using k-factor constant and higher in weak/moderate inversion than pure DG one [7] results in extremely low S values. However, pure DG biasing in weak/moderate inversion (i.e. k=5-10, according to [7]) results in S values close to the ideal (Fig.5a). However, nevertheless the origin, in this paper we show a benefit of such biasing regime in order to achieve steep subthreshold device characteristics. One can note, that application of a constant negative V_{sub} in a SG mode also allows S reduction, but firstly, it is not as efficient as a QDG mode, and moreover negative V_{sub} pushes V_{Th} to higher values, while QDG allows simultaneous reduction of V_{Th} and S.

B. Performance assessment

First of all, it is important to emphasize that due to constraints imposed by realistic limits on V_{sub} values reached at high k-factor in the case of high V_g, we focus our analysis predominantly on moderate inversion (i.e. around $V_{Th} \pm 0.2$ V).

1) Analog applications

Device performance was assessed using the methodology reported in [8] and focused on such parameters as: G_m/I_d, normalized drain current, $I_{d\_norm}=I_d/(W/L)$, taken at a constant $G_m/I_d = 10$ V^{-1} (which is independent on V_{Th}, thus assuring a fair comparison of different devices and operation modes [8]), Early voltage, $V_{EA}=I_d/G_d$, where G_d is output conductance and Intrinsic gain $A_v=G_m/G_d=G_m/I_d \cdot V_{EA}$.

Fig. 6 plots G_m/I_d as a function of I_{d_norm} for 50 nm-long device clearly demonstrating advantage of a QDG mode over SG. G_m/I_d maximum reaches 85 V^{-1} in QDG mode with high k, which is higher than theoretical limit (~38 V^{-1}) and directly correlates with S reduction (Fig. 5). Increase of G_m/I_d in weak-to-moderate inversion regime suggests that QDG mode is

efficient for high-precision applications, allowing higher gain for the same power consumption or lower bias current for the same gain-bandwidth. I_d was then extracted at a constant G_m/I_d (thus neglecting V_{Th} difference) of 10 V^{-1}, which is typically achieved at V_g of 0.1-0.2 V higher than V_{Th}. Such approach allows keeping V_{sub} within $\pm$ 5 V limit (inset in Fig. 7). Strong (up to ~4 times) I_d improvement provided by a QDG mode clearly shown in Fig. 7 is related to both μ and DIBL improvements in QDG mode.

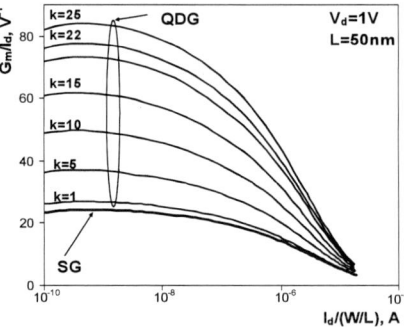

Figure 6. G_m/I_d versus normalized drain current for 50 nm-long n-MOSFET operating in SG and QDG modes with different k-factors. $V_d = 1$ V.

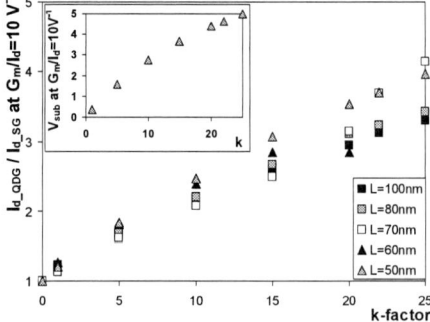

Figure 7. I_d improvement in QDG mode as a function of k-factor (k=0 corresponds to SG mode) for UTBB MOSFETs with different lengths extracted at G_m/I_d=10 V^{-1}. V_d=1 V. Inset gives V_{sub} reached at G_m/I_d=10 V^{-1} as a function of k-factor of 50 nm-long UTBB MOSFET.

Comparison of V_{EA} extracted in saturation at V_d=1 V in UTBB devices reveals 2 to 10 V increase in QDG mode over SG (depending on L and k), particularly in around-V_{Th} regime (i.e. where advantages of VI operation conditions dominate). Higher V_{EA} allows higher achievable A_v (5-6 dB) (Fig. 8a) and higher A_v combined with higher G_m are directly translated into strongly improved G_m vs. A_v FoM provided by QDG mode (Fig. 8b). On another hand, higher I_{d_norm} allows proportional device width reduction for same performance and hence reduction of gate and drain parasitics, allowing margin for accommodating the GP connection ones. It is also worth to point out that the obtained intrinsic A_v values would enable straightforward generation of k-factors up to 20-25 based on a simple inverting and differential amplifier stage using crossed outputs as input V_{sub} signals.

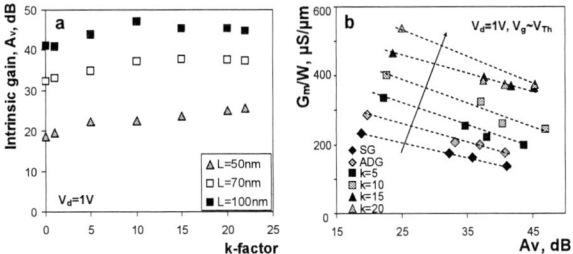 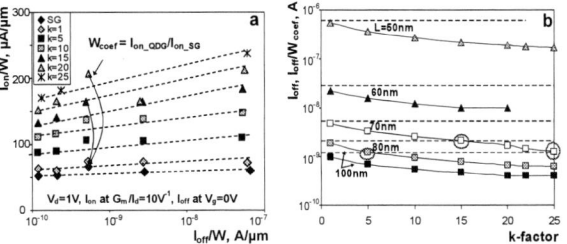

Figure 8. (a) Maximum achievable A_v vs. k-factor in QDG mode for devices with different lengths. V_d=1V; (b) G_m/W vs. A_v for devices in SG and QDG mode. L=50 nm…100 nm, V_d=1V, and $V_g = V_{Th}$. Lines indicate general trends.

Figure 9. (a) I_{on}/W vs. I_{off}/W for UTBB devices biased in SG and QDG modes with different k-factors. I_{on} is extracted at a constant G_m/I_d of 10V^{-1}, I_{off} at V_g=0V, V_d=1 V. L=50 to 100 nm. Lines indicate general trends; (b) I_{off} in SG mode (dashed lines) and I_{off} reduction allowed by a QDG mode (I_{off}/W_{coef}) (symbols+lines) vs. k-factor for devices with different L. Circles indicate points of possible L reduction in a QDG mode.

2) Digital applications.

As demonstrated above, QDG mode allows simultaneous I_d increase, V_{Th} reduction and DIBL lowering without degradation of I_{off}. This feature can be of interest for digital applications (e.g. sleep-mode switches), allowing the reduction of either the voltage needed to achieve specified on-current I_{on} while keeping the same I_{off}, or the I_{off} through W reduction needed to achieve a given I_{on} at constant voltage.

Fig. 9a summarizes I_{on}/W vs. I_{off}/W for UTBB MOSFETs operated in SG and QDG modes with different k-factors. I_{on} is extracted at a constant G_m/I_d=10 V^{-1}, I_{off} at V_g=0 V, showing a clear I_{on} improvement in QDG mode while I_{off} stays constant. In order to give a rough estimation of possible W (and hence I_{off}) reduction provided by a QDG mode, we fix I_{on} as in SG mode and extract W reduction coefficient, W_{coef}=I_{on_QDG}/I_{on_SG} allowed by the use of QDG mode. One can see that W_{coef} goes from ~1.2 (for an ADG mode) to 4.2 in a high k-factor case, meaning that we can reduce 4.2 times the W of our device to reach the same I_{on}. In turn, W reduction leads to I_{off} reduction (evidently with the same coefficient) (Fig. 9b). Moreover, I_{off} reduction provided by QDG mode means that we can probably go one step further and use shorter L device, which in turn will provide higher I_{on} (thus allowing either better performance or additional W and hence I_{off} reduction). Fig. 9b illustrates such approach plotting I_{off} constant level in SG mode and I_{off} reduction achieved by the use of QDG mode, i.e. I_{off}/W_{coef} for the devices with different L. One can see for instance that 70 nm-long device in QDG mode with k=25 reaches the same I_{off} as 100 nm-long device in SG mode. Simple calculations lead to consider and compare for example, a device 1 in SG mode with L=100nm, W=10µm, I_{off}=1.2nA, I_{on}=520µA, versus a device 2 in QDG mode (k=25) with L=70nm, W=2.4µm, I_{off}=1.2nA, I_{on}=660µA. Device 2 provides either ~30% better performance (in terms of I_{on}) than device 1 with 4 times lower W and 6 times lower gate area; or reciprocally, further ~30% W (and I_{off}) reduction at a given I_{on}. Hence, 'second iteration' of 70 nm-long device yields Device 3: L=70nm, W=1.9µm, I_{off}=1nA, I_{on}=520µA. Thus, for the same performance, total W reduction of ~5 times and L*W reduction of 7.5 times (0.13µm^2 and 1µm^2 for Devices 1 and 3, respectively) are allowed by QDG mode application. Furthermore, it is worth to point out that devices operated in QDG mode use 0.23 V lower V_g to reach the given I_{on}.

IV. CONCLUSIONS

Effects of a QDG operation mode (i.e. simultaneous gate and substrate (or GP contact) sweep as V_{sub}=k*V_g) on the behavior of UTBB MOSFET have been investigated and compared to a standard SG mode. We have shown that QDG mode allows simultaneous improvement of electrostatic features and performance enhancement. Contrarily to the constant V_{sub} biasing in SG mode, V_{Th} modulation achieved in QDG mode has been shown accompanied by S and DIBL reduction as well as constant I_{off}, thus of great interest for low-voltage, low-power applications. Improved performance in terms of I_d and G_m, accompanied by lower DIBL and higher A_v values can actually be exploited for analog and digital applications. Additional studies are evidently required to figure out the complexity of QDG mode realization as well as to tackle the impact of parasitic resistance and capacitance of such connection which would lead to dynamic performance reduction, but the potential gate area decrease we quantified opens significant margins for actual implementations.

ACKNOWLEDGMENT

The authors thank Olivier Faynot and Thierry Poiroux (CEA-Leti) and David Bol (UCL) for valuable discussions.

REFERENCES

[1] C. Fenouillet-Beranger, et al., "Efficient multi-Vt FDSOI technology with UTBOX for low power circuit design", VLSI Symp. 2010, pp.65-66.

[2] F. Andrieu, et al., "Low leakage and low variability ulita-thin body and buried oxide (UT2B) SOI technology for 20nm low power CMOS and beyond", VLSI Symp. 2010, pp. 57-58.

[3] F. Balestra, et al., "Double-gate silicon-on-insulator transistor with volume inversion: a new device with greatly enhanced performance", IEEE Electron Dev. Lett., vol. EDL-8, no.9, 1987, pp.410-412.

[4] T. Ernst et al., "Ultimately thin double-gate SOI MOSFETs," IEEE Trans. Electron Devices, vol. 50, no. 3, pp. 830-838, 2003.

[5] V. Kilchytska, et al. "Investigation of charge control-related perfor-mances in double-gate SOI MOSFETs," 11th Int. SOI symp.(ECS), 2003, pp. 225-230.

[6] M.K.Md.Arshad et al., "UTBB SOI MOSFETs analog figures of merit: effects of GP and asymmetric DG regime", EuroSOI 2012, pp.111-112.

[7] A. Ohata, et al., "Correct biasing rules for virtual DG mode operation in SOI MOSFETs", IEEE, Trans. on El. Dev., vol. 52, 2005, pp. 124-125.

[8] V. Kilchytska et al., "Influence of device engineering on analog and RF perfor-mances of SOI MOSFETs," IEEE Trans. El. Dev., vol. 50, 2003, pp. 577-588.

978-1-4673-1707-8/12 $31.00 © 2012 IEEE

An Integration Approach for Graphene Double-Gate Transistors

S. Vaziri, A.D. Smith, C. Henkel, M. Östling, M.C. Lemme
School of Information and Communication Technology
KTH Royal Institute of Technology
16440 Kista, Sweden
vaziri@kth.se

G. Lupina, G. Lippert, J. Dabrowski, W. Mehr
Leibniz-Institut für Innovative Mikroelektronik, IHP
Im Technologiepark 25
15236 Frankfurt (Oder), Germany

Abstract— **In this work, we propose an integration approach for double gate graphene field effect transistors. The approach includes a number of process steps that are key for microelectronics integration: bottom gates with ultra-thin (2nm) high-quality thermally grown SiO_2 dielectrics, shallow trench isolation between devices and atomic layer deposited Al_2O_3 top gate dielectrics. The complete process flow is demonstrated with fully functional GFET transistors and can be extended to wafer scale processing and other graphene-based devices.**

I. INTRODUCTION

The experimental discovery of graphene provided fertile ground for research in device physics and engineering. Among its unique properties, the exceptional electronic characteristics make it a promising material for ultra-fast electronic applications. Compared to other high-mobility semiconductor materials, graphene exhibits high charge carrier mobilities at room temperature, high velocity saturation at low carrier density [1] and is mostly compatible with silicon CMOS technology [2]. Other defining features of graphene are a linear energy-momentum dispersion relation and the absence of an energy band gap ($E_g \approx 0$ eV). This leads to ambipolar device behavior, but it also makes graphene unsuitable as a replacement material for transistor channels in logic applications: the current can be modulated, but not switched off, resulting in unacceptable on-off current ratios. Instead, the focus has switched to analog, radio-frequency (RF) applications, where switching off is not as critical. A recent theoretical comparison of 65-nm channel length silicon MOSFETs with 65-nm channel length graphene field effect transistors (GFETs) FET showed superior figures of merit for GFETs if the mobility is higher than 3.000 cm^2/Vs [3]. However, maximum operating speed and intrinsic gain require drain current saturation in the output characteristics, which is weak in conventional GFETs due to the lack of a band gap in single-layer graphene. This is particularly the case in the ballistic and near-ballistic transport regime, where scattering does not contribute to current saturation. Hence, a small bandgap would be favorable for RF GFETs. A recent study suggests an optimal E_g of ~100 meV [4]. Two options to tune the band gap of graphene are quantization through etching of nanoribbons [5] and applying a vertical electric field in bilayer graphene [6]. A manufacturable solution could be to combine these two methods: bilayer graphene nanoribbon FETs [7, 8].

The fabrication of appropriate graphene-double gate structures is challenging because it requires high quality and scalable dielectrics on both gates, as well as well-controlled transfer processes for graphene. Here, we present a CMOS compatible integration solution for graphene double gate structures with high-quality ultra-thin (2nm) bottom gate oxides (SiO_2) and atomic layer deposited (ALD) aluminum oxide (Al_2O_3) top gate dielectrics. We demonstrate the proposed process with well-behaved single-layer GFETs, including high breakdown fields of the oxides and tunability of the charge neutrality (Dirac) point. The process is wafer-size scalable and is adaptable to the fabrication of other graphene-based devices such as hot electron transistors [9] and micro-cavity-based optoelectronic devices [10, 11].

II. DEVICE FABRICATION

A. Substrate Preparation

Device fabrication started with the formation of the bottom gate. For this purpose, n-type Si wafers were covered with a silicon nitride layer, which served as a hard mask and a stop layer for chemical mechanical polishing (CMP). This was followed by the definition of active areas using photolithography and reactive ion etching of Si_3N_4 and Si. Subsequently the trenches were filled with about 400 nm of high-density plasma undoped silicon glass (HDP USG). After another CMP step, n-dopant (P) was implanted in the active regions (bottom gate electrodes). The substrate preparation was finalized by the formation of a bottom SiO_2 gate oxide down to a thickness of ~2 nm using dry-wet thermal oxidation.

B. Graphene Transfer

Deposition of graphene on copper using CVD is well known as a scalable and rather straightforward method of graphene production [12]. We chose poly(Bisphenol A) Carbonate (PC) as the polymer to transfer graphene from copper to the substrates, as this leads to less residue after removal compared to commonly used poly(methyl

978-1-4673-1707-8/12 $31.00 © 2012 IEEE

metacrylate) (PMMA) [13, 14]. After etching of copper in FeCl$_3$ solution and rinsing the graphene/PC stack, the sample was placed in dilute HCl solution for removal of iron ions. Then, the transferred graphene/PC sample was dried on a hot plate at 60°C. The samples were then submerged in chloroform to remove the PC from the surface of graphene. Afterwards, the samples were annealed in forming gas at 350°C for one hour to further remove polymer residues from the graphene surface. Next, a thin film of about 4 nm of aluminum (Al) was evaporated onto the graphene to prohibit photoresist touching the graphene surface in subsequent processing [15]. This prohibitory layer is removed during photoresist development in MICROPOSIT MF CD-26 DEVELOPER. Source and drain contacts were formed by photolithography followed by evaporation of 10/70 nm of titanium (Ti) and gold (Au) (compare schematic in Fig. 1). Micro-Raman spectroscopy confirmed the successful graphene transfer process.

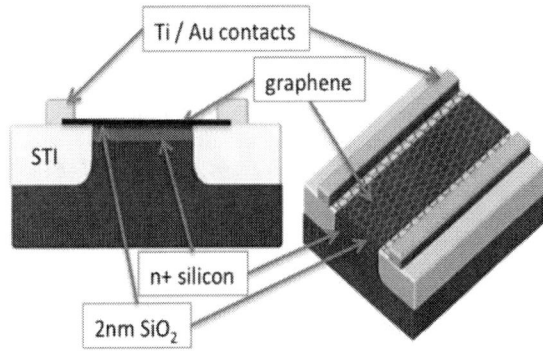

Figure 1. Schematic cross sectional and isometric views of the GFET structure with ultra-thin back gate dielectrics after contact deposition.

C. Top Gate Formation

The integration of top gate dielectrics typically degrade the the mobility in graphene FETs [2]. Among different approaches, ALD is one of the promising dielectric integration methods which induces minimum defects to the graphene crystal lattice [16] and deposits high quality pinhole-free high-k dielectrics [17]. Graphene surfaces are very similar to carbon nanotubes in terms of problems regarding ALD growth nucleation [18]. Hence, we deposited a 3-4nm thin film of Al as a nucleation layer prior to ALD. After oxidation in ambient air, high-k Al$_2$O$_3$ films of 10-30 nm were deposited using trimethylaluminum and water vapor as reactant gases in the ALD reactor. Raman spectra taken before and after top dielectric formation show that the PVD/ALD process does not induce significant defects to the graphene [17]. Finally, Ti and Au were evaporated as top gate electrodes. Fig. 2 shows schematics of the final device.

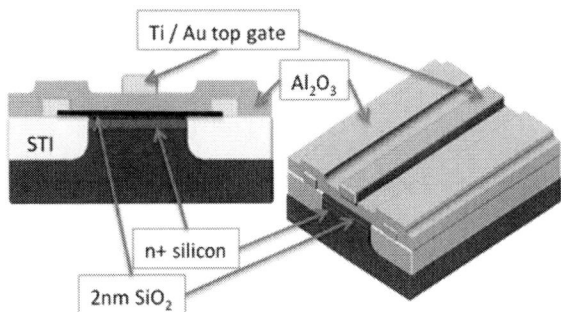

Figure 2. Schematic cross sectional and isometric views of the GFET double-gate structure with ultra-thin back gate dielectrics and ALD top-gate.

III. RESULTS AND DISSCUSSIONS

Fig. 3 shows an optical micrograph of a double gate GFET. The silicon active area, i.e. the bottom gate, is emphasized in the image for clarity. We note that the size and shape of the back gate can be adjusted to the particular application as a double-gate, hot electron or photo-transistor. In particular, it can be designed to enable local, individual back gate contacts. In this particular example the back gate is small compared to the top gate.

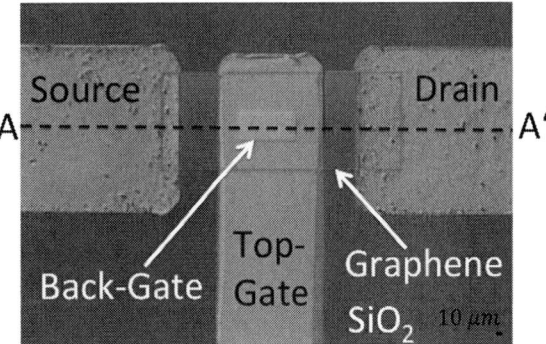

Figure 3. Optical micrograph of a double-gate GFET. The graphene and back gate regions are emphasized for clarity. A-A' indicates the cross section cut-line in Fig. 1 and Fig. 2.

The double gate GFETs were measured in standard back gate and top gate configurations separately. A top gate transfer characteristic is shown in Fig. 4. In this device, with a top gate Al$_2$O$_3$ layer of 28 nm, the bottom gate was biased at zero volts.

Fig. 5 shows the bottom gate transfer characteristics for different top gate voltages. In this case, the bottom gate and top gate insulators consist of 2 nm SiO$_2$ and 22 nm Al$_2$O$_3$, respectively. The Dirac point occurs in the negative region of bottom gate voltages, which implies n-type doping of the graphene, typical for Al$_2$O$_3$ gate oxides. Moreover, different top gate voltages result in a change of the conductivity and a shift of the Dirac point due to electrostatic doping. The typical top and the bottom gate current modulation ("V-shape") and the tunable Dirac point of our double gate GFETs demonstrate the feasibility of the proposed integration process.

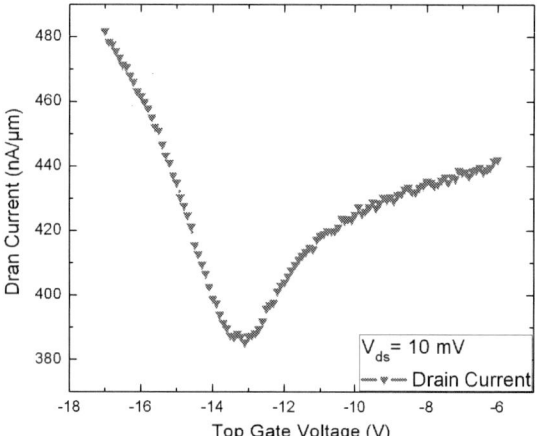

Figure 4. Top gate transfer characteristics of a GFET with 3 μm gate length and zero biased bottom gate.

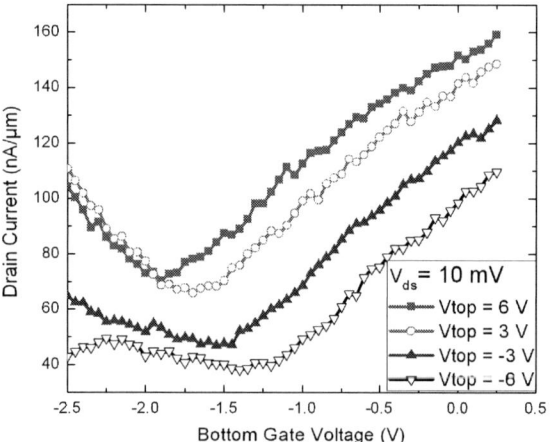

Figure 5. Bottom gate transfer characteristics at different top gate voltages for a GFET with 3 μm gate length. Conductivity and Dirac point position are modified by the top gate.

Hysteresis is a common issue in graphene FET technology. The ALD process used for Al_2O_3 effectively suppresses hysteresis. This is shown in Fig. 6, where transfer characteristics of a back gated GFET are compared before and after Al_2O_3 deposition. In contrast to the initial measurement (blue squares), hysteresis can no longer be observed after deposition (red triangles), which indicates a low density of oxide charge traps. The deposition of the Al_2O_3 film leads to a shift of the Dirac point towards negative gate voltages, which can be attributed to the introduction of positively charged impurities at the graphene/Al_2O_3 interface or in the Al_2O_3. The mobility is degraded roughly by a factor of two by the deposition process. We note that in order to rule out defects in the graphene films, this experiment was carried out using exfoliated graphene as the channel material and 90 nm thermally grown dry SiO_2 as the gate insulator.

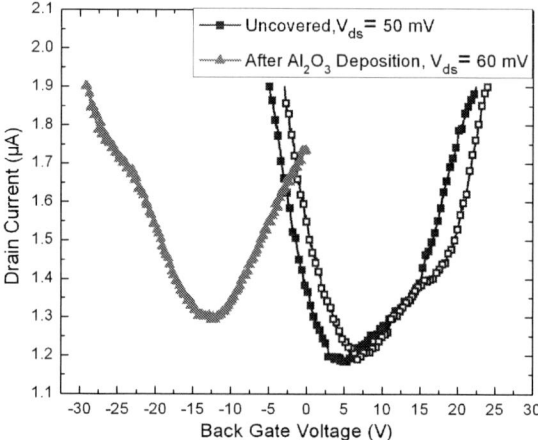

Figure 6. Double sweep transfer characteristics before (blue and open squares) and after (red triangles) Al_2O_3 deposition.

Dielectric breakdown measurements have been carried out to further assess the quality of the gate insulators. Fig. 7 shows the current-voltage characteristics of a ~2 nm SiO_2 in a double gate GFET. The bottom gate was set to $V_{bg} = 0$ V and the potential of the graphene film was varied via the source/drain contacts ("Voltage" in Fig. 7). The top gate was left floating. The total dielectric breakdown happens at around 7.6 V, corresponding to an electric field of 38 MV/cm.

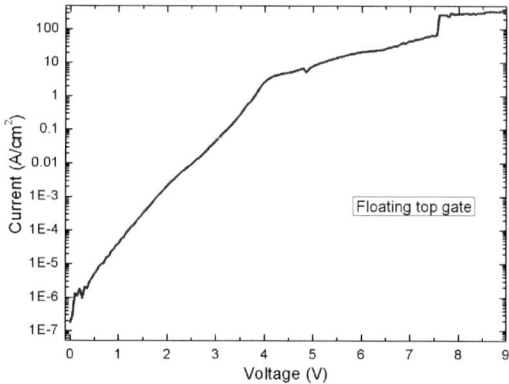

Figure 7: Current-voltage breakdown characteristics of a ~2 nm SiO_2 gate oxide.

While this high value seems to confirm the high quality of the bottom gate dielectric and graphene transfer, we also note that it is surprisingly close to the bond breaking energy in SiO_2. Hence, even though it is in line with recent findings on nanometer scale SiO_2 films [19], it requires further commenting. The breakdown field in this particular case has to be estimated considering the quantum capacitance of graphene, the band bending in the bottom silicon electrode, and the series resistance of the silicon wafer. The ultrathin oxide thickness results in a large oxide capacitance and due to the low density of states of graphene, the quantum capacitance is not negligible. In addition, the silicon electrode (wafer) has a doping level of $N_D = 5 \times 10^{15}$ cm^{-3}, which results both in a high series resistance as well as silicon band bending at the Si/SiO$_2$ interface. Even though a detailed model is beyond the scope of this article, we note that these effects lead to an overestimation of the breakdown field.

An identical measurement was carried out for a 10nm-Al$_2$O$_3$ top gate dielectric. The respective current-voltage characteristics in Fig. 8 shows dielectric breakdown at around 9.5 V, which translates to a breakdown electric field of 9.5 MV/cm, reasonably close to theoretical values [19]. While the hysteresis measurements imply a low density of charge traps in the high-k film, the breakdown electric field indicates a rather defect-free film with reasonably low roughness / inhomogeneity, even though a more detailed study with a larger, statistically relevant number of devices is needed to elaborate further.

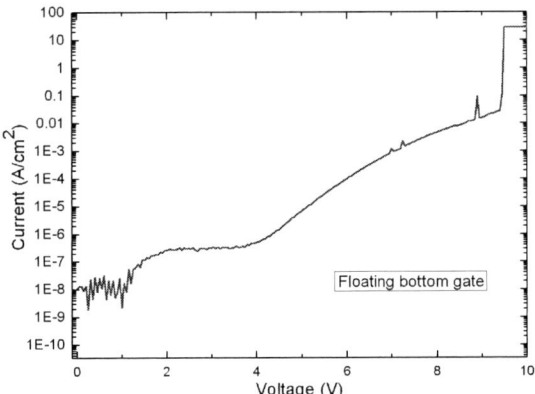

Figure 8. Current-voltage characteristics of 10 nm Al$_2$O$_3$.

IV. CONCLUSIONS

In this work, we have proposed an integration approach for double gate graphene field effect transistors. The approach includes a number of process steps that are key for microelectronics integration: bottom gates with ultra-thin (2nm) high-quality thermally grown SiO$_2$ dielectrics, shallow trench isolation between devices and atomic layer deposited Al$_2$O$_3$ top gate dielectrics. The complete process flow has been demonstrated with fully functional GFETs. Our approach will enable local control of vertical electric fields in double gate GFETs on a wafer scale. In addition, novel devices such as graphene base hot electron transistors (GBTs) [9] or optical micro-cavities for photodetection and graphene light emitters [10, 11] can be manufactured with this process on wafer scale.

ACKNOWLEDGMENTS

We thank L. Selmi (Univ. Udine), P. Palestri (Univ. Udine), F. Driussi (Univ. Udine) for fruitful discussions and A. Srinivasan (KTH), R. Sanatinia (KTH), H. Ruecker (IHP) and A. Trusch (IHP) for experimental support. The KTH authors acknowledge support through an Advanced Investigator Grant (OSIRIS, No. 228229) and a Consolidator Grant (InteGraDe, 307311) from the European Research Council.

REFERENCES

[1] V. E. Dorgan, *et al.*, "Mobility and Saturation Velocity in Graphene on SiO2," *Applied Physics Letters*, vol. 97, p. 082112, 2010.

[2] M. C. Lemme, *et al.*, "A Graphene Field-Effect Device," *IEEE Electron Device Letters*, vol. 28, pp. 282-284, 2007.

[3] S. V. S. Rodriguez, M. Ostling, A. Rusu, E. Alarcon, M.C. Lemme, "RF Performance Projections of Graphene FETs vs. Silicon MOSFETs," *arXiv:1110.0978v1*, 2011.

[4] S. Das and J. Appenzeller, "On the Importance of Bandgap Formation in Graphene for Analog Device Applications," *Nanotechnology, IEEE Transactions on*, vol. 10, pp. 1093-1098, 2011.

[5] M. Y. Han, *et al.*, "Energy Band-Gap Engineering of Graphene Nanoribbons," *Physical Review Letters*, vol. 98, pp. 206805-4, 2007.

[6] J. B. Oostinga, *et al.*, "Gate-induced insulating state in bilayer graphene devices," *Nat Mater*, vol. 7, pp. 151-157, 2008.

[7] F. Xia, *et al.*, "Graphene Field-Effect Transistors with High On/Off Current Ratio and Large Transport Band Gap at Room Temperature," *Nano Letters*, vol. 10, pp. 715-718, 2010.

[8] B. N. Szafranek, *et al.*, "Electrical observation of a tunable band gap in bilayer graphene nanoribbons at room temperature," *Applied Physics Letters*, vol. 96, pp. 112103-3, 2010.

[9] W. Mehr, *et al.*, "Vertical Transistor with a Graphene Base," *IEEE Electron Device Letters*, vol. 33, pp. 691-693, 2012.

[10] M. Engel, *et al.*, "Light-matter interaction in a microcavity-controlled graphene transistor," *Nature Communications*, 3, p. 906, 2012.

[11] M. Furchi, *et al.*, "Microcavity-integrated graphene photodetector," *arXiv:1112.1549v1*, 2011.

[12] Y. Lee, *et al.*, "Wafer-Scale Synthesis and Transfer of Graphene Films," *Nano Letters*, vol. 10, pp. 490-493, 2010.

[13] Y.-C. Lin, *et al.*, "Clean Transfer of Graphene for Isolation and Suspension," *ACS Nano*, vol. 5, pp. 2362-2368, 2011/03/22 2011.

[14] H. J. Park, *et al.*, "Growth and properties of few-layer graphene prepared by chemical vapor deposition," *Carbon*, vol. 48, pp. 1088-1094, 2010.

[15] A. Hsu, *et al.*, "Impact of Graphene Interface Quality on Contact Resistance and RF Device Performance," *IEEE Electron Device Letters*, vol. 32, pp. 1008-1010, 2011.

[16] L. Liao and X. Duan, "Graphene-dielectric integration for graphene transistors," *Materials Science and Engineering: R: Reports*, vol. 70, pp. 354-370, 2010.

[17] S. Vaziri, *et al.*, "A Hysteresis-Free High-k Dielectric and Contact Resistance Considerations for Graphene Field Effect Transistors," *ECS Transactions*, vol. 41, pp. 165-171, 2011.

[18] D. B. Farmer and R. G. Gordon, "Atomic Layer Deposition on Suspended Single-Walled Carbon Nanotubes via Gas-Phase Noncovalent Functionalization," *Nano Letters*, vol. 6, pp. 699-703, 2006.

[19] C. Sire, *et al.*, "Statistics of electrical breakdown field in HfO$_2$ and SiO$_2$ films from millimeter to nanometer length scales," *Applied Physics Letters*, vol. 91, p. 242905, 2007.

MTJ-based Implication Logic Gates and Circuit Architecture for Large-Scale Spintronic Stateful Logic Systems

Hiwa Mahmoudi, Viktor Sverdlov, and Siegfried Selberherr

Institute for Microelectronics, TU Wien, Gußhausstraße 27–29/E360, A–1040 Wien, Austria
E-mail: {mahmoudi|sverdlov|selberherr}@iue.tuwien.ac.at

Abstract—Because of the easy integration with CMOS, non-volatility, reconfiguration capability, and fast-switching speed of magnetic tunnel junctions (MTJs), this work proposes and investigates stateful IMP-based logic gates and circuit architecture for future reconfigurable and nonvolatile computing systems. Stateful logic uses the memory unit (MTJ device) as the main computing element (logic gate) unlike the previously proposed MTJ-based logic circuits, where MTJs are only ancillary devices in logical computations. Spintronic IMP logic gates are analyzed using a SPICE model for spin-transfer torque MTJs to demonstrate the reliability of the IMP operation. The realization of the spintronic stateful logic operations extends nonvolatile electronics from memory to logical computing applications and opens the door for more complex logic functions to be realized with MTJ-based devices.

I. INTRODUCTION

To enable stateful logic operations, the realization of a fundamental Boolean logic operation called material implication (IMP) has been demonstrated recently [1] using two equivalent TiO_2 memristive switches [2]. Stateful logic allows memory cells to serve simultaneously as logic gates and latches. This improves the conventional CMOS logic which combines logic circuits and memory elements to transfer back and forth information between them and also opens the door for a shift away from the Von Neumann architecture for innovation in computational paradigms.

By using two different circuit topologies of the magnetic tunnel junction (MTJ)-based IMP gates (Fig.1a and Fig.1b), we show the possible parallel execution of the same computational sequence in a large-scale logic system based on the spin-RAM architecture [3], which is near commercialization. The MTJ has the advantages of CMOS-compatibility, non-volatility, reconfigurable capability, and fast-switching speed [4], [5], [6] and has received great interest to overcome the significant increase in the leakage currents in CMOS circuits [7]. In addition, by utilizing the spin transfer torque (STT) effect [8] [9], the STT-MTJ gives pure electrical switching and better scalability than conventional MTJs switched by magnetic field.

Spintronic stateful logic architecture reduces the device counts and interconnection delay for which the memory elements acts also as logic gates. Using MTJ's non-volatility, spintronic stateful logic architecture is expected to reduce the static power dissipation similar to logic-in-memory [10] and non-volatile CMOS/MTJ hybrid [11] architectures. It reduces the device counts and interconnection delay by using the memory element as the main computation component (gate) as compared to the aforementioned architectures which use the MTJs as ancillary device embedded in the interconnection layer of the logic circuits.

II. SPINTRONIC IMP LOGIC GATES

Material implication, 'p IMP q' or 'if p, then q', is a fundamental Boolean logic operation which in combination with FALSE (Logic 0) forms a complete logic basis to compute any Boolean function. In contrast to [1], we use MTJs as the memory elements to build spintronic IMP gates. In addition to the conventional IMP circuit topology (Fig.1a), we consider a new topology (Fig.1b) driven by a current source, which offers a more energy-efficient and reliable implementation. The MTJ contains two ferromagnetic layers separated by a thin non-conductive tunneling barrier. The magnetization of one layer (fixed layer) is pinned, while the magnetization of the second one (free layer) can be switched freely using an external magnetic field or (spin) current passing through

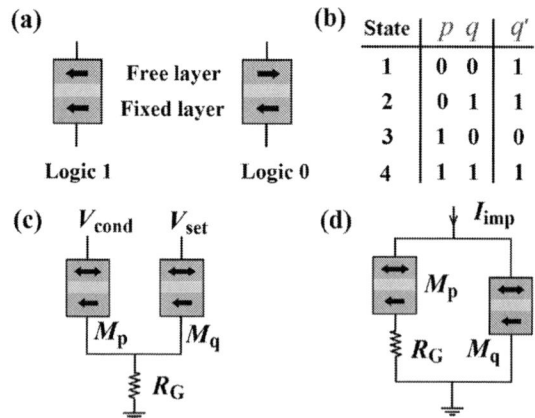

Fig. 1. (a) MTJ basic structure. (b) The IMP truth table. Conventional (c) [1] and proposed (d) IMP circuit topologies.

978-1-4673-1707-8/12 $31.00 © 2012 IEEE

the MTJ. The electrical resistance of the device depends on the relative orientation of the magnetization directions of the ferromagnetic layers. The parallel (P) magnetization state results in a low-resistance state (R_P; logic 1) across the barrier, while the antiparallel (AP) alignment places it in a high-resistance state (logic 0; R_{AP}). The resistance modulation is described by the tunnel magnetoresistance (TMR) ratio, defined as $(R_{AP} - R_P)/R_P$.

We analyze the IMP gates based on the SPICE model of the MTJ [12], which uses the equivalent circuit of the STT-MTJ shown in Fig.2. A curve-fitting circuit is used to model the TMR voltage(current) dependence, which is important to determine the $R - V$ characteristics of the MTJ and the voltage(current) division between the source and target MTJs in an IMP gate. The output signals of the decision circuit (V_1 and V_2) are used to determine when the device should switch states, based on the critical switching time and current (τ_0 and I_{C0}) characteristics of the device, which usually are defined corresponding to the 50% switching probability. In order to analyze the correct behavior and the reliability of the IMP gates, we use the theoretical expression of the MTJ switching probability as (1) (P_{sw}) [13], which is experimentally proved in [3].

$$P_{sw} = 1 - \exp\left\{-\frac{t}{\tau_0}\exp\left[-\Delta_0\left(1 - \frac{I}{I_{C0}}\right)\right]\right\} \quad (1)$$

Δ_0 is the magnetic memorizing energy without any current and magnetic field, t is the pulse width, and I is the current flowing through the MTJ. As shown in Fig.2, an error calculation circuit can be added to the SPICE model to calculate P_{sw}. Fig.3 shows how the proposed circuit can improve the SPICE model for direct calculation of P_{sw} and an IMP reliability analysis.

With the conventional topology the initial logic state of the source MTJ (M_p) provides a state dependent modulation (SDM) of the voltage across the target MTJ (M_q) through R_G. Due to this SDM, M_q switches (AP→P) in State 1, but remains unchanged in State 3. Thus, V_{cond} is chosen to leave

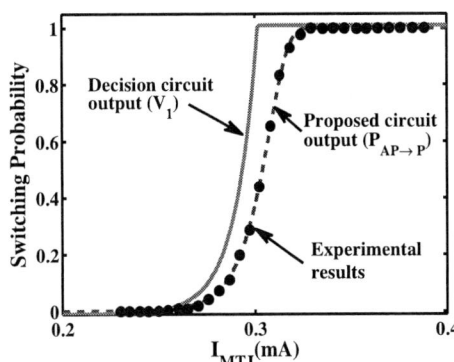

Fig. 3. Switching probability as a function of the applied current. The output of the proposed error calculation circuit reproduces the experimental data [3] as expected from the theory (1).

M_p unchanged. With our proposed topology the initial logic state of the M_p provides an SDM of the current through M_q, which results in a correct logic behavior of M_q as shown in Fig.4a. Because of R_G, the current through M_p is low enough to leave it unchanged. According to the IMP truth table (Fig.1c), we define the IMP error as

$$E_{imp} = (1 - P_{sw}^{q1}) + P_{sw}^{p1} + P_{sw}^{p2} + P_{sw}^{q3} \quad (2)$$

The value of the IMP gate circuit parameters can be optimized to minimize the error for fixed pulse duration and the TMR as shown in Fig.4b. Our results show that in the conventional topology, the optimal R_G is higher by a factor of $\times(2$ to $3)$ as compared to the new one. Therefore, the IMP energy consumption is about 60% lower than with the conventional gate topology (Fig.5a).

Robust IMP logic behavior requires a wide enough SDM window (Fig.4a). The width of the SDM window increases with the TMR ratio as shown in Fig.5b. It demonstrates that the IMP error decreases exponentially with increasing TMR ratio. At a fixed TMR the proposed topology provides a higher SDM, thus reducing the IMP error by about 60% as

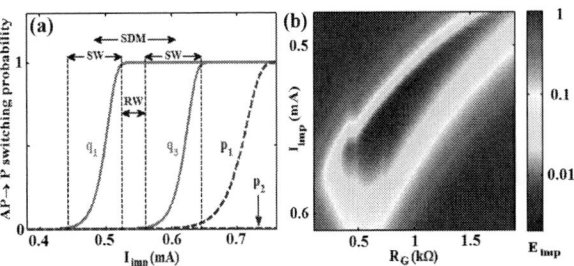

Fig. 4. Switching probabilities of M_q and M_p (a) and the total IMP error as a function of R_G and I_{imp} (b), plotted for a 50ns IMP execution in the new IMP circuit topology based on physical devices characterized in [3]. The SDM opens a reliable window (RW) between the switching windows (SWs) of the desired and disturbing AP→P switchings.

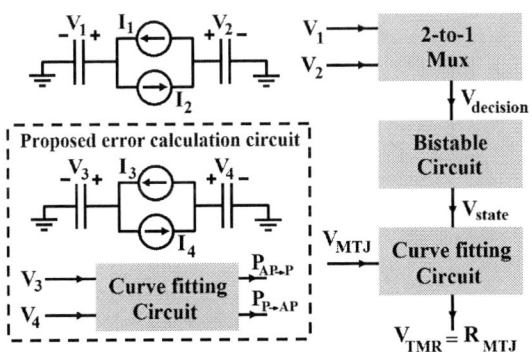

Fig. 2. The simplified equivalent circuit of the MTJ model in [12] and the proposed error calculation circuit. $I_3 = exp\left[-\Delta_0\left(1 - I/I_{C0(AP\rightarrow P)}\right)\right]$ and $I_4 = exp\left[-\Delta_0\left(1 - I/I_{C0(P\rightarrow AP)}\right)\right]$.

compared to the conventional. The record room temperature TMR of 604% [14] found in single-barrier MTJs is close to the theoretical maximum [15], [16]. This makes the MgO-based MTJs predominant candidates for STT magnetoresistive random access memories (STT-MRAMs) and promises highly reliable IMP gates.

III. SPINTRONIC IMP LOGIC CIRCUIT ARCHITECTURE

In conventional MRAM architecture the MTJ is connected to the crossing points of two perpendicular arrays of parallel conducting lines. The STT switching technique brought significant advantages and eliminates the difference between reading and writing in spin-RAM architecture [3]. A typical memory cell of the spin-RAM, which consist of an access transistor and an MTJ as its storage element (the 1T/1MTJ structure) is shown in Fig.6. Stateful IMP logic architecture offers the eliminate the need of using extra charge-based logic gates, while uses the memory cells simultaneously as logic gates and latches via IMP operation. Fig.7 shows a block IMP logic circuit architecture based on the 1T/1MTJ structure, which can realize the spintronic IMP gate shown in Fig.1d.

The IMP operation between two memory cells $C_{i,j}$ and $C_{i,j'}$ ($q_{i,j} \leftarrow p_{i,j'}$ IMP $q_{i,j}$) can be performed by simultaneous selection of the i-th word-lines (WLs) and the j-th and the j' source-line (SL) selectors which connect the SLs to ground directly and via R_G, respectively, and applying the current source I_{imp} to the j-th and j'-th bit-lines (BLs). Then the result of the IMP operation will be written in $C_{i,j}$.

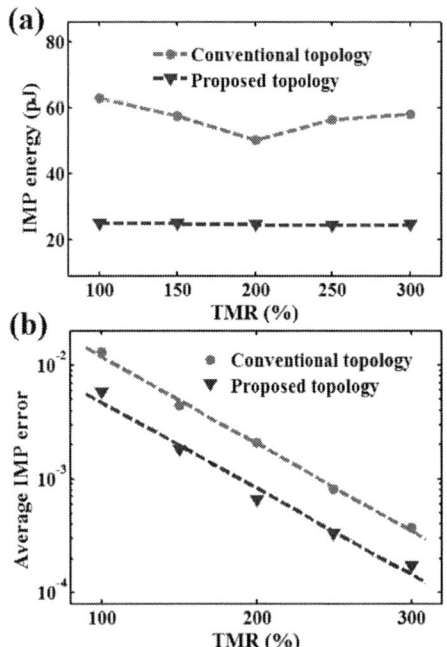

Fig. 5. The IMP energy consumption (a) and the average error (b) depends on the TMR ratio for both conventional and proposed topologies.

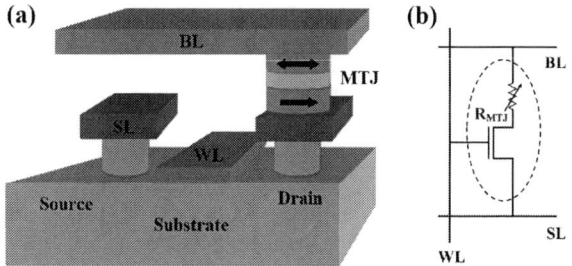

Fig. 6. 1T/1MTJ structure. Structural (a) and the equivalent circuit (b) diagrams [18].

As compared to the spin-RAM we have added two *work cells* to any WL, while it has been shown that with two additional memristors all Boolean functions on any number of memory cells can be performed [17]. These work cells can also be used to connect different WLs. In fact, in order to perform the IMP between memory cells from different WLs, one can copy the logic data stored in one memory cell to a work cell from the other WL.

It should be noted that the nonzero ON resistance of the access transistors (R_{on}) decreases the effective TMR of the 1T/1MTJ cells which can be defined as

$$TMR_{\mathrm{eff}} = \frac{R_{\mathrm{AP}} - R_{\mathrm{P}}}{R_{\mathrm{P}} + R_{\mathrm{on}}} \qquad (3)$$

Therefore, a robust IMP operation needs MTJs with sufficiently high TMR and electrical resistance. Our simulations show that the TMR of a 1T/1MTJ including the MTJ devices characterized in [3] and an access device with a width about 1-2μm at the 180-nm technology decreases about 10%-30%. Therefore, according to Fig.5, a 99.9% IMP correct logic behavior requires a TMR ratio higher than 250%.

IV. STATEFUL SPINTRONIC FULL ADDER

We consider a full adder which is a basic element of arithmetic circuits. As is well known, it adds three binary inputs (c_1-c_3) and produces two binary outputs, sum (S = c_1 XOR c_2 XOR c_3) and carry (C =[c_1 AND c_2] OR [c_3 AND {c_1 XOR c_3}]). Since IMP cannot fan out, two operations, FALSE ($c_j \leftarrow 0$) and IMP ($c_j \leftarrow c_i$ IMP 0), should be executed in subsequent steps to write $\overline{c_i}$ in an additional cell $\overline{c_j}$ ($j = 4-6$), in order to ensure that the logical value $\overline{c_i}$ (therefore c_i) is still available, when it is needed as an input for subsequent IMP operations. As 'p IMP 0' and 'p IMP q' are equivalent to 'NOT p' and '(NOT p) OR q', respectively, some operations can be eliminated to minimize the total effort. Our design involves only 27 subsequent FALSE and IMP operations on 3 input cells (c_1-c_3) and 3 additional cells (c_4-c_6), in contrast to the earlier proposed IMP-based scheme [19] with 19 and 18 operations (37 total) for generating S and C, respectively, and 4 additional cells. Therefore, our design requires less operations (delay) and devices (area).

978-1-4673-1707-8/12 $31.00 © 2012 IEEE

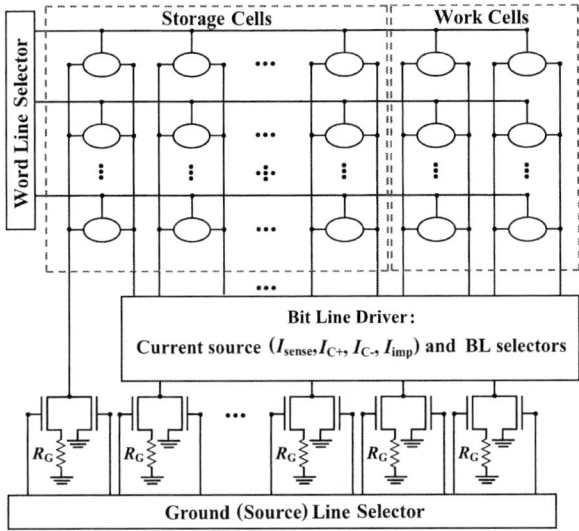

Fig. 7. A simplified spintronic IMP logic circuit architecture based on the spin-RAM architecture to realize the MTJ-based IMP gate with the proposed topology shown in Fig.1d. Controlling and programming the line drivers and selectors requires an external processor similar to the proposed circuit for the TiO$_2$-based architecture [1].

The logic-in-memory circuit presented in [10] uses 34 transistors and 4 MTJs for implementing a full-adder, while stateful architecture eliminates the need of using extra charge-based logic gates and offers superior logic density.

V. CONCLUSION

We have described MTJ-based IMP gates as basic elements to combine memory and logic computing in a stateful logic circuit based on existing spin-RAM architectures. This opens an alternative path towards reconfigurable and nonvolatile MTJ-based computing devices and systems [20]. The robustness of the IMP operation is based upon a state dependent modulation (SDM) of the voltage(current) division between the source and target MTJs. It has been demonstrated that the reliability increases exponentially with increasing TMR ratio.

Due to non-volatility and eliminating extra charge-based logic gates, the stateful IMP logic is expected to exhibit low power consumption, high logic density, and high speed operation simultaneously.

ACKNOWLEDGMENT

The work is supported by the European Research Council through the grant #247056 MOSILSPIN.

REFERENCES

[1] J. Borghetti, G. S. Snider, P. J. Kuekes, J. J. Yang, D. R. Stewart, and R. S. Williams, "Memristive Switches Enable Stateful Logic Operations via Material Implication," *Nature*, vol. 464, no. 7290, pp. 873–876, 2010.

[2] D. B. Strukov, G. S. Snider, D. R. Stewart, and R. S. Williams, "The Missing Memristor Found," *Nature*, vol. 453, no. 7191, pp. 80–83, 2008.

[3] M. Hosomi, H. Yamagishi, T. Yamamoto, K. Bessho, Y. Higo, K. Yamane, H. Yamada, M. Shoji, H. Hachinoa, C. Fukumoto, H. Nagao, and H. Kano, "A Novel Nonvolatile Memory with Spin Torque Transfer Magnetization Switching: Spin-RAM," *IEDM Tech. Dig.*, pp. 459–462, 2005.

[4] M. N. Baibich, J. M. Broto, F. N. V. D. A. Fert, F. Petroff, P. Etienne, G. Creuzet, A. Friederich, and J. Chazelas, "Giant Magnetoresistance of (001)Fe/(001)Cr Magnetic Superlattices," *Phys. Rev. Lett.*, vol. 61, no. 21, pp. 2472–2475, 1988.

[5] B. N. Engel, J. Akerman, B. Butcher, R. W. Dave, M. DeHerrera, M. Durlam, G. Grynkewich, J. Janesky, S. V. Pietambaram, N. D. Rizzo, J. M. Slaughter, K. Smith, J. J. Sun, and S. Tehrani, "A 4-Mb Toggle MRAM Based on a Novel Bit and Switching Method," *IEEE Trans. Magn.*, vol. 41, no. 1, pp. 132–136, 2005.

[6] C. Chappert, A. Fert, and F. N. V. Dau, "The Emergence of Spin Electronics in Data Storage," *Nat. Mater.*, vol. 6, pp. 813–823, 2007.

[7] N. S. Kim, T. Austin, D. Baauw, T. Mudge, K. Flautner, J. S. Hu, M. J. Irwin, M. Kandemir, and V. Narayanan, "Leakage Current: Moores Law Meets the Static Power," *Computer*, vol. 36, no. 12, pp. 68–75, 2010.

[8] J. C. Slonczewski, "Current-Driven Excitation of Magnetic Multilayers," *J. of Magn. and Magn. Mater.*, vol. 159, pp. L1–L7, 1996.

[9] L. Berger, "Emission of Spin Waves by a Magnetic Multilayer Traversed by a Current," *Phys. Rev. B, Condens. Matter*, vol. 54, pp. 9353–9358, 1996.

[10] S. Matsunaga, J. Hayakawa, S. Ikeda, K. Miura, T. Endoh, H. Ohno, and T. Hanyu, "MTJ-Based Nonvolatile Logic-in-Memory Circuit, Future Prospects and Issues," *Proc. Des. Autom. Test Eur. Conf. (DATE)*, pp. 433–435, 2009.

[11] Y. Gang, W. Zhao, J. O. Klein, C. Chappert, and P. Mazoyer, "A High-Reliability, Low-Power Magnetic Full Adder," *IEEE Trans. Magn.*, vol. 47, no. 11, pp. 4611–4616, 2011.

[12] J. D. Harms, F. Ebrahimi, X. F. Yao, and J. P. Wang, "SPICE Macromodel of Spin-Torque-Transfer-Operated Magnetic Tunnel Junctions," *IEEE Trans. Electron Devices*, vol. 57, no. 6, pp. 1425–1430, 2010.

[13] Y. Higo, K. Yamane, K. Ohba, H. Narisawa, K. Bessho, M. Hosomi, and H. Kano, "Thermal Activation Effect on Spin Transfer Switching in Magnetic Tunnel Junctions," *Appl. Phys. Lett.*, vol. 87, p. 082502, 2005.

[14] S. Ikeda, J. Hayakawa, Y. Ashizawa, Y. M. Lee, K. Miura, H. Hasegawa, M. Tsunoda, F. Matsukura, and H. Ohno, "Tunnel Magnetoresistance of 604% at 300 K by Suppression of Ta Diffusion in CoFeB/MgO/CoFeB Pseudo-Spin-Valves Annealed at High Temperature," *Appl. Phys. Lett.*, vol. 93, p. 082508, 2008.

[15] W. H. Butler, X.-G. Zhang, T. C. Schulthess, and J. M. MacLaren, "Spin-Dependent Tunneling Conductance of Fe|MgO|Fe Sandwiches," *Phys. Rev. B*, vol. 63, p. 054416, 2001.

[16] J. Mathon and A. Umersky, "Theory of Tunneling Magnetoresistance of an Epitaxial Fe/MgO/Fe(001) Junction," *Phys. Rev. B*, vol. 63, p. 220403, 2001.

[17] E. Lehtonen, J. H. Poikonen, and M. Laiho, "Two Memristors Suffice to Compute All Boolean Functions," *Electron. Lett.*, vol. 46, no. 3, pp. 239–240, 2010.

[18] H. Li and Y. Chen, "An Overview of Non-Volatile Memory Technology and the Implication for Tools and Architectures," *Proc. Des. Autom. Test Eur. Conf. (DATE)*, pp. 731–736, 2009.

[19] K. Bickerstaff and E. E. Swartzlander, "Memristor-Based Arithmetic," *Asilomar conf. on Sig., Sys., and Comp.*, pp. 1173–1177, 2010.

[20] X. Yao, J. Harms, A. Lyle, F. Ebrahimi, Y. Zhang, and J. P. Wang, "Magnetic Tunnel Junction-Based Spintronic Logic Units Operated by Spin Transfer Torque," *IEEE Trans. Nanotechnol.*, vol. 11, no. 1, pp. 120–126, 2012.

Resistive switching memory using titanium-oxide nanoparticle films

E. Verrelli, D. Tsoukalas

Department of Applied Sciences
National Technical University of Athens,
Heroon Politechniou 9, 15780 Athens, Greece
verrelli@central.ntua.gr

P. Normand, N. Boukos

NCSR "Demokritos",
Patriarchou Grigoriou, 15310 Aghia Paraskevi, Greece.

A.H. Kean

Mantis Deposition Ltd.,
Oxfordshire, OX9 3BX, United Kingdom.

Abstract— In this work we present symmetric metal-insulator-metal bipolar memristors based on room-temperature deposition of titanium-oxide nanoparticles (TiO NPs) formed in vacuum by a physical process. We report that deposition under substrate biasing conditions strongly affects the structural and electrical properties of the produced TiO-NP films including their bipolar switching behaviour. The application of an external electric field during deposition enhances the mean size and oxygen content of the TiO NPs as well as the high-to-low resistance (HLR) ratio of the memristive films. Under the substrate-biasing deposition conditions examined so far, we successfully achieved bistable devices with a HLR ratio increased by two orders of magnitude compared to devices using TiO-NP films formed without electric-field assisted NP deposition.

I. INTRODUCTION

In the last decade considerable efforts are underway for the development of new materials and alternative memory concepts. In that direction, non-volatile memories (NVMs) based on the electrically switchable resistance of materials located in between two-terminal electrodes, also called memristors and initially proposed by Chua [1], referred to as resistive random-access-memories (RRAMs or ReRAMs) have attracted considerable attention because of their potential application in crossbar resistive memories [2-4]. RRAMs makes use of binary and ternary metal oxides (but also demonstrative examples on perovskites [5] and organic materials [6] have been reported) in a metal-insulator-metal (MIM) device configuration where the low and high resistive states can be programmed by application of voltage pulses of same (unipolar switching) or opposite (bipolar switching) polarity to the metal electrodes [4].

Titanium oxide, which is a wide band gap semiconductor, has been already used in the form of nanoparticles to fabricate dye-sensitized solar cells [7]. Here, we introduce MIM bipolar memristors using titanium oxide (TiO) NPs formed in vacuum by a physical process and deposited on metal electrodes at room temperature. We demonstrate that application of an external electric field during NP deposition modulates the structural and electrical properties of the produced memristive films. The resulting devices exhibit reliable and reproducible

bipolar resistive switching without the need for an electroforming step. Under the deposition conditions examined so far, we successfully achieved bistable devices with a high-to-low resistance (HLR) ratio of ~ 1000 and switching voltages lower than 1V. This study allows at the same time to get insight on filament formation by correlating electrical and structural morphology of the obtained nanoparticle films.

II. EXPERIMENTAL

The NP fabrication is based on the terminated gas condensation method. Titanium NPs are created in a specific compartment attached to the main ultra-high-vacuum chamber where the NPs are swept through a small aperture. In the main chamber, an oxygen rich atmosphere oxidizes the NPs that deposit onto the sample at room temperature. At first, bottom contacts were formed on oxidized silicon substrates by successive e-beam evaporation of 30 nm of chromium and 200 nm of gold. Then, TiO NPs were deposited on the gold bottom contacts. Titanium NPs were fabricated by DC sputtering from a pure Ti target at 50W with Ar flow of 60 sccm. In the main chamber the oxygen flow was kept at 10 sccm. The final nominal thickness of the produced TiO NP films was kept constant at 40 nm. Transmission electron microscopy revealed that the deposited TiO NPs had a mean diameter of 5 nm and resulted to be mainly amorphous. No core/shell structure was observed indicating that the Ti NPs are well oxidized. Finally, top gold contacts of $2 \ 10^{-3} \ cm^2$ were deposited by e-beam evaporation of 200 nm of gold through a shadow mask. The NPs produced by the above technique are negatively charged and thus can be manipulated by electric fields [8]. The TiO NPs considered in this work are by about 90% negatively charged and are deposited onto the metal electrode under different sample holder bias voltage conditions. The two main MIM structures reported herein concern memristive films obtained (1) in the absence of any applied electric field with the NPs soft landing on the gold bottom contact and (2) following the application of a +5kV on the sample holder during deposition; they are referred to as the "as-deposited sample" and "high-voltage sample", respectively. SEM cross section images of the two TiO NP films revealed that the high porosity of the NP film resulted in a physical thickness of ~280

978-1-4673-1707-8/12 $31.00 © 2012 IEEE

nm and ~70 nm for the as-deposited and the high-voltage sample respectively, as shown in Figure 1c and 1d.

Figure 1. a) SEM plan view image of as deposited titanium oxide nanoparticle film b) SEM plan view image of high-voltage titanium oxide nanoparticle film, larger grains are visible compared with case (a). Images c) and d) represent cross section SEM images of the as-deposited and high voltage samples respectively. Although the nominal thickness of the two NP films is the same (40 nm), the different fabrication conditions used, produced films with different porosity and thus, physical thickness.

III. RESULTS

SEM cross section images as well as SEM plan view images taken on the as-deposited and high-voltage samples revealed a different NP-film morphology. The as-deposited sample (Figure 1a and 1c) has a NP film made of single particles with a mean diameter of 5 nm. Such a result is consistent with TEM analysis (not shown here). On the other hand, the high-voltage sample (Figure 1b and 1d) exhibits a NP film with grains much larger than the single NPs of the as-deposited film, having a mean diameter of about 25 nm. The above structural disparities between the TiO-NP films obtained in the absence and presence of an external accelerating electric field are reflected into differences between the electrical and physical properties of the films which present anyway a pronounced resistive switching memory effect. Current-Voltage (I-V) measurements performed on the as-deposited and high-voltage samples revealed a clear forming-free bipolar resistive switching phenomenon as shown in Figure 2. For the as deposited sample, the initial I-V sweeps in forward and backward direction between -0.8 V and 0.8 V (Figure 2a and 2b) reveal the existence of two states: the low-resistance-state (LRS) and the high-resistance-state (HRS) characterized by resistances of ~100 Ω and ~1 kΩ, respectively. The transition from the HRS to the LRS takes place at ~0.5 V while the opposite transition requires a voltage lower than ~ -0.5 V. It

should be mentioned that, in agreement with other observations related on thin 'bulk' films no forming process [9] is needed in order to achieve the resistance switching operation. The devices are initially found in the HRS and the first set pulse allows switching to the LRS. The operation of this resistive memory remains unaltered under pulsed regimes.

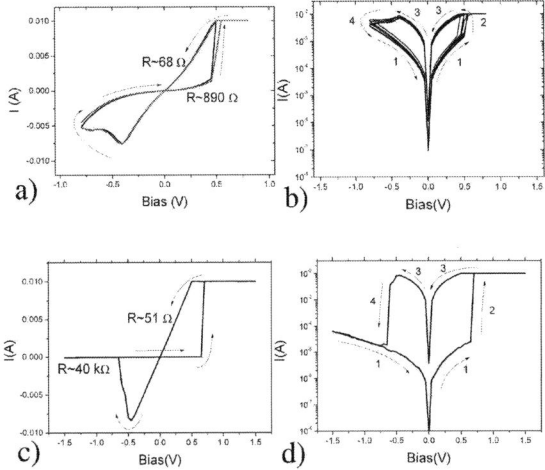

Figure 2. Characterization of the as-deposited (a and b) and high-volltage samples (c and d). I-V forward and backward sweeps showing the continuous transition between the HRS (forward sweep) denoted with 1 and the LRS (backward sweep) denoted with 3. The set and reset regions are denoted with 2 and 4.

Electrical characterization of the high-voltage sample (Figure 2) revealed a resistance switching operation very similar to the one observed for the as-deposited sample with the only exception that a much higher separation occurs between the LRS and the HRS. The initial forward and backward I-V sweeps, depicted in Figure 2c and 2d, show the existence of LRS and HRS characterized by resistances of ~100 Ω and ~100 kΩ, respectively. It should be pointed out that in pulsed regime no current compliance was considered during the application of the pulses while, on the contrary, in I-V sweep measurements a compliance of 0.01A was forced by the picoamperometer.

IV. DISCUSSION

The formation of large grains in the high-voltage sample can be understood considering the large amount of kinetic energy that the accelerated NPs carry and release in the form of heat when they impact on the sample, favouring the "bridging" of neighbouring NPs through oxygen atoms, i.e. meltdown of NPs by oxidation. Indeed, considering the simplified case of a single-charged oncoming NP accelerated by a voltage of 5 kV that impacts on a second NP laying on the sample surface, it is possible to estimate that the temperature increase during the deposition of accelerated NPs is of the order of hundreds of degrees. Rough calculations can be carried out quite quickly considering that the kinetic energy of the moving particle at the impact point is 5keV, $\sim 8 \cdot 10^{-16}$ J, which must be equal to the heat shared by the two NPs after the impact, i.e., $2 M_{NP} c_{TiO} \Delta T$ (M_{NP} is the mass of a single NP and c_{TiO} is the specific heat of TiO). From this calculation, a ΔT as high as 1000 K can be extracted. It is believed that such a kinetic-energy-transfer-induced-heating promotes NP sticking and oxidation.

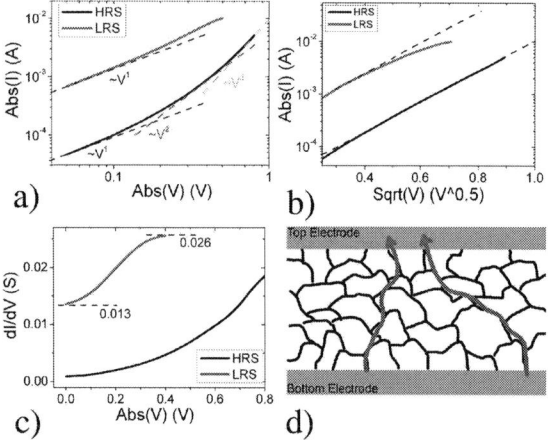

a)

b)

c)

d)

Figure 3. Analysis of the I-V characteristics of the as-deposited sample. a). Log-Log plot showing the nearly Ohmic behaviour of the LRS while the HRS deviate from any power-like dependence from V. b) Schottky plot demonstrating that HRS arises from the emission of carriers from the Au electrode. c) Differential conductance for the LRS and HRS I-V characteristics. d) Schematic of the conduction paths through grain boundaries that are responsible for the LRSs.

By comparing the resistive-switching results for the high-voltage and the as-deposited samples, it is clear that the structural modifications induced by the presence of an accelerating electric field have a direct effect on the HRS, which increased its resistance from ~ 1 kΩ, (as-deposited sample) to ~ 100 kΩ (high-voltage sample), while the LRS is left practically unaltered. While the origin of such a behaviour is difficult to explain, we suspect that it is related to the oxygen content of the NP films. Evidences in this direction have been recently presented by Chang et al. [10] who found that as the oxygen partial pressure increases, the HRS also increases by up to 2 orders of magnitude while the LRS remains unaffected. Furthermore, we should recall that in TiO resistive memories, oxygen vacancies are believed to be responsible for the LRS [11], while the HRS is believed to be mainly related to the dielectric properties of the oxide itself. A recent experimental evidence of the above mentioned statement was given by Kim et al. [12] who found that the I-V characteristics of the HRS depend strongly on the temperature, which is the signature of a defect related conduction through the dielectric, while the ones of the LRS show a clear metallic behaviour, i.e., ohmic I-V with a resistance that increases with the temperature. The fact

that the HRS depends on the dielectric properties of the oxide, when considered together with Chang's observations, might explain our results: the high voltage sample oxidizes more easily than the as-deposited one and therefore, should present a higher barrier to the conduction of electrons (TiO is an n-type semiconductor) leading to a higher HRS. Recent theoretical studies on the TiO$_x$ system, have shown that as x decreases to values lower than 2 (reduction of Ti), the band-gap of the film decreases while the conductivity increases [13,14] and thus could support our argument as well. Regarding the LRS, Chang results on thin films are already enough to support our results, but since we are dealing with a NP film, few further considerations are believed to be appropriate.

Oxygen vacancies are the key players in the resistance switching of titanium oxide films, but, in our case, also grain boundaries should play an important role in the formation of a conductive filament (CF) that determines the resistance of the LRS as depicted in Figure 3d. The important role of grains in the diffusion of oxygen vacancies has been already pointed out by Jeong et al. [15] and by Waser et al. [16]. Thus, we could advance the working hypothesis that following the set operation, the as-deposited sample might present multiple tiny filaments at the boundaries between neighbouring NPs while the high voltage sample, due to the larger grains, might present only few CFs or a single thick CF. According to Gao et al. [17], the different behaviour during reset operation might be related to the presence of multiple conductive filaments in the as-deposited sample and a single or few CF in the high-voltage sample. Regarding the former, accurate analysis of the I-V characteristics of the as-deposited sample revealed that the LRS and HRS are related to Ohmic conduction and Schottky emission [18] (or to quasi-Poole-Frenkel trap-assisted tunneling [19]) respectively, as shown in Figure 3. More in detail, while the HRS I-V characteristics do not show any power-like dependence on the applied voltage but fits very well with the Schottky emission model (Figure 3b) which subsequently is the evidence that the HRS depends upon the quality of the dielectric film, the LRS is ohmic (which is a characteristic of CFs) with a slight deviation at high currents (Figure 3a).

Considering the differential conductance shown in Figure 3c, it could be argued that the observed deviation from ohmic behaviour corresponds to a doubling (100% increase) of the conductance of the film at high currents and thus might be explained by Joule heating of the CFs which is responsible for the further growth of the filaments during the measurement itself [20]. On the other hand, the high-voltage sample shows ohmic conduction for both LRS and HRS as shown in Figure 4a and, similarly with the LRS of the as-deposited sample, both states exhibit an increase of the differential conductance with the applied bias (Figure 4b). Regarding the LRS the differential conductance increase is only 17% which might support the argument of a thicker filament in this sample. Indeed, only if the filament is thicker, Joule heating and further filament

978-1-4673-1707-8/12 $31.00 © 2012 IEEE

growth should have a smaller effect onto the filament conductance compared with the as-deposited sample.

The HRS, on the other hand, surprisingly shows an Ohmic behaviour with a 60% conductance increase during measurement, a behaviour that up to now was only observed for the LRSs. It might be explained by considering that a thicker filament is more difficult to disrupt and thus instead of the dissolution of the filament we might have to do simply with a thinning filament process. Given the inverse dependence of the resistance of a CF upon its cross section area, the observed three orders of magnitude increase of the HRS resistance compared to the LRS one, should be related to a 30 times decrease in filament diameter. The above argument drives to the conclusion that, assuming a LRS filament diameter of 10 nm, the HRS filament diameter should be in the sub-nm scale. Since the filament is so thin, its further growth during measurement should highly affect its resistance, explaining the 60% increase in conductance experimentally observed.

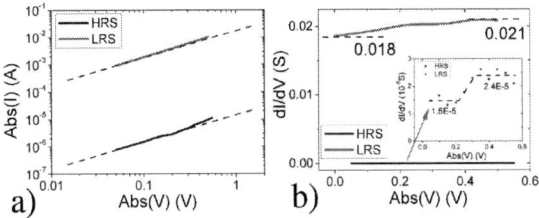

Figure 4. Analysis of the I-V characteristics of the high-voltage sample. a). Log-Log plot showing the Ohmic behaviour of the LRS and HRS. b) Differential conductance for the LRS and HRS I-V characteristics.

V. CONCLUSIONS

In this work we presented experimental evidence of forming-free bipolar memristive behaviour obtained from TiO_2 NPs deposited in vacuum at room temperature. The devices which were in the form of MIM structures demonstrated clearly distinct LRS and HRS, and did not require any electroforming process prior to operation. The two conditions studied which are achieved either by acceleration of the NPs via an externally applied voltage (high-voltage condition) or without any accelerating voltage (as deposited condition), present both very low set-reset voltages of $\sim\pm0.5$ V and an LRS of $\sim100\ \Omega$ while the HRS was ~1 kΩ for the as-deposited and ~100 kΩ for the high-voltage condition. The approach can be further investigated using smaller titanium oxide nanoparticles as well as different metal oxide materials.

ACKNOWLEDGMENT

The authors acknowledge the project IAPP 2007 – NANOSOURCE 218111 Marie-Curie actions for funding this research. We thank Miss G. Vatougia for technical support.

REFERENCES

[1] L. O. Chua 'Memristor-The missing circuit element' IEEE Trans. Circuit Theory, 18, 507 (1971)

[2] J.J. Yang et al., 'Memristive switching mechanism for metal/oxide/metal nanodevices', Nature Nanotechnol., 3, 429 (2008)

[3] Y. Fujisaki 'Current Status of Non-volatile Memory Technology' Jpn. J. Appl. Phys. 49, 100001 (2010)

[4] R. Waser et al., 'Redox-Based Resistive Switching Memories-Nanoionic Mechanisms, Prospects and Challenges', Adv. Mater. 21, 2632 (2009)

[5] S.S. Nonnenmann, E.M. Gallo and J.E Spamiera, 'Redox-based switching in perovskite nanotubes', Appl. Phys. Lett. 97, 102904 (2010)

[6] P.Y. Lai and J.S. Chen, 'Ultrahigh ON/OFF-Current ratio for Resistive Memory Devices with Poly(N-Vinylcarbazole)/Poly(3,4-thylenedioxythiophene)-Poly(Strenesulfonate) Stacking Bilayer' IEEE El. Dev. Lett 32, 3 (2011)

[7] B. O'Reagan and M. Gratzel 'A low cost high efficiency solar cell based on dye-sensitized TiO2 colloidal films' Nature 353, 737 (1991)

[8] J. Tang, E. Verrelli and D. Tsoukalas, 'Assembly of charged nanoparticles using self-electrodynamic focusing', Nanotechnology 20, 365605 (2009)

[9] J.J. Yang et al., 'The mechanism of electroforming of metal oxide memristive switches', Nanotechnology 20, 215201 (2009)

[10] W.Y. Chang et al., 'Influence of crystalline Constituent on Resistive Switching Properties of TiO2 Memory films' Electrochemical and Solid-State Letters 12, H135 (2009)

[11] S.G. Park et al., 'Impact of Oxygen Vacancy Ordering on the Formation of a Conductive Filament in TiO2 for Resistive Switching Memory', IEEE Elec. Dev. Lett. 132, 197 (2011)

[12] K.M. Kim et al., 'A detailed understanding of the electronic bipolar resistance prototypical memristive material', Nanotechnology 22, 254010 (2011)

[13] K. Szot, M. Rgala, W. Speier, Z. Klusek, A. Besmehn, R. Waiser' TiO2-a prototypical memristive material', Nanotechnology 22, 254001 (2011)

[14] K.M. Kim, D.S. Jeong, C.S. Hwang, 'Nanofilamentary resistive switching in binary oxide system; a review on the present status and outlook', Nanotechnology 22, 254002 (2011)

[15] D.S. Jeong et al., 'Characteristic electroforming behavior in Pt/TiO2/Pt resistive switching cells depending on atmosphere', Journal of Appl. Phys., 104, 123716 (2008)

[16] R. Waser, R. Diettmann, M. Salinga, M. Wuttig, 'Function by defects at the atomic scale-New concepts for non-volatile memories', Solid-State Electron. 4, 830 (2010)

[17] B. Gao et al., 'Inified Physical Model of Bipolar Oxide-Based Resistive Switching memory', IEEE EDL 30, 1326 (2009)

[18] J. Park et al., 'Resistive switching characteristics of ultra thin TiOx' Microel. Engin. 88, 1136 (2011)

[19] L. Zhang et al., 'Experimental investigation of the reliability issue of PRAM based on high resistance state conduction', Nanotechnology 22, 254016 (2011)

[20] K.L. Lin et al., 'Electrode dependence of filament formation in HfO2 resistive switching memory', J. Appl. Phys. 109, 084104 (2011)

An Array-Based Chip Lifetime Predictor Macro for Gate Dielectric Failures in Core and IO FETs

Pulkit Jain, *John Keane and Chris H. Kim

University of Minnesota, Minneapolis, MN 55455, USA
*Portland Technology Development, Intel Corporation, Hillsboro, OR 97124, USA

Abstract- **A comprehensive Chip LIfetime Predictor (CLIP) macro for automatically characterizing gate dielectric failure reduces the stress time and silicon area by a factor proportional to the number of FETs to be tested. A flexible DUT cell that can be stressed in isolation without thicker t_{ox} FETs to 4 times supply voltage, enables accurate lifetime prediction under different ON and OFF state dielectric breakdown modes for both low voltage core and high voltage IO devices.**

I. INTRODUCTION

Device reliability mechanisms such as bias temperature instability, hot carrier injection, and Time Dependent Dielectric Breakdown (TDDB) have become pressing concerns in scaled technologies. While parametric shifts due to the former two can be mitigated using frequency guard-banding or circuit adaptation [1-2], such techniques are ineffective against the more catastrophic TDDB where even a single instance in a chip can cause an outright system failure. Fig. 1 shows the different TDDB modes affecting common digital circuits. While the on-state TDDB is most severe and conventionally assumed to be critical due to the entire gate area being exposed to stress, High Drain, High Source (HDHS) and High Drain (HD) OFF-state modes [3-4] might lead to earlier failure in circuits such as SRAM access devices that are exposed to an off-state stress for most of their lifetime. As for Input-Output (IO) devices, Electrical OverStress (EOS) and ElectroStatic Discharge (ESD) are of particular concern. Reliability margin targets for them become an issue with extensive use of high-voltage IOs and high-power CMOS devices at interface circuits in system on chips.

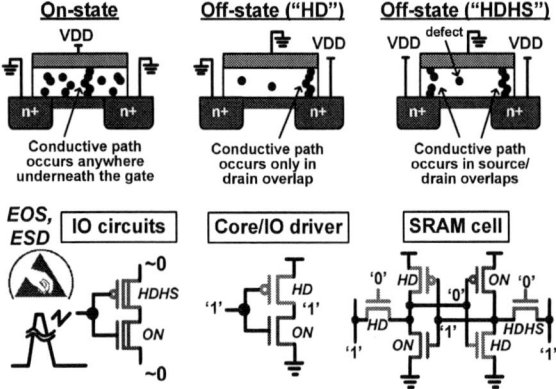

Fig. 1 Different occurrence of gate dielectric failure. While 'ON' and 'OFF-HD' cases are most prominent, 'OFF-HDHS' is also seen in certain cases such as SRAM access devices.

Particularly with TDDB, optimizing the fabrication process and using proper operating conditions based on

accurate lifetime predictions is the most practical and effective approach. The main challenge is in the collection of *massive* statistical data from accelerated tests, as TDDB is a function of a number of variables including voltage, temperature, area, dielectric thickness, and purity (Fig. 2). Given the need for up to thousands of samples to correctly define a *single* Mean-Time-To-Failure (MTTF) value, traditional device probing quickly becomes cumbersome (Fig. 3). A previous characterization array for TDDB [5] only considered ON-state stress in core transistors which is not enough to obtain an accurate picture of system lifetime. A combined lifetime prediction methodology is needed to take into account different modes in tandem with their predicted time to failures. In this paper, we propose an array-based Chip Lifetime Predictor (CLIP) macro for efficiently collecting failure statistics under various accelerated stress conditions including ON-state and OFF-state stress modes for both low voltage core and high voltage IO devices.

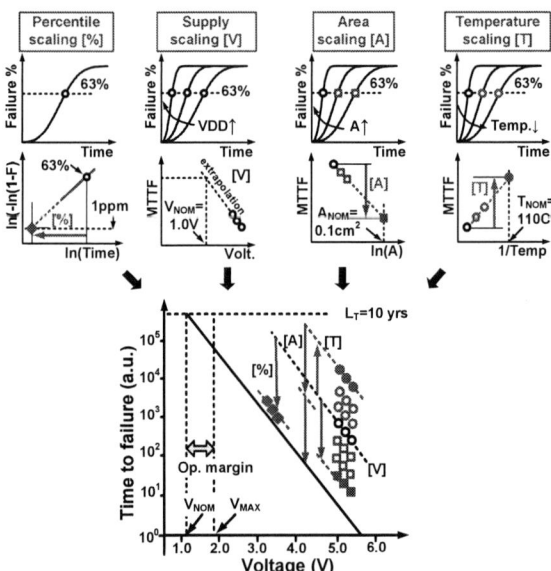

Fig. 2 Chip lifetime projection for TDDB based on accelerated stress involves mass data collection (e.g. up to 1000's of samples per MTTF data) to make voltage, percentile, temperature, and area projections to actual product usage conditions.

In the next section, we delve into the CLIP macro design and overall test strategy. Section III and IV describe the stress cell designs with measured statistics. In section V, the data is put together in perspective using the CLIP framework and finally we give a conclusion in section VI.

978-1-4673-1707-8/12 $31.00 © 2012 IEEE

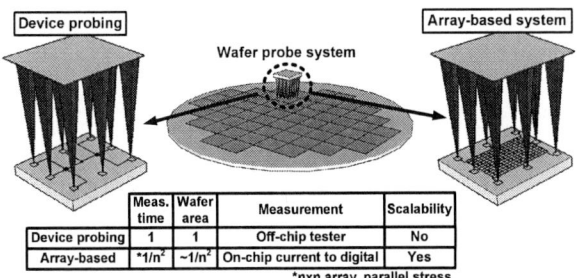

Fig. 3 Array based approach is an efficient way to carry out aging measurements compared to conventional probing.

II. MACRO DESIGN AND TEST STRATEGY

A. *CLIP macro design*

The basic framework of the proposed CLIP macro is an array based statistical collection setup that can stress the DUTs in parallel while taking fast serial measurements controlled by a convenient scan-based interface (Fig. 4). This feature reduces the test time and test silicon area by a factor proportional to the number of DUTs. The gate terminal of the selected DUT is connected to the shared BL for I_G measurements. The pre-charged BL gets discharged and the progressive TDDB in the form of I_G is converted to a count by an on-chip current-to-digital converter.

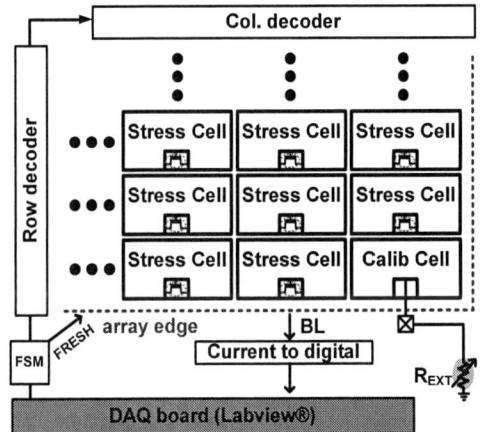

Fig. 4 Proposed array-based Chip LIfetime Predictor (CLIP) macro.

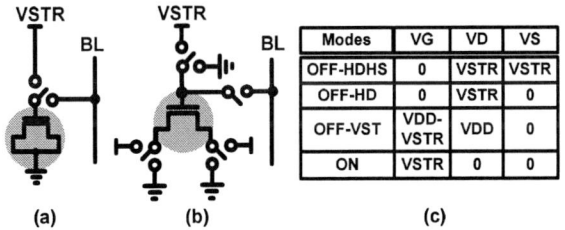

Fig. 5 Abstraction of different kinds of stress cells supported: (a) Conventional [7]; (b) Proposed flexible DUT; (c) Different flexible stress conditions.

The critical part is a flexible stress cell design that can be used for evaluation of the different OFF and ON-state TDDB modes with programmable control. Two different flavors of flexible stress cells are needed for IO and core cases as will be discussed in the next section. As shown in the abstraction in Fig. 5, the underlying principle is to connect each DUT terminal to a stress voltage using on-chip switches rather than a hardwired inflexible connection. Flexible stress conditions used for the DUT cells have been tabulated in Fig. 5(c). To avoid unrealistic GIDL behavior that may corrupt the stress data, a Voltage-Splitting technique (VST) was also implemented [3].

B. *Current to digital conversion*

Reliability engineers employ both 'hard' and 'soft' increase in dielectric conduction for characterizing TDDB. We therefore employ two variants for current to digital conversion in this work. Fig. 6(a) shows the Current to Count Converter (CCC) to facilitate soft breakdown evaluation for the core case similar to the one used in [5].

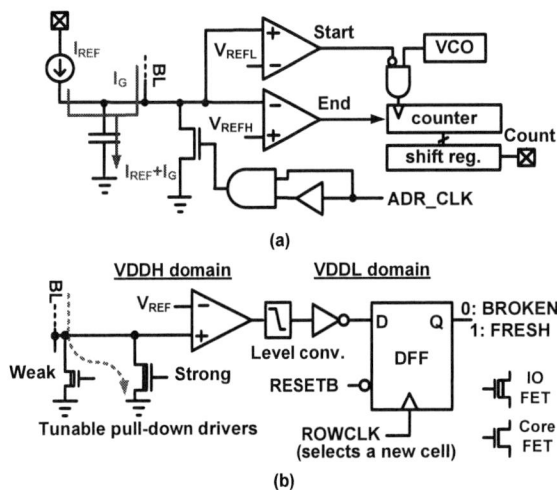

Fig. 6 Two flavors of current to digital blocks used a) CCC for soft breakdown in core FETs. b) CBC for hard breakdown in IO FETs.

Considering the high t_{ox} values in IO devices, as well as based on our preliminary findings on the core case, we did not expect to see progressive behavior in breakdown in our test setup. Thus, elaborate tracking using CCC was not needed. Therefore, a major simplification for higher timing resolution and ease of measurement is done in the form of Current to Binary Converter (CBC) scheme in Fig. 6(b). In this scheme, every time a cell is selected for measurement, BL voltage is decided by current balance of R_{DUT} and the programmable pull-down strength. This gets converted to a binary signal, FRESH/BROKEN by the comparator with a user defined reference, V_{REF}.

C. *Calibration*

We embed replica stress cells, called "calibration cells" directly in the CLIP array (Fig. 4). These calibration cells were identical to the stress cells, but they did not have DUTs. Instead, a metal interconnect path was routed from the DUT gate node out to a pad. During calibration, a known range of resistances, R_{EXT} were attached to that pad in order to mimic a range of R_{DUT}, and measurements were run in the calibration cell. This isolates the R_{DUT} from other extraneous resistances in the measurement path. The obtained results are shown in Fig. 7. The calibration cell also served useful as a marker cell during array operation.

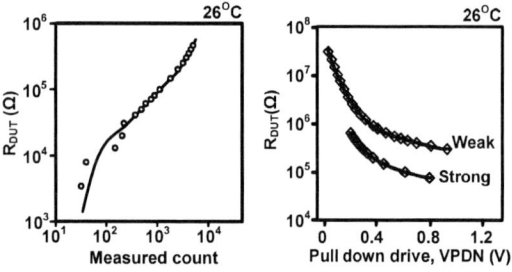

Fig. 7 Calibration curves using the two current to digital converters. (a) CCC case, and (b) CBC case

III. CORE DEVICE BREAKDOWN CELL

A. Stress cell design

Details of the core transistor stress cell are shown in Fig. 8. The terminal voltages of the DUT are separately controlled by the STR/MEASb and VS/D/G control signals to provide programmable control of different flexible stress modes. If 'FRESH'=0, the stress is gated off to prevent excess current during the long stress experiments. A timing logic selects cells in a manner that prevents over-shoot transients on the DUT and seepage of stress voltages to peripheral circuitry.

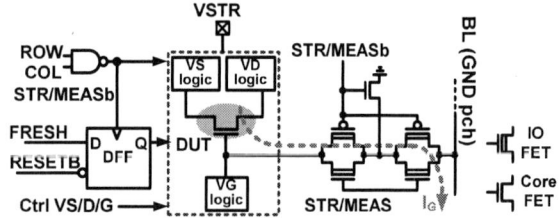

Fig. 8 Proposed DUT cell for core device breakdown. All FETs except DUT are thick t_{ox} devices.

B. Measured TDDB statistics

Adopting the methodology in Fig. 2, we first plot the results from different modes on a typical Weibull scale for straight line fits in Fig. 9(a-b). The rate of fails determined by the slope of this line, is related to [%] scaling. Fig. 9(a), shows the effect of pull down strength (or the breakdown threshold) used in CBC scheme. The 'harder' breakdown curves display a bend early in their evolution, and a lower slope.

Fig. 9(b) shows, that the two OFF modes have similar slopes and hence are projected to undergo almost identical [%] scaling. The MTTF versus voltage plot in Fig. 9(c) for different OFF and ON modes is used for [V] scaling. MTTFs for the different OFF modes were more than four orders of magnitude higher than ON mode at 4.5V. This difference is mainly attributed to the differences in effective gate area under stress. In ON mode, the entire gate area under the channel undergoes stress, while in the OFF mode it is limited to the

overlap region. The [V] and [%] slopes for different cases are summarized in Fig. 9(d)

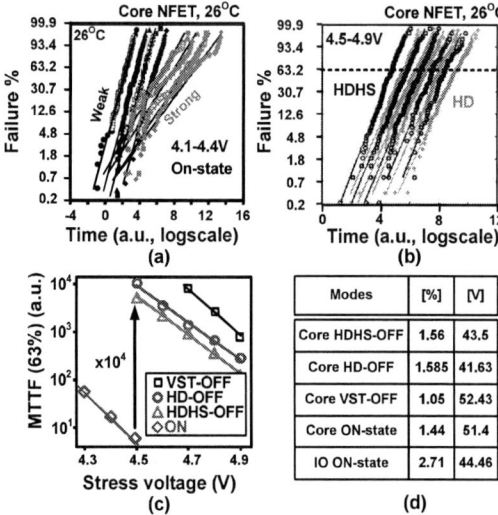

Modes	[%]	[V]
Core HDHS-OFF	1.56	43.5
Core HD-OFF	1.585	41.63
Core VST-OFF	1.05	52.43
Core ON-state	1.44	51.4
IO ON-state	2.71	44.46

Fig. 9 (a) Effect of pull down strength in CBC scheme with VPDN=0.35V (b) Comparison of different OFF-modes (c) Relative voltage scaling in different modes (d) Relative comparison of [V] and [%] in different cases.

IV. IO DEVICE BREAKDOWN CELL

A. Stress cell design

The higher stress voltage (3-4 times the IO supply) and lack of a thicker t_{ox} device complicate the design of the IO stress cell shown in Fig. 10. A blocking circuit with dynamic biasing was designed to distribute the high stress, post-breakdown. A stack of two blocking circuits (single stack shown in Fig. 10 for simplicity) was sufficient to stress the cell upto 4 times nominal supply. Two flexible stress configurations were employed to provide OFF-state mode control.

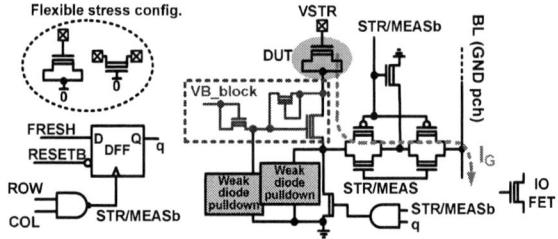

Fig. 10 Proposed DUT cell for IO breakdown (top). No thicker t_{ox} devices are available so a blocking circuit was used to protect non DUT devices.

B. Measured TDDB statistics

Measured TDDB for a range of stress voltages is shown in Fig. 11 (left). [%] scaling from 63% to 1ppm for the core case is projected to be 100X larger than the IO case. Fig. 11 (right) compares the relative MTTF with different VSTR. A steeper slope is observed for the core case which translates into a 20X MTTF difference due to [V] scaling. MTTF for different

978-1-4673-1707-8/12 $31.00 © 2012 IEEE   264

temperatures are shown in Fig. 12(a). Both core and IO FETs show Arrhenius trend in the measured regime. Spatial map shown in Fig. 12(b) of the individual cell's TTF shows no obvious correlation.

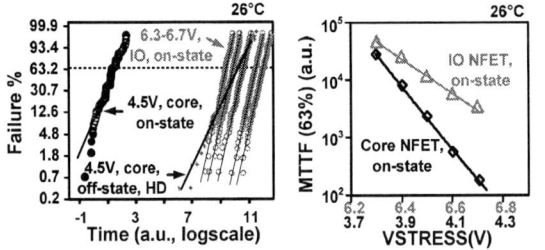

Fig. 11 Measured breakdown data at different stress voltages for IO case

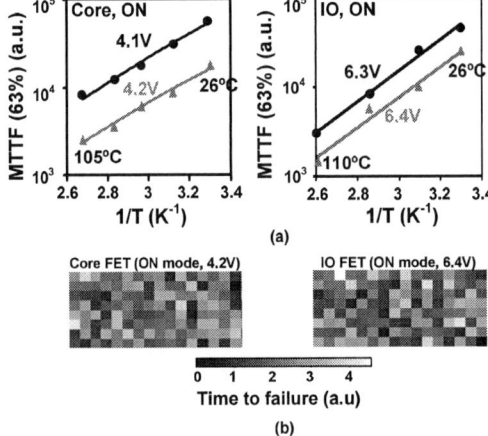

Fig. 12(a) MTTF for different temperatures. Both core and IO FETs show Arrhenius trend in the measured regime. (b) Spatial map of individual cell's time to breakdown.

V. LIFETIME ESTIMATION USING CLIP MACRO

Fig. 13 shows the applied CLIP methodology for different stress profiles and gate types in tandem. The total gate areas for core and IO transistors were assumed to be $0.1cm^2$ and $0.01cm^2$, respectively. Duty cycle between OFF-state and ON-state is assumed to be 50% and a 1ppm failure percentile criterion was used. We observe that IO devices meet the lifetime requirement by a sufficient margin while the core transistor barely meets it. The chip microphotographs and summary of the core and IO CLIP arrays are given in Fig. 14.

VI. CONCLUSION

Optimizing the fabrication process and using proper operating conditions based on accurate lifetime predictions are the most practical and effective ways of dealing TDDB. However, the main challenge with this approach is in the collection of massive statistical data from accelerated tests, as TDDB is a function of a number of variables including voltage, temperature, area, dielectric thickness, and purity. In this work, we propose a CLIP macro for gate dielectric breakdown to reduce the stress time and silicon area by a

factor proportional to the number of FETs to be tested. The essential part is a flexible DUT cell that can be stressed in isolation without thicker t_{ox} FETs to 4 times the VDD, enabling accurate lifetime prediction under different ON and OFF state TDDB modes for both low voltage core and high voltage IO devices.

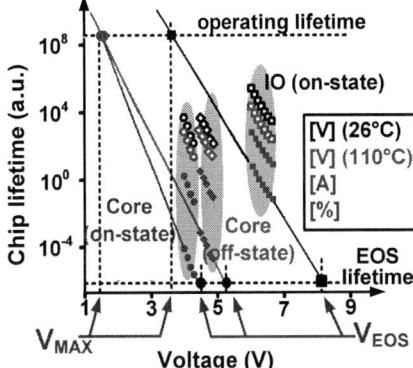

Fig. 13 Comparison of projected lifetimes for IO and core devices for ON and OFF (avg. of HD and HDHS) states. Voltage, area, percentile, and temperature extrapolations (solid) are performed from measured statistical data (open).

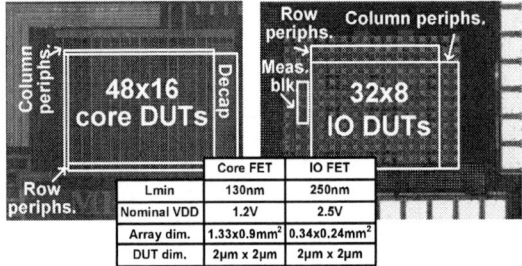

Fig. 14 Test chip microphotographs of core and IO CLIP macros with chip summary.

ACKNOWLEDGMENTS

The authors would like to thank SRC and the Texas Analog Center of Excellence (TxACE) for financial support, and Dr. Vijay Reddy at Texas Instruments for technical discussions.

REFERENCES

[1] E. Saneyoshi et al., "A precise-tracking NBTI-degradation monitor independent of NBTI recovery effect", International Solid State Circuits Conference, 2010

[2] E. Karl et al., "Compact in-situ sensors for monitoring negative-bias-temperature-instability effect and oxide degradation", International Solid State Circuits Conference, 2008

[3] E. Wu et al. ,"Off-state mode TDDB reliability for ultra-thin gate oxides: New methodology and the impact of oxide thickness scaling", International Reliability Physics Symposium, 2004

[4] S. Pae et al. ," Reliability characterization of 32nm high-k and metal-Gate logic transistor technology", International Reliability Physics Symposium, 2010

[5] J. Keane et al., "An array-based test circuit for fully automated gate dielectric breakdown characterization", Custom Integrated Circuits Conference, 2008

Unified characterization of RTN and BTI for circuit performance and variability simulation

N.Ayala, J. Martin-Martinez, R. Rodriguez, M. Nafria and X. Aymerich.

Departament d'Enginyeria Electrònica, Universitat Autònoma de Barcelona (UAB) 08193, Bellaterra, Spain.
Corresponding autor e-mail: Nuria.Ayala@uab.es

Abstract—**In small devices, Bias Temperature Instability (BTI) produces discrete threshold voltage (V_T) shifts, which are attributed to the charge and discharge of single defects. In this work, the voltage and temperature dependences of charging/discharging of individual defects, considering their stochastic behavior, have been analyzed. From the results, and considering a previously presented BTI physics-based model, the corresponding V_T shifts in the device have been obtained and included in a circuit simulator, to evaluate their effects on SRAM cells performance and variability.**

I. INTRODUCTION

In small MOSFETs, BTI reveals a stochastic behavior, which has been attributed to the charge/discharge of defects during stress/relaxation [1, 2].The analysis of these isolated defects provides very valuable information for the development of future BTI models [2, 3]. The behavior of these defects has been also related to the Random Telegraph Noise (RTN) observed in MOS structures [1] which is characterized by different and clearly distinguished conduction levels [4]. In this work, a characterization of defects, attending to their stochastic behavior, on pMOS is presented. The capture (τ_c) and emission (τ_e) times of the defects and their dependences with the gate voltage (V_G) and temperature have been analyzed. Based on these results and the probabilistic defect occupancy model for BTI [3], the corresponding V_T shifts have been evaluated and their impact on a SRAM cell behavior and variability analyzed.

II. EXPERIMENTAL

The samples used in this work were pMOSFET transistors with SiON as gate dielectric (EOT=1.7nm) and area of $0.15 \times 0.13 \mu m^2$. Since our aim is to characterize in detail the behavior of the BTI related defects, the data measured on a single device will be shown in the paper. Firstly, the fresh transistor drain current (I_D) versus V_G characteristic was measured to obtain the transistor V_{th}. Secondly, to obtain the τ_e and τ_c of the BTI related defects in the device, I_D was measured when applying -50mV at the drain, for different stress gate voltages and temperatures ($2.5°C \leftrightarrow 100°C$) for 140 seconds. For each temperature, V_G was sequentially changed starting from -0.6V to -1.4V in steps of -0.1V and τ_e

and τ_c were obtained from the I_D evolution for the different V_G. Before changing V_G, the sample was relaxed for 60 seconds (applying V_G= -0.5V and V_D= -50mV).

III. V_G AND T DEPENDENCES OF DEFECTS BEHAVIOUR

When a voltage is applied to the gate of the device, abrupt changes in the I_D-t traces (which imply abrupt changes in the device V_T) between clearly distinguished current levels can be observed (Fig. 1), which have been attributed to the charge/discharge of defects in the device [2]. Each defect has associated a fixed I_D increase/decrease (δI_D) which identifies the defect and the τ_e and τ_c times are the times elapsed between consecutive sudden I_D changes with the same δI_D (see Fig. 3b and c).

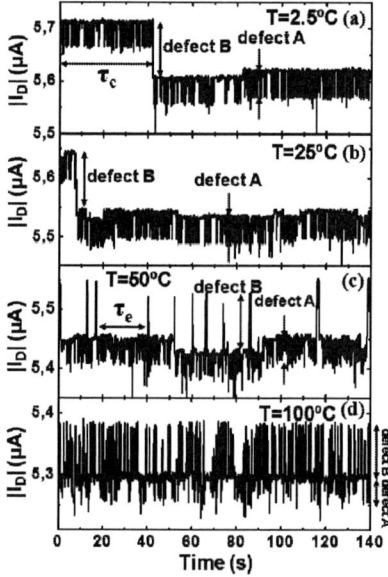

Figure 1. I_D evolution for V_G= -0.9V and 2.5°C (a), 25°C (b) 50 °C (c) and 100°C (d). τ_e and τ_c of defect B ($\delta I_D \approx 100nA$) decrease with temperature

First of all, the temperature dependence of τ_e and τ_c has been analyzed Fig. 1 shows the I_D evolution measured at V_G= -0.9V for different temperatures. Since defects can be identified by the magnitudes of the I_D changes, the charging/discharging of two defects are clearly distinguished in Fig. 1, which correspond to $\delta I_D \approx 50nA$ (defect A) and $\delta I_D \approx 100nA$ (defect B) Fig. 1 shows that the charging (current decrease) of defect B observed at ~45 s for T=2.5°C (Fig. 1a) is shifted to lower times (Fig. 1b and 1c) when the temperature increases and, for large enough T, its charging cannot be observed in this scale (Fig. 1d). Therefore, this result suggests that τ_c decreases with temperature. Moreover, when temperature increases, multiple discharging events (current increments) of defect B are observed (Fig. 1c and d), which also indicate that τ_e decreases with temperature. Fig. 2 shows the number of detected charge /discharge events for each defect, for V_G= -0.8V and different temperatures. For the lower temperatures, events related to defect A are dominant However, as temperature increases, the number of events related to defect B increases and dominate over those related to defect A.

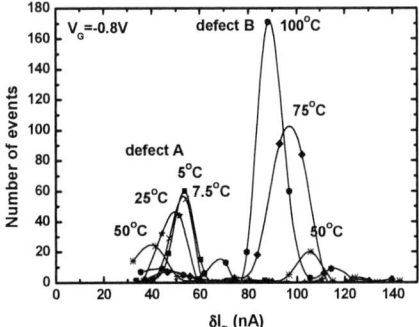

Figure 2. Number of charging/discharging events versus δI_D current change for different temperatures. For high temperatures defect B ($\delta I_D \approx 100nA$) dominates, while for the lower temperatures defect A ($\delta I_D \approx 50nA$) prevails.

The V_G dependence of τ_e and τ_c has also been studied. Fig. 3 shows the evolution of I_D for different gate voltages, for T= 7.5°C. For the lower gate voltage (-0.7V), defect B is charged at ~50 seconds. However, when V_G increases, this effect is shifted to lower times, which means a reduction of τ_c with V_G. In addition, for defect A, τ_c decreases and τ_e increases as V_G increases (Fig. 3b and 3c). For very high V_G, the defect B is charged immediately after the voltage is applied and it is not discharged again in this time window.

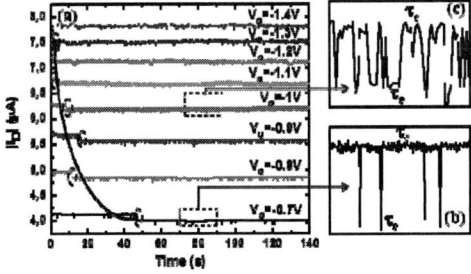

Figure 3. I_D evolution for different V_G (T=7.5°C). τ_c decreases and τ_e increases when V_G increases. The cycles and arrows correspond to charging events of defect B. Zooms of the traces are shown in (b) and (c) which correspond to charging/discharging of defect A.

The voltage and temperature dependences of the statistical distributions of the time constants measured for defects A and B have been analyzed, assuming an exponential distribution for τ_e and τ_c [3]. The cumulative probability function, F, satisfies (1).

$$Ln\big(1 - F(V,T)\big) = -\tau / \big\langle \tau(V,T) \big\rangle \qquad (1)$$

being $\langle \tau \rangle$, the mean value of τ_c or τ_e. As an example, Figure 4 shows Ln(1-F) versus τ_e for defect B for different temperatures and V_G= -0.7V. Good fittings of the experimental data (symbols) to equation 1 are obtained. The values of $\langle \tau_e \rangle$ can be determined from the slopes of these plots. Fig. 5 shows $\langle \tau \rangle$ versus V_G for defect B, for 2 temperatures. $\langle \tau_c \rangle$ and $\langle \tau_e \rangle$ exponentially decrease and increase, respectively, with V_G. In addition, lower values of $\langle \tau_c \rangle$ and $\langle \tau_e \rangle$ are observed for larger temperatures. In our device, the voltage and temperature dependences of $\langle \tau_c \rangle$ and $\langle \tau_e \rangle$ have been found to be given by (2)

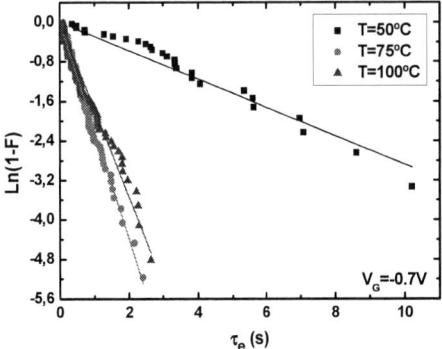

Figure 4. Symbols: experimental values of τ_e obtained for defect B at V_G=-0.7 and different temperatures. Lines: fittings of the experimental data to (1). $\langle \tau \rangle$ can be obtained from the plots slopes.

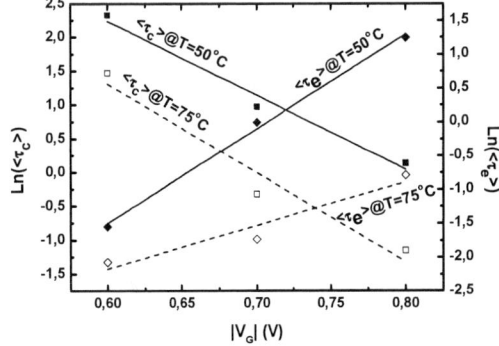

Figure 5. $\langle \tau_e \rangle$ and $\langle \tau_c \rangle$ versus V_G of defect B for different temperatures.

$$< \tau_{c,e}(V,T) >= K \cdot e^{\alpha|V|} \cdot e^{E_a / k_B T} \qquad (2)$$

where K, α and E_a are here empirical parameters which are related to the energy defect level and the structure band diagram [5]. k_B is the Boltzmann constant.

IV. BTI SIMULATION AND IMPLICATIONS ON SRAM CELLS

Based on the previous results, the effect of the charge/discharge of individual defects in pMOS transistors has been simulated. To do this, the probabilistic defect occupancy model for BTI has been considered [3]. According to this model, when the defect is charged, a V_T shift of value η is observed. η can be experimentally obtained from the sudden I_D changes (Fig. 1,3) and the initial I_D-V_G characteristic [6]. Moreover, in a device, η's related to different defects are exponentially distributed [3]. The probability of charging/discharging of the defect can be obtained from the values of τ_c and τ_e, which in a device are statistically distributed following a logarithmic law [6]. First of all, the effect on the threshold voltage of the charging/discharging of a **single defect** has been evaluated, as a function of the operation voltage and temperature. The occupancy probability of the defect has been computed from (1) by randomly generating a particular value of τ_c (τ_e). The average values of τ_c (τ_e) are evaluated from (2), with the experimentally determined parameter values for the considered V_G and T. If the defect is found to be occupied, V_{th} shifts by an amount η. An example of V_{th} shift related to the charging/discharging of a single defect is illustrated in Fig. 6, where a pulsed V_G waveform (Fig. 6a) is applied. In the example, we have considered a defect with $<\tau_c>$=1.2s and $<\tau_e>$=0.45s for V=-0.6V and T=25°C, that causes a change η = -3.81mV in V_{th} (defect A) when it is occupied. Fig. 6b shows the simulated ΔV_{th} (or η) caused by this defect when the gate voltage has an amplitude of V=-0.6V. Note that, as observed in Fig. 1, a typical RTN signal is obtained. Fig. 6c shows the ΔV_{th} trace for the same defect when the gate voltage is V=-1V. In this case, as in Fig. 1c, $<\tau_c>$ decreases and $<\tau_e>$ increases, so that the defect is rapidly occupied, and because its very low emission probability, the defect does not discharge during the high-state voltage and, consequently, the ΔV_{th} trace follows the gate voltage. In a device, however, **several defects** can coexist, with different values of $<\tau_c>$, $<\tau_e>$ and η, so that the contributions of all the defects have to be considered. Fig. 6d shows the ΔV_{th} trace obtained when 10 (small area device) and 1000 (big area device) are simultaneously considered. For these simulations the $<\tau_c>$, $<\tau_e>$ and η distributions shown in [3] have been used, which correspond to devices with similar characteristics. Note that the obtained ΔV_{th} evolution with time corresponds to the typical NBTI behavior during the stress (V=-1V) and relaxation (V=0) phases, being stochastic in small devices (blue line) and continuous (red line) in larger ones. This result confirms that RTN and BTI phenomenology have the same origin (they are caused by the same defects) and shows that their observation depends on the conditions (voltage, temperature, time) at which they are studied.

The effects of the BTI V_{th} shifts related to these defects on the performance and variability of SRAM cells, as circuit example, have been analyzed, as a function of the operation conditions. The device threshold voltage shift is considered in the circuit by adding a voltage source to the gate of each

pMOS transistor (Fig. 7). The value of each voltage source is determined by using the simulation procedure and the characterization data previously described, which lead to the results in Fig. 6, taking into account the gate voltage applied to each pMOS and the circuit temperature. Simulations have been carried out when the logic states change between '0' (V_Q=0, $V_{Q'}$=V_{DD}) and '1' (V_Q=V_{DD}, $V_{Q'}$=0) every 0.1s. 10 defects with distributed $<\tau_{c,e}>$ and η have been considered in each pMOS transistor.

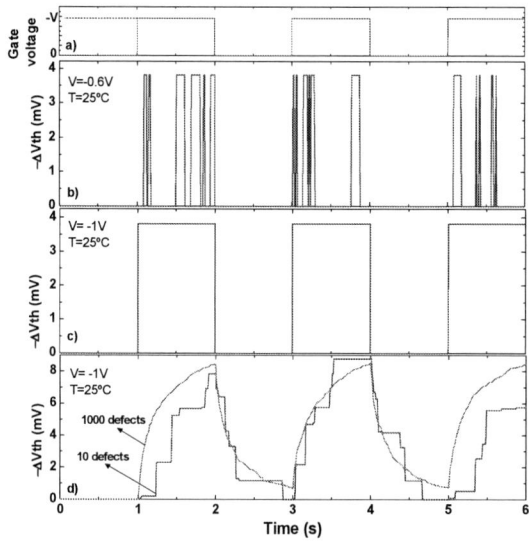

Figure 6. a) pulsed waveform used to evaluate the ΔV_{th} traces related to defects at different conditions. b) ΔV_{th} trace obtained for a defect with $<\tau_e>$=0.45s $<\tau_c>$= 1.2s and η=-3.81mV (defect A) when V=-0.6. (c) and V=-1V (d). ΔV_{th} traces obtained from the combination of several defects. The typical ΔV_{th} shifts of NBTI during stress and relaxation are well reproduced.

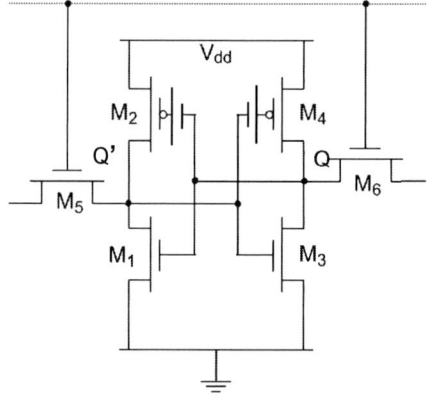

Figure 7. SRAM cell used to evaluate the effect of trapping/detrapping in defects. To account for the V_{th} variations, a voltage source is added at the gate of each pMOS transistor, whose values are obtained by performing simulations as those in Fig. 6.

Fig. 8 shows the corresponding butterfly plots at three different conditions, (a): V_{DD}=0.6V, T=25°C t=1s, (b): V_{DD}=1V, T=25°C, t=1s and (c): V_{DD}=1V, T=125°C and

$t=10^4$s. Clearly, the operation conditions of the SRAM cells have a strong influence on their performance variability.

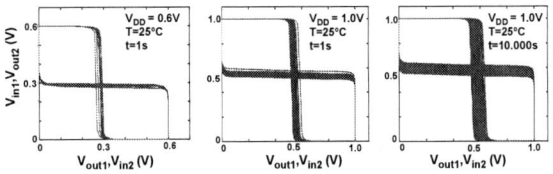

Figure 8. Butterfly curves obtained in SRAM cells at different conditions. 200 simulations have been performed for each case.

To carefully check this point, the Static Noise Margin (SNM) of the circuits have been calculated for each simulation, and their probability distributions plotted in Fig. 9. At V_{DD}=0.6V, T=25°C and t= 1s (Fig. 9a) a low spread of the SNM distribution is observed because only few defects are potentially capable to trap/detrap charge (RTN in the pMOS transistors, Fig, 6b). Increasing V_{DD} (Fig. 9b) increases the number of defects that can trap/detrap charge and, consequently, a larger SNM spread is observed. In the last case (Fig. 9c), for a higher temperature and operation time, several defects that were empty in cases (a) and (b) are now occupied (leading to a net shift of V_{th} ,typical of NBTI, Fig, 6d), and, consequently, a shift of the distribution to lower SNM values and an increase of the spread are obtained.

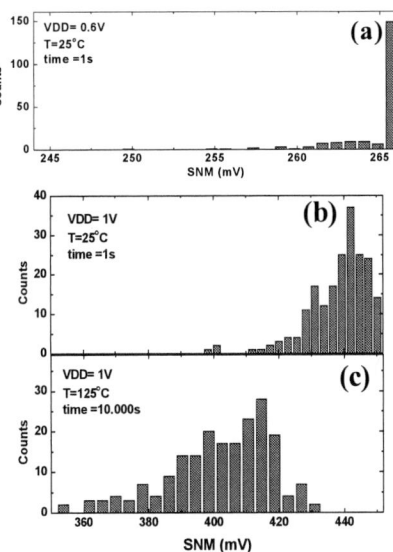

Figure 9. SNM distributions obtained for SRAM cells at different operation conditions.

The simulations show that if the behavior of individual defects are properly modeled and characterized, the impact of

RTN and NBTI on the circuit performance can be evaluated, taking into account the operation conditions of the transistors within the circuit.

V. CONCLUSIONS

In this work, the behavior of individual defects in pMOSFETs, which can trap/detrap charge during the operation of the device, has been characterized. The statistical distributions of their emission and capture times and their dependences with voltage and temperature have been experimentally obtained. From these data, the parameters of the previously presented probabilistic defect occupancy model for BTI have been determined, and the associated V_T shifts in the device evaluated. The results have shown that those defects that cause Random Telegraph Noise can induce NBTI degradation at higher voltages and operation times, so that the observation of RTN or BTI depends on the device operation conditions. Finally, the V_T shifts have been included in a circuit simulator, to study the RTN/BTI effects on SRAM cells. The results show that not only the performance, but also the variability of the cell depend on the operation conditions. In summary, using a physical-based model with experimentally obtained parameters values (which describe the underlying technology) the physical properties of defects can be directly translated into circuit response and variability.

ACKNOWLEDGMENT

This work was partially supported by the Spanish MINECO (TEC2010-16126) and by the Generalitat de Catalunya (2009 SGR-783).

REFERENCES

[1] B. Kaczer, T. Grasser, Ph. J. Roussel, J. Franco, R. Degraeve, L. –A. Ragnarsson, E. Simoen, G. Groeseneken, and H. Reisinger, " Origin of NBTI variability on deeply scaled pFETs," Proc. Int. Reliab. Phys. Symp., 2010, pp. 26-32.

[2] T. Grasser, H. Reisinger, P.-J. Wagner, F. Schanovsky, W. Goes, and B.Kaczer, "The Time Dependent Defect Spectroscopy (TDDS) for the Characterization of the Bias Temperature Instability", Proc. Int. Reliab.Phys. Symp., pp. 16-25, 2010.

[3] J. Martin-Marinez, B. Kaczer, M. Toledano-Luque, R. Rodriguez, M. Nafria, X. Aymerich, and, G. Groeseneken, "Probabilistic defect occupancy model for NBTI," Proc. Int. Reliab. Phys. Symp., 2011, pp. 920-925.

[4] K. Ito, T. Matsumoto, S. Nishizawa, H. Sunagawa, K. Kobayashi, and H. Onodera, "The impact of RTN on performance fluctuation in CMOS logic circuits," Proc. Int. Reliab. Phys. Symp., 2011, pp. 710-713.

[5] A. Avellan, D. Schroeder, and W. Krautschneider, "Modeling random telegraph signal in the gate current of metal-oxide-semiconductor field effect transistors after oxide breakdown", J. Appl. Phys., 94, 2003, pp. 703-708.

[6] B. Kaczer, T. Grasser, Ph. J. Roussel, J. Martin-Martinez, R. O'Connor, B. J. O'Sullivan, G. Groeseneken, "Ubiquitous relaxation in BTI stressing – new evaluation and insights", Proc. Int. Reliab. Phys. Symp., 2008, pp. 20-27.

Kink effect characterization in AlGaN/GaN HEMTs by DC and Drain Current Transient measurements

L. Brunel[1,2], N. Malbert, A. Curutchet, N. Labat
[1] IMS laboratory, UMR-CNRS 5218
Talence, France
Laurent.brunel@ims-bordeaux.fr

B. Lambert
[2] United Monolithic Semiconductors
Villebon-sur-Yvette, France
Benoit.lambert@ums-gaas.com

Abstract — This study reports on detection and characterization of parasitic effects of AlGaN/GaN HEMTs on SiC substrate. First, experimental conditions impact and temperature effects on static $I_{DS}(V_{DS})$ characteristics are studied between 160 and 390K. Then, dependences on temperature, electric field and integration time of the kink effect are demonstrated with DC and pulsed measurements. Two traps are identified by isothermal drain current transient spectroscopy : an emission process and a capture process with activation energy of 0.58eV and 0.64eV respectively.

I. INTRODUCTION

Since the 90's, new applications using wide band gap based devices came out to complete the overall technologies used in sensors, optoelectronics, power electronics and high frequency electronics. GaN material, with a direct band gap of 3.43eV, offers great intrinsic qualities for power amplifier : a high thermal conductivity, a high breakdown electric field and a high electron mobility. Consequently, GaN based technologies are promising in terms of electrical performances and they are quickly progressing [1-2].

Studied AlGaN/GaN HEMTs on SiC substrate are provided by United Monolithic Semiconductor and have a 2x75µm-finger gate topology and a 0.5µm gate length.

This paper presents a study of the kink effect through exhaustive $I_{DS}(V_{DS})$ characteristics measurements as a function of several parameters such as experimental measurement conditions and temperature ; traps characterization is also performed by drain current transient spectroscopy. Thus, the aim of this paper is among others to discriminate the influence of integration time, temperature and electric field on both trapping and de-trapping processes and to analyse their correlation with the kink effect.

II. STUDY OF STATIC $I_{DS}(V_{DS})$ CHARACTERISTICS

A. Experimental conditions impact

Static measurements were carried out with a HP4142B parameter analyser by the Kelvin measurement method. Voltage steps are applied on the gate and the drain and each biasing point is first held during 20ms before the beginning of the measurements and then during the integration time (IT). Short, Medium and Long measurement configurations

correspond to the respective IT values of 4ms, 26ms and 320ms. Figure 1 shows the static output characteristics of a HEMT measured at 300K for the different integration times and for V_{DS} varying from 0 to 15V.

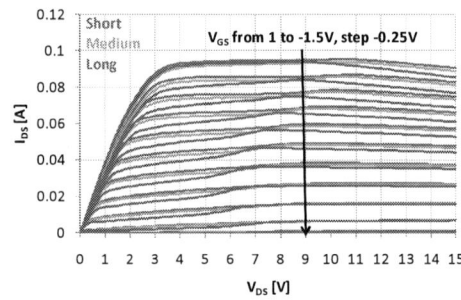

Figure 1. Static $I_{DS}(V_{DS})$ characteristic at 300K

According to figure 1, the kink effect observed near V_{DS}=6V becomes more pronounced with the increase of IT. Thermal effects are also visible for V_{DS}>10V and at open channel for long IT. It is important to notice that the first measurement is always free of kink in the case of fresh device.

Figure 2 presents the comparison between output characteristics obtained with upward and downward pumping in long IT configuration.

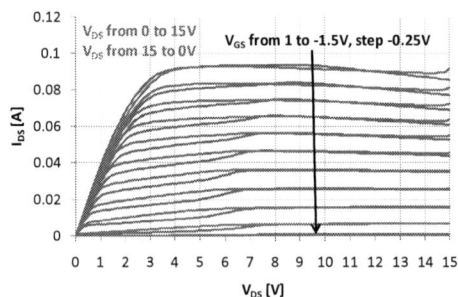

Figure 2. Comparison between upward (V_{DS} from 0 to 15V) and downward (V_{DS} from 15 to 0V) pumping in Long IT configuration

978-1-4673-1707-8/12 $31.00 © 2012 IEEE

According to figure 2, there is no more kink effect in the case of downward pumping and the change in drain current between $V_{DS} = 0V$ and 6V is due to trapping/de-trapping processes. Then the occurrence of the kink effect is possibly linked to the increase of the drain to source electric field while it disappears for V_{DS} decreasing during the measurement. We assume that the kink cannot be due to band to band impact-ionisation mechanisms, considering the large energy band gap of the GaN and drain voltage values at which the kink effect occurs.

Even if the mechanism responsible for the kink is still not well understood, several authors proposed some assumptions. According to M. Wang, a de-trapping process could take place during the measurement after application of a high drain bias ($V_{DS}=15V$) [3]. According to G. Meneghesso, a possible mechanism could be the following : when deep levels are negatively charged, electron de-trapping can occur due to the impact ionization of traps by hot electrons [4].

We propose in the next parts several experiments to bring to light the mechanisms responsible for the kink effect.

B. Output characteristics versus temperature

Output characteristics have been measured from 160K to 390K to determine the dependence of the kink effect versus temperature. Figure 3 shows the evolution of the output characteristics versus temperature measured at a fixed gate bias value in Long IT configuration.

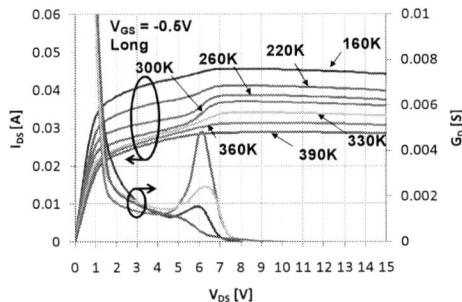

Figure 3. Output characteristics versus temperature at $V_{GS}=-0.5V$

According to figure 3, we demonstrate that the kink effect is thermally activated as its magnitude (represented by the magnitude of G_D) changes with temperature. Moreover, the kink effect seems to totally disappear for high temperature (Tc >360K) which suggests a more important de-trapping process at high temperature [5-6].

Figure 4 presents the evolution of V_{DSkink} (V_{DS} value where G_D is maximum) versus gate bias at a fixed temperature. V_{DSkink} decreases then increases with the gate bias V_{GS}, which has already been observed in [4] and reflects impact ionization of traps by hot electrons. Thus, the mechanism responsible for the increase of drain current cannot be only electric field assisted as the de-trapping process would be more efficient when the electric field is maximum (i.e. when the device is biased close to pinch-off condition) and thus V_{DSkink} would be minimum at $V_{GS}=-1.5V$.

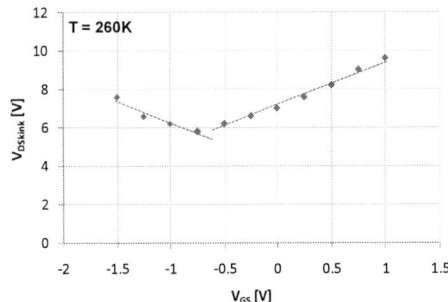

Figure 4. Evolution of V_{DSkink} versus gate bias V_{GS} at 260K

C. Transfer characteristic versus temperature

The transfer characteristics $I_{DS}(V_{GS})$ are also measured as a function of temperature. Figure 5 shows the evolution of the threshold voltage V_{Th} as a function of the drain bias value V_{DS} and at a fixed temperature.

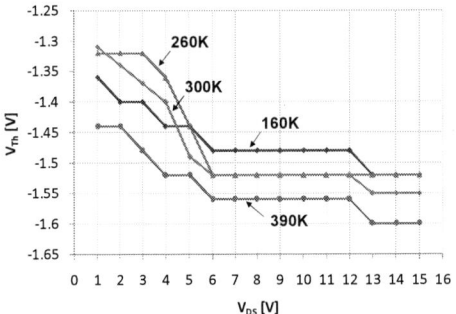

Figure 5. Evolution of V_{Th} as a function of V_{DS} and temperature

Figure 5 highlights that there is a large shift of V_{Th} only for low V_{DS} values (from 1 to 6V) which is reduced at high temperature (390K), i.e. when the kink effect magnitude is the lowest (see Figure 3). Moreover, there is no more shift of V_{Th} above $V_{DS}=6V$, which corresponds to the value of V_{DSkink}. Then, we assume that the kink effect is correlated with a shift of V_{Th}. A de-trapping process could take place at a high drain bias ($V_{DS}>6V$). It is then suggested that the traps responsible for the kink effect could be located in the GaN buffer layer below the gate [3, 7]. In this case, electrons trapped under the gate create a negative charge which shifts V_{Th} towards positive values.

III. KINK EFFECT STUDY

A. Impact of the temperature

To evaluate more precisely the effect of temperature, two identical consecutive measurements at $V_{GS}=0V$ with V_{DS} varying from 0 to 15V in Long IT configuration and at different temperatures have been compared. After the two consecutive measurements, the device is placed at room temperature for 24 hours to obtain a first measurement free of kink effect. Figure 6 shows that the kink effect does not totally disappear at high temperature, on the contrary to what is assumed in the previous part. However, the kink effect magnitude decreases with the increase of temperature which suggests a faster de-trapping process at high temperature.

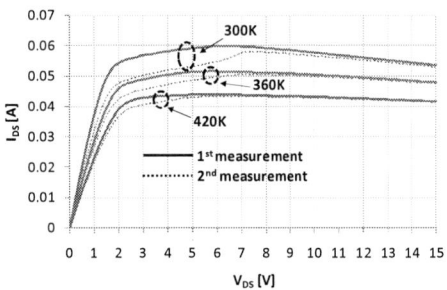

Figure 6. Comparison between two consecutive measurements for several temperatures at V_{GS}=0V

B. Study of the trapping process

A first measurement has been performed at V_{GS}=0V with V_{DS} from 0 to V_{DSmax} and then a second one from 0 to 15V. The device has been placed at room temperature for 24 hours before two others consecutive measurements to obtain a first measurement free of kink. Figure 7 shows the comparison between the second measurements for different values of V_{DSmax}.

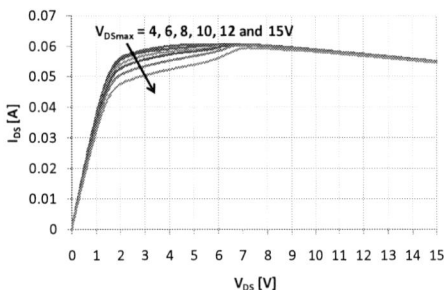

Figure 7. Output characteristics for different values of V_{DSmax} at V_{GS}=0V and 300K

According to these results, the more the maximum drain voltage of the first measurement is high, the more the kink effect is important.

Then, a first sweep with V_{GS}=-5V and V_{DS} from 0 to 15V (free of kink) has been performed and followed by a second one with V_{GS}=0V. A non negligible kink effect was observed on the second output characteristic (not shown). Thus, it is assumed that the trapping process highly depends on electric field [3, 4] and that a high drain current is not necessary to fill-in the traps inducing the kink effect to occur.

C. Study of the de-trapping process

A first sweep is applied (called first measurement) to fill-in the traps, then constant V_{DS} and V_{GS} biases are applied for 10s while the drain current is monitored and then a second sweep is performed. The same sequence is then repeated 24 hours later with different constant V_{DS} and V_{GS} bias conditions. As the drain current remains constant while applying constant biases, we assume that applying drain to gate voltage does not allow all traps to emit electrons. Figure 8

compares the second measurements obtained after applying constant biases for 10s with a consecutive sweep (used as reference) in order to evaluate their impact on kink effect.

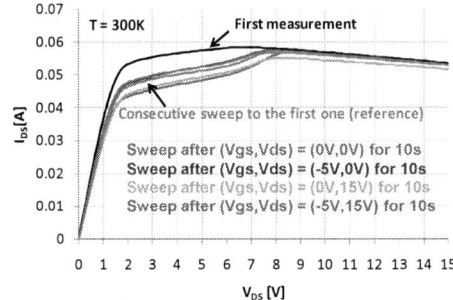

Figure 8. Influence of different gate and drain constant biases

According to figure 8, it is shown that once traps are charged, applying high constant electric field between gate and drain ((V_{GS};V_{DS}) = (0V;15V) and (-5V;15V)), increases the kink effect magnitude whereas a partial recovery of the drain current is obtained with a lower electric field ((V_{GS};V_{DS}) = (0V;0V) and (-5V;0V)). Thus we conclude that the kink effect depends on final electric field value applied between gate and drain on the device. This hypothesis is confirmed by the fact that replacing constant biases by a downward pumping (V_{GS}=0V and V_{DS} varying from 15 to 0V) leads also to a partial recovery of the drain current (not shown). It is then assumed that the de-trapping process becomes faster when the device is biased at low or decreasing electric field conditions.

D. Output characteristics in pulsed mode

The test bench used allows applying two short periodic voltage pulses on both gate and drain simultaneously. To localize traps responsible for the kink effect, the output characteristics have been measured for 4 different quiescent biases and with 2048 averages on each point. Once again, two identical and consecutive measurements have been performed at V_{GS}=0V with V_{DS} varying from 0 to 15V for the different quiescent biases with a pulse period of 30µs and a duty cycle of 1%.

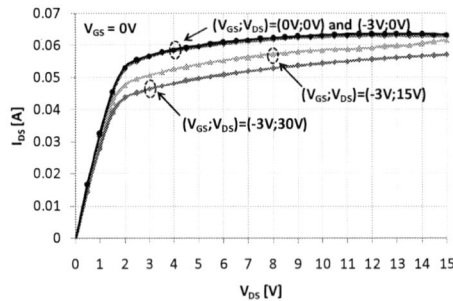

Figure 9. Output characteristics in pulsed mode at V_{GS}=0V and 300K for 4 quiescient points

According to figure 9, only drain lag effect is observed on the output characteristics. These results demonstrate that there are traps in the gate-drain region located in the GaN layer [8]. However, traps highlighted with DC and pulsed measurements cannot be the same due to their different time constant. Thus we conclude that the kink effect is closely related to slow traps effect since it is not observed in pulsed mode [6, 9].

IV. ISOTHERMAL DRAIN CURRENT TRANSIENT SPECTROSCOPY MEASUREMENTS

The principle of the measurements is to apply a voltage pulse with a period T (T=90s) on the gate and to measure the drain current transient response while the HEMT is biased in saturation mode (V_{DS}=3V). The test bench is composed of a pulse generator (HP8116A) and a numerical multimeter (HP34401A). Each measurement is performed 10 times then averaged to increase the signal to noise ratio. Measured drain current transients are then normalized to the final value of the current at equilibrium state. Figure 10 presents normalized drain current transients as a function of the temperature.

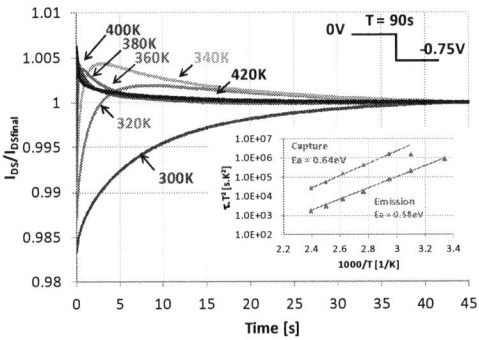

Figure 10. Normalized drain current transients with a period T=90s versus temperature and arrhenius plot for V_{GS} varying from 0 to -0.75V and V_{DS}=3V

Figure 10 highlights an electron emission mechanism followed by a capture one above 320K. Time constants encountered here become lower as the temperature increases which indicates faster electron de-trapping at high temperature. Then, drain current transients are split up into a sum of exponential components as follows:

$$F(t) = a_0 + \sum_{i=1}^{n} \left(a_i \times e^{\frac{-t}{\tau_i}} \right) \qquad (1)$$

with a_0 the normalized current value at equilibrium state, τ_i the time constant and a_i the amplitude associated to each detected trap. According to the Arrhenius plot in figure 10 obtained with a gate pulse varying from 0 to -0.75V, two trap activation energies have been identified at Ea = 0.64 ± 0.01eV and Ea = 0.58 ± 0.01eV; respectively corresponding to a capture process and an emission process. Similar activation energies were also obtained for V_{GS} varying from -0.5 to -1V and correspond to the same traps. Furthermore, many authors have reported similar activation energies [3, 9, 10 and 11] with different transient measurements techniques. As the time constant values are large, the traps identified by drain current transient measurement may be responsible for the kink effect

and are located under the gate either in the AlGaN or the GaN layer.

V. CONCLUSIONS

The impact of experimental conditions on the HEMT static output characteristics has been studied and two phenomena have been highlighted. The thermal effects which cause a decrease of the drain current at high electric field (V_{DS}>10V) and open channel conditions (V_{GS}>-0.5V). These effects are all the more important as the integration time is long and the temperature is low. The kink effect is characterized by an unusual increase of the drain conductance G_D. It has been proved that the kink effect depends on temperature, electric field and IT. Slow traps located under the gate have been found responsible for the kink effect since no kink effect was observed on pulsed measurements. Finally, these traps responsible for a capture process and an emission process have been identified with activation energy at Ea = 0.64 ± 0.01eV and Ea = 0.58 ± 0.01eV respectively.

ACKNOWLEDGMENT

This work was made possible by financial support from the French MoD (DGA-MI).

REFERENCES

[1] Nidhi, S. Dasgupta, S. Keller, JS. Speck and UK. Mishra, "N-polar GaN/AlN MIS-HEMT with f(MAX) of 204 GHz for Ka-band applications." IEEE EDL Vol. 32, No. 12, December 2011

[2] UK. Mishra, L. Shen, TE. Kazior and YF. Wu, "GaN-based RF power devices and amplifiers." Proceedings of the IEEE Vol. 96, No. 2, February 2008

[3] M. Wang and KJ. Chen, "Kink effect in AlGaN/GaN HEMTs induced by drain and gate pumping" IEEE EDL, Vol. 32, No. 4, April 2011

[4] G. Meneghesso, F. Zanon, MJ. Uren and E. Zanoni, "Anomalous kink effect in GaN high electron mobility transistors" IEEE EDL, Vol. 30, No. 2, February 2009

[5] G. Mouginot, R. Sommet, R. Quéré, Z. Ouarch, S. Heckmann and M. Camiade, "Thermal and trapping phenomena assessment on AlGaN/GaN microwave power transistor." Proceeding of the 5th EuMA, pp 110-113, September 2010

[6] R. Cuerdo, Y. Pei, Z. Chen, S. Keller, SP. Deenbaars, F.Calle, and UK. Mishra, "The kink effect at cryogenic temperatures in deep submicron AlGaN/GaN HEMTs." IEEE EDL Vol. 30, No. 3, March 2009

[7] G.Meneghesso, F. Rossi, G. Salviati, MJ. Uren, E. Muñoz and E. Zanoni, "Correlation between kink and cathodoluminescence spectra in AlGaN/GaN high electron mobility transistors." Applied physics letters 96, 263512, July 2010

[8] M. Meneghini, N. Ronchin A. Stocco, G. Menegheso, UK ishra, Y. Pei and E. Zanoni, "Investigation of trapping and hot-electron effects in GaN HEMTs by means of a combined electrooptical method", IEEE transactions on elecron devices, Vol. 58, No. 9, September 2011

[9] J. Joh and JA. del Alamo, "A current-transient methodology for trap analysis fr GaN hgh electron mobility transistors", IEEE transactions on elecron devices, Vol. 58, No. 1, January 2011

[10] JG. Tartarin, S. Karboyan, F. Olivié, G. Astre, L. Bary and B. Lambert "I-DTS, electrical lag and low frequency noise measurements of trapping efects inAlGaN/GaN HEMT for reliability studies", Proceeding of the 6th EuMA, pp 438-441, October 2011

[11] M. Gasoumi, MM. Ben Salem, S. Saadoui, B. Grimber, J. Fontaine, "The effect of he gate length variation and trapping effects on the transient response of AlGaN/GaN HEMTs on SiC substrate", Microelectronic Engineering 88, pp 370–372, October 2011

Random Telegraph Signal Noise
Properties of HfOx RRAM in High Resistive State

Francesco M. Puglisi, Paolo Pavan
Dipartimento di Ingegneria
dell'Informazione
Università di Modena e Reggio Emilia
Via Vignolese 905, 41125 Modena –
Italy
francescomaria.puglisi@unimore.it

Andrea Padovani, Luca Larcher
Dipartimento di Scienze e Metodi
dell'Ingegneria
Università di Modena e Reggio Emilia
Via Amendola 2, 42122 Reggio
Emilia – Italy

Gennadi Bersuker
SEMATECH
257 Fuller Road, 12203 Albany, New
York, and 207 Montopolis Drive,
78741 Austin, Texas - USA

Abstract— **In this paper we analyze Random Telegraph Signal (RTS) noise in hafnium-based RRAMs. RTS is measured in HRS, showing fast and slow multilevel switching events. RTS characteristics are examined through novel color-coded time-lag plots and Hidden Markov Model (HMM) time-series analyses. Noise is examined at different reset conditions to provide new insights on conduction mechanisms in HRS. Higher reset voltages result in an enhanced complexity in RTS due to a larger number of active traps.**

I. INTRODUCTION

Resistive switching memories (RRAMs) are one of the most important candidate to achieve reliable, fast and high-density NVMs (non-volatile memories). Among different alternatives to market-dominating Flash memories, i.e. Phase-Change Memories [1], RRAMs show good performances and optimal compatibility with standard CMOS processes since hafnium oxide is extensively used in ultimate commercial 45-nm and 32-nm CMOS nodes (High-K/Metal Gate). RRAMs based on this material have already been successfully implemented, even in sub-22-nm nodes [2], but further characterization is demanded. State-of-the-art understanding of RRAMs physics is based on the formation of a conductive filament (CF) during Forming / Set operations and its subsequent partial oxidation during Reset operation [3]. Current compliance determines CF equivalent cross-section and LRS resistance value [4]. While filamentary conduction in LRS exhibits ohmic behavior, HRS conduction is correctly described by a multi-phonon trap-assisted tunneling (TAT) process [3] via the traps in the oxidized portion of the filament.

This paper presents experimental results on Random Telegraph Signal (RTS) noise characterization in HfOx RRAMs. RTS fluctuations are clearly detected only in HRS. We believe that current fluctuations are related to the temporary unavailability/availability of one or more traps assisting the charge transport through the dielectric layer, in agreement with the interpretation that the dominant mechanism involved in HRS current conduction is trap-assisted tunneling [3]. Noise at reading voltage is analyzed

after reset in different conditions through powerful tools such as color-coded time-lag plots and Hidden Markov Model (HMM) analysis [6], suggesting that active traps are primarily located in the oxidized part of the filament. *This is, to our knowledge, the first time that RTS features are explored in HfOx RRAMs at different reset voltages.*

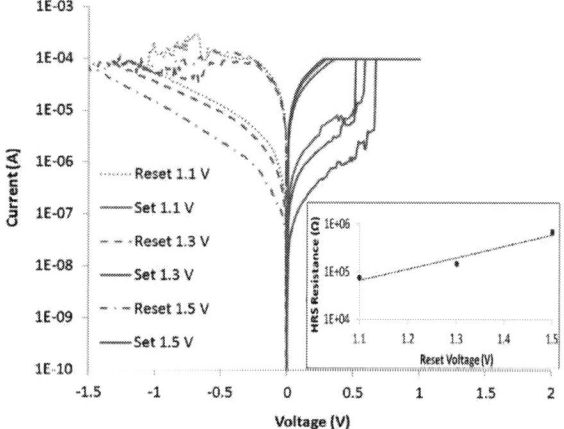

Fig. 1. I-V curves of MIM capacitors for different reset voltages. Current is limited by compliance (100 µA) during Set operation. Reset and Set operations are re-iterated at different Reset Voltages. Inset shows experimental HRS Resistance values at different Reset Voltages and exponential fitting.

A detailed description of the experimental procedure is available in Section II. Results are shown in Section III and discussed in Section IV. Conclusions follow.

II. EXPERIMENTS

Measurements are performed on a 49 µm^2 TiN/HfO$_x$/TiN RRAM (MIM capacitor) with a 5 nm thick oxide layer. Initial forming operation is performed to enable resistive switching, then LRS and HRS are reached through set and reset

978-1-4673-1707-8/12 $31.00 © 2012 IEEE

operations. An HP4155B Semiconductor Parameter Analyzer is used to acquire I-V curves during set and reset operations (see Fig. 1). Resistive switching to HRS and LRS is achieved by applying negative and positive (with a specific current compliance) voltage ramps, respectively. This process is repeated for different reset voltages, Fig. 1. Once the DUT is in the desired resistive state, noise is analyzed using the HP4155B as a biasing source (in reading conditions) and current monitor for RTS measurements. Then, collected data are processed by a custom-developed software based on HMM [6], allowing accurate RTS events detection. Moreover time-lag plots and probability density of current levels are combined and used to evidence traps activity. Measurement operations are controlled via a dedicated software, which allows fast and reliable data management.

III. RESULTS

Figure 1 reports I-V measurement for the investigated sample. Interestingly, HRS resistance value (V_{READ}= 100 mV) depends on reset voltage, as previously reported [5]. Conversely, LRS always shows the same value. This can enable multilevel operations: HRS resistance in reading conditions is found to be exponentially proportional to the applied reset voltage. The inset in Fig.1 reports experimental HRS resistance data vs. reset voltage and its exponential fitting.

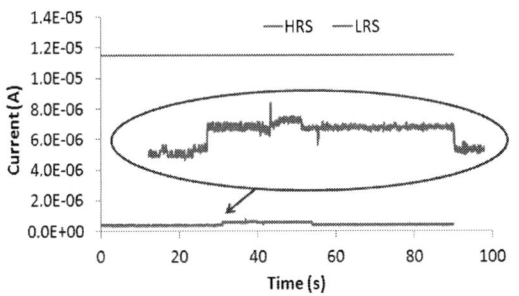

Fig. 2. Random Telegraph Signal is detected only in HRS. Slow and fast switching events are clearly visible.

RTS noise is measured in both LRS and HRS, Fig. 2: while in HRS we detect large RTS current fluctuations, no events are measured in LRS. HRS, moreover, is characterized by complex RTS, and both slow and fast switching are observed. Slow switching events are of great interest since they result in meta-stable states lasting several seconds and causing large current fluctuations while fast switching events are usually related to smaller current variations, see the zoomed curve in Fig. 2.

It is believed that reset voltage affects HRS resistance since it determines the re-oxidation length of the CF [5], hence the thickness of the dielectric barrier. Simulations show that trap-assisted tunnelling through this barrier is the main conduction mechanism in HRS: we expect an increasing number of active traps with increasing oxidation lengths, hence increasing reset voltages. RTS characterization is performed

applying a constant voltage (V_{READ}= 100 mV) and recording current fluctuations of a sample forced in HRS at given reset voltages: low (1.1V), medium (1.3 V) and high (1.5V). Data is processed through a custom-developed software based on discrete HMM analysis, a powerful tool commonly used in pattern recognition and statistical signal analysis in which the system is assumed to be a Markov process with hidden states, i.e. discrete current levels. An HMM is completely defined as a 5-tuple (N, M, A, B, π) in which N is the number of hidden states in the model (i.e. discrete current levels to be found), M is the number of distinct observable symbols per state (i.e. the possible current values), A is a N-by-N matrix defining the transition probabilities among states, B is a N-by-M matrix defining the observation probability of a each observable symbol in each state and π is a vector defining the initial state probability distribution [6]. The main task in HMM is to find, given an output sequence (i.e. acquired RTS data), the best set of hidden state transitions and output probabilities. This is achieved through a maximum likelihood estimate of the parameters of the HMM given the output sequence using the "Baum–Welch" or the "Viterbi Training" algorithm [7]. In our implementation we prefer to use "Viterbi Training" algorithm, characterized by a worse convergence but a faster response w.r.t. Baum-Welch method, resulting in an accurate and fast analysis of RTS switching events. Convergence is ensured by a good initial estimation of B matrix, extrapolated from acquired RTS data through multiple Gaussian fittings. HMM analysis can efficiently estimate discrete current levels and typical dwell times for n-levels RTS.

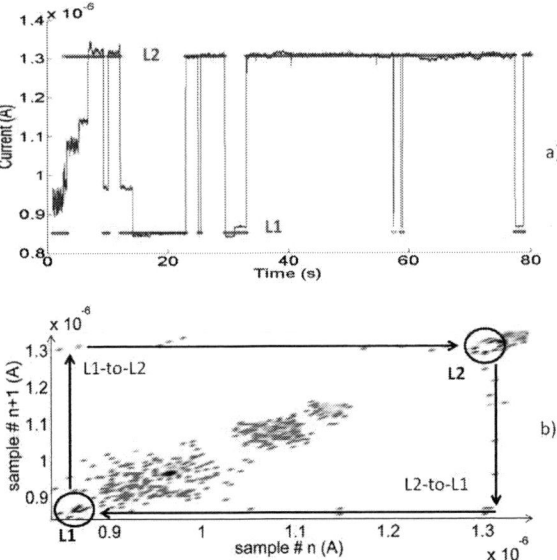

Fig. 3. a) RTS and b) colour-coded time-lag plot at 1.1V (low) Reset Voltage. Two levels, L1 and L2, are clearly recognizable in both representations, indicating the presence of one dominant trap and the good agreement between experimental data and 2-levels HMM approximation. Transitions among dominant levels in RTS trace are identified also in colour-coded time-lag plot. Spurious features can be observed both in the RTS trace and in colour-coded time-lag plot.

Fig.3 a) shows a typical acquired RTS trace (blue) and its approximation to a 2-levels HMM approximation (red). Along with this characterization method, another tool, based on time-lag plot, is employed to analyze RTS noise. Indeed any time-series can be represented in terms of time-lag plot (graphical data analysis technique for determining if an autocorrelation structure exists within the time series). A time-lag plot is built considering the n-th sample of the acquired time-series on the x-axis and the n+1-th sample on the y-axis: this results in big spots along the diagonal corresponding to discrete current levels and weak spots outside the diagonal representing transitions among levels. Moreover, we can enhance the time-lag plot visualization by means of a proper colour-code. Indeed, each point of the plot represents a couple of possible successive symbols in the observed time trace which can be associated to a value related to the probability density of finding the n+1-th symbol after the n-th symbol. An appropriate colour code based on the reciprocal of the aforementioned probability density is employed to properly reveal levels and transitions: black spots are associated with a high probability density; as probability density decreases spots colour tends to be yellow and finally red, see Fig. 3 b).

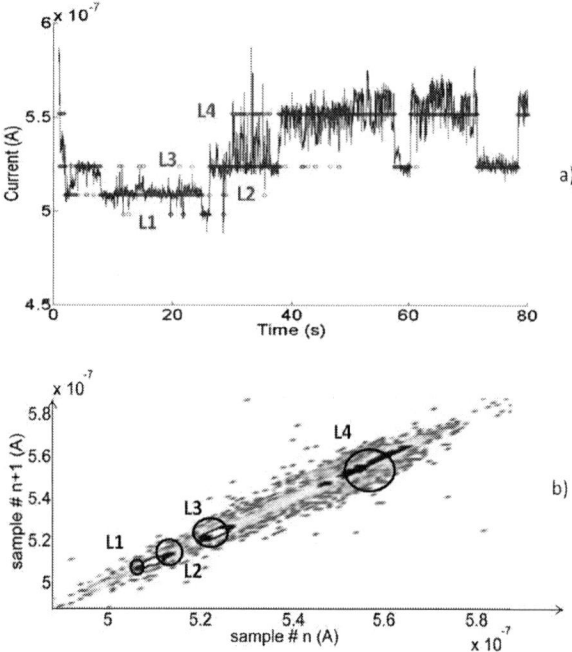

Fig. 4. a) RTS and b) colour-coded time-lag plot at 1.3V (medium) Reset Voltage. A 4-levels HMM approximation is used to describe RTS data. These levels are highlighted in the colour-coded time-lag plot, confirming that this approximation can be used to correctly represent data.

HMM analyses and colour-coded time-lag plots are exploited to reveal noise features in HRS at different reset conditions: Figs. 4 and 5 report acquired time-series (along with 4-levels HMM approximation) and colour-coded time-lag plots for the remaining two reset voltages, (medium, 1.3V, and high, 1.5V).

IV. DISCUSSION

The comparison of Figs. 3 to 5 helps in understanding the change in HRS conduction with reset voltage. In Fig. 3 a) a clear 2-levels fluctuation is recognizable, as confirmed by the HMM analysis: small fluctuations are visible but the dominant trend is given by a single switching mechanism related to a single trap activity. This is confirmed by colour coded time-lag plot in Fig. 3 b) in which two main black spots are visible along the diagonal, representing two discrete current levels, hence a single dominant trap. Transitions between those levels are also indicated. Transitions between intermediate levels (spurious) are clearly visible but their low probability density do not compromise the validity of this 2-levels approximation. As reset voltage is increased more traps are revealed: Fig. 4 a) shows acquired RTS trace at medium reset voltage and its HMM approximation: this time a 4-levels approximation has been used since time-series is more complex. With this increased complexity, the 4-levels representation (related to the activity of 2 main traps) is a good approximation, as confirmed by colour coded time-lag plot in Fig. 4 b). Indeed, four main black spots are visible along the diagonal. Notice that the black spot corresponding to L4 is wider, representing the large amount of noise around that level. Finally, the sample reset at high reset voltage (1.5V) shows a further increase in RTS complexity, see Fig. 5 a), where a 4-levels HMM approximation cannot properly fit the experimental data: more levels are visible in the time trace and they should be considered for a better HMM approximation but this leads to a critical time-to-solution. Colour coded time-lag plot confirms the presence of several black spots, associated to an increased number of active traps. Some black spots are evident even outside the main diagonal, revealing the existence of other levels and transitions among them.

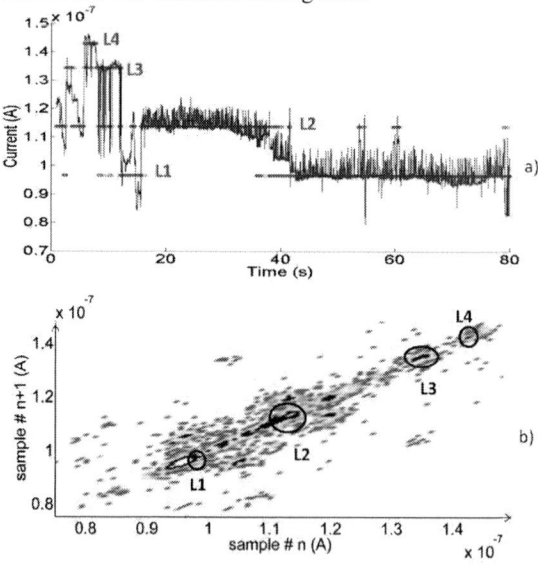

Fig. 5. a) RTS and b) colour-coded time-lag plot at 1.5V (high) Reset Voltage. Enhanced complexity is revealed by the color-coded time-lag plot: a four-level approximation cannot represent RTS noise.

These considerations confirm our description of HRS conduction: TAT occurs into the dielectric layer formed during reset and traps located in the barrier seem to be responsible of RTS fluctuations. As reset voltage increases also barrier thickness is increased and more traps are revealed.

In low reset conditions TAT involving one trap only is a good description of HRS conduction. Therefore, 2-levels HMM approximation can be used to extract capture (τ_c) and emission (τ_e) times. The TAT model in [8] describes τ_e and τ_c by compact formulae, allowing trap characterization. Indeed dwell times variations with reading voltage can be exploited to extract trap vertical position using the relation:

$$Z_{eff} = V_T \cdot \frac{\partial (\ln(\frac{\tau_c}{\tau_e}))}{\partial V_{READ}} \cdot \left(\frac{1}{2} \left[\alpha + \beta \cdot V_{READ} \right] \right) \quad (1)$$

where Z_{eff} is the relative trap position inside the dielectric barrier, V_T the thermal voltage and V_{READ} the reading voltage. The term in parentheses accounts for the structural relaxation occurring during the TAT process [8], thus providing a strong link with the physical nature of the defect at the origin of the RTS fluctuations. α and β coefficients depend on trap energy and relaxation energy (typical values in HfO$_x$ are used).

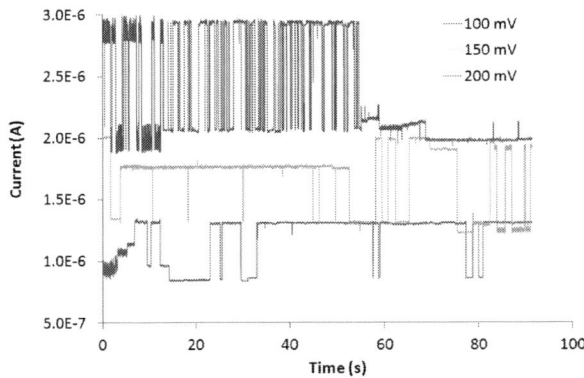

Fig. 6. Current traces at different reading voltages. DUT is reset at low reset voltage showing a clear 2-levels fluctuations. Dwell times can be extracted.

Fig. 6 shows low reset voltage RTS at different reading conditions. From these curves, dwell times can be extracted and the logarithms of their ratio are reported in Table I and shown in Fig. 7 as a function of reading voltage. Since their linearity, (1) can be simplified neglecting the variation of the trap position with the reading voltage used in this work. Indeed numerical calculations show that, in our experimental conditions, the term in parentheses can be approximated with α /2. The dominant trap causing RTS instabilities at low reset voltage is found to be located at ~2/5 of the dielectric barrier width (almost in the middle).

Reading Voltage (mV)	$\ln(\tau_c/\tau_e)$
100	-1.12559
150	-0.72222
200	-0.36103

Table 1. Reading voltage dependence of ln(τc/τe).

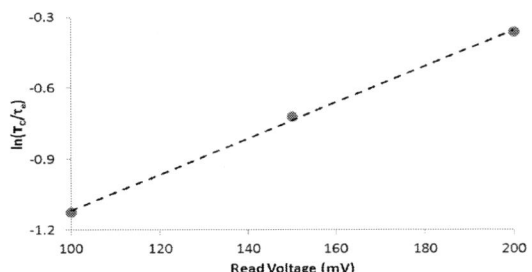

Fig. 7. Logarithm of the dwell times ratio vs. reading voltage and its linear fitting.

V. CONCLUSIONS

In this paper we presented experimental results on RTS noise characterization in HfOx RRAMs. Noise is investigated through powerful HMM analysis and colour coded time-lag plots. Noise behavior suggests that a higher reset voltage results in increased complexity in RTS, hence a larger number of active traps. A 2-levels approximation can be easily elaborated to infer trap characteristics, while it is still difficult to deal with multiple-levels oscillations.

ACKNOWLEDGMENT

The authors wish to thank Sematech for providing samples.

REFERENCES

[1] H.Y. Cheng, T.H. Hsu, S. Raoux, J.Y. Wu, P.Y. Du, M. Breitwisch, Y. Zhu, E.K. Lai, E. Joseph, S. Mittal, R. Cheek, A. Schrott, S.C. Lai, H.L. Lung, C. Lam, "A high performance phase-change memory with fast switching speed and high temperature retention by engineering the GexSbyTez phase change material", IEEE IEDM Tech. Digest, pp. 51-54, 2011.

[2] B. Govoreanu, G.S. Kar, Y.-Y. Chen, V. Paraschiv, S. Kubicek, A. Fantini, I.P. Radu, L. Goux, S. Clima, R. Degraeve, N. Jossart, O. Richard, T. Vandeweyer, K. Seo, P. Hendrickx, G. Pourtois, H. Bender, L. Altimime, D.J. Wouters, J.A. Kittl, M. Jurczak, "10x10 nm2 Hf/HfOx Crossbar Resistive RAM with Excellent Performance, Reliability and Low-Energy Operation", IEEE IEDM Tech. Digest, pp. 729-732, 2011.

[3] L. Vandelli, A. Padovani, L. Larcher, G. Broglia, G. Ori, M. Montorsi, G. Bersuker and P. Pavan, "Comprehensive physical modeling of forming and switching operations in HfO2 RRAM devices", IEEE IEDM Tech. Digest, pp. 421-424, 2011.

[4] G. Bersuker, D.C. Gilmer, D. Veksler, P. Kirsch, L. Vandelli, A. Padovani, L. Larcher, K. McKenna, A. Shluger, V. Iglesias, M. Porti and M. Nafria, "Metal oxide resistive memory switching mechanism based on conductive filament properties", Journal of Applied Physics 110, 124518, 2011.

[5] H.-L. Chang, H.-C. Li, C. W. Liu, F. Chen and M.-J. Tsai, "Physical Mechanism of HfO2-based Bipolar Resistive Random Access Memory", IEEE International Symposium on VLSI Technology, Systems and Applications (VLSI-TSA), pp. 1-2, 2011.

[6] L. R. Rabiner, "A Tutorial on Hidden Markov Models and Selected Applications in Speech Recognition", Proceedings of the IEEE, vol. 77 No.2, pp. 257-285, February 1989.

[7] S. Siddiqi, G. J. Gordon, and A. Moore, "Fast state discovery for HMM model selection and learning", Proc. AISTATS, 2007.

[8] L. Vandelli, A. Padovani, L. Larcher, R. G. Southwick, III, W. B. Knowlton, and G. Bersuker, "A Physical Model of the Temperature Dependence of the Current Through SiO2/HfO2 Stacks", IEEE Trans. Electron Devices, 58, 2878, 2011.

On the impact of Ag doping on performance and reliability of GeS$_2$-based Conductive Bridge Memories

E. Vianello, C. Cagli, G. Molas,
E. Souchier, P. Blaise, C. Carabasse, G. Rodriguez,
V. Jousseaume, B. De Salvo
CEA-LETI
17, rue des Martyrs 38054 Grenoble Cedex 9, France

F. Longnos, F. Dahmani, P. Verrier,
D. Bretegnier, J. Liebault
Altis Semiconductor
224 Bd John Kennedy F-91105 Corbeil Essonnes Cedex.
florian.longnos@cea.fr

Abstract — **In this work, we study the impact of Ag doping on GeS$_2$-based CBRAM devices employing Ag as active electrode. Several devices with Ag doping varying between 10% and 24% are extensively analyzed. First, we assess switching voltages and time-to-set as a function of Ag concentration in the electrolyte layer. Subsequently, we evaluate data retention at different temperatures. The results show that a Ag doping increase in the GeS$_2$ yields a strong improvement on data retention performance, increasing the 10-years data-ret temperature from 68°C for the 10% Ag doping to 100°C for the 24%, without any significant increase of the set voltage (50mV higher).**

I. INTRODUCTION

Resistive–switching memory technologies are attracting big interest and are undergoing extensive investigation as possible FLASH technology replacement for several applications. Among different emerging technologies, Conductive Bridge Random Access Memory (CBRAM) [1][2] appears as one of the most promising candidates thanks to its relatively simple structure, very low power consumption together with high programming speed. In particular, its low cost and BEOL (Back-End-Of-Line) compatibility make this technology particularly interesting for embedded applications.

As shown in Fig.1, the basic CBRAM cell is a resistor – i.e. it has two terminals – consisting of a thin film of solid electrolyte sandwiched between an electrochemically active electrode, and an electrochemically inert counter electrode. The CBRAM working principle is based on field-driven generation of metallic ions at the top active electrode and their deposition and neutralization at the counter electrode after electron-migration through the electrolyte. During the programming operation (SET) a positive voltage is applied to the anode, a conductive metal bridge is thus formed and the cell resistance drops. The process can be reversed by applying opposite bias (RESET operation), redepositing the metal back into the active electrode.

In the literature it has been shown that good solid electrolytes contain a large number of highly mobile positively charged metal ions [3]. The most widely studied materials are chalcogenide-based glasses, e.g. GeSe and GeS, doped with silver or copper. The doping can be either obtained in one step (with the co-sputtering of the chalcogenide-based glass and a metal target [4]) or in two steps (with the successive deposition of the electrolyte and the metal layers followed by a thermal- or photo-induced treatment in order to make the metal diffuse into the chalcogenide-based glass [5,6]). So far, several material studies have been reported on the microstructure of a huge number of metal ion-doped chalcogenide-based glasses, such as Ag-Ge-S [3,7,8,9], Ag-Ge-Se [7,8,10], Cu-Ge-S [3]. However, an extensive investigation of the impact of the metal ion doping concentration on the electrical performances of the CBRAM devices is still missing. In this work, we study performance and reliability of CBRAM cells based on GeS2 with different Ag doping concentrations. In particular, SET/RESET operation and data retention are assessed.

II. DEVICE FABRICATION AND STRUCTURE

The CBRAM cell used for this investigation consists of a Ag/GeS$_2$/W resistor integrated as a mesa structure.

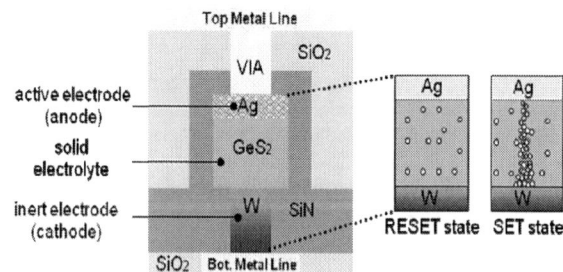

Figure 1 - 1R CBRAM memory structure and schematic view of the CBRAM resistor.

978-1-4673-1707-8/12 $31.00 © 2012 IEEE

As shown in Fig.1, in the CBRAM resistor, a Tungsten (W) plug, with a diameter of 200 nm, is used as inert electrode. The active storage layer consists of a 30nm thick Ag-doped GeS_2 layer. Then as a top electrode a 50nm-thick Ag layer is deposited. For the purpose of this work, three different samples with different Ag doping concentrations (namely S1: 10%, S2: 15% or S3: 24%) in the GeS_2 layer were integrated (see Table 1). The solid electrolytes of samples S1 and S2 were fabricated in a one-step process using co-sputtering of stoichiometric GeS_2 and Ag targets. The 24 at% Ag-GeS_2 film (sample S3) was obtained by a two step process, namely sputtering of a 10nm-thick Ag layer onto a 30nm-thick GeS_2 layer, followed by a UV photo-diffusion treatment. The composition and structure of the Ag-doped GeS_2 layers were studied by means of Rutherford Backscattering Spectroscopy (RBS) and X-Ray Reflectivity (XRR) measurements. On all the samples, RBS measurements show a constant Ge/S ratio of 0.64-0.66 with different Ag concentrations (see Table 1). XRR measurements show that S1 and S2 are structured as nanolaminates of Ag-doped GeS_2, while S3 shows a uniform Ag doping concentration across the layer. A probable consequence of the methodology of cell fabrication.

To get more insights into the structure of the Ag-doped GeS_2 layers, we performed *ab-initio* simulations. As shown in Fig.2, it appears that the Ag is always chemically bonded. Some S atoms are extracted from the initial GeS_2 backbone to react with the diffused Ag thus generating Ag_2S compounds. Therefore, after chemical reaction the remaining chalcogenide glass becomes deficient of Ge-S bonds and unbonded S atoms, hindering the dissolution of additional Ag into the GeS_2. These results could suggest that the Ag diffusion in Ag doped GeS_2 obey the Fick's law [11,12,13]. Consequently, increasing the Ag doping, the GeS_2 matrix becomes saturated thus limiting the Ag diffusion during the memory operations.

TABLE 1 – SPLIT-TABLE OF THE DIFFERENT CBRAM SAMPLES USED IN THIS STUDY.

Sample	%at Ag in the GeS₂ layer (RBS)	Fabrication methodology of the Ag-doped GeS₂
S1	10.7	Co-sputtering of stoichiometric GeS₂ and Ag targets
S2	15.2	
S3	24	30nm sputtered GeS2 layer + 10nm sputtered Ag layer + UV photo-diffusion treatment

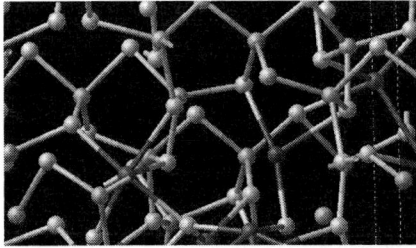

Figure 2 – Atomistic view of the Ag doped GeS_2 structure (Red=Ag, green=Ge, yellow=S) showing the Ag_2S compound.

III. RESULTS AND DISCUSSION

Program/Erase operations in quasi-static mode - Quasi-static (or DC) program/erase characteristics were performed applying to the anode a write voltage sweep from 0 to 1V followed by an erase voltage sweep from 0 to -1V. In order to study the influence of the compliance current flowing in the CBRAM resistor during the set operation, the compliance current was screened between 10^{-7} and 10^{-3}A. As shown in Fig. 3, the SET resistance decreases and the RESET current increases while increasing the SET compliance current [14,1]. It is worth noting that the latter ranges from 10^{-8} to 10^{-4} A, thus being smaller than the compliance current. Fig. 3 also indicates that the SET resistances and the RESET currents are almost completely independent of the Ag doping concentration.

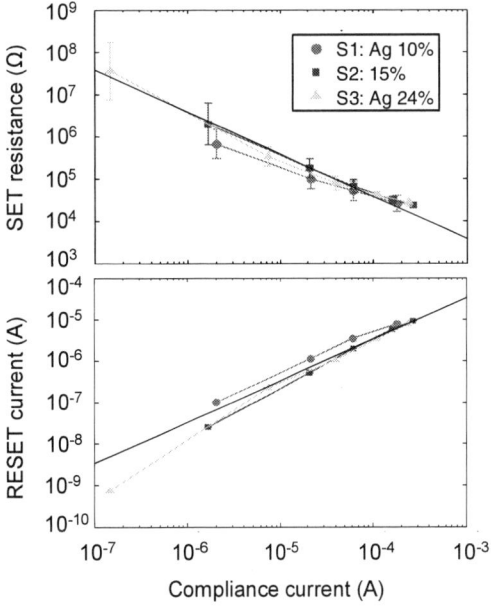

Figure 3 – SET resistance and RESET current vs compliance current passing through the CBRAM resistor, during quasi-static cycling. The two families of curves are respectively fitted to Rset.Ic=0.35V and Ireset=Ic/3.

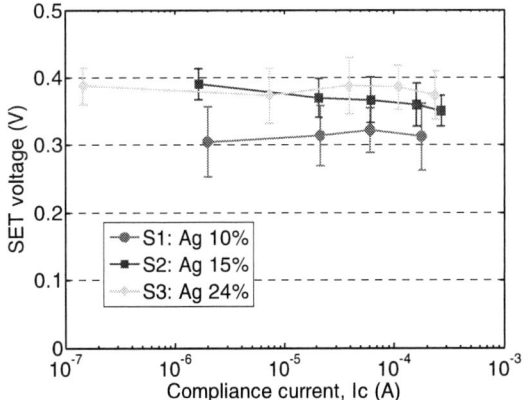

Figure 4 – SET voltage with respect to the compliance current through the CBRAM resistors.

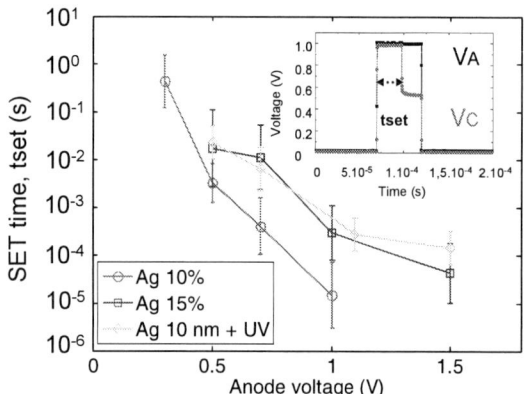

Figure 5 – SET time vs voltage applied across the CBRAM resistor during programming in the pulsed mode.

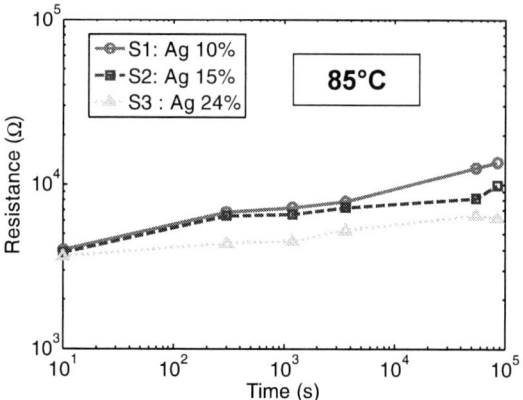

Figure 6 – Evolution of the CBRAM resistance vs time during a data retention test at 85°C.

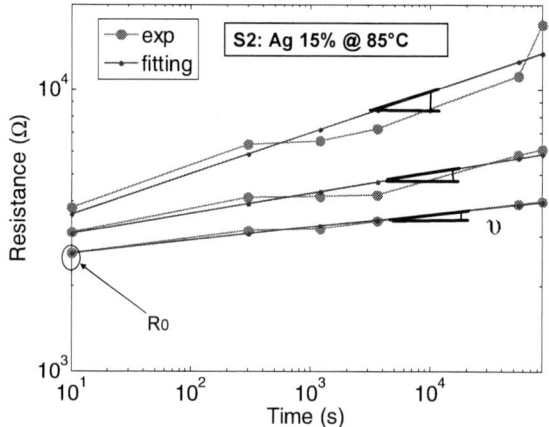

Figure 7 – Evolution of the CBRAM resistance vs time during a data-retention at 85°C starting from different initial resistances (R0) for S2 (Ag 15%). The resistance follows a power-low drift behaviour.

As expected, Fig. 4 shows that the SET voltage is independent of the compliance current, but increases slightly with the Ag concentration (V_{SET}=0.31V, 0.37V, 0.38V for S1, S2 and S3, respectively).

Program operation in pulsed mode – Fig.5 shows the SET time versus the applied voltage for the three samples with different Ag concentrations, as obtained during programming in the pulsed mode. Inset of Fig. 5 shows a typical programming pulse (in the μs-range). Different pulse amplitudes were used, and for each amplitude the time for switching the cell in the SET state was recorded by monitoring the voltage across the CBRAM resistor during the pulse with an oscilloscope. For a pulse amplitude of 1V the SET time is nearly: 15μs for the 10% Ag and 400 μs for higher Ag doping. Note that from these data it is possible to estimate the median energy per bit to SET the device: ~1 nJ for 10% Ag and ~30 nJ for 15% and above, thus very well illustrating the ultra low-power potentialities of the CBRAM technology.

Data retention behaviour - Fig. 6 reports the resistance evolution averaged on 30 cells during a 85°C bake for all the three samples, initially programmed in the low resistance state. It shows a slower resistance evolution (green curve) for the highly Ag doped sample (S3).

We note that the resistance-time curves obey a power law with an exponent ν. So that, quantitatively, the resistance evolution can be fitted by the following equation:

$$R(t) = R_0 \left(\frac{t}{t_0} \right)^{\nu} \qquad (1)$$

where R_0 is the resistance value at t = 0, and ν is the power low exponent, or the slope of the bi-logarithmic plot in Fig.7. We observe that ν, and thus the resistance evolution, depends on R_0. Fig.8 shows a scatter plot of the extracted ν as a function of the cell initial R_0 value. It appears that, as the resistance increases from about 2kΩ to 6kΩ, ν increases, eventually saturating at about 0.5. This is in good agreement with previously reported observations [1,15], as the lower resistance states can be related to larger conductive filaments which dissolution process is generally slower.

Data retention tests were then pursued on the three CBRAM samples at different bake temperatures, namely 50°C, 85°C and 130°C. Families of power law exponents' ν were then extracted for each sample at different temperatures. Fig.9 reports for example the Weibull plot of the power law exponent ν corresponding to the three temperatures retention for the S2 (15% Ag-doped GeS₂) sample. It appears that, as the temperature raises, the ν distribution shifts toward the right, meaning that the time dynamics of the resistance (i.e. the data-loss) accelerates.

Based on Eq. (1), as the initial resistance of the cells is known, the exponent ν allows us to determine the whole resistance evolution at a fixed temperature. From Eq. (1), it is then possible to extrapolate for each sample a time to failure (t_{fail}), here defined as the time at which the cell resistance has a value ten times larger than the initial one (namely $R(t_{fail})$=10.R_0). In this frame, we neglect the resistance dependence of ν, which only gives a second order contribution. So, from each sample distribution (equivalent to Fig.9), we extracted three values of ν, one for each temperature, reasonably choosing the 62% of the distribution (corresponding to the zero of the Weibull plot).

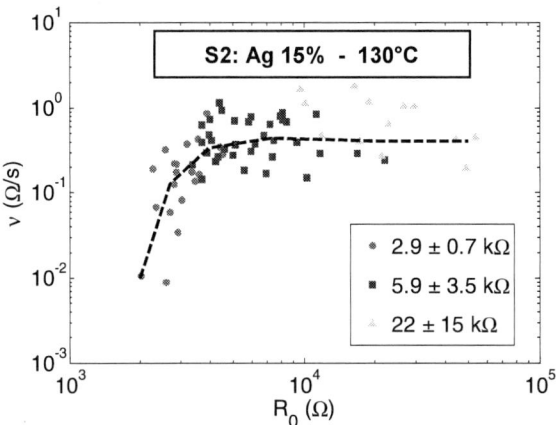

Figure 8 – Experimentally extracted power law exponent ν (see Eq.1) as a function of the initial resistance R0 for S2 (Ag 15%) during a data retention test at 85°C.

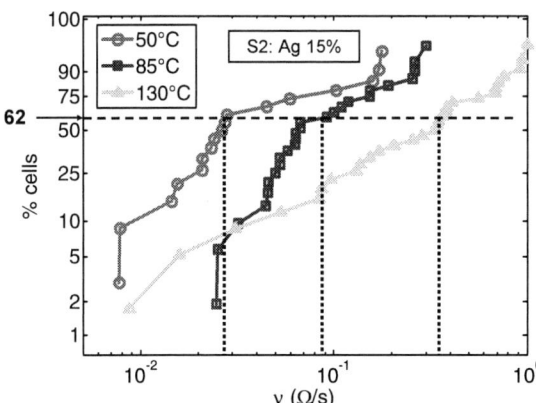

Figure 9 – Weibull plot of the power law exponent ν (see Eq.1) distribution corresponding to the three temperatures retention experiments of sample S2 (15% Ag).

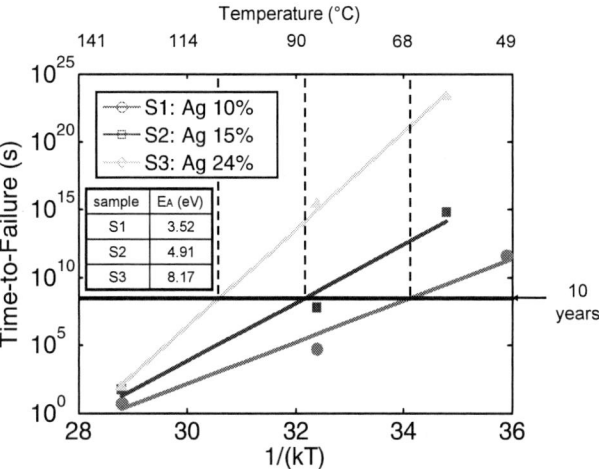

Figure 10 – Arrhenius plot of the time-to-failure (defined as the time at which the cell resistance has a value ten times larger than the mean initial one) vs the inverse of temperature for the three CBRAM samples. The extracted activation energies are indicated, as well as the extrapolation @ 10 years (extrapolated fail temperature @ 10 years is 100°C for 24%Ag, 87°C for 15%Ag, 67°C for 10%Ag).

IV. CONCLUSION

In this paper, we have studied the impact of Ag doping on GeS$_2$-based CBRAM performance, focusing on P/E conditions, switching kinetics and data-retention. The beneficial impact of the Ag doping on data retention has been clearly demonstrated, as the 24% of Ag doping allows for an activation energy of 8.17eV and a 106°C fail temperature for 10 years data-retention. Moreover, no significant trade-off on the switching speed has been put in evidence. Finally, *ab-initio* simulations put in evidence the generation of Ag$_2$S compounds in the Ag-doped GeS$_2$, suggesting that increasing Ag doping leads to a saturated GeS$_2$ matrix thus limiting the Ag diffusion during the memory data-retention.

Fig.10 reports the failure time t$_{fail}$ for the three samples as a function of 1/kT, (where k is the Bolzmann constant and T the bake temperature). As can be seen, the data-retention reliably obeys the Arrhenius law, thus allowing the extraction of an activation energy. Interestingly, we observe that as the silver doping concentration increases, from 10% to 24%, the activation energy becomes larger, going from 3.52eV to 8.17eV. This implies that for the 24% Ag-doped sample, the 10-years retention time specification is satisfied at 106°C, while for the 15% and 10% Ag-doping the same criterion is respected at 87°C and 67°C, respectively, thus strongly evidencing the beneficial impact of silver doping.

REFERENCES

[1] Michael Kund et al., *IEDM Tech. Dig.*, pp. 754–757, 2005.
[2] C. Gopalan et al., *Solid State Electronics*, vol. 58, pp. 54–61, 2011.
[3] M.N. Kozicki et al., *NVMTS*, pp. 83–89, 2005.
[4] Jiutao Li et al., U.S. Patent US 6.890,790 B210, May-2005.
[5] M. N. Kozicki et al., *J. Non-Cryst. Solids*, no. 352, pp. 567, 2006.
[6] Faiz Dahmani, U.S. Patent US 2008/0217670 A111, Sep-2008.
[7] V. Balan et al., *J. Opt. & Adv. Mater.*, vol. 8, no. 6, pp. 2112, 2006.
[8] M. Mitkova et al., *Thin Solid Film*, no. 449, pp. 248–253, 2003.
[9] M. Frumar and T. Wagner, *Curr Opin Solid State Mater Sci.*, no. 7, pp. 117–126.
[10] M.A. Ureña et al., *Solid State Ionics*, no. 176, pp. 505–512, 2005.
[11] A. Fick, Phil. Mag. **10**, 30, 1855.
[12] D. Brogioli and A. Vailati, *Phys. Rev., E* **63**, 012105/1-4, 2001
[13] E. Bychkov et al., *J. Non-Cryst. Solids*, no. 208, pp. 1–20, 1996.
[14] G. Palma and F. Longnos et al., *Proceeding of IMW*, 2012.
[15] S. Choi et al., *Proceeding of IMW*, 2012.

Analysis of the Effect of Cell Parameters on the Maximum RRAM Array Size Considering Both Read and Write

Leqi Zhang*, Stefan Cosemans, Dirk J. Wouters*, Guido Groeseneken*, Malgorzata Jurczak
imec, Kapeldreef 75, B-3001 Leuven, BELGIUM
*also with KULeuven, ESAT, Kasteelpark Arenberg 10, B-3001 Leuven, BELGIUM
Leqi.Zhang@imec.be

Abstract— **A numerical framework is developed to analyze the requirements of Self-Rectifying Resistive RAM cells for using in cross-point arrays. This paper analyzes the relation between maximum array size and cell characteristics, such as non-linearity, absolute current level and on/off ratio. Furthermore, optimal bias conditions are determined, and the advantage compared to a standard ½ voltage bias scheme is discussed.**

I. INTRODUCTION

Resistive RAM (RRAM) cross-point arrays are a very promising option for future storage memories, as they are expected to scale beyond the limits of flash technology due to their smallest cell footprint ($4F^2$) and simple fabrication structure. However, single 1R memory structure has its intrinsic drawbacks. The leakage current through the unselected cells degrades the accessibility to a specific device in the array. This reduces the output signal swing for distinguishing the different states of a selected cell and increases the power consumption of the application. To overcome these problems, concepts like using separate selectors (e.g. diode or transistor in series with a resistive element, 1S1R) or Self-Rectifying (selectorless) Cells have been proposed [1, 2]. Although they are conceptually effective, to what extent they improve the circuit performance remains questionable.

Several papers have studied the impact of cell behavior on the cross-point array performance using analytical approaches [3-5]. However, errors remain due to inexact assumptions (e.g. over-simplified device characteristics, over-look parasitics, etc.). Analysis based on circuit simulations will be useful but still not available at present. This work presents a comprehensive Matlab-level DC analysis that determines the required cell characteristics (e.g. non-linearity, current levels, on/off ratio) as a function of desired array size under the worst read and write scenarios.

The remainder of this paper is outlined as follows. The implemented constraints in the simulation are discussed in section II. In section III, the simplified cross-point memory architecture and simulation setup are described. The device

characteristic requirements using the optimized bias scheme are analyzed and the results are compared to a standard ½ voltage bias scheme in Section IV. Finally, in Section V the conclusions are presented.

II. IMPLEMENTED CONSTRAINTS

The maximum matrix size that can be constructed with a given cell is limited by several factors. This section details the constraints that are considered in this work.

A. Limitation from a System Perspective

The lowest energy values reported for single-level cell NAND flash products are 42pJ/bit for read, 410pJ/bit for program [6] (data from different products, values include the energy consumption of the peripheral circuits). As compared to NAND flash, Resistive RAM seems to be very promising for the purpose of energy saving since most of the energy consumption of NAND flash is due to the high voltages involved, requiring the use of charge pumps. Resistive RAM operates at much lower voltages. However, as the selectivity of the cells in two-terminal resistive memory cross-point arrays tends to be rather poor, the partially biased cells have fairly large leakage currents, and hence the leakage power is more pronounced in these cross-point arrays.

B. Limitation from the Techonology

Beyond 10nm technology node, the resistivity of Cu increases significantly due to the 'size effect' of Cu wire [7], which leads to a large voltage drop over the wire (Word-Line and Bit-Line). This effect degrades the accessibility to a specific cell in the array since the cell sees a smaller access voltage while cell disturbance may happen at the beginning of the line. Moreover, the current density in Cu wires increases as well. Once the current density reaches the Electro-Migration (EM) threshold, it may cause serious reliability problems. These two factors limit the maximum program current of a resistive memory as well as the acceptable leakage current on the selected WLs and BLs.

Part of this work is funded by the imec IIAP on RRAM

978-1-4673-1707-8/12 $31.00 © 2012 IEEE

C. Limitation from the Device

Cell disturbance voltage ($V_{disturb}$) is another important limitation in cross-point resistive memory array. Disturbance determines the maximum voltage that can be applied over the unselected cells in the write operation. Otherwise the state of the cell could be changed. Similarly, the read voltage cannot be larger than $V_{distrub}$ either to avoid destructive read.

III. RRAM CELL 'DEVICE' TEMPLATE AND SIMULATION SETUP

A. Array Configuration

To simulate large arrays, we are interested in the behavior of the RRAM cell under the worst case scenario. In our analysis, the wire resistances are not explicitly taken into account. Rather, we account for these resistances by imposing a maximum current through the wires. This way, all cells in the matrix can be categorized into four groups. Fig.1 shows the simplified array configuration. It is clear that the voltage over the selected cell is equal to the sum of the voltages over the WLHS cells, the NS cells and the BLHS cells. Notice that the NS cells are reverse biased compared to the other cells. One bit per matrix is assumed to be selected. Therefore, only one selected cell can be written or read. In our analysis, we optimize the biasing conditions of both the selected and unselected WLs and BLs.

B. RRAM Cell Device Template

Any cell element (i.e., combining selector and RRAM element as in a 1S1R cell, or using a Self-Rectifying Cell) I-V curves can be put into the framework. However, to investigate the requirements for a cell element, the I-V characteristics (i.e. in both on and off state) of the RRAM cell are abstracted into a device template. Fig.2(a) shows the device template as used in our simulations. The non-linearity in the region [0, $V_{disturb}$] is the most important because the unselected cells are biased in this range in any operation conditions. The cell non-linearity parameter S is defined as the ratio of the current read at $V_{disturb}$ and ½ $V_{disturb}$. We assume symmetrical I-V behavior with $|V_{set}|=|V_{reset}|=V_{write}$. (i.e. bipolar-switching). This template is more flexible and better suited to describe actual cell characteristics than previously used templates, such as a parabolic shape [5]: $I(V)=aV+bV^2$, Fig 2.(b) compares the template with an actual cell [1].

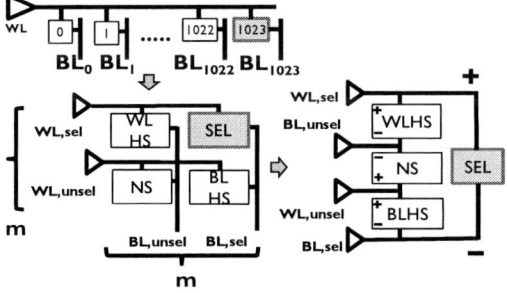

Fig 1. Simplified m x m (square) array. The four main elements are the selected cell (SEL, the cell farthest from the voltage source under the worst case operation condition), the Bit-Line Half Selected (BLHS) Cells, the Word-Line Half Selected (WLHS) Cells and the Non-Selected (NS).

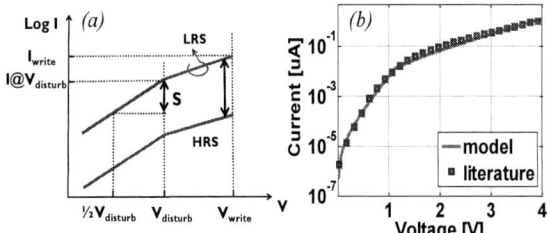

Fig 2. (a) Device template. V_{write}: program voltage, I_{write}: program current, $V_{disturb}$: disturbance voltage, $I@V_{disturb}$: current read at $V_{disturb}$, S: Non-linearity factor. Note that $V_{disturb}$ is also the maximum read voltage. (b)Template fit of an actual cell behavior (e.g. LRS I-V) [1].

To limit the complexity of the generated conclusions, we further simplify the template by fixing some of the parameters.
1) V_{write} is fixed at 1V.
2) I_{write} is set at 1uA, which is close to the max. current that can be supported for a 1024x1024 array in 10nm technology node due to the voltage drops across the wire[8] and EM limitation [9] (constraint II.B).
3) $V_{disturb}$ =1/2V_{write} =max. read voltage (constraint II.C).
4) As for HRS (off-state) current of this cell, all the current are scaled according to the cell on/off ratio.

C. Worst Case Patterns

Write operation: The worst case array pattern for the write operation is that all memory cells are in the LRS (on-state). This results in the highest leakage currents. On the selected word-line and bit-line, this results in the largest voltage drop. For the NS cells, this leads to the highest power consumption.

Read operation: The current on the selected BL is the sum of the read current of the selected cell and the current of the BLHS cells. Define 'read window' as the worst case current ratio between read of an LRS cell (with all the BLHS cells in HRS) and read of an HRS cell (with all the BLHS in LRS).

D. Write Operation Analysis

From Fig.1, an easy way to check whether a combination of cell properties and matrix size allows a functional write operation or not is to determine the max. voltage over each group of cells while taking into account all the constraints. Then, sum the values, and compare this to the cell program voltage. Fig.3 shows the applied constraints and the basic principle of determining the maximum voltage over each cell.

E. Read Operation Analysis

We optimize the read window, including optimization of the bias on the selected and the non-selected BLs and WLs while meeting the input constraints (i.e. maximum voltage can be applied across the WLHS, BLHS and NS cells). A successful read operation requires sufficient signal for the sense amplifier under the worst case conditions. To ensure this, we impose a minimum required read current (e.g. 10nA). After the read window optimization, we check whether a sufficient read window is achieved.

All constraints used in simulation are listed in Table.I.

TABLE I. Constraints Value Used in the Simulation

	$V_{disturb}$	$I_{constraint}$	Energy/bit	Access time	R_{wire}/sq[8]
Write	0.5V	≤1uA	50pJ/bit	1us	20ohm
Read	0.5V	≥10nA	25pJ/bit	1us	20ohm

Compared to NAND flash: $E_{program}$=410pJ/bit ,E_{read}=42pJ/bit ,$t_{program}$=100us,t_{read}=1us [6]
$I_{constraint}$:max. write current and min. read current
$V_{distirub}$=1/2 V_{write}

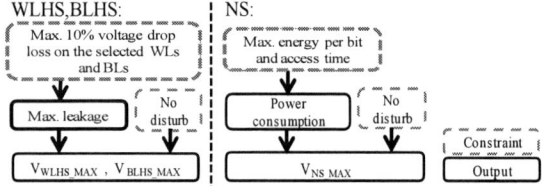

Fig 3. For both write and read, maximum voltage over the HS cells is determined by the voltage drop (constraint II.B) and disturbance (constraint II.C). Power consumption (constraint II.A) and disturbance (constraint II.C) determines maximum voltage over the NS cells.

IV. RESULTS AND DISCUSSION

A. Write Operation

1) Influence of Non-linearity S

Fix $I@V_{disturb}$ =0.5uA in our template and sweep the non-linearity parameter S to check the influence of non-linearity on the write performance. Fig.4(a) shows the calculated max. voltage over the selected cell as a function of array size. A larger array size can be achieved with larger values of S, as a higher non-linearity allows to apply a larger voltage over the NS and HS cells without exceeding the leakage constraints.

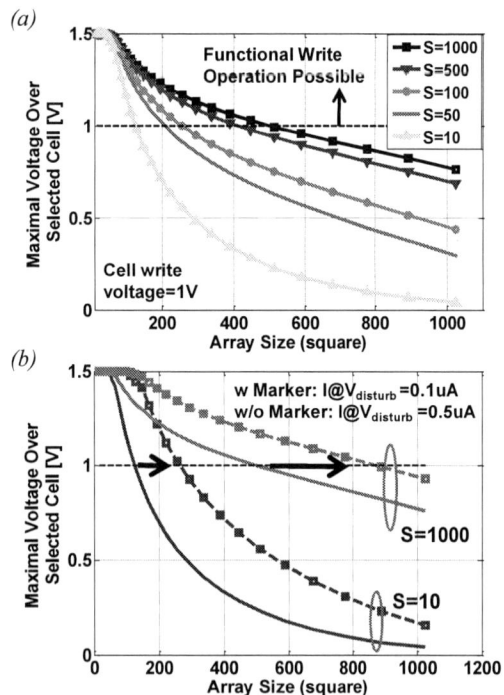

Fig. 4. (a) Max. voltage over the selected cell vs. array size for cell with different non-linearity S, curves intersect the required program voltage for our template device (i.e.1V), the cross-points indicate the maximal array size can be achieved (b) Lower $I@V_{disturb}$ also allows larger array size.

2) Influence of $I@V_{distrub}$

Fix the S parameter while changing the $I@V_{distrub}$, Fig.4(b) shows that larger array sizes can also be accomodated by decreasing $I@V_{disturb}$.

B. Read Operation

1) Influence of Non-linearity S

Fix $I@V_{disturb}$ =0.5uA, cell on/off ratio at 10 and sweep the S parameter. Assume that the minimum read window required to distinguish different states of the cell is two. Fig.5 (a) shows that increasing the non-linearity factor S also improves read performance. However, the benefit of increasing S is limited by the read window staircase behavior due to the leakage current constraints. Initially, optimized max. read window smoothly decreases as the array size increases. In this region, the read voltage is optimized at $V_{disturb}$ to allow maximum read current. The reduction of read window is entirely due to the increasing contribution of bit-line leakage current (more cells more leakage paths). Once the leakage current increases up to the applied constraints, the only way to increase further array size is to lower the voltage over the unselected cells, which means the read voltage applied to the selected cell has to be reduced as well.

2) Influence of $I@V_{disturb}$

Lowering $I@V_{disturb}$ directly improves the read window. Fig.5 (b) shows that the cells with larger $I@V_{disturb}$ require a

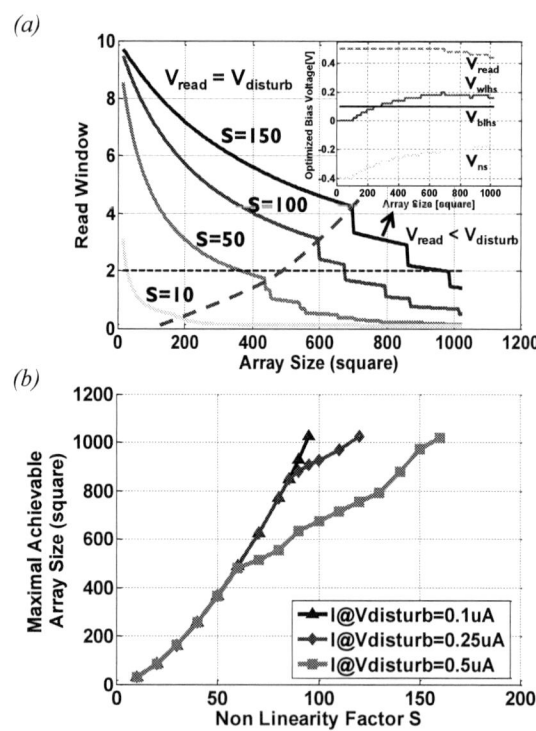

Fig 5. (a) Optimized read window vs. array size for cells with different non-linearity. Inset: example of optimized bias voltage vs. array size for cell with non-linearity S=150. Read window staircase behavior due to the discrete optimization step for the bias voltages. For sufficient sense amplifier signal: Vblhs≥100mV is required in the simulation (b) Cells with larger $I@V_{disturb}$ require a larger non-linearity to accomodate the same array size.

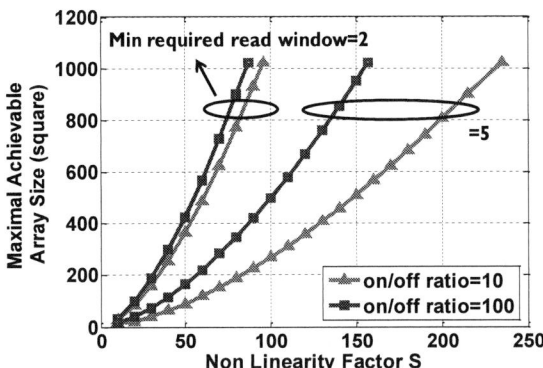

Fig 6. Influence of cell on/off ratio on maximal achievable array size for different minimal required read window (Fix $I@Idisturb$=0.1uA).

larger non-linearity to accomodate the same array size due to the high leakage currents. This indicates that it is more important to reduce the leakage currents than to achieve high read current.

3) Influence of Cell on/off Ratio

Fig.6 shows that, if we impose a minimum required read window of two, improving on/off ratio from 10 to 100 hardly improves the maximum array size. However, a higher ratio does help if we impose a higher required read window, e.g. five. This indicates that for single level cells, an on/off of 10 is good enough, while for multi-level cells, a larger on/off is desirable.

C. Optimized Bias Scheme vs. ½ Voltage Bias Scheme

In both read and write conditions, the above simulations determine the optimal voltages on the selected BL, WL, the non-selected BLs and WLs. However, most prior work relies on a fixed partial bias scheme (e.g. ½ voltage bias scheme), where half of the voltage is applied over WLHS cells and half of the voltage is applied over the BLHS cells. Fig. 7 shows that the max. achievable array size can be much larger for the optimized bias scheme, especially using cells with higher non-linearity. The reason is, that for more non-linear cells, we can allow a higher voltage over the NS cells while still meeting the current constraints. The benefit of increasing cell non-linearity vanishes using the ½ voltage bias scheme, because in our template, non-linearity only works in $V_{disturb}$

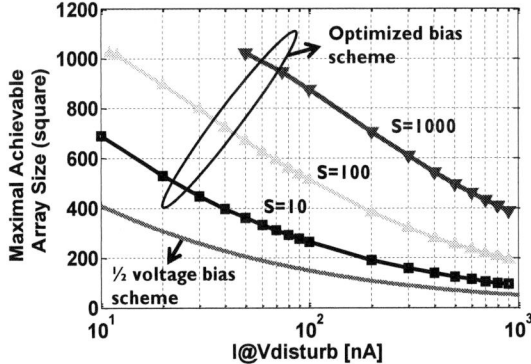

Fig 7. Non-zero bias on the non-selected cells using optimized bias scheme allows larger maximal achievable array size than using ½ voltage bias scheme in the write operation.

region. Moreover, the result indicates that $I@V_{disturb}$ should be around 10nA to achieve a 400x400 array if using the standard bias scheme. Cells with such low current may cause problems in the read operation (i.e. low read speed and small S/N ratio).

V. CONCLUSION

Cross-point memory arrays have been analyzed considering both read and write operations. A template-based exploration shows that larger non-linearity and low current level are important as they allow larger array size. LRS/HRS current ratio is not very critical when a small read window is required, such as for single-level cells, but it is important when a larger read window is required, e.g. for multi-level cells. Biasing the unselected bit-lines and word-lines to optimally distribute the voltage over the BLHS, WLHS and NS cells allows for much larger array size than a fixed ½ voltage bias scheme.

ACKNOWLEDGMENT

The authors wish to thank A.Fantini and Y.Y.Chen for the useful input. This work is partially funded by the IMEC IIAP on RRAM.

REFERENCES

[1] Christophe.J.Chevallier et.al., "A 0.13µm 64Mb Multi-Layered Conductive Metal-Oxide Memory", *ISSCC* 2010.

[2] Ching-Hua Wang et.al., "Three-dimensional $4F^2$ ReRAM cell with CMOS logic compatible process," *IEDM*, 2010 IEEE International , pp.29.6.1-29.6.4, 6-8 Dec. 2010.

[3] Jiale Liang et.al., "Size Limitation of Cross-Point Memory Array and Its Dependence on Data Storage Pattern and Device Parameters", *IITC*, 2010 International, pp.1-3, 6-9 June 2010.

[4] Flocke.A et.al., "A Fundamental Analysis of Nano-Crossbars with Non-Linear Switching Materials and its Impact on TiO2 as a Resistive Layer," *Nanotechnology*, pp.319-322, 18-21. 2008.

[5] An Chen, "Accessibility of nano-crossbar arrays of resistive switching devices," *Nanotechnology*, pp.1767-1771, 15-18 Aug. 2011.

[6] Grupp.L.M et.al., "Characterizing flash memory: Anomalies, observations, and applications," *Microarchitecture*, pp.24-33, 12-16 Dec. 2009.

[7] Flocke.A et.al., "Fundamental analysis of resistive nano-crossbars for the use in hybrid Nano/CMOS-memory," *ESSCIRC 2007*, pp.328-331, 11-13 Sept. 2007.

[8] ITRS website 2011 Report Interconnect.

[9] JEDEC

Carbon-doped Ge$_2$Sb$_2$Te$_5$ Phase-Change Memory Devices Featuring Reduced RESET Current and Power Consumption

Q. Hubert[1,3], C. Jahan[1], A. Toffoli[1], G. Navarro[1], S. Chandrashekar[1], P. Noé[1], V. Sousa[1], L. Perniola[1], J-F. Nodin[1], A. Persico[1], S. Maitrejean[1], A. Roule[1], E. Henaff[1], M. Tessaire[1], P. Zuliani[2], R. Annunziata[2], G. Reimbold[1], G. Pananakakis[3], B. De Salvo[1]

[1] CEA-Leti, Minatec Campus, 17, rue des martyrs, 38054 Grenoble Cedex 9, France
[2] STMicroelectronics, Technology R&D, via C. Olivetti 2, 20041 Agrate Brianza, Italy
[3] IMEP – LAHC, 3 parvis Louis Néel, BP 257, 38016, Grenoble Cedex 1, France
quentin.hubert@cea.fr

Abstract—In this paper, carbon-doped Ge$_2$Sb$_2$Te$_5$, integrating from 5% to 15% of carbon content, is studied as an alternative phase-change material. Accurate electrical characterizations were performed both on large and shrinked PCM devices. Compared to pure Ge$_2$Sb$_2$Te$_5$ based reference devices, a wide decrease of about 50% of the RESET current, which translates into a RESET power reduction of about 25%, is observed when 5% of carbon is added to Ge$_2$Sb$_2$Te$_5$. Moreover, an improved endurance up to 10^8 cycles is obtained while maintaining a programming window higher than 2 orders of magnitude. An increase of about 30% of the activation energy for the crystallization process is also observed. Therefore, this paper suggests that Ge$_2$Sb$_2$Te$_5$ doped with 5% of carbon is a promising phase-change material for future PCM technology.

I. INTRODUCTION

Thanks to a unique set of features such as short read and write times, multi-level capability and good data-retention, phase-change memory (PCM) is one of the most promising candidates for the next-generation of non-volatile memory (NVM) [1]. In PCM, a phase-change material, such as Ge$_2$Sb$_2$Te$_5$ (GST), is reversibly switched from a high-resistive amorphous state (RESET state) to a low-resistive crystalline state (SET state). However, the high current required to go from the SET to the RESET state, I_{RESET}, limits the minimum size of the selector element, and hence the maximum memory density [1]. Reduced I_{RESET} were observed in PCM devices integrating silicon-doped or oxygen-doped GST instead of pure GST [2], [3]. Carbon can also be an interesting doping solution for PCM [4], [5]. In this paper, the main electrical performances of carbon-doped GST (named GST-C) based PCM devices are studied and a significant lowering of I_{RESET} and of the power consumption is demonstrated.

II. PHASE-CHANGE MATERIAL DEPOSITION AND CHARACTERIZATION

A. Phase-change material deposition

To characterize carbon-doped GST as alternative phase-change material, blanket wafers with amorphous GST and GST-C, deposited at room temperature by plasma-assisted co-sputtering from one target of pure GST and one target of pure carbon, were fabricated. In order to screen the effect of carbon doping, various carbon percentages were incorporated into GST. Using Rutherford Backscattering Spectroscopy (RBS), Particle Induced X-ray Emission (PIXE) and Nuclear Reaction Analysis (NRA), the actual atomic percentages of Ge, Sb, Te and C elements in GST-C were measured and confirmed that the stoichiometry of Ge, Sb and Te is 2, 2 and 5, respectively. The atomic percentages of carbon into GST are 5% ± 1%, 8% + 1% and 15% + 1%. These materials are called GST-C5%, GST-C8% and GST-C15%, respectively.

B. Activation energy for the crystallization process

Using four-probe electrical resistivity measurements, the crystallization temperatures, T_C, of each phase-change material were measured with temperature ramps, β, varying from 1°C/min to 30°C/min. On Fig. 1, it is clear that for every temperature ramp, T_C widely increases of more than 100°C when 5% of carbon is added to GST. However, raising further the carbon content, a smaller T_C increase is observed. Indeed, only an increase between 10°C and 30°C is observed when the carbon content goes from 8% to 15%.

By applying the Kissinger method to the resistivity measurements, the activation energy for the crystallization process, E_A, is calculated (Fig. 1) [6]. We see that E_A increases from 2.5eV to 3.3eV for GST and GST-C5%, respectively. However, when the carbon content increases further the activation energy decreases and reaches 2.8V for GST-C15%.

978-1-4673-1707-8/12 $31.00 © 2012 IEEE

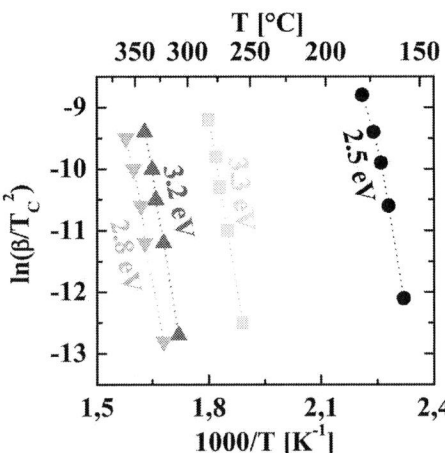

Fig. 1. Kissinger plot of crystallization temperatures of GST (black circles), GST-C5% (blue square), GST-C8% (red triangle), GST-C15% (green triangle) full-sheet layers for various temperature ramps and corresponding activation energy for the crystallization process.

III. DEVICE CHARACTERIZATION

To characterize the electrical performances of GST-C, all the materials were integrated on lance-type PCM devices having large 300nm-diameter tungsten heater on top of which a 100nm-thick phase-change layer is deposited (Fig. 2).

A. Electrical annealing of the PCM devices

Due to their high crystallization temperatures, the phase-change materials incorporating carbon are not crystallized at the end of the fabrication process. Therefore, an electrical procedure was developed in order to make them crystalline: a sequence of voltage pulses with increasing amplitude is applied to the PCM devices (inset of Fig. 3). At the beginning of the procedure, the device shows an initial high resistance due to the low conductivity of the amorphous GST-C (Fig. 3). Then at a voltage between 6V and 7V, the resistance of the device sharply decreases indicating the crystallization of the phase-change material. Due to these high switching voltages, a 1kΩ resistance is added in series with the device, in order to protect it by limiting the flowing current.

B. Reduction of the programming currents and the power consumption

Fig. 4 shows the programming characteristics of large GST-based PCM devices, i.e. the readout cell resistance, R_{CELL}, measured using four-probe method, as a function of the current flowing through the device, I_{CELL}, obtained with the setup described in [7]. We can see that I_{SET}, defined as the current where R_{CELL} reaches ten times the minimum resistance obtained for a given pulse width (Fig. 4), and I_{RESET} do not depend on the pulse width. Fig. 5 shows I_{SET} and I_{RESET} of large PCM devices for various carbon contents. It is worth noting that I_{RESET} is widely reduced for GST-C based PCM devices. Indeed, devices based on GST-C5%, GST-C8% and GST-C15% exhibits wide I_{RESEST} reductions of 53%, 23% and 51%, respectively. Moreover, I_{SET} is reduced by about 29% and 50% for GST-C5% and GST-C15% based devices respectively.

To further validate the I_{RESET} reduction all the phase-change materials were integrated on shrinked wall-like PCM devices, as in [9], with a doped-TiN heater which has a contact area reduced by at least two orders of magnitude compared to the one of the previous large test structure. On Fig. 6, one can clearly see that wide I_{RESET} reductions of 45%, 53% and 61% are obtained on GST-C5%, GST-C8% and GST-C15% based devices, respectively.

The power required to reset the large PCM cells, P_{RESET}, is also calculated using (1) where R_{PCM} is the resistance of the phase-change material during the application of the pulse while V_H is the holding voltage of the PCM device [8]. P_{RESET} is also reduced in GST-C based devices (inset of Fig. 5). Indeed, 26% and 24% of P_{RESET} reduction are achieved in GST-C5% and GST-C15% based devices respectively.

$$P_{RESET} = V_H \bullet I_{RESET} + R_{PCM} \bullet I_{RESET}^2 \qquad (1)$$

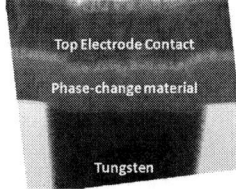 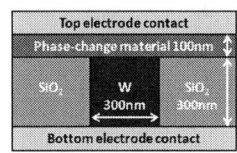

Fig. 2. Cross-section TEM image of the studied large PCM device (left) and corresponding scheme (right).

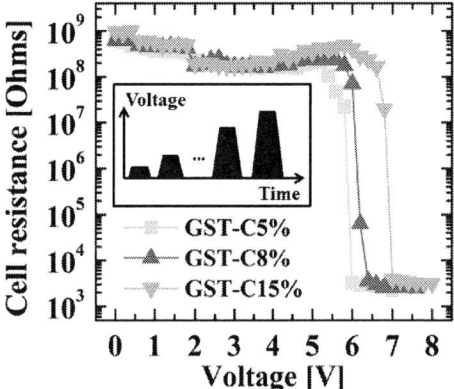

Fig. 3. Readout resistance of the large PCM devices versus the voltage at the output of the pulse generator during the electrical crystallization procedure. The inset shows the pulse scheme send to the device. Each curve is obtained by averaging data from more than 50 devices.

978-1-4673-1707-8/12 $31.00 © 2012 IEEE

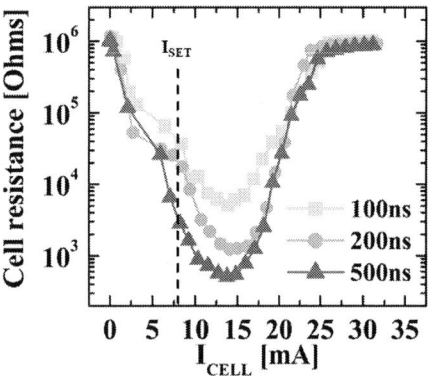

Fig. 4. Readout resistance of the large GST-based PCM devices versus the current flowing through the cell during the application of the pulse. Each curve is obtained by averaging the data from more than 10 cells.

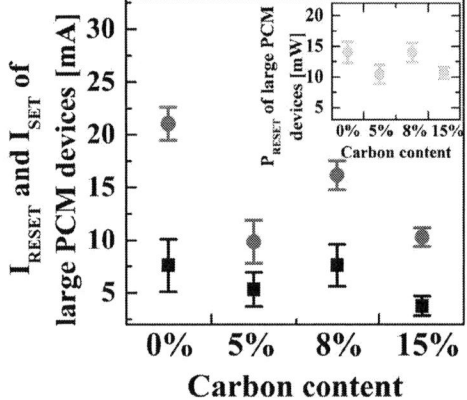

Fig. 5. RESET (red circle) and SET (black square) currents of large PCM devices for various carbon contents. The inset shows the RESET power of large PCM devices for various carbon contents. Each point is obtained by averaging the data extract from about 30 R-I curves and the error bars represent the standard deviation.

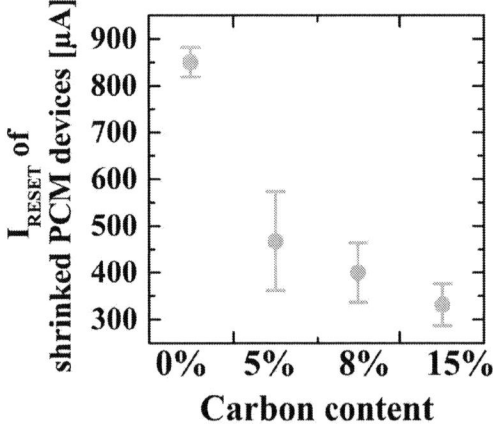

Fig. 6. RESET current of shrinked PCM devices for various carbon contents. Each point is obtained by averaging the data extract from about 30 R-I curves and the error bars represent the standard deviation.

C. Memory programming window and cycling endurance

Fig. 7 shows the memory programming window of the large PCM devices for the different carbon contents. We note that it decreases from about 3800 for GST reference devices to about 500, 100 and 40 for GST-C5%, GST-C8% and GST-C15% based devices, respectively.

Fig. 8 shows the cycling endurance of the large PCM devices for the different carbon contents, measured with the setup described elsewhere [10]. In GST reference devices, we see that the RESET state resistance sharply decreases after about 2×10^7 cycles and hence, closes the memory window. In GST-C5% and GST-C8% based devices, an improved cycling endurance of 10^8 cycles with a memory window higher than two orders of magnitude is obtained. In GST-C15% based devices, a memory window of one order of magnitude is available during the whole cycling endurance.

D. Amorphous phase-stability

The amorphous phase stability is one of the main reliability issues of the PCM technology. In the one hand, a spontaneous structural relaxation of the amorphous phase, which evolves into a phase with a higher thermodynamic stability, is correlated with a temporal increase of the RESET resistance of the PCM device. Such resistance drift phenomenon is undesirable for multi-level operation [11]. In the other hand, the thermal-induced spontaneous crystallization of the amorphous phase can reduce the retention of the data stored in the cell [12].

The spontaneous relaxation of each material was characterized by monitoring the RESET resistance drift of more than 50 large PCM devices at a constant temperature of 25°C. The data of each cell is fitted with an empirical power law (2) and the drift coefficient, ν, is calculated [11].

$$R = R_0 \bullet (t/t_0)^{\nu} \qquad (2)$$

R and R_0 are the current and initial resistances while t and t_0 are the current and initial times. We see on Fig. 9 that the ν value slightly decreases from 0.119 to about 0.113 for GST reference devices and GST-C5% and GST-C8% based devices, respectively. For GST-C15% based devices, the ν value is widely lower due to a lower RESET resistance [11].

For every phase-change material, the high-temperature data-retention of PCM devices was investigated on more than 60 large PCM devices, using the procedure described in [12]. No degradation is observed on the data-retention of GST-C based devices compared to GST reference devices. Indeed, in both cases, a 10years-data-retention temperature of about 125°C can be guaranteed. This data-retention measurement confirms that the activation energies for the crystallization process of GST-C are higher than the one of GST. Indeed, compared to the activation energy of GST, increases of about 28% and 22% are obtained for the activation energies of GST-C5% and GST-C8%, respectively. These increases are in agreement with the increases of about 32% and 28% obtained in section II.

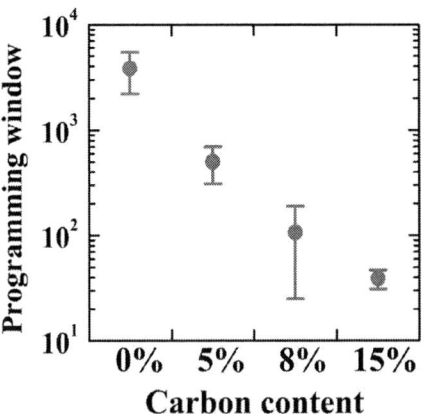

Fig. 7. Programming window of large PCM devices for various carbon contents. Each point is obtained by averaging the data extract from about 30 R-I curves and the error bars represent the standard deviation.

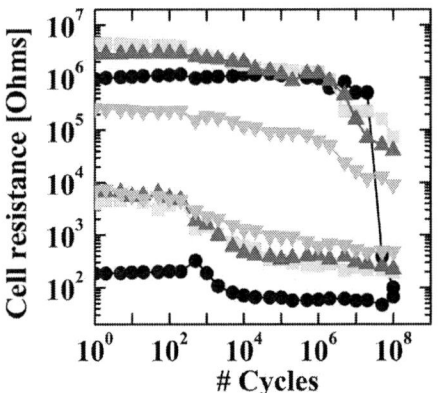

Fig. 8. Cycling endurance of large PCM devices integrating GST (black circle), GST-C5% (blue square), GST-C8% (red triangle) and GST-C15% (green triangle). Each curve is representative of about 5 devices.

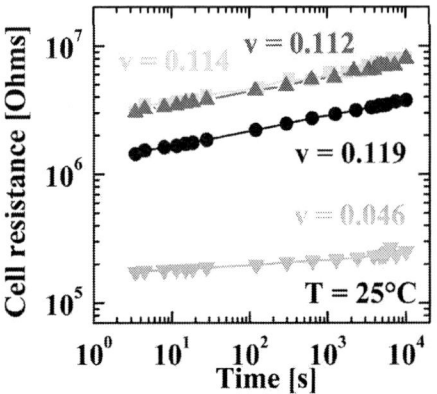

Fig. 9. Drift properties of large PCM devices integrating GST (black circle), GST-C5% (blue square), GST-C8% (red triangle) and GST-C15% (green triangle). Each curve is obtained by averaging the data from more than 50 devices.

IV. CONCLUSION

In this paper, we demonstrate that doped $Ge_2Sb_2Te_5$ with carbon is an efficient way to reduce the RESET current and the RESET power consumption of PCM devices. Especially, $Ge_2Sb_2Te_5$ doped with 5% of carbon shows the best trade-off between all the electrical properties required for the PCM technology. Indeed, a wide RESET current reduction of about 50% is obtained both on large and shrinked PCM devices, with an improved endurance up to 10^8 cycles maintaining a programming window of two orders of magnitude. Moreover, a RESET power consumption and a SET current reduced by 25% and 29% are also obtained, respectively. The activation energy for the crystallization process is increased by about 30%. Finally, no degradation in the drift properties or in the data-retention properties of carbon-doped GST based devices is observed. Therefore, the results here presented suggest that $Ge_2Sb_2Te_5$ doped with 5% of carbon is a promising phase-change material for the future PCM technology development.

REFERENCES

[1] H-S. Philip Wong, S. Raoux, S. Kim, J. Liang, "Phase-change memory", Proc. IEEE, vol. 98, pp. 2201-2227, 2010

[2] J. Feng, Y. Zhang, B. W. Qiao, "Si doping in Ge2Sb2Te5 film to reduce the writing current of phase-change memory", Applied Physics A: Materials science & processing, vol. 87, pp. 57-62, 2007

[3] N. Matsuzaki, K. Kurotsuchi, "Oxygen-doped GeSbTe phase-change memory cells featuring 1.5V/100μA standard 0.13μm CMOS operations", IEDM Tech. Digest., pp. 738-741, 2005

[4] W. Czubatyj, T. Lowrey, S. Kostylev, I. Asano, "Current reduction in ovonic memory devices", Proc. E*PCOS, 2006

[5] G. Betti Beneventi, E. Gourvest, A. Fantini, L. Perniola, V. Sousa, "On carbon-doping to improve GeTe-based phase-change memory data retention at high temperature", Proc. IMW, pp. 1-4, 2010

[6] I. Friedrich, V. Weidenhof, W. Njoroge, P. Franz, M. Wuttig, "Structural transformations of Ge2Sb2Te5 films studied by electrical resistance measurements", Journal of Applied Physics, vol. 87, pp. 4130-4134, 2000

[7] A. Fantini, L. Perniola, M. Armand, J-F. Nodin, V. Sousa, "Comparative assessment of GST and GeTe materials for application to embedded phase-change memory devices", Proc. IMW, pp. 1-2, 2009

[8] S. Braga, "Effects of alloy composition on multilevel operation in self-heating phase-change memories", Proc. IMW, pp. 1-4, 2011

[9] R. Annunziata, P. Zuliani, M. Borghi, "Phase-change memory technology for embedded non-volatile memory applications for 90nm and beyond", IEDM Tech. Digest., pp. 1-4, 2009

[10] A. Toffoli, A. Fantini, "Highly automated sequence for phase-change memory test structure characterization", Proc. ICMTS, pp. 38-42, 2010

[11] N. Papandreou, H. Pozidis, T. Mittelholzer, G. F. Close, "Drift-tolerant multilevel phase-change memory", Proc. IMW, pp. 1-4, 2011

[12] B. Gleixner, A. Pirovano, J. Sarkar, "Data retention characterization of phase-change memory arrays", Proc. IRPS, pp. 542-546, 2007

978-1-4673-1707-8/12 $31.00 © 2012 IEEE

Transport properties of strained silicon nanowires

(Invited Paper)

Yann-Michel Niquet[*], Christophe Delerue[†], Viet Hung Nguyen[*], Christophe Krzeminski[†] and François Triozon[‡]

[*] SP2M, UMR-E CEA/UJF-Grenoble 1, INAC, Grenoble, France
[†] IEMN, Dept ISEN, UMR CNRS 8520, Lille, France
[‡] CEA LETI-MINATEC, Grenoble, France

Abstract— **We discuss the effect of strains on the electron and hole mobilities in silicon nanowires with diameters near 10 nm. We show that silicon nanowires are very sensitive to strains, so that strain engineering shall be a highly efficient booster for nanowire technologies.**

I. INTRODUCTION

Silicon nanowires (Si NWs) are very attractive as building blocks for ultimate transistor devices. They are expected to show better gate control, hence reduced short channel effects [1]. Si NWs with diameters ranging from 4 to 10 nm have now been synthetized and electrically characterized [2]–[6]. Recent works [7]–[16] have however shown that the transport properties of ultimate Si NWs are, in general, degrading with decreasing diameter, thus raising concerns about the performances of 3D FinFET/NW devices beyond the 16 nm node. To address these challenges, it is therefore important to explore the potential of possible mobility "boosters". Mechanical strains have, in particular, been very efficient in enhancing the transport properties of planar transistors [17].

Band structure calculations in Si NWs [18]–[24] have suggested that strains could also improve the perfomances of Si NW devices. Quantitative data about carrier mobilities are, however, still missing. Enhancements of the mobility (or transconductance) under strain has been experimentally evidenced in Si NWs with diameters ranging from 150 nm down to quantum-confined < 10 nm NWs [25]–[30]. In this context, quantitative predictions about the transport properties of sub-10-nm Si NWs are highly desirable. In this paper, we report on a calculation of the low-field (phonon-limited) mobility of electrons and holes in Si NWs with diameters near 10 nm in a fully atomistic framework [31]. We show that Si NWs are very sensitive to strains, and that strain engineering shall allow unprecedented performances enhancements.

II. METHODS

We compute the phonon-limited mobility of electrons and holes in Si NWs within an atomistic framework. We apply a longitudinal strain $\varepsilon_\parallel$ along the nanowire axis, then relax the atomic position using the Valence Force Field (VFF) model of Ref. [32]. We next compute the phonon band structure of the strained Si NW with the same VFF model. The electron band structure of the Si NWs is computed with the $sp^3d^5s^*$ tight-binding (TB) model of Ref. [33], which reproduces all effective masses and deformation potentials of bulk Si. Spin-orbit coupling is taken into account in the valence bands.

The electron-phonon interactions are then computed from the derivatives of the TB model with respect to the atomic positions. All electrons (within at least 250 meV of the band edges) and phonons fulfilling the energy and momentum conservation rules are considered. Finally, Boltzmann's transport equation is solved exactly for the low-field mobility. Details can be found in Ref. [13] and [31].

III. RESULTS

The mobility of electrons and holes in $\langle 110 \rangle$, $\langle 001 \rangle$ and $\langle 111 \rangle$ oriented Si NWs with diameter $d = 8$ nm is given in Tables I and II for three different strains $\varepsilon_\parallel = -1.5\%$, $\varepsilon_\parallel = 0\%$ and $\varepsilon_\parallel = 1.5\%$. The conduction and valence band structures of these nanowires are plotted in Figs. 1 and 2.

A. Electrons

In unstrained Si NWs, the $\langle 001 \rangle$ Si NWs show the best electron mobilities, closely followed by the $\langle 110 \rangle$ Si NWs, the $\langle 111 \rangle$ Si NWs being much poorer. This hierachy follows from band structure effects [9], [16]: The lowest energy valleys at are light ($m^* \sim m_t^* = 0.19m_0$) in $\langle 110 \rangle$ and $\langle 001 \rangle$ Si NWs, and the subbands are well separated, especially in the higher energy (but heavier) valleys (which somewhat limits scattering to/from those valleys). On the contrary, the conduction band valleys are not split by quantum confinement in $\langle 111 \rangle$ Si NWs, and show quite heavy masses $m^* \sim 0.43m_0$.

When a tensile strain is applied along $\langle 001 \rangle$ and $\langle 110 \rangle$ Si NWs, the heavy valleys off- move up and are further split from the light valleys at , which move down. Consequently, the electron mobility is enhanced by up to $\sim 50\%$. Conversely, when a compressive strain is applied, the light valleys at are emptied in the heavy valleys off- , which decreases the mobility by a factor $2\times$ to $3\times$. As for $\langle 111 \rangle$-oriented Si NWs, both compressive and tensile strains tend to flatten the lowest conduction bands, hence decrease the electron velocity and mobility. This is due to the opening of mini gaps between the bands arising from opposite valleys (visible here at the point), and controlled by the shear conduction band deformation potential Ξ_u' [34]–[36]. In light of this beahvior, it appears that bottom-up (grown) $\langle 111 \rangle$ Si NWs are unsuitable for n-type device applications, while there are huge opportunities for strain engineering in (grown or etched) $\langle 110 \rangle$ and $\langle 001 \rangle$ Si NWs.

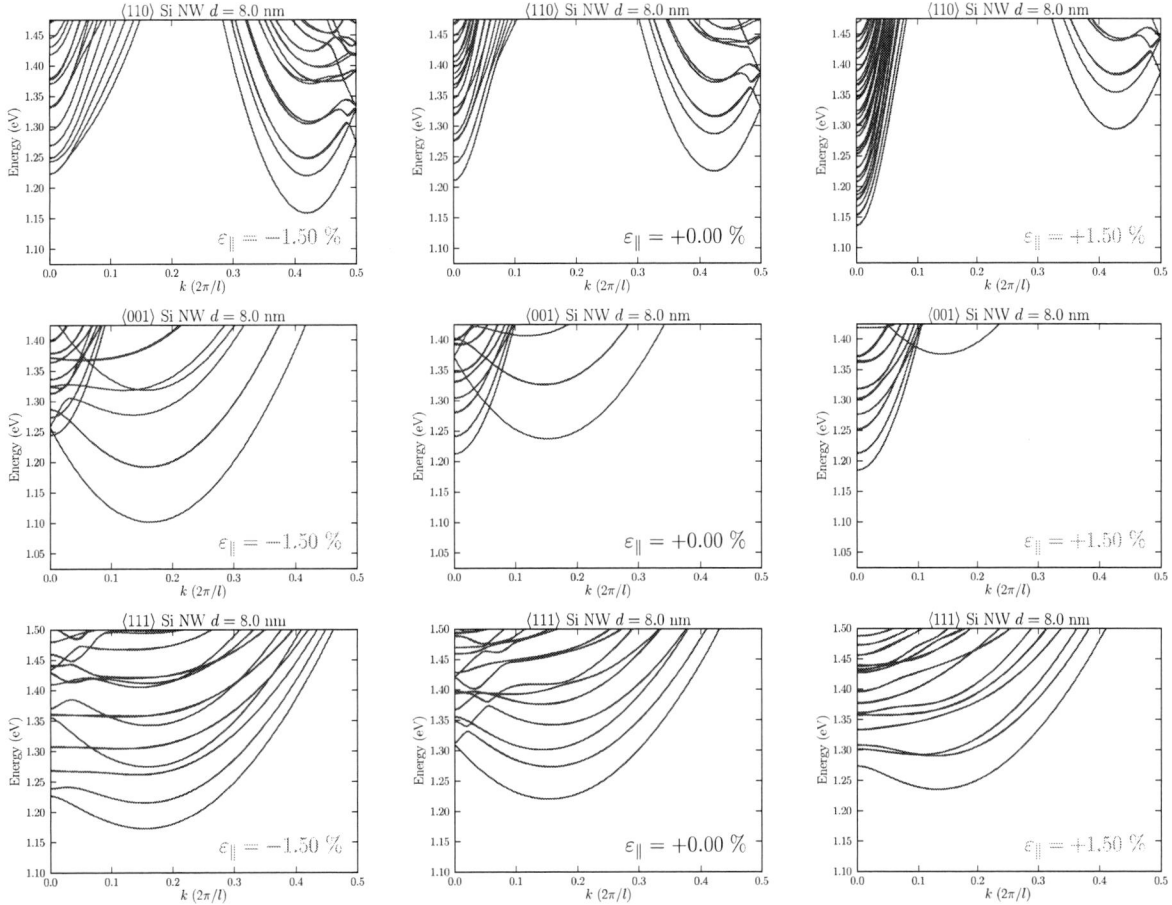

Fig. 1. Conduction band structure of $\langle 110 \rangle$ (top), $\langle 001 \rangle$ (middle) and $\langle 111 \rangle$ (bottom) oriented Si NWs with diameter $d = 8$ nm as a function of the longitudinal strain $\varepsilon_\parallel$.

	Electron mobility in Si NWs ($d = 8$ nm)		
	$\varepsilon_\parallel = -1.5\%$	$\varepsilon_\parallel = 0\%$	$\varepsilon_\parallel = 1.5\%$
$\langle 110 \rangle$	390	745	1163
$\langle 001 \rangle$	272	949	1297
$\langle 111 \rangle$	276	520	290

TABLE I

PHONON-LIMITED ELECTRON MOBILITY (CM^2/V/S) IN $\langle 110 \rangle$, $\langle 001 \rangle$ AND
$\langle 111 \rangle$ (BOTTOM) ORIENTED SI NWS WITH DIAMETER $d = 8$ NM AS A
FUNCTION OF THE LONGITUDINAL STRAIN $\varepsilon_\parallel$.

B. Holes

The situation is very different for holes. The $\langle 111 \rangle$, then $\langle 110 \rangle$ Si NWs show the best (unstrained) mobilities, while the $\langle 001 \rangle$ Si NWs perform much worse. This is due to the fact that quantum confinement brings light holes at the top of the valence bands in $\langle 111 \rangle$ and $\langle 110 \rangle$ Si NWs. The effects of quantum confinement on the mobility are still sizeable around $d = 10$ nm, and counteract the increase of the electron-phonon coupling expected in confined systems [10], [11], [15], [37].

The holes in $\langle 001 \rangle$ Si NWs remain, on the opposite, quite heavy, because "heavy" and "light" hole bands are almost degenerate and strongly mixed.

A compressive strain applied along $\langle 111 \rangle$ and $\langle 110 \rangle$ Si NWs reinforces this trend, pushing heavy hole bands further down, which increases the hole mobility (by up to $5\times$). On the contrary, a strong enough tensile strain brings back heavy holes at the top of the valence band, which effectively kills the mobility. The case of $\langle 001 \rangle$ Si NWs is more complex: Compressive as well as tensile strains tend to split the highest valence subbands, which decreases the hole mass and improves the mobility.

Extensive data and a complete discussion and can be found in Ref. [31]. Other issues, such as high field transport, will be discussed at the conference.

IV. CONCLUSION

We have shown the Si NWs with diameters in the 10 nm range are very sensitive to strains, and that electron mobilities can be improved by up to a factor 5 depending on the carrier and orientation. Electron mobilities in $\langle 001 \rangle$ and

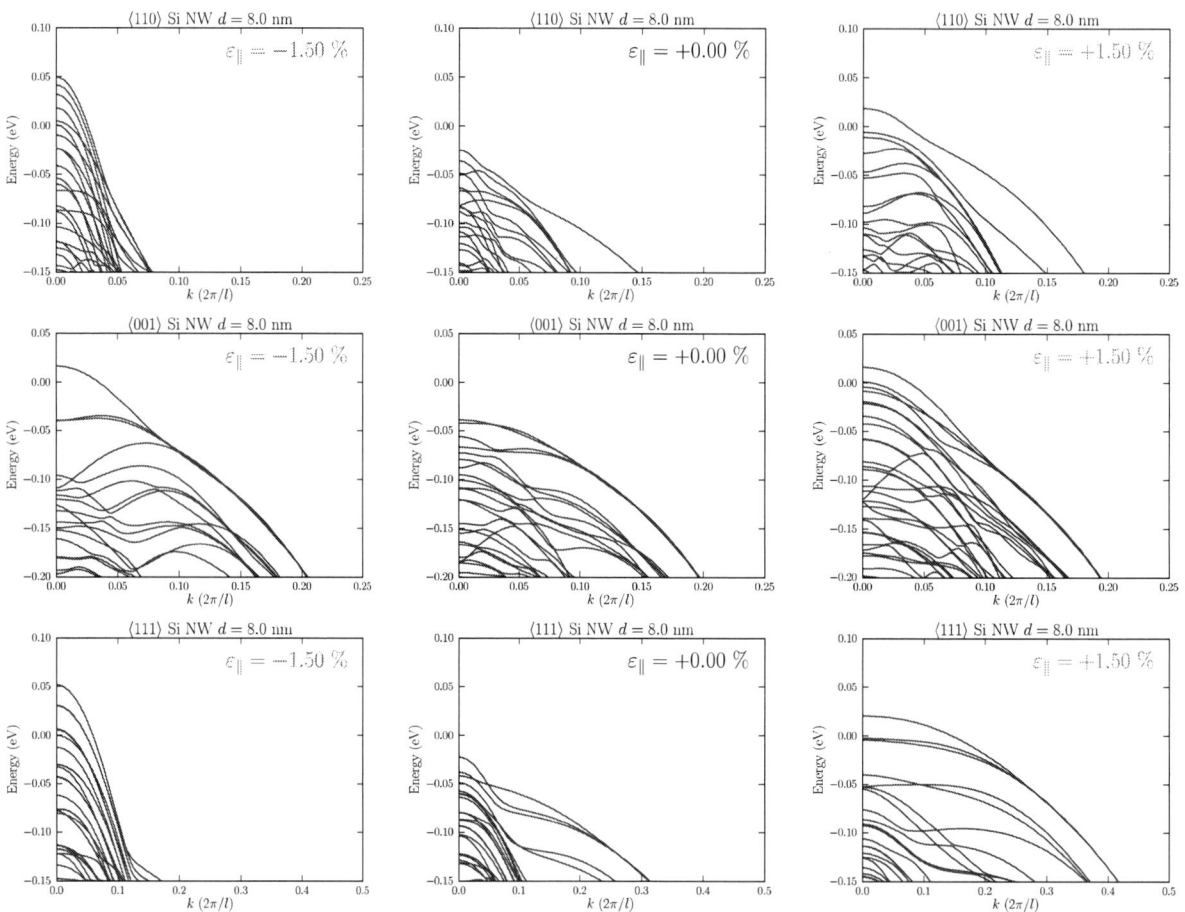

Fig. 2. Valence band structure of $\langle 110 \rangle$ (top), $\langle 001 \rangle$ (middle) and $\langle 111 \rangle$ (bottom) oriented Si NWs with diameter $d = 8$ nm as a function of the longitudinal strain $\varepsilon_{\parallel}$.

	Hole mobility in Si NWs ($d = 8$ nm)		
	$\varepsilon_{\parallel} = -1.5\%$	$\varepsilon_{\parallel} = 0\%$	$\varepsilon_{\parallel} = 1.5\%$
$\langle 110 \rangle$	2605	500	316
$\langle 001 \rangle$	1043	201	619
$\langle 111 \rangle$	3134	750	138

TABLE II

PHONON-LIMITED HOLE MOBILITY (CM^2/V/S) IN $\langle 110 \rangle$, $\langle 001 \rangle$ AND $\langle 111 \rangle$ (BOTTOM) ORIENTED SI NWS WITH DIAMETER $d = 8$ NM AS A FUNCTION OF THE LONGITUDINAL STRAIN $\varepsilon_{\parallel}$.

$\langle 110 \rangle$ Si NWs are typically enhanced by tensile strains, while hole mobilities in $\langle 111 \rangle$, then $\langle 110 \rangle$ Si NWs are enhanced by compressive strains. These results show that appropriate strain engineering can be a very efficient booster of NW (and FinFET) technologies. Thanks to their 1D geometry, NWs indeed provide unprecedented opportunities to apply large stresses to the channel. These data also draw attention to the adverse effects of non-intentional strains, which must be well controlled to achieve reproducible transport properties.

ACKNOWLEDGMENT

This work was supported by the French ANR "QUASANOVA". Part of the calculations were runt at the GENCI/CCRT supercomputing center.

REFERENCES

[1] J.-P. Colinge, "Multiple-gate soi mosfets," *Solid-State Electronics*, vol. 48, no. 6, pp. 897–905, 2004.

[2] N. Singh, A. Agarwal, L. Bera, T. Liow, R. Yang, S. Rustagi, C. Tung, R. Kumar, G. Lo, N. Balasubramanian, and D.-L. Kwong, "High-performance fully depleted silicon nanowire (diameter $\leq$ 5 nm) gate-all-around cmos devices," *Electron Device Letters, IEEE*, vol. 27, no. 5, pp. 383–386, 2006.

[3] J. Goldberger, A. I. Hochbaum, R. Fan, and P. Yang, "Silicon vertically integrated nanowire field effect transistors," *Nano Letters*, vol. 6, no. 5, pp. 973–977, 2006.

[4] S. Rustagi, N. Singh, Y. Lim, G. Zhang, S. Wang, G. Lo, N. Balasubramanian, and D.-L. Kwong, "Low-temperature transport characteristics and quantum-confinement effects in gate-all-around si-nanowire n-mosfet," *Electron Device Letters, IEEE*, vol. 28, no. 10, pp. 909–912, 2007.

[5] R. Ng, T. Wang, F. Liu, X. Zuo, J. He, and M. Chan, "Vertically stacked silicon nanowire transistors fabricated by inductive plasma etching and stress-limited oxidation," *Electron Device Letters, IEEE*, vol. 30, no. 5, pp. 520–522, 2009.

[6] K. Trivedi, H. Yuk, H. C. Floresca, M. J. Kim, and W. Hu, "Quantum confinement induced performance enhancement in sub-5-nm lithographic si nanowire transistors," *Nano Letters*, vol. 11, no. 4, pp. 1412–1417, 2011.

[7] S. Jin, M. V. Fischetti, and T.-W. Tang, "Modeling of electron mobility in gated silicon nanowires at room temperature: Surface roughness scattering, dielectric screening, and band nonparabolicity," *J. Appl. Phys.*, vol. 102, no. 8, p. 083715, 2007.

[8] M. Luisier, A. Schenk, and W. Fichtner, "Atomistic treatment of interface roughness in si nanowire transistors with different channel orientations," *Applied Physics Letters*, vol. 90, no. 10, p. 102103, 2007.

[9] M. P. Persson, A. Lherbier, Y.-M. Niquet, F. Triozon, and S. Roche, "Orientational dependence of charge transport in disordered silicon nanowires," *Nano Letters*, vol. 8, no. 12, pp. 4146–4150, 2008.

[10] E. B. Ramayya, D. Vasileska, S. M. Goodnick, and I. Knezevic, "Electron transport in silicon nanowires: The role of acoustic phonon confinement and surface roughness scattering," *J. Appl. Phys.*, vol. 104, no. 6, p. 063711, 2008.

[11] A. K. Buin, A. Verma, A. Svizhenko, and M. P. Anantram, "Significant enhancement of hole mobility in [110] silicon nanowires compared to electrons and bulk silicon," *Nano Letters*, vol. 8, no. 2, pp. 760–765, 2008.

[12] M. Luisier and G. Klimeck, "Atomistic full-band simulations of silicon nanowire transistors: Effects of electron-phonon scattering," *Phys. Rev. B*, vol. 80, p. 155430, 2009.

[13] W. Zhang, C. Delerue, Y.-M. Niquet, G. Allan, and E. Wang, "Atomistic modeling of electron-phonon coupling and transport properties in n-type [110] silicon nanowires," *Phys. Rev. B*, vol. 82, p. 115319, 2010.

[14] M. Luisier, "Phonon-limited and effective low-field mobility in n- and p-type [100]-, [110]-, and [111]-oriented si nanowire transistors," *Appl. Phys. Lett.*, vol. 98, no. 3, p. 032111, 2011.

[15] N. Neophytou and H. Kosina, "Atomistic simulations of low-field mobility in si nanowires: Influence of confinement and orientation," *Phys. Rev. B*, vol. 84, p. 085313, 2011.

[16] Y.-M. Niquet, C. Delerue, D. Rideau, and B. Videau, "Fully atomistic simulations of phonon-limited mobility of electrons and holes in $\langle 001 \rangle$-, $\langle 110 \rangle$-, and $\langle 111 \rangle$-oriented si nanowires," *Electron Devices, IEEE Transactions on*, vol. 59, no. 5, p. 1480, 2012.

[17] S. Thompson, G. Sun, Y. S. Choi, and T. Nishida, "Uniaxial-process-induced strained-si: extending the cmos roadmap," *Electron Devices, IEEE Transactions on*, vol. 53, pp. 1010–1020, 2006.

[18] K.-H. Hong, J. Kim, S.-H. Lee, and J. K. Shin, "Strain-driven electronic band structure modulation of si nanowires," *Nano Letters*, vol. 8, no. 5, pp. 1335–1340, 2008.

[19] D. Shiri, Y. Kong, A. Buin, and M. P. Anantram, "Strain induced change of bandgap and effective mass in silicon nanowires," *Applied Physics Letters*, vol. 93, no. 7, p. 073114, 2008.

[20] R. N. Sajjad and K. Alam, "Electronic properties of a strained $\langle 100 \rangle$ silicon nanowire," *Journal of Applied Physics*, vol. 105, no. 4, p. 044307, 2009.

[21] T. Maegawa, T. Yamauchi, T. Hara, H. Tsuchiya, and M. Ogawa, "Strain effects on electronic bandstructures in nanoscaled silicon: From bulk to nanowire," *Electron Devices, IEEE Transactions on*, vol. 56, no. 4, pp. 553–559, 2009.

[22] M. O. Baykan, S. E. Thompson, and T. Nishida, "Strain effects on three-dimensional, two-dimensional, and one-dimensional silicon logic devices: Predicting the future of strained silicon," *Journal of Applied Physics*, vol. 108, no. 9, p. 093716, 2010.

[23] C. Tuma and A. Curioni, "Large scale computer simulations of strain distribution and electron effective masses in silicon $\langle 100 \rangle$ nanowires," *Applied Physics Letters*, vol. 96, no. 19, p. 193106, 2010.

[24] L. Zhang, H. Lou, J. He, and M. Chan, "Uniaxial strain effects on electron ballistic transport in gate-all-around silicon nanowire mosfets," *Electron Devices, IEEE Transactions on*, vol. 58, no. 11, pp. 3829–3836, 2011.

[25] A. Seike, T. Tange, Y. Sugiura, I. Tsuchida, H. Ohta, T. Watanabe, D. Kosemura, A. Ogura, and I. Ohdomari, "Strain-induced transconductance enhancement by pattern dependent oxidation in silicon nanowire field-effect transistors," *Applied Physics Letters*, vol. 91, no. 20, p. 202117, 2007.

[26] K. Moselund, M. Najmzadeh, P. Dobrosz, S. Olsen, D. Bouvet, L. De Michielis, V. Pott, and A. Ionescu, "The high-mobility bended n-channel silicon nanowire transistor," *Electron Devices, IEEE Transactions on*, vol. 57, no. 4, pp. 866–876, 2010.

[27] P. Hashemi, M. Kim, J. Hennessy, L. Gomez, D. A. Antoniadis, and J. L. Hoyt, "Width-dependent hole mobility in top-down fabricated si-core/ge-shell nanowire metal-oxide-semiconductor-field-effect-transistors," *Applied Physics Letters*, vol. 96, no. 6, p. 063109, 2010.

[28] J. Koo, Y. Jeon, M. Lee, and S. Kim, "Strain-dependent characteristics of triangular silicon nanowire-based field-effect transistors on flexible plastics," *Japanese Journal of Applied Physics*, vol. 50, no. 6, p. 065001, 2011.

[29] P. Hashemi, L. Gomez, M. Canonico, and J. Hoyt, "Electron transport in gate-all-around uniaxial tensile strained-si nanowire n-mosfets," in *Electron Devices Meeting (IEDM)*, 2008, pp. 1–4, (DOI: 10.1109/IEDM.2008.4796835).

[30] P. Hashemi, L. Gomez, and J. Hoyt, "Gate-all-around n-mosfets with uniaxial tensile strain-induced performance enhancement scalable to sub-10-nm nanowire diameter," *Electron Device Letters, IEEE*, vol. 30, no. 4, pp. 401–403, 2009.

[31] Y.-M. Niquet, C. Delerue, and C. Krzeminski, "Effects of strain on the carrier mobility in silicon nanowires," *Nano Letters*, 2012, (DOI: 10.1021/nl3010995).

[32] D. Vanderbilt, S. H. Taole, and S. Narasimhan, "Anharmonic elastic and phonon properties of si," *Phys. Rev. B*, vol. 40, pp. 5657–5668, 1989.

[33] Y. M. Niquet, D. Rideau, C. Tavernier, H. Jaouen, and X. Blase, "Onsite matrix elements of the tight-binding hamiltonian of a strained crystal: Application to silicon, germanium, and their alloys," *Phys. Rev. B*, vol. 79, p. 245201, 2009.

[34] E. Ungersboeck, S. Dhar, G. Karlowatz, V. Sverdlov, H. Kosina, and S. Selberherr, "The effect of general strain on the band structure and electron mobility of silicon," *Electron Devices, IEEE Transactions on*, vol. 54, no. 9, pp. 2183–2190, 2007.

[35] V. Sverdlov, G. Karlowatz, S. Dhar, H. Kosina, and S. Selberherr, "Two-band k.p model for the conduction band in silicon: Impact of strain and confinement on band structure and mobility," *Solid-State Electronics*, vol. 52, no. 10, pp. 1563–1568, 2008.

[36] Z. Stanojevic, O. Baumgartner, V. Sverdlov, and H. Kosina, "Electronic band structure modeling in strained si-nanowires: Two band k.p versus tight binding," in *14th International Workshop on Computational Electronics (IWCE)*, 2010, pp. 1–4, (DOI: 10.1109/IWCE.2010.5677927).

[37] G. D. Sanders, C. J. Stanton, and Y. C. Chang, "Theory of transport in silicon quantum wires," *Phys. Rev. B*, vol. 48, pp. 11 067–11 076, 1993.

Tin Nanowire Field Effect Transistor

Lida Ansari, Giorgos Fagas and James C. Greer

Tyndall National Institute, University College Cork, Lee Maltings, Dyke Parade, Cork, Ireland

Lida.Ansari@tyndall.ie

Abstract—**Semimetal tin nanowires of sufficiently small diameters become semiconductors. Bandgap engineering based on this effect allows for the design of a confinement modulated gap field-effect transistor in which the need for doping in the source, channel or drain is eliminated. Functionality of a dopant-free, single-material field effect transistor is demonstrated through *ab initio* simulations. Drain-source current-voltage characteristic of the confinement modulated gap transistor shows that the subthreshold slope and the on/off ratio are 73 mV/dec and up to 10^4 at V_{DD}=400 mV, respectively.**

I. INTRODUCTION

In the past several decades, performance improvement in electronic devices was mainly achieved by reducing the size of the transistors. Further scaling will not be as straightforward as it has been in the past, because of the fundamental material and device limitations. For this reason, innovative structures are being explored to continue 'Moore's law'.

By engineering a material's geometry at the nanometer scale, it is possible to tailor bandgap energies [1] and to change, for example, an indirect gap semiconductor to a direct bandgap [2]. Exciting progress has been achieved toward the fabrication of electronic devices at scales approaching fundamental limits set by atom sizes. Ultrathin metal oxide semiconductor field effect transistors (MOSFETs) with multiple gates [3] and gate-all-around silicon nanowire devices with a mere 3 nm diameter [4] have demonstrated good electrostatic control of the gate over the channel.

Junctionless nanowire transistors (JNTs) [5] offers one of the few solutions to engineer out difficulties related to aggressive scaling of the traditional (bulk or nano-engineered) MOSFETs and in this regard the physical operation of these transistors with a gate length down to 3 nm has been studied [6]. In JNTs, issues due to the formation of ultrasharp p-n junctions are avoided [7] and short-channel effects are strongly suppressed owing to the absence of depletion regions associated with the junctions and related space charges in the channel region [8]. The variation of device performance due to the random doping fluctuations in the channel is one of the main concerns for the reproducible fabrication of nano-scale transistors. Therefore, in this study we proposed a dopant-free structure for FETs and investigate the performance of this new candidate for end-of-roadmap transistors.

Tin (Sn) crystals have a diamond cubic structure with a zero bandgap (gray tin or α-tin) [9,10]. In this work, we performed atomic scale simulation using DFT to study the electronic structure of Sn nano-wires with different crystallographic directions and assess the effects of nanowire diameter on the electronic structure. Based on the different electronic properties of Sn in both bulk and nanowire forms, we propose a new structure for FETs which is compatible with current silicon-based nanofabrication technology.

II. DEVICE DESIGN

The electronic structure of SnNWs derived from the diamond cubic crystal structure are studied with the axis of the nanowire oriented along the <100>, <110> or <111> crystallographic direction and for different nanowire diameters, using the density functional theory within the local density approximation with the OpenMX code [11]. The predicted band structure of bulk α-tin using these methods is in reasonable agreement with angular resolved photoemission spectroscopy [12] and GW calculations [13].

In contrast to the semimetal band structure of bulk tin, bandgap forms in nanowires, which increases in value with decreasing diameter [10]. The formation of the bandgap is attributable to the well known phenomena of quantum confinement. The variation of the bandgap energy in a tin nanowire is shown in Fig. 1 as a function of the nanowire diameter. The smaller the diameter of a wire, the larger the relative increase of bandgap energy would be. In addition to the diameter dependence, the Sn bandgap depends on the crystal orientation of the nanowire. As illustrated in Fig. 1, larger bandgap energies are found in nanowires with <100> orientation, followed by <111> and then by <110>. The bandgap energy vanishes for a tin nanowire with <110> orientation when the diameter is greater than approximately 4 nm.

978-1-4673-1707-8/12 $31.00 © 2012 IEEE

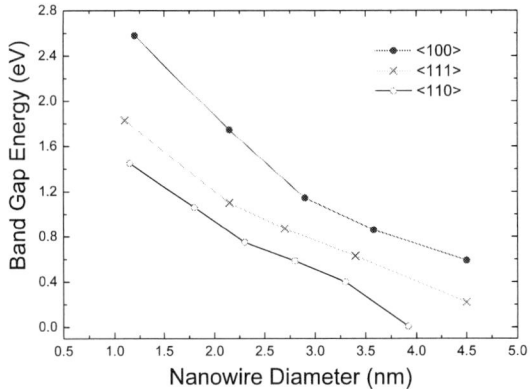

Figure 1. Bandgap energy as a function of nanowire diameter for wires grown along different crystallographic directions [10].

Based on electronic structure calculations for tin nanowires, a confinement modulated gap transistor (CMGT) is proposed in which the cross section of the central region is small enough to be a semiconductor and forms the channel region. The cross section is then increased on either side of the channel to form metallic source and drain. The channel region is surrounded by a gate stack to form a gate-all-around (GAA) MOSFET, as shown in Fig. 2. Gate bias is included as an external potential to the Kohn-Sham self-consistent equations, which are used to determine the charge density within a standard density functional theory (DFT) approach [14]. The GAA configuration offers optimum electrostatic control over the channel and reduces short channel effects [15].

Figure 2. Atomic scale illustration of the structure which has been used to simulate the confinement modulated gap transistor. The ring around the channel indicates an isopotential surface from the application of a gate bias [10].

A <110>-oriented SnNW channel with a 1nm diameter and with a gate length $L_g = 2.3$ nm is used for the channel. The oxide isolating the channel from the gate electrode is modeled as a continuous medium characterized by a dielectric constant of $\kappa_{HfO2} = 25$ corresponding to, for example, 1nm hafnium oxide.

III. RESULTS AND DISCUSSIONS

In order to study the CMGT structure the local density of states as projected onto the localized atomic orbital basis in the channel and source versus energy and displacement along the channel is illustrated in Fig. 3 (the plot is symmetrical along the channel axis at zero source drain bias). As can be seen, the source region toward the back of the plot is metallic and the channel region is semiconducting. The Fermi level is located at zero energy (E=0). Based on the density of state properties, the carriers for the channel are provided by the metal source and drain. Therefore, source, drain and channel are formed from a single material. It is worthy to note that the Fermi level is consistent with n-type doping due to the penetration of source and drain states into the channel and to the difference between the work functions different diameter wire cross sections.

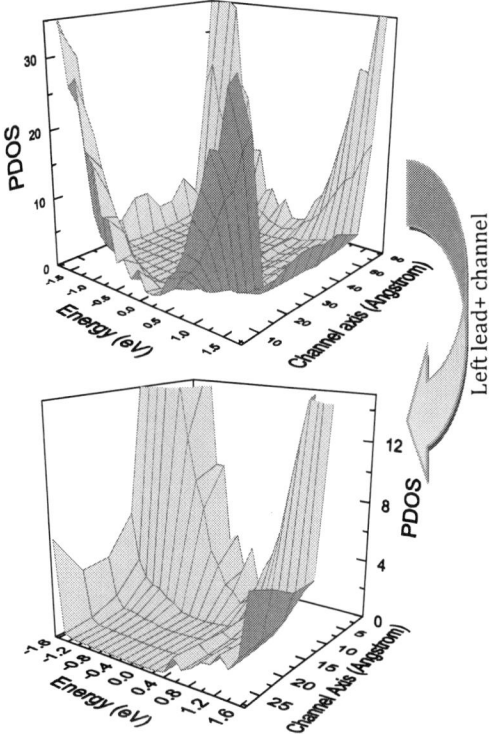

Figure 3. Local density of state along the channel axis at the left lead and the source (drain) – channel interface at zero bias.

The electronic transmission in the CMGT calculated using a self-consistent scattering formalism within DFT is shown in Fig. 4 for different gate voltages. The energies are referenced to the Fermi energy, E_F. The linear-response conductance is calculated from the following formula:

$$G(E_F) = 2e^2/h \ T(E_F) \qquad (1)$$

The inset in Fig. 4 shows the conductance in semi-log scale. The conductance is equal to zero at $V_G = 0$ V around the Fermi level and increases to ballistic values at higher gate voltages – indicating the gate electrode has excellent electrostatic control over the channel region.

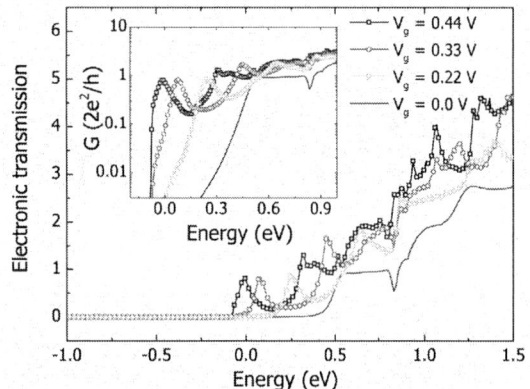

Figure 4. Electron transmission versus energy of injected carriers at different gating fields in the confinement modulated gap transistor. The inset shows the conductance in semi-log scale versus energy. Energies are referenced to the Fermi energy.

After calculating the transmission, the drain-source current is calculated using the Landauer formula as follows [16]:

$$I_{DS} - \int i(E)dE, $$
$$i(E) = \frac{e}{h} T(E,V)[f_L(E) - f_R(E)], \qquad (2)$$

where $f_L(E)$ and $f_R(E)$ are Fermi distribution functions in the left lead and right leads (i.e. source and drain), respectively:

$$f_L(E) = \left[1 + \exp\left(\frac{E - E_f}{kT}\right)\right]^{-1},$$
$$f_R(E) = \left[1 + \exp\left(\frac{E - E_f - qV_{ds}}{kT}\right)\right]^{-1}. \qquad (3)$$

The source-drain current is determined by the transmission probability within the bias window defined by the chemical potentials of the left and right leads.

The output characteristics ($I_D(V_{DS})$) of the CMGT are plotted in Fig. 5(a) for various gate voltages. As seen in the graph, the CMGT works as a conventional transistor in terms of response to the gate or bias voltage. The device is clearly turned 'off' at zero gate voltage and applying positive gate voltages turns it 'on'. A zero work function difference is used for the gate electrode but a simple recalibration of the gate voltage can readily be made for different gate metal choices.

The subthreshold slope extracted from Fig. 5(b) is 72.6mV/dec at room temperature - comparable to values commonly found in multi-gate silicon transistors [15]. This ideal value does not reflect any degradation which may arise from possible dielectric-channel interface states which would lead to gate tunneling.

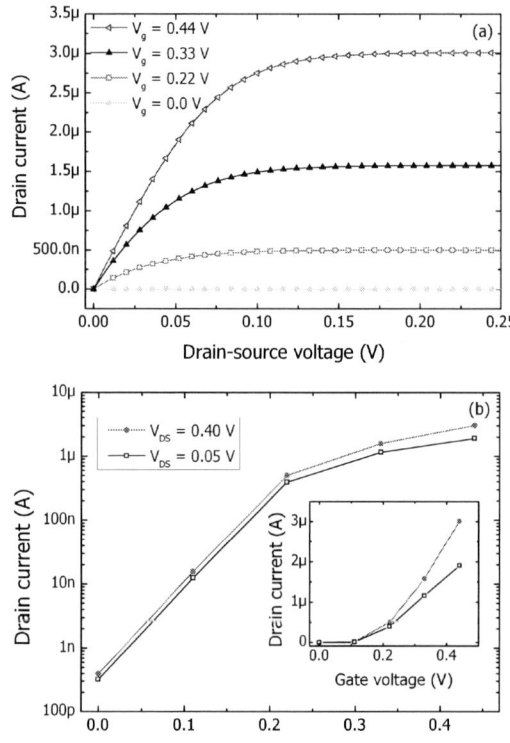

Figure 5. (a) Output characteristic of confinement modulated gap transistor. (b) I_D-V_{GS} characteristic of confinement modulated gap transistor in semi-log scale for drain voltages of 50 mV and 0.4 V. The inset shows the curves in linear scale [10].

ACKNOWLEDGMENT

This research was funded by Science Foundation Ireland under the Principal Investigator Grants No. 06/IN.1/I857. We thank the SFI/HEA Irish Centre for High-End Computing (ICHEC) for the provision of computational facilities and support.

978-1-4673-1707-8/12 $31.00 © 2012 IEEE

REFERENCES

[1] A. R. Guichard, D. N. Barsic, S. Sharma, T. I. Kamins, and M. L. Brongersma, "Tunable light emission from quantum-confined excitons in TiSi$_2$-catalyzed silicon nanowires," Nano Lett., vol. 6, pp. 2140–2144, 2006.

[2] X.Sun, J. Liu, L. C. Kimerling, and M. Jurgen, "Room-temperature direct bandgap electroluminesence from Ge-on-Si light-emitting diodes," Optics Lett., vol. 34, pp. 1198–1200, 2009.

[3] J. P. Colinge, "Multiple-gate SOI MOSFETs," Solid State Elec., vol. 48, pp. 897–905, 2004.

[4] N. Singh, et al., "SiGe nanowire devices by top-down technology and their applications," IEEE Trans. Electron. Dev., vol. 55, pp. 3107–3118, Nano Lett., vol. 6, 2008.

[5] J. P. Colinge, et al., "Nanowire transistors without junctions," Nature Nanotechnology, vol. 5, pp. 225–229, 2010.

[6] L. Ansari, B. Feldman, F. Fagas, J. P. Colinge, and J. C. Greer, "Simulation of junctionless Si nanowire transistors with 3 nm gate length," Appl. Phys. Lett., vol. 97, p. 062105, 2010.

[7] N. Dehdashti Akhavan, I. Ferain, P. Razavi, R. Yu, and J. P. Colinge, "Random dopant variation in junctionless nanowire transistors," Proc. IEEE International SOI Conference, USA, 2011.

[8] C. W. Lee, et al., "Short channel junctionless nanowire transistors," Proc. Solid-State Devices and Materials Conference (SSDM), Japan, pp. 1044–1045, 2010.

[9] W. Paul, "Band structure of the intermetallic semiconductors from pressure experiments", J. Appl. Phys., vol. 32, p. 2082, 1961.

[10] L. Ansari, F. Fagas, J. P. Colinge, and J. C. Greer, "A proposed confinement modulated gap nanowire transistor based on a metal (Tin)," Nano Lett., vol. 12, pp 2222–2227, 2012.

[11] OpenMX web site. T. Ozaki, H. Kino, J. Yu, M. J. Han, N. Kobayashi, M. Ohfuti, F. Ishii, T. Ohwaki, and H. Weng, http://www.openmx-square.org/.

[12] H. Höchst, and I. Hernández-Calderón, "Angular resolved photoemission of InSb(001) and heteroepitaxial films of α-Sn(001)," Surf. Sci., vol. 126, pp. 25–31, 1983.

[13] M. Rhohlfing, P. Krüger, and J. Pollmann, "Role of semicore d electrons in quasiparticle band-structure calculations," Phys. Rev. B., vol. 57, pp. 6485–6492, 1998.

[14] F. R. Zypmana, "Off-axis electric field of a ring of charge," Am. J. Phys., vol. 74, pp. 295–300, 2006.

[15] J. P. Colinge, FinFETs and Other Multi-Gate Transistors, Springer, New York, 2007.

[16] S. Datta, Electronic Conduction in Mesoscopic Systems, Cambridge University Press, Cambridge, 1996.

Effects of Disorder on Transport Properties of Extremely Scaled Graphene Nanoribbons

M. Poljak[1,2,*], E. B. Song[1], M. Wang[1], T. Suligoj[2], and K. L. Wang[1]

[1]DRL, University of California at Los Angeles, Los Angeles, CA 90095, USA
[2]FER-ZEMRIS, University of Zagreb, Zagreb, HR 10000, Croatia
[*]Corresponding author. E-mail address: mirko@ee.ucla.edu

Abstract—Influence of different disorders on the transport properties of extremely-scaled graphene nanoribbons is investigated using atomistic quantum transport simulations. We report the effects of different disorder strengths on quantum transmission, local density of states, and room-temperature ON- and OFF-state conductance. We find that transport gap could increase by up to 269% for realistic lattice defect densities. In the case of edge defects, ON-OFF conductance ratio increases up to 10^5 as defect density increases.

I. INTRODUCTION

Graphene is a promising material for future nanoelectronic devices because of its high carrier mobility that offers a possibility of ballistic transport at room temperature. Furthermore, due to its two-dimensional structure graphene is compatible with the planar CMOS technology. The problem of large OFF state current and low ON-OFF current ratios in large-area graphene, which is caused by zero bandgap, can be resolved by patterning graphene into graphene nanoribbons (GNRs) where a bandgap is formed by quantum confinement [1-2]. In this paper, we calculate the transport properties of realistic GNRs that include non-idealities caused by fabrication and impurities. The investigation of extremely-scaled GNRs ($L \leq 10$ nm, $W \leq 5$ nm) is necessitated by scaling demands and acceptable bandgaps. Namely, GNR-based field-effect transistors could be introduced into CMOS at the 12-nm node (channel length of 10 nm), according to the ITRS [3]. In addition, acceptable bandgaps are achievable in GNRs narrower than ~5 nm [2]. The effects of lattice defects have been studied previously in graphene and GNRs to a limited extent [4-8]. The goal of our study is to obtain the relative influence of three main disorders (edge defects, vacancies and potential fluctuations) for different disorder strengths, by averaging over an ensemble of randomly generated GNRs.

II. NUMERICAL MODELING

In this paper, we study GNR transport properties using atomistic quantum transport simulations. We focus on a 10.1 nm-long and 2.58 nm-wide armchair GNR and study the influence of edge defects, bulk vacancies and potential fluctuations separately. A tight-binding Hamiltonian with a

M.P. acknowledges support of the U.S. Department of State through the Fulbright Fellowship Program. M.P. and T.S. acknowledge support from the Ministry of Science, Education and Sports of the Republic of Croatia, under Contract No. 036-0361566-1567. E.B.S., M.W. and K.L.W. acknowledge financial support from the MARCO Focus Center on Functional Engineered Nano Architectonics (FENA).

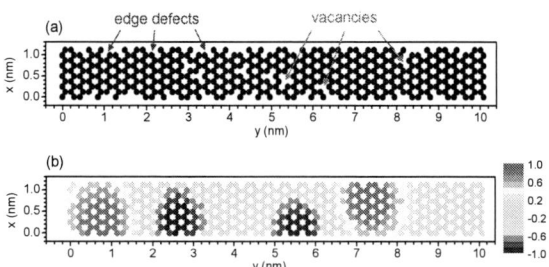

Fig. 1. (a) Illustration of a 1.1 nm-wide GNR with marked edge defects and vacancies. (b) A random realization of potential fluctuations (normalized).

single p_z orbital basis, which accounts for the first-, second- and third-nearest neighbor interaction, is employed:

$$H = \sum_i \varepsilon_i \, c_i^\dagger c_i + \sum_{k=1}^{3} t_k \sum_{i,j} c_i^\dagger c_j + H.c., \qquad (1)$$

where ε_i is the on-site energy, and t_k are the hopping parameters [9]. For edge C-C bonds we use a modified hopping parameter t_1' in order to cover edge-bond relaxation effect that increases the bandgap [10]. Non-equilibrium Green's function (NEGF) formalism is used to find the transmission that is given as $T(E) = \mathrm{Tr}\,(\Gamma_1 G_d \Gamma_2 G_d^\dagger)$, where G_d is the Green's function of the device, and $\Gamma_{1,2}$ are contact broadening functions. Surface Green's functions are calculated using an iterative procedure, which are then used to evaluate contact self energies $\Sigma_{1,2}$. Broadening functions are calculated as $\Gamma_{1,2} = (\Sigma_{1,2} - \Sigma_{1,2}^\dagger)$. Density of states (DOS) and local DOS (LDOS) are calculated from the diagonal elements of the spectral function $A(E) = i(G_d - G_d^\dagger)$. Lattice defects are realized by random removal of single carbon atoms from the lattice in the given percentage [see Fig. 1 (a)], either from the edge (edge defects, P_{ED}) or from the bulk (vacancies, P_V) of the GNR. Orbitals of the missing atoms are removed from the total Hamiltonian by setting the respective hopping parameters to zero. Reconstruction and relaxation of the defected lattice [11] is neglected in our work. Potential fluctuations are implemented as local on-site energy shifts, and are randomly generated as Gaussian potential profiles [Fig. 1 (b)] with an amplitude δV, see details in [12]. We simulate an ensemble of 50 randomly generated GNRs for each disorder case. All curves are obtained by averaging, unless stated otherwise.

III. RESULTS AND DISCUSSION

Impact of two different levels of vacancy density on DOS is presented in Fig. 2. Comparison between ideal and averaged

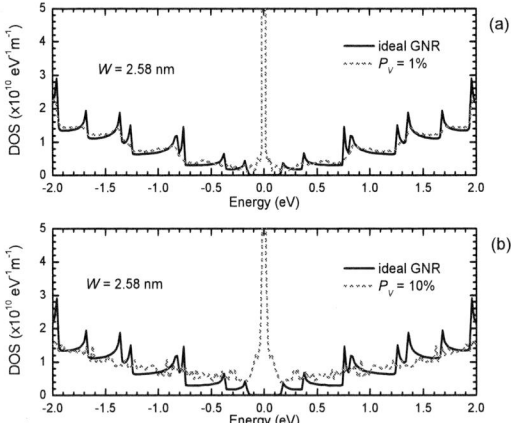

Fig. 2. Comparison of DOS between an ideal GNR and GNRs with vacancy density of (a) P_V = 1% and (b) P_V = 10%.

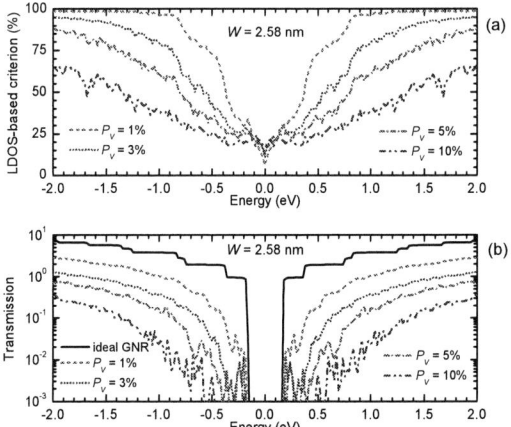

Fig. 3. (a) Dependence of the LDOS-based criterion on energy and vacancy density. (b) Transmission as a function of energy for different P_V.

DOS of defected GNRs reveals non-zero DOS in the bandgap. It has been reported previously for the case of edge defects that these states are strongly localized, which opens an effective transport gap (E_{TG}) despite the non-zero DOS [12]. Hence, an issue of extraction of the transport gap arises. We evaluate E_{TG} using two different approaches. First, transport gap is extracted directly from the transmission function as the energy range where the transmission is lower than 10^{-3}. Second, we define an LDOS-based criterion that can separate localized from extended states and use it to extract E_{TG} within the mobility edge theory [13]. This energy-dependent criterion is defined as $C_{LDOS} = N_\eta / (N_{TOT} - N_R) \cdot 100\%$, where N_{TOT} is the total number of atoms in the GNR, N_R is the number of removed atoms, and N_η is the number of atomic sites with an LDOS value higher than 5% of the maximum LDOS at the given energy. From the definition it follows that the criterion should decrease as localization effects become stronger, and it should increase in the energy range of extended states. Hence, C_{LDOS} indicates localization strength. Energy dependence of the criterion and transmission is shown in Fig. 3 (a) and (b),

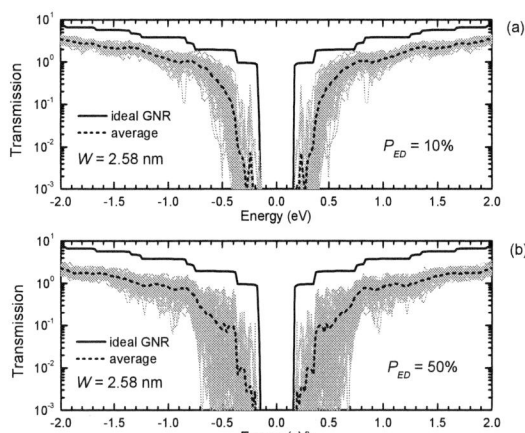

Fig. 4. Transmission functions of all randomly generated GNRs with edge defect density of (a) P_{ED} = 10% and (b) P_{ED} = 50%. Averaged transmission and the transmission of an ideal GNR is also shown.

respectively, for different vacancy densities. Higher number of vacancies induces more localized states, which is evident from the decrease of the criterion as P_V increases. Similarly, criterion decreases as energy decreases towards 0 eV, indicating increased localization at low energies. For lower P_V values, the criterion exhibits two distinct behaviors; it is almost constant in the high-energy range, and decreases sharply for lower energies. This property indicates extended states at high energies and localized states at low energies. Consequently, we define the criterion-based E_{TG} as an energy where C_{LDOS} = 50%, and from Fig. 3 (a) it is clear that the transport gap increases with increasing P_V. Fig. 3 (b) shows a strong suppression of transmission over the whole energy range. Transmission curves also reveal that E_{TG} increases as P_V increases. In the case of high vacancy density, we note higher transmission variability in sub-1 eV range, i.e. the occurrence of transmission peaks at certain energies.

A similar analysis is performed for the case of edge defects. Fig. 4 shows all the transmission functions and the averaged curve for two P_{ED} values. Variability from device to device is higher for larger edge defect density, as evidenced by a larger spread of transmission curves in Fig. 4 (b), and a larger variability is observed at low energies. Averaged LDOS-based criterion and transmission curves are shown in Fig. 5 for P_{ED} in the range from 10% to 50%. In contrast to the case of vacancies shown in Fig. 3 (a), criterion curves shown in Fig. 5 (a) are spaced more closely but they still allow us to distinguish a clear increase of the transport gap. On the other hand, influence of edge defects on the transmission curves saturates for higher P_{ED} values [Fig. 5 (b)]. For $T = 10^{-3}$, it is difficult to establish a trend in gap behavior as P_{ED} increases. We find that E_{TG} values are comparable despite a large difference of edge defect densities. The observed behavior of the transmission is in qualitative and quantitative contrast to that found in narrower edge-defected GNRs [12], i.e. sub-2.5 nm-wide GNRs exhibit strong transmission suppression and the enhancement of the transport gap is evident in transmission as it is in criterion curves. However, for the GNR with W = 2.58 nm the transmission is enhanced in the low-

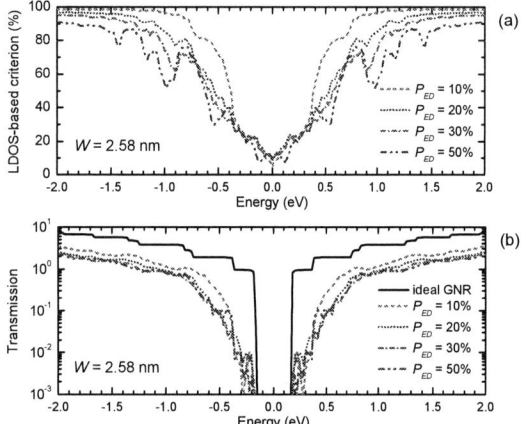

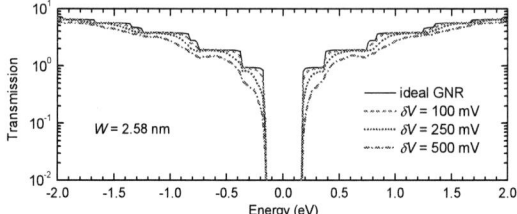

Fig. 7. Influence of potential fluctuations on transmission for different δV.

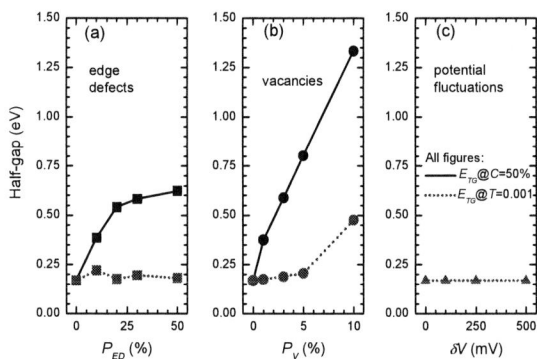

Fig. 8. Dependence of half-gap on disorder strength for (a) edge defects, (b) vacancies, and (c) potential fluctuations. Half-gap is extracted using two methods: from the LDOS-based criterion at C_{LDOS} = 50% (full line), and from transmission curves at T = 0.001 (dotted line).

Fig. 5. (a) Behavior of the LDOS-based criterion for different values of P_{ED}. (b) Dependence of transmission on energy and edge defect density.

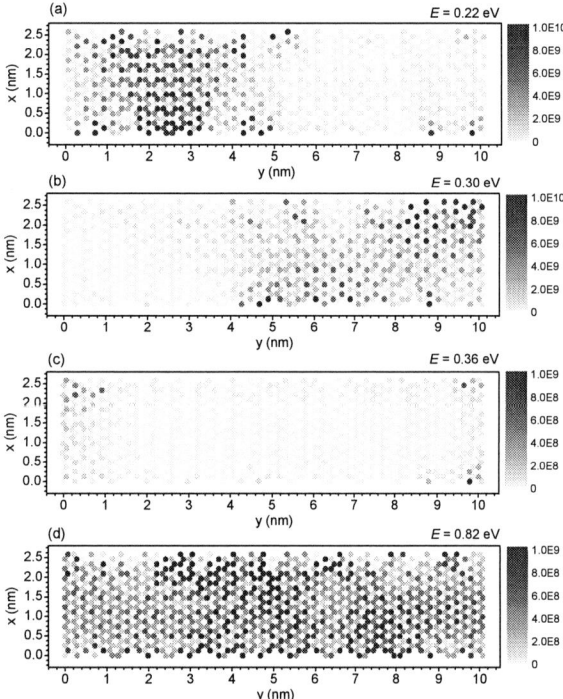

Fig. 6. LDOS of a randomly generated edge-defected GNR with P_{ED} = 50% at 4 different energies (a) 0.22 eV, (b) 0.30 eV, (c) 0.36 eV, and (d) 0.82 eV. LDOS magnitude at each atomic site is indicated by color, scale is linear and the legend is in the units of eV^{-1}m^{-1}.

Fig. 6. An extended state is shown in Fig. 6 (d), whereas localized states are observed in Fig. 6 (a), (b) and (c). Cases (a), (b) and (d) correspond to a transmission peak, while (c) exhibits low transmission (transmission curve is not shown). Therefore, despite being marked as strongly localized by the C_{LDOS} curve [see Fig. 5 (a) for $E < 0.4$ eV], states at energies of 0.22 eV and 0.30 eV exhibit high transmission. In other words, we find that the transmission function can exhibit a peak while at the same time the criterion can indicate a strong localization. In contrast to localized state at $E = 0.36$ eV in Fig. 6 (c), localized states in Fig. 6 (a) and (b) exhibit larger quantum transmission due to spatially long-reaching group of localized states. The effect of enhanced transmission is most likely related to variable range hopping that depends on the separation between the localized states [14], but further investigation is beyond the scope of this paper.

Potential fluctuations are found to have a negligible effect on the transport gap within the quantum transport approach, even for δV of 500 mV. However, transmission is reduced over the whole energy range, as demonstrated in Fig. 7. LDOS-based criterion never reaches 50% due to weak localization strength (not shown), which means that the criterion-based E_{TG} cannot be defined for W = 2.58 nm in the case of potential fluctuations.

Dependence of the transport gap on disorder strength is shown in Fig. 8. In the case of edge defects, half-gap increases up to 0.623 eV (+269%), according to the criterion-based method. We note a saturation of the gap increase for $P_{ED} > 20\%$. However, extraction from transmission curves

energy region (most probably due to increased quantum hopping between localized states [14]), which in turn causes a small difference in E_{TG} for different P_{ED}.

An advantage of using atomistic NEGF simulations is that we can investigate LDOS at any given energy and study the effects of localization. LDOS at 4 different energies of a random realization of GNR with P_{ED} = 50% is shown in

978-1-4673-1707-8/12 $31.00 © 2012 IEEE

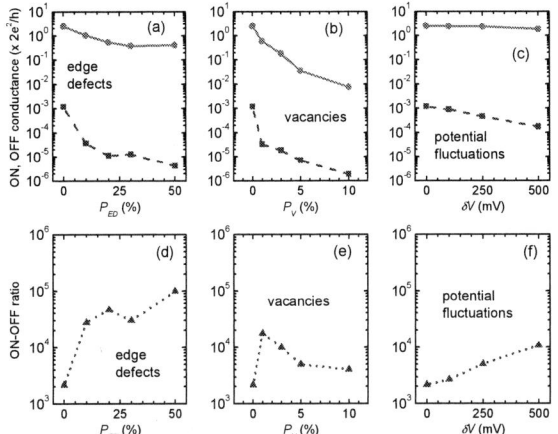

Fig. 9. (a)-(c) Influence of disorder strength on G_{ON} (full line) and G_{OFF} (dashed line) for different disorders. Dependence of G_{ON}/G_{OFF} ratio on P_{ED}, P_V and δV is shown in (d)-(f). Conductances are calculated at 300 K.

gives almost equal half-gap values (~0.2 eV) for all P_{ED}. The resulting E_{TG} (~0.4 eV) is close to experimentally observed energy gap in 2-nm to 3-nm-wide GNRs fabricated by unzipping of carbon nanotubes [2]. This agreement makes the further investigation of enhanced transmission due to quantum hopping indispensable. Regarding the influence of vacancy density, both methods give a similar trend, i.e. half-gap increases monotonically with increasing P_V. The results indicate that the increased number of vacancies effectively shuts off the transmission, without transmission-enhancing effects as in the case of edge defects, because of lack of atomic sites (and hence energy states) in the bulk of the nanoribbon. Criterion-defined half-gap increases to 0.375 eV and 1.333 eV for P_V of 1% and 10%, respectively. When the half-gap is extracted from the transmission curves, the increase equals 4% and 181% for P_V of 1% and 10% (half-gap equals 0.175 eV and 0.474 eV). We note that realistic P_V is expected to be <1%, and gap enhancement is low in that vacancy density range. Fig. 8 (c) summarizes the data from Fig. 7, showing a negligible impact of potential fluctuations on the transport gap.

Influence of disorder strength on the ON-state (G_{ON}) and OFF-state (G_{OFF}) conductance is presented in Fig. 9 (a)-(c), and the behavior of G_{ON}/G_{OFF} ratio is shown in Fig. 9 (d)-(f). G_{ON} and G_{OFF} are calculated at 300 K, using the expression:

$$G = \frac{2e^2}{h} \int_0^\infty dE\, T(E) \left(-\frac{\partial f}{\partial E} \right), \qquad (2)$$

where $T(E)$ is the transmission function, f is the Fermi-Dirac function, and h is the Planck's constant. G_{OFF} is defined at $E_F = 0$ eV, and G_{ON} at $E_F = 0.7$ eV (which equals supply voltage for the 12 nm CMOS node [3]). ON- and OFF-state conductances decrease with increasing disorder strength because of the suppressed transmission. The decrease is strongest in the case of vacancy disorder, due to strong increase of E_{TG}. However, dependences of G_{ON}/G_{OFF} ratio on P_{ED}, P_V and δV exhibit interesting features. In the case of edge defects, the ratio generally increases with increasing P_{ED}, since

the suppression of G_{OFF} is stronger than that of G_{ON}. In contrast, G_{ON}/G_{OFF} ratio decreases as vacancy density increases. Nevertheless, an interesting property can be seen in Fig. 9 (e), i.e. defected GNRs with $P_V = 1\%$ have a higher G_{ON}/G_{OFF} ratio than the ideal GNR. This effect is caused by a stronger decrease of G_{OFF} than that of G_{ON} when vacancies at 1% density are introduced [see Fig. 9 (b)]. Finally, in the case of potential fluctuations, ON-OFF conductance ratio increases with increasing δV because potential fluctuations cause a stronger transmission reduction at low than at high energies and, hence, G_{OFF} decreases more than G_{ON} [see Fig. 9 (c)]. The highest obtained G_{ON}/G_{OFF} ratio is 9.9×10^4, in the case of edge-defected GNRs with $P_{ED} = 50\%$. The obtained high G_{ON}/G_{OFF} ratios and the described G_{ON} and G_{OFF} behavior can be used to explain qualitatively the high ON-OFF current ratios of extremely-scaled GNRs reported in [2].

IV. CONCLUSION

Atomistic tight-binding NEGF simulations are employed to study the influence of three main sources of disorder, i.e. edge defects, vacancies and potential fluctuations, on transport properties of extremely-scaled GNRs. We report the dependence of transmission, transport gap, ON- and OFF-state conductance at 300 K, and ON-OFF conductance ratio on disorder strength. Edge defects and vacancies have a strong influence, whereas potential fluctuations have a negligible effect on the transport gap. In the case of edge defects and potential fluctuations, ON-OFF conductance ratio is enhanced as disorder strength increases due to differences in the behavior of ON- and OFF-state conductances. The reported results can be used to explain some recent experimental results, and provide valuable information for GNR-based CMOS device engineering.

ACKNOWLEDGMENT

The authors thank R. Lake (UC Riverside) and H. Dai (Stanford) for helpful discussions.

REFERENCES

[1] M. Y. Han, B. Ozyilmaz, Y. Zhang, and P. Kim, *Phys. Rev. Lett.*, vol. 98, p. 206805, 2007. [2] X. Li, X. Wang, L. Zhang, S. Lee, and H. Dai, *Science*, vol. 319, p. 1229, 2008. [3] International Technology Roadmap for Semiconductors (ITRS), 2011 Edition. Available online: http://www.itrs.net/ [4] V. M. Pereira, J. M. B. Lopes dos Santos, and A. H. Castro Neto, *Phys. Rev. B*, vol. 77, p. 115109, 2008. [5] D. Querlioz, Y. Apertet, A. Valentin, K. Huet, A. Bournel, S. Galdin-Retailleau, and P. Dollfus, *App. Phys. Lett.*, vol. 92, p. 042108, 2008. [6] D. Basu, M. J. Gilbert, L. F. Register, S. K. Banerjee, and A. H. MacDonald, *Appl. Phys. Lett.*, vol. 92, p. 042114, 2008. [7] Y. Yoon and J. Guo, *Appl. Phys. Lett.*, vol. 91, p. 073103, 2007. [8] J. J. Palacios, J. Fernández-Rossier, and L. Brey, *Phys. Rev. B*, vol. 77, p. 195428, 2008. [9] Y. Hancock, A. Uppstu, K. Saloriutta, A. Harju, and M. J. Puska, *Phys. Rev. B*, vol. 81, p. 245402, 2010. [10] R. Sako, H. Hosokawa, and H. Tsuchiya, *IEEE Electron Dev. Lett.*, vol. 32, p. 6, 2011. [11] S. M. M. Dubois, A. Lopez-Bezanilla, A. Cresti, F. Triozon, B. Biel, J.-C. Charlier, and S. Roche, *ACS Nano*, vol. 4, p. 1971, 2010. [12] M. Poljak, E. B. Song, M. Wang, T. Suligoj, and K. L. Wang, submitted to *IEEE Trans. Electron Dev.*, 2012. (unpublished) [13] M. Bresciani, P. Palestri, D. Esseni, and L. Selmi, *Solid-State Electron.*, vol. 54, p. 1015, 2010. [14] M. Y. Han, J. C. Brant, and P. Kim, *Phys. Rev. Lett.*, vol. 104, p. 056801, 2010.

978-1-4673-1707-8/12 $31.00 © 2012 IEEE

High Temperature Behaviour of GaN-on-Si High Power MISHEMT Devices

Dirk Wellekens, Rafael Venegas, Xuanwu Kang, Mohammed Zahid, Tian-Li Wu, Denis Marcon, Puneet Srivastava,
Marleen Van Hove, Stefaan Decoutere

Imec, Kapeldreef 75, 3001 Leuven, Belgium
Email: Dirk.Wellekens@imec.be

Abstract—The device performance of GaN-on-Si AlGaN/GaN MISHEMT devices with a Si_3N_4/Al_2O_3 bi-layer gate dielectric is studied as a function of temperature. In addition to the temperature dependence of the key DC parameters, which are also benchmarked against a silicon VDMOS device, special attention is paid to the behaviour under operating conditions, including thermal stability, switching behaviour and reliability.

I. INTRODUCTION

AlGaN/GaN high-electron mobility transistors (HEMTs) have been increasingly gaining interest for high-frequency and high-power applications. Because of their intrinsic properties, such as their wide band gap, good conductivity and high thermal stability, GaN-on-Si devices are extremely attractive for high temperature operation. Although several papers have reported on the high temperature behaviour of GaN-based HEMT devices with a Schottky gate on various substrates [1-8], very limited temperature data is available for GaN-on-Si devices [9,10]. Nevertheless, a Si substrate is most appealing with respect to cost, maturity and quality. In addition, the insertion of a gate dielectric between the Schottky gate and the AlGaN barrier, forming a MISHEMT, is known to result in a drastic decrease of the gate leakage current. Therefore, this work explores the temperature dependence of GaN-on-Si depletion-mode MISHEMT devices in the temperature range 25°C - 200°C. An in-depth investigation of DC parameters is performed with benchmarking of the data against a silicon VDMOS device. Furthermore, special attention is given to the behaviour under operating conditions, including thermal stability, switching behaviour and reliability.

II. PROCESS AND DEVICE DESCRIPTION

The device processing starts from an epi-stack grown by Metal Organic Chemical Vapor Deposition (MOCVD) on a Czochralski 6-inch Si (111) substrate. The buffer layer is grown at 1130°C and consists of a stack of $Al_{0.70}Ga_{0.30}N$, $Al_{0.40}Ga_{0.60}N$, and $Al_{0.18}Ga_{0.82}N$ on top of an AlN nucleation layer, also grown at 1130°C. A 10nm GaN channel layer is then sandwiched between this buffer layer and a 10nm $Al_{0.25}Ga_{0.75}N$ barrier layer to form a double heterostructure

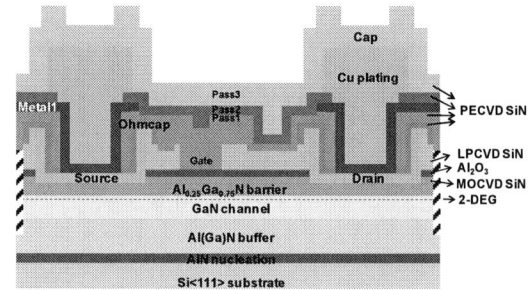

Figure 1. Schematic cross-section of the MISHEMT device

FET (DHFET). The epi-layer is terminated by a 5nm thick in-situ Si_3N_4 layer. Next, 5nm Al_2O_3 is deposited by ALD, followed by a 120nm high temperature low pressure chemical vapor deposited (LPCVD) nitride. The in-situ Si_3N_4/Al_2O_3 bi-layer serves as the gate dielectric and is effective in suppressing the gate and drain leakage current. Moreover, the use of an MOCVD-grown in-situ nitride below the ALD Al_2O_3 to form a bi-layer gate dielectric considerably enhances the breakdown voltage [11].

Ohmic contacts are formed by etching the triple dielectric stack with a low power SF_6 plasma for the Si_3N_4 layers and a Cl_2-based plasma for the Al_2O_3 layer. This is followed by Ti/Al/W deposition, dry etching of the metal stack and alloy at 600°C. The measured contact resistance was 0.65 Ω·mm. Before N-implant isolation, the ohmic metal is capped by a patterned plasma enhanced chemical vapor deposited (PECVD) nitride. The gate is formed by selective removal of the LPCVD nitride in an SF_6 plasma using Al_2O_3 as an etch stop layer, followed by deposition and dry etching of the W/Ti/Al gate metal stack. The process is completed by Al and Cu interconnect metallization layers. A schematic cross-section is shown in Fig. 1. Small devices with total gate width $W_g = 500\mu m$ were considered, as well as power transistors with a total gate width up to 40mm, while the gate length was $L_g = 1.5\mu m$, the gate-to-source distance $L_{sg} = 1.25\mu m$ and the gate-to-drain distance $L_{gd} = 5\mu m$ (500μm wide transistors) and $L_{gd} = 9.5\mu m$ (power transistors).

III. EXPERIMENTS

The devices were tested in the temperature range 25°C-200°C. Characteristics of drain current (I_{ds}) versus gate voltage (V_{gs}) and drain voltage (V_{ds}) were recorded at different temperatures to determine the thermal stable point and the main DC-parameters, such as the threshold voltage, transconductance, maximum current and on-resistance. Furthermore, the gate and drain leakage currents were measured at different temperatures in the voltage range V_{ds} = 0V-400V. Finally, the high temperature dispersion behaviour was assessed and compared to room temperature and the impact of off-state and on-state stress was studied at high and low temperature.

IV. RESULTS AND DISCUSSION

A. Thermal stable point

An important issue in the thermal behaviour of power devices is the risk of thermal runaway. Fig. 2 shows 2 sets of I_{ds}-V_{gs} characteristics at different temperatures, one for a VDMOS Si power MOSFET device, the other for a GaN power device. In the power MOSFET the curves at different temperatures show a crossing point, where the temperature coefficient changes from negative to positive. This crossing point is a result of a reduction in threshold voltage (V_t) and mobility degradation with temperature. Above this point the current at high temperature is reduced. This effect improves the thermal stability as the device heats up in this operating region, but the mobility reduction reduces the current at high temperature and hence, selfheating becomes self-limited. Therefore, the thermal stable point should be as low as possible. In case of a GaN device, it is evidently at a much lower current level than in the silicon device. This is an important asset of the GaN device over the power MOSFET device.

B. DC-parameters

The on-resistance (R_{on}) of the depletion-mode GaN transistors was determined from the I_{ds}-V_{ds} characteristics in the linear region (V_{gs} = 0V, V_{ds} = 1V). As the temperature

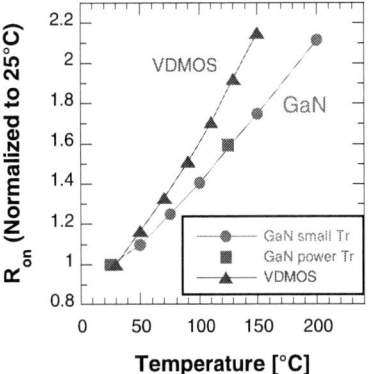

Figure 3. On-resistance change with temperature for GaN and Si (VDMOS) power device

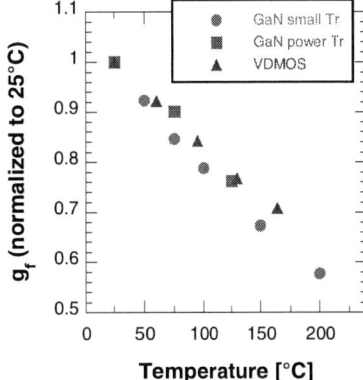

Figure 4. Transconductance change with temperature for GaN and Si (VDMOS) power device

increases the on-resistance increases almost linearly and roughly doubles between 25°C and 200°C as a result of mobility degradation. This is shown in Fig. 3, which represents the data normalized to the value at 25°C. The same increase in R_{on} is observed for small (500µm) devices (●) and 20mm power transistors (■). Fig. 3 also shows a comparison with the data for a VDMOS power MOSFET transistor (▲). Evidently, a much higher R_{on} increase with temperature is observed compared to the GaN device.

Next, Fig. 4 shows the variation of the transconductance (g_f) with temperature (normalized to the value at 25°C). For GaN devices the temperature evolution for small (500µm) transistors (●) and 20mm power transistors (■) is identical with a g_f reduction of approximately 40% between 25°C and 200°C. This reduction is actually equivalent to the reduction in the maximum (saturation) current of the GaN device. A comparison with the data for a VDMOS device (▲) shows the same trend with temperature. This is consistent, as the data reflect the change of the mobility with temperature.

Furthermore, the drain leakage currents ($I_{ds,leak}$) were measured in cut-off with V_{ds} swept between 0V and 400V. The increase of $I_{ds,leak}$ with temperature for 40mm power transistors is

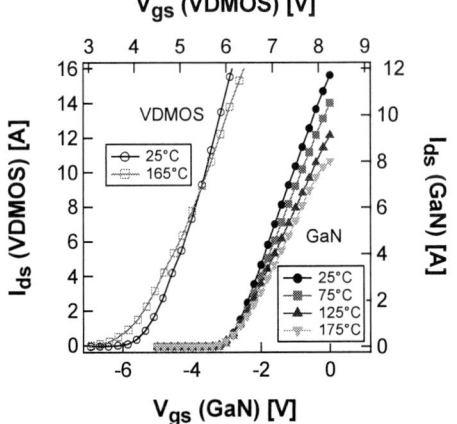

Figure 2. I_{ds}-V_{gs} for GaN and Si (VDMOS) power device at different temperatures

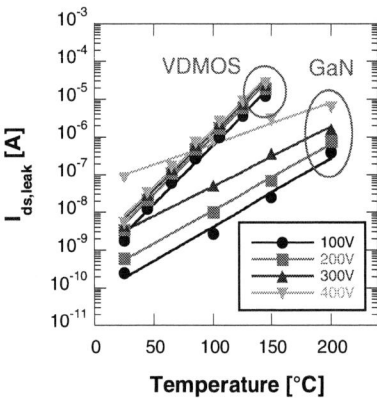

Figure 5. Drain leakage current versus temperature for different V_{ds}: comparison between GaN and VDMOS device

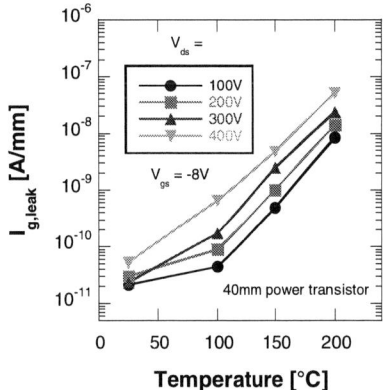

Figure 6. Gate leakage current versus temperature for different V_{ds}

shown in Fig. 5 for different V_{ds} and compared to the behaviour of a VDMOS device. For both devices the drain leakage current exhibits an exponential increase over the entire temperature range and for different V_{ds} values. Although the current increase with voltage is larger in the GaN device, its sensitivity to a temperature increase is much smaller than for a VDMOS device, which is another asset of the GaN power transistor.

Finally the gate leakage currents ($I_{g,leak}$) were measured in cut-off with V_{ds} swept between 0V and 400V and the temperature dependence of $I_{g,leak}$ for 40mm power transistors is shown in Fig. 6. The current increases roughly by 3 orders of magnitude between 25°C and 200°C, the increase being somewhat enhanced at higher temperatures. However, the gate leakage current is very small compared to a conventional HEMT (Schottky gate), but also much smaller than in case only a thin gate oxide is used [9]. For a VDMOS device gate leakage during high voltage operation is not an issue.

C. Dispersion

Gallium nitride devices are known to suffer from so-called dispersion effects, which are evidenced by current collapse and the associated dynamic on-resistance degradation. These

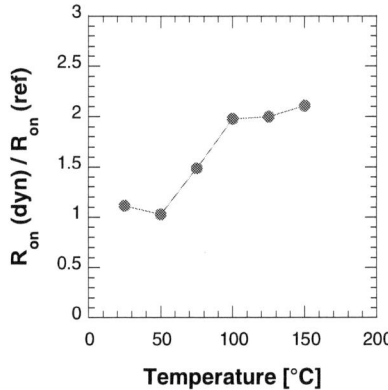

Figure 7. On-resistance degradation during pulsed measurements as a function of temperature

effects are caused by trapping of electrons in the buffer layer, the AlGaN barrier and in ionized donor states in the region between gate and drain. Modulation of the trapped charge density during pulsed operation modulates the 2DEG charge.

Drain lag measurements were performed under (V_{gs}, V_{ds}) = (-6V,50V) conditions and compared with (0V,0V) reference conditions. The ratio of the corresonding on-resistances is plotted versus temperature in Fig. 7. Although dispersion effects can be expected to improve at higher temperature, because of higher activity of the traps and higher electron energy [8], the on-resistance degradation in Fig. 7 shows a different picture with a twofold increase in degradation between 25°C and 150°C. This is because the dispersion behaviour is much more complex and governed by the presence of different hole and electron traps with different activation energies, which can even lead to an optimum temperature for dispersion effects [12].

D. Reliability

During electrical stress of a GaN transistor, its threshold voltage and current can be impacted by several mechanisms, such as trapping, detrapping and electron-hole recombination. These effects can occur during off-state as well as on-state stress and their impact can be expected to be different at different temperatures. A study was performed to assess the on-state and off-state stress effects in our GaN devices. The investigation was done on 60V devices with L_{gd} = 2µm, because the voltage range of the test set-up was limited to 100V.

Fig. 8 shows the shift in V_t and maximum current ($I_{ds,max}$) during off-state stress performed at 25°C and 200°C with -7V applied to the gate and 60V to the drain, while the source is grounded. During the off-state stress at 25°C, electrons are trapped in the Al_2O_3/Si_3N_4 under the gate, which shifts V_t in the positive direction. At longer times the V_t-shift shows a turn-around which is explained by electron discharging from the gate or emission of electrons from donor states. At 200°C, V_t shifts negative, which is attributed to thermal activation and discharging of traps in the gate region. As the stress time increases, trapping takes over as the dominant mechanism and

978-1-4673-1707-8/12 $31.00 © 2012 IEEE

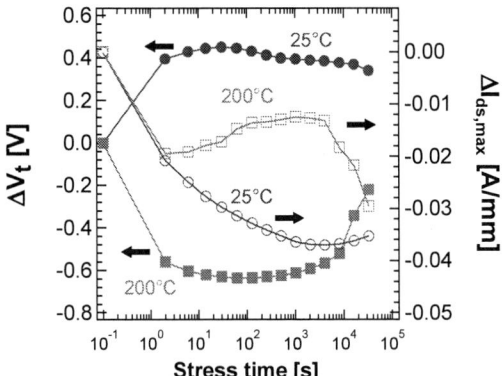

Figure 8. Shift in threshold voltage and maximum current during off-state stress at 25°C and 200°C

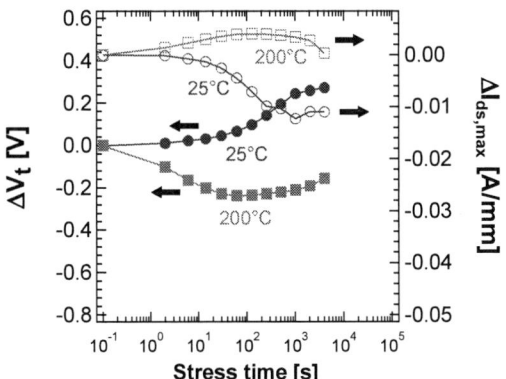

Figure 9. Shift in threshold voltage and maximum current during on-state stress at 25°C and 200°C

V_t starts to go up again. The current drop during the off-state stress at 25°C is due to trapping in the inter-metal dielectrics in the region between the gate and the drain, where the electric field is highest. At 200°C the initial drop (after 2s off-state stress) in $I_{ds,max}$ is the same as compared to 25°C due to trapping in the inter-metal dielectrics between gate and drain, but then the traps start to discharge with high temperature, resulting in a short time recovery of $I_{ds,max}$. However, at longer time trapping takes over again and $I_{ds,max}$ drops to the same level as for 25°C.

Also during on-state stress (V_{gs} = 0V, V_d = 4V) the shift in V_t and $I_{ds,max}$ is governed by an interaction between trapping and thermal activation of charges from traps or donor states (Fig. 9). Both at room temperature and high temperature, the shifts are very limited.

V.   CONCLUSIONS

The temperature dependence of GaN MISHEMTs with an in-situ Si_3N_4/ALD Al_2O_3 was investigated in the temperature range between 25°C and 200°C and compared to Si VDMOS power MOSFETs. The thermal stable point of the GaN

MISHEMT is at a much lower current level than that of the VDMOS device, which is an important asset with respect to thermal stability. Also, the on-resistance increase with temperature of the GaN device is significantly smaller than for VDMOS, while the transconductance variation is comparable. The drain and gate leakage currents of the GaN device are found to increase by roughly 3 orders of magnitude. However, the sensitivity of the drain current to temperature is much smaller in the GaN device. Dispersion measurements result in a doubling of the dynamic on-resistance degradation between 25°C and 150°C. Finally, off-state and on-state stress measurements were done, showing the impact of trapping/detrapping at different temperatures on the reliability of the GaN device.

REFERENCES

[1] O. Aktas, Z. F. Fan, S. N. Mohammad, A. E. Botchkarev and H.Morkoc, "High temperature characteristics of AlGaN/GaN modulation doped field effect transistors," Appl. Phys. Lett., vol. 69, pp. 3872-3874, 1996.

[2] N. Maeda, T.i Saitoh, K. Tsubaki, T. Nishida and N. Kobayashi, "Superior Pinch-Off Characteristics at 400±C in AlGaN/GaN Heterostructure Field Effect Transistors," Jpn. J. Appl. Phys. Vol. 38 pp. L 987–L 989, 1999.

[3] W.S. Tan, M.J. Uren, P.W. Fry, P.A. Houston, R.S. Balmer, and T. Martin, "High temperature performance of AlGaN/GaN HEMTs on Si substrates," Solid-State Electron, vol. 50(3), pp. 511-513, 2006.

[4] P. A. Houston, W. S. Tan, and R. T. Green, "Optimisation of high temperature performance and reliability of GaN HFETs," in Proc. 3rd EMRS DTC Tech. Conf., Edinburgh (U.K), A10, 2006.

[5] R. Cuerdo, F. Calle, A. F. Braña, Y. Cordier, M. Azize, N. Baron, S. Chenot, E. Muñoz, and E. Muñoz, "High temperature behaviour of GaN HEMT devices on Si(111) and sapphire substrates," Phys. Status Solidi C, vol. 5(6), 2008, pp. 971–973.

[6] M. Florovic, P. Kordos, D. Donoval, D. Gregusova, J. Kovac, "Performance of AlGaN/GaN heterostructure field effect transistors at higher ambient temperatures," J. Electr. Eng. vol. 59 (1), pp. 53–562, 2008,

[7] M. Hatano, N. Kunishio, H. Chikaoka, J. Yamazaki, Z. B. Makhzani, N. Yafune,K. Sakuno, S. Hashimoto, K. Akita, Y. Yamamoto, and M. Kuzuhara," Comparative high-temperature DC characterization of HEMTs with GaN and AlGaN channel layers," in Proc. CS MANTECH Conference, Portland (USA), pp. 101-104, 2010.

[8] L-Y. Yang, Y. Hao, Xi-H. Ma, J-C. Zhang, C-Y. Pan, J-G. Ma, K. Zhang, and P. Ma, "High temperature characteristics of AlGaN/GaN high electron mobility transistors," Chin. Phys. B Vol. 20 (11), 117302, 2011.

[9] C. Wang, N. Maeda, M. Hiroki, T. Kobayashi, and T. Enoki, "High Temperature Characteristics of Insulated-Gate AlGaN/GaN Heterostructure Field-Effect Transistors with Ultrathin Al_2O_3/Si_3N_4 Bilayer," Jpn. J. Appl. Phys., vol. 44, pp. 7889–7891, 2005.

[10] K. Son, A. Liao, G. Lung, M. Gallegos, T. Hatake, R. D. Harris, L. Z. Scheick, and W. D. Smythe, "GaN-Based High Temperature and Radiation-Hard Electronics for Harsh Environments," Nanosci. Nanotechnol. Lett., Vol. 2 (2), pp. 89-95, 2010.

[11] M. Van Hove, S. Boulay, S. R. Bahl, S. Stoffels, X. Kang, D. Wellekens, K. Geens, A. Delabie, and S. Decoutere, "CMOS-Compatible High-Power Low-Leakage AlGaN/GaN MISHEMT on Silicon Substrate," IEEE Electr. Dev. Lett., vol. 33 (5), 2012 (in print).

[12] S. Saadaoui, M. Mongi B. Salem, M. Gassoumi, H. Maaref, and C. Gaquière , "Anomaly and defects characterization by I-V and current deep level transient spectroscopy of $Al_{0.25}Ga_{0.75}N$/GaN/SiC high electron-mobility transistors, " J. Appl. Phys. Vol 111, 073713, 2012.

High Voltage Low R_{on} In-situ SiN/Al$_{0.35}$GaN$_{0.65}$/GaN-on-Si Power HEMTs Operation up to 300 °C

A.Fontserè[a,b], A.Pérez-Tomás[a], P.Godignon[a], J.Millán[a]

[a]IMB-CNM-CSIC, Campus UAB, 08193 and
[b]Dep. d'Enginyeria Electrònica, UPC, 08034,
Barcelona, Catalunya Spain
e-mail: abel.fontsere@imb-cnm.csic.es

J. M. Parsey[d] and P. Moens[c]

[c]ON Semiconductor, Power Technology Center, Westerring 15 B-9700 Oudenaarde, Belgium
[d]ON Semiconductor, 5005 East McDowell Road, Phoenix, AZ 85008, USA

Abstract—This paper reports some aspects of major practical interest now that GaN HEMTs are in the onset of commercialization for high voltage applications ($V_B > 300$ V). The device reproducibility of 900 V-class MIS-HEMTs with a thin in-situ grown Si$_3$N$_4$ gate insulator is investigated by means of wafer mapping and high-T stress. A remarkable result is that an exceptional yield of 99% across an entire 4-inch AlGaN/GaN-on-Si wafer when a reverse test of $V_{ds} = 100$ V was performed. In the same sense, it was found an impressive low dispersion of the threshold voltage, the saturation voltage and the on-resistance for 120 devices. Besides, under a DC reverse test, the typical gate and drain leakage currents are negligible ($I < 1$ µA/mm) up to 500 V and temperature independent (at the low drain bias) up to 250°C.

I. INTRODUCTION

AlGaN/GaN-based high electron mobility transistors (HEMTs) have demonstrated a great potential in the field of high-power, high-temperature and high-frequency electronics. In particular, GaN HEMTs have attracted great attention due to their low gate charge (Q_g) and reverse recovery charge (Q_{rr}), a low specific on-resistance (R_{on}) and high breakdown voltages [1]-[3]. The development of high-power AlGaN/GaN HEMT devices has progressed extraordinarily rapidly over the last few years and commercialization of devices and circuits based on this technology is now a reality. This is favored by the fact that GaN-on-Si wafers (up to 6-inch) are commercially available from several vendors, with buffers capable of withstanding high voltages.

The electrical performance of GaN HEMTs is superior to competing technologies, although the devices are still prone to long-term degradation due to the presence of defect-related charge traps. The relatively slow charging and discharging of these defect states, with time constants in the microsecond range, cause the HEMTs to display RF dispersion and current collapse behaviors. Mitigation of the RF dispersion has been demonstrated in the literature by passivating the AlGaN surface using a variety of dielectric films. The most commonly reported passivation being SiN deposited by chemical vapor deposition (CVD) methods. The SiN layers not only removed the current collapse phenomena but also improved the microwave noise performance and boosted the power density at high frequency [3]-[7].

A further step for improving the robustness of the passivated HEMT is to grow the SiN layer epitaxially *in-situ*. In-situ SiN deposition on AlGaN/GaN HEMT structures was recently shown to be feasible and advantageous mainly due to reduced AlGaN relaxation, increased sheet carrier concentration (n_s), improved Ohmic contacts and surface protection during processing [8]-[11].

In this paper, we summarize the results of a gold-free GaN-on-Si HEMT CMOS-compatible technology with reduced leakage currents and a reliable gate architecture. The MIS gate architecture is based on a thin in-situ grown Si$_3$N$_4$ layer. Focus will be put on the device reproducibility where 4-inch wafer-scale DC parametric mapping revealed minimal dispersion and remarkable stability of the MIS-HEMTs electrical parameters across the wafer. The SiN-isolated gate also allows operation at larger positive gate bias voltages and hence, reduces the on-resistance and increases the saturation current. The MIS-HEMT high-temperature device performance up to 300°C was also investigated.

II. EXPERIMENTAL DETAILS

Insulated gate MIS-HEMTs (Fig. 1(a)) were processed on 4-inch commercial AlGaN-GaN-on-silicon substrates capped with a thin in-situ silicon nitride layer. The in-situ passivation combined with the GaN buffer optimization yielded extremely low surface (gate) and bulk currents. Ideally, GaN-HEMT process and wafers should be capable of being processed in a standard CMOS line which requires low wafer bow and a gold-free metallization scheme. Ohmic contacts were formed after annealing a CMOS-compatible gold-free metal stack of Ti, Al and W. The gate metal was W-Al. The device isolation was achieved by nitrogen implantation.

978-1-4673-1707-8/12 $31.00 © 2012 IEEE

III. EXPERIMENTAL RESULTS

900V-class HEMTs were fabricated and characterized in this study. Extensive DC forward and reverse characterization (with wafer mapping) was performed. The wafer-level analysis was investigated with 120 test devices across the 4-inch wafer with a 3 μm source-gate (L_{gs}), 8-12 μm gate drain spacing (L_{gd}), and a gate length of 3 μm (L_g), respectively. The device width (W) was 150 μm in all cases unless noted. The fabrication process was shown to be stable with a large number of devices showing excellent on-state characteristics and breakdown voltages $V_B > 800$ V for gate-drain lengths greater than 8 μm (Fig. 1(c,b)). As shown in Fig. 1(b), the typical drain/gate reverse current characteristic showed a saturation of the breakdown voltage for $V_B \sim 900$ V ($L_{gd} > 8$-10 μm) and $V_B > 600$ V for gate-drain distance of $L_{gd} > 5$ μm. The drain and gate leakage current was below 1 μA/mm at 600 V.

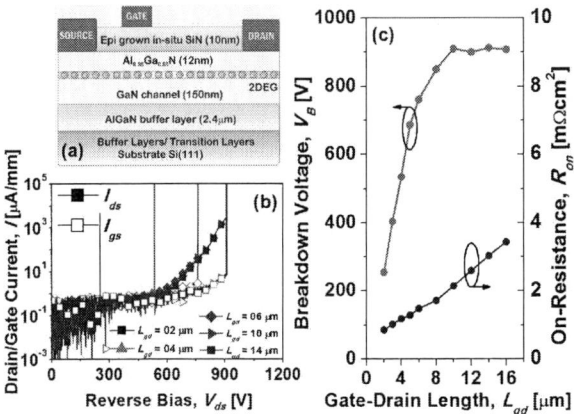

Figure 1. (a) Cross-sectional view of the MIS-HEMT. (b) Typical drain/gate current reverse characteristic showing the saturation of the breakdown voltage for $V_B \sim 900$ V and $V_B > 600$ V for gate-drain distance of $L_{gd} = 5$ μm. Gate and drain leakage current is maintained in any case below 1 μA/mm at $V_{ds} = 600$ V. (c) On-resistance and breakdown voltage dependence on the gate-drain length for MIS-HEMT with $L_{gs} = L_g = 3$ μm. R_{on} was extracted at $V_{gs} = 0$ V.

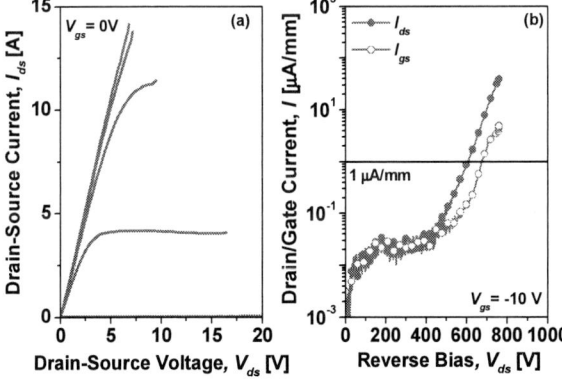

Figure 2. (a) Output DC characteristics of a large area AlGaN/GaN HEMT ($W = 31$ mm). A thicker second metal level was deposited on top of the gold-free Ohmic contacts. (b) Reverse characteristics (substrate floating) of the device showing leakage currents below 1 μA/mm up to 600 V.

The extracted on-resistance was found to be very stable across the wafer with $R_{on} = 2.5\pm0.3$ mΩ–cm^2 at $V_{gs} = 0$V. The ohmic contact area was included in the device pitch calculation and pad areas were excluded from the device area calculation. We believe that the relatively large value of the on-resistance is linked with the CMOS compatible gold-free technology. The current Al-based ohmic contact module yields a contact resistance of 2.6±0.6 Ω–mm, compared to 0.86±0.58 Ω–mm for a reference gold-based metallization scheme [12]. The higher contact resistance for the gold-free HEMT is consistent with literature data [13], [14] and clearly indicates the important role of gold in achieving low contact resistance to the 2D electron gas (2DEG) at the AlGaN/GaN interface. In spite of this slightly higher contact resistance gold-free large area HEMTs exhibited a large on-state forward current as shown in Fig. 2 (a). The gold-free contact solution did not affect the leakage current behavior where it was found to be remarkably low, less than 1 μA/mm up to 600 V as shown in Fig. 2(b).

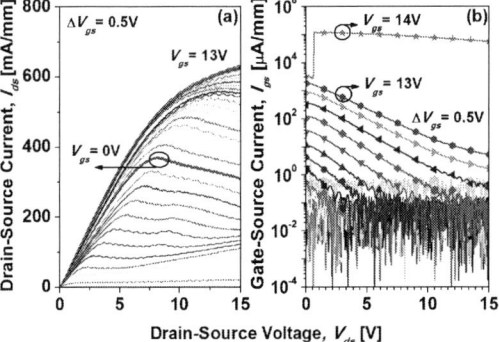

Figure 3. (a) Typical forward I-V drain-source current for the in-situ SiN MIS-HEMT for gate bias up to $V_{gs} = +14$ V. (b) Gate-source current during the same bias. It was observed negligible gate current up to $V_{gs} = +9$ V and the device gate degradation was observed at $V_{gs} = +13$ V.

The thin in-situ deposited SiN-based MIS approach also results in improved gate stability under large positive gate bias. The Schottky injection is mitigated when the thin insulator is introduced, with negligible gate-source current flow with a forward I-V sweep up to $V_{gs} = 9$ V. Irreversible device degradation (gate dielectric damage) was observed to occur at $V_{gs} = 13$-14 V. Both the saturation current and on-resistance are significantly enhanced when they are measured at a large positive gate bias. At $V_{gs} = 9$ V the on-resistance value drops to $R_{on} = 1.9\pm0.3$ mΩ–cm^2. In the same sense, the maximum saturation current (which was found to be very uniform across a wafer with $I_{d,sat} = 296\pm33$ mA/mm measured at $V_{gs} = 0$ V) increased significantly to $I_{d,sat} = 603\pm44$ mA/mm at $V_{gs} = 9$V. When full wafer transconductance maps were produced the threshold voltage value was found to be very homogeneous across the full wafer. The average V_{th} obtained for the population of devices was $V_{th} = -5.8\pm0.3$ V. Again, this indicated the good homogeneity of the epitaxial material with in-situ SiN deposition.

The introduction of the thin insulator has a marked beneficial role in reducing the gate and drain leakage currents all over the wafer. Figs 4 and 5 present the wafer maps of the reverse-biased gate and drain reverse currents at $V_{ds} = 100$ V.

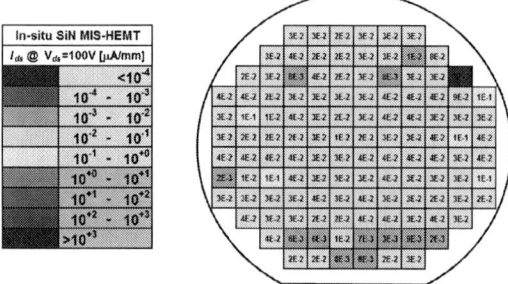

Figure 4. Reverse drain current wafer map ($V_{ds} = 100$ V) for the in-situ SiN MIS-HEMT with $L_{gs} = L_g = 3$ μm and $L_{gd} = 8$ μm measured at $V_{gs} = -10$ V. The yield of devices at $V_{ds} = 100$V with leakage currents lower than 1 μA/mm was as high as 99%. The drain leakage current for these devices was in the range of 2-50 nA/mm.

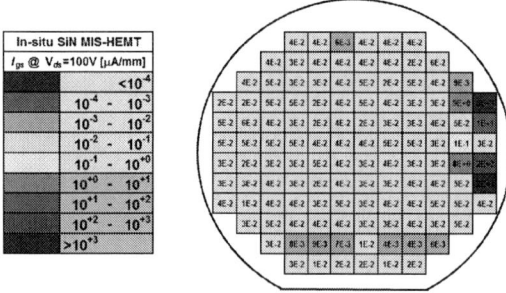

Figure 5. Reverse-bias gate current wafer map ($V_{ds} = 100$ V) for the in-situ SiN MIS-HEMT with $L_{gs} = L_g = 3$ μm and $L_{gd} = 8$ μm measured at $V_{gs} = -10$V. The yield at $I_{gs} < 1$ μA/mm at $V_{ds} = 100$ V in this case is 95%. The gate leakage current for these devices was in the range of 4-50 nA/mm. Only six devices on the wafer periphery exhibited a high gate leakage current which is an indication of the homogeneity and robustness of the in-situ SiN passivation.

The yield of devices exhibiting currents lower than 1μA/mm at $V_{ds} = 100$ V was as high as 99%. The drain leakage current was in the range of 2-50 nA/mm. In the same sense the reverse gate current wafer map showed a yield of 95% for a yield criteria of $I_{gs} < 1$ μA/mm at $V_{ds} = 100$ V. The gate leakage current for these devices was in the range of 4-50 nA/mm. It is worth noting that only six devices on the wafer periphery exhibited a high gate leakage current which is an indication of the homogeneity of the in-situ SiN passivation and the process uniformity.

Fig. 6 presents the temperature stability (up to 300°C) of the transconductance for a typical MIS-HEMT device. The MIS gate architecture was very effective in suppressing both the drain and gate off-state leakage up to ~250°C (Fig. 6 (c)). A negligible shift in the V_{th} was observed within experimental error. This indicates that the amount of mobile charge present in the gate insulator was also negligible. In any case, the forward current decreased with the temperature as can be seen in Fig.6 (b). The R_{on} increased by ~80% from room temperature to 300°C as shown in Fig.7 (a).

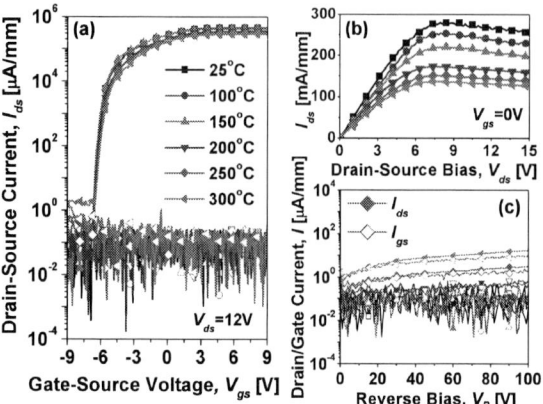

Figure 6. (a) In-situ SiN MIS-HEMT transconductance curve (drain/gate current vs gate-source) for different temperatures (25°C-300°C). A negligible shift in the threshold voltage was observed within the experimental error. (b) Drain current characteristics at varying temperatures. (c) The in-situ passivation along with the optimized GaN buffer resulted in HEMT drain/gate leakage currents with exceptionally temperature-independent behavior up to ~250°C.

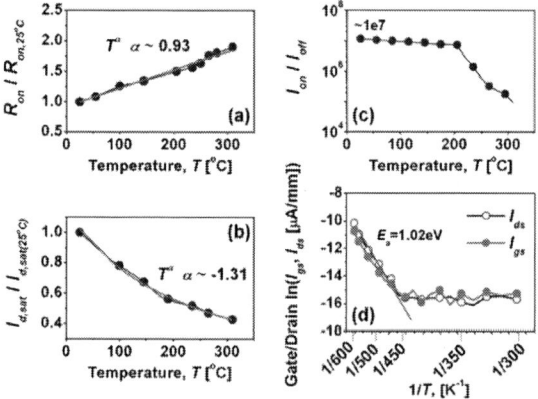

Figure 7. (a) On-resistance versus temperature for the in-situ SiN MIS-HEMT. R_{on} increased by ~80% from RT to 300 °C. The R_{on} increase ($R_{on}^{-1} \sim q\mu_n n_s$) is due to the reduction of the electron mobility (μ_n) with T although the 2DEG sheet concentration (n_s) is generally considered as virtually temperature independent. (b) Saturation current versus temperature where the self-heating effects showed a greater sensitivity to the temperature than R_{on}. (c) The transconductance I_{on}/I_{off} ratio is ~1e7 up to ~220°C where it started to decrease sharply down to ~1e5 at ~300°C. (d) An Arrhenius plot showing the activation energy for the in-situ SiN MIS-HEMT of $E_a = $ ~1.02eV for $T > 200$°C.

The R_{on} increase ($R_{on}^{-1} \sim q\mu_n n_s$) is attributable to the reduction of the electron mobility (μ_n) with temperature since the 2DEG sheet concentration (n_s) is generally considered as virtually temperature independent. In the AlGaN/GaN heterojunction, lattice vibrations due to polar-optical phonons in the non-intentionally doped GaN layer strongly increase with temperature and hence reduced the mobility at elevated temperatures [15]. The temperature dependence of the on-resistance can be fitted with a power law $R_{on} \propto T^\alpha$ with $\alpha = 0.97$ which is agreement with previous literature [12]. For the saturation current (Fig. 7(b)), the temperature coefficient can be fitted with $\alpha = -1.31$. We believe that the discrepancy in the temperature coefficient is due to the fact that in saturation the self-heating effects on the transport mechanisms become more dominant.

In previous works based on ALD HfO_2 MIS [12], it was observed that the HEMT leakage currents increase with the temperature which is typical of a rate-limited thermally activated process following an Arrhenius law ($I = I_0 \exp[-E_a/k_B T]$), up to a saturation regime. From the I vs. $1/k_B T$ (in the linear range), the activation energy of the gate/drain current was determined to be $E_a = 0.4$ eV. Differently, for the in-situ SiN MIS-HEMT it was observed that the leakage currents were negligible up to ~250 °C (using a $V_{ds} = 0$-100 V reverse sweep). The activation energy was determined to be as high as $E_a = 1.02$ eV. We suggest that this larger value of the thermal activation energy is related to the larger effective heterojunction electron barriers achieved by the in-situ passivation along with the optimized GaN buffer.

IV. CONCLUSIONS

The device reproducibility of 900 V-class MIS-HEMTs with a thin in-situ grown Si_3N_4 gate insulator was investigated by means of wafer-level mapping and high-T stress. The typical drain/gate reverse current characteristics exhibited breakdown voltage saturation for $V_B \sim 900$ V (for $L_{gd} > 8$-10 μm) and $V_B > 600$ V for gate-drain length of $L_{gd} > 5$μm. The gate leakage current was maintained in all cases below 1 μA/mm at $V_{ds} = 600$ V. The yield of devices at $V_{ds} = 100$ V exhibiting drain currents lower than 1 μA/mm was as high as 99%. The drain leakage current for these devices was in the range of 2-50 nA/mm. For the gate current, the yield with $I_{gs} < 1$ μA/mm at $V_{ds} = 100$ V was 95%. The gate leakage current for these devices was in the range of 4-50 nA/mm.

The thin in-situ based MIS approach also resulted in improved gate stability and robustness under positive gate bias. Negligible gate current flow was found at $V_{gs} = +9$ V. Irreversible degradation of the gate structure and device performance was observed at $V_{gs} = +13$ V. The on-resistance value was very uniform across the wafer with $R_{on} = 2.5 \pm 0.3$ mΩ−cm^2. At $V_{gs} = 9$ V this value dropped to $R_{on} = 1.9 \pm 0.3$ mΩ−cm^2. The maximum saturation current was also very uniform across the wafer with $I_{d,sat} = 296 \pm 33$ mA/mm measured at $V_{gs} = 0$V. At $V_{gs} = 9$ V this value increased significantly to $I_{d,sat} = 603 \pm 44$ mA/mm. In addition, a remarkably homogeneous threshold voltage, $V_{th} = -5.8 \pm 0.3$ V was determined. After temperature stress tests a negligible threshold shift was observed. R_{on} increased by ~80% from RT to 300 °C due to the reduction of the electron mobility. The in-situ SiN passivation, along with the optimized GaN buffer, also resulted in HEMT drain/gate leakage currents essentially temperature-independent up to ~250°C.

ACKNOWLEDGMENT

This work was sponsored by the IWT project GreenFETs (Flanders, Belgium) and the Spanish MICyN grants TEC2008-05577/TEC (THERMOS), and Ingenio-Consolider 2010 CSD2009-00046 (RUE).

REFERENCES

[1] P. Srivastava, J. Das, D. Visalli, M. V. Hove, P. E. Malinowski, D. Marcon, et al., "Record Breakdown Voltage (2200 V) of GaN DHFETs on Si With 2-μm Buffer Thickness by Local Substrate Removal," *IEEE Electron Device Lett.*, vol. 32, no. 1, pp. 30–32, Jan. 2011.

[2] E. Bahat-Treidel, F. Brunner, O. Hilt, E. Cho, J. Würfl, and G. Tränkle, "AlGaN/GaN/GaN:C Back-Barrier HFETs With Breakdown Voltage of Over 1 kV and Low $R_{ON} \times$ A," *IEEE Trans. Electron Devices*, vol. 57, no. 11, pp. 3050-3058, Nov. 2010.

[3] S. Hoshi, H. Okita, Y. Morino and M. Itoh, "Gallium Nitride High Electron Mobility Transistor (GaN-HEMT) Technology for High Gain and Highly Efficient Power Amplifiers," *Oki Technical Review*, vol. 174, no. 3, pp. 90-93, Oct. 2007; T. Kikkawa, T. Iwai, and T. Ohki, "Development of High-Efficiency GaN-HEMT Amplifier for Mobile WiMax," *Fujitsu Sci. Tech. J.*, vol. 44, no. 3, pp. 333-339, July 2008.

[4] J-C. Gerbedoen, A. Soltani, S. Joblot, J-C. De Jaeger, C. Gaquière, Y. Cordier, and F. Semond, "AlGaN/GaN HEMTs on (001) Silicon Substrate With Power Density Performance of 2.9 W/mm at 10 GHz," *IEEE Trans. Electron Devices*, vol. 57, no. 7, pp. 1497-1503, July 2010.

[5] H-K. Lin, H-L. Yu, and F.-H. Huang, "Performance Improvement of AlGaN/GaN HEMTs using Two-step Silicon Nitride Passivation," *Microwave and Optical Technology Letters*, vol. 52, no. 7, pp. 1614-1619, Sept. 2009.

[6] K. B. Lee, R. T. Green, P. A. Houston, W. S. Tan, M. J. Uren, D. J. Wallis and T. Martin, "Bi-layer SixNy passivation on AlGaN/GaN HEMTs to suppress current collapse and improve breakdown," *Semicond. Sci. Technol.*, vol. 25, pp. 125010 1-5, Sept. 2010.

[7] Z. H. Liu, S. Arulkumaran, and G. I. Ng, "Improved Microwave Noise Performance by SiN Passivation in AlGaN/GaN HEMTs on Si" *IEEE Microwave and Wireless Components Letterss*, vol. 19, no. 6, pp. 383-385, June 2009.

[8] B. Heying, I.P. Smorchkova, R. Coffie, V. Gambin, Y.C. Chen, W. Sutton, T. Lam, M.S. Kahr, K.S. Sikorski and M. Wojtowicz, "In situ SiN passivation of AlGaN/GaN HEMTs by molecular beam epitaxy," *Electron. Lett.*, vol. 43, no. 14, pp. 789-790, July 2007.

[9] H. Behmenburg, L. R. Khoshroo, C. Mgoldder, N. Ketteniss, K. H. Lee, M. Eickelkamp, M. Brast, D. Fahle, J. F. Woitok, A. Vescan, H. Kalisch, M. Heuken, and R. H. Jansen, "In situ SiN passivation of AlInN/GaN heterostructures by MOVPE," *Phys. Status Solidi C*, vol. 7, no. 7-8, pp. 2104-2106, May 2010.

[10] M. J. Tadjer, T. J. Anderson, K. D. Hobart, M. A. Mastro, J. K. Hite, J. D. Caldwell, Y. N. Picard, F. J. Kub, and C. R. Eddy JR., "Electrical and Optical Characterization of AlGaN/GaN HEMTs with In Situ and Ex Situ Deposited SiN_x Layers," *J. Electron. Matt.*, vol. 39, no. 11, pp. 2452-2458, Sep. 2010.

[11] M. Germain, K. Cheng, J. Derluyn, S. Degroote, J. Das, A. Lorenz, D. Marcon, M. Van Hove, M. Leys, and G. Borghs, "In-situ passivation combined with GaN buffer optimization for extremely low current dispersion and low gate leakage in Si_3N_4/AlGaN/GaN HEMT devices on Si (111)," *Phys. Status Solidi C*, vol. 5, no. 6, pp. 2010-2012, Apr. 2008.

[12] A. Fontserè, A. Pérez-Tomás, V. Banu, P. Godignon, J. Millán, H. De Vleeschouwer, J. M. Parsey and P. Moens, "A HfO_2 based 800V/300°C Gold-free AlGaN/GaN-on-Si HEMT Technology," *24th International Symposium on Power Semiconductor Devices & IC's, 2012. ISPSD 2012 Bruges, Belgium.*

[13] B. Van Daele, G. Van Tendeloo, J. Derluyn, P. Shrivastava, A. Lorenz, M. R. Leys, and M. Germain, "Mechanism for Ohmic contact formation on Si_3N_4 passivated AlGaN/GaN high-electron-mobility transistors)," *Appl. Phys. Lett.*, vol. 89, pp. 201908 1-3, Nov. 2006.

[14] H.-S. Lee, D. S. Lee, and T. Palacios, "AlGaN/GaN High-Electron-Mobility Transistors Fabricated Through a Gold-Free Technology," *IEEE Electron Device Lett.*, vol. 32, no. 5, pp. 623–625, May 2011.

[15] A. Pérez-Tomás, M. Placidi, N. Baron, S. Chenot, Y. Cordier, J. C. Moreno, et al.," GaN transistor characteristics at elevated temperatures," *J. Appl. Phys.*, vol. 106, no. 7, pp. 074519, Oct. 2009.

Critical Gate Module Process Enabling the Implementation of a 50A/600V AlGaN/GaN MOS-HEMT

S. G. Khalil, R. Chu, R. Li, D. Wong, S. Newell,
X. Chen, M. Chen, D. Zehnder, S. Kim,
A. Corrion, B. Hughes and K. Boutros
HRL Laboratories, LLC
Malibu, CA, USA
Email: sgkhalil@hrl.com

C. Namuduri
GM Global R&D Center
General Motors
Warren, MI, USA
Email: Chandra.s.namuduri@gm.com

Abstract—Two critical processes within the gate module of GaN-based MOS-HEMT with significant impact on device robustness and performance were identified and are presented in this paper. Specifically, data highlighting the impact of the number of cycles of the atomic layer etching of the AlGaN barrier to recess the gate region and the sequence of the gate dielectric anneal step on device performance are discussed. The optimization of these two critical steps enabled the implementation of a 50A/600V with an off-state leakage current of 455 μA at 600V and on-state resistance of 41mΩ at V_{GS}=2.5V.

I. INTRODUCTION

The recent increase in research and development activities of GaN-based power devices and materials is propelled by the high potential of these devices to lead the power electronics technology roadmaps in ever demanding energy efficient applications. The current efforts to bring GaN-based power electronics to market are consistent with a worldwide push to reduce CO_2 emission and pursue new alternatives to reduce energy consumption.

GaN-based lateral HEMT, implemented on group III-Nitride epitaxial stack and grown on a (111) Silicon substrate is the leading candidate for future energy efficient electronics.

GaN-based power devices bring the advantages associated with GaN material properties such as high critical electric field of 3MV/cm, high 2DEG mobility that ranges between 1600 and 2000 cm^2/V.sec and high temperature operation, together with the cost effectiveness of a large diameter Silicon wafer and the ability to utilize mainstream CMOS economies of scale [1].

Recent work on GaN-based lateral HEMT showed steady improvement of HV device performance [2-6]. However, little data have been published in the literature on the measures taken to implement a robust, low leakage device with high current carrying capability for high voltage applications.

In this paper we report on the results of a systematic study on the impact of two critical process steps in the gate module on key device parameters. The motivation for this work is to improve device robustness and performance to enable the implementation of a low leakage and large gate periphery 50A/600V GaN-based device.

II. DEVICE STRUCTURE AND PROCESS FLOW

A 2D cross-section view of the AlGaN/GaN MOS-HEMT of HRL's GaN-On-Silicon technology is shown in Fig. 1. The device is a hybrid between a HEMT structure and an MOS structure. A positive threshold voltage is achieved by introducing fluorine ions and forming a recess in the AlGaN barrier in the channel region and by the appropriate choices of the gate dielectric thickness, gate metallurgy and process conditions of the gate module.

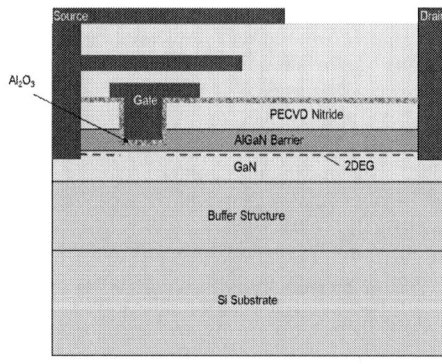

Fig. 1 2D Cross-section view of HRL's AlGaN/GaN MOS-HEMT

The starting material is a group III-Nitride epitaxial stack that consists of a nucleation layer, thick III-Nitride buffer layer and AlGaN barrier/GaN Cap layers grown on a 3" Silicon wafer by heteroepitaxy using the MOCVD method. The process starts with patterning the device isolation region followed by high dose ion implantation to remove the 2DEG charge outside the active region.

This is followed by source and drain Ohmic metallization using Ti/Al/Ni/Au stack followed by a brief 860°C anneal. Thereafter, a Si_3N_4 film is deposited for passivation using the PECVD method. Gate definition step followed by nitride etch, fluorine treatment and gate recess etch using a digital O_2-BCl_3 Atomic Layer Etching (ALE) scheme as described in [5], to implement a normally-off operation. This is immediately followed by Atomic Layer Deposition (ALD) of Al_2O_3 as the gate insulating film followed by gate metal deposition and patterning.

A key step is the rapid thermal anneal (RTA) process that is inserted within the gate module process and is performed at 400°C for 5 minutes to reduce the trap density at the interface between the Al_2O_3 film and AlGaN surface. Further back-end metallization and intermetallic dielectric evaporation/ deposition steps are performed to realize a low resistance interconnect network and complete the 3-step field plate structure for electric field shaping.

III. EXPERIMENTAL DETAILS AND RESULTS

A design-of-experiment (DOE) was constructed to study the impact of the following two critical steps on device parameters; 1) the number of ALE cycles used to etch the AlGaN barrier to recess the gate region and 2) the insertion point of the dielectric RTA step in the gate module process flow.

A total of 6 identical wafers were split into two main groups to study the impact of the RTA step on device parameters, one group had the RTA step inserted immediately after the Al_2O_3 ALD step while the other group had the RTA step done after gate metal deposition. Each of the two groups had in turn three variations of the ALE number of cycles which were chosen to be 5, 7 and 9 cycles as summarized in Table I below. The device parameters that are used to assess the response to these two steps are gate leakage (I_{g_leak}) at V_{GS}=2.5V, on-state resistance (R_{on}) at V_{GS}=2.5V, maximum drain current (I_{max}) at V_{GS}=3V and V_D=10V and threshold voltage (V_{th}) at I_d=10 μA/mm and V_d=0.1V.

TABLE I
Summary of the DOE

Wafer ID	ALE Cycles	RTA after ALD deposition
A	5	Yes
B	5	No
C	7	Yes
D	7	No
E	9	Yes
F	9	No

Statistical data of measured Process Control Monitors (PCMs) were generated for key device parameters by testing identical devices with gate periphery of 600 μm which have a unit cell design similar to the large gate periphery device (50A device). The testing involved a sample size of 30 reticles across the 3" wafers.

The aforementioned key device parameters were analyzed to study the impact of both the ALE number of cycles and the RTA process sequence on device performance and to arrive at optimized process conditions.

Threshold voltage for the 6 wafers, shown in Fig. 2, highlights the dependence on the number of ALE cycles. The data show that a minimum thickness of AlGaN barrier needs to be removed before the channel is switched off. It will require at least 7 ALE cycles (wafer C) to arrive at a positive V_{th}. The higher the number of cycles the more positive V_{th} is.

However, increasing the number of ALE cycles to move to more positive V_{th} values is not without constrains as it turns out there is an upper bound on the number of ALE cycles which is imposed by the extent of degradation of I_{max}, R_{on} and I_{g_leak} with the increase of ALE cycles as shown in Fig. 3, Fig. 4 and Fig. 5, respectively.

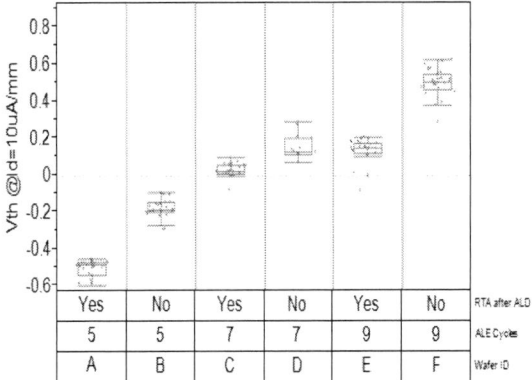

Fig. 2 Statistical data of V_{th} of 600 μm MOS-HEMT

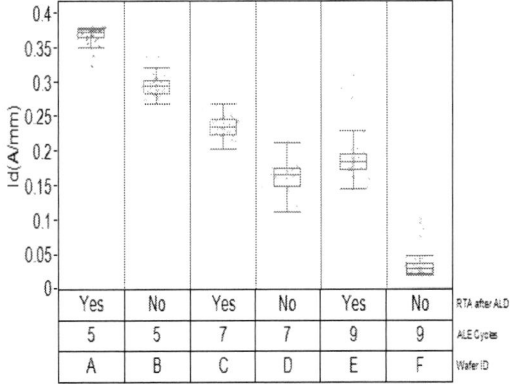

Fig. 3 Statistical data of I_{max} of 600 μm MOS-HEMT

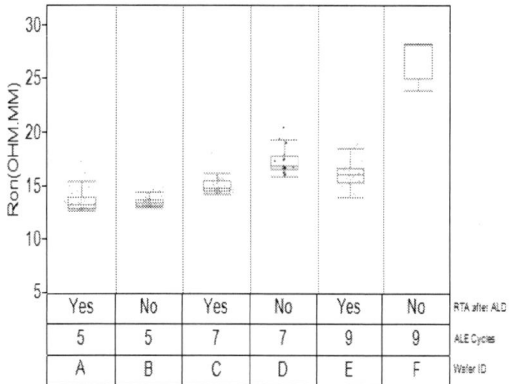

Fig. 4 Statistical data of R_{on} of 600 μm MOS-HEMT

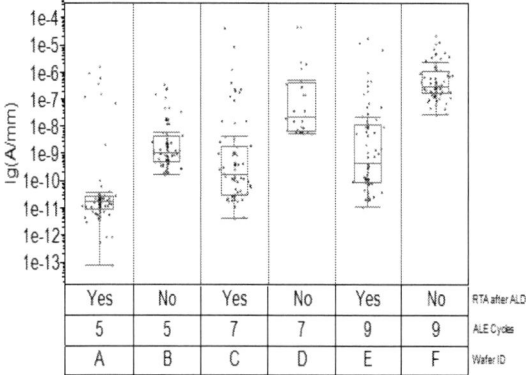

Fig. 5 Statistical data of I_{g_leak} of 600 μm MOS-HEMT

The on-state (I_{max} and R_{on}) degradation is mainly attributed to the rough interface at the bottom of the gate recess which is moved closer to the gate channel as the ALE number of cycles increases. The scattering of channel carriers at this rough interface is responsible for channel mobility degradation.

A low gate leakage current, which is a critical parameter for device robustness and reliability of GaN- based device and is a prerequisite to implement a large gate periphery device, is highly dependent on the gate stack composition and the process used to implement it.

The correlation between both the numbers of cycles, the gate dielectric anneal sequence and gate leakage current can be seen in Fig. 5. The gate leakage at positive gate voltage increases if the number of ALE cycles increases and if the RTA step was done after the gate metal deposition.

For a normally-off device, it is important to ensure that gate leakage remains low at a relatively high positive V_{GS} up to at least 3V since a higher gate voltage drive is required for lower R_{on} and faster switching time. This requirement, which can be met by thickening the gate dielectric, presents a design challenge since it has been

shown that increasing Al_2O_3 thickness to reduce gate leakage and to improve gate integrity results in shifting the flat-band voltage, and in turn the threshold voltage, to a more negative value [7]. Other gate dielectric such as the high-K HfO_2 deposited with ALD has a positive shift in the flat-band voltage with increasing film thickness and is a potential candidate but still suffers at the present time from high interface trap density of the order of 10^{13} cm^{-2}. Therefore, engineering the gate module to simultaneously achieve a positive V_{th} and low gate leakage at high positive gate voltage without degrading the on-state performance becomes a challenge.

For HRL's recessed gate structure, the insulating gate dielectric can be thought of as a composite stack made up of the Al_2O_3 film plus the portion of the AlGaN barrier that remains after the gate recess step, in other words the remaining AlGaN under the gate supports a portion of the voltage drop appearing between the gate electrode and the 2DEG layer, hence reducing the electric field in the Al_2O_3 film. The reduction of the gate leakage as a result of larger AlGaN thickness under the gate, as seen in Fig. 5, is attributed to enhanced immunity to Fowler Nordheim tunneling across the composite insulating gate stack.

The TEM images shown in Fig. 6 are taken for samples from an earlier but related experiment where two devices, one with 6 cycles of ALE and the RTA step performed prior to the gate metal deposition, and the second is for a device with 8 cycles and RTA performed after gate metal deposition. Key device parameters V_{th}, R_{on}, I_{max} and I_{g-leak} followed the same trends as the 6 wafer experiment. The images reveal that the 6 cycle wafer has a remaining layer of AlGaN under the Al_2O_3 whereas the 8 cycle wafer had almost no AlGaN remaining. The images also show that the gate metal/Al_2O_3 has a rough interface for the 8 cycles sample where the RTA was done after gate metal deposition whereas no roughness at that interface is observed on the 6 cycles sample where the RTA was done prior to metal deposition. The electrical results and the cross section differences between the two samples confirm the observations and conclusions extracted from the 6 wafer experiment.

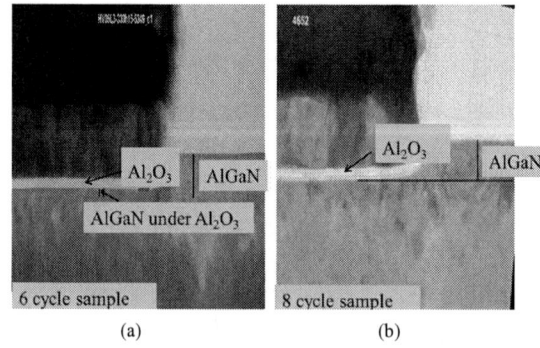

Fig. 6 TEM images of the gate region: (a) sample with 6 ALE cycles and (b) sample with 8 ALE cycles

In the 6 wafer DOE, wafer C which had 7 cycles of ALE and RTA process done prior to gate metal deposition represents an optimized process flow.

IV. IMPLEMENTATION OF A 50A/600V DEVICE

Arriving at process conditions that can meet key device parameters simultaneously and resolve the trade-offs described above enabled the implementation of a low leakage 50A device with a blocking voltage of 600V.

A large gate periphery device of 420 mm was packaged and tested. At 600V the drain leakage current was 455 μA (1.1 μA/mm) as shown in Fig 7(a). It should also be noted that the off-state measurements were done with the substrate connected to ground which represent the worst case scenario of the blocking capability of the lateral GaN-on-Si MOS-HEMT.

The output characteristics of the device are shown in Fig. 7(b) where the gate voltage is stepped from 0 to 2.5V in 0.5V steps. The measured R_{on} at V_{GS}=2.5V is 41mΩ which includes 5mΩ contribution from the external package and bond wire resistance.

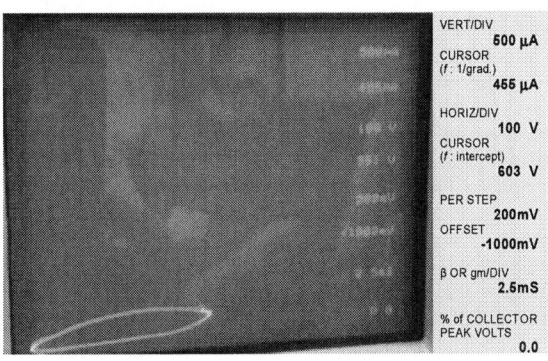

(a)

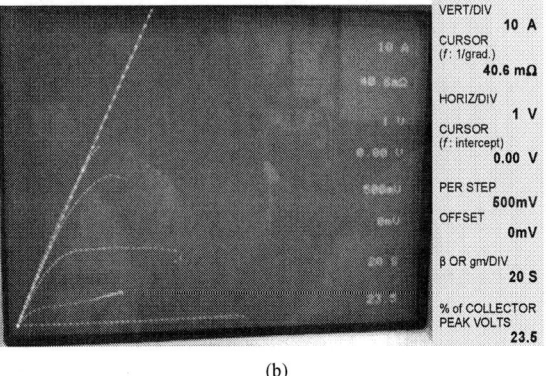

(b)

Fig. 7 Electrical characteristics of a 50A/600V device (Wg=420mm): (a) Off-state characteristics, (b) Output characteristics (V_{GS} stepped from 0V to to 2.5V in 0.5V steps)

V. CONCLUSIONS

Two processes of the gate module of an AlGaN/GaN MOS-HEMT were identified as critical to device robustness and performance. It was found that the gate recess process should be designed to allow a minimum thickness of the AlGaN barrier to reduce gate leakage current and reduce channel mobility degradation at the rough interface between the Al_2O_3 gate dielectric and AlGaN in the channel. It was also observed that it is more favorable to perform the dielectric anneal process prior to depositing the gate metal to reduce gate leakage currents and improve on-state performance for a given positive gate voltage. Optimizing the gate module processes has enabled the implementation of a low leakage 50A/600V MOS-HEMT with a positive V_{th}.

REFERENCES

[1] M.A. Briere, "The Next stage in the Commercialization of GaN-Based Power Devices," *Power Electronics Europe*, pp. 35-37, issue 3, 2010.

[2] K.S. Boutros, S. Burnham, D. Wong, K. Shinohara, B. Hughes, D. Zehnder and C. McGuire, "Normally-off 5A/1100V GaN-on-Silicon Device for High Voltage Applications," *in Proc. IEDM*, pp.161-163, 2009.

[3] A. L. Corrion, M. Chen, R. Chu, S. D. Burnham, S. Khalil, D. Zehnder, B. Hughes, and K. Boutros, "Normally-off Gate-Recessed AlGaN/GaN-on-Si Hybrid MOS-HFET with Al2O3 Gate Dielectric," *in Proc. DRC*, pp. 213-214, 2011.

[4] N. Ikeda, S. Kaya, J. Li, T. Kokawa, M. Masuda and S. Katoh, "High-power AlGaN/GaN MIS-HFETs with field-plates on Si substrates," *in Proc. ISPSD*, pp. 251-254, 2009.

[5] S. Burnham, K. Boutros, P. Hashimoto, C. Butler, D. Wong, M. Hu and M. Micovic, "Gate-recessed normally-off GaN-On-Si HEMT using a new O_2-BCl_3 digital etching technique," *Phys. Status Solidi* Volume 7, Issue 7-8, pp. 2010-2012, July 2010.

[6] R. Chu, A. Corrion, M. Chen, R. Li, D. Wong, D. Zehnder, B. Hughes and K. Boutros, "1200V Normally Off GaN-On-Si Field effect Transistors with Low Dynamic On-Resistance," IEEE Electron Device Letters, Vol. 32, No. 5, pp. 632-634, May 2011.

[7] M. Esposto, S. Krishnamoorthy, D. N. Nath, S. Bajaj, T.H. Hung and Siddharth Rajan, "Electrical Properties of Atomic Layer Deposited Aluminum Oxide on Gallium Nitride," *Applied Physics Letter* 99, 133503, 2011.

Scaling of InAlN/GaN Power Transistors

Daniel Piedra, Hyung-Seok Lee, Tomás Palacios
Department of Electrical Engineering and Computer Science
Massachusetts Institute of Technology
Cambridge, MA, USA
dpiedra@mit.edu

Xiang Gao, Shiping Guo
IQE RF LLC
Somerset, NJ, USA

Abstract—**This work presents the first use of InAlN/GaN high electron mobility transistors in power electronic applications. High voltage transistors of various gate widths (W_g) have been fabricated to observe how transistors scale. The high charge density enabled by the InAlN/GaN heterostructure has the potential to reduce the on-resistance of GaN power transistors, compared to the standard AlGaN/GaN structures. Transistors with maximum gate width of W_g=39.6 mm have been fabricated in this study and they showed a total current $I_{D,max}$ of 18.57 A and specific on-resistance of $R_{ON,sp}$=1.497mΩ·cm². In all these devices, the maximum drain voltage, transconductance, and gate-source capacitance scale well, however the off-state current and gate leakage show a super-linear increase with scaling gate width. The demonstration of transistors with large gate widths and high current levels is promising for use in actual circuit applications.**

I. INTRODUCTION

GaN-based transistors exhibit an unsurpassed combination of low on-resistance and high breakdown voltage. These features have made GaN transistors appealing for applications in the areas of RF-electronics as well as power electronics [1]. Traditionally, most GaN HEMTs have been fabricated with AlGaN as the top barrier, but the InAlN/GaN heterostructure gives a higher 2DEG charge density, thus enabling higher current density and lower on-resistance [2]. Previous work on InAlN/GaN HEMTs has focused on their use in high frequency applications and their performance as a function of gate length scaling [3, 4] and barrier thickness scaling [5], but to our knowledge there has not been a study of gate width scaling or a demonstration of wide-periphery (>2.5 mm) devices. This paper aims to investigate the scaling of transistor metrics as a function of gate width to understand the impact of epitaxial and processing defects on the device performance. Devices with a total gate periphery of 39.6 mm have been demonstrated and are being used in power circuits.

II. DEVICE FABRICATION

The devices studied in this work were processed on InAlN/GaN grown on SiC by IQE RF LLC. The epitaxial structure consisted of a 5.3 nm $In_{0.17}Al_{0.83}N$ top barrier, a 1 nm

AlN interfacial layer, a 20 nm GaN channel, and a 850 nm $Al_{0.04}Ga_{0.96}N$ back-barrier grown directly on the 4H-SiC substrate.

First, mesa isolation patterns were defined by photolithography and etched in BCl_3/Cl_2 ECR plasma. Then, a metal stack of Ti/Al/Ni/Au (20 nm/100 nm/25 nm/50nm) was deposited by electron beam evaporation and patterned by lift-off. Ohmic contacts were formed by performing a rapid thermal anneal of 870°C for 30 s in nitrogen atmosphere. An additional layer of Ti/Au (20 nm/500 nm) was deposited on the annealed ohmic contacts by electron beam evaporation and patterned by lift-off. This layer was to reduce the resistance in the contact and support high current levels. Next, the surface was cleaned by an ozone plasma descum step and a 12 nm layer of SiO_2 was deposited by atomic layer deposition to serve as the gate dielectric. Lift-off of an e-beam evaporated stack of Ni/Au/Ni (30 nm/200 nm/50 nm) on patterns defined by photolithography was done to make the gate electrodes. A 1.7 μm layer of SiO_2 was deposited by plasma enhanced chemical vapor deposition to serve as a passivation layer and interlayer for the multi-finger transistors. Pad openings were patterned in the SiO_2 layer and etched in CF_4/H_2 ECR plasma. Ti/Au (20 nm/700 nm) was deposited and patterned by lift-off to form the interconnects of the multi-finger transistors. A diagram of the cross-section of the device is shown in Fig.1.

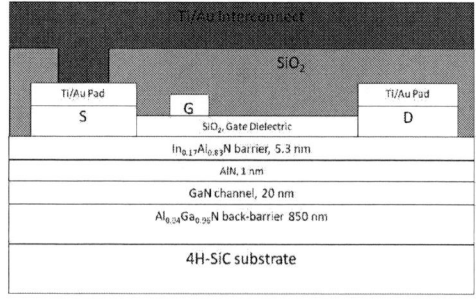

Figure 1. Cross-sectional diagram of the InAlN/GaN HEMT used in this study, showing both the epitaxial structure and the fabricated device.

This work was sponsored by the MARCO MSD and IFC programs, the DARPA MPC program and the ADEPT ARPA-E program.

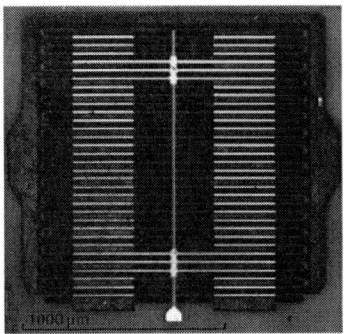

Figure 2.   Micrograph of wide-periphery multi-finger HEMT, W_g=39.6 mm

To study the scalability of these devices, the gate width of the transistors was varied while the other device dimensions remained constant. In all of the processed devices, the gate length (L_g) was 2 µm, the gate-to-source distance (L_{gs}) was 1.5 µm, and the gate-to-drain distance (L_{gd}) was 10 µm. Devices with the following total gate widths were fabricated: 25 µm, 50 µm, 100 µm, 150 µm, 200 µm, 300 µm, 400 µm, 22.8 mm and 39.6 mm. A micrograph of a completed wide periphery transistor (W_g=39.6mm) is shown in Fig 2.

III.   MEASUREMENT RESULTS AND DISCUSSION

Electrical characterization of the material properties of the processed sample was performed first. A sheet resistance of 295 Ω/sq, a two-dimensional electron gas charge density of 1.83×10^{13} cm^{-2}, and an electron mobility of 1147 cm^2/Vs were measured by Hall effect technique at room temperature. The device output and transfer characteristics were measured using an Agilent B1505 power semiconductor analyzer and a Cascade Tesla probe station. The output curves of each device were measured by sweeping V_{DS} from 0 V to 10 V at different V_{GS} levels while monitoring the drain current I_D and gate leakage I_G. Fig. 3 shows the output characteristics of the largest transistor used in this study (W_g=39.6 mm). From these curves, the maximum drain current $I_{D,max}$ of each device was extracted and plotted in Fig 4. The absolute value $I_{D,max}$ was divided by the W_g for each respective transistor to give the normalized $I_{D,max}$, shown in Fig 5.

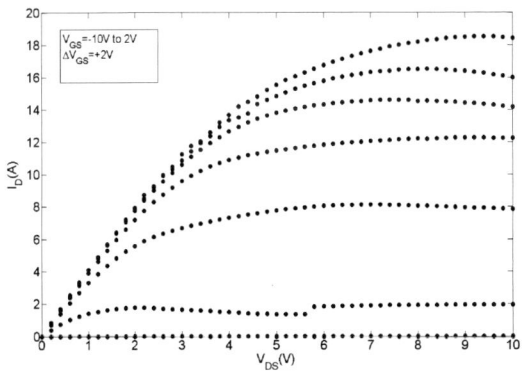

Figure 3.   Output curves of wide periphery HEMT, with W_g=39.6 mm

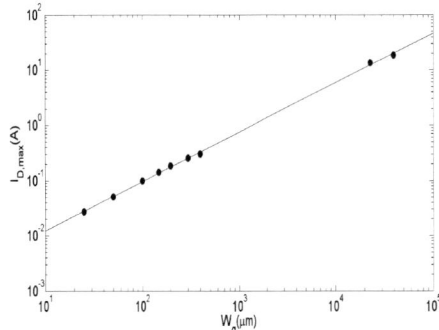

Figure 4.   Maximum drain current as a function of gate width (experimental data shown as dots and best fit line (power law by least means squares) $I_{D,max}=10^{-2.806}W_g^{0.8934}$).

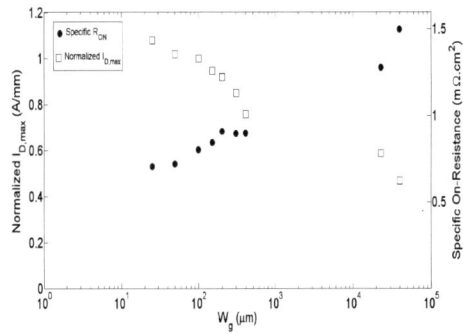

Figure 5.   Normalized Drain Current and Specific On-Resistance across various gate widths.

The transfer characteristics were measured by keeping V_{DS} constant at 8V and sweeping V_{GS} from -14 V to 0, while monitoring the drain current and the gate leakage I_g. The transconductance g_m was computed as the derivative of the drain current with respect to the gate voltage. Fig. 6 shows the transfer characteristics of the W_g=39.6 mm device. From these curves, the maximum transconductance $g_{m,max}$ of each device was extracted and plotted in Fig 7. The absolute value $g_{m,max}$ was divided by the W_g for each respective transistor to give the normalized $g_{m,max}$, shown in Fig 8. The off-state drain current $I_{D,off}$ and maximum gate leakage levels were also extracted and plotted in Fig 9.

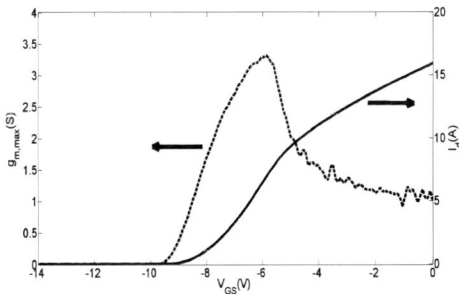

Figure 6.   Transfer curves of wide periphery HEMT, with W_g=39.6 mm.

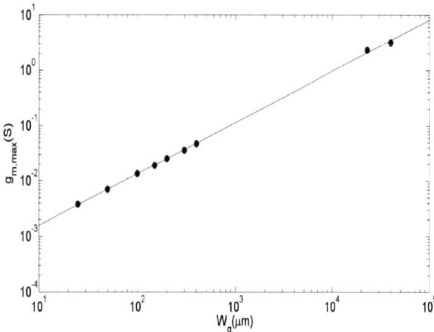

Figure 7. Maximum transconductance experimental data shown as dots and best fit line (power law by least means squares) $g_{m,max}=10^{-3.794}W_g^{0.9606}$.

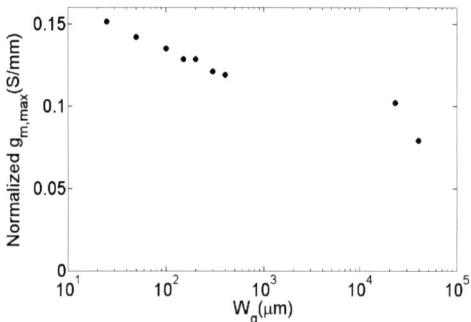

Figure 8. Normalized transconductance across various gate width

Capacitance-voltage measurements were performed using an Agilent 4292A Impedance Analyzer. The C_{gs} was measured with the oscillation voltage level set to 100 mV and the oscillation frequency set to 1 MHz. Fig 10 shows the resulting C_{gs} values for the various devices. The gate-to-source capacitance of each device was divided by the W_g for each respective transistor to give the normalized C_{gs}, shown in Fig 11.

The normalized values were plotted by taking the absolute value of the respective metric and dividing by the gate width (in mm). The maximum drain voltage, transconductance, and gate-source capacitance scaled well as seen by the fact that the normalized values did not decrease drastically over 3 orders of magnitude in gate width. Some ohmic loss in the contacts or slight non-uniformities across the sample may attribute to the lack of perfect scaling.

The leakage currents ($I_{g,max}$, $I_{D,off}$), however, did not scale as well in these devices. Rather, the leakage current levels remain fairly constant for the smaller gate widths (up to W_g=400 μm), but increased for the largest devices. It is suspected that this increase in the level of leakage current is due to non-uniformities in the epitaxial layer and/or the fabrication technology. Since the wide-periphery devices occupy such a large area, there is a greater probability that there may be a weak point in the GaN material or the gate dielectric and thus a higher probability that the dielectric

might break, resulting in higher I_g. This higher gate leakage in turn causes higher off-state leakage.

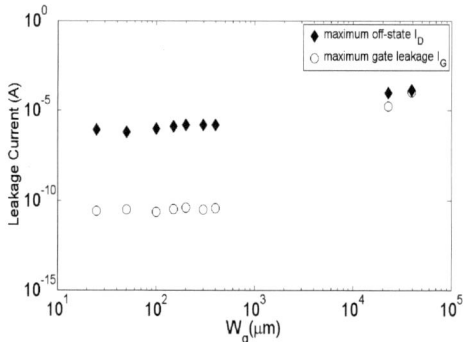

Figure 9. Off-state drain current and gate leakage current across various gate widths.

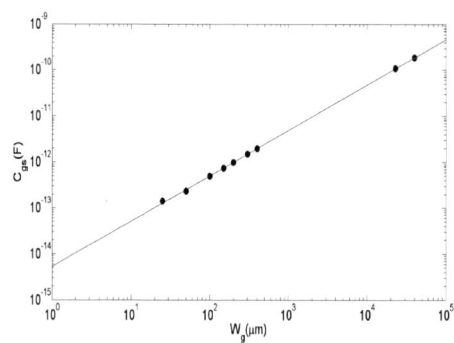

Figure 10. C_{gs} experimental data shown as dots and best fit line (power law by least means squares) $C_{gs}=10^{-14.28}W_g^{0.988}$

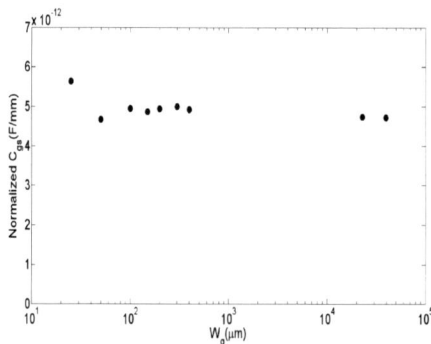

Figure 11. Normalized C_{gs} across various gate widths

IV. CONCLUSION

InAlN/GaN high electron mobility transistors of various gate widths were fabricated to study scaling in these devices. Wide periphery multi-finger transistors with a maximum gate width of W_g=39.6 mm have been fabricated as a part of this study. These devices showed a maximum drain current of $I_{D,max}$= 18.57 A and specific on-resistance of $R_{ON,sp}$=1.497mΩ·cm^2. In all these devices, the maximum drain voltage, transconductance, and gate-source capacitance scale well over 3 orders of magnitude in gate width. However the off-state current and gate leakage show a super-linear increase with scaling gate width. Demonstration of the wide-periphery devices is promising for use in power electronic circuit applications

ACKNOWLEDGMENT

The device fabrication was performed in the Microsystem Technology Laboratories at MIT and the Center for Nanoscale Systems at Harvard University.

REFERENCES

[1] U. K. Mishra, L. Shen, T. Kazior, Y-F. Wu; "GaN-Based RF Power Devices and Amplifiers," Proc. of the IEEE 96, 287 (2008)

[2] Medjdoub, F.; Carlin, J.-F.; Gonschorek, M.; Feltin, E.; Py, M.A.; Ducatteau, D.; Gaquiere, C.; Grandjean, N.; Kohn, E.; , "Can InAlN/GaN be an alternative to high power / high temperature AlGaN/GaN devices?," *Electron Devices Meeting, 2006. IEDM '06. International* , vol., no., pp.1-4, 11-13 Dec. 2006

[3] D. S. Lee; X. Gao; S. Guo; D. Kopp.; P. Fay; T. Palacios; "300-GHz InAlN/GaN HEMTs With InGaN Back Barrier," *Electron Device Letters, IEEE* , vol.32, no.11, pp.1525-1527, Nov. 2011

[4] Haifeng Sun; Alt, A.R.; Benedickter, H.; Feltin, E.; Carlin, J.-F.; Gonschorek, M.; Grandjean, N.R.; Bolognesi, C.R.; , "205-GHz (Al,In)N/GaN HEMTs," *Electron Device Letters, IEEE* , vol.31, no.9, pp.957-959, Sept. 2010

[5] Medjdoub, F.; Alomari, M.; Carlin, J.-F.; Gonschorek, M.; Feltin, E.; Py, M.A.; Grandjean, N.; Kohn, E.; , "Barrier-Layer Scaling of InAlN/GaN HEMTs," *Electron Device Letters, IEEE* , vol.29, no.5, pp.422-425, May 2008

Deterministic Simulation of 3D and Quasi-2D Electron and Hole Systems in SiGe Devices

Christoph Jungemann
Chair of Electromagnetic Theory
RWTH Aachen University
52056 Aachen, Germany
Email: cj@ithe.rwth-aachen.de

Anh-Tuan Pham
Synopsys Inc.
700 E Middlefiled Rd
Mountain View, CA 94043

Sung-Min Hong
Samsung Semic. Inc.
95 West Plumeria Drive
San Jose, CA 95134

Bernd Meinerzhagen
BST
TU Braunschweig
38023 Braunschweig, Germany

Abstract— **We present examples of deterministic solvers for the Boltzmann transport equation for electrons and holes in a 3D and quasi 2D $\vec{k}$-space. Compared to the standard approach, the Monte Carlo method, these deterministic solvers have certain advantages. They yield exact stationary solutions, which, for example, are required for the simulation of the floating body effect in SOI devices. They allow exact small-signal and noise analysis in the whole range of frequencies from 0 to THz. Inclusion of magnetic fields, the Pauli principle or rare events causes no problems. On the other hand, the deterministic solvers are more memory intensive and more difficult to code than the Monte Carlo method.**

I. Introduction

The Monte Carlo (MC) method is the standard approach to solve the Boltzmann transport equation (BTE), which describes the transport of electrons and holes in the semi-classical framework (e.g. [1]). The MC method itself is a numerical approach to integration and its accuracy is inversely proportional to the square root of the CPU time (e.g. [2]). In the case of the BTE the MC solver is inherently transient and the solution contains stochastic noise. If the physical MC approach is used, the stochastic noise is proportional to the physical noise of the carriers (e.g. [3]). The usual MC method for devices is charge conserving and in the case of sufficiently short time steps for the self-consistent solution with the Poisson equation (PE) relatively stable (e.g. [4]). The MC method is well understood and easy to code. Complex microscopic models can be included (e.g. full bands [5]). Due to its many advantages the MC approach is the method of choice for the solution of the BTE and almost universally used.

Problems arise for example, when the MC method is applied to processes, which are rare or evolve on a relatively long time scale. A typical example is the floating body effect in SOI devices, where rare events (impact ionization (II) or tunneling) lead to hysteresis effects in the millisecond range and low-frequency noise [6]. By now no successful simulation of this problem with an MC method has been demonstrated. Self-consistent MC device simulations require time step lengths in the order of femtoseconds to resolve the plasma oscillations [7] and simulations for milliseconds are not feasible. Even simulation times of nanoseconds are not feasible and the RF behavior of transistors at technically relevant frequencies (i.e in the lower GHz range) cannot be simulated by MC [8].

Furthermore, small-signal behavior is difficult to simulate, since no small-signal MC approach for devices is known and double randomization has to be used with its unfavorable stochastic properties. This makes it very difficult to calculate key figures of merit for RF transistors (e.g. cutoff frequency). Even standard stationary MC device simulations might require excessive CPU times [9]. Inclusion of the Pauli principle is possible [10], but requires in devices rather larger particle ensembles and restricts the simulations to very short durations.

The shortcomings of the MC method have let to a search for alternative methods and the most successful one is the deterministic spherical harmonics expansion (SHE), which was used already in the earliest days of solving the BTE (e.g. [11]).

II. Spherical harmonics expansion

The distribution function in the 3D $\vec{k}$-space is expanded with spherical harmonics based on spherical coordinates w.r.t. the angles [12], [13]

$$f(\vec{k},\vec{r}) = f(\varepsilon,\vartheta,\varphi,\vec{r}) \approx \sum_{l=0}^{l_{\max}} \sum_{m=-l}^{m=l} f_{l,m}(\varepsilon,\vec{r}) Y_{l,m}(\vartheta,\varphi) \,,$$

(1)

where $\vec{k}$ is the wave vector, $\vec{r}$ the position in real space, ε the energy, ϑ,φ the spherical angles in the $\vec{k}$-space , $Y_{l,m}$ the spherical harmonics and $l_{\max}$ the maximum order of the SHE. This leads to a projection of the BTE which results in balance equations for the coefficients of the expanded distribution function in energy and real space. Thus, the number of dimensions of the solution domain is reduced by two at the cost of multiple and coupled balance equations. These balance equations can be handled with numerical methods similar to the drift-diffusion model [14], [15] and easily integrated in a standard TCAD framework. The initial problems with numerical stability have been successfully solved (e.g. H-transform [12], maximum entropy dissipation scheme [16], [14], for details see [15]). With this approach stationary self-consistent solutions of the BTE and PE can be calculated for devices, where the final convergence is ensured by the Newton-Raphson method [17], even in the case where the Pauli principle is considered [18]. In addition, exact small-signal and noise analysis is possible directly in the frequency

978-1-4673-1707-8/12 $31.00 © 2012 IEEE

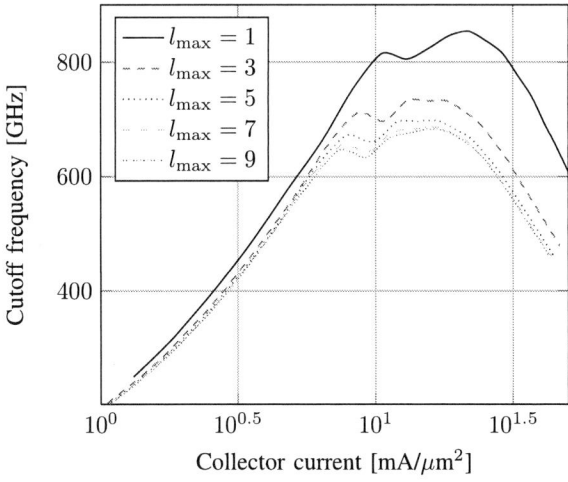

Fig. 1. Cutoff frequency of a 1D SiGe HBT at a collector/emitter bias of 1.0V.

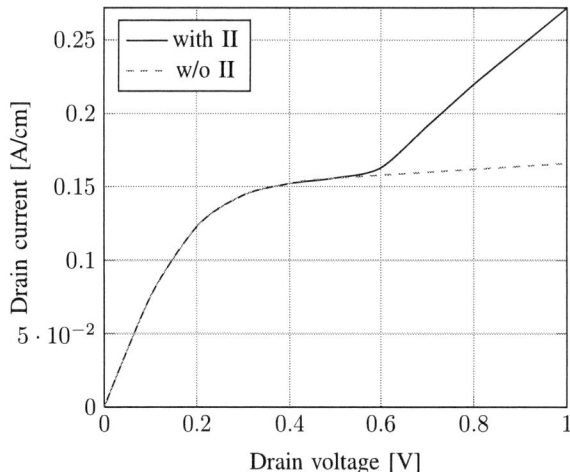

Fig. 2. Drain current of the SOI NMOSFET with and without impact ionization (II) at a gate bias of 1.0V.

domain [19], [15]. Cyclostationary simulations with the harmonic balance approach are possible [20]. Full-band structures can be included with different levels of approximation [21], [22], [15]. Exact inclusion of full-bands is possible [14], if the relation between energy and the modulus of the wave vector is monotonic [23]. Magnetic fields can be included and the small changes due to them can be accurately calculated [24]. Even quantum transport effects can be included [25].

A problem of the SHE approach is the huge memory requirement. About 100 GBytes of memory are required for a DC and small-signal simulation of a device, which is 2D in real space. Although memories of this size are available today, a method has been developed to reduce the memory requirements and a DC simulation of a FinFET, which is 3D in the real space, has been demonstrated recently [26].

Direct numerical solution of the BTE is not only possible in the case of a 3D $\vec{k}$-space, but can also be applied to systems with a lower number of dimensions. In the case of a 2D $\vec{k}$-space instead of the spherical harmonics expansion a Fourier harmonics expansion (FHE) in the polar angle can be used [27]. Otherwise the same numerical methods can be applied as in the case of SHE.

The expansion in (1) is only exact, if an infinite number of spherical harmonics is used. This is not possible in numerical calculations and the expansion has to be terminated at a certain maximum order $l_{\max}$. The first SHE solvers for devices were limited to a first order expansion ($l_{\max} = 1$) to keep the number of solution variables small (e.g. [12]). This restriction can introduce a rather large simulation error, if the transport is quasi-ballistic, and in Ref. [14] a variable order expansion was introduced. In Fig. 1 the cutoff frequency of a SiGe HBT is shown for different values of $l_{\max}$ [15]. The cutoff frequency converges for higher $l_{\max}$ and $l_{\max} = 5$ yields already a good result. On the other hand, $l_{\max} = 1$ results in a large error and the cutoff frequency is overestimated. Thus, if a sufficiently

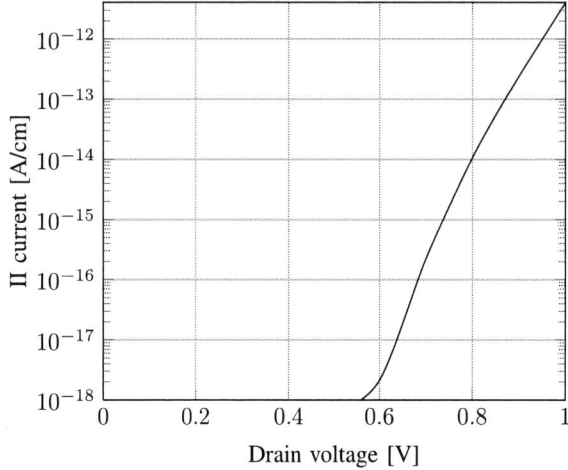

Fig. 3. Hole current generated in the SOI NMOSFET by impact ionization at a gate bias of 1.0V.

high $l_{\max}$ is chosen, the SHE solver can yield an accurate solution of the BTE similar to the MC method. The same holds for FHE [15].

III. SHE SIMULATION OF AN SOI-MOSFET

As mentioned in the introduction, the floating body effect in SOI devices cannot be simulated with the MC method, whereas it is possible with the SHE solver. In Fig. 2 the output characteristics of a partially depleted SOI NMOSFET is shown with and without the inclusion of II in the scattering integral of the BTE [15]. II leads to an accumulation of holes in the floating body of the SOI device and a kink occurs in the output characteristics. This hole generation current is very small and an II current of about 10^{-18}A/cm already modifies the drain current (Fig. 3). This current is more than

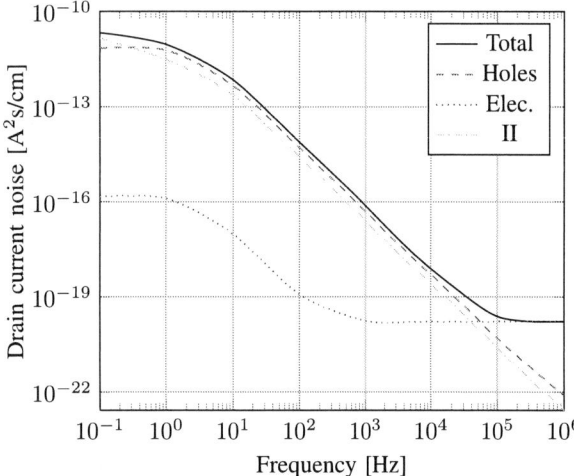

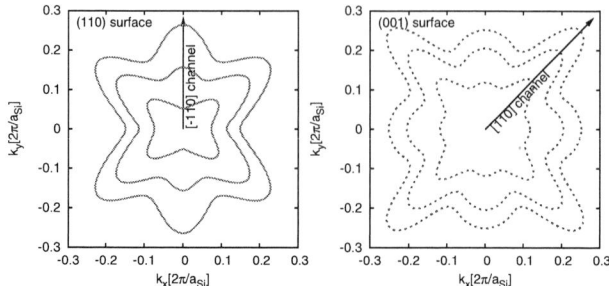

Fig. 5. Equi energy lines in multiples of 100meV of the first heavy-hole subband for the strained SiGe HOI (001) and (110) PMOS structures. For both devices the bottom of the subband is shifted to 0meV.

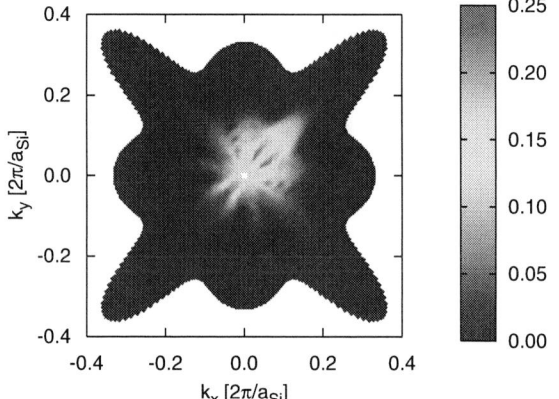

Fig. 6. Distribution function of the first heavy hole subband at the center of the channel in a double gate PMOSFET biased at $V_{DS} = -400$mV and $V_{GS} = -1$V.

Fig. 4. Power spectral density of the drain current of the SOI NMOSFET at a gate and drain bias of 1.0V.

ten orders of magnitude smaller than the drain current. Such a small current is very difficult to simulate by MC. Due to the low current value it takes about milliseconds or more, before a stationary state is reached, and transient solutions are not possible, whereas the stationary SHE solver causes no problems. Since the distribution function depends on energy, the many orders of magnitude involved in this problem are resolved in a numerically robust way and the SHE solver is more stable than a drift-diffusion model for this type of simulation.

Small-signal and noise analysis can be performed directly in the frequency domain for arbitrary frequencies. In Fig. 4 the power spectral density of the drain current noise of the SOI MOSFET is shown [15]. Due to the slow processes involved, the noise depends on frequency even below 1Hz. Such time scales are not accessible by MC. In addition, the SHE approach allows to calculate the contributions of the different noise sources separately (II, hole and electron scattering).

IV. QUASI-2D HOLE GAS

In the inversion layer of PMOSFETs the holes are quantized in the direction perpendicular to the channel. In addition to the BTE, which includes the Pauli principle, and the PE, the Schrödinger equation (SE) is solved, resulting in a quasi-2D hole gas (for details see: [15]). A 6×6 $\vec{k} \cdot \vec{p}$ Hamiltonian is used [28], which yields warped subbands. An example is shown in Fig. 5 for a SiGe heterostructure on insulator (HOI) for two surface orientations [29]. Obviously, the subband structure is not spherical and a full-subband approach is required and feasible by FHE [30], where the solution of the $\vec{k} \cdot \vec{p}$ SE takes much more CPU time than the solution of the BTE or PE and is performed in parallel based on an efficient discretization of the 2D $\vec{k}$-space [31]. Not only anisotropic subbands can be handled, but also the anisotropic distribution function itself, if a sufficient number of Fourier harmonics is considered

in the calculation. In Fig. 6 an example is shown, where the strong anisotropy of the distribution function for a DG PMOSFET with a (001) surface and [110] channel orientation is demonstrated [27].

Due to the complex dependency of the distribution function on the potential via the SE, the formulation of a complete Newton-Raphson approach is rather cumbersome in this case. Instead, the three equations (PE, BTE and SE) are solved under stationary bias conditions by a Gummel loop (Fig. 7). Convergence is achieved, when the change in potential is smaller than $10^{-8} \cdot kT/q$ ($kT/q = 26$mV). Since the deterministic solver for the BTE yields a solution that does not contain any stochastic error in contrast to MC, the Gummel loop converges linearly down to very low values and robust convergence is obtained. This convergence can be improved by an incomplete Newton-Raphson scheme for the BTE and PE [32].

V. CONCLUSION

We have demonstrated that SHE is a viable alternative to the MC method and that it yields accurate solutions of the BTE. SHE works in many cases, where the MC method fails. Its main advantages are that stationary solutions can be obtained directly, that small quantities can be calculated accurately, and that small-signal and noise analysis is possible

978-1-4673-1707-8/12 $31.00 © 2012 IEEE

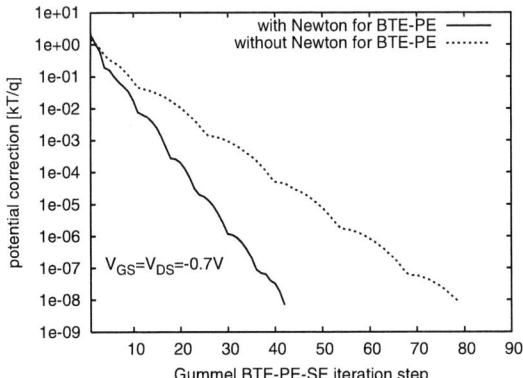

Fig. 7. Convergence of the BTE-PE-SE Gummel loop with and without the BTE-PE incomplete Newton-Raphson method for a PMOSFET biased at $V_{GS} = V_{DS} = -0.7$V.

in the frequency domain. The SHE approach fullfils much better the classical TCAD requirements of high convergence speed, applicability for all bias conditions and availability of all simulation domains (DC, AC, noise). This comes at the price of larger memory requirements, more cumbersome coding, and sometimes simplifications in the physical models (e.g. electron-electron scattering).

REFERENCES

[1] C. Jacoboni and P. Lugli, *The Monte Carlo Method for Semiconductor Device Simulation*. Wien: Springer, 1989.

[2] J. M. Hammarsley and D. C. Handscomb, *Monte Carlo Methods*. London: Methuen/Chapman and Hall, 1964.

[3] L. Varani, L. Reggiani, T. Kuhn, T. González, and D. Pardo, "Microscopic simulation of electronic noise in semiconductor materials and devices," *IEEE Trans. Electron Devices*, vol. 41, no. 11, pp. 1916–1925, 1994.

[4] P. W. Rambo and J. Denavit, "Time stability of Monte Carlo device simulation," *IEEE Trans. Computer–Aided Des.*, vol. 12, pp. 1734–1741, 1993.

[5] J. Bude and R. K. Smith, "Phase-space simplex Monte Carlo for semiconductor transport," *Semicond. Sci. Technol.*, vol. 9, pp. 840–843, 1994.

[6] W. Jin, P. C. Chan, S. K. H. Fung, and P. K. Ko, "Shot-noise-induced excess low-frequency noise in floating-body partially depleted SOI MOSFET's," *IEEE Trans. Electron Devices*, vol. 46, no. 7, pp. 1180–1185, 1999.

[7] R. Hockney and J. Eastwood, *Computer Simulation Using Particles*. Bristol, Philadelphia: Institute of Physics Publishing, 1988.

[8] C. Jungemann, B. Neinhüs, S. Decker, and B. Meinerzhagen, "Hierarchical 2–D DD and HD noise simulations of Si and SiGe devices: Part II—Results," *IEEE Trans. Electron Devices*, vol. 49, no. 7, pp. 1258–1264, 2002.

[9] C. Jungemann and B. Meinerzhagen, "In-advance CPU time analysis for stationary Monte Carlo device simulations," *IEICE Trans. on Electronics*, vol. E86-C, no. 3, pp. 314–319, 2003.

[10] P. Lugli and D. Ferry, "Degeneracy in the Ensemble Monte Carlo Method for High-Field Transport in Semiconductors," *IEEE Trans. Electron Devices*, vol. 32, no. 11, pp. 2431–2334, 1985.

[11] G. A. Baraff, "Maximum anisotropy approximation for calculating electron distributions; Application to high field transport in semiconductors," *Phys. Rev.*, vol. 133, no. 1A, pp. A26–A33, 1964.

[12] A. Gnudi, D. Ventura, G. Baccarani, and F. Odeh, "Two-dimensional MOSFET simulation by means of a multidimensional spherical harmonics expansion of the Boltzmann transport equation," *Solid–State Electron.*, vol. 36, no. 4, pp. 575 – 581, 1993.

[13] N. Goldsman, L. Henrickson, and J. Frey, "A physics-based analytical/numerical solution to the Boltzmann transport equation for use in device simulation," *Solid–State Electron.*, vol. 34, pp. 389–396, 1991.

[14] C. Jungemann, A.-T. Pham, B. Meinerzhagen, C. Ringhofer, and M. Bollhöfer, "Stable discretization of the Boltzmann equation based on spherical harmonics, box integration, and a maximum entropy dissipation principle," *J. Appl. Phys.*, vol. 100, pp. 024502–1–13, 2006.

[15] S.-M. Hong, A. T. Pham, and C. Jungemann, *Deterministic solvers for the Boltzmann transport equation*. Computational Microelectronics, Wien, New York: Springer, 2011.

[16] C. Ringhofer, "Numerical methods for the semiconductor Boltzmann equation based on spherical harmonics expansions and entropy discretizations," *Transport Theory and Statistical Physics*, vol. 31, no. 4-6, pp. 431–452, 2002.

[17] S.-M. Hong and C. Jungemann, "A fully coupled scheme for a Boltzmann-Poisson equation solver based on a spherical harmonics expansion," *J. Computational Electronics*, vol. 8, no. 3, pp. 225–241, 2009.

[18] S.-M. Hong and C. Jungemann, "Inclusion of the Pauli principle in a deterministic Boltzmann equation solver for semiconductor devices," in *Proc. SISPAD*, pp. 135–138, 2010.

[19] C.-K. Lin, N. Goldsman, Z. Han, I. Mayergoyz, S. Yu, M. Stettler, and S. Singh, "Frequency domain analysis of the distribution function by small signal solution of the Boltzmann and Poisson equations," in *Proc. SISPAD*, pp. 39–42, 1999.

[20] C. Jungemann and B. Meinerzhagen, "A frequency domain spherical harmonics solver for the Langevin Boltzmann equation," in *International Conference on Noise in Physical Systems and 1/f Fluctuations, AIP Conf. Proc.*, vol. 780, pp. 777–782, 2005.

[21] M. C. Vecchi and M. Rudan, "Modeling electron and hole transport with full-band structure effects by means of the spherical-harmonics expansion of the BTE," *IEEE Trans. Electron Devices*, vol. 45, no. 1, pp. 230–238, 1998.

[22] S. Jin, S.-M. Hong, and C. Jungemann, "An efficient approach to include full-band effects in deterministic boltzmann equation solver based on high-order spherical harmonics expansion," *IEEE Transactions on Electron Devices*, vol. 58, pp. 1287 –1294, may 2011.

[23] H. Kosina, M. Harrer, P. Vogl, and S. Selberherr, "A Monte Carlo transport model based on spherical harmonics expansion of the valence bands," in *Proc. SISDEP*, pp. 396–399, 1995.

[24] S.-M. Hong and C. Jungemann, "Simulation of magnetotransport in nanoscale devices," in *International Conference on Solid State and Integrated Circuits Technology*, pp. 377–380, 2008.

[25] N. Goldsman, C. Lin, Z. Han, and C. Huang, "Advances in the Spherical Harmonic-Boltzmann-Wigner approach to device simulation," *Superlattices and Microstructures*, vol. 27, pp. 159–175, 2000.

[26] K. Rupp, T. Grasser, and A. Jungel, "On the feasibility of spherical harmonics expansions of the boltzmann transport equation for three-dimensional device geometries," in *IEEE International Electron Devices Meeting (IEDM)*, pp. 34.1.1 –34.1.4, dec. 2011.

[27] A. Pham, C. Jungemann, and B. Meinerzhagen, "Deterministic multisubband device simulations for strained double gate pmosfets including magnetotransport," in *IEEE International Electron Devices Meeting (IEDM)*, pp. 1 –4, dec. 2010.

[28] R. Oberhuber, G. Zandler, and P. Vogl, "Subband structure and mobility of two-dimensional holes in strained Si/SiGe MOSFETs," *Phys. Rev. B*, vol. 58, pp. 9941–9948, 1998.

[29] A. Pham, C. Jungemann, and B. Meinerzhagen, "Comparison of strained sige heterostructure-on-insulator (001) and (110) pmosfets: C - v characteristics, mobility, and on current," in *Proceedings of the European Solid-State Device Research Conference (ESSDERC)*, pp. 230 –233, sept. 2010.

[30] A. Pham, C. Jungemann, and B. Meinerzhagen, "On the numerical aspects of deterministic multisubband device simulations for strained double gate pmosfets," *J. Computational Electronics*, vol. 8, no. 3, pp. 242–266, 2009.

[31] A.-T. Pham, B. Meinerzhagen, and C. Jungemann, "A fast k x p solver for hole inversion layers with an efficient 2d k-space discretization," *J. Computational Electronics*, vol. 7, pp. 99–102, 2008.

[32] A. T. Pham, C. Jungemann, and B. Meinerzhagen, "A convergence enhancement method for deterministic multisubband device simulations of double gate PMOSFETs," in *Proc. SISPAD*, pp. 115–118, 2009.

A Multi-Subband Monte Carlo study of electron transport in strained SiGe n-type FinFETs

D.Lizzit, P.Palestri, D.Esseni, F.Conzatti and L.Selmi

DIEGM, University of Udine, via delle Scienze 208, 33100 Udine, Italy, e-mail: david.esseni@uniud.it

Abstract— **This paper reports a simulation study investigating the drive current in the prototypical SiGe n-type FinFET depicted in Fig.1 and for different values of the Ge content x in the $\text{Si}_{(1-x)}\text{Ge}_x$ active layer. To this purpose we performed strain simulations, band-structure calculations and Multi-Subband Monte Carlo transport simulations accounting for the effects of the Ge content on both the band-structure and scattering rates in the transistor channel. Our results suggest that the largest on-current may be obtained with a simple Si active layer.**

I. INTRODUCTION

Nowadays the main hindrance to the performance growth of VLSI circuits is power consumption and the scaling of the supply voltage, V_{DD}, is the most powerful measure to reduce it. The degradation of dynamic performance for V_{DD} below 1V, however, has become particularly challenging. A promising route to boost the performance at low V_{DD} is the introduction of innovative channel materials, such as SiGe and Ge [1], [2], or III-V semiconductors [3], [4]. In particular, the use of new channel materials in FinFETs or Tri-Gate transistors and the strain engineering for alternative semiconductors represent the frontier of research for advanced CMOS technologies [5], [6].

In this framework, our paper investigates the drive current of the prototypical SiGe n-type FinFET depicted in Fig.1. In such a device structure the $\text{Si}_{(1-x)}\text{Ge}_x$ active layer could be grown on a relaxed $\text{Si}_{0.25}\text{Ge}_{0.75}$ virtual substrate. After etching the SiGe layers in a fin-shape, a thick insulator is formed next to the exposed side walls of the virtual substrate to limit the FET action to the channel layer. Adopting the FinFET architecture improves SCE control, but also has the effect of transforming the initial biaxial strain in the $\text{Si}_{(1-x)}\text{Ge}_x$ into predominantly uniaxial strain due to the small fin width [7], [8] (see also Sec.IV and Fig.5), which in turn has a positive impact on electron transport [9].

In the transistor of Fig.1 the Ge content x has a direct impact on the band-structure of the $\text{Si}_{(1-x)}\text{Ge}_x$ layer (through the percentage in the alloy), and also an indirect influence because the lattice mismatch with the $\text{Si}_{0.25}\text{Ge}_{0.75}$ contributes to determine the strain in the $\text{Si}_{(1-x)}\text{Ge}_x$ layer. Finally the Ge content affects the alloy scattering rates in the active layer [10]. The $\text{Si}_{0.25}\text{Ge}_{0.75}$ substrate considered in this paper has approximately the largest Ge content such that the material is still Si like (the gap reduces rapidly for larger Ge contents, see also Fig.2), and allows one to induce a tensile stress in the $\text{Si}_{(1-x)}\text{Ge}_x$ active layer in a large range of x values. In this sense the $\text{Si}_{0.25}\text{Ge}_{0.75}$ substrate appears a reasonable

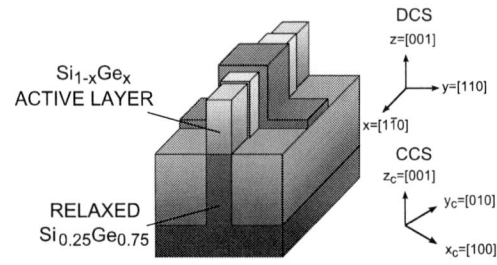

Fig. 1. Device structure and device coordinate system (DCS) for an (110)/[1$\bar{1}$0] FinFET. The correspondence between DCS and the crystal coordinate system (CCS) is also illustrated.

choice from the device design perspective; however the main conclusions of the paper would not change for a $\text{Si}_{(1-y)}\text{Ge}_y$ substrate with y somewhat different from 0.75.

This paper presents strain simulations, band-structure calculations and Multi-Subband Monte Carlo transport simulations that analyze the drive current in the device at study for different values of the Ge content in $\text{Si}_{(1-x)}\text{Ge}_x$ layer.

II. BAND-STRUCTURE CALCULATION AND PARAMETRIZATION

The band-structure of the $\text{Si}_{(1-x)}\text{Ge}_x$ active layer was calculated by employing a 30 bands $\mathbf{k}\cdot\mathbf{p}$ model as that described in [11]. It has been extensively demonstrated that the use of 30 bands in the $\mathbf{k}\cdot\mathbf{p}$ formulation allows to obtain a good agreement with pseudo-potential and DFT-GW calculations throughout the Brillouin zone of Si, SiGe and Ge [11]. Fig.2 illustrates the calculated valence and conduction band edges for relaxed $\text{Si}_{(1-x)}\text{Ge}_x$. Our results agree well with the pseudo-potential calculations of [10] and the experiments from [12]–[14].

We used the 30 bands $\mathbf{k}\cdot\mathbf{p}$ to extract, for the different $\text{Si}_{(1-x)}\text{Ge}_x$ layers, not only the band edges but also the effective longitudinal and transverse masses and the non parabolic corrections for the relevant minima of the conduction band; all these band-structure parameters were then used in Multi-Subband Monte Carlo transport simulations.

III. MULTI-SUBBAND MONTE CARLO METHODOLOGY

The n-type transistors were studied with a Multi-Subband Monte Carlo (MSMC) simulator accounting for the Δ, L and Γ valleys in the conduction band, for arbitrary crystal

978-1-4673-1707-8/12 $31.00 © 2012 IEEE

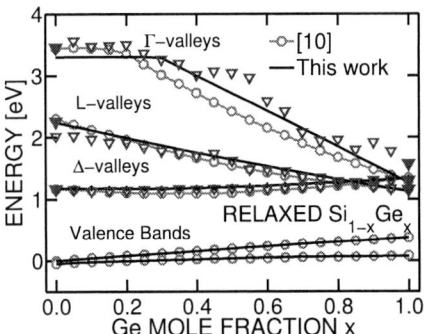

Fig. 2. Valence and conduction band edges for relaxed $Si_{(1-x)}Ge_x$ versus the Ge content x; the zero has been taken as the top of the unstrained silicon valence band. Black solids lines: 30 bands $\mathbf{k \cdot p}$ calculations. Open circles : pseudo-potential calculations [10]. Open triangles are experimental data from [12], [13], combined as explained in [10]. Solid symbols on the sides are the experimental values for the energies at symmetry points in Si and Ge [14].

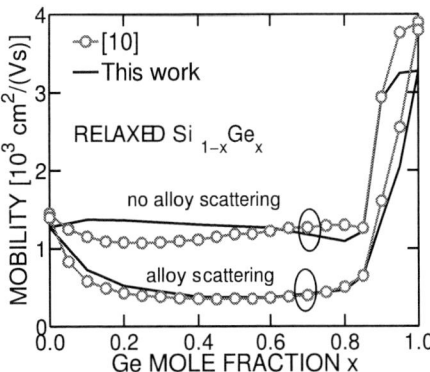

Fig. 3. Mobility in relaxed $Si_{(1-x)}Ge_x$ calculated in this work and compared with the results of [10]. Results obtained either including or neglecting the alloy scattering and plotted versus the Ge content x.

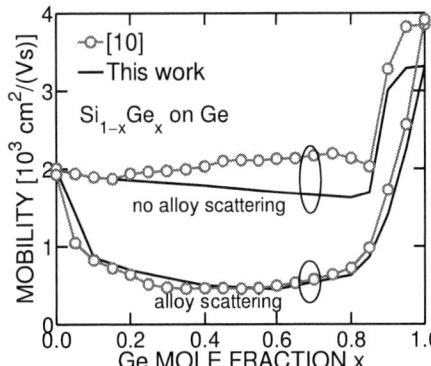

Fig. 4. Mobility in a $Si_{(1-x)}Ge_x$ active layer grown on an underlying Ge substrate calculated in this work and compared with the results of [10]. Results obtained either including or neglecting the alloy scattering and plotted versus the Ge content x.

orientations and for the most important scattering mechanisms [15]. As discussed in more detail below, for the purpose of this work the simulator was extended in order to account for alloy scattering. For any material and strain configuration the MSMC simulator uses the energy minimum and effective masses for each relevant conduction band valley, so that our approach is very flexible for the simulation of strained Si and Ge transistors [16], [17].

Both intra-valley and inter-valley phonon transitions are accounted for in our simulations. The deformation potentials for intra-valley, elastic phonons in Si and Ge are reported in Tab.I. The parameters for the inter-valley phonons in Si and Ge are taken from [18] (and references therein). For a $Si_{(1-x)}Ge_x$ alloy the squared value of the deformation potentials are obtained by linearly interpolating the corresponding values in Si and Ge according to the Ge content.

We verified that for the material and strain configurations considered in this work the contribution to the current of the conduction band minimum at the Γ point is essentially negligible, because both the density of states and the population is small. Thus for most of the results reported in the paper the Γ valley was not included in the calculations.

TABLE I

DEFORMATION POTENTIALS FOR INTRA-VALLEY PHONON SCATTERING IN SI AND GE; FROM [18] AND REFERENCES THEREIN.

Phonon type	Valleys	Def.pot [eV]
Si		
intra-valley	Δ	13
Ge		
intra-valley	Δ	13
intra-valley	Λ	10.83

For the analysis of SiGe FinFETs we extended the simulator used in [15]–[17] to account for the alloy scattering with

a formulation essentially consistent with [19]. In particular, we calculated the squared, unscreened matrix element for the transition between the bands n and n' in the valley ν as

$$|M_{n,n'}^{(\nu)}|^2 = \frac{\Omega_c}{A} \Delta U\, x(1-x) \int |\xi_{\nu,n}(y)|^2\, |\xi_{\nu,n'}(y)|^2\, dy \quad (1)$$

where y is the quantization direction normal to the transport plane, $\xi_{\nu,n}(y)$ is the envelope wave-function for the subband (ν,n), Ω_c is the volume of the unit cell and A is a normalization area. ΔU is the difference between the atoms average energies in the SiGe alloy.

For the alloy and the surface roughness scattering we accounted for the screening effect produced by the electrons in the inversion layer by employing the tensorial dielectric function, because the approach based on the scalar dielectric function is incorrect for double-gate MOSFETs or FinFETs, and in fact leads to artifacts in the simulation results [20]. The simulations of all the devices in this work were obtained by using a Gaussian spectrum of the surface roughness with a r.m.s. value Δ=0.62nm and a correlation length L_c=1.0nm, neglecting the still debated effect of strain on SR scattering.

The numerical value of ΔU was extracted as a fitting parameter to reproduce the mobility simulations for bulk $Si_{(1-x)}Ge_x$ systems reported in [10], which in turn were shown to be consistent with the available experimental data.

In this respect, Fig.3 reports the simulated mobility for relaxed $Si_{(1-x)}Ge_x$, while Fig.4 shows mobility for a strained $Si_{(1-x)}Ge_x$ grown on a Ge substrate. In these simulations the transport in a bulk $Si_{(1-x)}Ge_x$ material is approximated with our MSMC calculations by using a 50nm squared potential well. In such a structure the subband quantization effects are negligible, so that the system emulates a bulk $Si_{(1-x)}Ge_x$ material. Furthermore, for the simulations of Figs.3 and 4 the deformation potentials for intra-valley acoustic phonons were set to the values reported for the bulk materials, that is 9eV for the Δ subbands in Si, and 9eV and 7.5eV respectively for Δ and L subbands in Ge (see [18] and references therein).

For all the material configurations a fairly good agreement with the results of [10] was obtained by using a single $\Delta U = 1.3 eV$ value for both Δ- and L-valleys. It is worth noting that the agreement with [10] is good also when the alloy scattering is neglected in both simulation sets.

The inclusion of alloy scattering results in the well known degradation of mobility for intermediate x values, which is due to the fact that the scattering matrix elements are maximum for $x=0.5$, as it can be seen in Eq.1.

IV. DEVICES AND SIMULATION RESULTS

We designed the FinFET structures for a gate length $L_G = 14nm$, which is the L_G projected by the ITRS Roadmap for year 2017 [21], and used an equivalent oxide thickness of 0.7nm and a fin thickness $W_{fin} = 7nm$. The transport analysis is focussed on the lateral, $(110)/[1\bar{1}0]$ interfaces of the FinFETs (see Fig.1), simulated as double-gate SOI transistors. With a fin width $W_{fin} = 7nm$ this approximation seems reasonable for a fin height of 30nm or more; furthermore the main conclusions of our paper concerning the drive current for different Ge contents are not expected to be affected by such a simplified description of the device structure.

A simplified process was simulated using Sentaurus-Process, including epitaxial growth of the $Si_{(1-x)}Ge_x$ channel on the virtual substrate, fin etch and SiO_2 fill next to the sidewalls of the virtual substrate. Average values for the strain components ε_{xx}, ε_{yy}, ε_{zz} in the DCS were obtained by averaging in the fin height direction (see Fig.1). Fig.5 shows the simulated, average strain components versus the Ge mole fraction. As can be seen the component ε_{yy} in the fin width direction is always negligible and for small mole fractions we observe a predominantly uniaxial strain in the transport direction, consistently with [7], [8].

Then the ϵ_{xx}, ϵ_{yy}, ϵ_{zz} were converted to the strain components in the crystal coordinate system by using the appropriate linear transformations [18], and finally used in the 30 bands $\mathbf{k} \cdot \mathbf{p}$ model for the calculation of the band-structure. The energy minima of the Δ and Λ valleys and the corresponding effective masses were then extracted and used in the Multi-Subband Monte Carlo simulations.

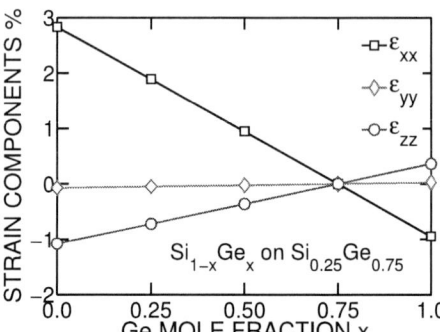

Fig. 5. Simulated strain components in the device coordinate system, DCS, for the FinFETs of Fig.1 versus the Ge content x. The strain component ε_{yy} in the fin width direction is essentially negligible with respect to ε_{xx} and ε_{zz}.

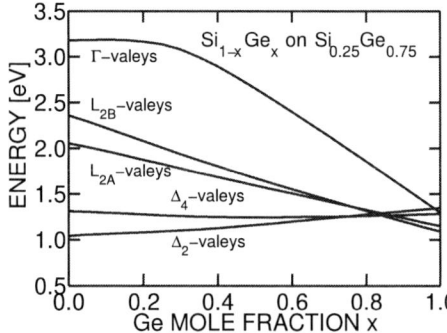

Fig. 6. Conduction band edges for the FinFETs of Fig.1 versus the Ge content x, and thus corresponding to the strain components of Fig.5; the zero has been taken as the maximum of the valence band of unstrained silicon. The strain results in a splitting of both the Δ and the L valleys. The Δ_2 valleys correspond to $\Delta_{[001]}$ and Δ_4 to $\Delta_{[100]}$ and $\Delta_{[010]}$ of the bulk crystal, while the L_{2A} valleys correspond to $L_{[111]}$, $L_{[11\bar{1}]}$ and L_{2B} to $L_{[1\bar{1}1]}$, $L_{[1\bar{1}\bar{1}]}$.

Fig.6 shows the band edges for the Δ, L and Γ conduction band minima versus the Ge content x. As can be seen, for all x values different from 0.75 the strain induced by the lattice mismatch with the $Si_{0.25}Ge_{0.75}$ layer results in a splitting of the L and Δ valleys, which in turn results in different valleys in the inversion layer. In particular, the valleys Δ_2 and Δ_4 in Fig.6 are named consistently with Fig.7 that illustrates the valleys in the inversion layer.

Fig.7 reports the effective transport mass m_x for the Δ_2 and Δ_4 valleys. For small x values the tensile shear strain reduces the transport mass for the Δ_2 valleys. Since the lowering of the Δ_2 valley reported in Fig.6 tends to repopulate the Δ_2 compared to the Δ_4 valleys, the overall effect of the strain is beneficial for the mobility and the drive current of the FinFETs [16].

This is confirmed by the I_{DS} versus V_{GS} characteristics reported in Fig.8 for different Ge content in the $Si_{(1-x)}Ge_x$ active layer and for a relatively small $V_{DS} = 0.5V$ (consistent with our focus on very low V_{DD} applications). We here assumed that the work-function of all the devices can be adjusted to obtain the same off-current of $100nA/\mu m$ [21].

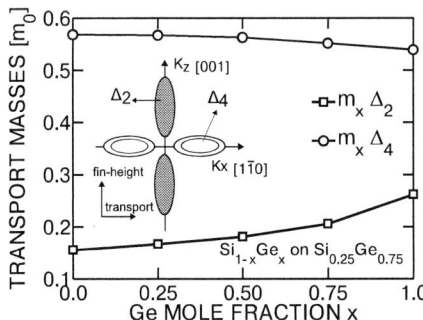

Fig. 7. Transport effective mass versus the Ge content x for the Δ_2 and Δ_4 valleys. For small x values the shear strain reduces the transport mass for the Δ_2 valleys. The inset shows a sketch of the equienergy curves for the Δ_2 and Δ_4 valleys in a (110) inversion layer corresponding to the lateral interfaces of the (110)/[1$\bar{1}$0] FinFETs in Fig.1.

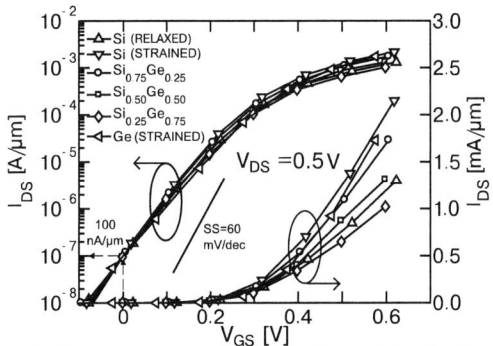

Fig. 8. Drain current versus gate voltage characteristics for $Si_{(1-x)}Ge_x$ over $Si_{0.25}Ge_{0.75}$ FinFETs and for different Ge content x. The corresponding curve for a relaxed silicon device is also shown. L_G=14nm.

As can be seen the I_{DS} for small Ge content is much better than in a *relaxed* Si channel, which clearly illustrates the strain induced I_{DS} enhancement. Furthermore the I_{DS} is reduced by increasing the Ge content because the beneficial effect of the strain is progressively lost. Then the current increases when we move from $Si_{(1-x)}Ge_x$ with x=0.75 to pure Ge because the L valleys become dominant in the transport (see Fig.6), which have more favorable transport masses. The Ge active layer, however, is subject to a compressive strain (see Fig.5), which is unfavorable for the transport [2], so that not even the pure Ge has an I_{DS} larger than the strained Si transistor (i.e. the device with x=0).

The results about the on current I_{on} are best illustrated by Fig.9, reporting I_{on} versus the Ge content and for a fixed I_{off}=100nA/μm; also shown are the I_{on} for x=0.25, 0.5 and 0.75 calculated by switching off the alloy scattering and the I_{on} for relaxed Si. As can be seen the I_{on} reduction with increasing x in the $Si_{(1-x)}Ge_x$ active layer is mainly due to the resulting reduction of the strain. In fact the trend is clearly the same when alloy scattering is switched off, namely the I_{on} is significantly deteriorated for increasing x with respect to the I_{on} for x=0, which is in turn remarkably larger than the I_{on} for relaxed Si.

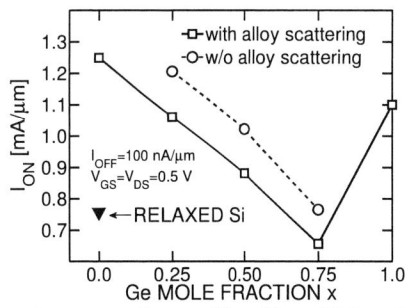

Fig. 9. On current I_{on} at $V_{GS}=V_{DS}$=0.5V and for a fixed I_{off}=100nA/μm versus the Ge mole fraction in the active layer. The simulation results obtained by switching off the alloy scattering are also shown for x=0.25, 0.5 and 0.75. The I_{on} for relaxed Si is also reported for reference.

V. CONCLUSIONS

This paper reported strain simulations, band-structure calculations and Multi-Subband Monte Carlo transport simulations that studied the drive current in the prototypical SiGe n-type FinFET depicted in Fig.1 for different values of the Ge content x in the $Si_{(1-x)}Ge_x$ active layer. The overall effect of x on the band-structure, the strain configuration and the intensity of the alloy scattering suggests that a simple Si active layer may outperform in terms of drive current a $Si_{(1-x)}Ge_x$ or even a pure Ge active layer.

Acknowledgments This work was supported in part by TSMC and by the MIUR FIRB RBFR 10XQZ8 project. Authors would like to acknowledge Gerben Doornbos and Edward Chen for valuable technical discussion. The authors would also like to thank to D.Rideau for useful discussions concerning band structure calculations with the 30 bands $\mathbf{k}\cdot\mathbf{p}$ method.

REFERENCES

[1] M.Kobayashi *et al.* IEEE TED, Vol.57, pp.1037, 2010.
[2] F.Conzatti *et al.* IEDM Proceedings., pp.363-366, 2010.
[3] G.Doornbos *et al.* IEEE Electron Device Letters, 31, pp.1110, 2010.
[4] J.Del Alamo, Nature, Vol.479, pp.317, 2011.
[5] M.Heyns *et al.*, IEDM, pp.299, 2011.
[6] M.Radosavljevic *et al.*, IEDM Proceedings, pp.765-768, 2011.
[7] G.Eneman *et al.*, VLSI Proceedings, pp.41-42, 2010.
[8] T.Mizuno *et al.*, IEDM Proceedings, pp.453-456, 2006.
[9] P.Hashemi *et al.*, IEDM Proceedings, pp.865-868, 2008.
[10] M.V.Fischetti and S.Laux Journal Appl. Phys., Vol.80, pp.2234, 1996.
[11] D.Rideau *et al.* Journal Appl. Phys., Vol.74, pp.195208, 2006.
[12] J.F.Morar *et al.*, J. Vac. Sci. Technol. B, Vol.10, pp.2022-2025, 1992.
[13] J.Weber *et al.*, Phys.Rev.B, Vol.40, pp.5683-93, 1989.
[14] J.R.Chelikowsky *et al.*, Phys.Rev.B, Vol.14, pp.556-582, 1976.
[15] L.Lucci *et al.* IEEE TED, Vol.54, pp.1156, May 2007.
[16] N.Serra and D.Esseni *et al.* IEEE TED, Vol.57, pp.482, 2010.
[17] F.Conzatti *et al.* IEEE TED, Vol.58, pp.1583, 2010.
[18] D.Esseni *et al.*, "Nanoscale MOS Transistors: Semi-Classical Transport and Applications", Cambridge, U.K.: Cambridge Univ. Press, 2011.
[19] A-T.Pham IEEE TED, Vol.54, pp.2174, 2007.
[20] P.Toniutti *et al.* IEEE TED, Vol.57, pp.3074, 2010.
[21] International Technology Roadmap for Semiconductors, 2011 Edition (http://www.itrs.net/Links/2011ITRS/Home2011.htm).

Electron Transport in Germanium Junctionless Nanowire Transistors

Pedram Razavi, Giorgos Fagas, Isabelle Ferain, Ran Yu, Samaresh Das
Tyndall National Institute, University College Cork, Cork, Ireland
e-mail: pedram.razavi@tyndall.ie

Abstract— The impact of channel dimension and crystallographic orientations on ballistic transport of germanium junctionless nanowire transistors is investigated using 3D quantum-mechanical simulations and is compared with those of the inversion-mode nanowire transistors. Analysis of the contribution of source-to-drain tunneling to the off-current shows suppressed direct tunneling in the junctionless design compared to the inversion mode devices.

I. INTRODUCTION

Due to the rapid shrinking of MOSFETs, semiconductor industries face drastic challenges in the formation of source and drain junctions in short-channel devices. Junctionless nanowire transistors do not need the formation of extremely abrupt junctions and can be promising candidates for future CMOS devices [1]. Furthermore, Ge as a high mobility material has been reported to be a promising candidate as channel material for future devices [2, 3]. Some studies on Si junctionless nanowire transistors can be found in the literature [4-15], but there are no 3D quantum mechanical studies of the Ge junctionless nanotransistors yet. In this work, we study ballistic transport of n-type Ge JNTs with different cross-section dimensions and different crystallographic orientations and compare them with those of inversion-mode (IM) nanowire transistors using an in-house 3D quantum-mechanical simulator based on the Non-Equilibrium Green's Function (NEGF) formalism and within the framework of effective-mass theory [16-18]. The source-to-drain tunneling in Ge JNTs is also studied and compared with those of IM devices. A brief overview of the simulation methodology is presented in the next section and followed by the presentation of the device structure and simulation parameters and results.

II. SIMULATION METHODOLOGY

For our simulations, we solve self-consistently the three-dimensional (3D) Poisson and Schrödinger equations with open boundary conditions. The 3D full stationary Schrödinger equation is given by:

$$H_{3D}\Psi(x,y,z) = E\Psi(x,y,z) \qquad (1)$$

where H_{3D} is the 3D device Hamiltonian, E is the energy and $\Psi(x,y,z)$ is the 3D wavefunction. Due to misalignment of the iso-energy surfaces of the conduction bands with the device coordinate system in arbitrarily oriented wires, the inverse effective-mass tensor has non-diagonal terms. Assuming an ellipsoidal parabolic energy band structure the H_{3D} is defined as:

$$H_{3D} = -\frac{\hbar^2}{2}\left(\frac{1}{m_{xx}}\frac{\partial^2}{\partial x^2} + \frac{1}{m_{yy}}\frac{\partial^2}{\partial y^2} + \frac{1}{m_{zz}}\frac{\partial^2}{\partial z^2} + \frac{2}{m_{xy}}\frac{\partial^2}{\partial x \partial y} + \frac{2}{m_{yz}}\frac{\partial^2}{\partial y \partial z} + \frac{2}{m_{xz}}\frac{\partial^2}{\partial x \partial z}\right) + V(x,y,z) \qquad (2)$$

where $1/m_{ij}$ is the reciprocal effective mass tensor (EMT) in the device coordinate system, and $V(x, y, z)$ is the potential energy. To obtain the electrostatic potential, we solve the 3D Poisson equation using COMSOL Multiphysics [19].

To solve the 3D Schrödinger, we use the mode space approach [17] which treats quantum confinement and transport separately. As a result the 3D Schrodinger equation is divided into: (i) a 2D Schrodinger equation solved for cross-section planes of the device to calculate the sub-band energies and the corresponding wave functions and (ii) a 1D transport equation solved using the NEGF formalism to obtain the electron density. The carrier transport is assumed to be ballistic. Using the method described in [20], energies associated with the 2D cross-section planes perpendicular to the transport (x) direction decouple from those associated with the longitudinal direction and the terms with m_{xy} and m_{xz} are cancelled. This respectively yields:

$$\left\{-\frac{\hbar^2}{2}\left(\frac{1}{m_{yy}}\frac{\partial^2}{\partial y^2} + \frac{1}{m_{zz}}\frac{\partial^2}{\partial z^2} + \frac{2}{m_{yz}}\frac{\partial^2}{\partial y \partial z}\right) + [V(x_i,y,z) - E_{sub}^n(x_i)]\right\}\xi^n(x_i,y,z) = 0 \qquad (3)$$

and

$$\left[-\frac{\hbar^2}{2m_{xx}}\frac{\partial^2}{\partial x^2} + E_{sub}^n(x) - E\right]\varphi^n(x) = 0 \qquad (4)$$

978-1-4673-1707-8/12 $31.00 © 2012 IEEE

for the 2D equation in the i-th plane and the 1D equation of the n-th mode along the channel; where, E_{sub}^n is the subband energy level, $\xi^n(x_i, y, z)$ is the corresponding transversal wave function at each slice $x=x_i$ and $\varphi^n(x)$ is the longitudinal component. The retarded Green's function of the active device is calculated using:

$$G = [EI - H_{1D} - \Sigma_1 - \Sigma_2]^{-1} \qquad (5)$$

where I is the identity matrix, $H_{1D} = -\frac{\hbar^2}{2m_{xx}}\frac{\partial^2}{\partial x^2} + E_{sub}^n(x)$ and Σ_1 and Σ_2 are the self-energy functions and account for the open boundary conditions [21]. Finally the electron density and current are obtained from G using the NEGF formalism [16, 17, 20]

III. DEVICE STRUCTURE AND PARAMETERS

Figs. 1 (a) and (b) show a schematic view of a gate-all-around junctionless and inversion-mode Ge nanowire as well as the doping profile in the longitudinal cross-sections of both junctionless and IM devices. The square cross-section dimensions of the Ge wire are $T_{Ge} \times T_{Ge}$ where T_{Ge} ranges from 7 nm down to 3 nm and the gate length is assumed to be 15 nm for all devices. The equivalent oxide thickness is 1 nm for all devices. Uniform doping concentration throughout the devices has been used. The doping concentration in JNTs is 2×10^{19} cm^{-3}. For the IM nanotransistors, the doping concentrations of the source/drain reservoir is 1×10^{20} cm^{-3} (n-type) and the channel is 1×10^{15} cm^{-3} (p-type), respectively. Abrupt junctions are considered for IM devices. These are typical doping concentrations in fabricated devices. The supply voltage (V_{dd}) is equal to 0.65 V and by tuning the gate workfunction all transistors are designed to have the same off-current of 10 pA/μm which is suitable for low standby power technologies [22]. The band alignment of the direct and indirect gaps of Ge at room temperature is shown in Fig. 2. In bulk semiconductor devices, valleys which are lower in energy have the largest contribution to transport. In bulk germanium the Λ-valleys are energetically lower than the other valleys and, as a result, most of the electron transport in the conduction band is through these valleys. However, in small dimension nanowires, quantum confinement becomes very important and effective masses in the cross-section plane perpendicular to the transport direction play an important role in determining the valleys that form the energetically lowest subbands. Table 1 shows the effective masses used in the simulations of Ge nanowires with different crystal orientations. In this study we consider the six-fold degenerate Δ-valleys, the four-fold degenerate Λ-valleys, and the non-degenerate Γ-valley to study their respective contributions to the electron transport as a function of the dimensions and crystallographic orientations.

IV. RESULTS AND DISCUSSION

Figure 3 shows the current-voltage characteristics of JNTs with <100>- and <110>-oriented wires fabricated on (010)-

wafer. As it can be seen in this figure, by decreasing the cross-section the subthreshold swing (SS) improves; from 65 mV/dec to 60 mV/dec and 66 mV/dec to 61 mV/dec for <100>- and <110>-channels, respectively. Since the gate length is constant in all devices, this is due to the fact that by decreasing the cross-section of the devices, the electrostatic control of the gate over the channel carriers increases and results in a better subthreshold swing.

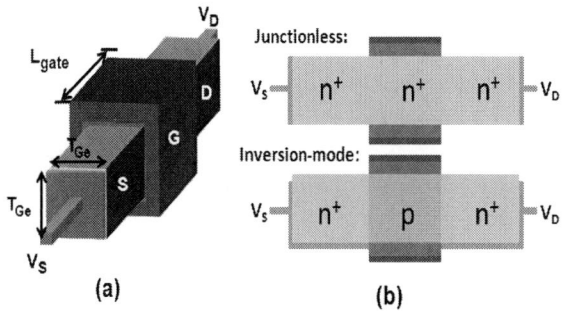

Figure. 1. (a) 3D schematic of JNT and IM device, (b) Longitudinal cross-section of JNT and IM along the gate.

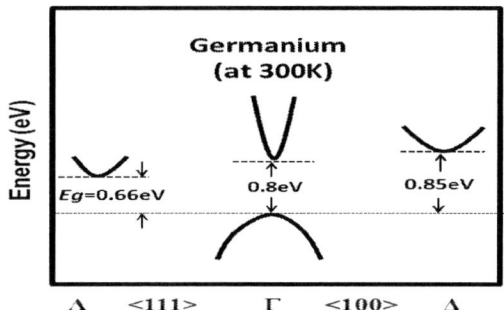

Figure. 2. The direct and indirect gap values used in the simulations of Ge nanotransistors.

Table. 1. Effective masses and subband degeneracy for <100>- and <110>- oriented Ge nanowires with (010) wafer orientation. The Γ-valley in Ge is non-degenerate and has an isotropic effective mass (0.038×m_0 where m_0 is the free electron mass).

wire	valley	m_{yy}	m_{zz}	m_{yz}	m_{xx}	Deg
<100>	Δ	0.2	0.2	inf	0.95	2
		0.95	0.2	inf	0.2	2
		0.2	0.95	inf	0.2	2
	Λ	0.12	0.12	-0.259	0.601	2
		0.12	0.12	0.259	0.601	2
<110>	Δ	0.95	0.2	inf	0.2	2
		0.33	0.2	inf	0.575	2
		0.2	0.33	inf	0.575	2
	Λ	0.12	0.224	-0.444	0.082	1
		0.12	0.224	0.444	0.082	1
		0.12	0.082	inf	1.12	2

978-1-4673-1707-8/12 $31.00 © 2012 IEEE

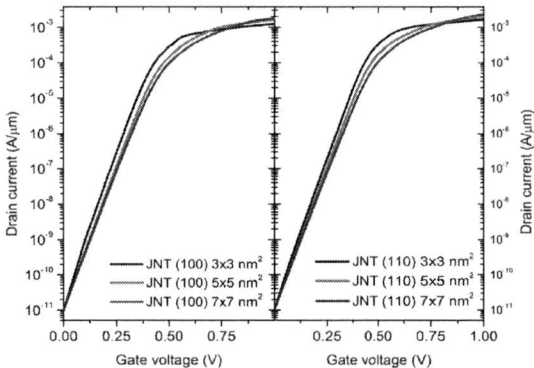

Figure. 3. I_d-V_g characteristics of JNTs with different wire orientations and cross-sections (V_{ds}=0.65V).

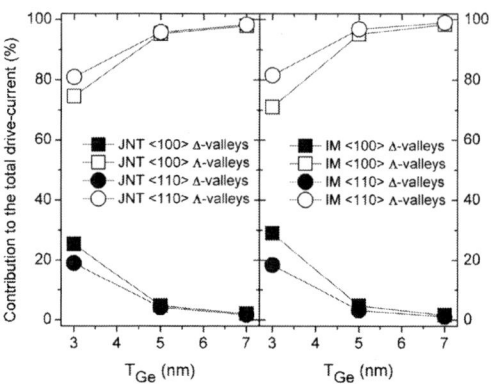

Figure. 4. Contribution of the various valleys to the total current for JNT and IM devices (drive-current extracted at V_{gs}=V_{ds}=0.65V).

Fig. 4 shows the contribution of different electron valleys to the total current. Since the contribution of the non-degenerate Γ-valley in total current is small due to a small effective mass in the transport direction, it is not presented in this figure. It is observed that by decreasing the cross-section the contribution of the Δ-valleys increases while the contribution of the Λ-valleys decreases. This is due to uplift of Λ-valleys owing to the smaller transverse effective masses compared to the Δ-valleys. Fig. 5 shows the transmission coefficient as a function of energy in Ge JNTs. It can be seen that with decreasing cross-section the Δ-valleys subbands shift lower in energy, thereby, contributing more to the total current. For a larger cross-section (7nm) the Δ-valley subbands are above the source Fermi level and their contribution to the total current vanishes. Notably, the subband separation of subbands is smaller for Δ-valleys which may result in more Δ-valley subbands contributing to the transport of small-thickness devices.

Fig. 6 illustrates the contribution of the source-to-drain tunneling to the off-current in JNTs and IM devices for different wire orientation and cross sections. As it can be seen in this figure, the tunneling current in JNTs and IM devices with <110>-wires is much larger than devices with <100>-wires. This is due to the small transport effective mass ($0.082 \times m_0$) of the Λ-valleys, which carry the largest contribution to the total current, compared to the much larger transport effective mass ($0.601 \times m_0$) of <100>-oriented Ge nanowires. The source-to-drain tunneling is smaller in JNTs compared to the IM devices. This is due the fact that JNTs do not have any junctions yielding different subbands profile than IM devices. These are obtained from the 3D quantum mechanical simulations and are shown in Fig. 7. A lower barrier for JNTs and extension of the potential barrier from the sides of the physical gate electrode into the source and drain regions produces a longer effective gate length than the physical gate length when the device is turned off and suppresses the tunneling component allowing for a dominant thermionic component.

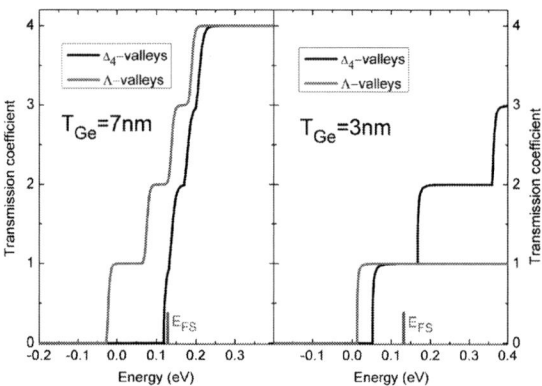

Figure. 5. Transmission contributions of the various valleys in Ge nanowire transistors with decreasing cross-section.The green line marks the source Fermi level. (<100>-wire, V_{gs}=V_{ds}=0.65V) ($Δ_4$-valleys are the four equivalent ellipsoids perpendicular to the transport direction and has more contribution to the total current than $Δ_2$-valleys)

V. CONCLUSION

The current characteristics, subthreshold swing, source-to-drain tunneling of Ge JNTs were investigated in the ballistic transport regime using 3D quantum mechanical simulations within the framework of effective-mass. We found that the electrical characteristics of the Ge JNTs and IM nanowire transistors strongly depend on the crystallographic orientation and the channel dimension. <100>-oriented nanowires yield a better performance and the contribution of Δ-valleys to the total current cannot be ignored in Ge nanowires with cross-sections smaller than 5nm. We also found that the source-to-drain tunneling in Ge JNTs is less than IM devices due to the longer effective gate length of JNTs compared to the IM devices.

978-1-4673-1707-8/12 $31.00 © 2012 IEEE

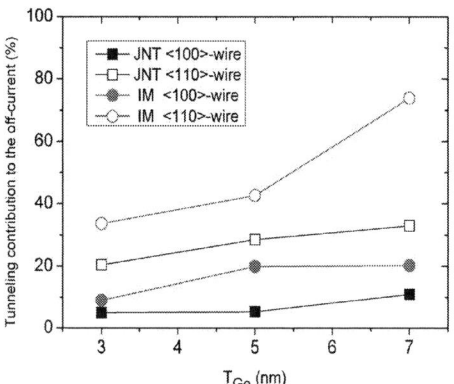

Figure. 6. Contribution of the tunneling components to the off-current (V_{gs}=0V, V_{ds}=0.65V).

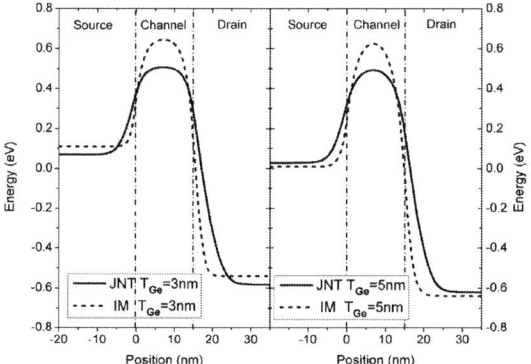

Figure. 7. Lowest-subband profiles of Λ-valleys in germanium JNTs and IM devices with <100>-wires fabricated on (010)-wafer which have the most contribution to the off-current (V_{gs}=0V, V_{ds}=0.65V).

ACKNOWLEDGMENT

This work was supported by the Science Foundation Ireland grants 05/IN/I888 and 10/IN.1/I2992 and the European project SQWIRE under Grant Agreement No. 257111. This work has also been enabled by the Programme for Research in Third-Level Institutions.

REFERENCES

[1] J.-P. Colinge, C.-W. Lee, A. Afzalian, N. D. Akhavan, R. Yan, I. Ferain, P. Razavi, B. O'Neill, A. Blake, M. White, A.-M. Kelleher, B. McCarthy, and R. Murphy, "Nanowire transistors without junctions," *Nat Nano,* vol. 5, pp. 225-229, 2010.

[2] K. Saraswat, C. O. Chui, T. Krishnamohan, D. Kim, A. Nayfeh, and A. Pethe, "High performance germanium MOSFETs," *Materials Science and Engineering: B,* vol. 135, pp. 242-249, 2006.

[3] K. C. Saraswat, C. O. Chui, T. Krishnamohan, A. Nayfeh, and P. McIntyre, "Ge based high performance nanoscale MOSFETs," *Microelectronic Engineering,* vol. 80, pp. 15-21, 2005.

[4] J.-P. Colinge, I. Ferain, A. Kranti, C.-W. Lee, N. D. Akhavan, P. Razavi, R. Yan, and R. Yu, "Junctionless Nanowire Transistor: Complementary Metal-Oxide-Semiconductor Without Junctions," *Science of Advanced Materials,* vol. 3, pp. 477-482, 2011.

[5] B. Sorée, W. Magnus, and G. Pourtois, "Analytical and self-consistent quantum mechanical model for a surrounding gate MOS nanowire operated in JFET mode," *Journal of Computational Electronics,* vol. 7, pp. 380-383, 2008.

[6] L. Ansari, B. Feldman, G. Fagas, J.-P. Colinge, and J. C. Greer, "Simulation of junctionless Si nanowire transistors with 3 nm gate length," *Applied Physics Letters,* vol. 97, pp. 062105-3, 2010.

[7] E. Gnani, A. Gnudi, S. Reggiani, G. Baccarani, N. Shen, N. Singh, G. Q. Lo, and D. L. Kwong, "Numerical investigation on the junctionless nanowire FET," in *Ultimate Integration on Silicon (ULIS), 2011 12th International Conference on,* 2011, pp. 1-4.

[8] J.-P. Colinge, A. Kranti, R. Yan, I. Ferain, N. D. Akhavan, P. Razavi, C.-W. Lee, R. Yu, and C. Colinge, "A Simulation Comparison between Junctionless and Inversion-Mode MuGFETs," *ECS Transactions,* vol. 35, pp. 63-72, 2011.

[9] C.-W. Lee, I. Ferain, A. Afzalian, R. Yan, N. D. Akhavan, P. Razavi, and J.-P. Colinge, "Performance estimation of junctionless multigate transistors," *Solid-State Electronics,* vol. 54, pp. 97-103, 2010.

[10] J.-P. Colinge, C.-W. Lee, I. Ferain, N. D. Akhavan, R. Yan, P. Razavi, R. Yu, A. N. Nazarov, and R. T. Doria, "Reduced electric field in junctionless transistors," *Applied Physics Letters,* vol. 96, pp. 073510-3, 2010.

[11] P. Razavi, N. D-Akhavan, R. Yu, G. Fagas, I. Ferain, and J.-P. Colinge, "Investigation of short-channel effects in junctionless nanowire transistors," presented at the proc. int. conf. Solid States Devices and Materials (SSDM), 2011.

[12] N. D. Akhavan, I. Ferain, P. Razavi, R. Yu, and J.-P. Colinge, "Improvement of carrier ballisticity in junctionless nanowire transistors," *Applied Physics Letters,* vol. 98, pp. 103510-3, 2011.

[13] P. Razavi, G. Fagas, I. Ferain, N. D. Akhavan, R. Yu, and J. P. Colinge, "Performance investigation of short-channel junctionless multigate transistors," in *Ultimate Integration on Silicon (ULIS), 2011 12th International Conference on,* 2011, pp. 1-3.

[14] C.-W. Lee, A. Borne, I. Ferain, A. Afzalian, R. Yan, N. Dehdashti Akhavan, P. Razavi, and J. P. Colinge, "High-Temperature Performance of Silicon Junctionless MOSFETs," *Electron Devices, IEEE Transactions on,* vol. 57, pp. 620-625, 2010.

[15] J.-P. Raskin, J.-P. Colinge, I. Ferain, A. Kranti, C.-W. Lee, N. D. Akhavan, R. Yan, P. Razavi, and R. Yu, "Mobility improvement in nanowire junctionless transistors by uniaxial strain," *Applied Physics Letters,* vol. 97, pp. 042114-3, 2010.

[16] S. Datta, "Nanoscale device modeling: the Green's function method," *Superlattices and Microstructures,* vol. 28, pp. 253-278, 2000.

[17] J. Wang, E. Polizzi, and M. Lundstrom, "A three-dimensional quantum simulation of silicon nanowire transistors with the effective-mass approximation," *Journal of Applied Physics,* vol. 96, pp. 2192-2203, 2004.

[18] A. Afzalian, N. Akhavan, C.-W. Lee, R. Yan, I. Ferain, P. Razavi, and J.-P. Colinge, "A new F(ast)-CMS NEGF algorithm for efficient 3D simulations of switching characteristics enhancement in constricted tunnel barrier silicon nanowire MuGFETs," *Journal of Computational Electronics,* vol. 8, pp. 287-306, 2009.

[19] http://comsol.com.

[20] M. Bescond, N. Cavassilas, and M. Lannoo, "Effective-mass approach for n-type semiconductor nanowire MOSFETs arbitrarily oriented," *Nanotechnology,* vol. 18, p. 255201, 2007.

[21] S. Datta, "Quantum Transport: Atom to Transistor," 2005.

[22] http://public.itrs.net/.

978-1-4673-1707-8/12 $31.00 © 2012 IEEE

Low-Frequency Noise Assessment of the Transport Mechanisms in SiGe Channel Bulk FinFETs

T. Romeo, L. Pantisano, E. Simoen, R. Krom, M.Togo,
N. Horiguchi, J. Mitard, A.Thean,
G. Groeseneken[a] and C. Claeys[a]

Imec
Kapeldreef 75, B-3001 Leuven, Belgium
[a]Also at EE Department, KU Leuven

F. Crupi
University of Calabria
Arcavacata di Rende (CS), Italy

Abstract—**This paper discusses the low-frequency noise behavior of SiGe-channel bulk FinFETs processed on (100) and (110) Si wafers. A comparison is also made with planar SiGe-channel pMOSFETs. It is shown that for devices with carriers confined in the quantum well, only 1/f noise is observed, dominated by mobility fluctuations. Surprisingly, SiGe pMOSFETs fabricated on (110) Si wafers exhibit the highest mobility but also the highest 1/f noise, corresponding with trapping/detrapping. This is also consistent with the density of interface traps extracted from charge pumping measurements.**

I. IMPORTANCE OF SiGe IN ADVANCED DEVICES

Further improving the logic CMOS device performance requires implementation of novel materials (high-k dielectrics, metal gate, high-mobility channels, stressor layers) and device architectures (vertical fin-type structures, different channel orientation). One of the more promising routes for achieving high p-channel performance is the use of thin strained (Si)Ge channel devices [1-6]. Moreover, it has been demonstrated that $Si_{1-x}Ge_x$-channel pFETs show advantageous scaling potential with up to 90% performance boost at narrow channels [7]. In addition, an intrinsic reliability improvement, with respect to NBTI and gate leakage current has been reported for biaxially strained SiGe channel pMOSFETs [8].

The next performance boost can be achieved by implementing SiGe-channel pMOSFETs in a bulk FinFET architecture: the combination of the higher hole mobility along the (110) sidewalls in the normal (100)<110> configuration with the strained SiGe layer should not only improve the on current but at the same time will assist in a better control of the Short-Channel Effects (SCEs) for narrow fins. The change from biaxial to uniaxial compressive stress in narrow fins can be assisted by adding the strain of other stressor techniques, like CESL caps, SiGe Source/Drains (S/Ds), etc.

One of the concerns of FinFETs is the potentially higher density of interface states (N_{it}) along the (110) sidewalls [9], which may reduce the mobility by Coulomb scattering.

Likewise, there can be an orientation dependence of the density of oxide traps (D_{ot}), which can become apparent in the low-frequency (LF) noise of the devices. So far, only few studies on the noise behavior of bulk FinFETs have been reported [10-12]. In the case of p-channel transistors, it was demonstrated that besides 1/f noise also Generation-Recombination (GR) noise was frequently present, which could be associated with defects in the fully depleted silicon fins [12]. The 1/f noise showed an intriguing behavior, namely, that on the same wafer devices could exhibit so-called number fluctuations (Δn) and mobility fluctuations ($\Delta \mu$) behavior, based on the dependence of the current noise spectral density S_{Id}, normalized by the drain current squared (I_d^2). The potential origin of the latter has been assigned to the roughness of the sidewalls, which could vary from fin to fin and device to device.

It is the aim of the present work to explore the LF noise behavior of SiGe-channel bulk FinFETs, fabricated in different orientations on the same (100) substrate and on (110) Si wafers and to relate this with the dc transport parameters with emphasis on the hole mobility. A comparison is also made with standard Si bulk FinFET references and with planar strained SiGe pMOSFETs. Overall, it is observed that the responsible noise mechanism depends heavily on the process parameters studied: the lowest noise power spectral density (PSD) is obtained for the (100) oriented SiGe BFFs, where the 1/f noise is governed by mobility fluctuations The highest mobility is found for the (110) BFF, whereby the PSD is enhanced by excess number-fluctuations and GR noise .

II. EXPERIMENTAL DETAILS

The Si and SiGe Bulk FinFETs (BFF) with gate lengths down to 40nm were realized on 300 mm diameter CZ Si wafers. The integration path was as following. The Si fin was initially formed by STI recess resulting in roughly a 30nm height and a minimum width of around 15nm. Afterwards 10nm SiGe (Ge 45%) was grown epitaxially (SiGe "wrap around") on top of the fin. After spacer definition S/D junctions were formed and activated (T>1000 °C). The process is completed by silicidation and BEOL for contacts. Additionally to FinFETs, also reference planar SiGe devices were considered [6].

The SiGe wraparound process has many advantages beyond its simplicity. Depending on the starting Si-fin orientation, the resulting SiGe-fin wraparound sidewalls can be tuned to be either <100> or <110>, as shown in the TEM insets of Fig. 1. This results in a dramatic boost of the SiGe mobility properties, especially for narrow fins, with a mobility increase up to a factor of 2.5.

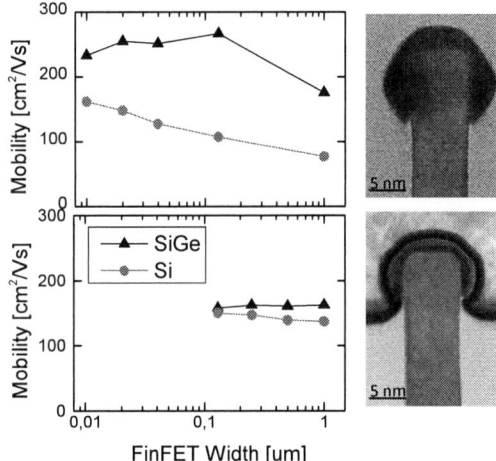

Figure 1: Mobility by split CV (L=10 µm) on SiGe <110> (top) and <100> (bottom), respectively. The TEM with the fin dimensions and orientations are also shown. The 10nm SiGe is wrapped around the Si BFF. Note the difference in the SiGe on the sidewall and dimensions.

LF noise has been measured on p-channel devices with a fixed length L=1 µm and various fin widths W_{fin}. A rather large device length was chosen in order to reduce the device-to-device variation in PSD. Measurements were performed at room temperature in linear operation and gate bias V_G from weak to strong inversion. Additional charge pumping measurements were done on the same samples to independently assess the level of interface states.

III. PLANAR SiGe TRANSISTORS AS A TOOL FOR A MOBILITY FLUCTUATIONS STUDY

The LF noise results of the different types of SiGe-channel pMOSFETs will be described, starting with planar transistors. Planar SiGe PFETs feature a width-dependent mobility [7]. For the same length the mobility increases up to a factor of 3 when the width is reduced to 80nm due to mechanical strain effects (Fig. 2). Therefore these devices are an ideal tool to study how noise is impacted by mobility fluctuations (see Fig. 2).

Generally, the spectra exhibit a $1/f^{\gamma}$ behavior, with $\gamma \approx 1$ (Fig. 3a). No GR bumps are found in most cases. From the continuous reduction of the normalized PSD with drain current in Fig. 3b, it is concluded that the flicker noise is dominated by $\Delta\mu$ fluctuations, especially at high inversion (i.e., high I_d). Following [8,13-17], due to the confinement of the holes in the buried channel the noise in SiGe channel

pMOSFETs is reduced with respect to regular Si counterparts. The buried channel further reduces the holes interaction with traps in the gate oxide (i.e., the number fluctuations) and thus emphasizes the impact of potential fluctuations and scattering on the hole transport.

The W-independent noise behavior is very surprising. A typical assumption is that the noise sources are uniformly distributed over the device area $L \times W$. In this case one expects a PSD which scales with the inverse of the total area, and therefore a $1/W$ dependence (L is fixed in Fig. 3b) in the PSD. While in Fig. 3b the narrower pMOSFETs do exhibit a higher normalized noise than the wider ones, it appears that the width dependence is weaker than expected from theory and seems to improve for narrower channels. This strongly suggests a better confinement and a more uniform channel for the narrower transistors.

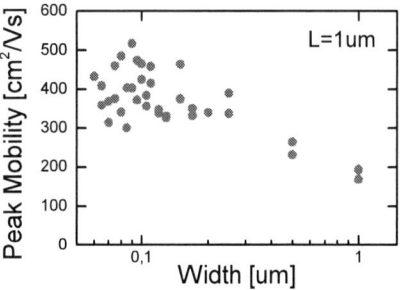

Figure 2: Mobility increases for narrow width due to strain effects. Mobility is estimated from the peak of the transconductance and the equivalent oxide thickness.

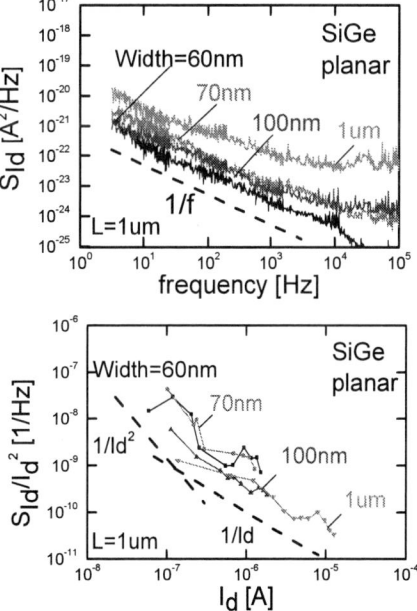

Figure 3: a) LF noise spectra of planar p-SiGe-channel transistors fabricated on a Si (100) wafer and with L=1 µm and different W_{fin}. Measurements have been performed in linear operation. b) Corresponding normalized current noise PSD versus drain current in linear operation at a frequency f=25 Hz.

IV. NOISE AS A TOOL TO STUDY DEFECTS IN BULK FINFETs

FinFET features a unique conduction mechanism where the transport occurs both on the top and lateral sidewalls and, therefore, changing the fin width does change the effective mobility, as shown already in Fig. 1. Note that mobility is sensitive to the hole effective mass, i.e., to the fin width as well as active defects in the inversion layer.

Surprisingly SiGe <110> has larger defect densities compared to <100>, as shown in Fig. 4 for different widths. For large widths on SiGe <110> (i.e., top-sidewall dominated) large G-R noise is observed, while for sidewall-conduction (narrow widths) a $1/f$ noise is generally observed. When one analyses the SiGe on <100> (i.e., more conformal) devices, as shown in Fig. 4b, these G-R features are not observed.

The PSD of Fig. 5 is the key figure demonstrating the very different noise mechanism found for different SiGe orientations. While the more conformal SiGe <100> shows the noise signature of mobility fluctuations (i.e., similar to the planar SiGe of Fig. 3), the SiGe <110> PSD dependence on I_d with a plateau in weak inversion and a $1/I_d^2$ roll-offstrongly suggests a *defect-mediated noise mechanism*.

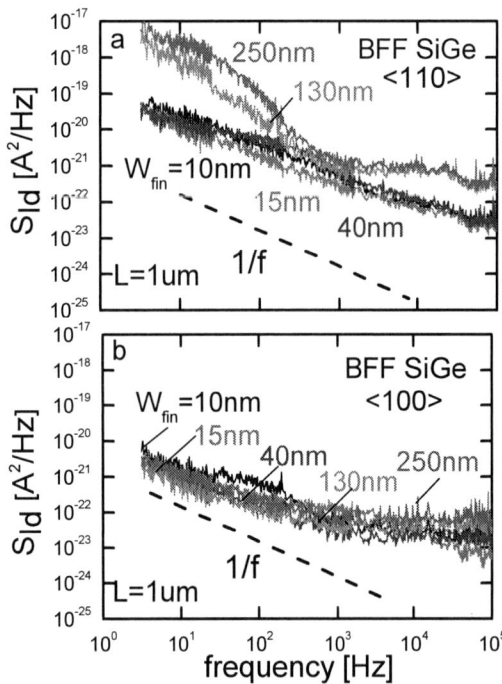

Figure 4: Normalized current noise PSD versus frequency, corresponding with p-type SiGe-channel BFFs fabricated on a <110> and <100> Si substrate respectively with L=1 μm and different W_{fin}.

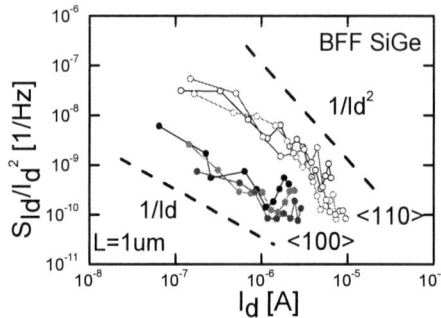

Figure 5. Normalized noise spectral density versus drain current for a number of p-type SiGe-channel BFFs fabricated on <100> and <110> Si substrates, at f=25 Hz.

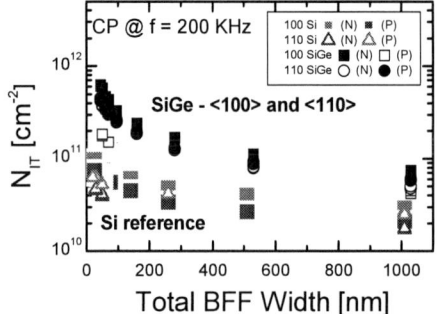

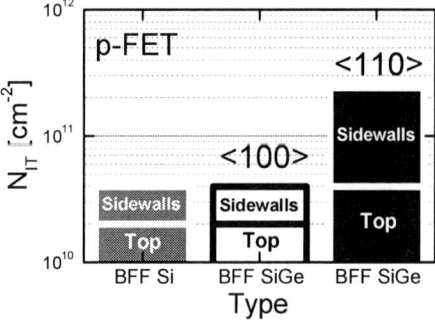

Figure 6: a) Interface states density as a function of the total device width (=2 x fin height + top width), as measured with charge pumping, for the Si reference and SiGe <110> and <100> n- and p-type devices. b) Schematic of the N_{it} contribution from top and lateral sidewall, as extracted from Fig. 6a.

V. HIGH NIT AND HIGH MOBILITY – A ROUTE FOR SIGE MATERIAL OPTIMIZATION?

The SiGe-orientation dependent interface state density (N_{it}) of Fig. 6 can explain the discrepancy in noise PSD behavior observed in Fig. 5.

Charge pumping measurements have been used to extract independently the interface states density for all the widths considered. The N_{it} has contributions from the top and the 2 lateral sidewalls. Large widths are dominated by the top interface, while narrow BFF widths are dominated by the sidewall defects. Figure 6 summarizes the defect densities found for different Si and SiGe orientations. The 10x larger defect densities for SiGe <110> compared to Si or SiGe <100> is also consistent with the noise PSD observed in Fig. 4 and Fig. 5 and strongly suggest that the SiGe crystal growth on <110> is more defective compared to the <100> case.

Despite the high defect density, SiGe <110> counter-intuitively shows a beneficial higher mobility (Fig. 1). Note that the high mobility is not a measurement artifact associated with the SiGe crystal growth / roughness as the peak mobility measured by split-CV in Fig. 1 is width-independent. The only explanation of such a high-mobility is a beneficial SiGe <110> orientation effect, somewhat similar to the case of hole transport in Si <110>.

VI. SUMMARY AND CONCLUSIONS

LF noise was studied on state-of-the-art planar and bulk FinFET SiGe devices with widths down to 40nm, enabling to study defects beyond the standard mobility and charge pumping analysis. Noise in optimized planar SiGe devices is dominated by mobility fluctuations. The lower noise in planar SiGe compared to reference Si (and its width-independence) suggests a beneficial effect of the hole buried channel conduction.

Bulk FinFETs feature noise sources on both the top and lateral sidewalls. The simple SiGe wraparound technique enables to study different SiGe growth conditions and orientations in a unique way. SiGe grown on a <110> direction features a very high mobility (250cm²/Vs) but also a high interface defect density, likely associated with the SiGe roughness. These results show the potential of SiGe <110> optimization for further device performance boost.

ACKNOWLEDGMENT

T. Romeo is indebted to the E.U. for granting him an ERASMUS scholarship. The imec Core Partners are gratefully acknowledged for their financial support within the frame of the Logic Program.

REFERENCES

[1] M.L. Lee, E.A. Fitzgerald, M.T. Bulsara, M.T. Currie, and A. Lochtefeld, "Strained Si, SiGe, and Ge channels for high-mobility metal-oxide-semiconductor field-effect transistors," *J. Appl. Phys.*, vol. 97, no. 1, pp. 1-27, Dec. 2005.

[2] L. Gomez, P. Hashemi, and J.L. Hoyt, "Enhanced hole transport in short-channel strained-SiGe p-MOSFETs," *IEEE Trans. Electron Devices*, vol. 56, no. 11, pp. 2644-2651, Nov. 2009.

[3] D.C. Gilmer, J.K. Schaeffer, W.J. Taylor, C. Capasso, K. Junker, J. Hildreth, D. Tekleab, B. Winstead, and S.B. Samavedam, "Strained SiGe channels for band-edge PMOS threshold voltages with metal gates and high-k dielectrics," *IEEE Trans. Electron Devices*, vol. 57, no. 4, pp. 898-904, Apr. 2010.

[4] S.-H. Lee, A. Nainini, J. Oh, K. Jeon, P.D. Kirsch, P. Majhi, L.F. Register, S.K. Banerjee, and R. Jammy, "ON-state performance

enhancement and channel-direction-dependent performance of a biaxial compressive strained $Si_{0.5}Ge_{0.5}$ quantum-well pMOSFET along <110> and <100> channel directions," *IEEE Trans. Electron Devices*, vol. 58, no. 4, pp. 985-995, Apr. 2011.

[5] L. Witters, S. Takeoka, S. Yamaguchi, A. Hikavyy, D. Shamiryan, M. Cho, T. Chiarella, L.-Å. Ragnarsson, R. Loo, C. Kerner, Y. Crabbe, J. Franco, J. Tseng, W.E. Wang, E. Rohr, T. Schram, O. Richard, H. Bender, S. Biesemans, P. Absil, and T. Hoffmann, "8Å Tinv gate-first dual channel technology achieving low-Vt high performance CMOS," in: 2010 *Symp. On VLSI Technol. Dig. Tech Papers*, pp. 181-182, Jun. 2010.

[6] J. Mitard, L. Witters, M. Garcia Bardon, P. Christie, J. Franco, A. Mercha, P. Magnone, M. Alioto, F. Crupi, L.-Å Ragnarsson, A. Hikavyy, B. Vincent, T. Chiarella, R. Loo, J. Tseng, S. Yamaguchi, S. Takeoka, W.-E. Wang, P. Absil, and T. Hoffmann, "High-mobility 0.85nm-EOT $Si_{0.45}Ge_{0.55}$-pFETs delivering high performance at scaled VDD," in: *IEDM Tech. Dig.*, pp. 249-252, Dec. 2010.

[7] G. Eneman, S. Yamaguchi, C. Ortolland, S. Takeoka, L. Witters, T. Chiarella, P. Favia, A. Hikavyy, J. Mitard, M. Kobayashi, R. Krom, H. Bender, J. Tseng, W.-E. Wang, W. Vandervorst, R. Loo, P.P. Absil, S. Biesemans, and T. Hoffmann, "High mobility $Si_{1-x}Ge_x$-channel PFETs: Layout dependence and enhanced scalability, demonstrating 90% performance boost at narrow widths," in: *Symp. On VLSI Technol. Dig. Tech. Papers*, pp. 41-42, Jun. 2010.

[8] S. Deora, A. Paul, R. Bijesh, J. Huang, G. Klimeck, G. Bersuker, P.D. Krisch, and R. Jammy, "Intrinsic reliability improvement in biaxially strained SiGe p-MOSFETs," *IEEE Electron Device Lett.*, vol. 32, no. 3, pp. 255-257, Mar. 2011.

[9] G. Kapila, B. Kaczer, A. Nackaerts, N. Collaert, and G.V. Groeseneken, "Direct measurement of top and sidewall interface trap density in SOI FinFETs," *IEEE Electron Device Lett.*, vol. 28, no. 3, pp. 232-234, Mar. 2007.

[10] E. Simoen, M. Aoulaiche, N. Collaert and C. Claeys, "Low-frequency noise study of p-channel bulk MuGFETs," in: the *Proc. of the 26th SBMICRO*, João Pessoa, Brazil, August 30-Sept. 2, ECS Trans., vol. 39 (1), pp. 53-60, Aug. 2011.

[11] M. G. Caño de Andrade, J.A. Martino, E. Simoen and C. Claeys, "Comparison of the low-frequency noise of bulk triple-gate FinFETs with and without dynamic threshold operation," *IEEE Electron Device Lett.*, vol. 32, no. 11, pp. 1597-1599, Nov. 2011.

[12] E. Simoen, G.M.C. Andrade, M. Aoulaiche, N. Collaert and C. Claeys, "Low-frequency noise investigation of n-channel bulk FinFETs for one-transistor memory cells," *IEEE Trans. Electron Devices*, May 2012.

[13] S. Okhonin, M.A. Py, B. Georgescu, H. Fischer, and L. Risch, "DC and low-frequency noise characteristics of SiGe p-channel FET's designed for 0.13-μm technology," *IEEE Trans. Electron Devices*, vol. 46, no. 7, pp. 1514-1517, Jul. 1999.

[14] S.J. Mathew, G. Niu, W.B. Dubbelday, and J.D. Cressler, "Characterization and profile optimization of SiGe pFET's on silicon-on-sapphire," *IEEE Trans. Electron Devices*, vol. 46, no. 12, pp. 2323-2332, Dec. 1999.

[15] J.A. Chroboczek and G. Ghibaudo, "Has SiGe lowered the noise in transistors?" *IEE Proc.-Circuits Devices Syst.*, vol. 149, no. 1, pp. 51-58, Feb. 2002.

[16] M. Myronov, O.A. Mironov, S. Durov, T.E. Whall, E.H.C. Parker, T. Hackbarth, G. Höck, H.-J. Herzog, and U. König, "Reduced 1/f noise in p-$Si_{0.3}Ge_{0.7}$ metamorphic metal-oxide-semiconductor field-effect transistor," *Appl. Phys. Lett.*, vol. 84, no. 4, pp. 610-612, Jan. 2004.

[17] M. von Haartman, B.G Malm, and M. Östling, "Comprehensive study on low-frequency noise and mobility in Si and SiGe pMOSFETs with high-k gate dielectrics and TiN gate," *IEEE Trans. Electron Devices*, vol. 53, no. 4, pp. 836-843, Apr. 2006.

Impact of front-back gate coupling on low frequency noise in 28 nm FDSOI MOSFETs

Christoforos G. Theodorou[1-2], Eleftherios G. Ioannidis[1-2], Sebastien Haendler[3], Nicolas Planes[3], Franck Arnaud[3], Jalal Jomaah[2], Charalabos A. Dimitriadis[1], Gerard Ghibaudo[2]

[1] Department of Physics, Aristotle University of Thessaloniki, 54124 Thessaloniki, Greece
[2] IMEP-LAHC, MINATEC, INPG, 3 Parvis Louis Neel, BP257, 38016 Grenoble, France
[3] STMicroelectronics, BP 16, 38921 Crolles, France
e-mail address: cgtheodo@auth.gr

Abstract--**Low-frequency (LF) noise has been studied on 28 nm FDSOI devices with ultra-thin silicon film (7 nm) and thin buried oxide (25 nm). A strong dependence of the noise level on the combination of the front and back biasing voltages was observed, and justified by the coupling effect of both Si/High-K dielectric and Si/SiO₂ interface noise sources (channel/front oxide and channel/buried oxide), combined with the change of the Remote Coulomb scattering. From comparisons of the experimental and simulation results, it is shown that the main reason of this dependence is the distance of the charge distribution centroid from the interfaces, which is controlled by both front and back-gate bias voltages, and the way this distance affects the Remote Coulomb scattering coefficient α. A new LF noise model approach is proposed to include all these effects. This also allows us to assess the oxide trap density values for both interfaces.**

I. INTRODUCTION

Fully Depleted (FD) Silicon On Insulator (SOI) technology is considered as one of the best candidate for the Short Channel Effect (SCE) control in future sub 28nm CMOS generations. The use of midgap/high-k metal gate stack with undoped SOI films allows for great improvement of variability as compared to bulk technology [1-3]. The use of ultra thin body and buried oxide thickness (UTBB) also enables to enhance the technology scalability, providing an ideal subthreshold slope and better drain-induced barrier lowering (DIBL) as well as larger back-to-front gate coupling effect useful for threshold voltage V_{th} control.

Besides, low frequency noise and RTS fluctuations, which scale as the reciprocal device area, become more important with technology scaling down. They are not only limiting the analog circuit operation but they should also jeopardize the digital circuit functioning. They could even appear as an ultimate variability source due to carrier dynamic trapping in undoped channel devices. For these reasons, the study of LF noise in UTTB FDSOI is a key issue for the technology evaluation. Previous works have shown that the LF noise in FDSOI devices should be influenced by the coupling effect between the back and front interfaces [4-8].

In this work, we present a thorough investigation of the impact of back/front interface coupling effect on the low frequency noise in FDSOI devices with very thin silicon film and thin buried oxide. We extend the existing carrier number fluctuations model established for fully depleted SOI to strongly coupled FDSOI structures, while including correlated mobility fluctuations. The studied devices are issued from a 28 nm FDSOI CMOS technology featuring high-k dielectric/metal front gate, 7 nm silicon film and a 25 nm BOX with channel width W=10 μm and gate length L=1 μm [9].

II. THEORETICAL APPROACH ON FD-SOI FLICKER NOISE

Based on the concept of flat band voltage fluctuations, it can be shown that the drain current fluctuations, δI_d, due to the carrier trapping-detrapping at front and back silicon/oxide interface can be expressed as [8],

$$\frac{\delta I_d}{I_d} = -\frac{g_{m1}}{I_d} \cdot \delta V_{fb1} - \frac{g_{m2}}{I_d} \cdot \delta V_{fb2} \qquad (1)$$

where $g_{m1}=dI_d/dV_{g1}$ and $g_{m2}=dI_d/dV_{g2}$ stand for the front and back gate transconductances, respectively.

Then, considering for each interface, carrier number fluctuations (CNF) due to trapping-detrapping by tunneling into the dielectric and additional correlated mobility fluctuations (CMF), yields for the normalized drain current noise,

$$\frac{S_{I_D}}{I_D^2} = S_{vfb1}\left(\frac{g_{m1}}{I_D}\right)^2\left(1 + \alpha_1 \mu_{eff} C_{ox1} \frac{I_D}{g_{m1}}\right)^2 +$$
$$S_{vfb2}\left(\frac{g_{m2}}{I_D}\right)^2\left(1 + \alpha_2 \mu_{eff} C_{ox2} \frac{I_D}{g_{m2}}\right)^2 \qquad (2)$$

with

$$S_{vfb1,2} = \frac{q^2 \lambda k T N_{t1,2}}{WLC_{ox1,2}^2 f} \qquad (3)$$

where $\alpha_{1,2}$ is the Coulomb scattering coefficient, λ is the tunneling constant in dielectric (≈ 0.1 nm), $N_{t1,2}$ denotes the front/back slow oxide trap densities, $C_{ox1,2}$ the front/back gate oxide capacitances, kT the thermal voltage and f is the frequency.

III. RESULTS AND DISCUSSION

A. Static Measurements

Typical transfer characteristics of the FDSOI devices operated in front and back gate modes illustrating the strong interface coupling are shown in Fig. 1. The drain voltage is $V_D=30$mV in all measurements. The front threshold voltage dependence on the back gate voltage V_{G2}, as well as the back threshold voltage dependence on the front gate voltage V_{G1}, are clearly emphasized in Fig. 2. Note that the front-to-back interface coupling can be significant when the device is biased in the strong accumulation regime ($V_{G2}<-10$V).

978-1-4673-1707-8/12 $31.00 © 2012 IEEE

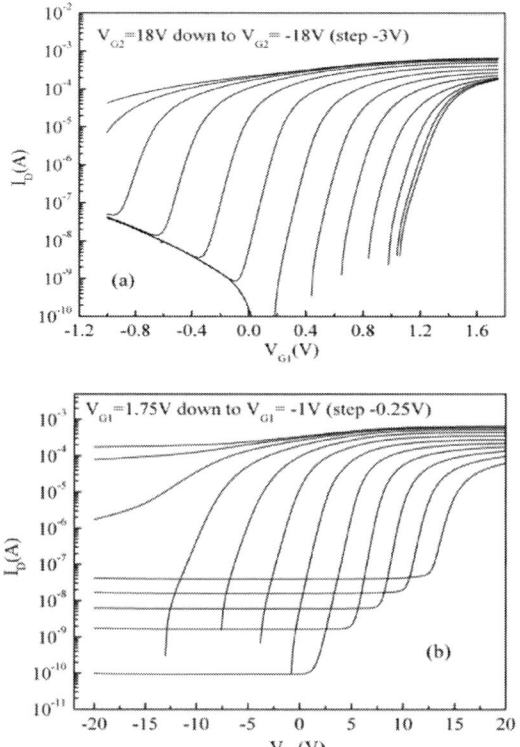

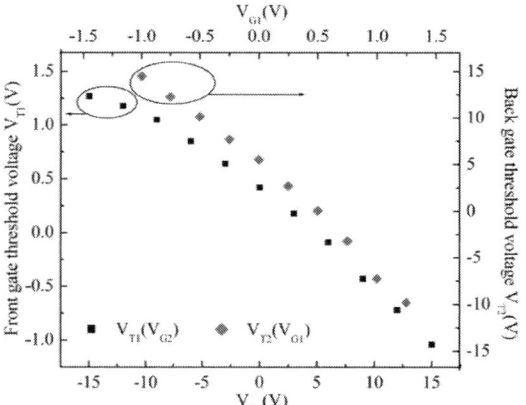

Fig.1. Transfer characteristics: (a) I_D-V_{G1} for various V_{G2}. (b) I_D-V_{G2} for various V_{G1}.

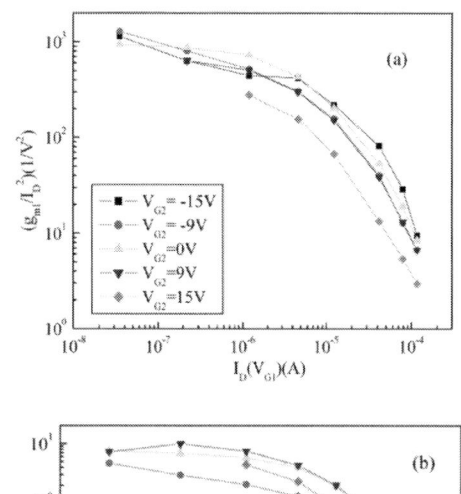

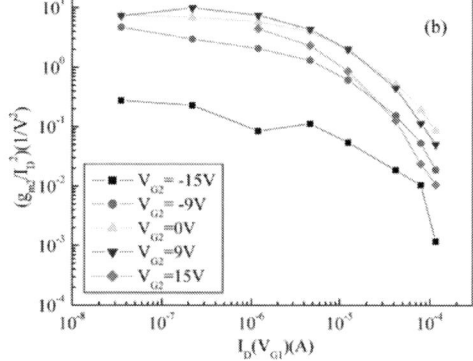

Fig. 3. $(g_{m1}/I_D)^2$ (a) and $(g_{m2}/I_D)^2$ (b) versus drain current for various V_{G2} values.

Fig. 2. Front and back threshold voltages versus gate voltages.

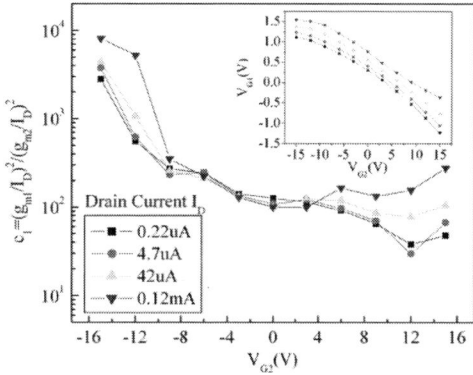

Fig. 4. Coupling factor c_1 and V_{G1} dependence on V_{G2} for various currents.

As indicated in Eq. 2, a very important parameter for flicker noise analysis in MOSFET is the squared transistor gain $(g_{m1,2}/I_D)^2$, measured here for front and back gate operations as a function of drain current (see Fig. 3). As it is usual, $(g_m/I_D)^2$ exhibits a plateau in weak inversion, before dropping above threshold in strong inversion.

But, in FDSOI devices, since there are two interfaces, a useful parameter that captures the relative impact of each interface on LF noise is the so called noise coupling factor, which can be more generally defined as:

$$c_1 = \left(\frac{g_{m1}}{I_D}\right)^2 \Big/ \left(\frac{g_{m2}}{I_D}\right)^2 \qquad (4)$$

As can be seen from Fig. 4, the coupling factor c_1 lies around ≈ 100 for the back interface in depletion and weak inversion regimes ($V_{G2}>0$), whereas it increases up to almost 10^4 for strongly accumulated back interface ($V_{G2}<<0$). As it is usually believed, this feature might indicate that the back interface LF noise contribution could be eliminated in the latter situation. One can also notice that c_1 is almost independent on I_D, especially at low drain currents.

B. Noise measurements

Power spectral density measurements were performed in the region of 5 Hz-20 kHz, with V_D at 30mV, using as variables both front and back gate voltages. Figure 5 shows typical 1/f noise spectra for different V_{G2} values at constant drain current.

978-1-4673-1707-8/12 $31.00 © 2012 IEEE

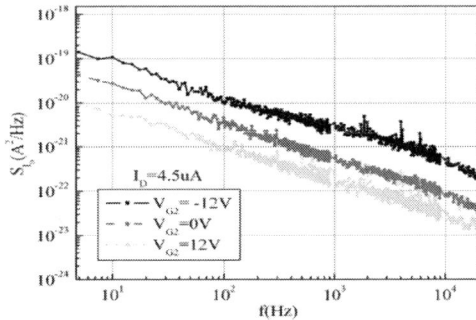

Fig. 5. PSD spectra for $I_D = 4.5\ \mu A$ measured at different V_{G2}.

Only few G-R or RTS components have been observed, since relatively large area devices were studied.

Typical normalized drain current noise versus drain current characteristics, measured in front gate mode, are illustrated in Fig. 6 for various back gate voltages. Note that the variations of the drain current noise follow the overall evolution of the squared transistor gain $(g_{m1,2}/I_d)^2$ with drain current as reported in Fig. 3(a), indicating that the LF noise can mostly be interpreted by the CNF noise model of Eq. 2.

The influence of the back interface coupling effect on the front gate operation LF noise can be better analyzed by plotting the normalized drain current noise, measured at a constant drain current, as a function of back gate voltage as shown in Fig. 7. As can be seen, the normalized drain current noise is significantly higher when V_{G2} takes high negative value. One can also notice from Fig. 7 that this dependence is stronger at higher drain currents. It should also be noted, by comparing Figs. 4 and 7, that the normalized drain current noise has the behaviour with V_{G2} as that of the noise coupling factor c_1.

The above finding seems peculiar and unexpected at first glance, because none of the quantities $(g_{m1}/I_D)^2$ and $(g_{m2}/I_D)^2$ is increasing that much when V_{G2} is varying from positive to negative values. Indeed, as can be seen from Fig. 8, the CNF LF noise model of Eq. 2 without correlated mobility fluctuations cannot explain this behaviour, since, for a strongly accumulated back interface situation, the noise should mostly stem from the front interface carrier trapping-detrapping contribution. Indeed, this is confirmed by numerical simulation made at constant drain current, which shows that the charge carrier distribution across the silicon film strongly depends on V_{G2} (see Fig. 10).

However, the CNF LF noise model applied in the positive V_{G2} range allows us to solve Eq. (2) considering α_1 and α_2 negligible for low drain currents and different bias conditions and extract constant values for S_{Vfb1} and S_{Vfb2}, from which the front and back gate oxide trap densities were determined using Eq. 3 ($N_{t1} = 9 \times 10^{17}$ cm^{-3}eV^{-1} and $N_{t2} = 2 \times 10^{17}$ cm^{-3}eV^{-1}). Note that the front interface shows higher oxide trap density as expected due to the high-k/metal gate stack, whereas the back interface trap density well reflects the quality of a pure thermal oxide in line with state of the art technology.

In order to interpret the abnormal LF noise behavior in the negative V_{G2} range, we have therefore taken into account the correlated mobility fluctuations and further

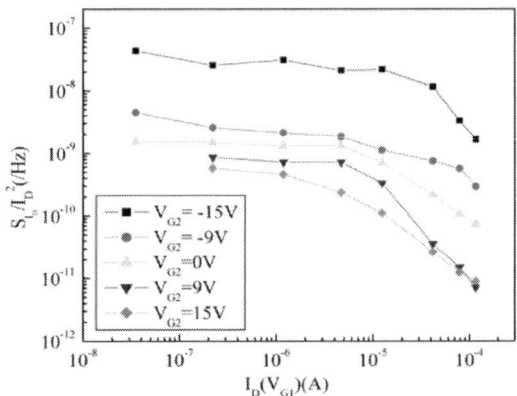

Fig. 6. Normalized PSD at f=10 Hz versus drain current for different V_{G2} values.

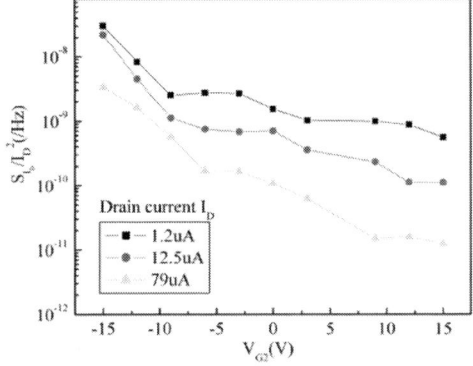

Fig. 7. Normalized PSD at f=10 Hz versus V_{G2} for different drain current.

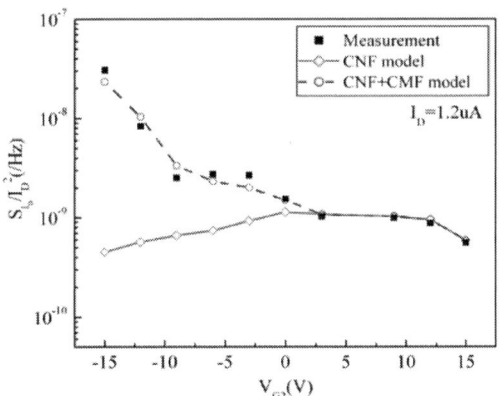

Fig.8. Comparison of experimental normalized drain current PSD versus V_{G2} results with the CNF model results for constant S_{Vfb1} and S_{Vfb2} and with CNF+CMF model results with the RCS parameter.

considered the influence of the remote Coulomb scattering (RCS) across the silicon film on the parameter α in Eq. 2. To this end, we used the following simple RCS model for the parameter α [9],

$$\alpha = \frac{\alpha_0}{\left(1 + \dfrac{x}{\lambda_C}\right)^2} \quad (5)$$

where $\alpha_0 = 10^5$ Vs/C approximately in our case, x is the distance of the charge distribution centroid from the

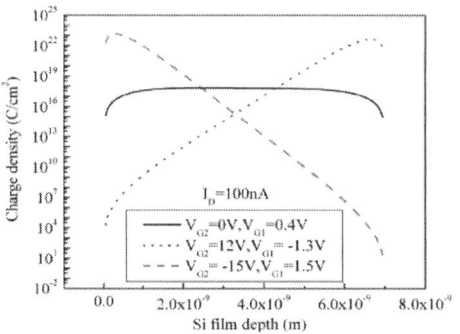

Fig. 9. Simulation results of charge density versus silicon film depth for different bias conditions, keeping the drain current value constant.

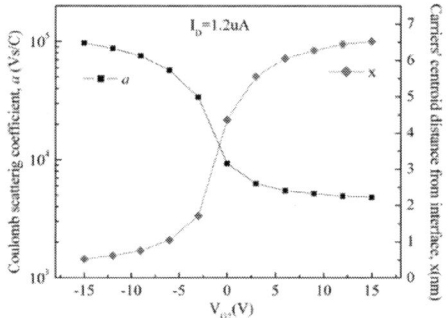

Fig. 10. Calculation results for carriers distribution centroid-front interface distance x and coulomb scattering coefficient α versus bias voltage V_{G2}, keeping the drain current constant.

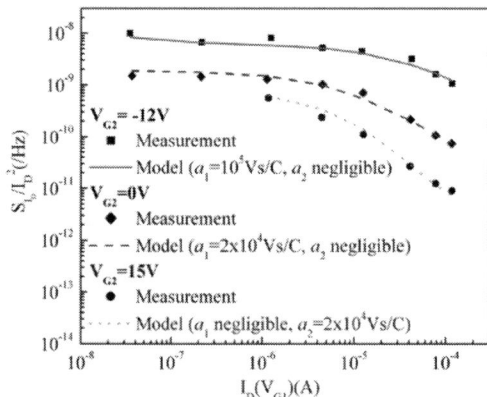

Fig. 11. Comparison of experimental normalized drain current PSD versus drain current results with CNF+CMF model results for different V_{G2} values.

interface and $\lambda_C = 1.2 \times 10^{-9}$ m. As obtained by the numerical simulation shown in Fig. 10 for a constant drain current, the RCS coefficient α strongly depends on the distance of the centroid from the interface, i.e. on the back gate bias condition.

Using this RCS formulation in the CNF LF noise model of Eq. 2 including CMF, the variation of S_{I_D} / I_D^2 with the back gate voltage is well described as shown in Fig. 8, especially the noise increase in the region of back gate accumulation ($V_{G2}<0$). It is noticed that for $V_{G2}<0$, the second term of Eq. (2) is negligible and the effective carrier mobility μ_{eff} corresponds to the low field mobility for I_D=1.2 μA, extracted from analysis of the transfer characteristic.

In addition, much higher values of the parameter α are required to explain the noise behavior at strong accumulation of the back interface, due to the closer proximity of the charge carriers at the interface (Fig. 9). Since in the measured drain current range the effect of the front gate interface on the coupling factor c_1 is negligible (Fig. 4), the normalized PSD as a function of the front gate biasing can be explained with a constant value for the scattering parameter α as shown in Fig. 11.

IV. CONCLUSIONS

The low frequency noise behaviour was studied systematically in ultra-thin body (7 nm) and thin buried oxide (25 nm) FD SOI devices, as a function of the front and back gate bias voltages. A peculiar behaviour of the normalized drain current power spectral density has been observed when the back interface is strongly accumulated. A simple model for the remote Coulomb scattering parameter has been proposed in terms of the distance of the charge distribution centroid from the interfaces, explaining the noise behaviour using the carrier number with correlated mobility fluctuations model. Although it is not possible to fully decouple the two interfaces, the coupling effect can be resolved to characterize both front and back interfaces in SOI devices.

ACKNOWLEDGEMENTS

This work has been performed under financial support from the Research Funding Program HERACLEITUS II, co-financed by the European Union (European Social Fund-ESF) and Greek national funds through the Operational Program "Education and Lifelong Learning" of the National Strategic Reference Framework (NSRF). This work has also been partly supported by Reaching 22 catrene project.

REFERENCES

[1] C. Fenouillet-Beranger et al, "Fully-depleted SOI technology using high-k and single-metal gate for 32 nm node LSTP applications featuring 0.179 μm^2 6T-SRAM bitcell," Proc. IEDM'07, p. 267.

[2] O. Weber et al, "High immunity to threshold voltage variability in undoped ultra-thin FDSOI MOSFETs and its physical understanding," Proc. IEDM'08, p. 245.

[3] N. Sugii et al, "Comprehensive study on Vth variability in silicon on Thin BOX (SOTB) CMOS with small random-dopant fluctuation: Finding a way to further reduce variation," Proc. IEDM'08, p. 249.

[4] E. Simoen, A. Mercha, C. Claeys, N. Lukyanchikova, "Critical Discussion of the Front–Back Gate Coupling Effect on the Low-Frequency Noise in Fully Depleted SOI MOSFETs," IEEE Trans on Electron Devices, 51, pp. 1008-1016 (2004).

[5] L. Zafari, J. Jomaah, G. Ghibaudo, "Low frequency noise in multi-gate SOI CMOS devices," Solid State Electronics, 51, pp.292-298 (2007)

[6] W. Cheng et al, "Suppression of 1/f Noise in Accumulation Mode FD-SOI MOSFETs on Si(100) and (110) Surfaces," Noise and fluctuations, AIP Conf. Proc., vol 1129, pp. 337-340 (2009).

[7] J. El Husseini et al, "New numerical low frequency noise model for front and buried oxide trap density characterization in FDSOI MOSFETs,"Microelectr. Eng. 88, 1286-1290 (2011).

[8] L. Zafari, J. Jomaah, G. Ghibaudo, "Modeling and simulation of coupling effect on low frequency noise in advanced SOI MOSFETs," Fluctuation and Noise Letters, 8, L87–L94 (2008).

[9] N. Planes et al, "28nm FDSOI Technology Platform for High-Speed Low-Voltage Digital Applications," VLSI Technology Symp., in press (2012).

[10] G. Ghibaudo, Chapter 3 in "Semiconductor-On-Insulator Materials for Nanoelectronics Applications", Springer, Berlin (2011).

On the correlation between the retention time of FBRAM and the low-frequency noise of UTBOX SOI nMOSFETs

E. Simoen, M. Aoulaiche, A. Veloso, M. Jurczak
and C. Claeys
Imec
Leuven, Belgium

A. Luque Rodríguez and J.A. Jiménez Tejada
Universidad de Granada : Departamento de Electrónica y
Tecnología de los Computadores, Facultad de Ciencias
Granada, Spain

L. Mendes Almeida and M.G.C. Andrade
University of São Paulo: LSI/PSI/USP
São Paulo, Brazil

C. Caillat and P. Fazan
Micron Technology
Leuven, Belgium

Abstract—**In this work, the low-frequency noise of UTBOX SOI nMOSFETs developed for Floating Body RAM (FBRAM) applications is reported and compared with the corresponding retention time. A clear trend is shown, relating a high retention time with a low noise spectral density S_{Id}. The one decade higher spread in S_{Id} compared with retention time indicates that other types of traps are responsible for both parameters. From the fact that for the same noise magnitude a different retention time can be observed in UTBOX nMOSFETs with a different channel processing strongly suggests that besides traps in the silicon film and at the interface, additional factors like the lateral electric field determine hole generation in the Si body.**

I. INTRODUCTION

One of the contenders for future scaled DRAM memory cells is the so-called capacitor-less, one-transistor (1T) floating-body cell, consisting of a single transistor fabricated on a Silicon-on-Insulator (SOI) wafer [1-3]. The use of an ultra-thin buried oxide (UTBOX) SOI wafer offers a better control of the short-channel effects [4] and provides a pathway to further scaling down to the 15-nm technology node of Floating-Body RAM memory cells [5]. One of the important performance parameters of a 1T FBRAM cell is the charge retention time, which is a measure for the time the charge of the 1-state can leak away from the body by carrier recombination or excess charge can be generated in the potential well to destroy the 0-state. Therefore, a clear correlation exists between the carrier lifetime in the body and the retention time [6]. While there exist several leakage mechanisms in a 1T FBRAM cell, it is clear that there is definitively a correlation with the defects in the structure, either at both interfaces or in the fully depleted film.

Assessing the carrier lifetime, or more generally, the deep-level Shockley-Read-Hall (SRH) generation-recombination (GR) centers in a Fully Depleted (FD) SOI transistor is not an easy task, either requiring a body contact for charge pumping measurements or a dedicated gated p-i-n diode structure [7]. Alternatively, low-frequency (LF) noise measurements on SOI transistors can be used as a tool for the investigation of defects in the gate oxide [8] or in the depletion region [9]. In the latter case, use is made of the excess Lorentzian noise contributions, associated with GR centers. It has been shown in the past that in the case these reside in the silicon depletion region, one expects that the parameters of the Lorentzian – the plateau amplitude A_0 and the corner frequency $f_0 = 1/2\pi\tau_0$ – are independent of the gate voltage [9], with τ_0 the trap characteristic time constant. It is the aim of the present work to investigate the LF noise behaviour of UTBOX SOI nMOSFETs, developed for 1T FBRAM memory cells and to compare it with the corresponding retention time. As will be shown, a general trend exists where a lower current noise spectral density S_{Id} corresponds with a higher retention time. The presence of excess GR noise may explain the device-to-device variability of the spectral density. A dedicated numerical model demonstrates that the GR components can be ascribed to defects in the fully depleted Si film. In addition, it will be shown that besides the trap density and energy level, another important parameter determining the retention time is the lateral electric field.

II. EXPERIMENTAL DETAILS

The nMOSFETs studied have been fabricated on 300 mm UTBOX SOI wafers with a BOX thickness of 10 nm and a nominal film thickness of 20 nm. The gate stack consists of 5 nm SiO_2 capped with 5 nm TiN as the gate metal and 100 nm polycrystalline silicon. A Transmission Electron Microscopy (TEM) cross section of an nMOSFET with 10 nm Si film thickness is given in Fig. 1. Devices with standard extensions have been compared with extensionless (also termed underlap) transistors. This results in a lower maximum electric field at the drain and a higher average retention time[10]. Charge retention time has been measured at 85 °C on transistors with W=1 µm and L=105 nm, with the front gate in accumulation and zero drain bias V_{DS}. As can be seen in Fig. 2, the charge retention is determined by the degradation of the 0-state by hole generation [10,11]. An activation energy of 0.32 eV has been derived for the retention time of the standard junction devices.

LF noise measurements have been performed at room temperature on the same n-channel transistors in linear operation (V_{DS}=0.05 V) and the front gate voltage V_{GS} stepped from weak to strong inversion in steps of 50 mV, typically. The back-gate bias V_{BS} was kept at 0 V.

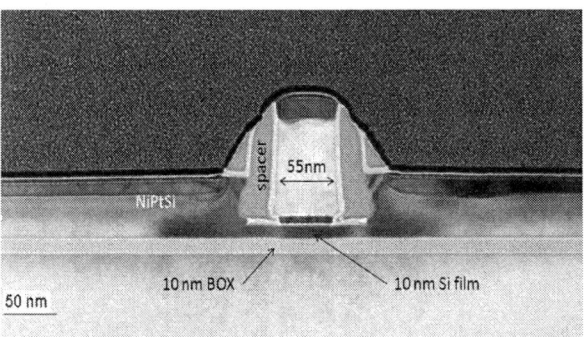

Figure 1. Cross-sectional TEM image of an UTBOX n-channel transistor, with 10 nm BOX and 10 nm Si film.

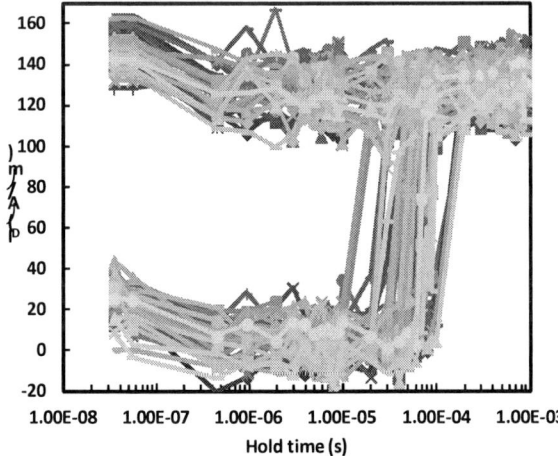

Figure 2. State-0 and 1 current as a function of the holding time measured for 36 different transistors over the wafer.

III. RESULTS AND DISCUSSION

As can be seen from Fig. 3, the spectra are in most cases dominated by flicker noise with a $1/f^\gamma$ dependence ($\gamma \sim 1$). From the behaviour of the normalized noise spectral density (S_{Id}/I_d^2) versus the drain current I_D, it can be derived that the 1/f-like noise is dominated by number fluctuations (Δn), related with trapping in the gate oxide [12,13]. This is also illustrated by Fig. 4, showing a proportionality between the normalized noise and $(g_m/I_d)^2$, with g_m the device transconductance.

On top of the 1/f-noise background there are Lorentzian noise components which are more clearly observable as peaks in Fig. 5, representing $f \times S_I$, with f the frequency. The trap time constant is calculated through $\tau_0 = 1/(2\pi f_0)$, with f_0 the peak frequency. Two different behaviors can be distinguished: in the first case, the corner frequency and peak amplitude change with V_{GS}, while for the second peak, there is no marked dependence. These GR noise components cause a significant variation of the noise spectral density over the wafer [13].

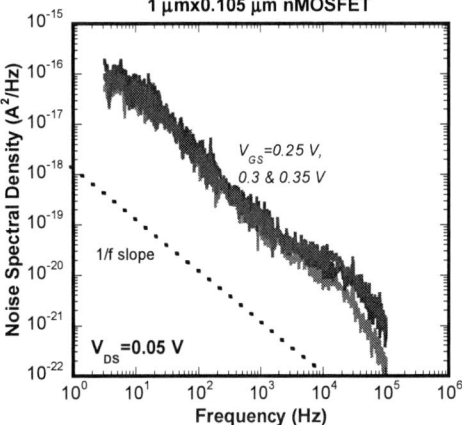

Fig. 3. LF noise spectrum for a 1 µmx0.105 µm nMOSFET in linear operation (V_{DS}=0.05 V) and corresponding with different gate voltages V_{GS}=0.25 V, 0.3 V and 0.35 V.

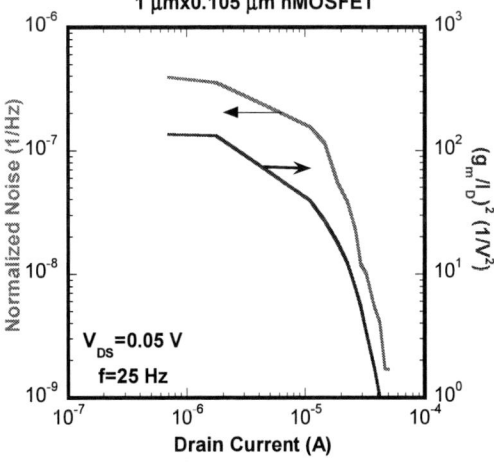

Fig. 4. Normalized current noise spectral density at f=25 Hz and $(g_m/I_d)^2$ versus drain current I_d in linear operation for a 1 µmx0.105 µm nMOSFET.

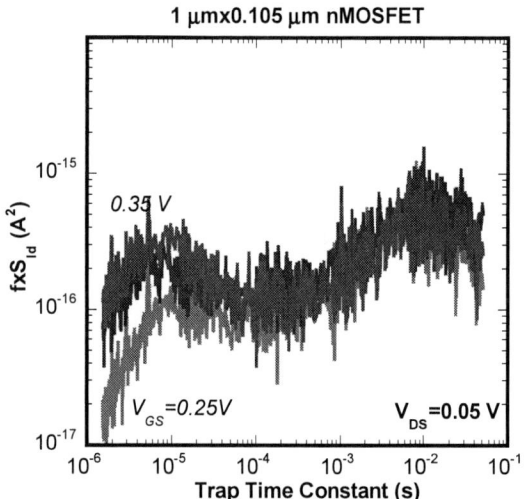

Fig. 5. Frequency normalized noise spectral density (fxS$_{Id}$) in linear operation and corresponding with a gate voltage of 0.25 V, 0.3 V and 0.35 V for a 1μmx0.105 μm UTBOX nMOSFET.

Finally, when comparing the noise spectral density with the retention time, it has been noted that there is an inverse relationship in Fig. 6: devices with a high retention exhibit a low noise magnitude and vice versa. Comparing the standard junction devices with the extentionless ones, a clearly higher retention time can be achieved in Fig. 7 for the latter [10]. It should be noted from Figs 6 and 7 that the range of noise spectral density amounts to about two decades while the retention time differs by one decade. At the same time, it is clear from Fig. 7 that if we consider the noise magnitude as a measure for the oxide trap density, a higher retention time can be obtained for the same S$_{VG}$, using extentionless junctions. From this, it is concluded that the traps responsible for the hole generation and degrading the 0-state are not the same as the ones giving rise to the 1/f and GR noise.

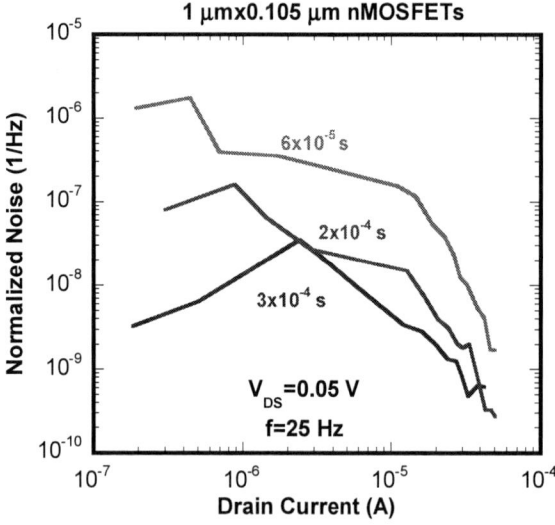

Fig. 6. Normalized noise spectral density at f=25 Hz versus drain current in linear operation for three 1 μmx0.105 μm nMOSFETs.

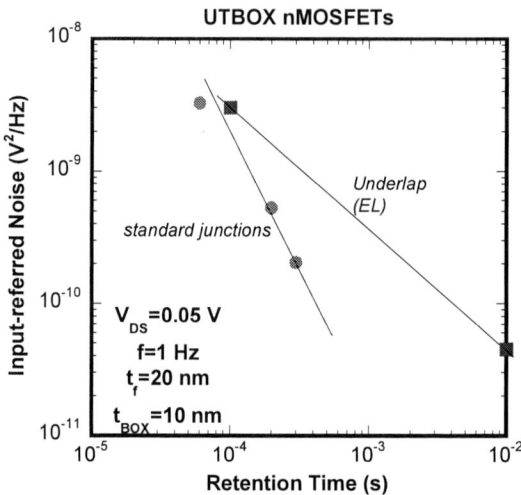

Fig. 7. Input-referred noise spectral density at flat-band and f=1 Hz versus retention time for UTBOX nMOSFETs with standard junctions (circles) or extensionless (squares) and with t$_f$=20 nm and t$_{BOX}$=10 nm.

In order to interpret the noise spectra, a dedicated model has been developed [12], which is an extension of the noise model developed for so-called four-gate SOI transistors [14]. Results of a fitting of the model to experimental noise spectra for standard HDD UTBOX nMOSFETs are represented in Fig. 8 for low retention time devices and in Fig. 9 for long retention counterparts. It can be concluded that the GR noise is generated by deep-level traps in the silicon film which are close to the conduction band edge (~E$_C$-0.12 eV).

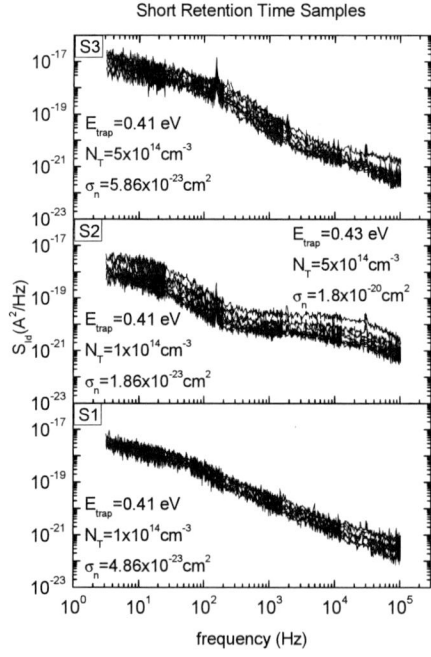

Fig. 8. Experimental S$_{Id}$ of the three samples which showed the shortest retention time and fitting parameters.

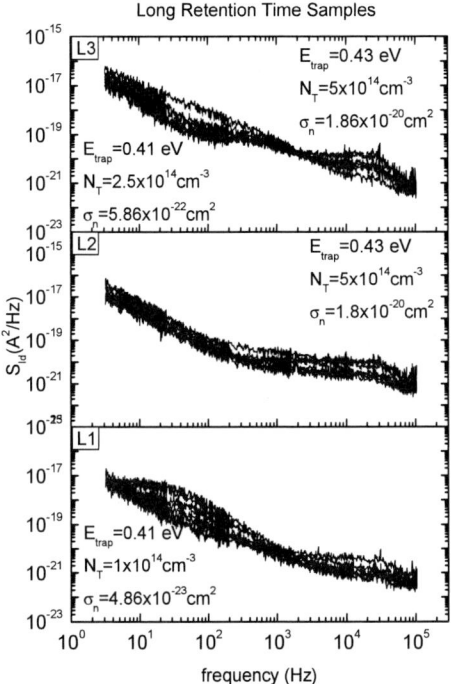

Long Retention Time Samples

L3: $E_{trap}=0.43$ eV, $N_T=5\times10^{14}$cm^{-3}, $\sigma_n=1.86\times10^{-20}$cm^2

$E_{trap}=0.41$ eV, $N_T=2.5\times10^{14}$cm^{-3}, $\sigma_n=5.86\times10^{-22}$cm^2

L2: $E_{trap}=0.43$ eV, $N_T=5\times10^{14}$cm^{-3}, $\sigma_n=1.8\times10^{-20}$cm^2

L1: $E_{trap}=0.41$ eV, $N_T=1\times10^{14}$cm^{-3}, $\sigma_n=4.86\times10^{-23}$cm^2

Fig. 9: Experimental S_{Id} of the three samples which showed the longest retention time and fitting parameters.

Whether the Lorentzian parameters shift with V_{GS} or not depends on the relative position of the Fermi level in the film with respect to the trap level. It is clear that the position of the GR centers observed here is too shallow to act as efficient (hole) generation centers. It should be remarked that the activation energy E_T has been defined with respect to the intrinsic Fermi level. In addition, the activation energy differs from the one derived for the retention time. Moreover, from the data in Fig. 8, one can derive that not only the trap concentration is determining the retention time – assuming for a moment that the noise is proportional to the relevant GR center concentration N_T – but also the electric field at the drain junction plays a crucial role [10].

IV. CONCLUSIONS

It has been observed that low retention UTBOX FBRAM nMOSFETs correspond with a high LF noise spectral density. At the same time, it has been shown that a different kind of traps is responsible for the hole generation, degrading the 0-state and the trapping/detrapping fluctuations responsible for the LF noise. While the observed GR noise components point to the presence of defects in the silicon film, most of them are not responsible for the hole generation degrading the 0-state.

REFERENCES

[1] S. Okhonin, M. Nagoga, J.M. Sallese, and P. Fazan, "A capacitor-less 1T-DRAM cell," *IEEE Electron Device Lett.*, vol. 23, no. 2, pp. 85-87, Feb. 2002.

[2] E. Yoshida and T. Tanaka, "A capacitorless 1T-DRAM technology using gate-induced drain-leakage (GIDL) current for low-power and high-speed embedded memory," *IEEE Trans. Electron Devices*, vol. 53, no. 4, pp. 692-697, Apr. 2006.

[3] S. Okhonin, M. Nagoga, E. Carman, R. Beffa, and E. Faraoni, "New generation of Z-RAM," in *IEDM Tech. Dig.*, 2007, pp. 925-928.

[4] V.P. Trivedi and J.G. Fossum, "Nanosacle FD/SOI CMOS: thick or thin BOX?" *IEEE Electron Device Lett.*, vol. 26, no. 1, pp. 26-28, Jan. 2005.

[5] I. Ban, U.E. Avci, D.L. Kencke, P. Tolchinsky, and P.L.D. Chang, "Integration of back-gate doping for 15-nm node floating body cell (FBC) memory," in *Symp. on VLSI Technol. Dig. of Tech. Papers*, pp. 159-160, Jun. 2010.

[6] S. Kim, S.-J. Choi, D.-I. Moon, and Y.-K. Choi, "Carrier lifetime engineering for floating-body cell memory," *IEEE Trans. Electron Devices*, vol. 59, no. 2, pp. 367-373, Feb. 2012.

[7] G. Kapila, B. Kaczer, A. Nackaerts, N. Collaert, and G.V. Groeseneken, "Direct measurement of top and sidewall interface trap density in SOI FinFETs," *IEEE Electron Device Lett.*, vol. 28, no. 3, pp. 232-234, Mar. 2007.

[8] E. Simoen, A. Mercha, C. Claeys, and N. Lukyanchikova, "Low-frequency noise in Silicon-on-Insulator devices and technologies," *Solid-State Electron.*, vol. 51, pp. 148-169, 2007.

[9] I. Lartigau, J.M. Routoure, W. Guo, B. Cretu, R. Carin, A. Mercha, C. Claeys, and E. Simoen, "Low temperature noise spectroscopy of 0.1 μm partially depleted silicon on insulator metal-oxide-semiconductor field effect transistors," *J. Appl. Phys.*, vol. 101, pp. 104511-1/5, 2007.

[10] M. Aoulaiche, E. Simoen, A. Veloso, L. Altimime, G. Groeseneken, M. Jurczak, L. Mendes Almeida and T. Nicoletti, "Junction field effect on the retention time of floating body RAM memory," *IEEE Trans. Electron Devices*, in print.

[11] M. Aoulaiche, L. Mendes Almeida, Ch. Caillat, N. Collaert, P. Blomme, E. Simoen, L. Altimime, G. Groeseneken and M. Jurczak, "Interface state impact on 1T-FBRAM cell retention," Paper to be published in the *Proc. of IRPS* 2012, April 2012.

[12] A. Luque Rodríguez, G. Cano de Andrade, M. Aoulaiche, L. Almeida, C. Claeys, J.A. Jiménez-Tejada, and E. Simoen, "Defect analysis in UTBOX SOI nMOSFETs by low-frequency noise," In: *Abstracts of EUROSOI* 2012, pp. 49-50, Jan. 2012

[13] E. Simoen, G.M.C. Andrade, L. Mendes Almeida, M. Aoulaiche, C. Caillat, M. Jurczak and C. Claeys, "On the variability of the low-frequency noise in UTBOX SOI nMOSFETs," Paper submitted for presentation at *SBMICRO 2012*, Brasilia, Brazil, Aug. 30- Sep. 2, 2012.

[14] J.A. Jiménez Tejada, A. Luque Rodríguez, A. Godoy, S. Rodríguez-Bolivar, J.A. López Villanueva, O. Marinov, and M.J. Deen, "Effects of gate oxide and junction nonuniformity on the DC and low-frequency noise performance of four-gate transistors," *IEEE Trans. Electron Devices*, vol. 60, no. 2, pp. 459-467, Feb. 2012.

Effect of Substrate Bias on Frequency Dependence of MOSFET Noise Intensity

Kenji Ohmori, Ranga Hettiarachchi, and Keisaku Yamada
Faculty of Pure and Applied Sciences, University of Tsukuba
Tsukuba, Ibaraki, 305-0051 Japan
ohmori.kenji.ff@u.tsukuba.ac.jp

Abstract—We have revealed that, using an originally-developed noise measurement system, the change in 1/f noise intensity due to a substrate bias mostly ranges in low frequencies, from 1 Hz to 100kHz. Above 100 kHz to 30 MHz, the 1/f component still continues. However, the noise intensity does not strongly depend on the substrate bias. The substrate bias alters the distance of inversion carriers from the SiO_2/Si interface. Therefore, these results indicate the possibility of separating several 1/f factors, such as the number fluctuation and the mobility fluctuation.

I. INTRODUCTION

Fluctuation of MOSFET properties is one of the most critical issues for achieving further shrinkage of ULSI devices. Characteristics of the drain-current noise depend upon a variety of conditions such as carrier types, voltages applied to gate (V_g), drain (V_d), and substrate (V_{sub}), dimension of transistors, trap densities, and gate-stack materials [1-10]. Two different models have been raised to explain the physical origins of the flicker (1/f) noise, i.e., the number fluctuation and the mobility fluctuation [11-18].

A number of study have been reported to elucidate the relationship which exists between the mobility fluctuation model on the one hand and the carrier number fluctuation model so called McWhorter model on the other to explain the noise in MOSFETs.

In this paper, we focus on the dependences of noise intensity upon frequency. An originally-developed measurement system for characterizing noise properties in a higher frequency region was carried out together with a conventional low-frequency measurement.

II. EXPERIMENTAL

A. Novel Measurement System for Noise Characterization of Higher Frequency Region

The conventional noise measurement system allows us to characterize MOSFET noise ranging from 1 Hz only up to about 100 kHz. The limit of this measurement range results from the bandwidth design of low-noise amplifier (LNA) and the distance between a DUT and the LNA [19-20]. In order to extend the frequency range, we have developed a novel measurement system. As shown in Fig. 1, a LNA with a frequency bandwidth from 14 kHz to 154 MHz was mounted

This work was partially supported by Core Research for Evolutional Science & Technology (CREST), Japan Science and Technology (JST).

on a probe so that the signal from a DUT can be amplified with lesser loss. The probe was equipped with 4 pins to contact on 4 pads (gate, source, drain, substrate) on a wafer. Several probes were fabricated with different LNA properties

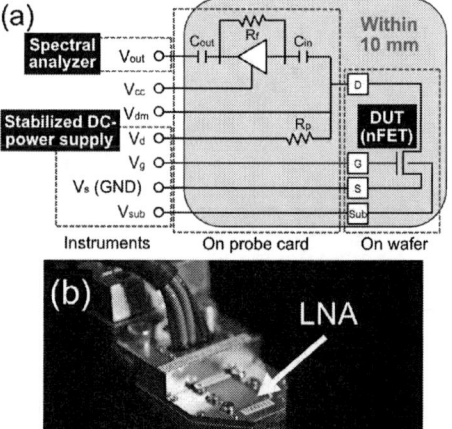

Fig. 1 (a) Block diagram of our developed high-frequency probe (HFP). A low-noise amplifier (LNA) on the probe was located as close as possible (< 10 mm) to DUT on a wafer. (b) Four pins for contacting to pads of gate, drain, source and substrate are equipped at the forefront of HFP. The LNA is double-shielded to avoid electromagnetic interference. (c) Several HFPs with different LNA properties are installed in a shielded probe station (only one of HFPs was used in this paper.)

as shown in Fig. 1(c), while we show data from only one of them in this paper. A spectrum analyzer (Agilent PXA N9030A) was used for monitoring the amplified signals.

B. Noise Measurement

We have characterized MOSFET noise with frequency ranges from 1 Hz to 100 MHz. For the lower frequencies (1 Hz to 100 kHz, a conventional system with fast IV units (Agilent B1530A) was carried out. The semiconductor parameter analyzer is equipped with four fast IV units so that the biases on the 4-terminals of MOSFETs are controlled individually.

The novel system with the probe card was used for the higher frequencies from 100 kHz to 100 MHz [19-20]. Since a capacitive coupling was adopted for extracting ac components from the drain current, it is difficult to characterize a noise intensity in a low-frequency region less than 10 kHz. In this paper, low-frequency measurement using the fast IV units is referred to LFM, while higher-frequency measurement using the probe card is referred to HFP.

N-type MOSFETs fabricated on a 12-inch wafer with a gate length L of 350 nm were used. Halo implantation was not applied to obtain a uniform impurity distribution in the channel. The gate stack structure is composed of poly-Si and SiO_2 films. The thickness of SiO_2 is 3.2 nm estimated from CV measurements. The applied substrate bias ranges from -1.0 to +0.4 V. The drain bias V_d used for noise characterization was 1.0 V, while for evaluating the threshold voltages a V_d value of 50 mV was used. Noise measurement was performed as a function of gate overdrive (V_g-V_t).

III. RESULTS AND DISCUSSION

A. Static Id-Vg properties

Fig. 2(a) demonstrates dependence of I_d-V_g properties on V_{sub} at V_d = 50 mV. Applying substrate biases from +0.4 to -1.0 V shifts the I_d-V_g curves, where the V_t values change from 0.33 to 0.70 V, respectively. Fig. 2(b) shows I_d-V_g curves at V_d = 1.0 V. The horizontal axis is a gate overdrive. After canceling the offset due to the different V_t values, the I_d-V_g curves for V_{sub} ranging from -1.0 to +0.4 V almost overlap.

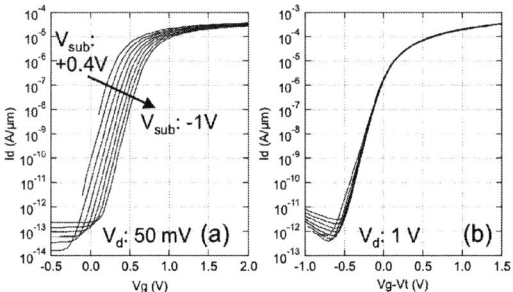

Fig. 2 (a) I_d-V_g characteristics dependence at V_d = 50 mV on substrate biases from -1.0 to +0.4 V, where the threshold voltage V_t estimated by the g_m method shifts from 0.70 to 0.33 V, respectively. (b) After offset by the V_t values, the I_d-V_g curves with substrate biases from -1.0 to +0.4 V almost overlap. Measurement was performed at V_d = 1 V.

B. Noise properties

Fig. 3(a) shows examples for noise spectral intensities of the MOSFET at V_d = 1.0 V and V_g-V_t = 0.6 V. The data shown is for V_{sub} = -0.8, -0.2, and +0.4 V. The noise floor of the system is shown in grey color, which was measured with the probe pins open (non-contact with a wafer). It clearly demonstrates that the intensity of noise floor is as large as the noise intensity from the MOSFET above 1 MHz where the evaluation of the noise intensity is difficult.

One of the physical origins of LFN is explained by the carrier number fluctuation theory, where the trapping phenomena between the inversion Si channel and oxide traps play an important role [21]. The trapping probability P is can

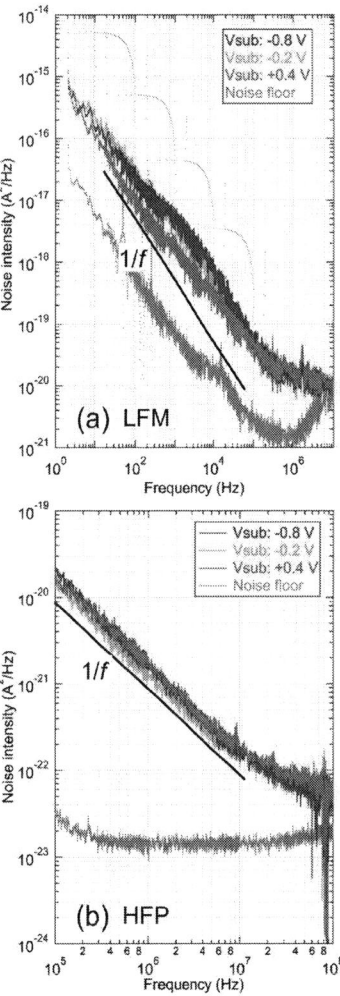

Fig. 3 (a) Noise intensities of n-MOSFETs for L = 350 nm with V_{sub} = -0.8, -0.2, +0.4 V. The noise floor of the measurement system is shown in a grey color, whose magnitude becomes as large as the noise from the MOSFET beyond 1 MHz. (b) Noise intensities of the MOSFET same with the one used in (a). The level of noise floor is low enough so that measurement up to 100 MHz is feasible.

be expressed as

$$P \propto \exp\left(-2d/\alpha - \Delta E/kT\right), \qquad (1)$$

where ΔE is Coulomb energy, α is attenuation length of wave-function, d is distance between a trap located in the SiO_2 layer and an electron in the channel, k is the Boltzmann constant, and T is absolute temperature [22]. For example, reduction of noise intensity by controlling the charge-centroid position has been demonstrated, in which a silicon-on-insulator (SOI) substrate or nanowire channel were utilized [9, 10, 23].

In the frequency region from 1 Hz to 10 kHz, the noise intensity shifts upwards as the V_{sub} value decreases from +0.4 to -0.8 V. The reverse bias shifts the charge-centroid position of inversion carriers close to the interface, so that the average distance between an electron and a trap decreases. Hence, the probability P of trapping/detrapping increases, resulting in the large the noise intensity. This behavior is different from random telegraph noise (RTN), which causes a Lorentz distribution. Here we emphasize that even though the $1/f$ property continue to exhibit up to 1 MHz, the increase in the noise intensity due to the substrate bias is observed only in the lower frequency region.

The noise properties in the higher frequency region up to 100 MHz are characterized by the novel high-frequency noise probe as shown in Fig 3(b). The unit of measured values by the spectrum analyzer (dBm) was converted to a unit of A^2/Hz using a value of transimpedance gain. The intensity of noise floor is small enough in this entire frequency region. The noise intensity increases when the V_g-V_t value increases (not shown) so that the characterization is properly performed. The slope of the intensities, β of $1/f$, becomes small beyond 10 MHz, suggesting the appearance of thermal noise. Note that in this frequency region, the noise intensity does not exhibit the significant dependence on V_{sub}.

The changes in noise intensity are shown in Fig. 4 as functions of V_g-V_t and V_{sub} for frequencies (a) 1 kHz, (b) 100 kHz, and (c) 30 MHz. The unit of noise intensity is in A^2/Hz, where the normalization by I_d^2 has not been adopted. For the comparison among different V_{sub} values, normalization by Id^2 is not necessary in this case because the actual I_d values for different V_{sub} is almost same after offset with V_g-V_t as shown in Fig. 2(b). It is clearly shown that the noise intensity increases as the V_g-V_t increases in each frequency region because of larger I_d. Although the dependence of the noise intensity on V_{sub} is remarkable in the low frequency region (10 kHz), the dependence on V_{sub} is diminished at the high frequency region (30 MHz).

Next, we focus on the slope value β of noise intensity. As shown in Fig. 3(a), the increase in noise intensity due to a

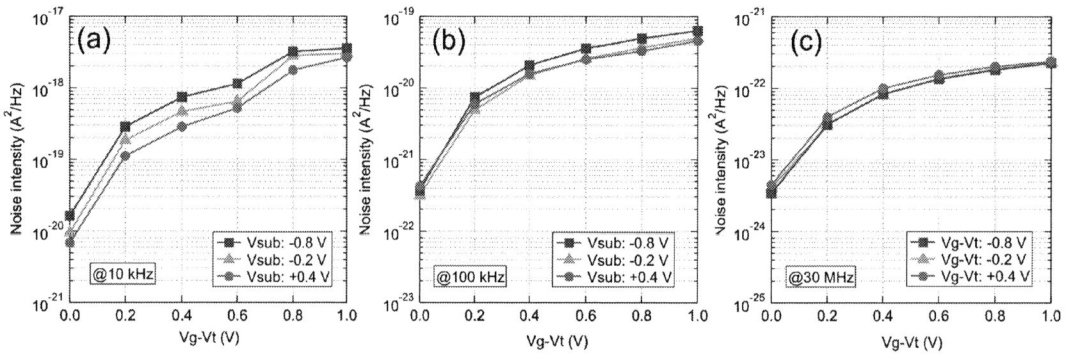

Fig. 4 Dependence of noise intensity on gate overdrive V_g-V_t for V_{sub} = -0.8, -0.2, +0.4 V. The noise intensity was evaluated at (a) 10 kHz, (b) 100 kHz, and (c) 30 MHz.

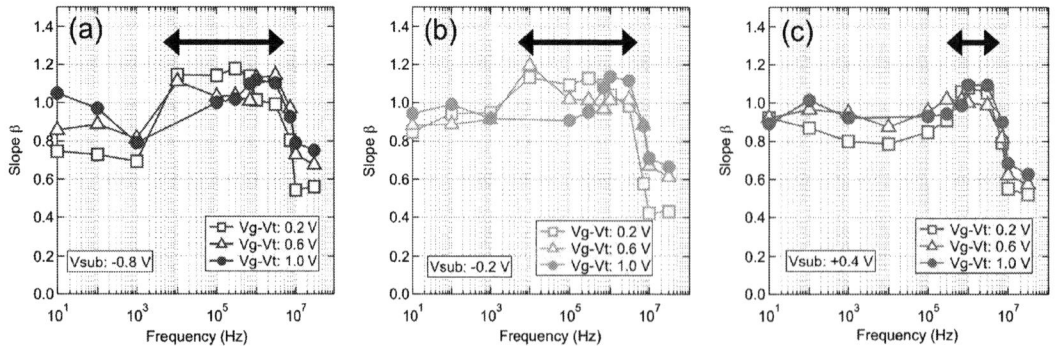

Fig. 5 The slope values β as a function of frequency for V_g-V_t = 0.2, 0.6, 1.0 V. The applied V_{sub} values are (a) -0.8, (b) -0.2, (c) +0.4 V. The bandwidths where large β values are observed change by applying the substrate bias.

978-1-4673-1707-8/12 $31.00 © 2012 IEEE

substrate bias was observed in the low frequency range from 10 Hz to 10 kHz. This increase results in the change in β values in the higher frequency region. Therefore, it is useful to estimate β values as functions of frequency and V_{sub}. Fig. 5 shows the β values for V_{sub} = (a) -0.8 V, (b) -0.2 V, and (c) +0.4 V. In the case that the charge-centroid position is far from the interface (V_{sub} = +0.4 V), the β values range almost from 0.8 to 1.0 in the low-frequency region (from 10 to 100 kHz), while having a broad peak around 1 MHz. Note that the maximum value of β is 1.2, which is much smaller than that of RTN ($\beta \approx 2$). The β values dramatically decreases to 0.6 beyond 10 MHz, indicating the appearance of thermal noise. This feature is observed regardless of V_g-V_t. As the V_{sub} is decreased to -0.2 and -0.8 V (under reverse biases), the increase in β values is observed at 10 kHz which is lower than that in the case of V_{sub} = +0.4 V. The extension in bandwidth of large β values corresponds to the large increase in noise intensity due to the substrate bias.

Fig. 6 summarizes the effect of substrate bias on the noise intensity of MOSFETs. $1/f$ low-frequency noise and thermal noise determines the corner frequency that is, in the MOSFETs we have measured in this study, beyond 100 MHz, although the dramatic decrease in β beyond 10 MHz probably corresponds to the appearance of thermal noise. By applying a negative substrate bias (reverse), the noise intensity specifically in the low frequency region (10 Hz to 10 kHz) increases, while the noise intensity beyond 100 kHz remains nearly unchanged. In addition, a peak in β values is observed around 10 kHz to 1 MHz, whose bandwidth depends on V_{sub}.

In order to clarify the physical origins for the noise components that show different behavior and frequency dependences, further investigation from many aspects, for example, using other gate-stack materials, temperatures,

dimension of MOSFETs, is necessary. However, we consider that our approach using a higher-frequency noise probe will allow us to have new insights on mechanisms of noise generation in nano-scaled MOSFETs.

IV. CONCLUSION

We have investigated the frequency dependence of noise intensity using n-MOSFETs with a gate length of 350 nm. Using a developed high-frequency noise measurement system, the frequency of noise characterization ranges from 10 Hz to 100 MHz. We applied substrate biases in order to change the effect of number fluctuation. The frequency region where transition of noise intensity from V_{sub}-dependent to V_{sub}-independent is observed. Our approach will give a key for further understanding of the noise mechanisms in MOSFETs.

REFERENCES

[1] L. K. J. Vandamme and F. N. Hooge: IEEE Trans. Electron Devices **55** (2008) 3070.

[2] F. Crupi, P. Srinivasan, P. Magnone, E. Simoen, C. Pace, D. Misra, and C. Claeys: IEEE Electron Device Lett. **27** (2006) 688.

[3] E. Simoen, G. Eneman, P. Verheyen, R. Loo, and K. D. Meyer: IEEE Trans. Electron Devices **53** (2006) 1039.

[4] M. V. Haartman, B. G. Malm, and M. Östling: IEEE Trans. Electron Devices **53** (2006) 836.

[5] N. Zanolla, D. Šiprak, M. Tiebout, P. Baumgartner, E. Sangiorgi, and C. Fiegna: IEEE Trans. Electron Devices **57** (2010) 119.

[6] N. B. Lukyanchikova, M.V. Petrichuk, N.P. Garbar, E. Simoen, and C. Claeys: Appl. Phys. A **70** (2000) 345.

[7] F. N. Hooge: Physica, **83B** (1976) 14.

[8] R. Hettiarachchi, T. Matsuki, W. Feng, K. Yamada, and K. Ohmori: Jpn. J. Appl. Phys. **50** (2011) 10PB04.

[9] W. Feng, R. Hettiarachchi, Y. Lee, S. Sato, K. Kakushima, M. Sato, K. Fukuda, M. Niwa, K. Yamabe, K. Shiraishi, H. Iwai, and K. Ohmori: Technical Digest of 2011 International Devices Meeting, p. 630.

[10] W. Feng, R. Hettiarachchi, S. Sato, K. Kakushima, M. Niwa, H. Iwai, K. Yamada, and K. Ohmori: Jpn. J. Appl. Phys. **51** (2012) 04DC06.

[11] F. N. Hooge: IEEE Trans. Electron. Devices, **41** (1994) 1926.

[12] L. K. J. Vandamme, X. Li, and D. Rigaud: IEEE Trans. Electron Devices **41** (1994) 1936.

[13] G. Ghibaudo and T. Boutchacha: Microelectron. Reliab. **42** (2002) 573.

[14] F. N. Hooge, T. G. M. Kleinpenning, and L. K. J. Vandamme: Rep. Prog. Phys. **44** (1981) 497.

[15] G. Ghibaudo, O. Roux, Ch. Nguyen-Duc, F. Balestra, and J. Brini: Phys. Status Solidi A **124** (1991) 571.

[16] H. Mikoshiba: IEEE Trans. Electron Devices **29** (1982) 965.

[17] K. K. Hung, P. K. Ko, C. Hu, and Y. C. Cheng: IEEE Trans. Electron Devices **37** (1990) 654.

[18] G. Ghibaudo, O. Roux, C. Nguyen-Duc, F. Balestra, and J. Brini: Phys. Status Solidi A **124** (1991) 571.

[19] K. Ohmori, R. Hasunuma, and K. Yamada: Proceedings of the 2012 International Conference on Microelectronic Test Structures (ICMTS), p. 169.

[20] K. Ohmori, R. Hasunuma, W. Feng, and K. Yamada: 2012 VLSI Symposium on Technology (*to be presented*).

[21] A. L. McWhorter: Semiconductor Surface Physics (1957) 207.

[22] N. Apsley, and H. P. Hughes: Philosophical Magazine **30** (1974) 963.

[23] M. von Haartman and Mikael Östling: J. Appl. Phys. **101** (2007) 034506.

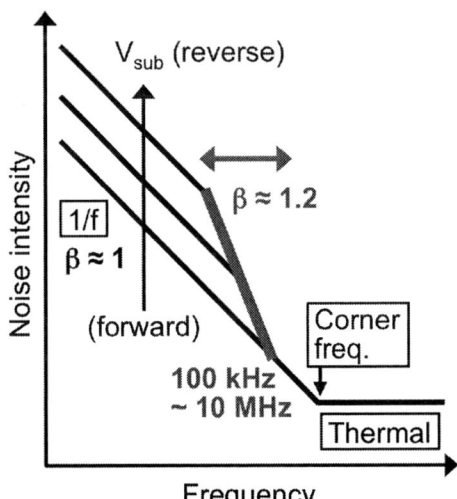

Fig. 6 Schematic diagram of the changes in noise intensity due to a substrate bias. The component of $1/f$ noise in the low frequency region, which was induced by a substrate bias, reduces at higher frequency (100 kHz to 10 MHz).

Author Index

A

Abdinia, Sahel 173
Abugharbieh, Khaldoon 18
Adamu-Lema, Fikru 109
Ahmet, Parhat 89
Alper, Cem 161
Al-Shahed, Muhammad 141
Amoroso, Salvatore Maria 109
An, Hokyun 149
Andrade, Maria Gloria C. 338
Andrieu, François 209, 246
Angelopoulos, Evangelos 141
Annunziata, Roberto 286
Ansari, Lida 294
Aoulaiche, Marc 338
Appel, Wolfgang 141
Ardouin, Bertrand 62
Arnaud, Franck 334
Asenov, Asen 109, 113, 205
Asenov, Plamen 205
Ashburn, Peter 137
Ashouei, Maryam 58
Ayala, Nuria 266
Aymerich, Xavier 266

B

Bablet, Jacqueline 173
Baccarani, Giorgio 105
Balestra, Francis 217
Banijamali, Bahareh 18
Barone, Gaetano 185
Barraud, Sylvain 73, 121
Behin-Aein, Behtash 36
Belleville, Marc 69
Bellisai, Simone 230
Ben Akkez, Imed 217
Benwadih, Mohammed 173
Bersuker, Gennadi 274
Blaise, Philippe 278
Boeuf, Frederic 217
Bonfiglio, Valentina 117
Boukos, Nikos 258
Bourdelle, Konstantin 153
Boutros, Karim 310
Bretegnier, Damien 278
Brockherde, Werner 230

Bronzi, Danilo 230
Brown, Andrew 113
Brunel, Laurent 270
Burghartz, Joachim 141
Bury, Erik 242
Buscemi, Fabrizio 97

C

Cagli, Carlo 278
Caillat, Christian 242, 338
Cantatore, Eugenio 173
Carabasse, Catherine 278
Cassé, Mikaël 73
Chalkiadaki, Maria-Anna 50
Chan, Mansun 101
Chandrashekar, Sandhya 286
Charbuillet, Clement 54
Chartier, Isabelle 173
Chauhan, Yogesh Singh 46, 50
Chen, Mary 310
Chen, Xu 310
Chen, Yin-Nien 157
Cheng, Binjie 113
Chiarella, Thomas 221
Cho, Youngseung 193
Choi, Byoungdeog 149
Chong, Harold 137
Chu, Rongming 310
Chuang, Ching-Te 77, 157
Chuang, Ming-Yeh 185
Chung, Chilhee 149, 193
Chung, Hyunwoo 193
Ciofi, Ivan 221
Claeys, Cor 330, 338
Clermidy, Fabien 69
Conzatti, Francesco 322
Coppard, Romain 173
Coquand, Remi 73, 121
Corrion, Andrea 310
Cosemans, Stefan 282
Cristoloveanu, Sorin 197, 209
Cros, Antoine 217
Crupi, Felice 330
Curutchet, Arnaud 270

D

Dabrowski, Jaroslaw 250

Author Index

Dagtekin, Nilay 161
Dahmani, Faiz 278
Danneville, Francois......................................54
Das, Samaresh 326
Datta, Supriyo 36
De Michielis, Luca 161
de Planque, Maurits 137
De Salvo, Barbara.................................278, 286
Decoutere, Stefaan..302
Delerue, Christophe.......................................290
Deshpande, Veeresh 121
Dimitriadis, Charalabos...................................334
Donetti, Luca 209
Dormieu, Benjamin 54
Dumania, Piotr...165
Dupuy, Jean-Yves 62
Durini, Daniel 230

E

Endler, Stefan 141
Enz, Christian C. 50
Esseni, David 322

F

Fagas, Giorgos.......................................294, 326
Fan, Ming-Long.......................................77, 157
Faynot, Olivier......................................121, 209
Fazan, Pierre..242, 338
Fenouillet-Beranger, Claire 217
Fenouillet-Berangerand, Claire 209
Ferain, Isabelle 326
Ferwana, Saleh 141
Flandre, Denis 246
Fontserè, Abel 306
Fregonese, Sébastien 189

G

Gamiz, Francisco 209
Gao, Xiang..314
Gardeniers, Han...169
Gaubert, Philippe...213
Gemmeke, Tobias 58
Gerrer, Louis 109
Ghibaudo, Gérard 73, 129, 217, 334
Ghosh, Sudip 62
Gnani, Elena 105, 185
Gnudi, Antonio.......................................105, 185

Godignon, Phillipe..306
Godin, Jean 62
Gong, Xiao.. 177
Grabiec, Piotr.. 165
Grandchamp, Brice.. 62
Grant, Lindsay A. 238
Greer, James C.. 294
Groenland, Alfons 169
Groeseneken, Guido 242, 282, 330
Guo, Hua Xin 177
Guo, Shiping 314
Gwoziecki, Romain 173

H

Haendler, Sebastien 334
Ham, Donhee 14
Han, Jaejong 149
Harendt, Christine.. 141
Hassan, Mahadi-Ul 141
Hattori, Takeo 89
He, Jr-Hau 234
Henaff, Ewen 286
Henderson, Robert 226, 238
Henkel, Christoph.. 93, 250
Hettiarachchi, Ranga...................................... 342
Hong, Hyeongsun 193
Hong, Soojin 193
Hong, Sung-Min 318
Horiguchi, Naoto 242, 330
Hu, Chenming.. 46, 50
Hu, Vita Pi-Ho 77, 157
Hubert, Quentin 286
Hueting, Raymond... 125
Hughes, Brian 310
Hwang, Yoosang 193

I

Iannaccone, Giuseppe...................................... 117
Ioannidis, Eleftherios.................................... 334
Ionescu, Adrian M. 161
Iwai, Hiroshi 73, 89

J

Jacob, Stephanie 173
Jacoboni, Carlo 97
Jahan, Carine 286
Jain, Pulkit 262

Author Index

Jan, Sebastien 54
Jehl, Xavier 121
Jeon, Dae-Young 129
Jeong, Gitae 149
Jeong, Seonghoon 149
Jiménez Tejada, Juan Antonio 338
Jin, Gyo-Young 193
Jomaah, Jalal 334
Joo, Min-Kyu 129
Jousseaume, V. 278
Jungemann, Christoph 318
Jurczak, Gosja 338
Jurczak, Malgorzata 282

K

Kakushima, Kuniyuki 89
Kang, Hokyu 149
Kang, Xuanwu 302
Kang, Yoongoo 149
Karim, Mohammed Ahosan Ul 46, 50
Kataoka, Yoshinori 89
Kawanago, Takamasa 89
Kean, Alistair H. 258
Keane, John 262
Kerner, Christoph 221
Khalil, Sameh 310
Khandelwal, Sourabh 46
Kilchytska, Valeriya 246
Kim, Bonghyun 149
Kim, Chris 262
Kim, Daeik 193
Kim, Gyu-Tae 129
Kim, Huijung 193
Kim, Jiyoung 193
Kim, Kinam 1
Kim, Namhoon 18
Kim, Samuel 310
Kim, Tony 201
Kim, Un Jeong 129
Kinkeldei, Thomas 133
Klauk, Hagen 41
Knoll, Lars 153
Koné, Gilles Amadou 62
Kovalgin, Alexey 169
Koyama, Masahiro 73
Krom, Raymond 330

Krzeminski, Christophe 290
Kwong, K. C. 101

L

Labat, Nathalie 270
Lambert, Benoit 270
Larcher, Luca 274
Lattanzio, Livio 161
Le Royer, Cyrille 197
Lee, Eunok 193
Lee, Hyung-Seok 314
Lee, Kongsoo 149
Lemme, Max C. 25, 250
Li, Qi 201
Li, Ray 310
Liebault, Jacques 278
Lien, Wei-Cheng 234
Lim, Hanjin 149
Lippert, Gunther 250
Liu, Yong 14
Lizzit, Daniel 322
Longnos, Florian 278
Lupina, Grzegorz 250
Luque Rodríguez, Abraham 338

M

Madden, Liam 18
Maddiona, Lidia 173
Mahmoudi, Hiwa 254
Mai, Andreas 181
Maiellaro, Giorgio 173
Maitrejean, Sylvain 286
Malbert, Nathalie 270
Malm, Gunnar B. 93
Maneux, Cristell 62, 189
Mangla, Anurag 50
Mantl, Siegfried 153
Marc, François 62
Marced, Maria .... No paper
Marcon, Denis 302
Mariucci, Luigi 173
Markov, Stanislav 109
Markovic, Bojan 230
Martin-Martinez, Javier 266
Masahara, Meishoku 81
Mehr, Wolfgang 250
Meinerzhagen, Bernd 318

Author Index

Mendes Almeida, Luciano....................338
Migita, Shinji....................81
Millán, Jose....................306
Millar, Campbell....................113, 205
Mitard, Jerome....................242, 330
Mizubayashi, Wataru....................81
Moens, Peter....................306
Mok, Philip K. T.....................101
Molas, Gabriel....................278
Morita, Yukinori....................81
Mouis, Mireille....................129
Münzenrieder, Niko....................133

N

Nafria, Montse....................266
Naiini, Maziar M.....................93
Nam, Seokwoo....................149
Namuduri, Chandra....................310
Nanver, Lis....................145
Natori, Kenji....................89
Navarro, Carlos....................209
Navarro, Gabriele....................286
Newell, Scott....................310
Nguyen, Viet Hung....................290
Niknejad, Ali....................46, 50
Niquet, Yann Michel....................290
Nishiyama, Akira....................89
Nodin, Jean-François....................286
Nodjiadjim, Virginie....................62
Noé, Pierre....................286
Normand, Pascal....................258

O

Oh, Yongchul....................193
Ohata, Akiko....................209
Ohmi, Tadahiro....................213
Ohmori, Kenji....................342
Östling, Mikael....................93, 250
Ota, Hiroyuki....................81

P

Padovani, Andrea....................274
Palacios, Tomas....................314
Palestri, Pierpaolo....................322
Palmisano, Giuseppe....................173
Pananakakis, Georges....................286
Pantisano, Luigi....................330

Pao, Chia-Hao....................157
Park, So Jeong....................129
Parsey, John M.....................306
Paschen, Uwe....................230
Pavan, Paolo....................274
Paydavosi, Navid....................46
Pérez-Tomás, Amador....................306
Perniola, Luca....................286
Perreau, Pierre....................121
Persico, Alain....................286
Petti, Luisa....................133
Peumans, Peter....................No paper
Pham, Anh-Tuan....................318
Piccinini, Enrico....................97
Piedra, Daniel....................314
Pisano, Albert P.....................234
Planes, Nicolas....................334
Poiroux, Thierry....................121
Poli, Stefano....................185
Poljak, Mirko....................298
Previtali, Bernard....................121
Puglisi, Francesco Maria....................274

Q

Qi, Lin....................145
Qin, Ling....................14

R

Rafhay, Quentin....................217
Ragnarsson, Lars-Ake....................242
Ramalingam, Suresh....................18
Razavi, Pedram....................326
Reggiani, Susanna....................105, 185
Reid, Dave....................205
Reimbold, Gilles....................73, 286
Rempp, Horst....................141
Riet, Muriel....................62
Ritzenthaler, Romain....................242
Roche, Benoit....................121
Rodriguez, Guillaume....................278
Rodriguez, Noel....................209
Rodriguez, Rosana....................266
Romeo, Tommaso....................330
Roule, Anne....................286
Roy, Scott....................205
Rücker, Holger....................181

Author Index

S

Sahoo, Amit Kumar 189
Sakic, Agata 145
Salvatore, Giovanni A. 133
Samukawa, Seiji 85
Sanquer, Marc 121
Santorelli, Marco 189
Sarkar, Angik 36
Schäfer, Anna 153
Scheer, Patrick 54
Schmitz, Jurriaan 169
Scholtes, Tom 145
Schram, Tom 242
Selberherr, Siegfried 254
Selmi, Luca 322
Senesky, Debbie G. 234
Simoen, Eddy 330, 338
Smith, Anderson 250
Song, Emil 298
Souchier, Emeline 278
Sousa, Véronique 286
Spessot, Alessio 242
Srivastava, Puneet 302
Srividya, Vidya 242
Su, Pin 77, 157
Sugawa, Shigetoshi 213
Sugii, Nobuyuki 89
Suligoj, Tomislav 298
Sultan, Suhana Mohamed 137
Sun, Kai 137
Sun, Nan 14
Sverdlov, Viktor 254

T

Takagi, Shinichi 85
Teramoto, Akinobu 213
Tessaire, Magali 286
Thakur, Pankaj 46
Thean, Aaron 242, 330
Theodorou, Christoforos 334
Thomas, Olivier 69
Tian, Weidong 185
Tisa, Simone 230
Tiwari, Sandip 28
Toffoli, Alain 286
Togo, Mitsuhiro 330

Tomaszewski, Daniel 165
Tosi, Alberto 230
Tosti, Lucie 121
Tramontana, Francesca 173
Trellenkamp, Stefan 153
Triozon, François 121, 290
Tröster, Gerhard 133
Tsai, Dung-Sheng 234
Tsai, Ming-Fu 157
Tsoukalas, Dimitris 258
Tsutsui, Kazuo 89

V

Valentian, Alexandre 69
van der Cingel, Johan 145
van Hemert, Tom 125
Van Hove, Marleen 302
Van Huylenbroeck, Stefaan 221
Vaziri, Sam 250
Veloso, Anabela 338
Venegas, Rafael 302
Venugopalan, Sriramkumar 46, 50
Vereshchagina, Elizaveta 169
Verrelli, Emanuele 258
Verrier, Pascal 278
Vianello, Elisa 278
Villa, Federica 230
Vinet, Maud 121
Vizioz, Christian 121
Voisin, Benoit 121

W

Wacquez, Romain 121
Wada, Akira 85
Walker, Richard J. 226, 238
Wan, Jing 197
Wang, Bo 201
Wang, Kang 298
Wang, Minsheng 298
Wang, Xingsheng 113
Webster, Eric A. G. 226, 238
Weiß, Mario 189
Wellekens, Dirk 302
Weyers, Sascha 230
Wise, Rick 185
Wolters, Rob 169
Wong, Danny 310

Author Index

Wouters, Dirk 282

Wu, Ephrem..18

Wu, Tian-Li 302

Wu, Xin 18

X

Xu, Guangyu 14

Y

Yamada, Keisaku..342

Yeo, Yee-Chia...177

Yoo, Wonseok 149

Yu, Ran...326

Z

Zaborowski, Michal..165

Zahid, Mohammed 302

Zappa, Franco 230

Zaslavsky, Alex 197

Zehnder, Daniel 310

Zhang, Leqi 282

Zhang, Rui..85

Zhang, Xingui...177

Zhao, Qing-Tai 153

Zimmer, Thomas 189

Zimmermann, Martin 141

Zuffada, Maurizio..7

Zuliani, Paola..286

Zysset, Christoph..133

9781467317078